ALBERT EINSTEIN

ALBERT EINSTEIN

A BIOGRAPHY

ALBRECHT FÖLSING

Translated from the German by
EWALD OSERS

VIKING

VIKING
Published by the Penguin Group
Penguin Books USA Inc., 375 Hudson Street, New York, New York 10014, U.S.A.
Penguin Books Ltd, 27 Wrights Lane, London W8 5TZ, England
Penguin Books Australia Ltd, Ringwood, Victoria, Australia
Penguin Books Canada Ltd, 10 Alcorn Avenue, Toronto, Ontario, Canada M4V 3B2
Penguin Books (N.Z.) Ltd, 182–190 Wairu Road, Auckland 10, New Zealand

Penguin Books Ltd, Registered Offices:
Harmondsworth, Middlesex, England

First published in 1997 by Viking Penguin,
a division of Penguin Books USA Inc.

1 3 5 7 9 10 8 6 4 2

Originally published in Germany as *Albert Einstein: Eine Biographie* by Suhrkamp Verlag.

PHOTOGRAPH CREDITS:
Lucien Aigner, The Institute for Advanced Study, Princeton: 29; American Institute of Physics, Emilio Segre Visual Archives: 7, 8, 12, 14, 18, 27, 34; Bibliothek der Eidgenössischen Technische: Hochschule, Zurich: 2, 3, 4, 6, 9; Bildarchiv Preussischer Kulturbesitz, Berlin: 11, 17; Bildarchiv Preussicher Kulturbesitz, Berlin and Süddentscher Verlag, Munich: 32, 33; Bundesarchiv Koblenz: 21; Einstein Archives, The Jewish National and University Library, Jerusalem: 5, 15, 16, 24; Lotte Jacobi, Dimond Library, Durham: 1, 25, 26, 28; Howard E. Schrader, Princeton University, Princeton: 31; Süddeutscher Verlag, Munich: 13, 30; Ullstein Bilderdienst, Berlin: 10, 19, 20, 22, 23.

LIBRARY OF CONGRESS CATALOGING IN PUBLICATION DATA
Fölsing, Albrecht, 1940–
[Albert Einstein. English]
Albert Einstein : a biography / by Albrecht Fölsing : translated from the German by Ewald Osers.
p. cm.
Includes bibliographical references and index.
ISBN 0-670-85545-6 (alk. paper)
1. Einstein, Albert. 1879–1955. 2. Physicists—Biography.
I. Title.
QC16.E5F5913 1997
530'.092—dc20 96-26341

This book is printed on acid-free paper.
∞

Printed in the United States of America
Set in Janson
Designed by Francesca Belanger

ACKNOWLEDGMENTS

THIS BOOK WOULD NOT HAVE BEEN POSSIBLE without the work of those who began early on to collect and deposit Albert Einstein's letters and manuscripts, as well as other documents relating to him. The central source is the Albert Einstein Archive, formerly in Princeton and now maintained in the Jewish National and University Library in Jerusalem. I would like to thank its curator, Ze'ev Rosenkranz, and Hanna Katzenstein for their generous support during my stay in Jerusalem, and the Einstein Archive, Hebrew University of Jerusalem, Israel, for its kind permission to reprint unpublished material. I was able to inspect sets of copies in the Mudd Library of Princeton University and in the Science and Engineering Library of Boston University. In Zurich Beat Glaus was an invariably helpful guide through the history of science collections of the Library of the Swiss Technical University, ETH. In the archive of the Max Planck Society in Berlin I enjoyed the kind assistance of Eckart Henning, Marion Kazemi, and Andreas K. Walter. Bernhardt Schell of the Anschütz company in Kiel was good enough to put the correspondence between Einstein and Hermann Anschütz-Kaempfe at my disposal before its publication.

I am grateful to Professor Yehuda Elkana and the Van Leer Foundation, Jerusalem, for enabling me to participate in a workshop on "Einstein in Context" in April 1990 in Jerusalem, and to its participants for many informative suggestions.

I greatly benefited from conversations with Anne J. Kox, Jürgen Renn, and Robert Schulmann of the project of *The Collected Papers of*

Albert Einstein; Robert Schulmann, especially, generously shared with me his knowledge about Albert Einstein's early years.

I owe a debt of gratitude to the publisher Siegfried Unseld for his great confidence in this difficult project, and for his patience.

My wife, Ulla, and our children, Philipp and Julia, had to bear with patience my prolonged preoccupation with this book; for this I not only thank them sincerely but also ask their forgiveness. This apology should also include our dog, Rufus, for whom it was most difficult to understand my changed lifestyle as I worked at my desk.

CONTENTS

VII THE PACIFIST AND THE BOMB

FOREWORD

Even four decades after his death, an exceptional fascination is evoked by the name of Albert Einstein. It far transcends the fact that he was indisputably the greatest physicist of our century, comparable to Isaac Newton—but Einstein is much more than a subject for specialists in the history of science.

Einstein's legendary greatness, which still touches many of us today, is based on a multitude of factors; but it is primarily linked to physics, in several respects. His concepts of space and time, of the "fourth dimension," and of a finite but unbounded universe in which light travels along a curved path, are regarded as revolutionary, comparable, in their effect on human understanding, only to those of Copernicus. However, the results of his profound reflections on nature are also—through his legendary formula $E = mc^2$—indirectly connected with the atom bomb and all that it has meant in terms of destruction, fear, and terror. If ever a theory born of an innocent search for knowledge became a material force, then it was in the mushroom cloud over Hiroshima.

The creator of this theory lived not in an ivory tower but in a time of wars and conflicts, and he faced this situation with a strong sense of humanitarian responsibility and a need to intervene in politics. His humanism, which assigned greater importance to what was common to all people than to what divided them, gave him a "left-wing" identity. However, he was not tied to any party doctrine but instead was stamped by the social ethics of Judaism, which include sympathy for the underdog. He put his fame—he was already a legend at the age of forty—at the service of social justice, democratic freedoms, pacifism, the welfare of the Jews, and a cosmopolitan internationalism, though rarely with success and frequently setting off controversy.

Einstein's kindness was often praised and his simplicity admired as if he were a saint. There was some justification for that, but he could also be rude and wounding, and below his modest surface there was unfathomable complexity. Although his attitude toward the country of his birth was complicated, his attitude toward Nazi Germany was unequivocal. It led him, despite his passionate pacifism, to write a letter to President Roosevelt suggesting the development of an atom bomb. After Hiroshima, when he warned against a nuclear arms race, many regarded him as wise old man, a personification of the world's conscience.

However, his "excursions into politics," as he called them, were just that—excursions. They were never nearly as important to him as physics. Physics was his passion and his life. No one else has ever enriched a science as Einstein enriched physics during the two decades between 1905 and 1925. If asking who was the greatest physicist of the century produces the answer "Einstein, for his theory of relativity," then asking who was the second-greatest physicist might justifiably produce the answer "Einstein, for all his other achievements." Over the last three decades of his life, he searched for the foundations of physics. His road led to no result, but he never gave up; and right to the end he remained addicted to physics.

In addition, he was a husband (twice), lover, and father (at least three times). He was a Jew. He was a citizen of four nations and was offered the presidency of a fifth, Israel—but he declined this honor even though his deepest loyalty was to those he called his "tribal brethren." He was born in Germany, and German remained his first language, the only language in which he wrote and in which he could adequately express his feelings and ideas. After the Holocaust he called German his "stepmother tongue," displaying a fine feeling for his language even at a distance. He never forgave the Germans.

The depth and variety of Einstein's thinking about nature, the scope and color of his life, and the complexity of his character have about them something alarming to a biographer; and in fact this book has turned out more voluminous than I intended. Wherever possible I have based my writing on Einstein's own testimony: his published work, his unpublished manuscripts, his countless letters, and his intermittent diaries, so far as they are at present accessible. In addition, I have used firsthand sources that seemed to me reliable. Some of the stories spread about Einstein are not mentioned in these pages, as there would be no point in discussing freely invented or unattested, fantastic assertions. Instead, I hope to offer much that is new, and in a manner which will cast new light on known facts. The most important aspect to me, always, was Einstein's physics. Physics was at the core of his identity, and only through physics can we get close to him as a seeker after truth, whose like we shall not see again.

PART I

CHILDHOOD, YOUTH, STUDENT YEARS

CHAPTER ONE

Family

He was born on March 14, 1879, in Ulm in southern Germany, on a cold but sunny Friday, half an hour before the church bells rang out midday. His parents and relatives, anxious to perpetuate the family name, were no doubt pleased that the first child was a boy. But as often happens with young couples who are facing parenthood for the first time, their joy was clouded by concern and even anxiety.

"When he was born"—his younger sister wrote many years later—"Mother was alarmed at the sight of his exceptionally large angular occiput and at first thought he was a monster."[1] The physician reassured the twenty-one-year-old mother, Pauline Einstein, that this peculiarity would soon disappear, and a few weeks later the size of the baby's skull was indeed quite normal, though a rather square occiput remained a lifelong characteristic.

The following morning the father, Hermann Einstein, put on his frock coat and went to the town hall to record the birth of his son. The boy was to be called Albert, only faintly echoing his grandfather's name, Abraham Einstein. Nothing, of course, suggested that the motto of Ulm, dating from its medieval prosperity, *Ulmenses sunt mathematici*—"The people of Ulm are mathematicians"—would be brilliantly confirmed by this Albert Einstein. In the column provided for religion, both parents and child were recorded as "Israelitic."[2]

In spite of their Jewish origin, Albert Einstein's ancestors could be described as true Swabians. On the paternal side the family had been settled in the region for more than two centuries—not in Ulm, but some forty miles to the south, in Buchau, a small township on Lake Feder in the foothills of the Alps. Under the patronage of an aristocratic abbey a small Jewish community had been in existence there from the sixteenth century; in 1665 it was joined by Baruch Moises Ainstein, who was originally from the area of Lake Constance, the large lake separating Germany from Austria and Switzerland. Ainstein's descendants later changed the first letter of their last name, producing the spelling familiar to us.

In the Jewish cemetery of Buchau, dozens of tombstones, now covered with overgrowth, are silent witnesses to the family history of the Einsteins over many generations. The last Jewish inhabitant of Buchau was Siegbert Einstein, a great-nephew of the physicist. He survived the Theresienstadt concentration camp and for a while after the Second World War was the mayor of Buchau, looking after the cemetery and opening its gates for its few occasional visitors. In 1968 he too was buried there—not only the last Jew in Buchau but also Albert Einstein's last relative in Germany.

An entry in the council records of the then Reich Town of Buchau, dated March 16, 1665, records the restrictions and conditions imposed on Jews settling in the town. Against payment of an annual "sitting charge" of twelve guilders, they were granted freedom to practice their religion and their trade—in Moises Ainstein's case, dealing in horses and cloth. Buying and selling were the only sources of livelihood permitted to Jews until the nineteenth century. In 1806, Buchau was assigned to the southern German kingdom of Württemberg, created under Napoléon's patronage. There, in 1828, a law was eventually enacted allowing Jews freedom in their choice of trade. This marked the first step in their emancipation as citizens, even though in Württemberg this was not fully attained until 1862.

Some of the Einsteins seized the new opportunities and became craftsmen—for instance, furriers and bookbinders. They lived in the old town in modest respectability, but its limitations and poverty were still reminiscent of the centuries following the Thirty Years' War, and conditions much too restrictive to allow any of them to excel in any way.

The tombstones in the Jewish cemetery also testify to the assimilation of the Buchau Jews and the Einstein family in the nineteenth century. The Hebrew inscriptions become less frequent and soon disappear altogether; and venerable biblical names, such as Samuel, David, and Abraham, come to be replaced by German names, such as August, Adolf, and Hermann. South German liberalism facilitated a gradual loosening of the formerly strong ties to the synagogue, the more so as the Buchau Jews—like other Jewish communities in southern Germany—were less strongly rooted in tradition than the Jews of Eastern Europe, with their shtetl culture. Moreover, the prosperity which the industrial revolution brought to the bigger cities tempted many to escape from the confines of the provincial towns.

The nearest such center was Ulm on the Danube, an ancient city whose walls were just then being pulled down to make room for a rapidly expanding "new city." The first member of the Einstein family to move to Ulm, in 1864, was Jette Dreyfus, née Einstein, with her husband Kosman Dreyfus, who also came from Buchau. She was followed after 1869 by several male Einsteins who were hoping to make their fortunes in Ulm. According to a census in 1875, the city then had thirty thousand inhabitants, including 692 Jews, who without much ado were accepted as fellow citizens.

In 1877, when the city festively observed the five-hundredth anniversary of the laying of the foundation stone of the cathedral—and the completion, at long last, of its southern tower—the Jewish community demonstrated its solidarity with the city and with its Protestant fellow citizens by a generous gift: a sculpture of the prophet Jeremiah by a local artist. Among the donors, the name Einstein appears six times; it seems likely that, including the Dreyfus and Moos families related to them by marriage, at least twelve Einsteins had by then moved from Buchau to Ulm. One of these was Hermann, Albert's father.

Hermann Einstein was born in Buchau, in 1847, the son of a merchant, Abraham Einstein. He was sent to Stuttgart, the capital of Württemberg, to attend its *Realschule*, a type of high school. Despite

Hermann's lively intelligence and some sign of mathematical talent, there could, given the family's financial position, be no thought of his going to a university. He therefore left school with a "medium maturity" certificate, which at any rate provided an entrée to the better classes of society and carried the privilege of having to serve only one year, instead of the usual three, of military service, and of serving as an officer cadet, with the prospect of a commission as a lieutenant of the reserve. However, Hermann evidently saw no point in participating in the two field exercises which were a condition of being commissioned, and thus spared the royal Württembergian army the problem of having to accept a Jew as a lieutenant of the reserve.

Albert Einstein's maternal ancestors also came from the Swabian Jewry. They lived in Jebenhausen, near Göppingen, on the northern spurs of the Swabian Alb. There Julius Dörzbacher, Albert's grandfather, supported his family with a small bakery. In 1842 the family name was changed to Koch, and in 1852 Julius Koch moved to Cannstatt, near Stuttgart. Together with his brother Heinrich he ran a profitable grain business, acquiring within a few years a considerable fortune and even becoming a "Supplier to the Royal Württembergian Court." Clearly, the business activities of the Koch family were in an entirely different class from the small trade of the Einsteins—not only more profitable, but also more extensive and worldly. When he married, Heinrich Einstein not only became the husband of a pretty young woman (she was eleven years younger than himself) who was regarded as efficient, well educated, and, because of her piano playing, musical; he also made what was called a "good match."

In Einstein's case, perhaps more than with anybody else, one is tempted to engage in the popular game of asking what he might have inherited from whom. One obvious answer would be that with his mathematical gifts he took after his father and with his love of music he took after his mother. There have, of course, been attempts to find the first indications of Albert Einstein's exceptional talents somewhere in his family tree. But he himself refused to go along with such speculations:

> First of all, I know virtually nothing about them, nor are there any people alive who could say a lot about them. If talents existed, then they could not emerge under their restricted living conditions. Besides, I know perfectly well that I myself have no special talents. It was curiosity, obsession, and sheer perseverance that brought me to my ideas. But as for any especially powerful thinking power ("cerebral muscles")—nothing like that is present, or only on a modest scale. Exploration of my ancestors therefore leads nowhere.[3]

More significant, without any doubt, is the fact that both on his father's and on his mother's side, Albert Einstein was born into a large, widely ramified family, whose members were soon settled in many cities and several countries of Europe. We will meet some of these relatives later. They include an aunt in Italy, who financed his studies; and his favorite uncle, Caesar Koch, a brother of his mother, whom the grain business had taken as far afield as St. Petersburg and Argentina, and who settled in Antwerp—where Albert, at age sixteen, sent him his first scientific essay.

These family connections were not only stimulating for young Einstein; they also helped him cope with many difficult phases in his life. And if in a city like Zurich there may have been no uncle, there would at least be a close friend of the family who looked after the young man. Much later, it was Professor Einstein, by then in America, who would try to help many of his relatives during the Nazis' persecution of the Jews.

After their marriage in 1876, Hermann Einstein and his young wife at first lived on Münsterplatz, the cathedral square, in the old part of Ulm. After two years, at the beginning of Pauline's first pregnancy, they moved into a bigger apartment. Early in 1879, with Pauline six months pregnant, they moved to the livelier Bahnhofstrasse 135B, to a comfortable apartment in a three-story building. We have already seen—thanks to his sister's notes—that Albert's birth was not without some alarm. From the same source, we learn that Grandmother Helene, on first seeing her grandson, exclaimed, "Much too fat! Much too fat!"[4] Little Albert seems to have been a quiet baby, causing no trouble to those charged with looking after him.

Albert Einstein did not develop any particular feeling for his birthplace, because a year later the family moved to Munich. When, on his fiftieth birthday, the owner of the building presented him with a photograph, he responded, not without some sarcasm: "For a place to be born in, the house is pleasant enough, because on that occasion one makes no great aesthetic demands yet; instead one first of all screams at one's dear ones, without bothering too much about reasons and circumstances."[5]

Still, even though Einstein spent only the first year of his life in Ulm—growing up in Bavaria, and later in Italy and Switzerland—something Swabian clung to him all his life. For one thing, there was the soft Swabian dialect, which the family never dropped after leaving Württemberg and which Einstein, if less markedly, kept to his old age. He himself became an object of its peculiar tendency toward diminutives: even as a grown man, he always remained, to his family and his second wife (his cousin Elsa), "der Albertl"—"Little Albert." Even during his final years in America, his English, which for him always remained a foreign language, seemed to have Swabian undertones.

In other respects, too, the Swabians would always have recognized him as one of their own: in his speculative brooding, in his often roguish and occasionally coarse humor, and in his pronounced, individualistic obstinacy. It was probably not just flattery when, as Ulm's most famous son, he was asked by the editor of the local paper for a comment, he readily came up with a compliment: "One's place of birth attaches to one's life as something just as unique as one's origin from one's mother. . . . I therefore think of Ulm with gratitude, because it combines artistic tradition with a simple and sound character."[6]

That Hermann Einstein planned another move so soon after the birth of his first child was due to the initiative of his youngest brother, Jakob. Jakob was the only one of the five brothers to have higher education. After leaving his *Realschule* he had attended the Polytechnic in Stuttgart, had qualified as an engineer, and as an engineer had served in the Franco-Prussian War of 1870–1871. In 1876 he had settled in Munich, where he ran a small firm that did water and gas installations.

No doubt Jakob convinced Hermann that there was little future in Hermann's business—dealing in goose feathers for bedding—and that the new industrial age held out greater promise in more appropriate fields.

At any rate, Hermann Einstein moved to Munich with his wife and one-year-old son in the summer of 1880 and became a partner in the firm Jakob Einstein & Cie. The family took an apartment at Müllerstrasse 3, close to the Sendlinger Tor, in the same building where Jakob, still a bachelor, was living and which was also the address of the firm. The division of labor was determined by the interests and abilities of the two brothers: Jakob dealt with technical matters whereas Hermann concerned himself with the commercial side. Two years later the brothers enlarged their business by acquiring two-thirds of the assets of Kiessling & Cie., a "mechanical-technical workshop and boilermaking firm" which had earned a name for itself making gas boilers. In this way, Hermann Einstein productively and profitably invested the major portion of his wife's dowry.

Jakob also saw to it that the activities of Einstein & Cie. were extended to the relatively new field of electrical engineering. In 1882, almost at the same time as they acquired their share of Kiessling & Cie., the two Einsteins took part in the International Electro-Technical Show organized in Munich by Oskar von Miller. They exhibited dynamos, arc lamps, and lightbulbs, as well as a complete telephone system. This side of the business developed so well that the brothers soon abandoned gas and water installations and boilermaking.

In 1885, they sold their shares in Kiessling & Cie. and invested their capital, along with loans from relatives, in a newly founded "Electrical engineering factory J. Einstein & Cie." To this end they had acquired a major piece of land in the suburb of Sendling, "property No. 14" on what was then Rengerweg but in 1887 would be given the unpronounceable name Adlzreiterstrasse. Facing the street was a residential building, which was immediately enlarged by a spacious addition, and behind which was a rather neglected but large garden with ancient trees. The factory was set up in buildings on a nearby property, Lindwurmstrasse 125, purchased for that purpose.

Thus the Einsteins had established themselves in an innovative industry with good prospects of growth. They were what we would now describe as high-tech venture entrepreneurs.

A photograph of Hermann Einstein from this time shows him as a typical patriarch of Germany's early industrial period: his hair is cropped short; he is clean-shaven, except for a precise mustache; he gazes severely through a monocle, demanding respect—in fact, he looks like a Prussian. But those who knew him remembered him differently—as a kind and friendly man, esteemed and loved by all of his family and friends, especially those of the female sex. He certainly was hardworking, but not to an extent that would have interfered with the pleasanter side of life. He made frequent excursions with his family to the surroundings of Munich, and he enjoyed the ancient Bavarian pastime of visiting beer cellars.

He was exceedingly fond of his wife, Pauline, and "the character of the couple harmonized so perfectly that throughout their whole lives the marriage was not only never clouded, but in fact proved the only solid and reliable element at all turns of fate."[7] This may also have been due to the fact that their religious views were in harmony. Both of them respected and declared their Jewish origins, and they probably never considered Christian baptism, either for themselves or for their children, as a way of assimilating further. However, the synagogue no longer played a role in their family life: they did not go to a temple, nor did they pray at home. The precepts of kosher cooking were ignored, and pork was eaten as a matter of course. With his freethinker's attitude Hermann even prided himself that Jewish rites and customs were not practiced in his house.[8] The writings of the Prophets were scarcely read, and the Talmud not at all. Instead, in the circle of his family Hermann Einstein recited Schiller and Heine[9]—Schiller as a Swabian national hero of the enlightened bourgeoisie and Heine as a popular Jewish poet writing in German. Comparing his own life with Heine's may indeed have buttressed his faith in the progress of civilization: Hermann Einstein—unlike Heine—did not have to be baptized to be accepted by his fellow citizens.

This, then, was the environment in which Albert Einstein grew up—at first to the pure joy of his parents and relatives. The earliest characterization of his personality comes from his grandmother Jette Koch, who visited Munich in the summer of 1881 and said of her two-year-old grandson: "Little Albert is so sweet and so good that it pains me already not to be able to see him for such a long time." A week later, she wrote to Munich: "Little Albert is fondly remembered by us; he was so sweet and good, and we have to repeat his amusing ideas again and again."[10] Unfortunately, the fond grandmother did not record any of those amusing ideas.

Little Albert's reaction to the birth of his sister Maria on November 18, 1881, was certainly amusing. No doubt the boy, then two years and eight months old, had been told of the arrival of a *Mädele*, a little girl, as a future playmate, because he promptly inquired where the *Rädele*, the wheels, of his new toy were.[11] This may have been an early hint of his later delight in making up rhymes, or it may have been no more than a little boy's mishearing and being disappointed to find that the screaming bundle was not a plaything. Actually, the second explanation is more probable, since Einstein's speech development was strikingly slow, as he himself would later confirm: "It is true that my parents were worried because I began to speak relatively late, so much so they consulted a doctor. I can't say how old I was then, certainly not less than three."[12] However, the delay seems to have been due to an early ambition to speak only in complete sentences. If someone asked him a question, he would first form the answer in his head, try it out in an undertone—deliberately, with obvious lip movements—and only after assuring himself that his formulation was correct would he repeat the sentence aloud. This often gave the impression that he was saying everything twice, and the maidservant therefore called him "stupid."[13] He gave up this habit only in his seventh year, or perhaps (according to some testimony) not until his ninth. One has the impression not only of particular thoroughness—the explanation his sister later gave for this peculiarity—but also of a boy's laborious and self-critical acquisition of language, in contrast to most children's natural, unproblematical learning.

Albert's younger sister—nicknamed Maja—recorded in her warm-hearted biographical notes that he was fondest of engrossing himself in all kinds of puzzles, making elaborate structures with building blocks and constructing houses of cards of breathtaking height. He was less interested in playing in the garden with young relatives who often came visiting, and he was totally averse to the fights of the boys in the street. These boys soon nicknamed him "the bore." If he could not avoid playing with other children, he deliberately sought the job of umpire, which, because of his instinctive sense of justice, was gladly assigned to him.

When Albert was five years old, a woman was engaged as a tutor to prepare him for the rigors of school life. She, however, found herself unequal to another trait in the boy's makeup—one that the family believed he had inherited from his grandfather Julius Koch. Whenever something was not to Albert's liking, he was seized by a sudden temper, his face paled, his nose turned white, and the consequences were terrible. On one occasion, when he did not like a lesson, "he grabbed a chair and with it struck the woman tutor, who was so terrified that she ran away in fear and was never seen again."[14] His little sister, too, had to suffer: "On another occasion he threw a large nine-pin bowl at [her] head, and yet another time he used a child's pickaxe to strike a hole in [her] head."[15] Fortunately, these tantrums receded during his seventh year and disappeared completely during his first years at school.

One might ask at this point how such a child—with conspicuously delayed speech development, averse to play and social behavior appropriate to his age, and moreover with an occasional total lack of self-control—would fare in the tests and examinations that now precede enrollment in school. Such a child, in a fit of temper, might attack a teacher or a psychologist with a chair, just as occurred a century ago with young Albert Einstein and his tutor. In the accepted view of child psychologists, a child like this should be diagnosed long before starting school and given some form of therapy or other, when, as with little Albert, there are speech problems suggesting defective development. The psychoanalyst Erik H. Erikson, who has ventured to make this remote diagnosis on the strength of the records, believes that cases of this kind deserve or even demand careful attention.[16] At the same time, he regards Albert Einstein's example as a warning against the present tendency to fit all children into the same mold; this could inhibit rather than promote the development of talent. In the event, Einstein grew up without the benefit of a therapist and developed his own distinctive character traits: his determination to apply his own yardstick, his intense brooding, and his profound way of wondering about things.

Einstein's receptiveness to "wonders" and "wondering" was of enormous importance to him throughout his life as a motivation for productive thought, especially in scientific matters. This was a trait which he felt he could not explain to himself, but he commended "wondering," and slowness, in a letter to a colleague, the Nobel Prize laureate James Franck:

> When I ask myself why it should have been me, rather than anyone else, who discovered the relativity theory, I think that this was due to the following circumstance: An adult does not reflect on space-time problems. Anything that needs reflection on this matter he believes he did in his early childhood. I, on the other hand, developed so slowly that I only began to reflect about space and time when I was grown up. Naturally I then penetrated more deeply into these problems than an ordinary child would.[17]

It is clear therefore that Einstein's notion of "wondering" is very different from the common meaning of that term—a noncommittal inability to understand. In his own view:

> It seems to occur whenever an experience comes into conflict with a conceptual world sufficiently fixated within us. If such a conflict is experienced strongly and intensively, then it reacts back in a decisive manner upon our mental world. In a certain sense, the development of that mental world is a continual flight from "wonder."—I experienced a wonder of just that kind as a child of 4 or 5, when my father showed me a compass.[18]

Thus in what Einstein facetiously called his "*Nekrolog*," his "Obituary"—published as *Autobiographisches*—he recalls an experience which he frequently related and which is recorded in several (basically agreeing) versions. He was sick in bed, when, no doubt to divert him, his father brought him a compass—not suspecting the lasting impression this instrument would make:

> The fact that the needle behaved in such a definite manner did not fit at all into the pattern of occurrences which had established itself in my subconscious conceptual world (effects being connected with "contact"). I remember to this day—or think I remember—the deep and lasting impression this experience made on me. There had to be something behind the objects, something that was hidden.[19]

Although the subject matter of Einstein's great accomplishment—the essay *Zur Elektrodynamik bewegter Körper* (*On the Electrodynamics of Moving Bodies*) of 1905, which contains the special theory of relativity—seems to be foreshadowed here, one should probably not read too much into this experience. A lot of children wonder about a rainbow, and some no doubt will have wondered about a compass needle, which seems to be moved by an invisible hand. A prism diffracting light or an apple dropping from a tree may evoke wonderment and clever questions. Altogether, as Sigmund Freud observed, the intelligence of adults pales against the brilliant intelligence of five-year-olds. Still, among all these children only one became an Isaac Newton and only one an Albert Einstein.

Einstein himself was unable to explain this powerful experience, because "a person has little insight into what goes on inside him. Seeing a compass for the first time may not produce a similar effect on a young dog, nor indeed on many a child. What then is it that determines a particular reaction from an individual? More or less plausible theories may be constructed about it, but one does not arrive at a deeper insight."[20] We will have to content ourselves with the suggestion that a productive result probably depends both on the "wonder" and on the person "wondering."

CHAPTER TWO

School

When Albert Einstein reached the statutory school age, six, his parents were spared the problem of choosing a school. The only Jewish private school in Munich had been closed in 1872 for lack of pupils,[1] a clear indication of the readiness of its Jews to assimilate. (One in fifty of Munich's population was Jewish, and this proportion had remained fairly constant during the city's growth over the last two decades of the nineteenth century. In the city center it was slightly higher, and in suburbs like Sendling it was distinctly lower.) In the absence of any alternatives, therefore, beginning on October 1, 1885, Albert attended the nearest school, the Petersschule on Blumenstrasse, a big Catholic elementary school with more than two thousand students from all strata of the population. At a brisk walking pace down Lindwurmstrasse it could be reached in about twenty minutes. Albert was accepted into the second grade: his private tuition, therefore, despite its disastrous end, cannot have been entirely in vain.

Albert was the only Jew among some seventy classmates. He participated in the Catholic religious studies and was in fact particularly liked by the teacher.[2] "The teaching staff in the elementary school were liberal and made no difference between denominations."[3] Such an attitude was a result of both the humanitarian educational reforms of the time and the progressive views of a large part of the Munich bourgeoisie.

Nevertheless, that same teacher of religious studies made Einstein clearly realize that among all those good Christians he must feel an outsider: "One day that teacher brought a long nail to the lesson and told the students that with just such nails Christ had been nailed to the Cross by the Jews."[4] This macabre method of teaching the Gospel was an indication that even liberal teachers were, as Christians, not free from an innate, if mild, anti-Semitism. Among the students this led to more outspoken aggression, as Einstein recollected: "Among the children at the elementary school anti-Semitism was prevalent. It was based on racial characteristics of which the children were strangely aware and on impressions from religious teaching. Physical attacks and insults on the way home from school were frequent, but for the most part not too vicious. But they were sufficient to consolidate even in a child a lively sense of being an outsider."[5] At the same time, though, we have no evidence that Einstein ever suffered from his "sense of being an outsider," either as a child or in later years. "Being a stranger" and "belonging" were both probably his most important personality traits from his earliest years.

Even in elementary school, therefore, Einstein never stepped out of his characteristic isolation. He rarely played with coevals, not even with the children who had meantime been born to Uncle Jakob or with his boy and girl cousins, who frequently came to visit. In his deliberate but usually reserved manner he got on reasonably well with them, but they gave him the nickname "Goody Goody." What probably helped to earn him that reputation was the fact that he always had to finish his homework before being allowed to play: "No excuse was accepted by our parents for any infringement of that rule."[6]

Success followed, for on August 1, 1886, at the end of Albert's first year at school in second grade, his mother wrote to her sister Fanny Einstein: "Albert got his grades yesterday, he was again top of the class, he brought home a brilliant report . . ."[7] Admittedly, he had an ingrained dislike of physical training and games, "as he easily got vertigo and got tired quickly."[8] Yet he did not get tired at all when he was engrossed with his beloved metal construction set, or with involved fretsaw work, or with manipulating a small, hissing steam engine which Uncle Caesar Koch had brought him as a present.

■ ■ ■

Albert Einstein the schoolboy would thus have appeared to his parents and teachers as a well-behaved child who had learned to submit to the inevitabilities of routine and obedience, as demanded by school and the world of adults. But behind that facade of adjustment there was still a determination to preserve his individuality, though this now manifested itself in a more sublimated, socially acceptable form: a dreamerlike, skeptical distancing from other people and things.

Now and again, however, his dislike of any coercion burst through the facade of the well-adjusted young man. Thus in November 1886 he was moved from class IIIa to IIIb, probably because of disciplinary problems in the wake of one of his last outbursts of anger.[9] Unusual also, for a boy, was the fact that he wanted neither to become a soldier nor to play with toy soldiers. In southern German states like Württemberg and Bavaria, the army did not quite enjoy the same overriding prestige as in Prussia; nevertheless, even in Munich young boys could see nothing more wonderful than the hope of one day wearing a uniform and serving king, emperor, and fatherland—as Uncle Jakob had served in the war with France—and most children were fascinated by military pomp. Albert Einstein, by contrast, displayed a definite dislike. On one occasion, when he was watching a parade and was told that someday he might march along with those men in uniform, he said to his parents: "When I grow up I don't want to be one of those poor people."[10]

Despite many positive features, schooling was pervaded by military drill and the principle of absolute obedience. There was an exaggerated emphasis on order and discipline. It seems that this was felt by the young Albert Einstein. At the age of eight or nine he was not ready to make any explicit criticism; in retrospect, however, he regarded his Munich schooldays with a mixture of anger and contempt: "The teachers at the elementary school seemed to me like drill sergeants, and the teachers at the *Gymnasium* [the high school] like lieutenants."[11] At the age of nine and a half he completed his three years of elementary school and moved over to the "lieutenants."

The Luitpold Gymnasium in Munich, where Albert Einstein was accepted on October 1, 1888, was by no means one of the worst institutions. It was not far from his old elementary school, on Müllerstrasse, the street where he had lived before the family moved to Sendling. Although here, too, Latin and Greek were at the center of education in the humanities, the school—under its principal, Dr. Wolfgang Markwälder—had gained a reputation as an enlightened, liberal institution, where mathematics and the sciences also had their place, however modest. At any rate, it evidently enjoyed great respect among parents: its steadily growing number of students cannot have been due solely to an increasing interest in education. When Einstein arrived, there were 684 students; by 1894, when he severed his (not altogether easy) relations with the school, it had increased to 1,330. Most of the students were Catholics, but five percent were of Jewish origin—two and a half times more than one might have expected on purely statistical grounds. Classes, at least the lower ones, were very crowded: a class photograph of Einstein's first year shows him with fifty classmates. Other than himself, only two were Jews.

The old story that Einstein was a bad pupil, or even failed altogether at school, has been repeated time and again, presumably to console poor students or their parents—though low marks at school are no guarantee of success in later life. In fact, the story is not true: at most it reflects the hope of parents that even the dumbest student might have a brilliant mind. Still, as early as the 1920s, Einstein was cited as a shining example of this thesis. As with so many other fanciful stories about him, he probably let it pass with a smile. Not so, however, a certain Dr. Wieleitner, the principal of the Neues Realgymnasium, the successor of Einstein's school.

When, in 1929, on the occasion of Einstein's fiftieth birthday, articles appeared in various journals, mentioning—if anything, with approval—Einstein's "total weakness in the ancient languages,"[12] Dr. Wieleitner was evidently worried that the boy's allegedly poor grades might damage the school's reputation more than that of the man who had meanwhile become a famous physicist. He therefore searched the school records and, in a letter to the editor of a Munich paper, saved the honor of the Luitpold Gymnasium by pointing out that Einstein had "always [received] at least a 2 in Latin, and in the sixth grade even a 1. In Greek he always had a 2 in his school reports. . . . Even in the 'secret reports' there is no complaint anywhere of a poor gift for the languages."[13] As the school records were destroyed in a bombing raid during the Second World War, the doughty principal's letter to the editor remains the only evidence that Albert Einstein was a good student in high school.

Despite his good reports, Albert Einstein would later remember his time at school as an almost traumatic experience.[14] While still a student he suffered in silence; he never voiced any criticism of the school and, presumably, without any comparable experience, was unaware of its shortcomings at the time. "According to the family, the taciturn child scarcely complained, nor did he seem too unhappy. Only much later did he identify the tone and atmosphere of the high school with those of the barrack-square, which in his eyes were the negation of everything human."[15] Even at the age of forty he told his first biographer that at the Gymnasium, "though he grew fond of some individual teachers, he felt himself harshly touched by the spirit of the institution."[16]

However, there exists some rather different testimony about the Luitpold Gymnasium. Thus Abraham Fraenkel, who attended it a short time after Einstein and later became a famous mathematician, has entirely pleasant memories—even though, as a practicing Orthodox Jew, he would have been far more of an outsider than Einstein. Indeed, Fraenkel referred to "nine happy years"[17] spent there. Einstein's experience cannot, therefore, have been the fault of the school alone; there must also have been powerful reasons within Einstein himself which prevented harmonious integration.

A glance at the syllabi[18] shows that they were not exactly to the taste of Albert Einstein's "intellectual stomach."[19] Eight hours of Latin a week—in some grades even ten hours—plus six hours of Greek from the fourth year on, did not leave much room for other subjects. This was not a favorable situation for someone who admitted: "My principal weakness was a bad memory, especially a bad memory for words and texts."[20] There were only three periods for German, as well as, in the upper grades, for French. Mathematics was taught only three or four times a week; geography and science were taught only twice a week. Physics appeared only in the seventh year; but by then this was no longer of any interest to Einstein, because "in mathematics and physics I was, through private study, far above the school requirements, also with regard to philosophy in so far as this has anything to do with the school program."[21] Thanks to his private study and the self-assurance this had given him, he eventually found it easy, despite good reports, to leave school early.

Conflicts also marked the development of young Albert Einstein's religious sentiments and beliefs. In his elementary school he had participated in the Catholic religious lessons, while at the same time being instructed in the Jewish tradition "at home by a distant relation,"[22] who was better versed in these matters than Albert's freethinker father—from whom Albert heard only "ironical and unfriendly remarks about dogmatic rituals."[23] At the Luitpold Gymnasium, unlike the elementary school, there were a few Jewish classmates, for whom a liberal school management provided their own Jewish religious instruction through Oberlehrer (senior teacher) Heinrich Friedmann. Friedmann's exegesis of the Prophets initially found a very receptive and grateful listener in young Albert. Like so many adolescents in search of a meaning to human existence, Albert Einstein

> acutely realized the vanity of hoping and striving, that drove most people restlessly through life. Everyone was condemned, by the existence of his stomach, to participate in this race. The stomach might well be satisfied by such participation, but not man as a thinking and sentient being. There the first way out is religion.[24]

He keenly studied the preacher Solomon, strictly adhered to ritual precepts, and in consequence no longer ate pork.[25] He even composed a few short hymns to the greater glory of God, which he sang with great fervor at home and as he was walking in the street.[26] Under the direction of Oberlehrer Friedmann and a rabbi he prepared to become a bar mitzvah, to be solemnly accepted into the Jewish community as a full member on the Sabbath following his thirteenth birthday. The reason this never took place was his encounter with the natural sciences.

One of the few Jewish customs still observed in the Einstein home was inviting a poor Talmudic scholar to lunch on the Sabbath. For the Einsteins, admittedly, the Sabbath had become Thursday, and the poor student did not want to become a rabbi: he was a medical student. Still, his last name was Talmud. Max Talmud was twenty-one when his weekly visits to the Einsteins in Sendling began, in 1889. For Albert, eleven years his junior, he seems to have been something like a substitute father in a spiritual or intellectual sense, or at least a substitute uncle. Max Talmud would bring popular science books for his hosts' son. In many respectable families such books would not have been considered suitable reading matter for a youngster—they presented a scientific, materialist picture of the world that reeked suspiciously of atheism and revolutionary attitudes. As no such misgivings existed in the Einstein household, the boy could engross himself, undisturbed, in Büchner's *Kraft und Stoff* (*Force and Matter*), which presented the philosophy of the French materialists to the German public in a somewhat diluted form. With great zeal he also studied the twenty volumes of Aaron Bernstein's *Naturwissenschaftliche Volksbücher* (*Science for the People* series); Bernstein was an author of educational works in a spirit of enlightenment, much read among emancipated Reform Jews. In addition, Albert browsed in Alexander von Humboldt's five-volume classic, *Kosmos—Entwurf einer physische Weltbeschreibung* (*The Cosmos—Attempt at a Description of the Physical World*), and read something by, or at least about, Charles Darwin.

These books soon convinced the boy "that a lot in the Bible stories could not be true. The result was downright fanatical freethinking, combined with the impression that young people were being deliberately lied to by the state: it was a shattering discovery."[27]

Albert Einstein therefore did not become a bar mitzvah, and by rabbinical standards was not a proper member of the Jewish community. His parents probably did not mind their son's freethinking any more than they had minded his earlier religious fervor. Religious disenchantment now also released what had been hidden for a few years under the mask of the well-adjusted schoolboy:

> From this experience grew a mistrust of any kind of authority, a skeptical approach to the convictions which were current in whatever social environment I found myself—an attitude which never left me, even though, with better insight into causal connections, it subsequently lost something of its original edge.[28]

For his expulsion from the "religious paradise of [his] youth"[29] Albert Einstein found more than compensation in mathematics. While he was still in elementary school, Uncle Jakob, the engineer, had introduced him to algebra, which was still far beyond the school curriculum and was yet in a sense appropriate to Albert's age, as "the art of lazy calculation. What one does not know one simply calls *x* and treats it as if the context were known; one writes that context down and determines *x* afterwards."[30] On another occasion Uncle Jakob drew his nephew's attention to Pythagoras' theorem. The boy felt that there was a need to prove it, "spent three weeks in strenuous reflection,"[31] and without help from anyone found a proof which his sister, fondly overestimating him, even claimed had been accomplished "in an entirely new way." Needless to say, Albert's proof, based on similar right-angle triangles, was new only to him, but he had worked it out for himself. Far more astonishing than even this achievement was surely his independent discovery that this was something that needed proving at all, and that it could be proved.

These mathematical preludes were the right tune-up for other intellectual stimulation, also coming from Max Talmud. Before Albert started his fourth year at the Gymnasium at age twelve, Talmud had brought him a textbook of plane geometry:[32] possibly Theodor Spieker's *Lehrbuch der ebenen Geometrie*.[33] It would be another two years before he came to this subject at school, and then with a much simpler textbook. Albert began to work his way through the book independently, and each Thursday he would show his mentor, Max Talmud, the problems he had solved during the week. The boy's astonishingly rapid progress left an indelible impression on Talmud: "After a short time, a few months, he had worked through the whole book of Spieker. He thereupon devoted himself to higher mathematics, studying all by himself Lübsen's excellent works on the subject.[34] These, too, I had recommended to him if memory serves me right. Soon the flight of his mathematical genius was so high that I could no longer follow."[35]

While Max Talmud was amazed mainly by the breathless speed with which his young friend absorbed his scientific reading, Einstein's own recollection, especially of his acquaintance with the "sacred little geometry book,"[36] had more to do with depth, with the mysterious regions of "wondering" that he had first experienced with the compass needle:

> At the age of twelve I experienced a second wonder of a totally different kind: over a little book of Euclidian geometry which came to my hands at the beginning of the school year. There were statements in it, such as for instance the intersection of the three altitudes of a triangle at a single point, which—though by no means self-evident—could be proved with such certainty that any doubt was ruled out. This clarity and certainty made an indescribable impression on me.[37]

Albert Einstein shared this awakening of an overwhelming love of geometry with other great intellects. Galileo Galilei at age seventeen made the chance acquaintance of this branch of mathematics, instantly dropped his medical studies, and from then on read nothing but Euclid. Bertrand Russell wrote about his studies of geometry, on which he embarked at age eleven under the guidance of his brother, as "one of the great events of my life, as dazzling as first love. I had not imagined that there was anything so delicious."[38] What impressed all three young enthusiasts was not so much the richness of geometry as the certainty and beauty of the axiomatic-deductive method described in canonical form by Euclid around 300 B.C.

Unlike eleven-year-old Bertrand Russell, twelve-year-old Albert Einstein was not bothered by the fact that Euclidian geometry rests on a foundation which cannot itself be questioned. The basic statements, known as "axioms," are unproven and unprovable and must simply be accepted. These axioms had been regarded as self-evident for more than two thousand years, so that the direct application of that geometry to real objects was considered the most natural thing in the world. Only in the course of developing his general theory of relativity did Einstein realize that the relationship between geometry and reality can be more complex and far-reaching—and he succeeded in elucidating this only after great effort and within a framework of non-Euclidian geometry. When he first encountered classical geometry, he thought it "sufficiently wonderful that man is able at all to attain such a degree of certainty and purity in thought, as the Greeks first demonstrated in geometry."[39]

These grand words—in which Albert Einstein in his old age recalled the "sacred little geometry book"—suggest that to him the encounter with mathematics was far more than a purely intellectual delight. He later interpreted his religious fervor as a boy's first attempt to free himself from "the fetters of the merely personal," from an existence dominated by desires, hopes, and primitive emotions."[40] After the disappointment of his experiment with traditional religion, he had found in mathematics another road to the same destination, one to which he could surrender himself with the same emotional intensity.

Through private study Einstein thus acquired the principal areas of higher mathematics, from analytical geometry through infinite series to differential and integral calculus. He found this occupation "truly fascinating: it contained high-points whose effect measured up entirely to that of elementary geometry."[41]

The fact that most people around him—Uncle Jakob was one of the few exceptions—were poor mathematicians and that most of his classmates and teachers tended to regard ignorance of mathematics as a virtue probably confirmed him in the belief that he had made the right choice for himself.

Although many of the other books given to him by Max Talmud made a deep impression on Einstein—especially by reinforcing his "fanatical freethinking"—their impact could not be compared to that of the "sacred little geometry book." He read Aaron Bernstein's *Naturwissenschaftliche Volksbücher* "with breathless suspense"[42] but found fault with it because the "presentation [was] almost entirely confined to the qualitative aspect." No doubt the theory of biological evolution and the wealth of the "description of the physical world," as presented by Alexander von Humboldt in his *Kosmos*, must have seemed interesting to young Albert, but here, too, the presentation was inevitably restricted to verbal argumentation. The "certainty and purity of thought" was less marked in this wealth of detail than it was in mathematics.

As Max Talmud was soon unable to follow Einstein's soaring mathematical flights, their conversations increasingly turned to philosophical problems. At the age of thirteen, with Talmud's recommendation and guidance, the boy studied Immanuel Kant's *Critique of Pure Reason*.[43] Talmud is probably correct in characterizing this work in his memoirs as "incomprehensible to ordinary mortals," and his statement that Kant's philosophy was instantly clear to the young Einstein is probably a glorified recollection. It seems likely, however, that Kant's attempt at a strict formulation of the "conditions of the possibility of cognition" generally was perceived by Einstein as the same striving for "certainty and purity" by which he himself was motivated.

Because of his feeling that others should share in this experience there emerged, in Einstein, for the first time, an inclination to preach to a small public if he could find one. One classmate recalled conversations of "such forcefulness that even today, after a good thirty years, the actual words come back to me whenever I happen to be in the neighborhood where we then strolled, you instructing, I receiving. At that time you were beginning to study the *Critique of Pure Reason*, of which I got quite a lot."[44] With all this seriousness, one is almost relieved to learn that the boy was also capable of behavior in line with his age. Alfred Einstein, who was two years younger than Albert and not related to him, and who became renowned as a musicologist and a reviewer for the *Berliner Tageblatt*, once reminded the famous scientist of their "old connection from 1894 or 1895, at the Luitpold Gymnasium, where, in our joint singing lessons, you were fond of pulling your younger namesake's hair."[45]

It was at about this time that Albert Einstein discovered his second great love, after mathematics—music. At the Einstein home "there was much & good music-making."[46] Hermann was not particularly keen on music; but Pauline, an excellent pianist, hoped to find a musical partner in her son. When he was six, she engaged a Herr Schmied as a violin teacher. But the technical practice and the boring études seemed to Albert to be merely a continuation of school drill, and no progress was made. Other teachers were engaged and dismissed: Einstein later believed that he had had no luck with them because for them "music did not go beyond the mechanical aspect."[47]

In a way which was beginning to be typical, Einstein's love of music awakened only when he himself became interested in certain pieces and replaced his lessons with self-teaching:

> I only began to learn something after I was thirteen, when I had fallen in love mainly with Mozart's sonatas. My wish to reproduce them to some degree in their artistic content and their unique gracefulness forced me to improve my technique; this I acquired with those sonatas, without ever practicing systematically. I believe altogether that love is a better teacher than a sense of duty—at least for me.[48]

Thus Pauline saw her hopes of playing duets with her son fulfilled. Their repertoire consisted primarily of Mozart and Beethoven sonatas for piano and violin: the mother probably preferring Beethoven and the son, quite certainly, Mozart.

Albert Einstein had grown up to be a strikingly handsome young man, with slightly wavy dark hair; full, sensuous lips, modified by the downturned corners of his mouth into something like skepticism; large dark-brown eyes; and a challengingly self-assured but often dreamy gaze. "In all these years," Max Talmud recalls, "I never saw him reading any light literature. Nor did I ever see him in the company of schoolmates or other boys of his age."[49] Even at age fourteen Einstein showed the beginning, at least, of those traits which made him describe himself as a "loner, who never belonged with his whole heart to the state, his country, his circle of friends, or even his closer family, but who felt with regard to all those ties a never overcome sense of being a stranger with a need for solitude."[50]

Such a young man would not always be popular, especially with people in authority, such as teachers at a German gymnasium. As mentioned above, he saw them as "lieutenants." But there were a few exceptions, like Dr. Ferdinand Ruess, who in Albert's fourth and sixth years was his ordinarius, or homeroom teacher. Dr. Ruess taught not only Latin and Greek but also history and German, and he "filled his students with great enthusiasm for the beauty of the classical world. . . . The boy's fondness [for] this teacher, who alone was able to satisfy his students' intellectual hunger, was so great that even being kept in after school, under his supervision, was a pleasure."[51] Einstein had such fond memories of Dr. Ruess that, as a man of thirty and a newly appointed professor, he paid Ruess a visit in Munich—presumably Einstein was en route from Bern to Salzburg, where in September 1909 he participated in his first physicists' conference. However, Ruess did not recognize his former student and was even afraid that this shabbily dressed man might want to borrow money from him. After a painfully long moment of embarrassment, Einstein had no choice but to leave in a hurry.[52]

It probably never occurred to Einstein to visit any of his other teachers. At sixty, he recollected that he had detested the "mindless and mechanical method of teaching, which, because of my poor memory for words, caused me great difficulties, which it seemed to me pointless to overcome. I would rather let all kinds of punishment descend upon me than learn to rattle something off by heart."[53] This recollection, however, seems hardly compatible with Dr. Wieleitner's report about his good to very good marks in the classics. Of course, all his teachers with the exception of Dr. Ruess might have been monsters—but that does not seem very likely in light of Abraham Fraenkel's testimony.

The conflict between Einstein and his schoolteachers was no doubt partly due to him. In his seventh year it had reached a point where his new ordinarius, Dr. Joseph Degenhart, informed him that "he would never get anywhere in life."[54] To Einstein's remark that surely he "had not committed any offence," he replied: "Your mere presence here undermines the class's respect for me."[55] Both sides must have wished to terminate this disagreeable relationship. That Albert Einstein was soon able to do so was due to a change in his parents' financial situation.

From its establishment in 1885, Electrotechnische Fabrik J. Einstein & Cie. speedily prospered. Within a year it had gained local renown by illuminating the Munich Oktoberfest on Theresenwiese by electricity for the first time.[56] Orders kept coming in for the installation of electric streetlights in Schwabing, a suburb which was then not yet part of Munich; and in the northern Italian cities of Varese and Susa. Favorable reports on Einstein & Cie. appeared in *Centralblatt für Elektrotechnik* and in *Elettricità*.[57] In retrospect, the Einstein brothers' high-tech firm looks like something that could well have developed into a giant of the electronics industry, or at least into a sound large-scale enterprise, successful both commercially and technologically.

The innovative head of the firm was Jakob Einstein, the engineer. Altogether, he held six patents. Three of them were for arc lamps (which were still in common use); two of these provided for an improved method of advancing the carbon electrode and the third was for an automatic circuit-breaker. The other three patents were for electric meters capable of measuring ampere-hours or watt-hours—a vital prerequisite for an electrified economy.[58]

In addition to lamps and electric meters, the Einstein firm manufactured dynamos of various sizes, from small workshop models to power-plant generators, along with cables and all the equipment needed for urban electrification, as well as electrolysis plants and gauges. The firm was important enough to be noticed at the spectacular International Show in Frankfurt in 1891, as well as at a symposium on urban lighting and power transmission which preceded the show. Twenty-one firms, including one from America, had been invited to present their concepts of an electrified future. Einstein & Cie. and Ingenieurbüro Oskar von Miller were the only firms from Munich.[59]

At its peak the firm had just under two hundred employees.[60] Its turnover and profits must have been considerable, and although, mindful of its persistent undercapitalization, the brothers withdrew only modest sums for themselves, these were enough to ensure a comfortable existence for both families.

In 1893, however, the fortunes of the firm changed dramatically. The Einsteins had directed all their efforts toward winning the contract for lighting the Munich city center, which would have kept the factory fully employed for several years.[61] After tough and bitter competition with Germany's three biggest electrical-engineering firms—Siemens and AEG from Berlin, and Schuckert from Nuremberg—the contract went to Schuckert in April 1893.

The modest volume of business left in Germany was not enough to cover the Einsteins' high overhead, and prospects for the future were represented only by a number of lesser projects in Italy. In March 1894 the two brothers, with their Italian representative as a partner, therefore founded the firm of Einstein, Garrone e C. in Pavia, in northern Italy; in July they liquidated their firm in Munich.

Leaving Munich was painful, especially for the children, who "right up to the moment of moving had to watch from their windows the destruction of their fondest memories."[62] An architect and a building contractor took possession of the fine properties on Adlzreiterstrasse, cut down the splendid old trees, and began to construct four-story residential blocks.[63] In the summer of 1894 the Einsteins moved to Milan, and the following year they went twenty miles farther south, to Pavia, where the new factory was built. Maja, the daughter, went to Italy with her parents. Albert was left in Munich under the care of some distant relations, because he was supposed to finish school there with the *Abitur*, the German high school graduation examination.

The liquidation of a firm in economic difficulties is always a sad affair, and its effects on the family were inevitable. For Albert, then fifteen, it must have been painful to find that the comfortable security of bourgeois life in Munich was now over for his family. Although he never referred to this upheaval at a formative time of his life, we may assume that his biting remarks on the vanity of restless striving, to which everyone was "condemned to participate by the existence of his stomach,"[64] had their origin then.

We may also assume that subliminal anti-Semitic sentiment had played a part in the contract for the Munich city center, since it was awarded to an outside firm rather than to a Munich firm—Munich's only major manufacturer of dynamos—which happened to be Jewish. Whatever caused the Einsteins to lose this crucial business, they must have been plagued by a feeling that the city's economic and social leaders regarded them as upstarts who should be cut down to the petty trade appropriate to Jews.

It is possible that in the young Albert Einstein, who had to watch helplessly as his father's firm folded, the conviction was then gaining ground that German society as a whole had robbed his family of its livelihood. His very reserved attitude to his native country, even before 1933, may well have begun to develop then. His negative, distorted recollections of the Luitpold Gymnasium and his own decisions when the firm closed—soon to be followed by his renunciation of German citizenship—may, at least in part, go back to the profound traumas of the year 1894.

No doubt the three years which, according to his parents' plans, he was to spend in Munich on his own must have seemed to him like an eternity. Moreover, he cannot have wanted to finish school—even though he hated it—because the gymnasium would then have been followed by the real barrack square, among people with whom, even as a small boy, he had not wanted to march in step. In this situation, the request by his ordinarius, Dr. Degenhart, to do him the favor of leaving the school must have looked like the benign hand of fate. For once, Einstein was ready to please his teachers. However, he was not to be provoked into a rash decision: instead, he proceeded with circumspection in order to limit the damage as far as possible.

First of all, he got a doctor—an elder brother of Max Talmud's—to give him a medical certificate attesting that he was suffering from "neurasthenic exhaustion" and demanding a suspension of his schooling. Next he persuaded his mathematics teacher, Joseph Ducrue, to confirm to him in writing that he had mastered mathematics up to *Abitur* level and that he was altogether quite an excellent mathematician.[65] Finally, on the strength of the medical certificate, he applied for release from his school. When these formalities were all completed, before the Christmas vacations,[66] he went straight to Munich's central railway station and the next day faced his startled parents in Milan—a young man of fifteen, his schooling cut short, with no plans or prospects for the future, but happy to have escaped the "lieutenants."

CHAPTER THREE

A "Child Prodigy"

MILAN AT THE TURN OF THE YEAR can be every bit as rainy and gloomy as any city north of the Alps. But for Albert Einstein the skies were brighter than they had been for him in Munich for a long time. Perhaps in Dr. Ruess's lessons he had learned of the traditional German longing for the south and now found himself in the land of his dreams, or perhaps he was simply happy to be with his family once more. Certainly he must have been glad to have escaped the restrictions of his high-school life in Munich and to have left behind many things he regarded as typically German. His parents, of course, were horrified at their son's decision: their hopes that he might graduate from school, move on to a university, and thus acquire status and reputation now seemed jeopardized. Young Albert, however, steadfastly declared that he never wanted to return to Munich. It seems probable that, in this situation, the family council was persuaded by Uncle Jakob's suggestion that the fugitive schoolboy be sent to the Eidgenössisches Polytechnikum (Federal Swiss Polytechnic) in Zurich, an advanced technological institution which did not insist on high-school graduation as a condition for admission.

Albert helped assuage his parents' misgivings by "assuring them most resolutely that, by the fall of that year, he would have prepared himself by private study for the Polytechnic's entrance examination."[1] His parents may also have been somewhat appeased by the unofficial testimonial he had brought with him from Munich about his exceptional knowledge and ability in mathematics. He even went one step further: in a university bookstore in Milan he purchased the first three volumes of the German edition of Jules Violle's demanding *Lehrbuch der Physik*.

Einstein's notes and glosses in the extant copies of Violle's physics texts show that he was entirely serious about the promises he was making to his parents. His method of private study, however, caused some astonishment:

> His working method was rather strange: even in . . . company, when there was quite a lot of noise, he could retire to the sofa, pick up pen and paper, precariously balance the inkwell on the backrest, and engross himself in a problem to such an extent that the many-voiced conversation stimulated rather than disturbed him.[2]

The sofa on which Albert Einstein balanced his inkwell originally stood in a large apartment on Via Berchet 2 in Milan, which was also the business address of Einstein, Garrone e C. In addition, offices had been established in Pavia and Turin. In view of their reasonable hope that the firm would be commissioned to set up a hydroelectric plant, along with the appropriate transmission lines and electric street lighting in Pavia, the Einsteins in the spring of 1895 moved to Pavia—an old city on the lower Ticino, just before it runs into the Po.

The two families moved into separate apartments; Hermann Einstein and his family occupied a floor with three reception rooms in a magnificent ancient building at Via Foscolo 11, formerly the house of the poet Ugo Foscolo, for whom the street was named.[3] A by no means small factory was built on the bank of the Naviglio di Pavia, a canal connecting Pavia with Milan. In addition to the money the Einsteins had saved from the liquidation in Munich and an investment by Signor Garrone, considerable sums on credit came from a cousin, Rudolf Einstein, the husband of Pauline's sister Fanny, whose affluence came from a textile mill at Hechingen in Württemberg. As the firm acquired a reputation in Italy, several of its craftsmen and technicians arrived from Munich to work with the Einsteins again.

For Albert Einstein, who until then had known only Munich and its immediate neighborhood, his flight to Italy—if we may so describe it—was his first major journey. His experience of the southern landscape, of a different culture and lifestyle, made an unforgettable impression on him, as on many Germans before him: "I was so surprised"—he would say four decades later—"when I crossed the Alps to Italy to see how the ordinary Italian, the ordinary man and woman, uses words and expressions of a high level of thought and cultural content, so different from the ordinary German. This is due to their long cultural history. . . . The people of northern Italy are the most civilized people I have ever met."[4]

This educational experience, however, was confined to northern Italy. Trips to Florence and Rome, or farther south, were not possible, and we do not even know if he would have greatly wished to undertake them. His only longer journey, in the early summer of 1895, took him to Genoa, to see his uncle Jakob Koch, a brother of his mother's. He traveled the first twelve miles by local train, via Casteggio to Voghera, and then crossed the Ligurian Alps to Genoa on foot, a hike of some sixty miles, which took several days.[5] At the height of summer he spent a vacation with his family at Airolo on the St. Gotthard pass.

As for his new hometown of Pavia, he described it in rather rude terms in a letter to a girlfriend in Switzerland: "The city's soul could be defined in mathematical terms as roughly (1) the sum total of the ramrods the various gentlemen & ladies have swallowed, (2) the mood created in the observer by the uniformly filthy walls & streets everywhere. The only beautiful aspects are the delightful, graceful little children."[6]

Nevertheless the land and its people, its culture, and its language left an indelible impression on him. When, two decades later, he corresponded with the mathematician Tullio Levi-Civita about the general theory of relativity, he asked Levi-Civita to "write me in Italian next time." Einstein profusely thanked him for the next letter "written in the familiar long-missed Italian. You can hardly imagine the pleasure it gives me to receive such a genuine Italian letter. It revives in me the most beautiful memories of my youth."[7] He could not rally enough courage to reply to the mathematician in Italian, "because that would come out rather too bumpily." But in his old age he tried his skill in that beautiful language when writing to a "Cara Ernestina," a friend of his sister's in their younger years: "The happy months of my sojourn in Italy are my most beautiful memories. . . . Days and weeks without anxiety and without worries."[8]

Along with his main occupation—preparing for the Zurich Polytechnic—Einstein seems to have done various jobs in the factory and occasionally even to have helped in Uncle Jakob's design office. "You know, this is quite fantastic about my nephew," Jakob Einstein once said to one of his assistants, "where I and my assistant engineer have racked our brains for days, this young fellow comes along and solves the whole business in a mere quarter-hour. He'll go far one day."[9] The chronicler unfortunately does not relate the nature of the "business."

As a spin-off of his private study Albert Einstein during the summer months of 1895 also wrote what is somewhat grandly described as his first physical essay, *Über die Untersuchung des Aetherzustandes im magnetischen Feld* (*Examination of the State of the Ether in the Magnetic Field*).[10] This was probably intended as a self-test and a warm-up for his entrance examination in Zurich, and perhaps also as evidence of his studies for the family. At any rate, he sent this first essay, together with a covering letter, to his favorite uncle, Caesar Koch, in Brussels, though he could hardly expect that grain merchant to understand it: "It deals with a rather specialized subject and moreover, as is natural for a young fellow like me, it is still somewhat naive and imperfect. If you don't read the stuff at all, I won't blame you in the least."[11]

The "stuff," five pages in the neat gothic handwriting he had learned at the gymnasium, is an examination of the relationships among electricity, magnetism, and the ether—that nonmaterial substance which was then being postulated as filling all space and being the medium in which electromagnetic waves, discovered in 1888 by Heinrich Hertz, were thought to propagate themselves. The author announced his essay as "the first modest utterance of a few simple thoughts on this difficult subject," more of "a program than a dissertation."[12]

Albert Einstein argued entirely on the lines of the ether theory of his day, according to which the propagation of waves was understood as analogous to the mechanical theory of waves—which he had come across in Violle's textbook. He must have heard something about "Hertz's wonderful experiments," and on that basis he now developed his ideas for "measuring the elastic deformations [occurring in the ether] and the acting forces." He certainly, by this rather strange route, derived the phenomenon of "self-induction"—purely qualitatively—without, however, using that term.

Some enthusiasts see this first essay as a "harbinger of what was to come,"[13] but that is overinterpreting it. In the wake of Heinrich Hertz's epoch-making discovery a great many popular accounts of electromagnetic theory appeared in Germany, and Einstein would have read at least some of these. In fact, striking parallels have been found between passages in Einstein's text and an article, *Die Umwälzungen unserer Anschauungen vom Wesen der elektrischen Wirkungen* (*The Revolution in Our Concepts of the Nature of Electrical Effects*), in a popular-science monthly.[14]

As the fine Italian summer was drawing to a close, the entrance exam in Zurich inexorably approached. Meanwhile it had been discovered that this examination stipulated a minimum age of eighteen. A special exemption had therefore to be requested for young Albert, then only sixteen. To this end a friend of the family, Gustav Maier, a resident of Zurich, was approached. Like the Einsteins, Maier came from Ulm, where he had been branch manager of the local Deutsche Reichsbank. A successful career had led him via Frankfurt to Zurich, where he managed a bank and a department store, and where he was part of the liberal, freethinking elite.

Maier must have recommended Albert Einstein to the principal of the Polytechnic as a "child prodigy" deserving of special provisions. But the principal, Professor Albin Herzog, in his (still extant) reply was against "taking even so-called 'child prodigies' away from the institution in which they began their studies before these studies are completed."[15] Only on condition that his "information on the talents and the mental maturity of the applicant were confirmed in full in writing by the principal of the institution concerned" would he waive the minimum-age rule for Einstein. It appears that Professor Herzog was satisfied with the unofficial testimonial of the mathematics teacher in Munich, who praised Albert Einstein's "mathematical knowledge and abilities," describing them as equivalent to graduation level.[16] Anyway, at the beginning of October, Albert traveled by train to Zurich, where the Maier family put him up. "With a sense of well-founded diffidence"[17] he reported for the exam, choosing the engineering department on the strength of his father's and uncle's area of interest.

The examination began on October 8 and probably extended over several days.[18] It covered many subjects, some general and others specific to the candidate's chosen field of study. For Albert Einstein it was a disappointing experience, "for it made me realize painfully the gappy character of my previous schooling, even though the examiners were patient and understanding. That they failed me seemed to me entirely just."[19] The negative outcome was due mainly to the verbal descriptive subjects; otherwise, the examiners' suspicion that they were dealing with a "child prodigy" was confirmed. Professor Heinrich Friedrich Weber was so impressed by the mathematical and physical knowledge of the young candidate that he invited Einstein—against all regulations—to attend his physics lectures for second-year students, provided he remained in Zurich. Einstein, however, followed the sensible advice of the principal that he should spend a year at the cantonal school in Aarau in order to qualify for study at the Polytechnic.[20]

The cantonal school in Aarau—about thirty miles west of Zurich—had a great reputation as a liberal, forward-looking institution. While originally a classical gymnasium, it had been enlarged to include modern languages and science. It was famous for its "physical cabinet," which was extended, during Einstein's time there, into a superbly equipped laboratory with a dynamo, an AC motor, batteries, switchboards, and electrical metering instruments.[21] The principal of this remarkable school, to which Albert Einstein was admitted on October 26, 1895, was the physicist Dr. August Tuchschmid, formerly an assistant to Professor Weber at the Polytechnic.

Though Einstein was placed in the third year, his admission report noted "great gaps" in his knowledge of French, as well as a need to "catch up" on chemistry. A little later, a teachers' conference urged him to "take private coaching in French, natural science, and chemistry."[22] On the other hand, he was exempted from singing, from physical training, and—as a foreigner—from military instruction.

Although he was obliged to take additional instruction and was no longer receiving good grades in all his subjects, Albert Einstein's recollections of the cantonal school are very different from his memories of the Luitpold Gymnasium:

> By its liberal spirit and by the simple seriousness of its teachers, unsupported as they were by any outward authority, this school has left unforgettable impressions on me. Comparison with six years' schooling at a German authoritarian *Gymnasium*, made me clearly realize how much superior an education towards free action and personal responsibility is to one relying on outward authority and ambition. True democracy is no empty illusion.[23]

Of equal importance were Einstein's domestic circumstances in Aarau. Gustav Maier had arranged for him to be a paying guest in the home of Jost Winteler. What he found there was far more than bed and board—he found a second home, one which probably molded him more than his own family had. All his life Einstein remained close to the Wintelers.

Winteler taught Greek and Latin at the cantonal school, so Einstein was not one of his students. He had studied first in Zurich and then in Jena, Germany, where he obtained his doctorate with a linguistic study of the Kerenz dialect, the speech of his native Toggenburg region.[24] In Jena he had met his future wife, Pauline Eckart, who was later always called Rosa. With their seven children—three daughters and four sons—plus one or two paying guests, they must have seemed to Albert Einstein like a family idyll straight out of Swiss literature. Before long the two Wintelers were "Mama" and "Papa" to him.

Albert's cousin Robert Koch from Hechingen was living in the house next door. The same age as Albert, Robert was still in the second class at the gymnasium. Gustav Maier, who had arranged for both boys to be placed in Jost Winteler's care, had informed him that Albert

Einstein "is much more mature than his cousin and therefore less in need of supervision."[25]

Anna, the eldest of the Wintelers' daughters, recalls that Einstein was a "pleasant, very respectable member of the household, and never a spoilsport. He was fond of conducting scientific conversations, yet he had a great sense of humor and at times could laugh heartily. In the evenings he very rarely went out, he often worked, but more often he would sit with the family around the table, where something was being read aloud or discussed."[26]

One classmate claims to have realized even then that Albert Einstein did not fit "into any mold even as a young man,"[27] but that the "sharp wind of skepticism" at the cantonal school suited him. "The cheeky Swabian fitted quite well into that atmosphere, his original self-assurance setting him apart from the rest." This classmate painted a romantically exaggerated portrait of Einstein:

> The grey felt hat pushed back over the silky mass of dark hair, he strode along with vigor and assurance, at the rapid—I am tempted to say sweeping—tempo of the restless mind that carried a world within it. Nothing escaped the acute gaze of the large sun-bright eyes. Whoever approached him was captivated by his superior personality. A mocking trait around the fleshy mouth with its protruding lower lip did not encourage the Philistine to tangle with him. Unconfined by conventional restrictions, he confronted the world spirit as a laughing philosopher, and his witty sarcasm mercilessly castigated all vanity and artificiality.

That may sound like dubious, overliterary idealization; yet if one studies a group photograph of ten graduates taken in Aarau, one easily spots the cheeky, exotic type to whom the description would apply.

Albert Einstein's discovery of his own identity first of all entailed the surrender of two identities almost universally regarded as matters of course: citizenship and, albeit to a lesser degree, religious denomination. It is no longer possible to discover exactly when he decided that he no longer wished to be a German. Perhaps this happened in Munich, when he realized that it was the only way of avoiding military service, or perhaps it did not happen until later, in Aarau, under the influence of "Papa" Winteler.

The records[28] show that a "petition by the merchant Hermann Einstein in Pavia for the release of his son Albert Einstein from Württemberg citizenship" was granted on January 28, 1896. The reason for the application was given as "for the purpose of his emigration to Italy." The actual date on which Hermann Einstein had taken this step—as the legal representative of his son Albert, then still a minor—is unknown, but the motive is fairly certain: avoidance of military service. Conscription in the German Reich began only at age twenty; however, under the law then in force, a male applicant over seventeen would not be released from citizenship, to make sure he would be there when the army wanted him. Albert Einstein must therefore have been happy when the hoped-for document releasing him arrived at his parents' house in Pavia six weeks before his seventeenth birthday. In the register of persons released from Württemberg citizenship Albert Einstein's name appears in the column headed "trade and business assistants and factory workers."[29] In another column we find, for the first time, the entry "no religious denomination."

Initially, Albert Einstein's decision to renounce German nationality may have been emotional, but "Papa" Winteler must have supplied him with rational arguments. While still a student in Jena, Winteler had watched the rise of German nationalism, especially after the Franco-Prussian War of 1870–71. In his native Switzerland he missed no opportunity to warn against pan-German expansionism.

Winteler drew Einstein so intensively into his own political thought that in later years Einstein frequently recalled his mentor's amazing political farsightedness. In a letter to his sister Maja, written in 1933 during a summer vacation in Old Lyme, Connecticut, he said: "I am often reminded of Papa Winteler and of the prophetic correctness of his political views. I have always felt this, but not with this purity and intensity."[30] Jost Winteler died in 1929 and thus did not see the Nazis' seizure of power in Germany. But from Einstein's remark about the "prophetic correctness" of his views it is clear that to Winteler, a Swiss republican, the gathering clouds of the Nazi dictatorship were less an incomprehensible disaster than an almost inevitable consequence of German political pathology. At any rate, in 1936 Einstein wrote from Princeton to his friend Michele Besso, a son-in-law of Jost Winteler's: "human affairs in our age are less than agreeable, not to mention the clowns in Germany. Now it is obvious what a prophetic mind Prof. Winteler had when he perceived this grave danger so early in its full magnitude."[31]

Einstein's decision, after having renounced German nationality, to apply for Swiss citizenship was surely due largely to the example of this upright Swiss. The prescribed five-year waiting period was no particular problem. Before World War I people were not yet tied to their passports. With a confirmation from the authorities in Pavia or the Zurich "residents' control," Einstein was allowed to travel wherever he wished and as often as he wished.

In addition to renouncing his old nationality, Einstein also stripped off his religious identity. This step had been foreshadowed ever since his brief phase of religious fervor in adolescence:

> The religion of the fathers, as I encountered it in Munich during religious instruction and in the synagogue, repelled rather than attracted me. Nor did I feel anything like national community or community of destiny. The Jewish bourgeois circles, which I came to know in my younger years, with their affluence and lack of a sense of community, offered me nothing that seemed to be of value. Loneliness, at first painful, then productive and strengthening, was the result.[32]

A visible result was that, in the application for release from Württemberg citizenship, his father—certainly with Albert's agreement if not at his request—had entered "no religious denomination." In the records of the cantonal school, he was listed as "Israelitic," but this is no doubt because the school believed that every student had to be assigned to some religion or other. In a questionnaire for "right-of-residence applicants" in Zurich,[33] which Einstein had to complete in October 1900, he entered "no religious denomination," and he would keep this status for more than two decades.

In view of persistent assertions that at about age sixteen Albert Einstein had left the Jewish community[34]—assertions which seem to be borne out by the records—it is worth quoting Einstein's correction of this account, made a year before his death:

> At that time I would not even have understood what leaving Judaism could possibly mean. Traditional religion had no place at all in my consciousness. But I was fully aware of my Jewish origin, even though the full significance of belonging to Jewry was not realized by me until later.[35]

The youthful skeptic and freethinker thus distanced himself from the religion of his forebears, but not from his forebears themselves.

Albert Einstein's maturing personality also underwent the experience of first love in Aarau. He did not have far to look: it was Marie, the eighteen-year-old daughter of the house. She had just graduated from the local teacher-training college and was still living with her parents before accepting her first post in a small village in the canton of Aarau. From Pavia, where he spent the Easter vacation, Albert Einstein wrote to his "Beloved darling" in the fullness of his emotional experience: "I have now, my angel, had to learn the full meaning of nostalgia and longing. But love gives much more happiness than longing gives pain. I only now realize how indispensable my dear little sunshine has become to my happiness."[36] The two did not have to hide their feelings from either the Einsteins or the Wintelers. Greetings were exchanged between Pavia and Aarau, and everything seemed to indicate something like an unofficial engagement.

But when Albert Einstein embarked on his studies in Zurich, he changed his mind. True, the two young people continued to meet at the Wintelers' house whenever he visited Aarau and she was able to come home from her teaching job. But six months later, in May 1897, he communicated his determination to break off the relationship, not to Marie herself but to her mother:

> So as not to continue fighting a mental conflict whose outcome to me is unshakable: I cannot come to you for Pentecost. It would be unworthy of me to buy a few days' pleasure with new pain—already I have inflicted too much on the dear child through my

> own fault. It fills me with a strange kind of satisfaction to have to savor myself some of the pain that my frivolousness & my ignorance of such a delicate nature has caused the dear girl. Strenuous intellectual work & observation of God's nature are the angels that will guide me, reconciling, strengthening, & yet implacably severe, through all the troubles of this life.[37]

"Mama" Winteler did not blame Albert. Maybe she was even relieved to find that his youthful wisdom was sparing her daughter the risk of a premature liaison. The end of the romance certainly did not affect Einstein's relations with the Wintelers. Indeed, he soon became related to them. His sister Maja attended the Aarau teacher-training college from 1899 to 1902 and eventually married the Wintelers' son, Paul. And Michele Besso, Einstein's friend from the Zurich Polytechnic, married Anna, the eldest daughter. Marie, who later married a watchmaker, wrote about her relationship with Einstein: "We loved each other sincerely, but it was an entirely ideal love."[38]

When he was not personally involved, the forward young man could talk about love very differently. Consoling a woman friend who was unhappily in love with an older man, the twenty-one-year-old gave her this rather presumptuous advice:

> Do you really believe you can find lasting happiness in life through others, even through the only man you love? Oh, I know that animal personally, from personal acquaintance, being one myself. There is not all that much to be expected from them, that I know for sure. Today we are grouchy, tomorrow frivolous, the next day cold, and then again irritable & half-tired of life. . . . So it continues, but I nearly forgot to mention infidelity and ingratitude—matters in which we do a lot better than the good girls.[39]

Young Albert Einstein's repertoire for dealing with women was not without refinement. From the enriching love of one "angel" one could escape by invoking other angels—"strenuous intellectual work & observation of God's nature." And rejecting the illusion of finding "lasting happiness in life through others" looks like a reference to himself. Yet he always felt comfortable, and even happy, in female company, the more so as this feeling was often reciprocated. Many a young or elderly woman was enchanted not only by his violin playing, but also by his appearance, which suggested a passionate Latin virtuoso rather than a stolid student of the sciences. "He had the kind of male beauty that, especially at the beginning of the century, caused such havoc"[40] is how a friend of his second wife's described Albert Einstein's effect.

School, as far as his grades were concerned, remained on an even keel. The "reproof"[41] about his unsatisfactory performance in French recurred throughout all the intermediate reports: despite his private lessons he was unable to catch up with his Swiss classmates. His father bore this blemish on his report of Christmas 1895 with equanimity: "I have always been used to Albert bringing home, alongside some very good grades, also some poorer ones, & I am not disconsolate."[42] Einstein made a bad start also in his other modern language, Italian, but by graduation had improved, earning the second-best grade.[43] His best was a 6, in algebra and geometry. In physics he had a 5–6, and the rest were 5s or 4s. Only in French did he get a 3.

With this school report, dated September 5, 1895, confirming his successful completion of the "fourth technical class," Einstein was now able to take the *Maturitätsprüfung*, the final exam.[44] This lasted several days, with a written and an oral part. For the latter it was customary for some professors of the Zurich Polytechnic to come over to Aarau to take a look at their future students. This time Professor Albin Herzog—the man who a year earlier had recommended the cantonal school to Einstein—was present.

On September 18 at seven o'clock Albert Einstein got down to his first task—outlining the plot of Goethe's play *Götz von Berlichingen*. This he managed in two and a half hours, unenthusiastically and unoriginally. His German teacher, Adolf Frey, out of the goodness of his heart, marked the outline "mostly 5." Next—handled in a masterly way, completed rapidly, and earning high grades—came geometry, algebra, and physics, though his work here revealed a certain carelessness. Thus one mathematical term was called "irrational" when it should have been "imaginary," and "Wheatstone bridge" appeared as "Watston bridge." No doubt his teachers had long realized that this student did not waste time over such trifles. They were marked as mistakes but evidently not allowed to affect his grades.

As he had also done well in chemistry and nature study, Einstein achieved an average of 5½—the best grade among the nine examinees, and a commendable result for the student who was by far the youngest of them. It should be pointed out, however, that this examination was relatively easy. A German—and no doubt a Swiss—traditional gymnasium would have expected more, not only in the German essay but certainly also in mathematics.

Einstein's French paper, the worst of the lot and marked 3–4, is also the most interesting—not because of the mistakes in every other line, but because of the subject of the essay, *Mes Projets d'avenir* (*My Plans for the Future*). Despite its execrable French, it shows that Einstein had found his objectives:

> A happy person is too content with the present to think much about the future. On the other hand, young people in particular are fond of making bold plans. Besides, it is natural for a serious young man to form as precise an idea of the goal of his strivings as possible.
>
> If I am lucky enough to pass my examinations, I will attend the Polytechnic in Zurich. I will stay there four years to study mathematics and physics. My idea is to become a teacher in these fields of natural science and I will choose the theoretical part of these sciences.
>
> These are the reasons which have led me to this plan. It is primarily a personal gift for abstract and mathematical thought and a lack of fantasy and practical talent. Moreover, my hopes lead me to the same resolution. This is quite natural: one always wishes to do the things one has the most talent for. Moreover, there is a certain independence in the profession of science that greatly appeals to me.[45]

Quite apart from the form of these hopes and dreams, they also contain something like a third renunciation of identity, a breaking away from a life pattern linked to his family. Originally he had been sent to

Zurich to study electrical engineering in order to be useful to his father's and uncle's firm and eventually to take it over and carry it on. But as he explained to a friend two decades later, Aarau had marked his decisive renunciation of the profession of a technologist:

> . . . because the thoughts of applying my inventiveness to things that would make everyday life even more sophisticated, with the aim of piling up capital, was unbearable to me. Thought for its own sake, like music![46]

This kind of thought had already found its subjects during his year in Aarau. The propagation of electromagnetic waves in the ether continued to occupy him productively, as he confirmed in a letter three years later: "In Aarau I had a good idea for investigating the way in which a body's relative motion with respect to the luminiferous ether affects the velocity of the propagation of light in transparent bodies."[47] This "good idea" appears to have been a variant of Fizeau's famous experiment of 1853 for determining the velocity of light in moving matter. In fact, this experiment is one of the two results from optics mentioned by Einstein six years after this letter in his special theory of relativity.

Far more important was another question, also from the range of problems of electromagnetic waves—a question which seemed to Einstein "worth asking" that year in Aarau. At the time it cannot have been much more than a puzzle, and from a distance of sixty years Einstein recalled it with forbearance: "If one were to run behind a light-wave with the velocity of light, one would have before one a time-independent wave-field. But it does not seem that something like that can exist! This was the first childish mental experiment to do with the special theory of relativity."[48]

However, the cantonal schoolboy Albert Einstein was no longer all that childish, and the problem he recognized at age sixteen or seventeen was then not perceived as a problem by even the greatest scientists. Einstein's mental experiment was precisely what Goethe had defined as the key to scientific knowledge: "Everything in science depends on what is called an *aperçu*, a realization of what lies behind the phenomena. And such a realization is infinitely fruitful."[49] This applies even more to Einstein's theory of relativity: it is not so much a case of understanding "what lies behind the phenomena" as an analysis of what should be understood by the concept "phenomenon." It would take ten years for the "infinitely fruitful" character of that aperçu to emerge.

CHAPTER FOUR

"Vagabond and Loner": Student Days in Zurich

In the second week of October 1896 Albert Einstein enrolled in Department VI, the "School for Specialized Teachers in the Mathematical and Science Subjects" of the Polytechnic in Zurich. He was still six months short of the official minimum age—eighteen—and must therefore have been one of the youngest students ever to have entered that venerable institution.

The impressive main building at the foot of the Zürichberg was designed by the Polytechnic's first professor of architecture, Gottfried Semper.[1] One of the most striking edifices in the city, it offers from its terrace a splendid view of the historic city center in the valley of the Limmat. The Polytechnic, founded in 1855, was the first university-type school of the Swiss Confederation (created in 1848). Unlike the later universities of Basel, Zurich, and Geneva, which were financed and supervised by the cantons, the Polytechnic was subject directly to the Swiss government in Bern. Compared with these universities, the Polytechnic was of slightly inferior status, but only in that it could not award doctoral degrees. This was changed in 1911, when it was upgraded to Swiss Technical University with all academic privileges. But the people of Zurich to this day refer to it as the "Poly."

At the turn of the century the institution had just under a thousand students,[2] the great majority of them in the engineering fields. Science came under Department VI, the "School for Specialized Teachers"; in addition to providing basic mathematical and scientific training for engineers, the department was also concerned with fundamental research. Mathematical Section VIA, which comprised mathematics, physics, and astronomy, was the one Albert Einstein attended in 1896, along with ten other freshmen, including Mileva Marić, the only woman student in the Mathematical Section, which then, with the freshmen, numbered only twenty-three students.

When in 1855 the first forty professorial appointments were made to the new Polytechnic, a journalist with a sense of history wrote: "Since the foundation of the University of Berlin no scholarly institution has set out on its activity with such a wealth of talent."[3] This applied not only to architecture, but also to mathematics, in which the renowned Rudolf Dedekind was the first professor. In Adolf Hurwitz and Hermann Minkowski, Einstein had two outstanding mathematicians as his professors, men from whom he might have received first-rate training, but he let this opportunity slip by more or less unused: "I saw that mathematics was split into many specialized areas, each of which could take up the short lifetime that is granted us. I thus found myself in the position of Buridan's ass, which could not make up its mind between one bundle of hay and another."[4] Added to this was a certain subject-specific arrogance: Albert Einstein, in his innocence, believed that

> it is sufficient for a physicist to have clearly understood the elementary mathematical concepts and to have them ready for application, while the rest consists of subtleties unprofitable to the physicist—a mistake I realized only later, with regret. My mathematical talent was evidently not sufficient to enable me to distinguish the central and fundamental from the peripheral, from what was not fundamentally important.[5]

At any rate he remembered Professor Carl Friedrich Geiser's lectures on infinitesimal geometry in his second year as "veritable masterpieces of the pedagogic art"[6]—possibly because he was able to make use of what he had learnt from Geiser while working on his own general theory of relativity.

Every beginner in Department VI had a study plan worked out for him by his professors. This consisted of core subjects, in which grades were given, and of useful optional subjects which were not assessed.

After the prescribed subjects of the first three semesters Albert Einstein found that he could now enjoy the academic freedom customary at universities. He soon discovered that "I had to content myself with being a mediocre student."[7] Perhaps with some coyness, he enumerated all that he lacked for being a "good student"—ease of comprehension, willingness to concentrate on what was being offered in lectures, and tidiness in making and processing lecture notes. But gradually he learned "to arrange my studies to suit my intellectual stomach and my interests. Some lectures I would follow with intense interest. Otherwise I 'played hookey' a lot and studied the masters of theoretical physics with a holy zeal at home."[8]

"At home" was, initially, the apartment of Frau Henriette Hägi at Unionstrasse 4, not far from the Polytechnic or from the Physical Institute—a separate building set up in 1890. In his third year he stayed at a small *pension* run by Stephanie Markwalder at Klosbachstrasse 87, where he also had his midday meal. In his last year he returned to Frau Hägi, with whom he later moved to Dolderstrasse 17. All three student lodgings were in the bourgeois district of Hottingen, favorably situated both for the "Poly" and for the Hotel Bellevue by the lake.

His monthly draft of 100 Swiss francs came not from his parents but from his wealthy relatives in Genoa. The reason was that Einstein, Garrone e C. in Pavia had come to grief. In an attempt to get licenses for power stations, the Einstein brothers and their Italian partner had tried to overbid local interests and had failed so badly that in the summer of 1896, barely more than a year after the foundation of the firm, they had to go into liquidation, losing their entire capital as well as some loans from relatives.[9] Uncle Jakob gave up his entrepreneurial ambitions and became an employee. Later, as the manager of an instrument-making firm in Vienna, he led a comfortable life. Hermann, on the other hand, wanted to try his luck as an entrepreneur one more time, with help from the family. He immediately set up a new firm in Milan, on Via Manzoni, for the "production of dynamos and electric motors." Jakob Einstein's position as technical manager was taken by an efficient foreman named Sebastian Kornprobst,[10] who had followed the Einsteins from Munich to Pavia. Despite financial difficulties, the family's lifestyle in Milan was still bourgeois—the Einsteins occupied a floor with eleven rooms at Via Bigli 21.[11]

When the firm in Pavia went into liquidation, Albert Einstein had been busy preparing for his examination in Aarau. He had tried to persuade his father to seek employment, like Uncle Jakob: "This would have saved him and us from the worst,"[12] as two years later the Milan firm also went bankrupt. To his sister Maja he wrote sadly:

> What depresses me most, of course, is the misfortune of my poor parents who have not had a happy minute for many years. It also pains me that, as an adult person, I have to stand idly by, without being able to do anything at all. I am nothing but a burden to my family. . . . It would really be better if I did not live at all. Only the thought that I have always done everything my feeble strength allowed & that year-in year-out I never permitted myself any pleasure, any diversion, except that which my studies offer, keeps me upright and must protect me from despair.[13]

But things soon improved for his parents, and in any case Einstein was not given to prolonged sorrow, knowing only too well "that I am a jolly Joe and, unless my stomach is upset or something similar, have no talent at all for melancholy moods."[14]

The stateless young student Albert Einstein soon felt at home in Zurich. The Winteler boys and his cousin Robert Koch often came over from Aarau, and the house of Gustav Maier, also from Ulm, was always open to him. Maier—a wealthy man of liberal views—together with a few like-minded friends, had just founded a "Swiss Society for Ethical Culture," where social reform, educational problems, and the danger to peace from militarism and chauvinism were discussed.[15] The founding members included "Papa" Winteler and Robert Saitchick, professor of literature at the Polytechnic. Einstein's entrée to this circle ensured that his republican views, and the skeptical attitude toward Bismarck's Germany first aroused by Jost Winteler, would be confirmed.

Through friends in Milan, Einstein also made the acquaintance of the family of Alfred Stern, a notable historian of German-Jewish origin, who was teaching at the Polytechnic. Einstein visited the Sterns regularly every week and enjoyed their cultivated hospitality. For his personal problems, too, he always found a sympathetic ear with Professor Stern. After completing his studies, he summed up his gratitude to Stern in these touching words: "But what can I say about all the kindness and fatherly friendship which you have always bestowed on me whenever I had the pleasure of being with you? . . . One thing is certain: no one has met me the way you have, and that I more than once came to you in a dejected or bitter mood and there invariably found joy and an inner equilibrium again."[16]

In addition, Einstein was invited for Sunday lunch every week by a family named Fleischmann, in much the same way as Max Talmud had been a guest in the Einsteins' house in Munich every Thursday. Michael Fleischmann had originally managed the Zurich branch of the Koch family's grain business in Genoa, but had then set up his own firm on Bahnhofstrasse, representing the Koch interests as an agent. After his Sunday roast, Einstein would usually be seen in a Bahnhofstrasse café, deep in thought while smoking a pipe—a newly discovered passion that he would maintain for a long time.

The 100 Swiss francs which the Kochs remitted to him each month should have been enough for a typical student lifestyle—modest but not poor. Of course, when Albert Einstein had to pay a ten-franc fine to the Zurich Residents' Control[17] because he had carelessly omitted to deposit certain documents, money would be tight, especially as he was saving 20 francs every month for the fee that would be payable when he received Swiss citizenship. Now and again he would earn a little money by coaching private students—for instance, Dora, the daughter of his fatherly patron Professor Stern.

The traditional social life of a student was not to Einstein's taste: in retrospect he described himself as "something of a vagabond and loner."[18] That does not mean that he was lonely at the Polytechnic. He developed a "genuine friendship" with Marcel Grossmann, a mathematics student a year older than himself. "Once every week I would solemnly go with him to the Café Metropol on the Limmat Embankment and talk to him not only about our studies but also about anything that might interest young people whose eyes were open."[19] The "vagabond" Einstein felt strangely fascinated by Grossmann's firm roots in the stolid, yet liberal, Swiss environment that he came to know when visiting the Grossmann home in Thalwil on Lake Zurich. Grossmann in turn was impressed by his friend's intellectual profundity; he is reported to have told his parents: "That Einstein will one day be someone really great."[20] Einstein regarded Grossmann as a "model student, close to his teachers . . . myself apart and unsatisfied, not too popular."[21] Grossmann not only zealously attended all lectures, but also wrote them up so neatly that they could have gone straight to the printer. These notebooks were Albert Einstein's "lifesaver" as soon as examinations approached: "I would rather not speculate how I might have fared without them."[22] In later years, too, Grossmann was a lifesaver—once in connection with Einstein's first post at the Patent Office in Bern and later with the first mathematical calculations of the general theory of relativity. After Grossmann's untimely death in 1936, Einstein wrote to his widow: "But one thing is great: we were and remained friends all our lives."[23]

Another lifelong friend was Michele Besso, six years older than Einstein. Besso had been born in the canton of Zurich but had grown up in Italy. He had qualified as a mechanical engineer at the Polytechnic and was working for a firm in Winterthur. Besso advised Einstein to read the works of Ernst Mach; he discussed with him endlessly the philosophical foundations of physics, and soon became a sounding board for Einstein's ideas.

The acquaintance with Besso came about through Einstein's unswerving love of music. During Einstein's first semester, they had met at the Zurich home of a woman named Selina Caprotti, where people would meet to make music on Saturday afternoons.[24] Young Einstein's considerable skill as a violinist is attested to by the objective record of a school inspector who had examined the musical skills of seventeen students in Aarau. He mentioned only one of the examinees by name: "One student, [named] Einstein, actually sparkled [in] his emotional performance of an adagio from a Beethoven sonata."[25]

At that time, long before radio and other advanced means of reproducing music, good amateurs like Albert Einstein were sought-after guests at domestic musical entertainments. Hardly any woman pianist was immune to his musical passion—not a young one like Susanne Markwalder, his landlady's daughter, whom he presented with the Mozart sonatas inscribed to her "In loyalty and devotion"; and not an elderly "charming old maid," like the piano teacher, Fräulein Wegelin.[26] At times one even has the impression that he preferred older woman as duet partners, as there was no risk of complications: "I am spending the Pentecost days & nights with musical pleasures, which God sends me through one of those angels who do not, with their two-edged sword, threaten impressionable hearts. It is a lady who is already a grandmother."[27]

As numerous as the women are the anecdotes connected with Einstein's violin playing, often testifying to his considerable self-assurance. When, at the Markwalders' house, he had a small audience of women, some of whom began to click their knitting needles during his recital, he simply put his violin back into its case: "We must not disturb you in your work."[28]

The extent to which his violin had become indispensable to him as an instrument of intimate dialogue became clear to Einstein toward the end of his third year in Zurich, when he injured his hand so seriously in the physics laboratory that he had to have stitches put in at the clinic. In a letter to a "Dear lady," with whom he had played duets in Aarau, he regretted that, as a result of that mishap, he could not play his violin: "I greatly miss my old friend, through whom I say & sing to myself everything that, in sober thoughts, I often do not even admit to myself, or at most laugh at when I see it in others."[29]

Alongside that "old friend" there was now also a new one of the kind which, "with their two-edged sword, threaten impressionable hearts." This one had nothing to do with music but was a student of physics—Mileva Marić, the one woman student in Section VIA. She came from Vojvodina, then part of the Hungarian half of the Habsburg empire and later part of Yugoslavia, an ethnically colorful region. Mileva came from a family of Serbian landowning farmers in the village of Titel and had grown up in what was then Neusatz and later became Novi Sad.

Though not exactly supported by her family or the school system of the day, Mileva had set her heart on going to the university.[30] Switzerland being the only German-speaking country where women could study, she went to Zurich, which attracted scholarly young women from all over the world. She had had to pass her "school-leaving" exam at the Young Ladies' College in Zurich before starting on her medical studies at Zurich University in the summer semester of 1896. In the winter semester, she switched to the Polytechnic and, simultaneously with Albert Einstein, enrolled in the program that would lead to a teacher's diploma.

There is some evidence that by the second semester, if not before, Albert Einstein the loner and Mileva Marić, three years his senior, had become closer than just fellow students. Perhaps it was Mileva who was behind Einstein's decision to break off his relationship with Marie Winteler in the spring of 1897. During the summer vacation, at any rate, Einstein had written to her and she had told her parents about him and excited their interest: "Papa has given me some tobacco and I am to give it to you myself, he wanted to make your mouth water for our little bandit country."[31] This letter, however, was posted from Heidelberg, where Mileva was walking "under German oaks in the charming Neckar valley,"[32] having enrolled as a guest student for the third physics semester at the local university. Whether this change of locale had been planned some time ahead or put into effect to gain some distance from Einstein is a matter of surmise. But the fact that Mileva regarded the move as a kind of test of her feelings emerges between the lines of her letter to Einstein.

The physics taught in the first few semesters at the Polytechnic tended to be an appendix to the mathematical training of future engineers rather than a discipline in its own right. In the first semester there was only mathematics; in the second and third semesters there was mechanics, taught by Albin Herzog and attended by 140 students, also attuned to the needs of the engineering students. Einstein's account for Mileva Marić was: "Herzog dynamics and strength of materials, the latter very clear and good—in dynamics a little superficial, as is quite natural in a mass course."[33]

Not until the third semester was there even a lecture announced as "physics." This was given by Professor Heinrich Friedrich Weber and met with Einstein's applause: "Weber lectured on heat . . . and with great mastery. I look forward from one of his lectures to the next."[34] Einstein had in fact taken down and written up this lecture,[35] so that he was able to give the "little runaway" the—perhaps not quite unselfish—"advice to get back here as soon as possible, because here you will find everything you need closely recorded in our notebooks."[36] Mileva Marić followed this advice and from April 1898 was back in Zurich. However, it was not all that simple to make up for the semester she had spent in Heidelberg, so that, despite Einstein's and Grossmann's notes, she had to postpone the "intermediate diploma examination"—due after the second semester—by a year.

Einstein, in his own eyes a "mediocre" student, had reported in time for the exam fixed for October 1898 and spent the summer working hard, mostly with Grossmann—which was not always a pleasure: "When one takes such an exam one feels responsible for everything one thinks and does, as if one were in a penal institution."[37] The grind, however, was crowned by success: in his five subjects Einstein twice received the maximum grade of 6, and otherwise 5.5. This gave an average of 5.7, which made him the best student of his year.[38] Marcel Grossmann came second, scoring 5.6.

After his brilliant work on the intermediate examination, Albert Einstein spent his third year at the Polytechnic working "with zeal and application . . . in Professor H. F. Weber's physical laboratory."[39] For this course—described in the catalogue as "Electrotechnical Laboratory"—Zurich had the best possible conditions. With support from Werner von Siemens, Weber had created a magnificently equipped Physical Institute, meeting the requirements both of experimental fundamental research and of scientifically based electrical engineering.

The building on Gloriastrasse had been completed in 1890, almost in rivalry with Semper's main building, and towering above it, being situated higher up on the slope of the Zürichberg. An American physicist who visited Zurich a year before Einstein began his studies was amazed by its opulence: "They not only have the most complete instrumental outfit I have ever seen, but also the largest building I have ever seen used as a physical laboratory. . . . Tier on tier of storage cells, dozens and dozens of the most expensive tangent and high resistance galvanometers. . . . The apparatus in this building cost 400,000 francs ($80,000), the building—the Phys. Laby.—alone cost one million francs."[40] These perfect working conditions may have reminded Einstein of the homemade equipment in Uncle Jakob's workshop, but his grateful recollection of the fascination of "contact with experience"[41] proves that he was not the narrow-chested thinker that the public imagined a theoretician to be, but a full-blooded physicist with practical interests and abilities.

Einstein was less lucky in "Physical Exercises for Beginners" under Professor Jean Pernet, which he also took in his third year. Whether he did not like his professor or the experiments, he played hookey a great deal and, "upon written request by Herr Pernet, citing neglect of the practical work," received a "reprimand from the director for lack of application" in March 1899. Small wonder that many a sharp clash followed between Einstein and Pernet. When the professor asked his "neglectful" student why, instead of studying the difficult subject of physics, he did not prefer to study medicine, law, or philology, Einstein is reported to have answered: "Because, Herr Professor, I have even less talent for those subjects. Why shouldn't I at least try my luck with physics?"[42] Pernet had his revenge when he gave the cheeky student a 1, the lowest possible grade.

With regard to theoretical physics, the situation at the Polytechnic did not meet Einstein's range of interests either, though for different reasons: "Physics was not greatly favored," complained Adolf Fisch, a student who, like Einstein, had come to the "Poly" from the cantonal school in Aarau. Weber, to both of them, was "a typical representative of classical physics, who simply ignored anything since Helmholtz. At the conclusion of one's studies one was acquainted with the history of physics, but not with its present or future. We were therefore dependent on studying the newer literature privately."[43]

Only during his final semester, when Einstein was already working on his diploma essay, did the mathematician Hermann Minkowski give a lecture on analytical mechanics, as a supplement to Professor Herzog's standard class on mechanics. Minkowski, who had just written a study on capillarity for the *Enzyklopädie der mathematischen Wissenschaften* (*Encyclopedia of Mathematical Sciences*), supplied his few listeners with a reprint and discussed the subject within the framework of his lecture. At the end of it Einstein enthusiastically, but also sadly, told a fellow student: "This is the first lecture on mathematical physics we have heard at the Poly."[44] At least the lecture stimulated Einstein to write his first scientific publication,[45] which he submitted to the *Annalen der Physik* six months after receiving his diploma, even though the subject was not exactly at the center of scientific discussion, nor indeed of Einstein's own interests.

It seems probable that at one of the better universities in Germany—in Göttingen, Bonn, or Berlin—Albert Einstein would have found training more in line with his "intellectual stomach," but his graduation from the school in Aarau entitled him to study only at the Zurich Polytechnic. Besides, Germany to him was a closed chapter, and switching to Vienna or Paris was out of the question, as the Swiss citizenship that he hoped to be granted required a prolonged stay in Switzerland.

The intellectual state of physics at the end of the nineteenth century did not, in retrospect, greatly impress Albert Einstein: "There was dogmatic rigidity on matters of principle. In the beginning (if there was a beginning) God created Newton's laws of motion, together with the necessary masses and forces. This is the lot: everything else derives by deduction from the development of suitable mathematical methods."[46] This observation probably contains at least a grain of truth. Eloquent propagandists of a "mechanical world picture" had put forward a similar formulation—only more expansively than Einstein—and the major part of physics teaching at the Polytechnic was probably no more than pedagogical application of those maxims. But in fact, physics just before the turn of the century was a lot more lively than Einstein's characterization suggests, and he himself was aware of this, was fascinated by physics, and was developing his problems for the future. "I soon learned to ferret out that which might lead to the bottom of things, to disregard everything else, to disregard the multitude of things that fill the mind but detract from the essential."[47]

Despite his sarcastic view of the dry, rigid program of universal mechanics, Einstein, like "any receptive person,"[48] was full of admiration for what had been achieved within that framework. Yet he was less impressed by the solution of even the most tricky problems than he was by the efficiency of the mechanical program when applied to areas which, at first glance, bore no relation to mechanics. This meant, above all, the kinetic theory of gases, some of whose essential theorems could be plausibly derived only by the assumption that gases consisted of minute globules of matter, whose movements and impacts obeyed the laws of mechanics.

This part of the theory of gases had been dealt with by Professor Weber in his lectures. The other, more profound aspect—that the statistical theory of the mechanical treatment of an ensemble of mechanical particles was capable of deducing the basic laws of thermodynamics—Einstein had learned about by private study of the recently published fundamental books of Ludwig Boltzmann,[49] which he is known to have read in the summer of 1899[50] and which show anything but "dogmatic rigidity." If one considers that the very existence of atoms was still being questioned at the time but that, on the other hand, X-rays, cathode rays, and other types of radiation had just opened a new world of microphysics, one is left with the impression of an exceedingly interesting phase of scientific research.

This applies equally to the other major development in physics in the second half of the nineteenth century: the electromagnetic field theory developed by James Clerk Maxwell and brilliantly confirmed in 1888 by Heinrich Hertz. Of course, Maxwell and Hertz had regarded mechanics as the sure foundation of all physics and had therefore tried to anchor electromagnetic field theory in mechanics, or indeed to derive it from mechanics. This was one of the reasons for the invention of the "ether," a strange immaterial substance which initially served as a substratum for the so-called "polarization states" of the electromagnetic field and which was subsequently drawn on to provide a basis for the propagation of electromagnetic waves, such as light or Hertzian waves, since these, in mechanical concepts, could not proceed in "nothing" but needed an oscillating medium, more or less like sound waves in air.

The price paid by physicists for tying electromagnetic fields to mechanics was, of course, rather high, as the "ether" proved to be a conceptual construct full of internal contradictions. As an elastic, all-pervading body, though weightless and noncompressible, it led a kind of "ghost existence" alongside normal matter; it was not permitted to offer resistance to the motion of normal matter or interact with it in any other way. In order to explain the velocity of light in moving matter, it had to be assumed that the ether, independently of matter, was in a state of permanent rest. However, when an attempt was made to measure the Earth's motion through the ether in the famous experiments of Albert Michelson and Edward Williams Morley, it had to be acknowledged, sadly, that apparently there was no such motion. Here, then, were enough problems capable of "leading to the bottom of things," and, with his sure instinct for the essential, Einstein felt himself drawn to them.

His thinking probably proceeded from what he had called his "first childish mental experiment to do with the special relativity theory."[51] Our earliest information on these reflections comes from a conversation which Max Wertheimer, one of the founders of Gestalt psychology (*Gestaltpsychologie*), had with Einstein in 1916 in Berlin. Wertheimer has represented Einstein's "problem horizon" during that first probing phase as follows:

> What if one were to run after a ray of light? What if one were riding on the beam? . . . If one were to run fast enough, would it no longer move at all? . . . What is "the velocity of light"? If I have it in relation to something, this value does not hold in relation to something else which itself is in motion.[52]

In all probability Einstein turned these problems over and over in his mind, first in Aarau and then during his first year in Zurich; certainly we know of no partners in conversation or studies in literature that could have led him to such ideas. In his second year at the Polytechnic he surprised Professor Weber by proposing an experiment to determine the velocity of the Earth relative to the ether—evidently in ignorance of the Michelson-Morley experiments, as "he learned only later that physicists had already conducted such experiments."[53]

As a matter of fact, Einstein's proposal was no more than a variant of the Michelson-Morley experiment: "I thought of the following experiment using two thermocouples. Set up mirrors so that the light from a single source is to be reflected in two different directions, one parallel to the motion of the Earth and the other anti-parallel. If we assume that there is an energy difference between the two reflected beams, we can measure the difference in the generated heat using two thermocouples."[54] Needless to say, this remained just a suggestion, because "there was no way of building that apparatus. The skepticism of his professors was too great and the persuasive force of the project too small."[55] We do not know whether Professor Weber rejected the experiment because he regarded it as uninteresting, or as the crazy idea of a generally odd student, or as technically impossible. In the last case one would, in retrospect, have to agree with him.

It was chiefly because electromagnetic field theory and associated problems were not dealt with at the Polytechnic that Einstein cut classes and "studied the masters of theoretical physics with a holy zeal at home."[56] For him this was "simply the continuation of an earlier practice."[57]

We owe our first clue to which books on field theory he was studying in his second year to his habit of forgetting his key. When he found himself once more keyless at Frau Hägi's locked front door, he went straight to the Pension Bächthold, where Mileva Marić was staying, and in a note requested the absent "Dear young lady" not to "be too angry with me if in my need I abduct your Drude, so that I can study a little."[58] The purloined object was a book by the young German physicist Paul Drude, *Physik des Aethers* (*Physics of the Ether*),[59] a work written in the conviction that Maxwell's theory of the electromagnetic field was more likely than the old mechanical approach to do justice to light and to Hertzian waves.

By the following year Einstein had worked his way through the most important publications of Hermann von Helmholtz[60] and Heinrich Hertz.[61] This fact emerges from a letter to Mileva Marić written during the summer vacation of 1899: "I returned the Helmholtz volume and am now rereading Hertz's *Ausbreitung der elektrischen Kraft* (*Propagation of Electrical Force*) with great care because I didn't understand Helmholtz's treatise on the principle of the least action in electrodynamics."[62] Whatever it was he failed to understand, he had certainly gained an entirely independent critical attitude toward Heinrich Hertz's views:

> I'm more and more convinced that the electrodynamics of moving bodies as it is presented today, doesn't correspond to reality, and that it will be possible to present it in a simpler way. The introduction of the term "ether" into theories of electricity has led to the conception of a medium whose motion can be described without, I believe, being able to ascribe physical meaning to it.[63]

Here we have a first hint of the abolition of the ether—to be accomplished six years later in the special theory of relativity. Even the title of that later study, *On the Electrodynamics of Moving Bodies*, is foreshadowed in this letter to Mileva, though admittedly as an echo of the title which Heinrich Hertz had given his treatise of 1890. A few of Einstein's remarks, such as the concept of "electrical currents . . . as the motion of true electrical masses"[64] and the future definition of electrodynamics as "the theory of the movements of moving electricities & magnetisms in empty space,"[65] suggest that he was also familiar with the work of the great Dutch theoretician Hendrik Antoon Lorentz.[66]

Albert Einstein, still a student in Zurich, had thus turned to the same subject as, less than a year earlier, the Gesellschaft deutscher Naturforscher und Ärzte (Society of German Scientists and Physicians) at its annual convention in Düsseldorf in September 1898. The ether and its behavior in moving media had been chosen as the topic of a special session, to which all the leading figures had been invited, including Lorentz from the Netherlands, and for which the young Aachen assistant professor Wilhelm Wien had prepared an introductory "overview" paper.

Wien's very first sentence revealed the confusion of the situation: "The question whether or not the ether participates in the motion of bodies, and whether it is altogether to be credited with mobility, has been agitating physicists for a long time, and there is no end to presumptions and assumptions which it has been thought necessary to make of the properties of the carrier of electromagnetic phenomena."[67] Wien's observations were essentially a juxtaposition of contradictions arising from conflicting concepts of the ether, as well as a careful listing of, all together, thirteen experiments that might yield information on the Earth's movement relative to the ether. In its theoretical part the discussion was somewhat reminiscent of the debate of astronomers in the late Middle Ages, when epicycles and other auxiliary constructs were piled on top of each other for the "salvation of phenomena" and hence for defending the concept of the Earth as resting at the center of the universe. These efforts were all ultimately doomed to failure because they produced new complications, which were resolved only by the radical new start represented by the "Copernican revolution."

In Düsseldorf, though, there were no signs of a comparable revolution in the matter of the ether. Even an authority like Professor Lorentz, who gave the co-report to Wien's survey, could not imagine the future of physics without the ether: "The ether, ponderable (that is: heavy) matter, and let us add electricity, are the building blocks making up the material world, and if we knew whether or not matter, on its motion, carried the ether with it, we would have a way of penetrating a little further into the nature of these building blocks and their mutual relations."[68] None of the physicists present in Düsseldorf asked, as the twenty-year-old Albert Einstein had, whether any physical meaning could be ascribed to that statement—in other words, whether the ether was not perhaps a superfluous concept for physics, not to be investigated but to be abolished.

The student Einstein had not been present at the learned convention in Düsseldorf, but he was informed about it, probably better informed than anyone else in Zurich. From Milan, where, after a stay in Mettmenstetten, he was spending the rest of the summer vacation of 1899 with his parents, he consoled Mileva Marić, cramming for her intermediate examination, saying that he too had been "quite a bookworm myself lately, trying to work out several ideas, some of them very interesting."[69] "I also wrote to Professor Wien in Aachen about my paper on the relative motion of the luminiferous ether against ponderable matter, the paper which the 'boss' handled in such an offhanded fashion. I read a very interesting paper by Wien from 1898 on this subject. He'll write me via the Polytechnic."[70] The "interesting treatise" was Wien's paper at the Düsseldorf convention of the previous year; Einstein had probably discovered it in *Annalen der Physik und Chemie*.[71] Unfortunately, Einstein's letter is lost, and we do not even know whether Wien replied to it.

Extant documents, however, show that Albert Einstein in his second and third year at the Polytechnic developed, or at least consolidated, self-assurance as a physicist. Through intelligently chosen reading he involved himself directly in the mainstream of topical research, and he was strong enough not to be carried away by the vortices of that stream but to choose his own bank. But no one then realized any of this, except perhaps Marcel Grossmann and Mileva Marić.

These years were a happy time for Albert Einstein, and not only in physics. His father had been lucky with his newly founded firm in Milan, winning contracts for the streetlights in the small towns of Cannetto sull'Oglio and Isola della Scala, both near Mantua.[72] Even though these contracts had to be immediately pledged to his main creditor—his cousin Rudolf Einstein in Hechingen—and although his debts exceeded his assets, there was no hardship at Via Bigli 21 in Milan now. This change of fortune was also a great relief to Albert, who admitted to his sister that "I now sometimes find time to stroll for an hour or so in Zurich's beautiful surroundings. I am happy at the thought that the worst anxieties are over for my parents. If everyone lived like me, romance-writing would never have been invented."[73]

Meanwhile he had started on his own romance. Private study was mostly done *à deux*, and from his 1899 spring break he wrote to Mileva

Marić from Milan: "Your photograph had quite an effect on my old lady. While she studied it carefully, I said with the deepest sympathy: Yes, yes, she certainly is a clever one."[74]

During the summer vacation the two were separated. She had to prepare for her intermediate examination, first in Novi Sad and then in Zurich, while he was in Milan, having spent August with his mother and sister in Mettmenstetten, a small village between Zurich and Zug. From the Pension "Paradies" he reported to Mileva about his "nice, quiet, philistine life, just as the pious & the upright imagine paradise to be."[75] He then went on: "When I read Helmholtz for the first time I could not—and still cannot—believe that I was doing so without you sitting next to me. I enjoy working together & I find it soothing & also less boring."[76]

With his sister, he climbed the 8,200-foot Säntis, where, three years previously, as a "badly shod tourist" on a three-day outing with the cantonal school, he would have crashed to his death if a classmate "had not, as he was beginning to slip down the steep slope, quickly extended his mountaineer's cane to pull him up again."[77] Einstein, however, sent Mileva some news from his "paradise" that was not so cheerful: sometimes he felt disturbed by "unpleasant visits from Mama's acquaintances. . . . I can usually escape their mindless prattle by slipping away if we don't happen to be at the dinner table. At the end of our stay my aunt from Genoa is coming, a veritable monster of arrogance & insensitive formalism."[78]

His escape, of course, was straight into physics—and to Mileva Marić, whom in his letters he still addressed with the formal *Sie*, but along with a southern German dialect expression meaning something like "Dollie." He was generous with his advice for her exam, which he himself had passed the previous year, and sincerely wished that "you could be here with me for a while. We understand one another's dark souls & also drinking coffee and eating sausage etc."[79] Finally he announced that he would bring his sister along to Switzerland and take her to Aarau, where she would attend the teacher training college and live with the Wintelers. That the mention of Aarau would arouse divided feelings in Mileva was clear to Albert Einstein, because that was where the "critical daughter with whom I was so madly in love 4 years ago" lived.[80] He went on: "For the most part I feel quite secure in my high fortress of calm. But I know that if I saw her a few more times I would certainly go mad. Of that I am certain & fear it like fire."[81] This was neither the first nor the last time that Albert Einstein's remarks on personal matters were guided less by tact or sensitivity than by ruthless frankness.

In his fourth and last year at the Zurich Polytechnic Einstein evidently set aside his reflections on the ether and the electrodynamics of moving bodies—presumably because he had found no new point of attack, rather than because of the necessities of preparing for exams and writing his diploma essay. His own interests coincided with those of the principal, Professor Weber, in the field of heat, so that it may have seemed advisable to him to deal with that subject in greater detail. "Scientific work in the physical laboratory" was the most time-consuming of his subjects; in addition he attended the rather more technical lectures on alternating current, also by Professor Weber,[82] as well as a minimum of mathematics, and "analytical mechanics" by Hermann Minkowski. Although this lecture suited his "intellectual stomach," he stayed away from Minkowski's advanced mathematics class, for "no one could make him attend the mathematical seminars."[83]

All students had to attend at least one class each year outside their special field. Einstein chose, and evidently enjoyed, the general studies of Department VII, the "General and Economics Department," which were often taught by original minds attracted by the amenities of a faculty without regular syllabi.[84] Einstein in fact enrolled for considerably more of these lectures than the mandatory minimum, covering a wide spectrum of subjects, such as "Man's Prehistory" and "Geology of Mountain Ranges," both given by Albert Heim, who started at seven in the morning yet always had a crowded lecture room. Even in old age Einstein commended the "magic" of Heim's way of lecturing.[85] He heard a lecture on Goethe by Robert Saitschick, whom he had already met in his patron Gustav Maier's Swiss Society for Ethical Culture, and he heard August Stadler on Kant's philosophy and on the "Theory of Scientific Thought." There were also some practical subjects, such as "Banking and Stock Exchange Dealing" and "Statistics and Life Insurance."

It was probably his friendship with Friedrich Adler that led Einstein in his final year to attend lectures on "Fundamentals of Economics" and on "Income Distribution and the Social Consequences of Free Competition." Adler was a socialist in his family's tradition: his father, Victor Adler, was the unchallenged leader of the Austrian Social Democrats and one of the spiritual fathers of Austro-Marxism.[86]

Friedrich Adler had come to Zurich a year after Einstein. He had first studied chemistry, and later also physics, though not at the Polytechnic but at the university. Einstein revered Adler as "the purest and most fervent idealistic character"[87] he ever met. The two shared an admiration for Ernst Mach's empiricist philosophy and, with their girlfriends—Einstein with Mileva Marić and Adler with a Russian woman, Katya Germanishskaya—would sit in the small lecture room where Hermann Minkowski taught analytical mechanics. Adler probably did not have to work hard to win Einstein over to his socialist ideas, but he could not persuade him to join the party. He concluded that "Einstein was a typical emotional socialist."[88] This characterization, given Einstein's sense of justice, was probably quite correct, and it applied throughout his life. His social attitudes may have been socialistically colored, but he did not feel greatly attracted to the dogmas of any political party.

Zurich at that time was a haven for socialist and anarchist student groups, especially from the Slavic countries, but Einstein's occasional contacts did not lead to more than a confirmation of his "emotional socialism." It was only the horrors of World War I that turned him into a political person. But he may well have felt in Zurich that not only the foundations of physics could be fragile, but those of society as well.[89]

What Albert Einstein called his "household"[90] with Mileva Marić seems to have been somewhat unconventional, if not by Zurich student standards, then certainly by those of respectable bourgeois morality. Mileva lived within a few minutes of Einstein's lodgings, in the *pension* of Fräulein Engelbrecht at Plattenstrasse 50, and Einstein appears to have been a frequent and popular visitor among the group of foreign women students. Whenever parcels of delicacies arrived for him from his mother, he would take them straight to Plattenstrasse.[91] When, on the occasion of his twenty-first birthday, a particularly generous package arrived, Mileva described to a friend the "tremendous effect" which it had on Einstein: "radiantly he walked down Plattenstrasse, carrying the box in both hands, so pleased that he did not look right or left."[92] Sometime in his fourth year, the two must have decided to get married.[93]

Einstein and Mileva Marić also spent the spring break of 1900 together in Zurich, working on their diploma essays. In March Professor Weber had accepted the suggestions of the two candidates, and Mileva was looking to the future full of hope: "I am looking forward to the research I will have to do. E. also has chosen a very interesting subject."[94] The essays had to be completed within three months, so no real flashes of genius were, or could be, expected.[95] Einstein's comment in his old age was: "My and my first wife's diploma essays dealt with heat conduction; they were of no interest to me and are not worth mentioning."[96] In the assessment Mileva's essay was given a 4 (6 being the top grade), while Einstein's received 4.5—neither of them a brilliant result.[97]

In the actual subject exams Einstein would seem to have relied too much on Grossmann's lecture notes, because he did not repeat his success in the intermediate examination. Of the five candidates of Section VIA the three mathematicians did better than the two physicists, Einstein and Marić. After various calculations and weightings of individual grades, Einstein finished fourth with an average of 4.91 and Mileva Marić achieved only an average of 4.0. The "conference of examiners" therefore decided "to award the diploma to the candidates Ehrat, Grossmann, Kollros, Einstein, but not to Fräulein Marić."[98]

The inevitable "cramming" for the exam, even though it probably went on for only a few months, left traces of trauma in Albert Einstein. In his old age he related, full of horror, "that one had to stuff all that jumble into oneself for the exams, whether one liked it or not. This compulsion had such a deterrent effect that, having passed the final exam, I lost all taste for any reflection on scientific problems for a whole year."[99] However, the letters he wrote immediately after the exam reveal nothing of what he called his "intellectual depression."[100] He took some physics books with him on his vacation, and a mere three days after finishing his exam he was writing to Mileva: "My nerves have calmed down enough, so that I'm able to work happily again."[101] Moreover, he was firmly counting on a position as assistant at the Polytechnic, in order to establish himself in both the scientific and the bourgeois world, so that he could marry Mileva. However, he was soon to discover that "there are worse things in life than exams,"[102] because once more there were long and tortuous routes to be covered before he arrived at his goal.

CHAPTER FIVE

Looking for a Job

EINSTEIN SPENT THE SUMMER OF 1900 in Melchtal, south of Lake Lucerne, in the company, as usual, of his "ghastly aunt" from Genoa, his sister, and his mother. Tension began on the very first evening, when his mother with affected casualness asked what was now going to become of "Dollie," to which Einstein, just as casually, replied, "She'll become my wife." There then unrolled a family drama, often to be repeated: "Mama threw herself onto the bed, buried her head in the pillow, and wept like a child."[1] As the mountains were drenched in rain, Einstein had no other escape than his books, "mainly Kirchhoff's famous investigations of the motion of the rigid body.[2] I can't stop marveling at this great work."[3]

What the Einsteins soon called the "Dollie affair" continued to give rise over the next few years to fierce clashes between parents and son. Needless to say, neither his mother's histrionics nor his father's more moderate arguments could shake Albert in his determination, especially as he believed he had already won the battle: "Mama and Papa are phlegmatic types and have less stubbornness in their entire bodies than I have in my little finger."[4] But he was mistaken: his mother's obstinacy was quite equal to his own.

Trying to put himself in his parents' position, Albert Einstein wrote to Mileva: "I understand my parents quite well. They think of a wife as a man's luxury which he can afford only when he is making a comfortable living. I have a low opinion of that view of the relationship between man and wife, because it makes the wife and the prostitute distinguishable only insofar as the former is able to secure a lifelong contract from the man because of her more favorable social rank."[5] Such theorizing on bourgeois sexual morality, however, is rare in his letters to Mileva, which are pervaded by feelings of love and longing for his "little darling": "I can go anywhere I want—but I belong nowhere, and I miss your two little arms and the glowing mouth full of tenderness and kisses."[6] And to cheer her up during that separation he would occasionally include a few rhymed ditties in dialect.[7]

While his mother was horrified and his father worried about this liaison, Einstein's fellow students were merely astonished that a man so successful with women should so firmly and inseparably tie himself to Mileva Marić. She spread around her an aura of Slavic melancholy, if not indeed gloom, and soon also—perhaps not without reason—of jealousy. Besides, she was by no means a stately beauty. She was very short, barely coming up to Einstein's shoulder; and tuberculosis in childhood had done lasting damage to her joints, with the result that she had a noticeable limp. But Einstein did not let these matters worry him. When a fellow student remarked to him that "he would never have the courage to marry a woman who was not entirely healthy," Einstein is reported to have replied: "Why not? She has a sweet voice."[8] Important as this acoustic component of love may be, it is hardly sufficient. Nor was it sufficient to Einstein. His letters, in which declarations of love are oddly intermingled with scientific matters, instead suggest that as a bohemian kind of physicist he was interested in his future wife as a colleague. Even though Mileva still had to repeat her diploma examination the following year, Einstein already saw her working for her doctorate at the university: "I'm also looking forward to working on our new paper. You must continue with your investigations—how proud I will be to have a little Ph.D. for a sweetheart while I remain a completely ordinary person."[9] With a cheerfulness which seemed still totally unaware of life's difficulties, he painted to her a future in which the two of them, arm in arm, would bestride the scientific stage and amaze the world: "I can hardly wait to be able to hug you and squeeze you and to live with you again. We'll happily get down to work right away, and money will be as plentiful as manure."[10]

▪ ▪ ▪

The economic foundation of their hoped-for life together over the next few years was to be the salary of an assistant at the Polytechnic, and during the first few weeks of his vacation Einstein was convinced that such a post would be his as a matter of course. The professors of Department VI needed several assistants because of the large number of engineering students; and, as only a few students enrolled in the less lucrative fields such as mathematics and physics, virtually any graduate after his exam could, if he wanted, become an assistant for a few years. This, however, did not apply to Einstein, as he was soon to discover.

With Professor Pernet, who had initiated his reproof for "lack of application," he had no prospects in any case. But Professor Weber, too, was not particularly fond of him. Weber, after all, was a German and therefore unlikely to give a post to someone who, even as a student, had persistently addressed him as "Herr Weber"[11] instead of "Herr Professor" and had generally shown scant respect. As Einstein's diploma essay had been rather mediocre, Weber felt no obligation toward the rebel graduate and was inclined instead to engage two mechanical engineers as his assistants.

Einstein probably surmised all that, and therefore turned his hopes to the mathematicians. On August 9, barely two weeks after his exam, he was in Zurich again "to straighten out my 'business and political' affairs."[12] The conditions he found there seemed to him so favorable that he summed them up for Mileva to the effect that he would be "provided for in any case."[13] He felt so confident that he actually turned down a temporary job one of his fellow students offered to arrange for him with an insurance firm: "an 8-hour day of mindless drudgery. . . . One must avoid such stultifying affairs."[14]

Einstein spent most of the vacation with his parents in Milan, where discussions of the "Dollie affair" were poisoning the atmosphere. In bed at night he wrote to Mileva: "Mama often cries bitterly & I don't have a single moment of peace. My parents weep for me almost as if I had died. Again and again they complain that I have brought misfortune upon myself by my devotion to you, they think you aren't healthy . . . oh Dollie, it's enough to drive one mad."[15] By way of consolation he was able to pass on to Mileva the latest news from Zurich, in a letter from his fellow student Jakob Ehrat, that one of Professor Hurwitz's assistants would be moving into the educational service, which made Einstein conclude that "I'll become Hurwitz's servant, God willing."[16]

But the vacation was not all lamentation over Mileva. Einstein met his old friend Michele Besso, who had been married to Anna Winteler for a couple of years and had settled in Milan as a technical consultant to the Società per lo Sviluppo delle Industrie Elettriche in Italia (Society for the Development of the Electrical Industry in Italy). Einstein spent many of his evenings with the Bessos and came to know the joys of family life and fatherhood: "I like him a great deal because of his sharp mind and his simplicity. I also like Anna, and especially their little kid."[17] A spinoff was a small—but presumably paid—job for Einstein; he examined for Besso's society the "interesting question: how does electric energy radiation through space occur in the case of a sinusoidal alternating current?"[18] He accompanied his father on a trip to his power plants in Cannetto and Isola della Scala, because he was anxious to "learn a little bit about the administration of the business so I can take Papa's place in an emergency."[19] As promised by his father, a detour was made to Venice. And toward the end of September he once more escaped to the mountains,[20] traveled to Lake Maggiore, and visited Isola Bella, the largest of its Borromean Islands.

By the end of September Einstein finally thought it necessary to write to Hurwitz with "the humble inquiry whether I have any prospect of becoming an assistant in your department."[21] The Herr Professor must have been rather surprised, as this student had never shown up at the mathematical seminars. In response to a courteous inquiry, Einstein had to confess his omissions. He did so in a letter to Hurwitz without any attempt at whitewash and without any remorse: "As, because of lack of time, I was unable to take part in the mathematical seminars . . . I can say nothing in my favor except that I attended most of the lectures on offer to me."[22] Even that, as Hurwitz could easily discover, was a bit of a tall story as far as mathematics was concerned, so that one cannot really blame the professor if he simply shelved this rather cheeky application. But when Einstein traveled to

Zurich at the beginning of October, he was still confident of getting a post at the Polytechnic: "Hurwitz still hasn't written me, but I have hardly any doubt."[23]

When, shortly after his arrival in Zurich, Einstein had to face the unpleasant truth, it must have been a serious blow to him, especially as his financial problems were becoming more acute. His relatives in Italy had discontinued their monthly remittance when he received his diploma, and his parents were in no position to give him regular support. Unsuspecting of the many disappointments still in store for him, he soon adjusted to the new situation and a few days later wrote to a woman friend of Mileva's: "We have neither of us landed a job and live by private lessons—if only we could find some, which is very doubtful. Isn't this a journeyman's life, or even a gypsy existence? But I think we'll be cheerful about it, as usual."[24]

Mileva Marić still had to repeat her fourth year and her diploma exam at the Polytechnic but was intending to work simultaneously on a doctoral thesis for the university. Albert was planning to develop some earlier reflections on the thermoelectric Thomson effect into a doctoral thesis and, because he needed to conduct the experiments in Professor Weber's laboratory, remained enrolled as a student at the Polytechnic. But he probably realized that, given his not too brilliant diploma grades, some further scientific qualifications would not come amiss. He therefore first of all developed his ideas on capillarity, intended for early publication in a reputable scientific journal.

Interface phenomena, which include the rise of water in fine glass tubes, were thought by most physicists to be due to the interaction of individual atoms and molecules, and it was in this area that Einstein believed he was on the verge of new discoveries. During the vacation, full of enthusiasm, he had written to Mileva: "The results on capillarity I recently obtained in Zurich seem to be entirely new despite their simplicity. . . . If this yields a law of nature, we'll send the results to Wiedemann's *Annalen.*"[25] On December 13 his manuscript was received by the editors of the *Annalen der Physik;* it was accepted and appeared on March 1 of the following year.[26] Einstein was now able to enclose with his job applications reprints from the most renowned

German physical journal, and to that extent the enterprise was a success. But he was less lucky with the natural law he was hoping to discover, and even his hypothesis about the forces between two molecules, which he rather superficially connected with Newton's law of gravitation,[27] soon proved to be untenable.

Meanwhile, however, he was entirely happy with his hypothesis and tried to develop it further. Six months after his first publication, Einstein wrote to Marcel Grossmann that he had

> conceived a few marvelous ideas, which now only have to be properly hatched. I now firmly believe that my theory of attraction forces between atoms can be extended also to gases, and that the characteristic constants for nearly all elements will be determined without major difficulties. Then the question of the inner relationship of molecular forces and Newtonian forces will move a big step closer to solution.[28]

In the event that his expectations were fulfilled, he would "use everything achieved so far about molecular attraction for the doctoral thesis."[29] This suggests that there must have been problems with the thesis about the Thomson effect begun under Professor Weber. He concluded his reflections with a lyrical hymn to the glorious perspectives of properly understood scientific research, entirely in the spirit of Alexander von Humboldt, whose *Kosmos* he had read as a schoolboy in Munich: "It is a wonderful feeling to realize the unity of a complex of phenomena which, to immediate sensory perception, appear to be totally separate things."[30]

The effusiveness, however, did not yield much, at least not in the field of molecular forces of attraction. He applied his hypothesis to saline solutions, which indeed yielded a second publication in the *Annalen*, but nothing of lasting value.[31] Einstein also intended to pursue this subject for his doctoral thesis at the University of Zurich. His method was now to be applied, after surfaces and solutions, also to gases. He submitted his work to Professor Alfred Kleiner sometime in November 1901, convinced "that he won't dare reject my dissertation"[32]—but precisely that seems to have happened. With a receipt dated February 1, 1902, Albert Einstein confirmed that he had "received back, in cash, the doctoral fee of Frcs. 230 (Two hundred and thirty) paid on Nov. 23, 1901."[33] No doubt Kleiner took a closer look at Einstein's ideas on the molecular forces than the editors of *Annalen*, and did not think them tenable. At least he gave Einstein the opportunity to withdraw his thesis before it was formally turned down, so that the "rejected" doctoral candidate could at least save his high financial stake. This receipt is the only extant trace of his second attempt to gain a doctorate.

The two articles published in *Annalen* would have vanished forever in the deep pit of the history of science if their author had not been Albert Einstein. He himself had no illusions about their shortcomings, and a mere five years later called them "my two worthless firstling works."[34]

Before it was even certain that the first paper would appear in *Annalen* Mileva wrote to a woman friend how proud she was of her "sweetheart" because of this "very significant" article. "We have also privately sent it to Boltzmann and would like to know what he thinks of it, we hope he'll write us."[35] Ludwig Boltzmann, the unchallenged master of statistical physics and the kinetic theory of gases, held the chair of theoretical physics—one of the few then existing—at the University of Leipzig. No record exists of any reply to Einstein.

As soon as the article appeared, Einstein also wrote off to other professors in Leipzig. On March 9, 1901, he inquired of the experimental physicist Otto Wiener "whether perhaps you require an assistant";[36] in his letter he referred to his "small treatise" which had appeared a few days previously and which the Herr Professor was presumably expected to track down for himself in the library. Ten days later Einstein enclosed a reprint to the famous professor of physical chemistry Wilhelm Ostwald, with the remark that "it was your work on general chemistry that encouraged me to produce the enclosed treatise." He inquired "if maybe you might have some use . . . for a mathematical physicist" and confessed that "I am impecunious and only that kind of post would enable me to continue my studies."[37]

When there was no reply, Einstein—a few weeks later, during his spring break in Milan—wrote again, with the excuse that "I am not sure whether I then enclosed my address."[38] The ploy did not work.

This camouflaged cry for help was followed ten days later by another, this time more explicit, from Hermann Einstein. Without his son's knowledge he wrote to Ostwald, after appropriate introductory courtesies:

> My son, lacking a post at present, feels deeply unhappy & each day the thought gains strength in him that his career has been derailed & he cannot find a connection any longer. He is moreover depressed at the thought that he is a burden to us, who are not very well off.
>
> As my son probably reveres and esteems you, dear Herr Professor, more than any other physical scientist working today, I am taking the liberty of turning to you with the respectful request that you will read the treatise published by him in *Annalen der Physik* & perhaps send him a few lines of encouragement, so he might regain his vitality and working vigor. If, moreover, you were to find it possible to find for him an assistant's post, either for now or for next autumn, my gratitude to you would be boundless.[39]

This letter from a father suffering along with his son likewise remained unanswered by Ostwald, and it is unlikely that Albert Einstein ever knew of his father's desperate act. It would be amusing to speculate whether Ostwald remembered those letters when, as the first of those entitled to propose names for the Nobel Prize,[40] he nominated Albert Einstein in 1909 for the 1910 prize and twice more repeated his proposal.[41]

Depressed though Einstein may have been at times, he certainly was not inactive. On April 12, 1901, he seems to have bought a stack of postage-paid reply postcards and to have sent them to a dozen or so professors whose names he happened to know. Two samples of this scattershot approach have been preserved—one addressed to Heike

Kamerlingh Onnes, the founder of low-temperature physics, in Leyden, the Netherlands, and the other to Professor Carl Paalzow, at the Technische Hochschule in Berlin, the capital of the German Reich—a country he was not at all fond of. Whereas in writing to Kamerlingh Onnes he at least referred to information received from a fellow student "that there is a vacant post in your department,"[42] he came straight to the point with Professor Paalzow: "I take the liberty of inquiring from you whether there is an assistant's post vacant in your department, and, if so, of applying for it."[43] Otherwise the two postcards are identical to the last comma. Another thing they have in common is the fact that both reply cards, to be returned to "Herr Albert Einstein, Via Bigli 21, Milano, Italia," have come down to us unused. It might perhaps be stated here, in anticipation of later events, that though Einstein did not then obtain an assistant's post in Berlin or in Leyden, he later became a professor in both cities.

As he reported to Mileva in his regular letters from Milan, other hopes, and indeed less outlandish ones, were also dashed. Thus Professor Eduard Riecke in Göttingen had actually advertised two assistants' posts in *Physikalische Zeitschrift.* Einstein applied but had soon "given up on the position. I can't believe that Weber would let such a nice opportunity pass without meddling in things."[44] Einstein had by then become convinced that his former professor must be behind his lack of success; Weber was bound to put further obstacles in his way by giving him poor references. He was, therefore, not surprised when Riecke turned him down, being "absolutely convinced that Weber is to blame."[45] And to Marcel Grossmann he wrote: "For the past three weeks I have been here with my parents and am trying from here to find a post as an assistant at some university. And I would have long found one if Weber wasn't double-crossing me." But he was not discouraged: "God created the donkey and gave him a thick hide."[46]

Whatever opinion Weber may have had of his student, he probably would not even have had an opportunity of "double-crossing" him. Most of the professors simply shelved Einstein's letters or postcards; there is no evidence that any of them even took the trouble to make inquiries in Zurich about this odd applicant. A much more plausible reason for Riecke's refusal is that in his advertisement he had specifically stipulated a doctoral degree; Einstein's mediocre diploma and not very impressive first publication were surely no substitute. Besides, the time between the publication of the advertisement and Riecke's rejection was far too short for any intervention by Weber.[47] There simply was no reason why any professor should have considered appointing a candidate who was unknown to him and whose qualifications, on paper, were indifferent.

His rejection by the academic world—due, as he believed, to Weber's intrigues—was very painful to Einstein. It is this deep sense of injury which, ten years later, when he was a full professor in Prague and shortly to be appointed to a professorship at the Zurich Polytechnic, made him comment on Professor Weber's death in a letter to a Zurich friend and colleague that "Weber's death is a good thing for the Polytechnic."[48] The old wounds were still smarting when, in 1918, the Polytechnic tried to bring Einstein back from Berlin by making him an exceptionally generous offer: "How happy I would have been 18 years ago if I could have become a humble assistant at the Polytechnic! But I couldn't bring it off! The world is a madhouse, reputation is everything!"[49]

Expecting nothing but bad references from Weber, Einstein conceived the unusual idea of asking his former teachers in Munich and Aarau for references in order to apply for posts in Italy: "To begin with, one of the main obstacles in getting a position doesn't exist here, namely anti-Semitism, which in German-speaking counties is as unpleasant as it is a hindrance. And in [the] second place, I have very good connections here."[50] This referred to the fact (not very important) that a friend of the Einstein family was acquainted with a professor of chemistry in Milan and that, in addition, an uncle of Michele Besso, the mathematician Giuseppe Jung, had been asked to make inquiries among his Italian colleagues. "I will soon have graced every physicist from the North Sea to the southern tip of Italy with my offer,"[51] Einstein boasted to Mileva, assuring her that he would not leave a single stone unturned.

Meanwhile a glimmer of hope appeared from a totally unexpected direction. On the very day that his father had described him to Professor Ostwald as "deeply unhappy," Albert Einstein received a letter from Marcel Grossmann, "in which he informed me that I'll be getting a permanent position at the Swiss Patent Office in Bern."[52] He went on, to Mileva: "Just think what a wonderful job this would be for me! I'd be overjoyed if something came of it."

Marcel Grossmann's father was a friend of Friedrich Haller, the director of the "Federal Swiss Office for Intellectual Property"—the Patent Office—and had recommended his son's unemployed colleague for the next vacancy in Bern. Work at the Patent Office had nothing to do with an academic career, the date of the next vacancy was uncertain, and the prospect of being appointed to it was still rather vague; but Einstein thanked Marcel Grossmann effusively: "I am truly touched by your loyalty and humanity which did not let you forget your old luckless friend. . . . I need hardly tell you that I would be happy to be granted such a fine field of activity, and that I would do everything in my power not to disgrace your recommendation."[53] He also knew how he would spend the first few months of the waiting period. Jakob Rebstein, formerly an assistant to Professor Herzog at the Polytechnic and now teaching at the Technical College in Winterthur, had to enlist for his two-month mandatory military service in mid-May and had asked Einstein if he could stand in for him. "You can imagine how happily I would,"[54] Einstein wrote to Mileva.

One of the conditions of employment at the Office for Intellectual Property was Swiss citizenship, but this condition, after more than a year of bureaucratic delays, had now been met. Einstein's wish to become Swiss was due not only to the inconvenience of being stateless. Actually, later, recalling the good old days before World War I, Einstein wrote: "An ordinary person then would not even know what a passport was, because none was needed for traveling. Besides, I was stateless for five years without any difficulties arising from that circumstance."[55] Nevertheless it had seemed useful to him, after the conclusion of his studies, to become a Swiss citizen in order not to exclude himself from state-controlled jobs, such as teaching posts. An equally important reason, according to his sister Maja, was the "agreement between his political convictions and Switzerland's democratic constitution."[56]

Swiss citizenship is an automatic consequence of cantonal or municipal "civic rights," but these require the permission of the Federal Council in Bern. On October 19, 1899, Albert Einstein therefore submitted to the "Illustrious Federal Council of the Swiss Confederation in Bern" his application for "permission to acquire Swiss cantonal & municipal civic rights." He enclosed a good-conduct testimonial from the Zurich police and his certificate of release from Württemberg citizenship. Being still a minor, he also needed the written agreement of his father; this he supplied subsequently.[57] The Zurich cantonal police sent a "favorable" report to the Federal Attorney's Office in Bern, and on March 10, 1900, the Federal Council approved Einstein's application. At the end of June, therefore, during his final exams at the Polytechnic, he submitted his petition to the Zurich City Council. This Council then involved various other departments and immediately charged Detective Hedinger to report on the applicant. Hedinger's report stated that Einstein was "a very zealous, hard-working, and exceedingly respectable person (teetotaler)."[58]

Einstein confirmed his dislike of alcohol when summoned on December 14, 1900, before the Immigration Section of the City Council,[59] which then recommended his application to the Plenary City Council. (Einstein remained a teetotaler in his later years. Thus when, with much hullabaloo, he arrived at San Diego, California, on December 30, 1930, on board the *Oakland*, and a newsreel reporter asked him what he thought of Prohibition, Einstein replied, cheerfully laughing at the camera: "I don't drink, therefore I couldn't care less."[60])

Among his answers to a "questionnaire for civic rights applicants,"[61] it is worth noting that he described himself as being of "no religious denomination" and, under the heading "occupation," declared: "I am giving private lessons in mathematics until I find a permanent post." It is also interesting that nowhere in these documents are there any questions about patriotic avowals or basic knowledge of civics. An applicant's wish and the satisfaction of the legal requirements were sufficient for the authorities.

The records also disclose that the "directorate of the interior" of the canton of Zurich gave instructions for Einstein's financial situation to be examined: "with regard to his economic circumstances, in so far as we endeavored to investigate the same, he offered no indication of any kind, except that it was possible to establish that he has no income from his own work. Concerning the information received from Milan this day, it emerges that he likewise has no resources from the side of his parents."[62] As the "Swiss Information Bureau," a kind of civil secret service, was not allowed to operate outside Switzerland, a private detective was hired to investigate Einstein's family. After a brief note on the bankruptcy of the firm of Einstein e Garrone in Pavia, the detective concluded: "In Milan, Einstein senior seems to be somewhat better off, but there is no real estate property and Einstein junior could certainly not count on pecuniary support from his father."[63] This information, however, did not harm the applicant at all; on the contrary, it had a beneficial effect: the authorities reduced his fee for the cantonal civic rights to 200 francs—he had had to pay twice that amount for the civic rights of the city of Zurich. On February 21, 1901, with a badly depleted savings account, the unemployed and impecunious Albert Einstein became a Swiss citizen.

The new citizen instantly became an object of interest to the military authorities, and on March 13, one day before his twenty-second birthday, the Zurich District Command summoned him for a medical examination. Under the heading "Diseases and Complaints" his service book[64] listed: "*Varices, Pes planus, Hyperidrosis ped.*," in plain language varicose veins, flat feet, and sweating feet. He was exempted from recruit training because of his classification of "Unfit A," according to which he had to perform only "auxiliary services, local service," though in fact he was never called on to do even these. Up to his forty-second birthday he regularly paid his military taxes—by way of compensation for not serving—though this did not prevent him later, when he had become a consistent pacifist, from publicly arguing against the Swiss militia system[65] and from encouraging a Swiss conscientious objector in a letter, telling him that through his example "the machines of war [would be] destroyed or at least the unworthy compulsory service abolished."[66]

Whatever other citizenship Einstein was to acquire in later years, whether through necessity or voluntarily, he always held his Swiss citizenship dearest of all. He invariably traveled on the red passport of the Swiss Confederation, and even when he was in America and an American, he emphasized in a brief account of his citizenship career that he had always remained Swiss: "I am still Swiss from Switzerland's point of view, as I have never renounced that citizenship. In addition, I was for a while an Austrian citizen (Prague), and as a member of the Berlin Academy from 1919 to 1933 I even was (horrible thought) a Prussian. The latter came to a dramatic end. . . . Now I am also an American. This country generously disregards the fact that one may also be a citizen elsewhere, even though it does not officially recognize it."[67]

He always had his own ideas about national labels. In 1919, when he explained his theory to the English in *The Times* (London), he startled them with an "application of the theory of relativity to the taste of readers," while at the same time poking fun at misunderstood popularizations and national claims to his person:

> Today in Germany I am called a German man of science, and in England I am represented as a Swiss Jew. If I come to be regarded as a *bête noire*, the descriptions will be reversed, and I shall become a Swiss Jew for the Germans and a German man of science for the English.[68]

This he wrote as a man who had suddenly become world-famous; for the moment, however, he was no more than a modest Swiss who had to reconcile himself to the thought that a university career would remain barred to him and for whom the principal advantage of Swiss citizenship was simply the fact that at least one formal obstacle to a post in the Swiss public service had been removed.

Einstein's job as a locum at Winterthur Technical College, which would keep him above water for two months, was a challenge for a young beginner in the profession. He had to teach thirty hours a week; moreover, he was teaching descriptive geometry, a subject he had frequently cut at the "Poly"; but, as he repeatedly assured Mileva, quoting a line from the poet Uhland, "the valiant Swabian is not afraid."[69]

Before starting the job he had met Mileva for a few days together at Lake Como. They sailed across the lake to Cadenabbia, admired the lush gardens of the Villa Carlotta, and crossed the Italian-Swiss frontier by the still snow-covered Splügen pass in a small horse-drawn sleigh. "It was merrily snowing all the time," Mileva shortly afterward wrote to a friend,[70] "as we drove one moment through long galleries and the next on the open road, where, all the way to the remotest distance, our eyes could see nothing but snow and more snow, so that at times I shuddered at this cold white infinity and firmly kept my arm round my sweetheart under the coats and blankets which covered us. . . . I was so happy to have my lover for myself again for a while, the more so as I saw that he was just as happy." Soon after this, Mileva must have told him that she was pregnant, for by the end of May he inquired: "How's the boy?"[71] Of course he assured his "dear kitten" of a better future. "Be happy and don't fret, darling. I won't leave you and will bring everything to a happy conclusion. You just have to be patient! You'll see that my arms aren't so bad to rest in, even if things are beginning a little awkwardly."

There is a mention also of happiness, indeed at the very beginning of the letter, but it refers to physics. That would scarcely have surprised Mileva, who by then had known her Albert for a few years: "I just read a wonderful paper by Lenard on the generation of cathode rays by ultraviolet light. Under the influence of this beautiful piece I am filled with such happiness and joy that I absolutely must share some of it with you."[72] His happiness and pleasure were to bear fruit: four years later his theoretical interpretation of the "photoelectric effect" would become the foundation of quantum theory, and two decades later it would earn him the Nobel Prize.

Einstein found teaching, and life, in Winterthur more agreeable than he had expected. He met Hans Wohlwend again—a friend from the Aarau cantonal school—and rented a room from Wohlwend's landlady at Schaffhauser Strasse 38, at the edge of town. He made music with his friend and—as he informed Mileva—"an older lady. I really enjoyed it."[73] Sundays he took the train to Zurich, twelve miles away, was in turn cheerful and sad with Mileva, and supplied himself with scientific literature. Teaching, as he wrote to Papa Winteler, was pleasing him "quite extraordinarily. I never suspected that I would enjoy teaching as much as in fact I did."[74] In order to accumulate some savings he also gave private lessons.

Of vital importance for Albert Einstein and his future plans was the surprising and pleasant discovery that the practice of a regular occupation did not exhaust him at all: "Having taught 5 or 6 hours in the morning, I am still quite fresh and work on my further education at the library in the afternoon, or on interesting problems at home."[75] Failure to find a university post clearly did not mean the end of his love of physics: "I have entirely abandoned the ambition of getting into a university now that I realize that, even as it is, I have enough strength and inclination left for scientific endeavor."[76]

As the Patent Office was, at best, a long-term prospect, Einstein applied for a couple of teaching posts, first at the Technical College in Burgdorf in Canton Bern, and next at the cantonal school in Frauenfeld, Canton Thurgau—both times without success. The fact that he applied for the Burgdorf job twice in short succession and that he sent his application to the wrong authority instead of to the address stated in the advertisement might suggest to some psychopathologists that deep down he was not all that keen on becoming a schoolteacher.[77]

The Frauenfeld post went to his friend Marcel Grossmann, whom Einstein warmly congratulated on his success: "[it] will provide for you a pleasant activity and a secure future. I too applied for that post, but only so I do not have to tell myself that I was too timid to apply. For I was firmly convinced that I had no prospect of getting this or any similar post."[78]

Einstein's self-assurance had not suffered from the many refusals and setbacks, and he was as busy as ever pursuing his scientific work in Winterthur. He reported to Mileva about a long letter he had written to Paul Drude, "with two objections to his electron theory. He'll hardly be able to offer a reasonable refutation, as my objections are very straightforward. I'm terribly curious to see if he replies, and to what effect. I mentioned of course that I'm without a position."[79] Drude, however, who was the director of the Physical Institute of the University of Giessen and also the editor of *Annalen der Physik*, had no post to offer and, besides, was by no means convinced by Einstein's objections—which Einstein saw as such "manifest proof of the wretchedness of its author that no further comment by me is necessary."[80] And to Papa Winteler he wrote that his dubiousness about the eerie neighbor in the north, and especially about German professors, had by no means been exaggerated. "I have again made the acquaintance of a sorry example of that species—one of the leading physicists of Germany. . . I will shortly give that man a kick up his backside with a hefty publication. Befuddled authority is the greatest enemy of truth."[81] As both Einstein's letter to Drude and Drude's reply have been lost, the factual background of the controversy can no longer be elucidated. But as Drude's electronic theory of metals was also criticized by other physicists and was not tenable in the long run—a fact which cannot have escaped so eager a reader of the journals as Einstein was—it is not impossible that Einstein did in fact touch on a few sore spots in Drude's theory.

Having finished his temporary teacher's job on July 15, Einstein once more joined his mother and sister in Mettmenstetten while Mileva was cramming for her exam in Zurich. On July 26 it was clear that she had failed again.[82] Deeply depressed, she traveled to see her parents in Novi Sad. Einstein returned to Winterthur, where he had retained his room, worked on his doctoral thesis, and kept his eyes skinned for new sources of income. In the *Schweizerische Lehrerzeitung* (*Swiss Teachers' Gazette*) he saw an advertisement for a "private tutor," to coach a young Englishman for the Swiss "school-leaving" exam at the private school of a Dr. Nüesch in Schaffhausen. In order to "be rid of the worries about where the next meal is coming from,"[83] he accepted the offer. "You can imagine how happy I am about it," a more modest

Albert Einstein wrote to Marcel Grossmann, "even if such a post is not exactly ideal for an independent character." In mid-September he therefore moved from Winterthur to Schaffhausen on the Rhine.

Einstein initially stayed at the house of his employer, Dr. Jakob Nüesch, who, in addition to being a teacher in a secondary school, ran a "Teaching and Education Institute." Einstein seems to have gotten on reasonably well with his young English student, Louis Cahen, who intended to study architecture at the Polytechnic and therefore needed the Swiss "school-leaving" examination. Neither of them seems to have shown excessive zeal.[84] But he did not feel at all happy in the house of his employer, who under their contract was to provide free bed and board, in addition to a monthly salary of 150 francs. Einstein would gladly have done without the domestic setting and the meals with the large Nüesch family. When he discovered that Cahen's mother had to pay 4,000 francs for one year of instruction for her son, and that Nüesch therefore was pocketing a huge profit, Einstein provoked one row after another, with the result that he was allowed, first, to live on his own and, next, to take his meals in a restaurant—all at Nüesch's expense. To Mileva he wrote that "the Nüesches are in a vicious rage against me, but now I'm just as free as any other man. . . . Long live impudence! It is my guardian angel in this world."[85]

His only other hope was the Patent Office, but more than six months had elapsed since the Grossmanns' intervention. "I've given up on the position in Bern, as no notice has appeared in the newspaper yet,"[86] he wrote to Mileva toward the end of November. She in turn, writing to a friend, bewailed "the misfortune of Albert not finding a post. . . . You know that my sweetheart has a sharp tongue and moreover he's a Jew."[87]

Albert Einstein now placed his hopes in the effect which his doctorate, once he received it, would have on his search for a post: "As soon as I receive my doctorate I'll apply for a secure position. Someday fate will smile on us."[88] When he had submitted his thesis to Professor Kleiner on November 23 and paid his examination fee, he boasted to Mileva that not one of his former colleagues, who had landed assistants' posts, had yet completed a thesis: "See, your Johnnie finished his paper first, despite being hounded in the process."[89] But the degree procedure, too, brought him nothing but annoyance.

Professor Kleiner, in Einstein's eyes, was a useless "shortsighted person." "If I had to be at his beck and call in order to become a professor, I'd rather remain a poor private tutor."[90] Much to Einstein's chagrin Kleiner was in no hurry to read his thesis, and he aroused Einstein's fury by refusing to keep the university library open especially for him during the Christmas vacation, when he intended to come over from Winterthur: "It's really terrible, all the things these old philistines put in the path of people who aren't of their kind. They instinctively view every intelligent youth as a danger to their fragile dignity,"[91] he ranted. Even six years later Friedrich Adler mentioned this episode as an illustration that Einstein was treated "rather contemptuously by the professors at the Polytechnic"; he was "locked out of the library, etc."[92] Einstein, who presumably had included a few cheekily formulated objections to Drude's theory of electron conduction in metals, was curious to see how not only Kleiner but also Drude would react: "a fine bunch, all of them. If Diogenes were alive today, he'd be looking in vain for an honest person with his lantern."[93] It was predictable that this emotionally overcharged degree procedure would end in disaster, and Einstein was lucky to recover at least his fee of 230 francs when he withdrew his doctoral thesis.[94]

Mileva spent these months with her parents. After some outrage demanded by bourgeois propriety they eventually reconciled themselves to Mileva's pregnancy and, like her, were hoping that Einstein would establish himself in a sound occupation and as head of a family as soon as possible. But that was the very thing Einstein's parents, especially his mother, were dreading. Although they were unaware of Mileva's condition, they feared that Albert would not give her up and therefore resorted to drastic measures. "They felt no compunctions"—Mileva complained to a friend—"about writing a letter to my parents, reviling me in a manner that was a disgrace."[95] Toward the end of October, Mileva went to Switzerland. But in order not to compromise the young private tutor by her then "funny shape," she stayed at a safe distance from Schaffhausen, at the Hotel Steinerhof in Stein am Rhein, some twelve miles along the river. Soon everything was just as in the old days: "When we are together we are merrier than anyone else. . . . And, do you know, in spite of everything bad, I must love him so much, quite frightfully much, especially when I see that he loves me like that too."[96] When Mileva returned to Novi Sad again, Einstein showered her with good advice and seemed to look forward to the birth of their child without misgivings: "Just take good care of yourself and keep your spirits up, and be happy about our dear Lieserl, whom I secretly (so Dollie doesn't notice) prefer to imagine as a Hanserl."[97]

At last there was cause for optimism. On December 11 Einstein received news from Marcel Grossmann: "a very sweet letter in which he said that the position in Bern will be advertised within the next few weeks and that he's certain I'll get it. In two months' time we could find our lives brilliantly changed for the better, and the struggles would be over. I'm dizzy with joy when I think about it. I'm even happier for you than for myself."[98] Grossmann after all had not promised too much. The same day an advertisement appeared in the *Federal Gazette*, almost as if it had been tailored for Einstein. It demanded "thorough university education of a mechanical-technical or specifically physical direction."[99] The head of the Patent Office, Friedrich Haller, had added the (not normally customary) "physical direction." Until then there had been no physicists in the Patent Office.[100]

In polite bureaucratic German, Einstein on December 18 composed his application for the "Engineer II Class vacancy in the Swiss Federal Office for Intellectual Property,"[101] casually mentioning his thesis "on an aspect of the kinetic theory of gases."[102] The following day he went to see Professor Kleiner in Zurich. Kleiner had not yet read Einstein's work, but they talked about all kinds of physical problems, and Einstein concluded that the professor was "not quite as stupid as I'd thought, and moreover, he's a good fellow. He said I can count on him for a recommendation anytime."[103] Small wonder that Einstein was "absolutely crazy with happiness."[104] Over the Christmas vacation he stayed in Schaffhausen, full of expectation of the "opportunities that beckon to us in the near future."[105] Only on Christmas Day and the next day did he allow himself a special treat: with his sister, who had come over from Aarau, he spent these two days at the Hotel-Pension Paradies in Mettmenstetten, known to the family from past vacations.

Nothing is known about the events of January 1902 which led to Einstein's premature departure from Schaffhausen. We know neither the doctoral thesis nor Professor Kleiner's reservations which induced Einstein to withdraw it. But in any case the prospect of the post in Bern would have made it easy for him to pack his bags at the end of January. By way of Zurich, where he recovered his doctoral fee from the university chancellery, he probably went straight to Bern. Certainly, as he wrote to a friend, he had "cast off with a bang"[106] from Schaffhausen and Dr. Nüesch's institute.

Meanwhile, he had become a father—but this he learned only from a letter from Novi Sad, forwarded to him in Bern. The baby, as Mileva had hoped, was a Lieserl. It was not until the middle of the year that Einstein could at last start on his job. Toward the end of the year Mileva came to Switzerland—but without Lieserl.

Einstein may have arrived in Bern with very little baggage in the conventional sense, but his mind by then was well stocked with creative physical thought. Unusual for a young man was his focus on the "fundamentally important," as well as the broad spectrum of his interests. Alongside the "great" themes of thermodynamics and the kinetic theory of gases, and Maxwell's theory of the electromagnetic field, he concerned himself with capillarity, thermoelectricity, and the electronic theory of conductivity in metals.

Although his studies proceeded on the margin of topical physical research and largely in an autodidactic manner, he seemed to aim, almost with a sleepwalker's assurance, straight at the central problems. He was a regular and thorough reader of *Annalen der Physik*, though he could not afford to subscribe to the journal himself. In his letters to Mileva Marić he would regularly comment on articles which had especially interested him. These included not only Wilhelm Wien's article on the ether, Drude's work on metal conductivity, and Lenard's experiments on the photoelectric effect, but above all Max Planck's work, published toward the end of 1900, about his new radiation formula which contained his quantum, subsequently to be known as "Planck's constant." Einstein commented on this article quite critically as early as April in a letter to Mileva,[107] and he realized, "shortly after Planck's pioneering work, that neither mechanics nor electrodynamics can (except in limiting cases) claim exact validity."[108]

As a way out of the dilemma in which physics found itself at the turn of the century, young Einstein had in mind a fundamental theory along the lines of thermodynamics: "the discovery of a general formal principle might lead us to reliable results."[109] Such a theory was presented by Einstein a few years later in the form of the special theory of relativity, but his letters from his "apprenticeship period" are already pervaded by this theme. It is possible that the crucial insight had already come to him while he was at school in Aarau and that since 1898, with a physicist's tools, he had further pursued the idea. "I'll be so happy and proud"—he wrote to Mileva in the spring of 1901—"when we are together and can bring our work on relative motion to a successful conclusion!"[110] Toward the end of 1901, in Schaffhausen, he was "busily at work on an electrodynamics of moving bodies, which promises to be quite a capital piece of work."[111]

Unfortunately it is impossible to conclude from such remarks, in those letters that have come down to us, what direction his thinking took at that time, or indeed what wrong turnings he must have taken. His "splendid treatise" still needed a lot of work and one flash of inspiration—and that would come to him only in the spring of 1905, when he was an Expert III Class at the Patent Office in Bern. Nevertheless, his earlier letters already reveal an unshakable conviction that he would succeed in solving the riddle, no matter what job he was holding down. There probably never was a young man about to enter a modest post with, at the same time, such high-flying plans as Albert Einstein, when he arrived in Bern in February 1902. And the most astonishing thing is that his hopes in fact came true.

PART II

THE PATENT OFFICE

CHAPTER SIX

Expert III Class

ALBERT EINSTEIN'S MOVE TO BERN had all the characteristics of an escape by an angry young man—it was rash and not free of risk. He would have to live hand to mouth, subsisting on hope, because all that was known at that point about the post at the Patent Office was that his application had been received and that the director, Friedrich Haller, would consider it sympathetically as soon as any new posts were authorized. But he was glad not to have to be a private tutor in Schaffhausen any longer. "It does me a world of good to have escaped from those unpleasant surroundings."[1] Besides, if he did have any worries, he was not going to show them.

"It's wonderful here in Bern,"[2] he wrote to Mileva immediately after his arrival at the beginning of February 1902. "An ancient, thoroughly pleasant city, in which one can live exactly as in Zurich." What Einstein liked best about this architectural gem of a city, more than the massive ramparts or the impressive baroque towers, were the arcades along both sides of the old streets. He found a room on Gerechtigkeitsgasse 32 at the lower end of the fine line of streets crossing the city, quite close to the Nydegg Bridge and the bear pit, where Bern's heraldic animals are kept.

There was still the problem of a livelihood. Since the matter at the Patent Office was making no visible progress, Einstein, as a holder of the Swiss specialist teacher diploma, offered his "exceedingly thorough" services as a private tutor in mathematics and physics in the local advertising paper—with "free trial lessons."[3] He received a few replies, including one from an engineer and one from an architect, and already saw himself as the professor of a small private college, with enough earnings to cover the waiting period for the Patent Office.

One of his first students was Louis Chavan, a French-speaking Swiss technician in the Swiss Postal and Telegraph Service, and before long one of Einstein's most loyal friends in Bern. In his meticulously kept notebooks Chavan not only recorded Einstein's lessons but also left us a thumbnail sketch of his youthful teacher: "His short skull seems unusually broad. His complexion is a matt light brown. Above his large sensuous mouth is a thin black moustache. The nose is slightly aquiline. His striking brown eyes radiate deeply and softly. His voice is attractive, like the vibrant note of a cello. Einstein speaks correct French, with a slight foreign accent."[4]

Immediately on his arrival in Bern, Einstein had received a letter from Mileva's father, addressed to Schaffhausen but forwarded to him, which, when he read it, "frightened [him] out of his wits."[5] He was quite prepared, as an unemployed man in his early twenties, to be accused, in terms of bourgeois morality, of having become, scandalously prematurely, a father, and without the blessing of a rabbi or at least a registry office clerk. What he was not prepared for was to learn about the serious complications of the birth. Mileva was actually so exhausted that she could not write to him herself. Einstein's concern about Mileva's health, however, was tempered by his concern for and joy over the baby: "Is she healthy and does she cry properly?" he inquired. "What are her eyes like? Which one of us does she more resemble? . . . I love her so much and don't even know her yet."[6] With a researcher's curiosity he continued: "Is she looking at things yet? Now you can make observations." And finally he made a frank admission of men's inferiority compared with women's reproductive ability: "I would like to make such a Lieserl myself one day, it must be most interesting!" This wish, almost too explicit on a psychoanalytic interpretation, was to be echoed on many occasions, when Einstein described his mental efforts as "hatching" and sometimes as "laying eggs."

Despite Einstein's protestations—"I long for you every day" and

"I'd rather be with you in some provincial backwater than without you in Bern"[7]—there was no mention in his letters of an imminent marriage. The reason probably was not so much the fact that without a post at the Patent Office he would not be able to support a family but rather the vigorous opposition of his parents. Pauline Einstein reacted to what to her was anything but a happy event, and to Albert's suggestion of at least an official engagement, by declaring, "We are resolutely against Albert's relationship with Fräulein Marić, and we don't ever wish to have anything to do with her. . . . That Fräulein Marić is causing me the bitterest hours of my life; if it were in my power, I'd do anything to banish her from our sight, I have a veritable antipathy toward her."[8] She complained of "having lost any influence on Albert." Given his scorn for "Philistines" and his angry resolution not to give a damn about the world, this sounds entirely credible. Yet basically he was an obedient son, reluctant to rebel against his mother's fierce determination—with the result that for the time being there was no engagement, let alone a wedding. Mileva, therefore, having to bear the precarious consequences of premarital motherhood on her own, stayed with her parents in Novi Sad, while Einstein was trying to survive in Bern until a decision came through about the Patent Office.

Max Talmud, his mentor from his school days in Munich, had the impression that Einstein was only just about managing to survive. Traveling through northern Italy in April 1902, Talmud remembered that the Einsteins were now living in Milan and visited them. He found them depressed, and, in response to his inquiries about their son, was told only that he was now living in Bern. Unaware of the cause of this estrangement, he went to Switzerland specifically to see his former pupil, whom he found in conditions "testifying to great poverty." The lodging on Gerechtigkeitsgasse, which Einstein had described to Mileva as though it were a small palace, struck Talmud as a "small, poorly furnished room."[9] In conversation Einstein blamed his sorry situation—no assistant's post and his failure to get a doctorate—on obstacles "laid in his way by people who were jealous of him."

But Einstein had a knack for making the best of adversity. Although he had arrived in Bern totally unknown and with no contacts either socially or in university circles, he did not remain alone long.

Paul Winteler, one of the sons of Einstein's "parents" in Aarau, had just begun to study law there. And from his time in Schaffhausen Einstein knew Conrad Habicht, who was working on a doctoral thesis in mathematics. He also met Hans Frösch again, a classmate from Aarau, now studying medicine. Einstein explained some of his physical problems to Habicht, the mathematician, who was "very enthusiastic about my good ideas."[10] He accompanied Frösch to a class on forensic pathology with "demonstrations ad oculos." He was "so fascinated" by the drunkards, arsonists, and megalomaniacs presented to the students "that from now on I'm going to go every Saturday."[11]

In his own special subject of physics, however, Einstein did not discover anything of comparable interest. Aimé Forster, the head of the department, lived in the astronomical observatory, converted into a laboratory, on the Grosse Schanze, the old city ramparts, not far from the huge site where the new main building of the university was then under construction. Forster had gained some renown, but as a local meteorologist rather than a physicist, and his lectures did not go beyond an elementary level; he made only derogatory remarks about theory.[12] The self-assured young Einstein could not take Forster quite seriously and had to find intellectual contacts elsewhere.

In response to Einstein's advertisement in the *Berner Anzeiger* a young Jew from Romania, Maurice Solovine, turned up one day. Solovine had come to the University of Bern with a thirst for knowledge but no clear idea of what to study. He had enrolled for both philosophy and physics but had found Professor Forster's lectures theoretically and philosophically shallow. So he turned to the private tutor on Gerechtigkeitsgasse. In old age, Solovine recalled climbing the stairs to Einstein's room, hearing his forceful "Come in," and being impressed by the brilliant clarity of Einstein's eyes.

After a few hours of regular, paid teaching, Einstein found it more interesting to discuss with Solovine the general philosophical foundations of physics, for which purpose Solovine was to visit him whenever he felt like it. At the beginning of the summer semester, Conrad Habicht also joined these conversations, and the three decided to set up a kind of club, a discussion circle with a firm agenda and an absurdly grandiloquent name, Akademie Olympia. The three would meet regularly in the evening for a frugal meal of sausage, some Gruyère cheese, a little fruit, honey, and tea. That, according to Solovine's recollections, was enough for them to "brim over with merriment."[13]

Regardless of the fun and games, the cheerful "Academy" was based on a serious, systematic reading program. Solovine has noted the books the three members studied and discussed—an impressive list, with a clear emphasis on theoretical works touching on the foundations of physics. Einstein was probably the driving force, putting on the program Ernst Mach's *Analysis of Perception* and *Mechanics and Its Development*—antimetaphysical works which he knew from his student days. Under the same heading came Karl Pearson's *Die Grammatik der Wissenschaft* (*Grammar of Science*) and Richard Avenarius's two-volume *Kritik der reinen Erfahrung* (*Critique of Pure Experience*), though only one chapter of this work was discussed. Several weeks were devoted to *La Science et l'hypothèse* (*Science and Hypothesis*) by the great Frenchman Henri Poincaré. But older works were studied as well, such as John Stuart Mill's reflections on the problem of induction in the third volume of his *Logic*, and David Hume's subtle critique of causality. Five decades later Einstein still remembered that "we chiefly concerned ourselves with D. Hume (in quite a good German edition). His work was of considerable effect on my development—along with Poincaré and Mach."[14]

Now and again the "Academy" meetings were enriched by a little violin recital by Einstein, and fine literature was read in a kind of "general studies" program—Sophocles' *Antigone*, Racine's tragedies, and *Don Quixote*. These works were probably more in line with Solovine and Habicht's interests than Einstein's. In reply to a question from a literary man, Einstein once confessed that "as a young man (and also later) I did not concern myself much with poetical literature or novels."[15] He explained this, however, as resulting from too much empathy rather than too little: "It is partly due to the fact that the artistic aspect was easily lost on me because the fate of the characters as such gripped me too strongly." Thus some plays by Gerhart Hauptmann had "enormously inflamed" him, and when a friend read *Hanneles Himmelfahrt* aloud "I had to cry like a child, half from bliss, half from pain."[16] His sensibility evidently made it necessary for him to shape his preferences differently: "What I loved more were books of ideological character, and especially philosophy."[17] At any rate, Einstein's preferences were served by Baruch Spinoza's *Ethics*, which he continued to read frequently, long after the little "Academy" had been scattered to the winds.

Although he had to abandon his hope of an assistant's post at a university, Einstein had not given up his scientific interests or even the idea of a scientific career. One prerequisite, at the turn of the century as today, was for his name to become known through publications. Einstein therefore used his ample leisure time to make his reflections on electrolysis, begun in Schaffhausen or possibly even in Zurich, ready for publication.

His second essay, like his first, dealt with the conclusions deriving from the hypothetical assumption of molecular forces and, more particularly, with the experimental verification of these conclusions. Toward the end of April the manuscript was sent to the editor of *Annalen*, and ten weeks later, without any problems, it was published under a long, involved title.[18] More significant than his somewhat turgid and ultimately unproductive physical reflections is Einstein's concluding apology for "only setting out a meager plan for a demanding investigation without contributing anything to an experimental solution." Because he was in no position to perform the necessary experiments himself, he believed, rather grandly for an unknown twenty-three-year-old, that his paper would have accomplished its objective "if it induces some researcher to attack the problem of the molecular forces from this angle."[19]

Nothing of the sort happened, and five years later the author himself would describe this paper as one of his "two worthless firstling works."[20] But his pattern of throwing out an idea and inducing his colleagues to pursue it was to recur frequently, and within a few years with every justification.

Because it did not yield much in terms of science, the study of molecular forces had lost its attraction for Einstein. There were plenty of problems to be pondered, but for the next two years only one field ripened into publications—the foundations of thermodynamics. As early as June 1902, Einstein was able to send off a paper titled *The Kinetic Theory of the Heat Equilibrium and of the Second Law of Thermodynamics.* This was the first of a series of three publications which will be described in context later. At this point it should merely be said that with them Einstein not only established himself as an original researcher but also developed the foundation of his later magnificent contributions to statistical physics. To him, however, just as important as the completion of his first thermodynamics treatise was the fact that, after a long and nerve-racking waiting period, he at last got into the Patent Office.

Spring was well advanced when the Swiss bureaucracy moved into gear again to fill two vacancies in the Patent Office. Sometime toward the end of May, Friedrich Haller invited the candidate Einstein, recommended to him by Grossmann *père*, for a "thorough oral examination."[21] This time Einstein's hopes were not disappointed: a proposal soon went to the Swiss Federal Council that the mechanical engineer Heinrich Schenk and Albert Einstein be "provisionally elected Technical Experts III Class at the Federal Office for Intellectual Property, at an annual salary of 3,500 francs each."[22] Two weeks later came their appointment by the Federal Council, and on June 23, 1902, Einstein reported for work. The "annoying business of starving"[23] had come to an end. Marcel Grossmann had again proved to be a "lifesaver," and Einstein would remain eternally grateful to him. But for that opportunity, he observed, "I might not have died, but I would have been intellectually stunted."[24]

Henceforth, every morning at eight o'clock, Einstein went to "the office." The Patent Office was then on the upper floor of the new, somewhat pompous building of the Postal and Telegraph Administration on Genfergasse, near the railroad station. His first impressions were clearly favorable: "I very much like my work at the office," he soon reported to a friend, "because it is enormously varied and calls for much thought. What I like even better is my handsome pay."[25] He got on well with the director and the dozen or so colleagues ranked as "experts," but he complained about the workload: "I have a frightful lot of work. Eight hours at the office each day and at least one private lesson, and then I have my scientific work."[26] Once he had settled in, though, he found his forty-eight hours per week at the office tolerable. When his friend Habicht was not entirely satisfied with the school service, in which he had landed after completing his studies, Einstein suggested that one day he would smuggle Habicht in among the "patent slaves" and tried to commend the work to him by observing that "along with the eight hours of work there are also eight hours of fun in the day, and then there is also Sunday."[27]

Even though the Patent Office had been Einstein's second choice, he never felt that the job was merely a matter of survival. For more than seven years he was a good "patent slave,"and for that reason, perhaps, the Patent Office even seemed in retrospect to have been an ideal refuge: "Working on the final formulation of technological patents was a veritable blessing for me. It enforced many-sided thinking and also provided important stimuli to physical thought."[28] In old age he even considered himself lucky to have escaped the academic treadmill by having a practical occupation. Academia, he believed, "places a young person under a kind of compulsion to produce impressive quantities of scientific publications—a temptation to superficiality, which only strong characters can resist."[29] He further argued that "most practical occupations are moreover of a character which allows a person of normal gifts to achieve whatever is expected of him. In his civil existence he is not dependent on special illuminations. If he has deeper scientific interests, let him engross himself in his favorite problems alongside his regular work."[30] Einstein was lucky; in the seclusion of the Patent Office he managed to produce "impressive quantities" of publications, which also happened to be of outstanding importance. When he had attained his goal and become a famous professor, he looked back nostalgically, in a letter to his friend and sometime colleague Michele Besso, to "that temporal monastery, where I hatched my most beautiful ideas and where we spent such a pleasant time together."[31]

▪ ▪ ▪

Albert Einstein may have found himself quite at home among patents. As a boy he had watched Uncle Jakob, the busy development engineer of the family firm, applying for six patents. In Professor Weber's laboratory he must also have come into contact with recently patented inventions. Nevertheless, the duties of an Expert III Class at the Swiss Patent Office were something a novice had to be instructed in. This was done personally by way of individual meetings with the boss. By his strict, almost schoollike instruction of new staff members, Professor Halder ensured that the patent examiners would all judge submitted inventions according to objective, verifiable, uniform criteria that could, if necessary, stand up in a court of law. Einstein, moreover, as a physicist with only slight experience in reading and interpreting technical drawings, lagged behind his colleagues with engineering training; this called for private instruction from the director. Einstein seems to have accepted Haller's severe regime without demur: he regarded him as "a splendid character and a clever mind. One soon gets used to his rough manner. I hold him in very high esteem."[32]

The reason why the job called for "much thought" was that a patent officer's central role was as a bookkeeper of technical progress on a scientific basis. As a rule, an inventor—in Einstein's days rarely represented by a patent attorney—would formulate his claims; and the officer, in addition to verifying the formal criteria, had to decide whether the submitted invention was in fact new and deserving of patent protection, whether it infringed on existing patents, and, in the case of more elaborate machinery, whether it actually worked. All that had to be done on the basis of drawings and specifications.

It turned out that for young Albert Einstein, examining patents was more than just a livelihood. In fact, it agreed quite strikingly with his characteristic approach to his favorite problems in physics. His virtuosity with "mental experiments" was not all that far removed from intellectual penetration of an invention, and his typical way of thinking in images involved visual more than conceptual operations. Einstein, in short, had almost providentially landed in a job which was so much in line with his own way of thinking that he might have experienced it as an agreeable exercise in technological and scientific imagination.

Even the procedures favored Einstein's inclination toward criticism and contradiction. Of course, he could not oppose Professor Haller's authority, but he could criticize and contradict the applicants—moreover, on his boss's instructions. "When you pick up an application, think that anything the inventor says is wrong,"[33] Haller advised his experts; otherwise, "you will follow the inventor's way of thinking, and that will prejudice you. You have to remain critically vigilant." This procedure of "brushing an argument against the grain" and, if possible, refuting it by a counterexample, greatly sharpened one's thinking and was entirely to Einstein's taste.

It would be interesting to prove this point with Einstein's expert opinions, but we have too few traces of his activity at the Patent Office. Under the rules, all papers were destroyed after eighteen years of patent protection. Even in the 1920s, when it was realized that no other employee of the Bern Patent Office, or any patent office anywhere, would ever rise to Einstein's heights, neither Friedrich Haller nor his successor wished to make an exception from that rule for the benefit of future biographers. Thus the last papers processed by Albert Einstein probably went into the shredder in 1927. Only one of his expert opinions has come down to us, because it found its way into a court record and there survived. It was compiled in 1907, when in the official judgment Einstein was "one of the most highly esteemed experts at the Office."[34] It rejected a patent claim by the AEG Company of Berlin for an alternating-current collector as "incorrect, imprecise, and not clearly drafted." "As for the various shortcomings of the design, we can deal with those only when the subject of the patent has been clarified by a correctly drafted claim."[35] That was a sovereign, curt judgment, entirely in the spirit of Einstein's chief, who was anxious to teach inventors, especially big firms, who was the boss.

This critical comment suggests that Einstein was concerned mainly with processing electrical-engineering patents. The original advertisement for the post had specified "thorough university education of a mechanical-technical or specifically physical direction." As physics had not previously figured in these advertisements, Einstein believed that "Haller put this in for my sake."[36] But this probably says more about young Einstein than about Haller, who, having headed the Patent

Office since 1887 with great propriety, would hardly have indulged in such a maneuver *ad personam.*

Actually, Haller's reason for enlarging his staff of predominantly mechanical engineers by adding a physicist was the rapid development of the electrical industry and a resultant flood of patent applications in that field. The first decade of the nineteenth century was characterized by the development of advanced alternating-current and polyphase-current machines, by telephone technology, and especially by wireless transmission of information by means of electrical oscillations. To a far greater extent than the pioneering achievements of the self-taught geniuses Werner Siemens and Thomas Alva Edison, the inventions of this later period were based on a theoretical understanding of electromagnetic phenomena—which was why Einstein, as a physicist familiar with Maxwell's theory, found a rich and interesting field of activity at the Patent Office.

Also, he enjoyed working with patents, and he continued to do so long after he left Bern. In later years he often served as an expert or consultant on patents, and he kept up the connection with his "temporal monastery" by having a few of his own inventions patented in Switzerland. When anyone expressed surprise that such a famous scholar should stoop to technology, he would say: "I . . . never ceased to concern myself with technical matters. This was of benefit also to my scientific research."[37] His first biographer even noted that for Einstein there was "a definite connection between the knowledge he acquired at the Patent Office and the theoretical results which, at that same time, emerged as examples of the acuteness of his thinking."[38]

The "handsome pay" of 3,500 francs a year was about double what Einstein could have expected from an assistant's post—by no means grand but sufficient for a modest bourgeois family existence. Nevertheless, there was still no question of marriage. His parents' opposition was too great, and so, despite all his protestations, was his emotional dependence on them. He finally received their consent in the saddest possible circumstance—when his father was dying.

Hermann Einstein's health had been prematurely undermined by the ceaseless worries of his life as an entrepreneur. Although his two power plants in northern Italy did not, for once, go bankrupt, they did not yield enough profit to repay his loans, chiefly from relatives. "The poor things have been constantly aggravated and worried about the damned money,"[39] is how his son described the situation. "My dear uncle Rudolf ("The Rich") has been nagging them terribly." In the fall of 1902, just after Hermann Einstein had celebrated his fifty-fifth birthday, his heart proved no longer up to the stress. His son arrived in Milan just in time to see his father on his deathbed. At that painful hour of parting, Einstein received his parents' consent to his marriage. Hermann Einstein died on October 10, 1902, and was buried in Milan. "When the end came, Hermann asked all of them to leave the room, so he could die on his own. His son never recalls that moment without a sense of guilt."[40]

Less than three months after his father's death, about the turn of 1902–1903, Einstein summoned Mileva to Bern. He had given up his modest room on Gerechtigkeitsgasse in August and had moved to Kirchenfeld, an area newly developed after the construction of a bridge over the Aare. In a pretty house of typical Bernese style on Tillierstrasse 18 he had rented a small attic apartment with a big balcony and a splendid view of the Bernese Alps. This became the first home of Albert and Mileva Einstein.

On January 6, without much ado, the wedding took place at the registry office in the old city. No wedding guests had arrived either from Einstein's or from Mileva's family. The witnesses were the other two members of the Akademie Olympia, Conrad Habicht and Maurice Solovine. Then the small party went to a photographer, and in the evening they celebrated a little.

When the newlyweds arrived at Tillierstrasse, Einstein had to rouse the landlord: he had forgotten the key to his apartment.[41] That tells us something.

Einstein, in fact, drawing up the balance sheet of an unhappy marriage that eventually ended in divorce, would recall that he had married primarily from a "sense of duty." "I had, with an inner resistance, embarked on something that simply exceeded my strength."[42] The fact that he had wrested consent to that marriage from his father under tragic circumstances must have been a greater trauma than he at first admitted to himself. And the fact that he was the first of his family to marry outside the faith must also, despite their assimilation and emancipation, have hung as a grave if silent reproach between them. Many years later, in Princeton, when he was asked by Jewish students in a discussion whether marriage outside the tradition was permissible, his answer reflected his personal experience: "That is dangerous—but then, any marriage is dangerous."[43]

However, he never mentioned the greatest stress to which his marriage with Mileva must have been subject from the outset—the fate of their daughter. In a letter to Mileva shortly before the child's birth, while Einstein was still waiting for his job in Bern, the only question yet to be resolved was "how to keep our Lieserl with us; I wouldn't want to have to give her up."[44] Despite these initial intentions, the girl at first remained with Mileva's parents in Novi Sad.

None of these problems or any other problem is reflected in his first accounts of his new marital status. "So I'm a married man now,"[45] he reported to Besso, "and lead a very pleasant comfortable life with my wife. She looks after everything splendidly, is a good cook, and is always cheerful."

Einstein also had every reason to be satisfied with his scientific work. Immediately before assuming his post at the Patent Office he had sent off his first study on the molecular-kinetic explanation of the theory of heat,[46] and two weeks after his wedding a second treatise, developing this subject in greater depth, was ready.[47] He thus took up again a subject which had already "made a very great impression on the student":[48] the explanation of many superficially disparate properties of matter, especially of gases, solely on the basis of the mechanical movement of countless minute particles—the atoms or molecules—as well as the derivation of the laws of thermodynamics through the statistical treatment of mechanical systems.

In Professor Weber's lectures at the Polytechnic no mention had been made of the latest advances by James Clerk Maxwell or Ludwig Boltzmann in this field. Einstein had acquainted himself with the theory through private study. By September 1902 he believed he had achieved something suitable for *Annalen*: "I have lately concerned myself thoroughly with Boltzmann's work on the kinetic theory of gases & over the past few days have written a small piece of my own, supplying the final brick to a chain of proof started by him."[49]

He did not publish it immediately, though, presumably because he wanted to use the idea in his doctoral thesis. But five months after withdrawing his thesis he produced a treatise "deriving the laws on heat balance and the second law of thermodynamics solely by the use of mechanics and probability calculus."[50]

Einstein began by declaring that while "Maxwell's and Boltzmann's theories had already come close to that objective," his observations would "fill a gap"—without, however, revealing to the (to this day) somewhat bewildered reader what exactly that gap was. On the other hand, the gaps in Einstein's own acquaintance with Boltzmann's work are more obvious: although he had studied Boltzmann's two-volume *Vorlesungen zur Gastheorie* (*Lectures on the Theory of Gases*), he knew nothing of the subtle investigations which Boltzmann had published in the *Accounts of the Imperial Academy of Sciences in Vienna*, to which Boltzmann primarily owed his reputation as the "unchallenged head of theoretical physics in Germany."[51] Thus Einstein made some discoveries in his paper that he might have found in the literature. Admittedly, Boltzmann's acute reflections were far from being the common property of physicists at the time, or even of those interested in theory; but even so the novelty value of the paper must have been greater for its author than for the few experts in the field.

Nevertheless, Einstein had reason to be proud of his interpretation of some central concepts of thermodynamics, such as temperature and entropy, and in deducing these he formulated essential aspects of statistical mechanics which were to remain valid as the foundations of that discipline. Most noteworthy, however, is his endeavor to manage as far as possible without specific assumptions about mechanics and to base his arguments on general laws alone. Now and again the reader is crudely reminded that only very slight use has been made of mechanics;[52] this justifies the assumption "that our results are more general than the mechanical presentation commonly used." Summing up,

Einstein claims that "no assumptions had to be made . . . about the nature of the forces, nor even that such forces occur in nature."[53]

Einstein's next steps were obvious: for one thing, to further reduce the assumptions about mechanics; for another, to include the difficult problem of irreversibility, considering that in the past he had confined himself to states of equilibrium. Somehow statistical mechanics had to resolve the inherent contradiction that the basic equations for the atoms are reversible in time, but not the macroscopic processes—which, though ultimately reduced to these mechanics, do have a direction in time. Thus milk is stirred into coffee a million times each day, resulting in light coffee, but no amount of stirring will ever reverse that process, i.e., separate out the few drops of white milk and restore the black coffee.

This next paper was completed only "after a lot of rewriting and amending," which kept Einstein busy, in addition to his official duties, during his first few weeks of married life. "But now it is perfectly clear and simple, so that I am quite satisfied with it,"[54] he wrote with relief. "On the assumption of the energy principle and atomic theory, the concepts of temperature and entropy, as well as . . . the second law of thermodynamics in its most general form follow logically." If certain assumptions on the structure of mechanics were to be found correct, he saw "the generalization achieved by my last paper in the elimination of the concept of force"[55] as being entirely in the spirit of Heinrich Hertz's theoretical program.

This paper too contained a few things the author might have read in Boltzmann, especially on the subject of irreversibility, and on this point there was even a false conclusion. The fact that Einstein referred in his footnotes to Boltzmann's *Vorlesungen* demonstrates his isolation from the mainstream of scientific discussion; at the same time, however, the paper is a brilliant testimony to his creative treatment of even the most complex problems.

Both papers would have deserved to become milestones of statistical mechanics. That this did not happen—that they were largely ignored—was due to the fact that an epoch-making study by the American Josiah Willard Gibbs, *Elementary Principles in Statistical*

Mechanics, which included the topics also treated by Einstein, had been published in 1902. "The resemblance is downright startling"[56] was Max Born's comment. Discussion in the field was henceforward determined by Gibbs's comprehensive treatise. Thus Einstein, at the very beginning of his career, was a victim of that by no means rare phenomenon, parallel discovery. Einstein actually did not come across Gibbs's book until 1905, in its German translation. He later called it a "masterpiece, though tough reading, and most of it between the lines,"[57] adding that "many have read it, verified it, and *not* understood it."[58]

One person who had read and understood Gibbs was Paul Hertz, a privatdozent at Heidelberg and a distant relative of the great Heinrich Hertz. Referring to Gibbs in 1910—eight years after Einstein's publication—Paul Hertz criticized his derivation of the second law of thermodynamics: "If one assumes, like Einstein, that more probable distributions follow upon less probable ones, one is thereby introducing a special assumption which is by no means evident and which certainly demands special proof."[59] Such proof, as Hertz realized, was of course not possible. Clearly, in 1902 young Einstein was unaware of the pitfalls besetting any proof of the second law of thermodynamics; nor did he know Boltzmann's analysis, according to which more probable distributions follow on less probable ones not inevitably but merely with overwhelming probability. Therefore, it is not certain but only overwhelmingly probable that entropy increases. After Paul Hertz had visited Einstein—by then Professor Einstein—in Zurich and their conversation had produced complete agreement, Einstein unhesitatingly announced in *Annalen* that he regarded "this criticism as entirely correct."[60] Besides, he added, had he known of Gibbs's book earlier, he would "not have published those papers at all, but confined myself to the treatment of some few points."[61] At the time of publication, however, he viewed the papers in a different light: they were to ensure his entree, albeit by a side door, to an academic career.

Among his colleagues at the Patent Office Einstein discovered one with similar scientific interests—Dr. Josef Sauter, a French-Swiss, who had also studied at the Polytechnic and who had been Professor Weber's chief assistant for a while. Since Sauter was eight years older than Einstein, they had not met before. Sauter, like Einstein, tried to fill the gaps in the Polytechnic's syllabus by private study, so that Einstein was able to discuss with him Maxwell's thermodynamics and Helmholtz's and Hertz's theoretical concepts. To the astonishment of his older colleague, Einstein frequently declared: "I am a heretic."[62] The two also discussed Einstein's publications on thermodynamics, with the result that Sauter discovered a mistake in them, which Einstein accepted "without being in the least upset."[63] Fifty years later Einstein recalled "that I had a lot of discussions with Sauter about . . . my thermal-statistical papers,"[64] but he had forgotten "what aspects were then being discussed."

At least as important as his help with the "rewriting and amending" were Sauter's connections with scientific circles in Bern, to which he soon introduced his new colleague. Thus, shortly after starting work at the Patent Office, Einstein was invited, as Sauter's guest, to the meetings of the Naturforschende Gesellschaft (Natural Science Society) in Bern, an association of professors, high school teachers, and the inevitable prominent figures in medicine and pharmacology. In the absence of a Swiss Academy, the Bern Society played the part, if on a modest scale, of the great scholarly institutions in other countries. It was there, if not earlier, that Einstein made the acquaintance of a friend of Sauter's, Dr. Paul Gruner, a high school teacher and simultaneously a privatdozent in physics at the university.

Gruner was a man with predominantly theoretical interests and may have seemed to Einstein an ideal combination of practical livelihood and academic career. The title privatdozent, of course, yielded more prestige than money, as a privatdozent received no salary from the university but was entitled only to teach and collect the lecture fees of his few students. But, after the doctorate, it represented the first rung on the academic ladder and was the customary prerequisite to a professorship after the ponderous *Habilitation* procedure—an impressive original thesis plus a trial lecture. Einstein now hoped to become a privatdozent like Gruner, even though Gruner had to spend nine years in that academic limbo before he was granted the humble position of titular professor—really no more than a privatdozent, unsalaried, but with the right to call himself professor—in 1903 and thus personified the risks of an academic career based chiefly on theory.[65]

Normally a doctorate was a prerequisite for *Habilitation*, but someone, possibly Gruner, must have drawn Einstein's attention to an exceptional regulation for Bern,[66] according to which doctorate and *Habilitation* thesis could be dispensed with in the case of "other outstanding achievements"—and these Einstein believed he had accomplished by his two papers on thermodynamics. He therefore wrote in January 1903: "I have now decided to become a *Privatdozent*, provided of course that I can get away with it. I won't, on the other hand, take my doctorate, as this doesn't help me much and the whole comedy has begun to bore me."[67]

He did not get away with it. He must have failed at an early stage, as we have no indications even of a properly initiated, though subsequently disallowed, procedure.

It is not hard to imagine that the professors at the university would regard Einstein's demand as a piece of youthful impertinence. The exceptional regulation was intended for "other achievements" by considerable scholars, not for two papers in *Annalen* by a greenhorn. No doubt any number of physicists whose published work fell far short of Einstein's papers had been appointed to professorships, but no one in the department in Bern was able to recognize the importance of Einstein's work—and this included the head of the department, Aimé Forster. Einstein, refusing to acknowledge that, ranted: "The university here is a pigsty. I won't lecture there, it would be a waste of time."[68] Thus ended Einstein's first attempt to become a "great professor."

Nevertheless, the episode was not entirely in vain. Einstein had made contact in Bern with circles interested in science. At its meeting of May 2, 1903, held as usual in the assembly room of the Storchen Hotel, "Hr. Alb. Einstein, mathematician at the Patent Office" was accepted as a member of the Naturforschende Gesellschaft.[69] That evening Gruner gave the customary lecture: he spoke about atmospheric electricity. On December 5, 1903, Einstein, for the first time, was the speaker: his subject was "The Theory of Electromagnetic Waves." A second lecture that same evening was on veterinary medicine—evidence of the broad spectrum of interests in the society.

Einstein became a fairly regular visitor to the meetings, but not exactly a zealous speaker. Only once more, on March 23, 1907, did he mount the rostrum: to report "On the Nature of Microscopic Particles Suspended in Liquids," a subject which he had discovered for statistical physics in his *annus mirabilis*, 1905. More often than in the Storchen, Einstein would give informal lectures at Paul Gruner's home; apart from Sauter, Gruner was probably the only person to recognize Einstein's outstanding talent.

In 1936, when the Society observed its 150th anniversary, and long after he had left Bern, Einstein was elected an honorary member. He was obviously touched: "It was, in a sense, a message from my long vanished youth,"[70] he wrote to its president, thanking him for the scroll. "The pleasant and stimulating evenings emerge again in my memory. . . . I had the charter framed straight away and hung it up in my study—the only one among similar acknowledgments where I have done this—as a memento of my time in Bern and my friends there."

Sometime in the summer of 1903, Einstein and his wife must have reached a decision to part for good from their daughter, Lieserl. Even before she was born, when Mileva visited Einstein in Schaffhausen with her "funny shape,"[71] they must have discussed the option of having the child adopted. There are a few hints in a letter from Mileva to Einstein, in which some kind of role seems to have been intended for Mileva's friend Helene Savić, who was living in Belgrade with her husband and had just become a mother herself. Although Mileva did not then wish to say anything to her friend about her baby, she asked Einstein to write to her every now and then: "We must treat her well"[72] was the explanation, "because she can help us with something important." In the circumstances, the "something important" can only have concerned their child's future.

Meantime Lieserl had remained in Novi Sad with Mileva. Einstein had initially been delighted with his daughter, but after their marriage

Lieserl was evidently not wanted in Bern. Judging by what we know, her existence was carefully concealed from their friends in Bern.

The reasons for this remain uncertain. It might be supposed that Einstein, after the disappointments of two years of job-hunting, did not wish to jeopardize his position in the Patent Office. After all, he was only "provisionally elected"—in other words, on trial—and a premarital child might have offended the Swiss authorities' sense of propriety, especially as Einstein, though naturalized, was still a foreign Jew. But it is equally possible that on his deathbed Hermann Einstein had consented to Albert's marriage, but not to the legitimation of a child "born in shame." Whatever the real reason, Lieserl was eventually handed over to strangers. In August 1903 Mileva went to her parents in Novi Sad, most probably in order to take the child to Belgrade—a possibility envisaged two years earlier. On a postcard Einstein asked her how the child was registered, voicing some concern lest disadvantages might accrue to her.[73] During her journey, meanwhile, Mileva had discovered that she was pregnant again. Einstein did not mind and recommended careful "hatching." He never saw his first child.

Evidence of the first two years of Lieserl's life is scant enough; beyond that, the fate of Albert Einstein's first child is totally unknown. The daughter was never again mentioned in a letter, and despite intensive searches no entries have been found in parish registers, registry office documents, or anywhere else.

If it was Einstein who regarded this parting with their daughter as the best solution and enforced it—an assumption supported by at least a few indications—then it is not surprising that Mileva did not regain in Bern the cheerfulness of her student days. She spread around her an aura of melancholia, mistaken for a Slavic phenomenon, which contrasted dangerously with her husband's jovial, extroverted nature. He later described her basic attitude as "depressive or gloomy,"[74] attributing this, with a then common overestimation of genetic factors, to an inherited schizophrenic disposition in the family of Mileva's mother, as well as to Mileva's limp. These were probably oversimplifications.

"Generally she was very cool and suspicious toward anybody who, in some way or other, came close to me,"[75] he complained, but there were some exceptions, such as his friends of the Akademie Olympia, especially the "kind Solovine." Solovine reports that Einstein's marriage had made no difference to their meetings, which were now usually held in the young couple's apartment, when Mileva, "clever and reserved, listened to us attentively without ever intervening in our discussions."[76] As for Einstein's conversations on physics with Paul Gruner, Josef Sauter, or Michele Besso, who came to Bern later, she apparently did not take part in those at all; at least we have no indication that she did. Nothing, therefore, seems to have come of the joint work so often referred to in earlier years, when Einstein would be "proud and happy" when "we are together and can bring our work on relative motion to a successful conclusion."[77]

After the disappointment of two failed exams it would not be surprising if Mileva had lost all interest in physics and withdrawn into a domestic role, leaving her husband to pursue his scientific endeavors on his own. Certainly there is no indication of any close collaboration or even that Mileva helped in a more modest way.[78]

In November 1903 the Einsteins moved from the Kirchenfeld neighborhood back into the city, renting a third-floor apartment at Kramgasse 49. This apartment was more modest than the exterior of this fine house would lead one to expect. It was reached by a steep narrow staircase and consisted of two rooms, one of them with large windows onto the fine street. This was where Einstein's second child was to be born. He announced the event in his usual boisterous manner: "We'll have a pup in a few weeks."[79] The child, a boy, was born on May 14, 1904, and was named Hans Albert.

In the summer of 1904 Michele Besso joined Einstein as a colleague at the Patent Office. In Trieste, Besso had experienced difficulties earning a living as a freelance engineer. Thus, when a vacancy for a "Technical Expert II Class" was advertised toward the end of 1903, Einstein drew his friend's attention to it.

Needless to say, the Expert III Class Albert Einstein had also applied for this higher position, but the director judged that, while he had "displayed some quite good achievements," it would be wiser "to wait with his promotion until he has become fully familiar with mechanical engineering, because by his qualifications he is a physicist."[80] Einstein is unlikely to have blamed his boss for rejecting his rather premature application, the less so as the post, which carried a salary of 4,800 francs, went to his friend Michele out of a field of thirteen applicants. Einstein himself, in line with regulations, was "made definitive" on September 16, 1904, after more than two years' employment.[81] His salary was increased to 3,900 francs, but his status continued to be Expert III Class.

In Besso Einstein now had an ideal friend, both at work and often also during their leisure hours. To Einstein their "conversations on our joint way home . . . [were] of unforgettable charm."[82] Although Besso had studied mechanical engineering, his quick, acute intellect was not satisfied with that; he was passionately interested in nearly all questions in the exact sciences, both philosophical issues and the more prosaic aspects of research.

During his student days in Zurich Einstein had received much stimulation from Besso, who was his senior by eight years, but in Bern the emphasis now shifted. Einstein was no longer a student but an active researcher with some brilliant publications and, more important, an acute awareness of the problems of contemporary physics. The former mentor was hardly able to offer stimulation, but he was a valuable critic: not exactly a collaborator but an ideal sounding board.[83]

Einstein's only publication in 1904 had been completed toward the end of March, before Besso's arrival. It should not be overlooked because, for the first time in the pages of *Annalen*, it shows the young Einstein on a creative, original path, in a way that can be seen as preparation for the strokes of genius soon to come. He mapped out his objective in a postcard to Marcel Grossmann, who had written to him about the joys of parenthood and sent him a mathematical treatise: "There is a strange similarity between us. We too will have a child next month. And you too shall receive a paper which I sent to Wiedemann's *Annalen* a week ago (*General Molecular Theory of Heat*). You deal with geometry without the axiom on parallels, I with the atomistic theory of heat without the kinetic hypothesis."[84]

This paper, a mere eight pages, is rather disparate. Since his problematical derivation of the second law of thermodynamics of January 1903 had not satisfied him even then,[85] he now begins by presenting an alternative, which, however, does not stand up to criticism either. He then analyzes the constant later named after Boltzmann, which occurs in such a variety of connections in the kinetic theory that Einstein may have assumed that hidden behind it was the crux of that concept. Einstein first of all finds a new relation between this fundamental constant and the equally important Avogadro number N, which for any substance gives the number of molecules in one mole. The constant thus becomes a kind of yardstick in the microcosm of molecules.

But Einstein had more to offer. He uncovers a surprising new meaning of these constants, which emerges in the analysis of fluctuation phenomena—in a sense, a thermodynamics on a small scale. The basic laws of thermodynamics naturally refer to "large" systems, consisting, like all objects in our everyday experience, of an enormous number of molecules. Even though, strictly speaking, these laws are merely statements about mean values, they are nevertheless regarded as strictly valid because all irregularities are evened out by the colossal number of molecules. However, in "small" systems, which of course may still consist of many thousands or millions of molecules, the chaotic confusion of molecular movements is no longer evened out, so that thermodynamic magnitudes should reveal deviations from mean values. Admittedly, no one had yet observed such fluctuations, and the theoreticians, occasionally running ahead of experiment, had discussed this concept only sporadically and controversially. In this situation Einstein briefly and tersely develops a simple theory of these fluctuations for the energy of a system and derives a condition of the thermal stability of a system, in which Boltzmann's constant appears as a yardstick for the magnitude of the fluctuations. Einstein greatly liked this relation because "it no longer includes any quantity that is reminiscent of the assumptions underlying the theory."[86]

At the time, it seemed out of the question to apply these reflections to specific systems consisting of molecules. Boltzmann and Gibbs, the two giants of statistical theory, had discussed the observability of fluctuations and ruled it out, and Einstein too, "at the present state of our knowledge," sees no possibility of this. But he does not leave it at that. He assumes, of a totally different kind of physical system, "that energy fluctuations attach to it: this is vacant space filled with temperature radiation."[87] In a daring step, justified by nothing except a kind of primal confidence in methods he had worked out by himself, he applies the formulas developed for material molecules to immaterial electromagnetic radiation. Perhaps to his own surprise, he obtains an empirically verifiable relation between the energy maximum of the radiation and the temperature, a relation which agrees with Wien's displacement law. This, he concludes, "given the great generality of our assumptions, cannot be attributed to coincidence."[88] The seminal nature of the treatment of fluctuation phenomena first tested in this treatise would be shown the very next year, in Einstein's theory of Brownian movement and in his radiation theory.

Before the year 1904 was out, Einstein had become a collaborator of the *Beiblätter zu den Annalen der Physik* (*Supplements to the Annals of Physics*), an early "journal about journals" founded in 1877. In it were published not original papers, but reviews of papers in other journals—especially foreign-language journals—and in rare cases also reviews of books. We do not know how Einstein came to be one of the sixty-two referees of *Beiblätter*,[89] but we might not be wrong in assuming that the editor had noticed Einstein's five publications in *Annalen* and had therefore invited him to referee papers on the "theory of heat." The subjects were laid down by the editor, who, whenever necessary, supplied the referees with offprints of articles to be refereed, or with review copies. At the end of the year there was even a modest honorarium.

All together, Einstein wrote twenty-three reviews, twenty-one of which appeared in 1905. Over the next two years he wrote only one book review each year but one of these two was in 1906, his noteworthy review of Max Planck's *Vorlesungen über die Theorie der Wärmestrahlung* (*Lectures on the Theory of Heat Radiation*) in 1906. Einstein had sent to him articles from the most varied journals, from the *Philosophical Magazine* of the British Royal Society to the *Schweizerische*

Bauzeitung (*Swiss Construction News*). In addition to German publications, he also refereed French and Italian papers, being familiar with both languages. He also reviewed four articles written in English, a language he had not learned; it is probable that someone helped him with them—possibly Mileva, who knew a little English, or a colleague at the Patent Office.

Depending on the fees paid for his reviews, and on his own range of interests, Einstein's reviews differed a good deal. Sometimes he would seem apathetic, writing a mere five lines berating the author for some awful mistake;[90] at other times he went into such detail that his review might have replaced the original article. He was generous when a work's usefulness outweighed its shortcomings, as in the case of a little book called *Die Grundzüge der mechanischen Wärmelehre* (*Fundamentals of the Mechanical Theory of Heat*), which despite "some inaccuracies" he recommended to any polytechnician facing an exam with incomplete lecture notes.[91]

His work for the *Beiblätter* provided Einstein with an opportunity to acquaint himself more thoroughly with the topical literature than would otherwise have been possible, given his official duties at the Patent Office. Without it he might easily have missed the *Festschrift* on the occasion of Ludwig Boltzmann's sixtieth birthday, which included 117 contributions by prominent authors and thus offered an exceptionally broad panorama of physics at the beginning of the twentieth century. Einstein discussed three papers from this volume, and he probably read the rest. Needless to say, his reviews also enhanced his reputation—but by the time most of them appeared in print he no longer had any need of that.

If, toward the end of 1904, Albert Einstein had decided to concentrate on a career in the Swiss public service and to abandon his scientific work, this would probably not have been considered a serious loss to science. His contemporaries would scarcely have noticed that a colleague who had published a few papers but otherwise was quite unknown had stopped writing. Professor Weber in Zurich might have found his judgment confirmed that this impertinent young man would never achieve anything worthwhile. And many decades later, some historian might have been surprised to discover that an outsider had, all by himself, developed an equivalent to Gibbs's statistical physics.

Actually, Einstein's publications up to this point were only the surface of his ceaseless wrestling with the most difficult problems of physics. But no one—perhaps not even Einstein himself—suspected that these ruminations would come together in 1905, his *annus mirabilis*, in an explosion of creativity.

It was probably in late May of 1905 that Conrad Habicht received a letter from Einstein in Bern—undated, as usual—which may well be the most remarkable letter in the history of science. After a boisterous and jocular opening, Einstein promised to send Habicht

> four papers, the first of which I could send off soon, as I am to receive my free copies very shortly. It deals with radiation and the energetic properties of light and is very revolutionary, as you will see provided you send me *your* paper first. The second paper is a determination of the true size of atoms by way of the diffusion and internal friction of diluted liquid solutions of neutral substances. The third proves that, on the assumption of the molecular theory of heat, particles of the order of magnitude of 1/1000 millimeters suspended in liquids must already perform an observable disordered movement, caused by thermal motion. Movements of small inanimate suspended bodies have in fact been observed by the physiologists and called by them "Brownian molecular movement." The fourth paper is at the draft stage and is an electrodynamics of moving bodies, applying a modification of the theory of space and time; the purely kinematic part of this paper is certain to interest you.[92]

These four papers would transform physics. They were completed in the brief span between March 17 and June 30, 1905—a little more than three months. The first—"very revolutionary"—publication was far ahead of its time but sixteen years later would earn Einstein the Nobel Prize. The second, which soon earned him a doctorate from Zurich University, is one of the most frequently quoted works of the century. The third established him as the founder of modern statistical mechanics. The fourth contains in fundamental form what would soon come to be called the special theory of relativity.

Never before and never since has a single person enriched science by so much in such a short time as Einstein did in this *annus mirabilis*. And his creative vigor continued: over the next two years his publications came thick and fast. The man at the Patent Office in Bern would place twentieth-century physics on a new foundation and open up perspectives that would influence research well into the next millennium.

To enable the reader to understand this unique climax of scientific creativity, its external conditions, and its inner connections, Chapters 7 through 11 will present Einstein's contributions to physics between 1905 and 1907, with their initial reception and their consequences. Einstein's own words may serve as an introduction to this material:

> What is essential in the life of a man of my kind lies in *what* he thinks and *how* he thinks, and not in what he does or suffers.[93]

CHAPTER SEVEN

"Herr Doktor Einstein" and the Reality of Atoms

ONE OF THE MORE ORIGINAL CONTRIBUTIONS to the observances in 1979 of the hundredth anniversary of Albert Einstein's birth was a list of papers in all areas of the exact sciences, from physics through chemistry to physiology, published before 1912 and most frequently cited between 1961 and 1975,[1] in other words papers which, after half a century, still had a major influence on ongoing research. Of the eleven "classics" at the top of the list, Einstein had written four (the other seven had seven different authors). Of the four works by Einstein, topping the list was *A New Determination of Molecular Dimensions*,[2] followed by his paper on the Brownian movement.[3] Both deal with the reality of molecules.

Counting citations or footnote references, however, is not necessarily the best way to measure a work's scientific value. Einstein's epoch-making papers of 1905, on light quanta and on relativity theory, were not included in this list, and that was because they had too much influence on scientific progress. They are the prerequisites of all modern physics and have become so integrated into physics that virtually no one quotes them any longer. In fact, hardly any physicists today have read the original papers: everyone has learned about them in classes or from textbooks.

To return to Einstein's top-ranking papers on the list, these have of course affected an unusually wide range of investigations. Both of them deal with the movement of large molecules or colloidal particles in liquids and were therefore quoted, for instance, in ecological studies of the dispersal of aerosols in the atmosphere and in dairy research papers dealing with the behavior of casein particles in milk during cheesemaking.[4] It is interesting to note that it was this counting of footnotes that led to the posthumous discovery of Einstein's doctoral thesis, which had been dismissed by his biographers as an insignificant mandatory academic exercise, not to be compared to the three famous papers of 1905.[5] Physicists and historians of science had also ignored it[6] when writing about his *annus mirabilis*—possibly because it did not appear in that famous Volume 17 of *Annalen* but was published later, in 1906. It had been completed on April 30, 1905, however, in close connection with his work on the Brownian movement.

In the summer of 1905, Einstein, who had given up the idea of a doctorate "as this doesn't help me much and the whole comedy has begun to bore me,"[7] decided after all to get his Ph.D. The reasons for his change of mind are obvious: at the Patent Office it could well help him, and for an academic career it was a prerequisite. Should he try for it in Bern or in Zurich, he asked his colleague Dr. Sauter, who was experienced in such matters. Sauter's reply was: "Zurich—and for you it'll be a cinch."[8]

It would have been the custom for Einstein to agree on the subject of his thesis with the head of the department. Instead, though—according to his sister Maja—Einstein first submitted his recently completed *Electrodynamics of Moving Bodies*, in other words the theory of relativity, which "seemed a little uncanny to the decision-making professors"[9] and was rejected. Einstein thereupon—having, as Professor Kleiner records, "chosen his own problem"[10]—simply picked from his "work in progress" whatever he thought would least upset the department: nothing too revolutionary or too speculative, but solid assumptions, conventional mathematics, and (since pure theory was still in bad odor in Zurich as something rather exotic) an investigation based on experiment. These criteria were best met by his investigation of the movement of large molecules in aqueous solution.

On July 20 Einstein addressed his degree application to the dean and, together with his treatise, sent it to Zurich. As the semester was drawing to a close, everything moved very quickly: within four days the paper, with comment, was circulating among the faculty. Kleiner emphasized that "the arguments and calculations are among the most difficult in hydrodynamics and could be approached only by someone who possesses understanding and talent for the treatment of mathematical and physical problems, and it seems to me that Herr Einstein has provided evidence that he is capable of occupying himself successfully with scientific problems."[11] Because of the tricky mathematics, Kleiner had brought in the head of the mathematics department, Professor Heinrich Burkhardt, who had thereupon "examined the most important part of the calculations, especially the passages marked by my colleague Herr Kleiner. What I have examined I found correct in every respect, and the manner of the treatment testifies to *a thorough mastery of the mathematical methods concerned.*"[12] In fact, the mathematics professor had missed one mistake—which had some consequences, but not until four years later. Like Kleiner, Burkhardt recommended accepting the thesis, though he criticized it for a lack of fastidiousness in detail: "Stylistic infelicities and slips of the pen in the formulas will have to be, and can be, eliminated for publication in print." Einstein was now free to take his emended paper to the printer. That did not cost much, as it was only seventeen pages long. As in less happy days, it was dedicated to "my friend Dr. Marcel Grossmann." After handing in the prescribed copy to the university, Einstein now was Herr Doktor Einstein.

The dissertation belonged to a range of subjects where—according to his own later characterization—Einstein was concerned chiefly with "discovering facts which would establish with certainty the existence of atoms of definite finite magnitude."[13] It may seem strange to us that at the beginning of the twentieth century the existence of atoms was still in contention. Even stranger is the passion which characterized the dispute, especially among German scientists. Radioactivity and the electron had already been discovered; moreover, ever since the first decades of the nineteenth century chemists had been regarding the transformation of substances as combinations of atoms into molecules or as reactions between molecules. In the second half of the nineteenth century this view became universally accepted, and chemists—who were not plagued by metaphysical questioning—were in no way bothered by the fact that they had never actually seen an atom and that, considering what was being discovered about the dimensions of atoms and molecules, they were not likely ever to see one.

The situation was entirely different in physics. Although the success of the atomistic hypothesis in the kinetic theory of gases and solid bodies was perhaps even more impressive, some influential scholars, mainly those priding themselves on methodological strictness and philosophical acumen, regarded the atom as a superfluous, if not indeed harmful, invention—partly because they had not yet seen one and partly because they refused to accept the fictions of chemists as an acceptable basis of physics. Thus the great physical chemist Wilhelm Ostwald headed a school of thought which hoped to base all scientific research on the concept of energy; and another, shorter-lived school believed that electromagnetism was the basis of all physics. The influential Ernst Mach is reported to have asked anyone who mentioned atoms to him: "Ever seen one?" Mach's statement "I do not believe that atoms exist"[14] positively alarmed his colleague Ludwig Boltzmann in Vienna. At the turn of the century Boltzmann, aware of "how powerless an individual is against the trends of the day," lamented "the damage to science if the theory of gases were to be relegated to temporary oblivion by the prevailing hostile mood."[15] Boltzmann's suicide on September 15, 1906, should not, of course, be seen as a direct reaction to this dispute—but perhaps he would have borne his life longer if science had brought him more joy and recognition.

Einstein later regarded Mach's and Ostwald's rejection of atomic theory as "an interesting illustration of how even researchers of bold intellect and subtle instinct can be prevented by philosophical prejudice from an interpretation of facts."[16] In this instance it had been the positivist belief "that facts alone without free conceptual constructs should and could lead to scientific knowledge." More than anyone else, the man from the Patent Office helped make the skeptical positivists eventually accept the atom.

Ever since his student days, Einstein had as a matter of course regarded the existence of atoms as unquestionable. All five of his publications had, in a sense, been variations on the atomic theory of natural phenomena. His dissertation now aimed at providing evidence for atoms and molecules. Naturally, he was not able to make an atom or even a molecule visible—that would become possible only in the 1950s, with the field-ion microscope. But Einstein invented his own kind of "microscope": an elegant theory which allowed the size of sugar molecules to be determined from something as ordinary as the viscosity of an aqueous sugar solution.

Einstein had worked out the basic method two years earlier. "Have you already calculated the absolute size of the ions, on the assumption that they are spheres and large enough for the equations of the hydrodynamics of viscous liquids to be applicable to them?"[17] he had then asked Besso. "I would have done it myself, but I lack the literature and the time; you might also draw upon diffusion to obtain information on neutral salt molecules. If you don't know what I mean, I'll be glad to write you again in greater detail." It seems that Einstein did have to explain himself in greater detail, and the dissertation therefore looks like a direct fulfillment of his promise given to Besso, as well as an attempt to convince Ernst Mach—whom he otherwise revered for his *Mechanik*—that atoms were not fictional.

Einstein's argument proceeded not (as usual) from the theory of gases but, for the first time, from the behavior of liquids. Because—in contrast to a similar theory of gases—a molecular-kinetic theory of liquids would, in Einstein's view, be faced with "insuperable difficulties"[18] he confined himself to a simpler model. His model was an aqueous solution of a substance whose molecules are large compared with molecules of water, so that (in a reasonable approximation) the water can be treated as an unstructured homogeneous medium in its effect on the dissolved molecules, which for the sake of simplicity are assumed to be spherical. If a substance is dissolved in water, viscosity increases. This can be measured. As a first step, Einstein was able to establish a relation between this change in viscosity and the total volume of the dissolved molecules. Despite his simplified assumption, this called for involved calculations and represented the most demanding part of the investigation. As a second step, mathematically simpler but more demanding in terms of physics, Einstein dealt with the diffusion of a swarm of molecules dissolved in the water, obtaining a diffusion coefficient which, with experimentally determinable values, again gave information on the size of the molecules. He had thus developed a method which, surprisingly, combined experimentally measurable properties of solutions, such as viscosity and diffusion, to create his ingenious "microscope." Though one could not, of course, "see" the molecules, Einstein's "microscope" made it possible to determine their size.

Such a theory demands practical application, and Einstein provided it for sugar water because experimental data were available for this. The radius of the sugar molecules he found in this way—one ten-millionth of a centimeter—was new. Added to this, by way of a countercheck, was the determination of the Avogadro number, and Einstein's result in fact agreed "satisfactorily, as for order of magnitude, with the values found for that quantity by other methods,"[19] This confirmed both the reliability of the method and the reality of molecules.

The judges at the university in Zurich were satisfied with Einstein's results, but Paul Drude, the editor of *Annalen*, was not. Einstein had submitted his treatise to Drude in August 1905, after the conclusion of the degree procedure; however, it was published not within the customary eight weeks, but only about six months later. This had never before happened with any of Einstein's papers, nor did it ever happen afterward. Drude evidently knew of better data for sugar solutions and must have asked for a small addendum.[20] Einstein supplied it at the beginning of the following year, with a substantially improved result for the Avogadro constant.[21]

Nothing happened for the next four years. With the sensation caused by Einstein's paper on the Brownian movement, his dissertation was scarcely noticed. This applied also to Jean Perrin, a young professor at the Sorbonne in Paris, who with superb experimental skill was investigating the Brownian movement in 1909 and corresponded with Einstein on the subject. Einstein, who had just become a professor in Zurich, used the opportunity to draw Perrin's attention to his dissertation in the autumn of 1909. Thereupon one of Perrin's colleagues, Jacques Bancelin, took the subject up experimentally.[22] He did not dissolve molecules, but instead suspended accurately prepared microscopic mastic globules of known dimensions in water. Much of Einstein's theory was confirmed by Bancelin's experiments, but on one point there was a major discrepancy. When Einstein repeated his calculations, he was unable to find a mistake. Although he did not rule out some experimental error, he nevertheless requested Ludwig Hopf, the assistant at the Zurich institute, to have another good look at the dissertation: "I have now re-examined my earlier calculations and arguments and not found a mistake in them,"[23] he wrote to Hopf during the Christmas vacation. "You would be doing a great service to the subject if you could seriously check my investigations." Hopf found the mistake, which the mathematician Burkhardt had previously also failed to spot—a rather trivial slipup, but it threw off the numerical result. Einstein sent a correction to *Annalen*, with an acknowledgment of Bancelin's and Hopf's work and with an even better result for Avogadro's number.[24]

Nine days after completing his dissertation—if not sooner—Einstein had his next paper ready; it was received by the editor of *Annalen* on May 11, 1905. Its title is of almost baroque convolution: *On the Movement, Demanded by Molecular-Kinetic Theory, of Particles Suspended in Liquids at Rest.* He might have called it, more succinctly, *On the Brownian Movement*, but at the very beginning of his paper he admitted that he could only assume "that the movements to be here dealt with are identical with the 'Brownian molecular movement.' However, the data available to me are so inaccurate that I cannot form a definite opinion on this."[25]

In 1828 the botanist Robert Brown privately published a paper in London, entitled *A brief Account of Microscopical Observations, conducted in the months of June, July and August 1827, on the particles contained in the Pollen of Plants; and on the general existence of active molecules in organic and inorganic bodies.* Brown described how he had, under his microscope, seen pollen grains in a permanent trembling movement. He regarded this as a typical characteristic of male sex cells, similar to spermatic filaments. But he had the brilliant idea of testing this assumption by observing minute particles of inanimate matter in water. He found that the same permanent erratic movement was present in very finely ground splinters of glass and granite, as well as in smoke particles. Hence the cause could not be the vitality of living matter, and the "Brownian movement" therefore passed from the hands of botanists or physiologists into the hands of physicists.

In the second half of the nineteenth century some physicists suggested a molecular-kinetic model to explain the Brownian movement. The zigzag movements of the suspended particles, they believed, were due to impacts from the molecules of the liquids.[26] This idea was basically sound, but all the theories had serious flaws[27] and failed to stand up to experimental testing. The theoretical situation remained confused and controversial.

Even if Einstein had been familiar with the work of his predecessors, and even if he had known everything there was to know about the Brownian movement, his theoretical explanation would still have been a brilliant achievement. Unburdened by any previous knowledge, however, he chose an entirely different and more fundamental approach. He asked himself if the irregularities in the movement of molecules, demanded by the molecular-kinetic theory, might not after all cause observable effects. To his own surprise he discovered that the theory in fact predicted fluctuations observable under a conventional microscope, and that the measurement of these fluctuations represented a kind of penetration into the microcosm of atoms. This was an original, theoretically founded concept of the Brownian movement and of its characteristic properties as a fluctuation phenomenon—a phenomenon that had been observed for nearly a hundred years without being understood.

Einstein therefore observed not molecules in solution, but suspended particles about a thousandth of a millimeter in diameter—still clearly visible under a microscope and actually, in kinetic theory, gigantic macroscopic structures. Unlike his predecessors (who—to repeat—were unknown to him) Einstein evidently realized from the outset that the velocity of the particles could not be observed directly. According to simple calculations, their velocity would amount to about one-tenth of a millimeter per second: in other words, a particle would travel a distance about one hundred times its own diameter in one second. Under a microscope, such a particle would flit through the viewing field like a wraith. However, the particle is also greatly slowed down by the liquid and simultaneously struck by individual molecules. The result of these two effects is an extremely irregular trembling—a movement whose track and velocity cannot be measured directly. At the same time, though, a kind of mean value, the mean square displacement, should be observable under the microscope, and this would be enough.

Einstein first of all demonstrated—and this was a very bold innovation—that "osmotic pressure," which according to classical thermodynamics should exist only in solutions, was present also in suspensions of "gigantic" spheres or globules. Next, in much the same way as in his dissertation, but more elegantly, he worked out a diffusion formula for the spherules and examined the interplay of diffusion and ceaseless impacts from the molecules of the liquid as a statistical process. He finally obtained an expression for the mean displacement of the particles, depending only on measurable or otherwise familiar values. From this he was able to calculate that his standard particle of one-thousandth of a millimeter, suspended in water, must after one second have moved by just under one-thousandth of a millimeter, and after one minute by six-thousandths of a millimeter. Conversely, the Avogadro number could be determined from that expression, provided the displacement and the time were measured.

That suggestion was surprising to experimental physicists: with minute spherules, a microscope, and a clock they were to count atoms. Einstein, moreover, had formulated his argument as a yes-or-no experiment. If his prediction was not correct, "this would mean a weighty argument against the molecular-kinetic concept of heat."[28] He therefore concluded his article with an exclamation mark: "May some researcher soon succeed in deciding the question here posed, a question vital to the theory of heat!"[29]

This time Einstein could not complain that he got no reaction. Soon after the appearance of the paper on July 18, 1905, Henry Siedentopf wrote to him from Jena, Germany, confirming that the predicted phenomenon probably was the Brownian movement.[30] Siedentopf, at the Carl Zeiss Works, was engaged in improving the ultramicroscope invented in 1903 by Richard Zsigmondy. This instrument illuminates objects by light projected from the side and, by intercepting the light scattered by them, makes it possible to view particles which are substantially smaller than the wavelength of light. As the Brownian movement is even more erratic for smaller and therefore lighter particles than for larger ones, this new ultramicroscope made it possible to study particularly hectic trembling. Zsigmondy compared what he saw in colloidal gold suspensions to a swarm of midges dancing in a sunbeam. But he had not, any more than Siedentopf, carried out any measurements that might have permitted comparison with Einstein's detailed predictions.

Einstein quite obviously enjoyed this subject. Even before Christmas of 1905 he dispatched to *Annalen* a further paper, this time with an appropriate title, *On the Theory of the Brownian Movement.*[31] In it he presented the theory in a substantially more elegant form, developed it further, and in particular discussed the limits of its validity for short periods: less than a ten-millionth of a second. As a bonus he calculated something no one had seen yet—a Brownian rotation, that is, a trembling rotational movement of the suspended particles. If this could be measured, it too would be suitable for determining the Avogadro number.

Interest in the Brownian movement increased and brought Einstein numerous letters from scientists, as well as one visitor. Heinrich Zangger, professor of forensic medicine at the University of Zurich, later to become famous as the founder of emergency medicine and director of spectacular rescue actions in the mining industry, was interested as a researcher in the Brownian movement. When he ran into difficulties with his counting under the microscope, the professor of mechanics, Aurel Stodola, had told him to "go and see Einstein in Bern."[32] The meeting was the beginning of a lifelong friendship, but it did not yield any new insights into the Brownian movement. These came from elsewhere.

In Uppsala, Sweden, a young physicist, The Svedberg, was experimenting with the ultramicroscope. Unfortunately he had failed to observe the difference between velocity and mean displacement (the only observable quantity), so that Einstein had to make a slight correction,[33] "which corrected only the worst mistakes, because I could not bring myself to impair Herr S.'s enjoyment of his work."[34] Other studies, including some by the Frenchman Victor Henri, using cinematographic pictures, likewise failed to provide unequivocal proof of Einstein's theory. Proof came only in 1908, when Jean Perrin in his laboratory at the Sorbonne in Paris studied the Brownian movement and, in a series of excellent experiments, confirmed all aspects of the theory. Einstein was delighted: "I wouldn't have thought it possible for the Brownian movement to be investigated with such precision; it is a piece of good luck for this subject that you undertook to study it."[35] By means of a sophisticated method of tagging minute mastic spherules Perrin was even able to measure the Brownian rotation calculated by Einstein, which surprised him greatly: "I wouldn't have thought a measurement of the rotation possible. To me it was merely an amusing pastime."[36] This was the final proof.

Einstein meanwhile had been concerned more with popularization. On March 23, 1907, he gave a lecture at the Natural Science Society in Bern[37] on the Brownian movement, and the following year, at the suggestion of Richard Lorenz, the professor of chemistry at the Polytechnic, he wrote *Elementary Theory of the Brownian Movement*,[38] to be comprehensible also to chemists. In addition, he was on the lookout for other macroscopically observable fluctuation phenomena. As early as his second paper he had considered an electrical circuit[39] and briefly discussed what subsequently came to be known as "noise." This gave rise to his own experimental study of the Brownian movement, though in the field of electricity, in voltage fluctuations in condensers. He initially published a theoretical concept,[40] and then, together with the Habicht brothers, began to build an apparatus for measuring very small amounts of charges. More about this "little machine" will come later.

Einstein must have been exceedingly gratified by a letter from Wilhelm Conrad Röntgen, the first Nobel laureate for physics, even though Röntgen objected that the Brownian movement "will be difficult to reconcile with the second law of thermodynamics."[41] Einstein's reply is lost—a great pity, as Einstein had never thoroughly examined this tricky question. However, in the second paragraph of his first paper he had pointed out that in the observation of the Brownian movement "along with the regularities to be expected . . . classical thermodynamics can no longer be regarded as absolutely valid even for microscopically distinguishable spaces."[42] This was something he had surmised anyway, and it fit into his overall ideas of the fundamental problems of physics. But it took another quarter-century before Leo Szilard satisfactorily proved that it was impossible to build a machine that would utilize the kinetic energy of the suspended particles for work, or to withdraw energy from the solvent.

Einstein never disclosed when he had himself first seen the Brownian movement he had predicted. He would have had an opportunity to see it in Bern, but he must have seen it, if not before, at the annual meeting of the Deutsche Gesellschaft der Naturforscher und Ärzte (German Society of Natural Scientists and Physicians) in Salzburg in September 1909, when Henry Siedentopf gave a lecture with demonstrations. A few years later, Einstein would write, with restrained emotion, that the significance of the Brownian movement was "that in it the disordered elemental processes are made directly visible. In a manner of speaking, one can see, directly under the microscope, a part of the thermal energy in the form of the mechanical energy of moving particles."[43] This was a spectacular assertion, and the agreement of the theory with Perrin's accurate measurements played a major part in convincing even the last skeptics of the reality of the atoms.

In 1913, when Einstein was to be brought to Berlin, Max Planck in an expert opinion emphasized, among many other points, that Einstein's contribution to the kinetic theory of matter "had a seminal effect on experimental research in different directions, above all the beautiful measurements of the Brownian molecular movement, which acquired their real value primarily through Einstein's work."[44]

In 1926, three protagonists of the research on the Brownian movement met in Stockholm. Jean Perrin was awarded the Nobel Prize for

Physics. The Svedberg and Richard Zsigmondy received the Nobel Prize for Chemistry—Svedberg for 1926 and Zsigmondy, retroactively, for 1925. As early as 1910, when Einstein was first proposed by Ostwald, the Nobel committee had pointed out in its internal report that the theory of the Brownian movement had earned Einstein great recognition. This achievement was mentioned on several subsequent occasions in his nominations; but when he was eventually awarded the prize in 1922, it was for a different paper, though from the same legendary year, 1905. That paper had been completed in March and was actually the first of his magnificent series. It dealt with light quanta.

CHAPTER EIGHT

The "Very Revolutionary" Light Quanta

ALBERT EINSTEIN DID NOT SEE PHYSICS as a sequence of scientific revolutions, nor did he see himself as a revolutionary. Indeed, he was extremely cautious about describing discoveries or theories as revolutionary. In his references to his own contributions to physics, I have come across only one use of the word—in his workshop report to Conrad Habicht in the spring of 1905, when he commended to Habicht's attention the first of the four promised papers: "It deals with radiation and the energetic properties of light and is very revolutionary, as you will see."[1] This confident assessment was not youthful hyperbole; it was accurate at the time and is even more so in retrospect. In this paper, Einstein questioned the universally recognized model of light as waves, and with it the unlimited validity of Maxwellian electrodynamics; instead, he "invented" a granular structure for light—the light quantum, the particle associated with electromagnetic radiation. This radical and immensely bold proposal made its young author a father of quantum physics.

In his title, Einstein did not promise anything like a theory, but rather "a heuristic viewpoint concerning the generation and transformation of light."[2] This may have seemed a bit frivolous to some readers of *Annalen*: "heuristic viewpoints" did not form part of theoretical physics at the turn of the century. A concept was either not yet confirmed and therefore open to future verification or falsification, in which case it was regarded as a hypothesis; or else it had proved its worth in practice, in which case it would be elevated to the rank of theory. Einstein had presumably encountered "heuristic viewpoints" in the course of his philosophical studies, perhaps as early as in his schooldays, when he read Immanuel Kant, who frequently used "heuristic principles."[3] The purpose of Einstein's "heuristic viewpoint," like that of Kant's "heuristic principle," was to state, or perhaps invent, an assertion from which familiar facts could then be deduced. From the outset, therefore, Einstein was focusing on something that would emerge only at the end of the paper. Experimentally observed oddities of the photoelectric effect and other phenomena that posed a riddle within the framework of Maxwell's theory of electromagnetic waves were to be effortlessly explained by reference to the "heuristic viewpoint" of the light quanta. Einstein thus took seriously the quantum hypothesis introduced into physics by Max Planck—unlike all his contemporaries and indeed unlike Planck himself, who for many years would remain reluctant to accept Einstein's radical step.

On December 14, 1900, at a meeting of the German Physical Society in Berlin, Max Planck had presented his famous radiation formula, which contained the quantum of action later to be named for him. This was a crucial innovation and, in retrospect, constituted the birth of the modern quantum theory of the microcosm, the theory which gave twentieth-century physics an entirely different appearance from nineteenth-century physics. All physics not involving the quantum now became "classical" physics.

At the time, Planck was scarcely aware of the radical nature of his work. At forty-two he was at the peak of his vigor, but by nature he was "peaceable and averse to risky adventures" in science.[4] He had become an unwilling revolutionary, anxious, almost at any cost, to avoid any split between his own research and "classical" physics—though this term was not yet being used.

What Planck was after was not a revolution in physics, but the solution of an old problem expressed in 1860 by Rudolf Kirchhoff. This concerned heat radiation. Everyone knows that heated metals glow—red at first, turning yellow at higher temperatures, and eventually turning almost white. In each case, this radiation is a mixture of different frequencies, with its spectrum extending far beyond visible light: into ultraviolet at high frequencies and into infrared, invisible heat radiation, at low frequencies. On the basis of abstract thermodynamics, Kirchhoff derived a number of statements on emission and absorption, valid for all materials. In these, a central role was played by an ideal object, the "black body," which completely absorbs all radiation striking it. The ideal case of "black radiation," totally independent of the properties of materials, was postulated as a cavity in which the radiation is in a state of equilibrium, determined solely by temperature, with the material of the walls. If a small part of this "black-body radiation" is allowed to escape through a minute opening, it may be observed and its frequency spectrum analyzed.

Kirchhoff brilliantly summed up all that was known and surmised about "black radiation" by claiming that for the emission capacity of "black bodies" there must exist a function that depends solely on temperature and frequency. "It is a task of great importance to discover that function. Its experimental determination is faced with great difficulties, yet there seems to be justified hope that it may be determined by experiment, as undoubtedly it is of a simple form, as indeed are all functions discovered so far that are not dependent on the properties of individual bodies."[5] Every part of Kirchhoff's statement was correct, including the great experimental difficulties. Not until after his death in 1887 did experimental skill or measuring techniques reach a level which made it possible to compare theoretically derived radiation formulas with actual measurements. However, all formulas sooner or later revealed major shortcomings: the best of them, Wien's formula, failed at low frequencies in the infrared range; and Lord Rayleigh's resulted at high frequencies in what Paul Ehrenfels later called "the disaster in the ultraviolet."

Max Planck, Kirchhoff's successor at Berlin University, firmly believed that the frequency distribution of "black cavity radiation" was "something absolute. And as the search for the absolute always seemed to me the finest research task, I tackled it with zeal."[6] When Planck concentrated on this problem in 1894, he was helped by the fact that some outstanding experimental physicists were equally fascinated by it. Friedrich Paschen, for instance, regarded Kirchhoff's problem as important enough to "decline a professorship for its sake."[7] He remained in his laboratory at the Polytechnic in Hanover and produced a graph with which he was able to improve the numerical factors of Wien's formula. For two or three years it looked as if this was the solution—but that proved to be wrong.

At the Physical-Technical Reich Institute in Berlin, then probably the world's best-equipped laboratory, Otto Lummer and Ernst Pringsheim had greatly refined the measuring techniques, especially in infrared, in the range of long wavelengths. Heinrich Rubens and Ferdinand Kurlbaum achieved a new degree of precision by Rubens's "rest radiation" method, whereby the rays of shorter wavelengths were faded out, so that very reliable measurements were made possible in the extreme longwave infrared at high temperatures. All the results in that range contradicted Wien's formula. It was the resolution of this contradiction that provided the key to the new physics.

The date when the new quantum theory was born can be stated very precisely. On October 7, 1900, a Sunday, Rubens and his wife were visiting with the Plancks. The men were unable to refrain from talking shop, and Rubens informed Planck that the latest measurements at the Reich Institute had shown that at very long wavelengths the energy density of radiation was proportional to temperature. This information must have excited Planck greatly, because that same evening, as soon as the guests had left, he got down to work. Thanks to his efforts over many years, the information he had received from Rubens bore fruit. That same night Planck developed a radiation formula which accurately matched all data. The formula had two constants: one was interpreted by Planck as a gas constant for a single molecule and was later named Boltzmann's constant (though, strictly speaking, Boltzmann had little to do with it); the other was a quantity until then unknown in physics, having the dimension of action.

However, a formula arrived at by trial and error, no matter how accurate, needs theoretical interpretation. This was what Planck concerned himself with during the next few weeks. According to "classical" physics the total energy would pass from the walls into the cavity, with no equilibrium being established. Overcoming his past rejection of the atomic view, Planck was eventually compelled to interpret the radiation as the emission of individual atoms; he conceptualized them as "harmonic oscillators," the simplest model for periodic processes. He now treated these "resonators" with Ludwig Boltzmann's statistical methods—which until recently he had found unacceptable. He was later to describe "the whole business as an act of desperation,"[8] meaning, evidently, what to him seemed an illegitimate use of atomic concepts. With these methods, however, it followed from the equilibrium between matter and radiation "that energy is compelled from the outset to keep together in certain quanta."[9]

On December 14, 1900, Planck presented his theoretical interpretation of the radiation formula to a meeting of the Physical Society, laying great emphasis on its novelty: "We therefore regard—and this is the most essential point of the entire calculation—energy to be composed of a very definite number of equal finite packages, making use for that purpose of a natural constant $h = 6.55 \times 10^{-27}$ ergsec."[10] This unimaginably small, though finite, "magnitude" (it is written with twenty-six zeros after the decimal point) represented the abandonment of the continuity-based conceptual apparatus of "classical" physics and the foundation of a new physics. But this was not realized until later. To Planck, the quantum was initially "a purely formal assumption and I didn't give it much thought, except only that, under all circumstances and at whatever cost, I had to produce a positive result."[11] He was, moreover, able to placate his "peaceable" nature because the energy quanta would play a role only in statistical counting procedures by the resonators, while radiation would continue to be understood, in line with Maxwell's theory, as a continuous wave in the ether.

Neither Planck nor his listeners suspected that a *terra incognita*—a new microphysics—was opening up before them. Indeed, for a whole decade Planck endeavored "somehow to harness the quantum *h* into the framework of classical physics,"[12] and other physicists, like Lord Rayleigh and James Jeans in England and Hendrik Antoon Lorentz in Leyden, the Netherlands, were doing the same. These clever men were examining such delicate problems as the interaction of the material resonators with the ether, never for a minute questioning the strict validity of Maxwell's theory and thus never questioning the wave nature of light. Only one man thought differently, recognizing the revolutionary element which had come to physics with energy quanta—the "heretic" at the Patent Office in Bern.

Albert Einstein had first interested himself in heat radiation while still a student. He came across Kirchhoff's work in Ernst Mach's *Wärmelehre*, which he studied, along with his assigned reading, in his second year at the Polytechnic. In his third year Professor Weber presented his own measurements of the energy spectrum of heat radiation, together with an empirical formula, and it may well have been this lecture that led Einstein to further reflection. After the end of the winter semester he wrote to Mileva Marić: "My musings on radiation are beginning to take on more substance—I myself am curious if anything will come of it."[13]

Two years later, during the depressing period of job-hunting, he studied at least one of Planck's papers in *Annalen*—which came immediately before the discovery of the correct radiation formula—but he had "reservations of a fundamental nature, so much so that I'm reading his paper with mixed feelings."[14] Einstein may have missed the report in *Proceedings of the Physical Society*, but he is sure to have read Planck's comprehensive article in the March 1901 issue of *Annalen*, the same issue in which his own "firstling" publication on capillarity appeared. At the beginning of April Einstein was intending to "have a go at it now,"[15] but we have no record of his immediate reaction to Planck's radiation formula. Only in his *Nekrolog* did he refer back to that period. According to the *Nekrolog*, it had quite early struck Einstein that Planck's derivation of the radiation formula "is in conflict with the mechanical and electrodynamic basis on which that deduction otherwise rests."[16] It is true that Planck's thermodynamic arguments, and especially his abstract subdivision of total energy into separate elements, seemed like an attempt to avoid an explicit discussion of the role of energy quanta. "In reality," Einstein said, summarizing his earlier reflections, "the deduction implicitly assumes that the energy can be absorbed and emitted by an individual resonator only in 'quanta' of magnitude $h\nu$, that therefore the energy of an oscillating mechanical structure, as well as the energy of radiation, can only be converted into such quanta—in contrast to the laws of mechanics and electrodynamics. . . . All this I realized a short time after the publication of Planck's fundamental paper."[17]

This realization had been helped along also by Einstein's interest in the photoelectric effect, which was then being investigated quite separately from the problem of the radiation formula. Heinrich Hertz had discovered this effect about 1888 in the course of his experiments on the propagation of electromagnetic waves. A fortunate coincidence made him notice that in a spark gap illuminated by ultraviolet light, a spark gains in brightness. The significance of that observation, however, was not at first realized. It was only by the discovery of X-rays in 1895 and of the electron two years later that this matter was clarified. Soon it was assumed that the cause of the photoelectric effect was the release of electrons from gas molecules or metal surfaces irradiated with ultraviolet light—those molecules which had just been identified as the corpuscular components of the so-called cathode rays. Maxwell's theories would have led one to expect that with increasing intensity of light both the number and the energy of the electrons would increase. But this was not so.

Sophisticated experiments, especially by Hertz's former assistant Philipp Lenard, showed that the energy of the electrons was not governed at all by the intensity of light, but only by its frequency, in other words its "color"—this term being understood to apply also to the invisible ultraviolet or X-ray radiation. The yield of electrons certainly increases with the intensity of the light, in normal conditions, but for every metal there is a definite frequency below which no electrons are observed at all, no matter how long or intensively they are irradiated. Above this threshold frequency, on the other hand, electrons are emitted even at exceedingly weak irradiation—all this in contradiction to accepted theory.

This was very much to the taste of young Einstein, as can be gauged from the opening of a letter to Mileva Marić: "I just read a wonderful paper by Lenard on the generation of cathode rays by ultraviolet light. Under the influence of this beautiful piece I am filled with such happiness and joy that I must absolutely share some of it with you."[18] Although his "dear kitten" had just informed him that she was pregnant, he came to that topic only in a later passage of his letter.

Some of Einstein's letters suggest that he also concerned himself with the photoelectric effect as an experimenter. Thus he intended, after his third year of study at the university, "to work scientifically with a gentleman from Aarau."[19] This was Conrad Wüst, principal of the Aarau district school, a physicist who was experimenting with X-rays in his school laboratory. No details are known about their cooperation, but "radiation experiments" were at least intended.[20] Even as a student, Einstein had regarded the ether as superfluous, had intended to deprive electromagnetic waves of their substrate, and had believed that "electric forces can be directly defined only for empty space."[21] Thus it seems reasonable to assume that as early as 1901, after studying Lenard's and Planck's papers, he had been toying with the idea that light could propagate not as a wave in a medium such as the ether, but as a stream of corpuscles—"light quanta"—through empty space.

Einstein begins his article by highlighting a contradiction to which the supporters of atomic theory, at any rate, had become so accustomed that they scarcely saw it as a contradiction. This was the kind of opening Einstein liked. His own style of reflection was fired by a hidden contradiction, and in his fundamental treatises he would use that device whenever possible. In this particular case it was the "deep-going formal difference"[22] between the atomistic structure of matter and the description of all electromagnetic phenomena, including light, by continuous mathematical functions in space. Thus the energy of a material body is understood as the sum over its atoms and electrons, which therefore cannot be subdivided into just "any number and any size of small parts," whereas according to the wave theory of light the energy of a ray of light "is continuously distributed over a steadily increasing volume." It soon emerges that Einstein is proposing to resolve that contradiction by ascribing a corpuscular structure to light.

Naturally, Einstein concedes that the wave theory has "superbly proved its worth for the description of purely optical phenomena and will probably never be replaced by another theory." He points out, however, that optical observations relate to mean values over time for a multitude of waves, so that it is at least conceivable "that the theory of light, operating with continuous spatial functions, will clash with experience when applied to phenomena of light generation and light transformation." After this preparation, and an announcement that entire "groups of phenomena will appear more readily comprehensible on the assumption that the energy of light is distributed in space discontinuously," the "heuristic viewpoint" comes as a thunderbolt:

> On the assumption here to be considered, energy during the propagation of a ray of light is not continuously distributed over steadily increasing spaces, but it consists of a finite number of energy quanta localized at points in space, moving without dividing and capable of being absorbed or generated only as entities.[23]

This is the most "revolutionary" sentence written by a physicist of the twentieth century.

Though formulated apodictically and programmatically, Einstein's statement is still provisional, merely a "heuristic" assumption. The test of its value and usefulness will be the extent to which it helps explain physical phenomena. It will be shown that it does explain those phenomena to a very great extent. The next fifteen pages of Einstein's paper endeavor to make this viewpoint and its significance comprehensible or at least plausible.

As Einstein has no compelling theory to offer, in the first few paragraphs he presents a panorama of what at first glance are rather disparate arguments. He begins with a critique of Planck's formula for "black radiation" by pointing out that in deriving it Planck used two other formulas which contradict each other. Next follows an elegant determination of the Avogadro number from Planck's black-radiation formula. (A few weeks later, in his statistical papers, Einstein would offer two further methods of determining Planck's important constant.) All he is trying to demonstrate here is "that the determination of the elementary quantum set out by Herr Planck is independent of his theory of 'black radiation' "[24]—in other words, that energy quanta and Planck's black-radiation formula may be used without having to accept what Einstein regarded as a mistaken derivation and interpretation.

Next come thermodynamic reflections on the entropy of radiation. Einstein confines himself to the range of high frequencies, for which Wien's formula is valid, and derives an expression for the volume-dependence of the entropy of monochromatic radiation at a slight radiation density. The reason for this is that in a paper he wrote the previous year, volume-dependence played an important role in his investigation of the energy fluctuations of radiation. Einstein next interprets the expression he obtained in light of what, for the first time, he calls the "Boltzmann principle," according to which the entropy of a system is related to the probability of the system's state. Next he considers the same situation in gases and in dilute solutions, obtaining formally identical formulas for the volume-dependence of entropy, on the one hand for gases and on the other for radiation.

The purpose of Einstein's disparate argumentation now emerges: he concludes, by analogy, that just as a gas consists of atoms, so radiation should be seen as consisting of independent particles. "Monochromatic radiation of slight density (within the validity range of Wien's radiation formula) behaves in a heat-theory respect as if it consisted of mutually independent energy quanta of a magnitude $h\nu$."[25]

That much his contemporaries might accept as a theoretical exercise—crazy, perhaps, but harmless. After all, it is irrelevant for observation what is thought about radiation enclosed in a cavity. But now follows the "heuristic viewpoint" announced in the title. To Einstein, and only to him, "it seems reasonable now to examine if the laws on the generation and transformation of light are of a nature as if light consisted of such energy quanta." Whereas Planck's energy quanta had been postulated only in connection with involved argumentation in order to derive the radiation formula, Einstein, in a manner of speaking, has liberated the quantum from its cavity and made it useful for a whole range of other phenomena. He illustrates the consequences by a number of examples.

The most interesting consequence of Einstein's "heuristic viewpoint" was his law on the photoelectric effect, and not only because it won him a Nobel Prize in 1922. (The Royal Swedish Academy of Sciences awarded the Nobel Prize to Albert Einstein for "his services to theoretical physics, especially for his discovery of the law of the photoelectric effect.") According to Einstein's explanation of this effect, a light quantum—later to be called a "photon"—penetrates like a minute missile into a metal, there encounters an electron, and transfers its whole energy to that electron. On its way to the surface, the electron can lose some of the energy that was transferred to it by the light quantum, and additional work has to be done to escape from the surface. These relations can be rather involved in detail, but the maximum energy of a photoelectrically ejected electron depends solely on the frequency of the incident light, and in the simplest imaginable manner: $E = h\nu\ 2\ P$, where P is the exit (or photoelectric) work function.

This was the "second appearance"[26] of the quantum of action, but its first appearance outside the black cavity, and it constituted the basis of an unequivocal prediction: energy, plotted against frequency, must be a straight line, whose gradient is represented by a constant that is identical with the quantum in the radiation equation. Here was a program for experimenters.

The only conclusion which it was then possible to draw from Lenard's measurements was that the energy of photoelectrically emitted electrons depended solely on the frequency of the incident light, not on the intensity of irradiation—that is, not on the "quantity" of the incident light. Quantitative investigations of the photoelectric effect, however, were exceedingly delicate because there were all kinds of interference, especially electrostatic. The best measurements were obtained for the "threshold frequency" at which the irradiated metal did not yet lose any electrons but with a minimal increase in frequency electrons would be ejected. For this threshold frequency Einstein obtained a result which "as for order of magnitude agrees with Herr Lenard's results." This was about all that Lenard's treatise, praised by Einstein as "pioneering," had to offer; in particular, his data were not nearly adequate to verify the linear dependence of energy on frequency. Einstein therefore had to confine himself to the statement that his concept, "as far as I can see, does not contradict" Lenard's observations.

Further "groups of phenomena" discussed by Einstein were Stokes's law of photoluminescence, according to which the frequency of luminescence, or re-radiation, is less than that of the incident light; and the ionization of gases by ultraviolet light. Both of these phenomena conflict with the wave theory of light but are easily explained by light quanta.

Almost exactly a year later, Einstein also derived a relation between the so-called Volta effect and photoelectric diffusion. This was done in a marginal note in a paper entitled *On the Theory of Light Generation and Light Absorption*, in which Einstein continued his argument with Planck's theory of radiation. Although Planck's theory had initially seemed to him a "counterpart"[27] to his own "heuristic viewpoint," he was now able to show that the "theoretical basis on which Herr Planck's radiation theory rests differs from the basis that would follow from Maxwell's theory and from electron theory—specifically in that Planck's theory makes implicit use of the above-mentioned light quantum hypothesis."[28]

According to Einstein's argument, Planck's formula presupposed that the energy of an elementary resonator could assume only values which were integral multiples of $h\nu$. It followed, therefore, that "the energy of a resonator changes by absorption and emission only by leaps, specifically by an integral multiple of $h\nu$."[29] The quanta had thus acquired a new meaning, and Planck's radiation formula (which was undoubtedly correct but had until then lacked justification) was now provided with at least a provisional theoretical foundation—admittedly not in classical physics, as Planck would have wished, but in the newly emerging quantum physics. Planck did not like it at all.

If a hallmark of "revolutionary" work in science is that one's contemporaries refuse to follow it and that it takes a great many years to be accepted, then Einstein's "heuristic viewpoint" was indeed "very revolutionary." The fact that his paper was evidently accepted by *Annalen* without any quibbles[30] testifies to the liberal attitude of Max Planck, the coeditor responsible for theoretical papers, since initially he does not seem to have regarded light quanta as even worthy of discussion. In the summer of 1906, a year after publication of the paper, Planck's assistant Max von Laue wrote Einstein in Bern: "Incidentally, I never discussed your heuristic viewpoint with my chief. Maybe there are differences of opinion on it between him and me."[31] The very different opinion of his "chief" emerges from the earliest extant letter from Planck to Einstein, in the summer of 1907: "I look for the meaning of the elementary quantum of action (light quantum) not in the vacuum, but at the points of absorption and emission, and I believe that the processes in the vacuum are *accurately* described by the Maxwellian equations. At least, I don't as yet see any compelling reason for departing from this assumption, which, for the moment, seems to me the simplest one, and one which characteristically expresses the contrast between ether and matter."[32]

In 1909 Hendrik Antoon Lorentz, who for a long time had been reluctant to accept even Planck's radiation formula, attempted to connect the quantum of action only "with a limitation of the degrees of freedom of the ether."[33] And so it continued, all the way to that curious apology which Planck, in the summer of 1913, inserted into his otherwise overgenerous nomination of Einstein for membership in the Prussian Academy of Sciences in Berlin:

> That sometimes, as for instance in his hypothesis on light quanta, he may have gone overboard in his speculations should not be held against him too much, for without occasional venture or risk no genuine innovation can be accomplished even in the most exact sciences.[34]

Einstein of course did not see this nomination, but he was aware of Planck's views. At almost the same time, using a kind of imaginary dialogue, Einstein in a tribute to Planck referred to the difficulties of the radiation formula with which everything had started. In a cheerfully flippant tone, he observed: "It would be uplifting if we could place on a balance the amount of brain substance sacrificed by theoretical physicists on the altar of this universal function—and there is no end in sight yet of these cruel sacrifices!"[35] It would take ten years more before light quanta were eventually accepted, on the eve of the new quantum mechanics.

Why did it take two decades for Einstein's "very revolutionary" concept to be generally accepted? There has never been a comparable situation in twentieth-century physics. For one thing, there is no doubt that the stubborn opposition to the new idea of light quanta was primarily due to the fact that the wave theory of light had proved itself in a thousand different ways and that discarding it was almost unthinkable. But there were other factors, some of them having to do with experiments and others with Einstein himself. Let us look first at the experiments.

Although Einstein's equation for the photoelectric effect was simple, it was difficult to confirm or disprove experimentally. It remained an object of contention for a whole decade. For example, Lenard, who received the Nobel Prize for his cathode-ray experiments in the fall of 1905, corresponded with Einstein soon after the publication of the "heuristic viewpoint" and even sent an offprint of one of his own papers, which Einstein "studied with the same sense of admiration as your earlier work."[36] But Lenard, as an experimental physicist, clung to a resonance theory based on Maxwellian electrodynamics and saw no reason to take account of light quanta. Rudolf Ladenburg, three years younger than Einstein, in a thorough, sixty-page overview in 1909,[37] juxtaposed the two views, clearly emphasizing the advantages of Einstein's light quanta. His experimental data, however, were insufficient to let him decide for or against a linear relation between energy and frequency.

From about 1905 Robert Andrews Millikan at the University of Chicago was working on the photoelectric effect, at first along with other problems and unaware of Einstein's equation. After 1912, he devoted a great deal of effort to an attempt at refuting that equation. In 1915, contrary to all his expectations, he found himself compelled "to assert its unambiguous experimental verification in spite of its unreasonableness since it seemed to violate everything that we knew about the interference of light."[38] In his publications of 1916, though, Millikan did not hesitate to attack the assumptions behind the experimentally confirmed equation as if they had come from a fantasizing outsider rather than from a scientist who by then was famous. In a comprehensive article intended for publication in Germany he first gave Einstein the good news "that the Einstein equation accurately represents the energy of electron emission under irradiation with light,"[39] only to continue by saying that he considered "the physical theory upon which the equation is based to be totally untenable." At any rate, his result for the numerical value of the quantum of action was in close agreement with Planck's, and from this he concluded that his findings were "the most direct and most striking evidence so far obtained for the physical reality of Planck's *h*."[40]

Millikan's measurements had confirmed Einstein's equation for the photoelectric effect, but by no means the "heuristic viewpoint" on light quanta. As late as 1922, the Swedish Academy emphasized this distinction when, avoiding the suspect terminology, it awarded the Nobel Prize to Einstein solely for "the discovery of the law of the photoelectric effect." It chose the same cautious language again the following year, when it awarded the prize to Millikan. However, the breakthrough occurred almost simultaneously, in 1923, when the diffusion of light on electrons demonstrated that light did in fact consist of discrete energy packets.

The exceedingly hesitant acceptance of the light quanta may also have been partly due to Einstein's own language. He never actually asserted the existence of light quanta but preferred a form of words between the conditional and unreality. Monochromatic radiation, he said, behaved "as if it consisted of independent energy quanta of magnitude $h\nu$." Such "as if" formulations were not likely to persuade physicists to abandon their faith in the proven wave theory of light. If in later years Einstein referred to the "light quanta hypothesis" and once even to his "theory of light quanta,"[41] these were concessions to common usage or else just stylistic slips. Basically, he stuck to his "heuristic viewpoint" and even emphasized the "provisional character of this auxiliary concept"[42]—all of which made his colleagues feel entitled to reject light quanta.

Einstein's choice of words had nothing to do with excessive caution, let alone insecurity; it was based on what he expected of a genuine theory. So far, quanta tended to highlight cracks in the established theory rather than fitting into a new concept based on first principles. That was why Einstein stayed with his provisional "heuristic viewpoint" even when, before the end of 1906, he presented a further, exceedingly important, application of the quantum concept in an entirely different field, one that was moreover supported by experimental results. This was the third appearance of quanta: this time not in radiation but, for the first time, in the behavior of matter—in a theory of specific heat. It was the first quantum theory of solid bodies.

In 1820, two Frenchmen, Pierre Dulong and Alexis Petit, had made an interesting observation during an investigation of the thermal behavior of solid bodies. The amount of heat needed to raise the temperature of a body by one degree was virtually constant if it was related to atomic weight. For many metals, from copper through nickel to gold, as well as for sulfur, they invariably found the same value of "specific heat."[43] Their surprising discovery indicated an atomic structure of matter and suggested that "the atoms of all simple bodies have exactly the same capacity for heat."[44] This "law of Dulong and Petit" did not receive its theoretical foundation until half a century later, when Ludwig Boltzmann, in 1876, with his fundamental "equipartition theorem"[45] of statistical physics, firmly based the empirical regularity found by the two Frenchmen on the kinetic theory of matter.

But as so often happens in physics, no sooner was the theory established than experimental results began to accumulate which would not fit into that neat concept at all. As a young man in Berlin in 1870, Weber, subsequently Einstein's teacher at the Zurich Polytechnic, had investigated specific heat at various temperatures, first for diamond and then for boron and silicon, and had found marked deviations from the Dulong-Petit law. Only at high temperatures did specific heat agree with expectations; at lower temperatures it diminished, and for these three substances it was much too low even at room temperature. While still a student under Weber, Einstein became familiar with this "specific heat anomaly," partly through Weber's lectures, partly through his own laboratory work, and possibly also through his diploma thesis, which dealt with heat conduction.

The first inspiration had come to Einstein on a train. As he reported to Mileva, while traveling to see his parents in Milan in the spring of 1901, he "came up with an interesting idea. It seems to me that it is not out of the question that the latent kinetic energy of heat in solids and liquids can be thought of as the energy of electrical resonators."[46] This idea may have been inspired by his study of Planck's work of a year earlier, immediately before the quantum of action. In line with his microphysical concept of matter, Einstein was linking the thermal and optical properties of matter, because "if this is the case, then the specific heat and the absorption spectrum of solids would have to be related." He immediately commanded his "little devil" to go to the library, because he was interested in these relations for glass: "See if you can find some literature on this!"

But whatever Mileva may have discovered then (if anything), the subject required a few more years of "hatching" and pondering, especially on the basis of Planck's radiation formula, before Einstein was able to bring it all together in the fall of 1906, under the title *Planck's Theory of Radiation and the Theory of Specific Heat*.[47]

In this article, Einstein again starts with a new examination of the mean energy of Planck's oscillator, "which clearly reveals its relation to molecular mechanics." And again he presents a further variation of the "Boltzmann method," which differs from Planck's procedure. From these reflections—which are by no means mere preliminaries and which actually contain an interesting mathematical innovation[48] that would be rediscovered two decades later—Einstein derives a profound transformation of mechanics in the interaction of atoms or molecules with electromagnetic radiation. This is the first explicit reminder that in the microcosm everything will be different from everyday experience based on the senses, and hence also from the "classical" physics based on that everyday experience:

> Whereas until now the molecular movements had been regarded as subject to the same regularities as those valid for the motion of bodies in the world of our senses, we now find ourselves compelled . . . to make the assumption that the variety of states

which they are capable of assuming is less than for bodies of our experience.[49]

Transfer of energy proceeds not continuously, but only in discrete packets of magnitude $h\nu$. Einstein asks if this might perhaps apply also to "the other oscillating systems suggested by the molecular theory of heat," that is, to the atoms of solid bodies. His answer is not long in coming.

Einstein confines his observation to a simple model: a homogeneous isotropic crystal whose atoms oscillate with a single frequency about their equilibrium. Initially Einstein thought only of electromagnetic forces as the cause of the oscillations, so that his model would represent only electrically conducting substances, whose atoms are separated into heavy ions and light electrons. Only after his paper had gone to the printer did he realize that there was no reason for this limitation. He amended his error in a "Correction,"[50] which is important because in it, for the first time, quanta appear in purely mechanical oscillations, and thus become totally independent of electromagnetic radiation. Applying his quantum formulas he derives, in a few steps, a formula for specific heat, which, just as it should, leads at high temperatures to the Dulong-Petit law, and at low temperatures results in a steady diminution of specific heat down to absolute zero. The temperature above which the old rule remains valid would later be called "Einstein temperature." It is closely connected with the oscillation frequency of the atoms, which in turn characterizes the optical properties of the crystals. For light atoms such as carbon, the Einstein temperature is fairly high—for diamond around 1,000° C—so that deviation from the Dulong-Petit law already appears at room temperature. That noble crystal on a beauty's neck thus displays quantum characteristics.

Einstein compared his quantum-theoretical formula with the data found for diamond by his teacher H. F. Weber in 1875. The agreement of his theoretical graph with the measurements is so excellent that Einstein felt justified in thinking it "probable that the new view . . . will prove its worth in principle."[51] At the same time he realized that "there can of course be no thought" of the new theory's "exactly matching the facts." He had been using a model which he realized had too many simplifications; in particular he regarded the assumption of a single temperature-independent oscillation frequency as "undoubtedly inadmissible."[52] Nevertheless, the theory, quite apart from its pioneering character, would certainly "prove its worth in principle."

Unlike Einstein's "heuristic viewpoint" on light quanta, his quantum theory of solid bodies soon moved into the scientific spotlight. This was due not to Max Planck, who preferred to keep silent on this new application of the quantum concept, but to his colleague Walther Nernst, the professor of physical chemistry at Berlin. In 1905 Nernst had not so much derived as postulated a general thermodynamic law according to which entropy rather than energy, and in consequence also specific heat, disappears at absolute zero. Henceforward Nernst devoted himself with the passion of a lover to the experimental consequences of his law—now raised to the status of the third law of thermodynamics—at really low temperatures.

In 1875, Weber had to suspend his measurements when the snow thawed; but liquefaction of air and even hydrogen had meanwhile become possible, providing a close approach to absolute zero, −273°C. Nernst had built a plant for the liquefaction of hydrogen, and an entire army of young coworkers at his institute on Bunsenstrasse were busy measuring the specific heat of the most varied substances over wide ranges of temperature.

For this research program Nernst could not disregard Einstein's new quantum theory of specific heat. Three years after Einstein's publication it was clear that his quantum formula was substantially correct in describing the actual conditions; and in 1911 Nernst, in a comprehensive article, stated: "It is obvious that the observations in their totality provide a brilliant confirmation of Planck's and Einstein's quantum theory."[53] After that, quantum theory was no longer to be excluded from discussions of the behavior of solid bodies—and, as we shall see, Einstein had gained an influential supporter and patron. But although he was the most vigorous propagandist of quantum theory, Nernst was never really sure whether to regard Einstein's concept as a kind of mathematical tool or the foundation of a totally new physics.

Having visited Einstein in Switzerland in the spring break of 1910 and discussed with him this question and many others, Nernst reported to a colleague about Einstein's theory, sounding as if he had let himself in for a very enticing but somehow illegitimate affair:

> Einstein's quantum hypothesis is probably the strangest thing ever thought up. If correct, it opens entirely new roads both for so-called ether physics and for all molecular theories. If false, it will remain "a beautiful memory" for all times.[54]

The theory was correct—at least "in principle"—but it nevertheless did remain "a beautiful memory." As for the "ether physics" mentioned by Nernst, Einstein no longer had anything to do with that. He had discarded the ether implicitly in March 1905 with his invention of light quanta; and he discarded it explicitly three months later in his theory of relativity.

CHAPTER NINE

Relative Movement: "My Life for Seven Years"

It happened about the middle of May 1905, shortly after he had sent in his paper on the Brownian movement. Einstein could not remember the exact date, only that it was a "beautiful day."[1] He had visited his friend and colleague Michele Besso to discuss a difficult question with him. "We discussed every aspect of the problem," Einstein reported seventeen years later. "Then suddenly I understood where the key to this problem lay." He must have spent an exciting night, with some mathematics and a lot of concentrated thinking. The key he had found in conversation with Besso magically opened a door to a new understanding of the fundamental concepts of all physics. The following day, "without even saying hello," Einstein pounced on his friend with the explanation:

> Thank you. I've completely solved the problem. An analysis of the concept of time was my solution. Time cannot be absolutely defined, and there is an inseparable relation between time and signal velocity.[2]

Einstein's analysis started off with the question of what is meant by simultaneity in two different places. He was observed gesticulating to friends and colleagues as he pointed to one of Bern's bell towers and then to one in the neighboring village of Muri. Michele Besso was the first person and Josef Sauter the second[3] to whom he explained in this manner that the synchronization of spatially separated clocks represented a problem which, properly understood, must lead to profound changes in the concept of time. If the customary concept of space, and a lot else in physics, had to undergo a transformation, then these, no matter how exciting they seemed, were simply logical consequences.

"Five or six weeks elapsed before the completion of the publication in question."[4] Toward the end of June it was all written up, and on June 30 receipt of the manuscript was recorded at the editorial office of *Annalen* in Berlin. The thirty-page article, published three months later, was titled *On the Electrodynamics of Moving Bodies*.[5] It was a treatise beyond compare and without precedent, one of the greatest scientific achievements in content and one of the most brilliant in style. Of course, there were later additions, some from Einstein himself and some from others, but these were mere addenda to a theory which had appeared before the world all ready and complete, valid for all time. A few years later it was called the "theory of relativity" or "relativity theory"—not by Einstein but by others—and after a few more years it became known as the "special theory of relativity."

At the age of twenty-six Einstein had successfully brought to fruition an intellectual adventure which had occupied him for a decade (if we include his schooldays in Aarau) and which had held him in thrall for at least seven years: the relativity principle in electrodynamics. He was later to declare that this had been his "life for over seven years and this was the main thing,"[6] but he found it difficult to retrace the development of his ideas over that period.[7] These ideas involved a complete rethinking of the entire conceptual tradition of modern physics from its beginning. Let us therefore briefly consider its first 250 years.

The principle of relativity had been discussed at the very beginning of modern science. It was formulated by Galileo in his *Dialogue on the Two Chief World Systems*, published in 1632. Galileo gave it the kind of colorful setting that was then fashionable. On the "Second Day" of the *Dialogue* Galileo's alter ego Salviati invites his friends to assemble in a spacious room inside a ship, a room containing midges and butterflies as well as a fish tank. Then comes an instruction: "Suspend a bucket from the ceiling, from which water drips into a second, narrow-necked vessel."[8] Finally, also in the service of science, the friends are instructed to leap forward and backward, and see what distance they cover.

The stage, of course, is set for a comparison between a ship at rest and a ship in motion. "Now let the vessel move at any speed whatever: provided only its movement is uniform and does not fluctuate one way or the other, you will observe that there is not the slightest change in any of the observed phenomena."[9] The midges and butterflies will fly about as before, and no difference will be noticed in the movements of the fish. The distance achieved by leaping remains the same regardless of whether one leaps in the direction of the ship's movement or in the opposite direction. Perhaps most convincing of all is the fact that all the drops continue to be caught in the lower container with the narrow neck: "Not one of them will fall on its rear part, even though the ship covers many spans while the drop is in midair."

With this argument Galileo intended to dismiss objections to the assumption of a revolving Earth. At the same time he demonstrated—entirely correctly and precisely, and not in the dry style of a modern textbook—that there is no way in mechanics of distinguishing, in the case of two referential systems moving rectilinearly and uniformly (i.e., with constant velocity) relative to each other, which of them is in motion and which at rest.

Fifty years later this idea was included by Isaac Newton in his monumental *Principia Mathematica*, the fundamental presentation of modern mechanics. It appears, however, not in a prominent position as an axiom, but only as an addition, as Corollary V: "The position[s] of bodies included in a given space are the same among themselves, whether that space is at rest, or moves uniformly forwards in a right line without any circular motion."[10] However, it is clear from this corollary that Newton was not concerned with equivalent referential systems, but that he postulated an "absolute space" relative to which the "given space" was either at rest or in motion. Moreover, the "absolute space" in Newton's concept, "in its own nature, without regard to any thing external remains always similar and immovable,"[11] and he needed it in order to explain inertia as well as the objective character of rotational movement. It was a kind of all-embracing container within which the privileged status of uniformly moving referential systems could be defined.

Newton's "absolute space" was by no means accepted uncritically, but the overwhelming success of his mechanics eventually silenced all objections, with the result that this concept was soon elevated to a prerequisite of thought. In the philosophical analysis of Immanuel Kant, who revered Newton, his concept of space was confirmed as "pure a priori experience" and hence a prerequisite of knowledge altogether.[12] Restating Newton's Corollary V, Kant declared: "The space that is itself movable is called the material, or also the relative, space; that in which all motion must ultimately be imagined (which therefore is immovable) is called the pure, or also the absolute, space."[13] This was so evident, to physicists as well as philosophers, that the equivalence of referential systems or "relative spaces" moving rectilinearly and uniformly relative to each other was not seen as a problem at all. The principle of relativity was so securely anchored in mechanics that it did not even have a name.

However, toward the end of the nineteenth century, when it seemed advisable to take a closer look at the foundations of mechanics, referential systems moving uniformly with regard to each other were given their modern name, "inertial systems,"[14] and the simple rules for calculating from one inertial system to another became known as the "Galileo transformations."[15] The principle of relativity could therefore be briefly formulated to the effect that the laws of physics have the same form in all inertial systems, and that they must therefore be invariant with regard to Galileo transformations.

That was evident in mechanics, but the emergence of the new terminology indicated enhanced awareness. This was due to the use of electrodynamic theory as a uniform way of describing all electrical, magnetic, and optical phenomena. Maxwell's theory, however, did not satisfy the relativity principle of mechanics, because its basic equations are clearly not Galileo-invariant. The ether, moreover, represented a privileged referential system.

After Einstein, the ether was so thoroughly swept out of physics that today a lot of historical empathy is required to appreciate the importance attached to it by nineteenth-century physicists. The fact is that less than a hundred years ago the ether was as real as air, light, or an ocean liner. In interpreting their optical experiments even the cleverest researchers felt that they were "virtually touching the ether with their fingers."[16]

Maxwellian electrodynamics, the finest and most impressive contribution to nineteenth-century physics, had actually moved the ether to the center of all physical thought. Thus Heinrich Hertz observed:

> Take electricity out of the world, and light vanishes; take the luminiferous ether out of the world, and electric and magnetic forces can no longer travel through space.[17]

Electromagnetic fields and waves, as well as light waves identified as such, were transversal oscillations of the ether, perpendicular to their direction of propagation; and for Hertz and most of his colleagues they made proper sense only once they were reduced to mechanical models of that medium. When Wilhelm Conrad Röntgen in 1895 discovered a "new kind of ray," experimental and theoretical physicists had an interpretation ready: X-rays could be nothing other than the long-suspected longitudinal oscillations of the ether, in the direction of propagation.

Ether physics[18] was fascinating and intellectually demanding, but it found itself in conflict, in a variety of ways, with the mechanics which was to continue as the foundation of all physics and which therefore should be able to describe the ether as well. A medium which did not share in the movement of matter and which pervaded all space offered itself as a natural, and simultaneously preferred, referential system for the propagation of light, as the only system in which the velocity of light has its value of 300,000 kilometers per second.[19] As the Earth could hardly be at rest in the ether—that assumption could at best be made for the sun—the Earth's movement through the ether would have to be observable through optical effects. At an orbital speed around the sun of thirty kilometers per second these effects could be expected to be in the readily measurable first order of magnitude of v/c (v = velocity; c = speed of light), i.e., one ten-thousandth; but nothing of the sort was observed.

In 1881 the American Albert Abraham Michelson, then not yet

thirty, achieved a fantastic increase in precision. With the benevolent support of Helmholtz, he constructed in Berlin a two-arm interferometer for determining the second order of magnitude v^2/c^2, i.e., one hundred-millionth. His apparatus was so sensitive that even the horse cabs passing outside the Physical Institute impaired its operation; it was therefore moved to the solitude of the Astrophysical Observatory in Potsdam. The outcome, published in 1881, was disappointing: no movement of the Earth relative to the ether could be proved. Six years later the experiment was repeated, with still greater accuracy, by Michelson and his colleague Edward W. Morley in Cleveland, but it merely confirmed the surprising Potsdam findings. "I think it will be admitted"—Michelson consoled himself in 1907, when he was the first American to receive the Nobel Prize (not for his ether experiments but for similar optical precision measurements)—"that the problem, by leading to the invention of the interferometer, more than compensated us for the fact that this particular experiment gave a negative result."[20]

But no one wanted to go back to before Copernicus, to a geocentric view, or conclude from the Michelson–Morley experiment that the Earth was resting motionless in the ether. Instead, brilliant theories were designed to prove that it was impossible to observe a movement relative to the ether. In these endeavors the theory of Hendrik Antoon Lorentz offered the most valuable insights: while it remained wedded to the ether, it came fairly close to the relativity theory. To Einstein it was important because a reconstruction of Lorentz's theory from general principles represented one of the touchstones of his own new concept of space and time.

In 1877, when he was not yet twenty-five, Lorentz was invited to take the newly created chair of theoretical physics at the University of Leyden in the Netherlands. In the 1890s, after fifteen years of hard work, he developed a new version of electrodynamics, his "electron theory." His terminology reflects the increase in knowledge during the final decade of the nineteenth century: in 1892 he referred simply to "charged particles"; in 1895, in his comprehensive *Attempt at a Theory of Electrical and Optical Phenomena in Moving Bodies*, carriers of charge were called "ions"; and after 1899, two years after the discovery of charged light particles, he called them "electrons." The fields exist independently of the charge carriers in the ether and react back on matter by exerting a force on the charge carriers. This force subsequently came to be known as "Lorentz force." On this basis he created a theory embracing all electromagnetic and optical phenomena known at the time, a theory that proved its worth even with new phenomena.

In 1896 Lorentz's assistant, the privatdozent Pieter Zeemann, succeeded in observing the splitting of spectrum lines in a magnetic field—an effect that had been sought in vain since Faraday. This was such a spectacular triumph for Lorentz's theory that he and his assistant were honored in 1902 with the second Nobel Prize for physics. Even in his old age, Einstein was enthusiastic whenever Lorentz's theory was mentioned: "It is a work of such consistent logic, clarity, and beauty as has been rarely achieved in a science based on empiricism."[21]

It is true that in Lorentz's theory the relativity principle could not find expression through the "Galileo invariance" of mechanics. A kind of relativity principle was upheld in Galileo's sense—that experimenters have no way of distinguishing whether they are at rest or in uniform rectilinear motion. But even that statement entailed considerable effort. Lorentz solved the problem of motion relative to the ether with his theorem of "corresponding states." By this ingenious device he was able to relate the electromagnetic values in moving inertial systems to the one system for which the Maxwellian equations are strictly valid—the ether at rest. To that end he invented for the moving system an auxiliary construct which he called "local time," in which "true time" appeared linked with the spatial coordinates. This was the first manipulation of the parameter of time in a physical theory; but bold as it was, "local time" to Lorentz was no more than a mathematical artifice without any consequence for the traditional understanding of time. After all, he had merely invented a trick to make the theory agree with observation, to "explain away" the first-order effects.

For second-order effects, such as the Michelson-Morley experiment, Lorentz had to resort to a further hypothesis, which had simultaneously and independently been introduced by George Fitzgerald. According to this hypothesis, the dimensions of a body moving through the ether are shortened in the direction of movement by a characteristic factor dependent on velocity:

$$\frac{1}{\sqrt{1-\frac{v^2}{c^2}}}$$

This contraction was understood by Lorentz as an effect based on real dynamic forces, caused by a compression (not specifiable in detail) of the charge carriers through their interaction with the ether.

Lorentz was thus able to develop a consistent theory for *electromagnetic phenomena in a system moving at any velocity not reaching the velocity of light*—the title of his comprehensive treatise of 1904, in which the approximations of earlier versions were overcome and the theory was valid for all orders in v/c. The success of Lorentz's extension of Maxwell's theory was so impressive that physicists, especially in Germany, already saw it as the overthrow of traditional mechanics-based physics. All physics, including mechanics, was now to be rebuilt within the framework of an "electromagnetic picture of the world."

Henri Poincaré was then the world's most famous mathematician, with epoch-making contributions to both the fundamentals of his discipline and its applications, especially in physics. Poincaré was sympathetic to Lorentz's theory but at the same time criticized Lorentz for creating more and more hypotheses for every new experimental result. Not only did he encourage Lorentz to improve his theory—and after 1904 Lorentz in fact made some allowance for Poincaré's objections[22]—but he himself made substantial contributions to a critical analysis and mathematical structuring of the theory.

A good opportunity was provided in 1900 by a *Festschrift* in honor of the twenty-fifth anniversary of Lorentz's doctorate. Poincaré in his contribution[23] showed, among other things, that Lorentz's "local time" could be interpreted as equivalent to a procedure whereby clocks were synchronized by light signals. This interesting link between Lorentz's "local time" and the problem of time measurement was presented by Poincaré four years later at the International Congress on the Arts and Sciences held, in true American style, on the occasion of the World Exhibition in St. Louis in 1904. As an internationally acknowledged authority he was entrusted with the key lecture, *On the Present State and the Future of Mathematical Physics.*[24]

In a section devoted to the relativity principle—the first text in which not only the subject but the name appears—Poincaré explained his synchronization procedure, along with the consequence that clocks in different inertial systems also show a different time. However, he followed Lorentz in the concept of "true time": "The clocks synchronized in that manner do not therefore show the true time, but what one might call 'local time,' so that one of them is slow with regard to another. This does not matter much, as we have no way of determining it."[25] This is hardly surprising, as everything had been shrewdly arranged in such a way that experimenters could not possibly tell whether they were at rest or in motion, just as was called for by the relativity principle.

This lecture contained some other hints at the future development of physics. Actual calculations and experiments showed that the mass of electrons was apparently not constant—as would have been expected as a matter of course from the time of Newton and Lavoisier—but depended on their velocity. "From all these results," Poincaré summed up,

> provided they are confirmed, an entirely new mechanics would arise, characterized mainly by the fact that no velocity can exceed the velocity of light, just as no temperature can drop below absolute zero. For an observer himself in a translational motion, which he does not suspect, no velocity whatever can any longer exceed that of light.[26]

This is the first indication that the velocity of light could play a major role, structuring theory not only in optics and electrodynamics but also in mechanics.

Poincaré, however, made no use of this far-reaching assumption when in the following year he published *Dynamics of the Electron.*[27] This did not go beyond Lorentz's theory in terms of physics, but moved its mathematical structure into a new light. Poincaré combined the three spatial coordinates and time into a "quadruple vector" and operated with these structures as in conventional Euclidian geometry.

The transformations of Lorentz's theory, which mediate between "local time" and true time on the one hand, and—because of contraction—between the spatial coordinates on the other, would then correspond to a rotation in a four-dimensional space. Poincaré also proved that the transformations, which he called "Lorentz transformations," displayed the kind of structure mathematicians call a "group," its important characteristic being that two consecutive Lorentz transformations for their part represent an admissible transformation. Specifically this means that, against all expectation, "addition" must not simply be performed arithmetically, but that the combination of two velocities always produces a result that is smaller than the velocity of light.

This was probably as much as could be achieved by Lorentz and Poincaré for electrodynamics as a physics of the ether. But two types of theory were now suddenly standing side by side: first, the physics of matter, mechanics, with a powerful relativity principle, realized in invariance toward the Galileo transformations; and second, the physics of the ether, electrodynamics, in which the Lorentz transformations had to be valid in order that any relative movement with regard to the ether, as the privileged system at rest, could not be observed. This conflict was certainly realized by the theoreticians, but they also reconciled themselves to it. It was certainly never clearly formulated in discussions at the beginning of the century, except in a few prophetic aperçus by Henri Poincaré.

The conflict, however, was at the center of the scientific endeavors of the Expert III Class at the Patent Office in Bern. When Poincaré's *On the Dynamics of the Electron* appeared in Paris on June 5, 1905, Einstein was just getting his solutions of this and other problems ready for the press. It was something no one had expected: the theory of relativity.

The axiomatic structure of Einstein's paper betrays little of the intellectual efforts, the roads and the wrong turnings, which eventually led him to the solution. Footnotes are no help here: the paper is unique—among other ways—in that it does not contain a single bibliographical reference.[28] The few extant letters between 1903 and 1905 contain nothing about the ideas that led Einstein to his relativity theory, which means that any reconstruction of his thinking must rest on the scant recollections of others and on the equally scant hints in his own scientific papers—and, of course, on his recollections. But his recollections were not recorded until very much later, and Einstein himself, in his *Nekrolog* written in 1946, points out that "the present man of sixty-seven is not the same as the one of fifty, thirty, or twenty. All memories are colored by what I am at present, in other words by a deceptive viewing angle."[29] Two weeks before his death he told a young historian of science that he himself had "always found himself a very poor source of information concerning the genesis of his own ideas."[30] But even if they are not always consistent, and at times even contradictory, Einstein's recollections are still our best and certainly our most interesting source.

Einstein often described the beginning of the ten-year incubation period—a mental experiment in his schooldays in Aarau. He had pictured an observer running after a ray of light. What he would perceive would be similar to the impression of a surfer riding ahead of a wave, at rest relative to the water while between two crests. In the case of light this situation corresponds to a electromagnetic field spatially oscillating but at rest:

> But such a thing does not seem to exist, either on the grounds of experience or according to the Maxwellian equations. But intuitively it seemed to me clear from the outset that, judged by such an observer, everything would have to unroll according to the same laws as for an observer at rest relative to the earth. For how could that first observer know, or be able to discover, that he is in a state of rapid uniform motion?[31]

This paradox clearly revealed a crack in the foundations of physics. On the one hand there was mechanics, in which an observer traveling with the speed of light or even faster was entirely thinkable; on the other hand there was electrodynamics, according to which such an observer would have to see something that evidently does not exist. That was why an observer, or indeed any material body, could never attain the velocity of light—which is therefore a limiting velocity, for any observer, no matter how fast the observer might move in any inertial system. In retrospect, Einstein believed "that the germ of the special relativity theory was already present in that paradox.[32] He later stated that this mental experiment had always been with him.[33] And like an inner compass, it seems to have led the student straight to what was "fundamentally important"[34] in physics.

During the summer vacation after Einstein's third year at the Polytechnic the old problem emerged in a letter to Mileva, now against the background of his study of Helmholtz and Hertz, and with remarkable self-assurance for a young man:

> I'm more and more convinced that the electrodynamics of moving bodies as it is presented today doesn't correspond to reality, and that it will be possible to present it in a simpler way. The introduction of the term "ether" into theories of electricity has led to the conception of a medium whose motion can be described without, I believe, being able to ascribe physical meaning to it.[35]

A month later, evidently in connection with Fizeau's famous experiment, Einstein "had a good idea for investigating the way in which a body's relative motion with respect to the luminiferous ether affects the velocity of the propagation of light in transparent bodies. I even came up with a theory about it that seems quite plausible to me."[36] Einstein, in a manner of speaking, had imbibed the ether with his mother's milk, and despite his doubts that statements about its movement had any meaning, he intended, as a good empiricist, to tackle it with the tools of observation.

At about that time he conceived another experiment, analogous to the Michelson-Morley experiment though not based on any detailed knowledge of it;[37] but this was not performed because the "boss," Weber, was skeptical. Disappointed, Einstein turned to Wilhelm Wien in Aachen.[38] We do not know if he received a reply, or what came of these efforts generally, but it is obvious that during his last year at the Polytechnic he was very much concerned with relative movement.

After his exam, while he was job-hunting, Einstein placed much hope in the completion of a paper on relative movement.[39] We have no idea what this paper contained. Nor do we know anything about another experiment which Einstein had thought up: "I have now thought of a very much simpler method of investigating the relative movement of matter against the luminiferous ether," he informed his friend Grossmann, "one that is based on ordinary interference experiments. If only inexorable fate would grant me the time and tranquillity necessary for them."[40] Toward the end of 1901 he was back with theory: "I am busily at work on an electrodynamics of moving bodies, which promises to be quite a capital piece of work."[41] At one time he had doubted his ideas, but when he discovered that only a "simple calculation error" had slipped in and spoiled everything, he was jubilant: "I now believe in them more than ever."[42] But we still do not know what he believed in.

He certainly explained his ideas to Professor Kleiner of the University of Zurich, and this experienced physicist had liked them: "He advised me to publish my ideas on the electromagnetic theory of light of moving bodies along with the experimental method," is how he summed up the conversation for Mileva. "He found the method I've proposed to be the simplest and most expedient. I was quite happy about the success. I'll write that paper in the next few weeks for sure."[43] However, he probably was, again, overoptimistic.

He must have encountered difficulties in putting his ideas on paper, because instead of publishing he decided to "get down to business now and read what Lorentz and Drude have written about the electrodynamics of moving bodies."[44] Jakob Ehrat, a former fellow student and now an assistant, "will have to get the literature for me."[45] No matter how much Einstein may have read, his reading did not produce anything that went into print. A year later, by then at the Patent Office, he was once more "engaging in comprehensive studies in electron theory,"[46] making use, probably not for the first time, of Lorentz's *Versuch* of 1895, now available in German.

■ ■ ■

Einstein by then had totally lost his initial belief in the ether. Even as a student he had considered the possibility that electrodynamics might become "the theory of the movements of moving electricities & magnetisms in empty space."[47] Poincaré in *La Science et l'hypothèse*—which, according to Solovine's account[48] "for weeks on end captured and fascinated" the members of the Akademie Olympia—had reduced the ether to a hypothesis which was "convenient for the explanation of phenomena" and even predicted that "one day the ether will undoubtedly be discarded as unnecessary."[49] Einstein may have understood this as a programmatic invitation, but by then he was himself much closer to that opinion than the great mathematician.

This must have been recognized by his Patent Office colleague Josef Sauter when he wanted to discuss his own ideas on the mechanical ether models of Maxwellian theory, which had been published in *Annalen*.[50] Einstein was not interested at all and once more declared: "I am a heretic."[51] His intensive study of radiation theory had convinced him "that Maxwell's theory does not describe the microstructure of radiation and therefore is not universally tenable."[52] This view eventually culminated in Einstein's "heuristic viewpoint" on light quanta, which no longer had any use for the ether.

Two months before his death, Einstein replied to a question on his state of knowledge at the time he developed his relativity theory: he said that in 1905 "I only knew Lorentz's important treatise of 1895, but not Lorentz's later work, nor Poincaré's follow-up work. In that sense my work in 1905 was independent."[53] Most probably Lorentz's publication of 1904, which for the first time presented the "Lorentz transformations" in generally valid form, was not available in Bern, having been published in the *Proceedings* of the Amsterdam Academy, which was not widely distributed. But he must have known a lot more than only Lorentz's *Versuch* of 1895.

Einstein's passionate interest in *Annalen* is unlikely to have diminished during his time in Bern, especially as he was now a regular contributor to it. Although he once complained that "during my free time the library is closed,"[54] he should have had no problem keeping up with publications. His official duties would have occasionally involved visits to the city and university libraries, so that, in addition to electrical engineering publications, he could have studied physical journals.[55] If that was not enough, Professor Gruner would surely have lent him some publication for excerpting.

It therefore seems virtually impossible that Einstein could have missed the seventy-five-page treatise *Principles of the Dynamics of the Electron*[56] by the Göttingen privatdozent Max Abraham. He certainly had Abraham's paper, published in 1904, *On the Theory of Radiation and Radiation Pressure*,[57] or at least notes on it, available when he wrote his own paper. That same year Einstein would have been able to read in *Annalen* Wilhelm Wien's *Differential Equations of the Electrodynamics of Moving Bodies*,[58] which contained many references to the latest literature, as well as a subsequent polemic between Wien and Abraham, in which Wien not only quoted Lorentz's work of the same year but actually provided an outline of it.[59] The outstanding Strasbourg theoretician Emil Cohn likewise published his phenomenological reflections *On the Equations of the Electromagnetic Field for Moving Bodies*[60] in *Annalen*, as an alternative to Lorentz's theory.

The most important experimental contribution of those years, Walter Kaufmann's measurements of the deflection of electrons in electric and magnetic fields, appeared not in *Annalen* but in *Physikalische Zeitschrift*,[61] founded in 1900, which was also available in Bern. Indeed Einstein must have positively had his nose rubbed in it by Abraham's and Wien's theoretical treatment of that question.

The only aspect of this discussion of immediate interest in this context is the fact that the velocity of light was emerging as a limit for the movement of the electron, with Wien, for instance, declaring that "by exceeding the velocity of light an infinite amount of work would be performed."[62] More particularly, the increment in the mass of the electron at increasing velocity was seen as an indication of the electromagnetic origin of all mass, the more so as this view fitted well into endeavors for an "electromagnetic world picture" as the foundation of physics and as an alternative to mechanics.

▪ ▪ ▪

Despite their physical sophistication and mathematical virtuosity, the theories put forward and discussed at the beginning of the century must have seemed imperfect to Einstein because of their adherence to the ether as a referential system at rest. Also, he had long been unhappy about the customary view of electrodynamics, if only because of its asymmetry. This concerned Michael Faraday's classic experiment, a simple school experiment watched by millions of students without setting off much thought in them. To Einstein, however, it was of crucial importance. This we can read between the lines of the opening paragraph of one of his publications, and we know about it from a manuscript, completed in 1920, for a special issue of the English periodical *Nature* devoted to relativity theory. When it was ready he complained to his translator that "unfortunately it has grown so long that I very much doubt if it can be published in *Nature*."[63] In fact, a greatly abridged version[64] was published, leaving out (among other things) some personal reminiscences which, fortunately, have come down to us with the thirty-one-page original version.[65]

After nineteen pages of objective didactic exposition, Einstein abandoned his dry scientific style and offered a surprising insight into the subjective aspects of his reflections:

> In the creation of the special relativity theory the following, not previously mentioned, idea about Faraday's magnetoelectric induction played a leading role.
>
> During the relative movement of a magnet with regard to an electric circuit an electric current is, according to Faraday, induced in the latter. It makes no difference whether the magnet is moved or the conductor, what matters is the relative movement only. According to the Maxwell-Lorentz theory, however, the theoretical interpretation of the phenomenon is very different for the two situations.
>
> If the magnet is moved, a time-variable field exists in space, which, according to Maxwell, gives rise to closed electric lines of force, i.e. a physically real electric field; this electric field then sets in motion the movable electric masses within the conductor.

> If, however, the magnet is at rest and the electric circuit is moved, then no electric field is created. Instead, the current is caused in the conductor through the fact that the electricities moved with the conductor are subject, through their (mechanically enforced) motion relative to the magnetic field, to an electromotive force hypothetically introduced by Lorentz.

What Einstein described here had been discovered by Michael Faraday in 1831 as one of the milestones on the road to a uniform understanding of electricity and magnetism. Quite apart from its theoretical importance, Faraday's demonstration had enormous practical consequences: it gave rise to the development of generators to produce an electrical current and motors to use it. As a youngster in the family's electrical engineering firm, Einstein must have learned something about electrical machines from his uncle Jakob. As for the pitfalls of theoretical interpretation, he had probably encountered those as a student at the Polytechnic in August Föppl's[66] book on Maxwell's theory. In its fifth chapter, "The Electrodynamics of Moving Conductors," Föppl analyzed the arrangement of magnet and conductor in terms of relative movement—without, however, noting any profound conflict of principle.

Einstein's friend and colleague Besso believed that it was he who had introduced the topic into their conversations, and "thereby having participated in the relativity theory, realizing as an electrical engineer that what in the framework of Maxwellian theory appears in the induced part as an electromotive force or as an electric force, according to whether the inductor of an alternator is at rest or in rotation" must be viewed "as a peculiar practical anticipation of the relativity concept."[67] Einstein, of course, had been aware of that problem since his student days, so that his talks with Besso merely gave it sharper outline.

Let us see, then, what conclusions Einstein drew from the asymmetric description of Faraday's experiment:

> The idea that these were two disparate situations was intolerable to me. I was convinced that the difference between the two was

> merely a difference in the choice of the station of the observer. Viewed from the magnet there certainly was no electric field present, viewed from the electric circuit, such a field certainly existed. The existence of the electric field, therefore, was a relative one, according to the state of motion of the system of coordinates used, and only the electric and magnetic fields *jointly*—regardless of the state of motion of the observer, or the system of coordinates—could be adjudged a kind of objective reality. This phenomenon of magnetoelectric induction compelled me to postulate the [special] relativity theory.[68]

From a footnote to the above account we may conclude that Einstein for a while was toying with alternatives to the Lorentz-Maxwell theory: "The difficulty to be overcome was in the constant nature of the velocity of light in a vacuum, which initially I thought I would have to discard." This is one of the few indications of endeavors in the course of which he intended to do without the universally constant velocity of light, inherent in the theory. The velocity of light was to be constant only for an observer stationed next to the light source, whereas all observers moving relative to that source would measure a different value, depending on their own relative velocity with regard to the source. That was not only plausible but also in line with Einstein's ideas on light quanta. In fact, he later wrote of one such "emission theory" that "prior to the rel. theory it was also mine."[69]

But Einstein returned penitently to Lorentz's theory, now richer by the knowledge that the independence of light's velocity from the state of motion of the light's source must be a crucial part of any future theory, even if it openly contradicted the relativity principle in mechanics. This brought him back to the same contradictory situation as before. Later he would sum up that dilemma as follows:

> Mechanically all inertial systems are equal. According to experience, this equality extends also to optics, or electrodynamics. This equality, however, seemed unattainable in the theory of the latter. At an early stage I gained the conviction that this was due to a profound imperfection of the theoretical system. The wish to discover and to eliminate this created in me a state of

> psychological tension which, after seven years of vain searching, was resolved by the relativization of the concepts of time and dimension.[70]

These "profound imperfections," however, were not to be eliminated by the well-tested methods of "normal science," even at the exalted level of *Annalen*. In this *Nekrolog* Einstein described how intensely he was aware of those difficulties and in which direction he eventually looked for a solution:

> I was more and more in despair about the possibility of discovering the true laws by means of constructive efforts based on known facts. The longer and the more desperately I tried, the more I gained the conviction that only the discovery of a general formal principle could lead us to safe results. I regarded thermodynamics as a model. There, the general principle was stated in the theorem: The laws of nature are of such a character that it is impossible to construct a perpetuum mobile (of the first or second type). But how was such a principle to be discovered?[71]

Not until Einstein had actually discovered the principle did he understand that his earlier efforts to resolve these painful paradoxes had been "doomed to failure so long as the axiom of the absolute character of time and simultaneity were anchored, albeit unrecognized, in the subconscious."[72] What was it about time that Einstein had to raise from the subconscious to the conscious mind before he could "relativize" it?

"What is time?" This question was asked long ago by St. Augustine, who was not the first to find himself perplexed by it: "If no one asks me about it I know it, but if I am to explain it to a questioner I do not know it."[73] Not so the fathers of modern physics. Newton, for one, was less concerned about any internal experience of time than he was impressed by the regularity of the planetary system and the logic of mechanics. He saw an "absolute, true, and mathematical time" which, he said, "of itself, and from its own nature flows equably without regard to any thing external and by another name is called duration."[74] Newton himself must have been aware that this explanation was circular—and that this absolute time, unchanging and independent of the material world, evidently could not be measured or read off anywhere. What was measurable was something different: "relative, apparent, and common time," which according to Newton "is some sensible and external (whether accurate or unequable measure of duration by the means of motion, which is commonly used instead of true time, such as an hour, a day, a month, a year." This distinction between absolute and common time was forced on Newton, not only by fluctuating human perceptions of time and by the inaccuracy of the clocks of his day, but also by small irregularities in that best of all natural timepieces, the rotation of the Earth. Newton even thought it possible that there was no motion by which common time could be measured accurately. But this did not affect the existence or unchangeability of true time; and getting close to true time was a task of science: "The true, or equable, progress of absolute time is liable to no change."

True time, though not a substance like the ether, was regarded as an objective "something," present throughout space, but independent of space and of matter, or of their states of motion. If someone in London defined a moment as "now," then this "now" would also be valid not only for Hamburg or Beijing, but also for the moon or for Sirius. This is in line not only with human perceptions of time but also—and more significantly for physicists—with Newtonian mechanics. Gravity, according to Newton's law of gravity, propagates instantaneously throughout space, so that a stone falling to the ground on Earth must, "at the same moment," cause effects—albeit immeasurably small ones—on the moon. That was why everybody, especially mathematicians and physicists, had been happy for centuries with Newtonian time. Einstein later caricatured this idea as an "eternally uniform tic-tac perceptible only to ghosts, but to them everywhere."[75]

As a student Einstein had already learned from Mach that "absolute time" made no sense. In his *Mechanik*, Mach pointed out that absolute time could not be measured anywhere and that therefore it "was of no practical and also of no scientific value; no one is entitled to say that he knows anything about it, it is a useless 'metaphysical' concept."[76] Mach's strong language was confined to criticism and initially was of no importance to physics. But it must have been to the taste of the rebellious student, who probably remembered that, apart from the definition of time spans with clocks of whatever kind, there was no "time in itself."

From his study of Poincaré, Einstein gathered that time should be regarded merely as a convenient convention. In *La Science et l'hypothèse* Poincaré not only rejected "absolute time" but widened his critique to other intuitive certainties such as simultaneity at different locations: "Not only do we have no direct experience of the equality of two times, but we do not even have one of the simultaneity of two events occurring in different places."[77] For details Poincaré referred to his essay *The Measure of Time*, published in 1898 in a philosophical journal[78] not read by physicists.

Given the eagerness with which, according to Solovine's account, the members of the Akademie Olympia studied Poincaré's book, it would be surprising if they had not also gotten hold of that essay. Poincaré concluded that, before it could be measured, time would first of all have to be defined—not, however, in an arbitrary manner but in line with the simplest possible form of the laws of nature: "Simultaneity of two events or their sequence, and the equality of two spaces of time must be defined in such a way as to ensure the simplest possible formulation of natural laws. In other words, all these rules, all these definitions, are merely the fruits of unconscious optimism."[79]

This unconscious optimism had worked as well and gained such power over human minds because a naive sense of time and Newton's theory of gravitation had converged in it. A good opportunity for elevating it to enlightened optimism was provided by Lorentz's electrodynamics through the introduction of a transformed "local time" $t' = (t - vx/c^2)$ for referential systems moving with a velocity v relative to the ether. Lorentz, however, clung to the idea of "true time" and regarded "local time" as merely a mathematical device.

A physical interpretation of Lorentz's local time was provided by Poincaré in his outline of a method of synchronizing clocks by light signals—the same procedure Einstein would subsequently use as a vital element in constructing his relativity theory. This would suggest that Poincaré came very close to the relativity theory himself, but in fact he never intended any further modifications, believing with Lorentz that clocks thus regulated "show not the true time, but what could be called 'local time.' "[80] He thus remained within the conceptual framework of Lorentz's theory, and relativity theory was discovered by someone else.

Einstein's paper of 1905 reveals nothing of the background to his stroke of genius, and in later years he always referred to it with regrettable brevity. The first such reference was in a major overview article written toward the end of 1907 for the *Jahrbuch der Radioaktivität und Elektronik*[81] (to be referred to later as his "*Jahrbuch* article"). "It turned out, surprisingly, that it was only necessary to define the time concept precisely enough to overcome the . . . difficulty. All it needed was the realization that an auxiliary term introduced by H. A. Lorentz and called by him 'local time' could be defined as 'time' purely and simply."[82]

His later comments were, if anything, even more laconic. In 1920 he wrote: "Only after years of probing did I become aware that the difficulty was due to the arbitrary nature of the basic kinematic concepts."[83] And four years later: "By means of a revision of the concept of simultaneity in a shapable form I arrived at the special relativity theory."[84] Further details are nowhere to be found. Whenever the genesis of the theory came up in conversation, Einstein expressed himself "in a strangely impersonal manner. He called the hour of birth of the relativity theory 'the step.' "[85] Thus we know that the hour of birth was that fine May evening when Einstein discussed his "difficult problem" with Besso, and we also know that Einstein found his solution by "an analysis of the time concept."[86] But just what the two of them discussed, and how the crucial idea emerged, and how Einstein, immediately afterward, "completely solved" his problem—all that remains concealed in the darkness of a night in May.

What probably happened during that conversation with Besso? It is likely that the friends had before them one of Poincaré's papers in which he presented his method for the synchronization of clocks as being equivalent to Lorentz's "local time"—either his 1904 lecture in St. Louis or his contribution to the Lorentz *Festschrift*. The latter was quoted by Einstein a year later, in a different context,[87] so it must have been available to him in Bern. As for Poincaré's lecture *The State and Future of Mathematical Physics*, Einstein could have found that either in a widely read journal[88] or in a copy, hot off the press, of a collection of essays called *The Value of Science*.[89] It also seems likely that in their conversation Einstein and Besso discovered some aspects of Poincaré's synchronization procedure that may have escaped Poincaré himself. How would it be—the two friends, by then skeptical about "true time," might have asked—if the time defined by Poincaré's experiment was not just a mathematical device for Lorentz's "local time" but in fact everything that a physicist could expect of a meaningful concept? Admittedly this would give a different "time" for every inertial system, but the constancy of the velocity of light for any observer would in that case be inherent in Poincaré's definition of simultaneity and would not, as with Lorentz, have to be forcibly brought about by a laborious adjustment to theory.

The fruitfulness of this exceedingly daring idea might have later struck Einstein at home, when he easily succeeded in deriving the "Lorentz-Fitzgerald contraction"—introduced by Lorentz into his theory as an independent hypothesis—from this modified time concept, without any further assumptions, and in thus obtaining a transformation of the local coordinates. As a skilled electrodynamicist he would then have examined the behavior of the Maxwell-Lorentz equations under these transformations. When it emerged in the course of his nocturnal calculations that these equations were invariant and that, moreover, the "Lorentz force," introduced as an independent hypothesis into electron theory, also resulted readily from the transformation behavior, virtually everything was accomplished. Relativity principle and universal constancy of the velocity of light, Maxwellian theory and Lorentz transformations: everything came together in the most wonderful way, and the following morning Einstein jubilantly informed his friend Besso that he had "completely solved" the problem.

This discovery was undoubtedly Einstein's most intense experience. The five weeks he needed to prepare it for publication were a happy time for him. In fact, he was speechless with happiness. To his colleague Sauter he merely said: "My joy is indescribable."[90]

CHAPTER TEN

The Theory of Relativity: "A Modification of the Theory of Space and Time"

IT IS OBVIOUS from the style and structure of Einstein's paper *On the Electrodynamics of Moving Bodies* that its author expected it to live not only in *Annalen der Physik*, but also in the annals of history. It is written with great care, and its axiomatic structure reflects not only its novelty but also its definitive character.

The programmatic introduction is followed by the "Kinematic Part" which in turn is succeeded by the "Electrodynamic Part." Each of these parts is further subdivided into five sections. Einstein had promised his friend Conrad Habicht: "the purely kinematic part is certain . . . to interest you"[1]; and in fact the first two sections of that part, on a mere five printed pages, contain all the essentials of the new concept of time and space. In the next three sections of the first part the consequences affecting the kinematic space-time structure are purely deductively derived. In the second part some central problems of electrodynamics are presented in their new garb, as a practical application of what has just been set out.

The structure of the paper, giving preference to kinematics over dynamics, indicates that it will be not just a new contribution to an ongoing debate, but a revision that should transform the conceptual foundations of physics.

"Kinematics" is the theory of the purely geometrical movements of bodies without any consideration of forces; once forces are included, physicists speak of "dynamics." Einstein's famous contemporaries were then all concentrating on dynamics; it was expected that kinematics would at most undergo certain modifications through the theoretical development of dynamics. The title of Einstein's paper—*On the Electrodynamics of Moving Bodies*—is entirely in line with that tradition. But its content is not: by presenting a new kinematics, Einstein simultaneously obtains a new dynamics, and thus a new physics.

As in the paper on his "heuristic viewpoint" of light quanta, Einstein again begins with a contradiction: "It is well known that Maxwell's electrodynamics—as usually understood at present—when applied to moving bodies, leads to asymmetries that do not seem to attach to the phenomena."[2] In this first sentence Einstein suggests not that there is anything wrong with the theory, but only that it is wrongly interpreted. Nevertheless, this will make all the difference, because in an appropriate interpretation, he implies, the symmetry of the phenomena would also have to be reflected in the theory. This will call for a profound change in fundamental concepts.

Actually, no one else had regarded these asymmetries as a problem worth discussing, so that the opening sentence—beginning "It is well known . . ."—urgently needs an explanation. Einstein provides this by using the example of Faraday's induction experiment, one of the main themes of his years of "pondering." Although the magnitude and direction of the induced current depend solely on the relative movement of conductor and magnet, the theory, "as usually understood at present," makes a strict distinction between the situation of moving magnet and stationary conductor versus stationary magnet and moving conductor. The fact that Einstein found this asymmetry "intolerable"[3] does not of course belong in a scientific publication.

Einstein next establishes a surprising link between this "intolerable" asymmetry and a class of experiments which seem to represent a totally different problem:

> Examples of a similar kind, as well as the unsuccessful attempts to confirm a motion of the Earth relative to the "light medium," lead to the assumption that not only in mechanics, but also in electrodynamics, there are no properties of the phenomena that are in accordance with the concept of absolute rest.

At first glance, there is no connection between the theoretical description of the induction experiment (involving conductor and magnet) and the experiments on ether drift. To have recognized how closely both are connected with the problem of relative movement was a stroke of genius. It brought together the difficulties of interpreting Maxwellian theory and led straight to the relativity principle as a signpost on the road to resolving them.

A system involving absolute rest results in "intolerable" asymmetries and, moreover, is not observable by ether-drift experiments or any other methods. Such a metaphysical monster makes no physical sense and should therefore be banished from physical theory. For that reason, Einstein continues:

> . . . in all coordinate systems in which the mechanical equations are valid, also the same electrodynamic and optical laws are valid, as has already been shown for quantities of the first order.[4]

This proposal is a kind of update, made possible by Poincaré, of Galileo's ship, moving uniformly and rectilinearly, whose occupants are unable to determine whether they are at rest or in motion, no matter what mechanical or electrodynamic-optical experiments they may perform.

Immediately afterward, Einstein compresses the entire range of his thinking—from the mental experiment of his schooldays in Aarau to the complexities of the Maxwell-Lorentz theory—into a single sentence, which already contains a hint of the solution:

> We shall raise this conjecture (whose content will be called "the principle of relativity") to the status of a postulate and shall introduce, in addition, the postulate, only seemingly incompatible with the former one, that in empty space light is always propagated with a definite velocity c which is independent of the state of motion of the emitting body.[5]

The "principle of relativity" implies (though it does not explicitly state) that the velocity of light is constant not only for a single observer—for instance, one sitting at the light source—but for any observer.

These two postulates are mutually incompatible in Newtonian mechanics. Einstein's main task, therefore, will be to prove that this incompatibility is only apparent. At the same time Einstein announces firmly and bluntly that the "luminiferous ether" will "prove superfluous." Thus the concept that governed throughout the nineteenth century will be discarded. Toward the end of his introduction Einstein points out that his theory will base itself on the kinematics of rigid bodies, "since assertions of each and any theory concern the relations between rigid bodies (coordinate systems), clocks, and electromagnetic processes."[6] He thereby indicates that the "new thinking" will represent a reinterpretation of space and time in concretizing the measurement of spatial and time distances—in fact, a theory of measuring as a new prerequisite of all physics.

No one had considered anything so elementary to be necessary or even to make sense. Despite the complexities of Lorentz's theory, the procedures for measuring space and time were so much a matter of course in physicists' prescientific, unconscious "old thinking" that even Poincaré, though he made a profound critique in several directions, did not achieve a decisive revision.

Before turning to Einstein's new ideas of space and time, it may be useful to examine the methodological status of his two presuppositions: the "principle of relativity" and the universal constancy of the velocity of light.

Both principles are, of course, related to experience, but they are not a direct consequence of experience. The "principle of relativity," though inherent in mechanics, is not inherent in electrodynamics based on the ether. Only with major efforts had it been possible to "rescue" it as a generalization up to magnitudes of first order in v/c. For Einstein to elevate it to a general postulate, based at least as much on philosophical reflection as on experience, was therefore a significant step beyond empirical knowledge.

Einstein's procedure, based on postulated principles, emerges even more clearly with his second presupposition. The fact that the velocity of light is independent of the state of motion of its source is purely empirical and might equally well have turned out otherwise. In his *Jahrbuch* article, two and a half years later, Einstein would return to this question: whether or not this presupposition "is in fact really fulfilled in nature is anything but a matter of course, though it is rendered probable—at least for a coordinate system of a definite state of motion—by the confirmations which Lorentz's theory, based upon the assumption of an ether at absolute rest, has received from experiment."[7] Direct experimental evidence of Einstein's postulate was unobtainable at the beginning of the century.[8] Thus his presupposition does not follow inevitably from experience but can be justified only by success. According to Einstein's announcement, such success consists in the fact that "these two postulates suffice for arriving at a simple and consistent electrodynamics of moving bodies on the basis of Maxwell's theory for bodies at rest."[9] In his two principles, therefore, Einstein reached that firm ground which, on the model of thermodynamics, "could lead to reliable results."[10]

Einstein's decisive step was the "relativization" of space and time, or, more accurately, his careful examination of what measurement of time intervals means in physics. His definitions, set down in conjunction with these principles, jointly provide the foundation from which everything comprised by the "special relativity theory" can then be derived.

The formulation in which the theory was born is actually scarcely known today. Subsequent presentations, some by Einstein himself, placed the emphases somewhat differently,[11] and most textbooks, as well as the more demanding popularizations, have followed suit.[12] But the "new thinking" still emerges more clearly than anywhere else in Einstein's paper of 1905, especially in the first few sections of the "kinematic part." A fairly full account of that "prelude" is as tempting as it is indispensable, although the contents of the later sections can be sketched out only in their essentials.

The first section is headed "Definition of Simultaneity." This may sound trivial, as we all believe we know what simultaneity means. But Einstein had to know precisely and to formulate his definition precisely. To that end he proceeds from a "system at rest," the customary three-dimensional Euclidian space with Cartesian coordinates, in which the movement of a body is described by its coordinates as a function of time. This is so conventional that many readers must have asked themselves why it was even mentioned. Einstein, however, continues by stating that "for such a mathematical description to have a physical meaning, we first have to clarify what is to be understood by 'time.' "[13] Here we have a suggestion that this has not been the case in the past, and, evidently using Poincaré's critique of the customary understanding of time, Einstein makes it clear "that all our propositions involving time are always propositions about simultaneous events." To ensure that no one will miss the importance of this assertion, Einstein illustrates it with what is perhaps the most unassuming sentence ever printed in *Annalen*: "If, for example, I say that 'the train arrives here at 7 o'clock,' that means more or less, 'the pointing of the small hand of my clock to 7 and the arrival of the train are simultaneous events.' "[14]

This sufficiently defines a "time," though initially only for the location of the clock. Anyone who has ever reflected on time, whether a physicist or not, will readily extend this plausible understanding of "time" to the whole universe, in the sense that the position of the hands of the Bern station clock shows the "true" time not only for Geneva or Zurich, but also for the moon and for Sirius. This could even be correct if signals propagated instantaneously or, in other words, at infinite velocity.

Einstein, however, is concerned with the electrodynamics of moving bodies, and that is why it makes sense to introduce here the velocity of light—which plays such an overwhelming role in the Maxwell-Lorentz theory. This velocity is enormously high, but it is finite. That is why Einstein insists that his definition of simultaneity is valid only for the location of the clock. "It becomes insufficient as soon as series of events occurring at different locations have to be linked temporally, or—what amounts to the same—events occurring at places remote from the clock have to be evaluated temporally."[15] This would require a definition of the assignment of times to spatially separated locations, to be effected by specific clocks and real physical processes.

Einstein accomplishes this assignment by proposing a synchronization procedure which is exactly the same as Poincaré's. Let two identical clocks at spatially separated locations A and B indicate "A time" and "B time." The time common to A and B is not, for Einstein, something that exists and can somehow be discovered, but something that has first to be appropriately established by the physicist. It

> can now be determined by establishing by definition that the "time" needed for the light to travel from A to B is equal to the "time" it needs to travel from B to A. For, suppose a ray of light leaves A toward B at "A time" t_A, is reflected from B toward A at "B time" t_B, and arrives back at A at "A time" t'_A. The two clocks are synchronous by definition if[16]
>
> $$t_B - t_A = t'_B - t'_A$$

With this definition all clocks in any spatial location B can be set to the "A time" of a clock at location A in accordance with[17]

$$t_B = 1/2\,(t'_A + t_A)$$

For the "system at rest" for which these observations were initially made, it may be stated "in accordance with experience"—i.e., in line with Maxwell-Lorentz theory—that the velocity of light in a vacuum is a universal constant. Einstein thus obtained a time concept which differed fundamentally from that of Newtonian theory, where any "A time," just like any "B time," was valid throughout the universe. This new definition of time would have been useful even for Lorentz's theory, with its ether at rest, as ultimately equivalent to Lorentz's local time. Poincaré had noticed that but had failed to draw the implied conclusions.

Einstein, on the other hand, would draw important conclusions from these seemingly pedantic reflections—in the next section, where he considers rigid bodies, clocks, and observer in motion relative to each other.

At the beginning of the second section, "On the Relativity of Dimensions and Times," Einstein once more formulates his two principles as precisely as possible. Then, abruptly and without any lead-in, he says: "Suppose a rod at rest; measured with a measuring rod likewise at rest; let it have a length *l*." Now suppose the rod is put into a state of motion with a constant velocity *v*. Einstein, asking about the length of the moving rod, describes how this can be determined by two fundamentally different operations.

To be sure, Einstein is using almost "prerelativist" terminology by referring, throughout this section, to a system "at rest" in which the rod, either at rest or in motion, is observed. While this formulation lets the background of Lorentzian theory—a motionless ether—shine through, it also leads to complications in which even an attentive reader can lose the thread. For that reason I shall use two referential systems: this will deviate from Einstein's text but will not change his argument. In fact, in his next section Einstein himself goes over to this clearer presentation.

Consider, then, two referential systems *k* and *K*, both furnished with measuring rods and synchronized clocks, initially at relative rest and congruent. The length of the rod arranged parallel to the *x*-axis is, of course, the same in both systems: *l*. Let the rod and system *k* be accelerated until they move along the *x*-axis at a constant velocity *v* relative to *K*, with the rod at rest in *k* but moving in *K* at velocity *v*. The length of the rod in *k* can now be "thought to be obtained by two operations."[18]

First let an observer determine the length of the rod at rest in *k*. "According to the relativity principle" this length must be the same as the length *l* of the rod at rest in *K*. Any deviation would annul the equivalence of the two systems and thereby violate the relativity principle. While no one at this point—even without the relativity principle—would expect anything different, the second operation immediately brings a surprise.

An observer in system *K*, relative to which the rod is moving at velocity *v*, now determines by means of the synchronized clocks where ends A and B of the rod are at a definite moment. If the distance between these two points is measured with the measuring rods from *K*, one obtains "also a length which one might call 'the length of the rod.' "[19] This is the length r_{AB} of the moving rod in system *K*. Einstein announces that he will now determine this length "on the basis of our two principles, and will find it to be different from *l*."[20]

This is in striking contrast to the commonsense idea—which had been the well-tested basis of physics for several centuries—that the length of the rod should be entirely independent of whether it is at rest or in motion relative to the observer. In his *Jahrbuch* article Einstein will use the term "geometric shape" for the length which an observer at rest next to the rod determines by the application of measuring rods; the length of the moving rod is its "kinematic shape." "It is obvious that an observer at rest relative to an inertial system *K* can determine only the kinematic shape, related to *K*, of a body moving relative to *K*, but not its geometrical shape."[21]

To determine the kinematic properties of the rod at rest in *k* but moving in *K*, imagine clocks fitted to ends A and B of the rod, these clocks being synchronized with those in system *K* and hence showing the same time as the clocks in *K*. "Suppose also," Einstein continues his thought experiment, "that each clock has an observer co-moving with it, and that these observers apply to the two clocks the criterion for synchronization formulated in §1."[22] At time t_A a ray of light is emitted from A; this is reflected at B at time t_B and returns to A at time t'_A. All times, Einstein points out in a footnote, are those of system *K*. Allowing for the constant velocity of light in *K*, the time differences on the forward and return travel are:

$$c\,(t_B - t_A) = r_{AB} + v\,(t_B - t_A)$$
$$c\,(t'_A - t_B) = r_{AB} - v\,(t'_A - t_B) \text{ and hence}$$
$$t'_A - t_A = r_{AB} \,/\, (c + v) + r_{AB} \,/\, (c - v)$$

If the clocks were synchronous in *k*—that is, for observers moving along with the rod—the equation $t'_A - t_A = 2\,r_{AB}/c$ would apply. The clocks at A and B are therefore synchronous in *K* but not for observers in *k*. Einstein thereby demonstrates "that we must not ascribe absolute meaning to the concept of simultaneity, instead two events that are simultaneous when observed from some particular coordinate system can no longer be considered simultaneous when observed from a system that is moving relative to that system." There are as many "times" as there are inertial systems: i.e., an infinite number. This is the burden of Einstein's critical reflections on time and also the proof of the relative nature of simultaneity.

In the heading of his second section Einstein also mentions the relativity of dimensions, and in the text he actually proposes to determine the length r_{AB} "and find that it is different from l." But for this discovery the reader waits in vain. The text strongly suggests that a paragraph or two may have been lost at the proof stage. But if Einstein intended to let the reader perform that exercise—which would not be unthinkable—he certainly failed to make that clear. The gap remains unexplained, even though after what has been said in the demonstration of the relativity of time, the relativity of dimensions is entirely plausible.

In conclusion, it can be stated that in his first two sections Einstein used no more "mathematics" than I have used in this account.

The recognition that every inertial system has its own time and that kinematic dimensions in different inertial systems differ from each other cannot, of course, benefit physics until the relationship between those times and dimensions has been discovered. This is the task Einstein addresses in his third section, "Theory of Coordinate and Time Transformations." He presents a purely kinematic deduction, not drawing on any other physical assumptions or theories, but based solely on the two principles and on the definition of time. This reduction of assumptions underlines Einstein's endeavor to establish his transformations as the structure of space and time—and hence as affecting physics as a whole, not merely as a feature of a special theory such as Lorentz's electrodynamics.

Consider, therefore, two coordinate systems k and K with a common x-axis. Let k move relative to K at velocity v; thus K moves relative to k at velocity $-v$. In "traditional kinematics," in line with Galileo's and Newton's ideas of space and time, the Galileo transformations would be the simple relations $x' = x - vt$ and $t' = t$, coordinates with a prime sign ($'$) representing k and coordinates without it representing K. If in K the velocity of light has the value c, then, according to the Galileo transformations, in k it has the value $c - v$. This would be in conflict with Einstein's second principle of the universally constant velocity of light—therefore, other transformations must be the correct ones.

Einstein thus proceeds from the postulate that clocks synchronized by light signals and at rest in system K record time t, whereas identically synchronized clocks at rest in system k record time t'. A relation has therefore to be found between t' and t and between x' and x. After a rather lengthy derivation of the transformations (this takes up four pages and incidentally, like the whole of the "Kinematic Part," manages with quite elementary mathematics), Einstein obtains

$$x' = \frac{x - vt}{\sqrt{1 - v^2/c^2}}$$

and

$$t' = \frac{t - v/c^2 x}{\sqrt{1 - v^2/c^2}}$$

These are the relations which Lorentz had presented a year earlier and which Poincaré had meanwhile named "Lorentz transformations."[23] However, Einstein was acquainted neither with Lorentz's paper of 1904 nor with Poincaré's of June 1905. "In that sense," he was therefore able to claim, "my 1905 paper was independent."[24] Above all else, it was independent in its fundamentally different justification and interpretation of these transformation equations.

Einstein must have begun with some idea of what he wanted to deduce. At an opaque point in his deduction he introduces, without any warning or explanation, a slight mathematical operation whose purpose becomes obvious only if the desired result is already known.[25] This underhand device, by means of which he rather forcibly "computes his way" to the Lorentz transformations, deprives the deduction of some of its elegance and stringency. Thus it is scarcely credible that Einstein actually arrived at his formulas in the way he presents them in his paper. In fact, in later years he never again used this rather awkward method.

Whether or not this derivation was conclusive, Einstein was able, without any difficulties or tricks, to demonstrate that his equations were the correct ones and, particularly, that they were compatible with a constant velocity of light. When the Lorentz transformations are applied for time and space coordinates, an electromagnetic spherical wave propagating from the origin of the coordinate system in K is again a spherical wave in system *k*—which is in motion relative to *K* at a velocity *v*—propagating in *k* at the velocity of light.[26] He has thereby resolved the seeming contradiction and demonstrated "that our two fundamental principles are compatible."

In his *Jahrbuch* article Einstein would proceed in the opposite direction and derive the Lorentz transformations from the postulate that a spherical wave in one inertial system invariably results in a spherical wave in transition to any other inertial system. This is the procedure that would prevail and, within a few years, would enter the textbooks as a "classical" method.

The transformation equation for time

$$t' = \frac{t - vx/c^2}{\sqrt{1 - v^2/c^2}}$$

highlights Einstein's departure from Newtonian kinematics more clearly than the equation for the spatial coordinates. Because it links time with the spatial coordinates, it is the physical basis of the four-dimensional presentation of relativity theory, with time as the fourth "dimension." Three years later, Einstein's former mathematics professor Hermann Minkowski was to introduce it with rather more bombast: "Henceforward space on its own and time on its own will decline into mere shadows, and only a kind of union between the two will preserve its independence."[27] Minkowski's presentation not only is very elegant but would soon also prove very useful—with the result that his formulation has long been common in textbooks. On the other hand, his ebullient rhetoric and especially his reference to time as the "fourth dimension" have encouraged the erroneous popular belief that relativity theory is exceedingly complicated in its mathematics, and that as an intellectual construct it is so abstruse that some special, higher enlightenment is required to understand it. Needless to say, this is not so, and we may take solace from the fact that Einstein himself formulated his ideas much more simply.

With the Lorentz transformations as equipment, it is now easy for Einstein, in the fourth section of his paper, to deduce a few consequences "concerning moving rigid bodies and moving clocks." For an observer "at rest," a measuring rod moving at velocity v appears shortened by the factor $\sqrt{1-v^2/c^2}$ in the direction of the movement. This already applied in Lorentzian theory, except that there it was a dynamic effect produced by interaction with the ether, so that the "Lorentz contraction" was an asymmetrical phenomenon, valid only for a measuring rod moving relative to the ether. In Einstein's theory, however, it is clear "that the same results apply for bodies at rest in a system 'at rest,' " when viewed from a uniformly moving system.[28] This is enough for Einstein to highlight the difference between his theory and Lorentz's—a difference that could hardly be more fundamental.

To Einstein this contraction has nothing to do with any forces; it is a purely kinematic effect, a consequence of the finite velocity of light. Moreover, the contraction is symmetrical: if two observers A and B move relative to each other, then the measuring rods at rest for B seem shortened to observer A, and the measuring rods at rest for A seem shortened to observer B. To that extent the question, often asked uncomprehendingly, whether the contraction is "real" or "apparent" misses the point: the only thing that can be measured is the kinematic shape, and that is shortened for any measuring rod in motion relative to an observer.

Up-to-date physicists, already familiar with the Lorentz contraction, "merely" had to get used to Einstein's purely kinematic symmetrical interpretation; but even Einstein himself described the results he obtained for moving clocks as "peculiar." To begin with, he found, by simple reflection and even simpler calculation on the basis of the transformation formula for time, that a moving clock is slow compared with time measured in a system assumed as being "at rest." The amount of that time lag is $(1-\sqrt{1-v^2/c^2})$.

Such a "time dilatation" was totally alien to Lorentzian theory: there, only one time existed—"true" time—and transformation of time was understood purely as a mathematical device. That was why Einstein's result, along with his proof of the relativity of simultaneity, was bound to shock, or at least surprise, his contemporaries. To make that surprise complete, Einstein escalates the consequences.

Imagine a system with time defined by synchronized clocks at rest. If a clock at a point A is moved at a velocity v relative to some point B, with this movement taking time t, then this clock after its arrival at B will no longer be synchronous with the clock at B, but will be slow by the amount $(1-\sqrt{1-v^2/c^2})t$. As any number of polygonal lines can be produced by joining straight lines, all the way to a closed figure in which points A and B would therefore coincide, the same considerations that apply to the transportation of a clock along a single straight line also apply to transporting it along such a polygon. By approximating polygons to evenly curved lines, Einstein arrived at the remarkable statement: "If there are two synchronous clocks in A, and one of them is moved along a closed curve with a constant velocity until it has returned to A, which takes, say, t sec, then this clock will lag on its arrival at A by $^1/_2\ tv^2c^2$ sec."[29] As a good physicist, he adds an experimental point: "A balance-wheel clock that is located at the Earth's equator must be very slightly slower than an absolutely identical clock . . . at one of the Earth's poles."[30]

Simple and convincing as time dilatation was for rectilinear uniform motion, once Einstein's definition of time and his view of the Lorentz transformations were accepted, this was not necessarily the case for closed paths. With a closed path, the two clocks were evidently not equal, because acceleration along a closed path violates the symmetry otherwise present in relativity theory. This difficulty, as well as Einstein's paradoxical result, led to widespread controversy, not only among people who refused to accept Einstein's theory but also among the "relativists" themselves.

In this regard, the "twin paradox" became very popular. The twin paradox is based on the fact that biological processes, despite their insufficient "regularity," can be accepted as clocks in the physical sense. This idea was first mooted by Einstein in a lecture to the Naturforschende Gesellschaft in Zurich on January 16, 1911,[31] though it attracted only slight attention. Much more sensational was a striking elaboration of Einstein's deduction by the French physicist Paul Langevin at a philosophical congress in Bologna a few months later, in April 1911.[32] In the audience was the famous Henri Bergson, who was as impressed by Einstein's ideas as he was challenged to oppose them.

Langevin discussed a thought experiment: a pair of twins, one of whom remains on Earth while the other, in a Jules Verne cannonball, goes off on a journey through the universe. At a velocity sufficiently approaching that of light, the traveling twin might return to Earth aged by a mere two years, while his brother would have long since died—two hundred years having elapsed on Earth.

The twin paradox, running counter to our everyday concept of time, was as difficult to grasp then as it is now. Small wonder that opposition to relativity theory focused primarily on the relativization of time and on its counterintuitive consequences—the more so because no direct experimental evidence of time dilatation was then attainable. Half a century would pass before time dilatation was observed as an isolated effect, first in certain elementary particles of cosmic radiation. Not until 1971 was the effect convincingly demonstrated by accurate "atomic clocks" carried by passenger aircraft around the world.[33] But by then relativity theory had long been a pillar of physics, so that no one expected anything but a confirmation that Einstein had been right in 1905.

In the fifth and final section of the "Kinematic Part" Einstein derives the "theorem of Addition of Velocities." In Galilean-Newtonian mechanics the combination of two equidirectional velocities v and w would amount to a simple arithmetical addition $u = v + w$. But in relativity theory this operation must have a different form, if only because of the special role played by the velocity of light. After some simple and transparent arguments Einstein arrives at the somewhat more complicated expression

$$u = \frac{v+w}{1+vw/c^2}$$

In a kind of consistency test, this formula now confirms one of the starting concepts of the entire theory: the overwhelming importance of the velocity of light, which had been used to synchronize the clocks. While still working on his physical interpretation of the Lorentz transformations, Einstein had referred to the velocity of light as a limit that could not be exceeded.[34] Now he demonstrates explicitly that for two velocities v and w, which are smaller than c, the relativistic "sum" u can never be greater than c. Moreover, the velocity of light is not changed by combination with another velocity. In that case, invariably,

$$u = \frac{c+w}{1+cw/c^2} = c$$

Finally, Einstein expands the addition procedure for three velocities, and thereby for any number of velocities. He concludes the section with an outline of a proof that, thanks to his theorem of addition, the transformations for spatial and time coordinates exhibit the mathematical structure of a group—"as they indeed must."[35]

Thus, on the first sixteen pages of his treatise, Einstein derives the essential elements of "a kinematics that corresponds to our two principles,"[36] and with them the complete foundation of the theory of relativity. The rest of the paper consists of applications showing the efficacy, elegance, and profundity of the new relativistic viewpoint.

In the second part of his paper, the "Electrodynamic Part," Einstein first examines the transformation behavior of Maxwell's equations for a vacuum. To do so he uses a frightful, but then still customary, way of writing the formulas;[37] behind the awkward notation, however, lie clear thinking and simple calculations. The essence of this exercise is not only the Lorentz invariance of the equations for the electromagnetic field, but also a demonstration that a force acting on a charge is arrived at effortlessly if the field is transformed into a system at rest relative to the charge. From this follows a ready explanation of the "Lorentz force," postulated by Lorentz as an independent axiom added to the Maxwellian equations.

Reducing the number of axioms of a theory has always been considered an intellectual triumph, and Einstein savored his triumph, though in the restrained language of a physics paper. In the "theory developed,"[38] as he now calls his achievement, this force merely plays "the role of an auxiliary concept whose introduction is due to the circumstance that the electric and magnetic forces do not have an existence independent of the state of motion of the coordinate system." In the same breath, it follows "that, when considering the currents produced by the relative motion of a magnet and a conductor, the asymmetry, mentioned in the Introduction, disappears." For the same reason the fiercely debated questions about the "location" of the electromotive force in unipolar machines are declared to be "irrelevant."

In the next section Einstein applies the same methods to two problems from optics: the Doppler effect and the aberration of starlight. He devotes a whole section to the "Transformation of the Energy of Light Rays"; in it he simultaneously develops a theory of "light pressure exerted on the reflecting surface." Here the treatment of electromagnetic radiation is entirely "classical," just as if his paper on light quanta, completed three months earlier, had never been written. Probably he did not wish to encumber the first appearance of his relativity theory with the "very revolutionary" light quanta. Even so, it is a triumph of classical physics in the new garb of relativity theory, and a brilliant example of the strength of relativity as a computational tool.

For light pressure acting on a mirror, Einstein obtains an expression "in agreement with experience and with other theories."[39] The other theories, which he neither identifies nor quotes, must have been those of Lorentz and, more especially, the mathematical virtuoso Max Abraham. A year previously, Abraham had unfolded the problem in *Annalen*,[40] using established methods and requiring more than forty printed pages. To arrive at the same result, Einstein needed not quite three pages.

As with most of his results on electrodynamics, Einstein did not claim to have discovered anything new; what was new was his relativist method, whose efficiency he emphasized: "This reduces every problem in the optics of moving bodies to a series of problems in the optics of bodies at rest."[41] And with this method, the problems could be treated far more comprehensibly and elegantly than had been possible so far.

In the ninth and penultimate section Einstein sketches out the Lorentz invariance of the Maxwellian equations with regard to moving charges, i.e., the foundation of the so-called electronic theory. This seemed appropriate not only for completeness but also because the "important proposition can be deduced" that electric charges do not change under Lorentz transformations: their amount and sign have the same value in any coordinate system.

This is an important result not only in its own right, but also as a preliminary for the final section, which deals with the dynamics of the electron[42] in the electromagnetic field, thus extending relativity to mechanics. Here Einstein makes a clumsy slip, which was soon to be corrected by Max Planck,[43] but it does not affect the arguments which follow. First of all, mass and kinetic energy are found to depend on velocity in such a way that the speed of light once again emerges as a limit: "As in our previous results, superluminal velocities have no possibility of existence."[44]

In conclusion Einstein lists three "properties of the motion of the electron . . . that are accessible to experiment": a relationship between velocity and the ratio of electrical to magnetic deflection; a relationship between velocity and voltage traversed in an electrostatic field; and the radius of curvature in a magnetic field. The formulas he arrived at are declared to be "a complete expression of the laws by which the electron must move according to the theory presented here."[45] Experiments with fast electrons had been conducted for a number of years, and Einstein must have known at least some of the results. In his paper, however, he ignores them. The reason was simple: they contradicted the "theory presented here."

At the very end, there is a novelty: an acknowledgment. Einstein concluded his treatise—a paper without bibliographical references, citing no names other than those of Maxwell, Hertz, and Lorentz, and these only as labels for theories and formulations—with the remark that "my friend and colleague M. Besso steadfastly stood by me in my work on the problem here discussed and . . . I am indebted to him for many a valuable suggestion."

By the time the foundation of relativity theory was presented to the world of physics, in the September 28, 1905, issue of *Annalen*, the editors had already received a supplement from Einstein. Under the rather unusual interrogative title *Is the Inertia of a Body Dependent on Its Energy Content?* Einstein presented what was subsequently considered his most famous and most spectacular conclusion—the equivalence of mass and energy. He answered the question posed in the title, which most of his colleagues would have regarded as rather abstruse, with an enthusiastic yes.

Einstein's initial ideas about this could have arisen from the velocity-dependence of mass in the final section of his great treatise. Perhaps they did arise, but he preferred to keep silent just then. A few weeks later, perhaps during his reading of the proofs, the idea was fully developed. "I have thought of yet another consequence of the electrodynamic paper," he reported to his friend Conrad Habicht:

> The relativity principle in connection with the basic Maxwellian equations demands that the mass should be a direct measure of the energy contained in a body; light transfers mass. With radium there should be a noticeable diminution of mass. The idea is amusing and enticing; but whether the Almighty is laughing at it and is leading me up the garden path—that I cannot know.[46]

This was the first but not the last time that Einstein brought in the Almighty. The reference to God, meant less piously than metaphorically, as the creator of the world *more geometrico*, whose construction plans needed to be discovered, pleased Einstein and would later appear in many variations.

To return to the relation between mass and energy, this was then a subject of research in the Bern Patent Office and elsewhere. The velocity-dependence of the electron's mass, experimentally confirmed at least qualitatively and integrated into the prevailing theories, seemed to support the idea. Moreover, the champions of an "electromagnetic world picture" tried to interpret mass as electromagnetic "innate" energy of the electron and hence as an effect of the electromagnetic field. In 1904, the Vienna Academy of Sciences had awarded a prize to Friedrich Hasenöhrl for research in that area. Hasenöhrl had shown that radiation enclosed in a vacuum has to be credited with an apparent mass, proportional to the energy of the enclosed radiation. It is scarcely credible that Einstein would have missed Hasenörhl's prize-winning paper, which was published in *Annalen*,[47] but he certainly did not refer to it.

Einstein addressed the problem quite differently, and his result is much more elegant and of incomparably greater universal validity. In his supplement he considers any body emitting radiation in two opposite conditions—first in a system in which the body is at rest, and second in a system in which the body is moving at constant velocity. Referring to the formula he had derived in his treatise on the transformational characteristics of radiation energy, Einstein succeeds, in less than two pages, in deducing the following theorem: "If a body releases the energy E in the form of radiation, its mass decreases by E/c^2."[48] Einstein follows this with the rather cryptic remark that it "obviously is here inessential that the energy withdrawn from the body happens to turn into radiation" in order to proceed to "the more general conclusion": "The mass of a body is a measure of its energy content; if the energy changes by E, then the mass changes in the same sense by E/c^2."[49]

He also has an idea about the experimental implications of the formula $E = mc^2$: "Perhaps it will prove possible to test this theory using bodies whose energy content is variable to a high degree (e.g., salts of radium)." Einstein, however, did not concern himself with the practicality of such a test for the decay of radium. This was taken up by a colleague the following year, with the result that the effect "for the time being probably lies beyond the realm of possible experience."[50]

To Einstein the relation between inertial mass and energy was clearly of great importance. His first follow-up paper on relativity theory, completed in May 1906, was devoted to the theoretical aspects of this subject. In it the "theorem of the constancy of the mass" was understood "as a special case of the energy principle."[51] After another year, in May 1907, he endeavored to test "the necessity and justification of these assumptions [made in 1905] in a more general way."[52]

Eventually, in the fall of 1907, in his extensive *Jahrbuch* article, Einstein returned in detail to the "dependence of mass on energy,"[53] calling his formula a "result of exceptional theoretical importance" and also thoroughly discussing its experimental side. But whichever way he approached it, the requisite accuracy of measurements was "of course impossible."

It was understandable that, given the state of knowledge at the time, Einstein had focused his attention on radioactive disintegration. The atomic nucleus made its first appearance in physics in 1911, and the binding energy of nuclei was soon interpreted as a "mass defect." But it took first the invention and subsequent improvement of the mass spectrometer, and next—in 1932—the discovery of the neutron as the second building block of the atomic nucleus, before a reliable test of Einstein's formula through nuclear binding energy became possible. Within a few years, it was confirmed with fantastic accuracy by a multitude of nuclear reactions, and by 1937 it was regarded as an empirically confirmed "fundamental law of physics."[54]

When on August 6, 1945, Einstein learned of the destruction of the Japanese city of Hiroshima by an atom bomb, he may well have been reminded of what, nearly four decades earlier, he had written in his *Jahrbuch* article:

> However, it is possible that radioactive processes will be detected in which a significantly higher percentage of the mass of the original atom will be converted into the energy of a variety of radiations than in the case of radium.[55]

CHAPTER ELEVEN

Acceptance, Opposition, Tributes

IT IS WIDELY BELIEVED that remaining unrecognized by one's contemporaries is a hallmark of genius, not only in the arts but also in science. As Max Planck once observed: "A new scientific truth does not as a rule prevail because its opponents declare themselves persuaded or convinced, but because the opponents gradually die out and the younger generation is made familiar with the truth from the start."[1] If this was the experience of Max Planck, a scientist with a straightforward career within the academic establishment, how much more must it be true of a youthful outsider at the Patent Office in Bern?

However, Einstein had no reason to complain—and never did complain—of any lack of response to or acknowledgment of the papers of his *annus mirabilis.* Of course, his publications did not exactly have the effect of bombshells: physicists are too reserved, conservative, and skeptical for that. And of course there was opposition, some of it blinkered. But the scientific community on the whole accepted the ideas of this genius who had suddenly emerged, some physicists with enthusiasm and others more hesitantly. Still others resolutely rejected them—but no one ignored them.

Einstein's "very revolutionary" light quanta had already attracted much notice and discussion. Thus the winner of the 1905 Nobel Prize, the experimental physicist Philipp Lenard, whose measurements of the photoelectric effect had been a basis of Einstein's "heuristic viewpoint" paper, honored the unknown man at the Patent Office with an offprint in the autumn of 1905. Einstein thanked Lenard, assuring him (as we saw earlier, in Chapter 8) that he had studied the article "with the same sense of admiration as your previous work."[2]

A letter to Einstein on the quantum problem from Planck's assistant Max von Laue dated June 2, 1906, suggests that this was not the first letter; and it seems probable that Planck had also written to Einstein, with the result that soon after the publication of his paper Einstein was in correspondence about light quanta with the most famous experimenter and the most respected theoretician in the German-speaking world. As we have seen, this earned him sympathetic attention but by no means acceptance. With regard to light quanta, Einstein actually stood alone for nearly two decades. His statistical interpretation of the Brownian movement, on the other hand, soon gained him the recognition of his colleagues. What, then, was the response of leading physicists to the theory of relativity?

The cliché of the misunderstood innovator was perpetuated by Maja Einstein in her sketch, written two decades after the event, of her brother's state of mind following the publication of his paper on relativity: "The young scientist had believed that his publication in the respected and widely-read journal would be noticed immediately. . . . But he was bitterly disappointed. Icy silence followed the publication. The next issues of the journal did not mention his paper in a single word."[3] But in fact, the situation was quite different.

Certainly Einstein was a temperamental, impatient young man. But as a regular contributor to *Annalen*, he would have realized that, given the usual lapse of two months from submission to publication, any comment on his paper, published on September 28, 1905, could hardly appear before Christmas, even if some colleagues did find something to say within four weeks. Actually, only two months later, toward the end of November, Walter Kaufmann, in an account of his experiments with electron beams,[4] first mentioned Einstein's treatise—though not in *Annalen* but in the *Sitzungsberichte* (*Proceedings*) of the Prussian Academy of Sciences in Berlin, a publication that Einstein might easily have missed. We may also assume that Einstein was not immediately informed that several leading physicists were concerning themselves intensively with his concept immediately after its publication.

In any case, by the beginning of May 1906 Einstein was able to report, not without some complacency: "My papers are meeting with much acknowledgement and are giving rise to further investigations. Prof. Planck (Berlin) wrote me about it recently."[5] That was a mere seven months after publication. Seven years later, in a tribute to Planck, Einstein recorded his gratitude: "It is largely due to the determined and cordial manner in which he supported this theory that it attracted notice so quickly among my colleagues in the field."[6]

We do not know when Planck first wrote to Einstein, or what he wrote. Einstein evidently believed that he had better things to do than keep letters,[7] even if they were from the top men in the field. And most of Einstein's letters to Planck perished along with Planck's house under a hail of bombs in World War II. Nevertheless, there is no doubt that Planck's role as an advocate of relativity theory was of decisive importance.

As the coeditor of *Annalen* responsible for theory, Planck held a key position in the information network of physics. He discharged his tasks with a very open mind. Only rarely were submitted papers rejected; with "established" authors—among whom, after five articles and his refereeing for *Beiblätter* (the supplement to *Annalen*), Einstein could by then count himself—this happened only in exceptional cases of patent nonsense.[8] Thus Planck had unhesitatingly accepted Einstein's "heuristic" paper on light quanta, even though it ran counter to his own concepts. The treatise on relativity received toward the end of June, however, was highly unusual, even in appearance: many pages of prose, more suitable to a philosophical journal, including some sentences of almost offensive triviality alongside ideas of staggering audacity; and finally an elegant, albeit rather opaque, exposition of electrodynamic problems without any new results beyond Lorentzian theory. No one could have blamed the editor if he had had second thoughts.

Max Planck's greatness as a physicist, however, is reflected also in the fact that this paper did not in the least strain his tolerance, but instead "immediately aroused my lively attention."[9] When Max von Laue came to Berlin as Planck's assistant in the fall of 1905, the first lecture he heard was one by Planck on Einstein's newly published paper *On the Electrodynamics of Moving Bodies.*[10] Much the same happened in Würzburg: immediately after the publication of Einstein's paper, Professor Wilhelm Wien entered the room of the students working on their doctoral theses one morning and instructed Jakob Johann Laub—of whom we shall hear more later—to give a colloquium on it. "There was a lively discussion," Laub recalls, "from which it was clear that it was not too easy to get inside the new concepts of time and space."[11]

Despite these difficulties, interest in the relativity theory spread, not exactly like wildfire, but steadily and persistently. The great Röntgen in September 1906 honored Einstein with a request for an offprint, presumably because he was preparing a lecture on the movement equations of the electron. Paul Drude, the editor of *Annalen*, mentioned Einstein's paper in 1906, both in a new edition of his authoritative book on optics and in an article on the same subject in *Handbuch der Physik*. In 1907 Hermann Minkowski, Einstein's former mathematics professor at the Polytechnic but since 1902 a professor at Göttingen, requested an offprint because he and David Hilbert were planning a seminar on the electrodynamics of moving bodies. This, as will be seen, was to have consequences for the mathematical formulation of the theory.

Einstein's ideas also met with lively interest among the younger generation of physicists. When Fritz Reiche, having obtained his doctorate under Planck in Berlin, brought news of the new theory to Breslau, Max Born, Rudolf Ladenburg, and Stanislas Loria there formed a circle of enthusiastic young relativists. When Ladenburg had asked for an offprint, Einstein was "highly delighted over your interest in that paper" and immediately sent off three copies, "one for yourself and two for the other two gentlemen."[12]

In Munich, Arnold Sommerfeld, one of the few full professors of theoretical physics, had likewise done his homework: "I have now studied Einstein, who impresses me greatly," he wrote to Wien toward the end of 1906. A year later his reflections had given rise to some curious conclusions, which he unashamedly communicated to Lorentz:

> But now we are all longing for you to comment on that whole complex of Einstein's treatises. Works of genius though they are, this unconstruable and unvisualizable dogmatism seems to me to contain something almost unhealthy. An Englishman would scarcely have produced this theory; perhaps it reflects, similarly as with Cohn, the abstract-conceptual character of the Semite. I hope you will succeed in imbuing this inspired conceptual skeleton with real physical life.[13]

No answer from Lorentz is known, but Sommerfeld was much too good a physicist to indulge in this attack of "sound common sense" for long. In fact he seems to have written to Einstein directly a few days later and sent him some offprints, naturally without the anti-Semitic remarks, as in January 1908 the two were keeping up a correspondence which in cordiality left nothing to be desired.

Perhaps this incident had some repercussions many years later, when Einstein wrote about Sommerfeld "that this person, for God knows what subconscious reason, did not ring entirely true to me."[14] Had Lorentz shown him Sommerfeld's letter or told him about it?

The most important figure in establishing relativity theory after 1905 was Max Planck. He not only commended it to the attention of his assistant Max von Laue and his predoctoral student Kurd von Mosengeil, but was the first to publish a paper linking up with Einstein's concept and developing it further.

In this paper Planck proved that the "principle of least action," a foundation of physics, remains correct in Einstein's concept; this ensured the connection of the relativity theory with advanced formulations of theoretical physics.[15] With this proof Planck made a substantial contribution to the shaping of the theory, and his personal engagement greatly enhanced its respectability.

Respectability was something Einstein's theory needed, for two reasons. First, it did not initially differ from Lorentzian theory in its electrodynamic consequences, so that several years of explanatory work were needed before even well-disposed physicists could grasp the central difference between the two views. Second, Einstein's conclusions concerning the movement of an electron ran counter to experimental findings at the time.

Although Einstein, like probably every physicist, subscribed to the principle that experience is the supreme judge of the usefulness of a theory, he had some reservations. With regard to the relationship between experiment and theory, he remarked in his *Nekrolog* that "Obvious as this postulate may at first appear, its application is very subtle."[16] He may have been reminded of the conflict that confronted him in the development of relativity theory. In the final section of his treatise he had firmly declared that three formulas he had derived represented "a complete expression of the laws by which the electron must move according to the theory presented here."[17] That was a bold statement, because when he wrote it, electrons were apparently moving according to entirely different laws.

Experiments on the behavior of fast electrons in electric and magnetic fields had been performed since 1897 by Walter Kaufmann, first as an assistant in Berlin, later as a privatdozent in Göttingen, and finally as a professor in Bonn. In 1902 he demonstrated that the mass of an electron increases with its velocity. This was a striking refutation of a sacred Newtonian principle, conservation of mass; and for many theoreticians it was a cogent argument for an "electromagnetic world picture"—the more so as Kaufmann's measurements were entirely in line with the world picture of his Göttingen colleague Max Abraham, who interpreted mass as a consequence of inherent electromagnetic energy.

In 1904 Lorentz published his magnificent completion of his electronic theory, which in terms of physics differed from Abraham's. And before the end of the year a third theory was put forward by Alfred Heinrich Bucherer, representing a kind of mediation between Lorentz's and Abraham's concepts. As all three authors had obtained different laws for the movement of electrons, Kaufmann had the fascinating task of deciding between the rival theories on the strength of his own experiments. While he was still engaged in delicate measurements at the very limits of what was observable, Einstein put forward his theory of relativity with formulas for the tracks of electrons, which were identical with those in Lorentz's theory.

In November 1905 Kaufmann published his tensely awaited results in a preliminary report in the *Sitzungsberichte* of the Prussian Academy (this is the paper in which Einstein's relativity theory is first mentioned) and followed this up in January 1906 with a full report in *Annalen*.[18] However Kaufmann interpreted his results, they fit Abraham's theory best, Bucherer's a little less well, and Lorentz's and Einstein's theories worst. Kaufmann concluded that his measurements were "not compatible with the fundamental assumptions of Lorentz and Einstein."[19] He concluded that the endeavors to base all physics on the relativity principle had failed, and he called for further experiments to prove the existence of an ether absolutely at rest. If experiment were truly the supreme judge, as empiricist philosophers of science like to portray it, both Lorentz's and Einstein's theories would have met a sudden end—at least for the time being. And in fact, when Lorentz learned about Kaufmann's experiments, he found himself compelled to give up his theory: "I am at my wits' end,"[20] he wrote to Poincaré, crushed.

In September 1906 the problem was on the agenda of the annual general meeting of the German Society of Scientists and Physicians in Stuttgart. What mattered was not just the correct formulas for the movement of electrons, but a decision between "world pictures": the "electromagnetic" picture on the one hand, and on the other a picture based on the relativity principle. Max Planck, cautious as ever, in his lecture analyzed Kaufmann's data; he did not find their interpretation wrong, but neither was it beyond all doubt, and he suggested that "in the theoretical interpretation of the magnitudes measured there is still some substantial gap that will first have to be filled before the results . . . can be used for a definitive decision."[21] In the discussion, Kaufmann still insisted that "unless there is a fundamental error in the observations, the Lorentzian theory is liquidated"[22]—and with it, needless to say, Einstein's. Planck once more advised waiting and continuing research "until the experiments eventually supply the decision." In view of Kaufmann's experiments this was all he could do: the fact that the relativity principle seemed to him "really more attractive"[23] was not a sufficient argument.

How did Einstein react to this controversy? The Expert III Class was presumably never asked for his opinion; nor did he offer it on his own initiative, at least not publicly. In the fall of 1907, however, when he was writing his comprehensive article for the *Jahrbuch der Radioaktivität und Elektronik*, he had to declare his position, and this he did with deep-rooted self-assurance.

To begin with, Einstein with great fairness described Kaufmann's experimental setup, emphasized the "admirable care" of the measurements, and compared the graph obtained by Kaufmann with the conclusions of relativity theory.[24] Never having had a high opinion of exaggerated accuracy, Einstein would have been inclined "to regard the agreement as sufficient" if the deviations had not been outside the range of error and, moreover, systematic. He commented on this situation in much the same way as Planck, by pleading for a postponement of a verdict and for further experiments: "Whether the systematic deviations are due to some not yet identified source of error or to the fact that the foundations of the relativity theory are not in line with the facts, will only be determined with certainty when a more copious experimental material is available."

What he really thought emerges between the lines of his subsequent assessment of the rival theories. He frankly admits that "Abraham's and Bucherer's theories of electron movement present graphs which are considerably closer to the graph observed than is the graph derived from the relativity theory." But while the famous Lorentz was "at his wits' end," the man at the Patent Office—who in 1905 had simply ignored Abraham's theory and Kaufmann's experimental results relating to it—refused to be rattled: "However, the probability that their theories are correct is rather small, in my opinion, because their basic assumptions concerning the dimensions of the moving electron are not suggested by theoretical systems that encompass larger complexes of phenomena."[25]

Einstein therefore does not even attempt to offer proof (which would scarcely be possible anyway) that Abraham's and Bucherer's theories are wrong. He simply does not consider them "probable" by metatheoretical criteria, because some of their basic assumptions are isolated and arbitrary. Basic assumptions, he believes, must not be invented ad hoc for specific cases but must cover wider areas—ideally, physics as a whole. Only then does one encounter that "marvelous feeling of realizing the uniformity of a complex of phenomena."[26] In Einstein's scientific credo this cannot be expected of ad hoc assumptions; it can be expected only of principles.

Less than a year passed before Bucherer, in a greatly improved version of Kaufmann's experiments, confirmed not his own theory but the formulas of Lorentz and Einstein. Bucherer wrote to Bern "that by careful experiments I have proved beyond any doubt the validity of the relativity principle."[27] Einstein thanked him by return of post in a "friendly letter."[28] He is unlikely to have been head over heels with joy, but he probably took satisfaction in having known all along that he was right.

These measurements of electron tracks, however, could not lead to a decision between Einstein's and Lorentz's theories, which had the same laws of motion for fast electrons. Further experiments were therefore needed to examine effects predicted by Einstein's theory but not by Lorentz's. One such possibility, as Einstein had suggested in his "$E = mc^2$ supplement," concerned the transformation of mass into energy in radioactive processes. Another possibility was the measurement of time dilatation, postulated only by relativity theory. In March 1907 Einstein published a proposal along those lines. It was obvious, of course, that the differences between moving clocks could not be measured by anything from a pocket watch to a precision chronometer. But atoms emitting spectral lines are very accurate clocks and can, moreover, be accelerated to very high velocities. Einstein therefore proposed experiments with electrically charged accelerated atoms, then known as "canal rays," whose frequencies of oscillation should, according to his theory, change.[29]

Although canal ray experiments were then being conducted, especially by Johannes Stark, their accuracy was not sufficient to prove time dilatation. Einstein did not let go: "The main thing now," he declared in 1911, "is to conduct the most accurate experiments possible to test the fundamentals. There is nothing much to be gained from a lot of pondering at the moment."[30] But he had to be patient. Not until the early 1930s was the conversion of mass into energy confirmed in the study of nuclear reactions, and time dilatation was not directly proved until 1938. Until then any physicist choosing a theory on the basis of experiment alone could have opted as well for Lorentz's theory as for Einstein's. If, nevertheless, the theory of relativity gained support so quickly, this was due not to conclusive experiments but to the fact that most physicists responded to its axiomatic, fundamental character and its beauty.

There probably is no other theory in modern physics that, like the special theory of relativity, had to wait a quarter-century for direct experimental evidence in its favor. And there is no other theory whose eventual experimental confirmation was received with greater indifference—simply because no one had expected anything else.

Later in his life, Einstein clearly formulated the difference between "theories based on principles" and "constructive theories." Constructive theories "endeavor, from a relatively simple fundamental formalism, to construe a picture of a more complex phenomena,"[31] as, for instance, the macroscopic properties of matter can be constructively explained by assuming molecular movements. Theories of principle, on the other hand, are based on "empirically found general properties of natural processes, on principles from which mathematically formulated criteria follow, which individual processes, or their theoretical representations, must observe,"[32] as is the case with thermodynamics and, of course, with relativity theory. Each type has its advantages: constructive theories are characterized by "completeness, adaptability, and clarity"; theories of principle have "logical perfection and secure foundations." Indeed Einstein regarded constructive theories as "the more important category,"[33] presumably because of their many applications. But it is obvious that his great love was for theories of principle, in line with his intention, dating back to his student days, of "sniffing out what might lead to the root of things."[34]

At this point it should be mentioned that Einstein, though he was the creator of relativity theory, was not the creator of its name. In his paper *On the Electrodynamics of Moving Bodies*, with which everything began (and almost reached completion) he referred only to the "relativity principle," and he kept to this formulation for the next few years. It was Planck who, at the meeting in Stuttgart discussed above, first spoke of *Relativtheorie*,[35] "relative theory." With Bucherer and others, this soon became *Relativitätstheorie*, "relativity theory." In his arguments with other physicists and his comments on their work, Einstein was progressively, and reluctantly, drawn into this new terminology, though in headlines and in the text of his own publications he continued to speak of the "relativity principle." In 1911 he eventually gave in and used what had meanwhile become the common term for the first time in a title—*Die Relativitäts-Theorie*[36]—though not without distancing himself from it by quotation marks. In this form, within quotation marks, he continued his ultimately futile resistance for a few more years.[37]

Einstein's uneasiness with this terminology was justified, since a principle is not a theory. A principle is something that has to be borne in mind in the formulation of any theory, something that may provide a useful hint on how to find the correct theory, but certainly not the theory itself. It would have been in line with his own understanding of the methodological status of the relativity principle[38] if he had introduced it in the title of his great treatise as a "heuristic principle"—except that he had already used that term in his paper on light quanta three months earlier.

When, after World War I, Einstein's name and his relativity theory were catapulted into the public limelight, it was probably unavoidable that the theory soon became shortened to the formula "Everything is relative," and that this formula was then applied also to morals, political institutions, and so on. To obscurantist minds, relativity theory appeared as some particularly reprehensible Jewish contribution to social decay.

Needless to say, even in physics it is arrant nonsense to claim that "everything is relative." Indeed, Max Planck was instantly fascinated by Einstein's paper because, on the contrary, it revealed ways of "finding the absolute, the universally valid, the invariant"[39] in natural laws—such as the universally constant velocity of light. In Minkowski's "four-dimensional" representation of 1908, the invariance of natural laws was developed with regard to the group of Lorentz transformations, and "the term relativity postulate" seemed to Minkowski "very feeble . . . for the postulate of an invariance."[40] When, in the early 1920s, nonsensical controversies broke out over relativity theory, some physicists believed that they could pull it out of the line of fire by renaming it "invariance theory." Einstein agreed only to the extent that this name would describe the method, not the physical content, of the theory. But he pointed out that even if the new name was possibly better, it "would only give rise to confusion if after such a long time the generally accepted name were now changed."[41] Relativity theory therefore retained its name.

Whether it is a "principle" or a "theory," relativity is widely believed to represent a scientific revolution, perhaps even *the* scientific revolution. Those who believe this may be perfectly correct, and moreover in very good company—except that they cannot call on Einstein as a witness. As with a great many other things, he had his own views also on revolutions in science.

Certainly Einstein referred to revolutions, though only sparingly and only for very significant events, such as the transition from forces acting at a distance to the fields of Maxwellian theory.[42] Nor had he shrunk from describing his light quanta as "very revolutionary,"[43] though not in print but in a private letter. A "revolution" in science evidently meant to him a break with tradition and a radical new beginning, such as the field concept in the nineteenth century and quantum physics in the twentieth.

On the other hand, he did not see relativity theory as coming under that heading. In the same letter to Conrad Habicht in which he called his light quanta "very revolutionary" he characterized what would later be called relativity theory as "an electrodynamics of moving bodies, making use of a modification of the theory of space and time"—and a modification is certainly no revolution. In later years, Einstein never used the term "revolutionary" or any of its synonyms in connection with his theory of relativity, and at times he would laugh when others used it.

This was certainly not due to modesty, a virtue toward which Einstein had no inclination anyway. Like Isaac Newton, he might have said: "If I have seen further, it is by standing upon the shoulders of giants."[44] For Einstein there was

> in the development of the sciences only a building-up, never a pulling-down. . . . Unless one generation can build on what earlier ones have achieved there can be no science. It would be a sad thing if the relativity theory had to overthrow the earlier mechanics, something like one tyrannical ruler overthrowing another. The theory of relativity is nothing but a further step in the centuries-old evolution of our science, one which preserves the connections discovered in the past, deepening them and adding new ones.[45]

When he paid his first visit to the United States he began right away by instructing a sensation-hungry public:

> There has been a false opinion widely spread among the general public that the theory of relativity is to be taken as differing radically from the previous developments in physics from the time of Galileo and Newton, that it is violently opposed to their deductions. The contrary is true. . . . The four men who laid the foundations of physics on which I have been able to construct my theory are Galileo, Newton, Maxwell and Lorentz.[46]

Against this historical background Einstein saw relativity theory as "simply a systematic development of the electrodynamics of Maxwell and Lorentz.[47] And in his Nobel Prize speech he described it as "an adaptation of the foundations of physics to Maxwell-Lorentzian electrodynamics."[48] There are numerous quotations along those lines: one, from his later years, was made in connection with a series of articles in *The New York Times*: "The reader gets the impression that every five minutes there is a revolution in science, somewhat like a *coup d'état* in some of the smaller unstable republics."[49]

But what did his colleagues think? "This is a revolutionary" was Max von Laue's impression when he was the first German physicist to visit the man at the Patent Office during the summer of 1907. "During the first two hours of our conversation he overthrew the entire mechanics and electrodynamics, doing so on statistical grounds."[50] In this deconstruction of the classical foundations, Einstein's radiation theory probably held center stage, and this Einstein himself regarded as "very revolutionary."

Soon, however, his relativity theory, too, was viewed as a revolution. The conservative Max Planck, of all people, set the tone in the spring of 1908, even though he replaced the politically objectionable "revolution" with a German equivalent (*Kühnheit*, boldness) when he referred to Einstein's definition of time:

> In boldness it exceeds anything so far achieved in speculative natural science, in philosophical cognition theory; non-Euclidian geometry is child's play by comparison.[51]

To Planck, "the revolution in the physical world picture" brought about by the relativity principle was "in extent and depth comparable only to that caused by the introduction of the Copernican world system." Sometime later, Einstein's old teacher and mentor, Professor Kleiner of Zurich, evidently reflected the majority view of physicists when, as a matter of course, he remarked that the relativity principle was being "described as revolutionary"[52]—and this view had as much justification as Einstein's own.

In fact, relativity theory—both the special theory of relativity of 1905 and, if we may anticipate, the general theory of relativity of 1915—is a deepening rather than a revolution, and Einstein is a perfecter of "classical" physics rather than a revolutionary. Nevertheless, the reshaping of the fundamental concepts of time and space, for centuries regarded as a priori concepts, certainly was a revolution, one of the greatest in the history of science, even according to Einstein's own criteria, although he preferred to see it merely as a modification.

As rapidly as relativity theory became known among physicists and mathematicians and was accepted by most of them, two of the greatest figures, men who had contributed a good deal to the establishment and analysis of the relativity principle, remained aloof—Poincaré and Lorentz.

As for Poincaré, one cannot even say that he rejected Einstein's theory, because he quite simply ignored it. It is scarcely credible that Poincaré, who was familiar with German, should not have read Einstein's work in *Annalen* at some time or other. Did some of it, such as the synchronization of clocks by light signals or the Lorentz transformations, seem to him too familiar? Was he looking for tributes in footnotes, and was he put off by their absence? Did he think he had himself presented all that was necessary about the relativity principle in his papers of 1905 and 1906, in a more complete and incomparably more elegant mathematical form? Or was he not greatly interested in its further development, since he was not a physicist but a mathematician, indeed the most famous mathematician in the world?

Poincaré preserved such total silence on these matters that we know nothing of his attitude or motives. The one thing that is certain is that he avoided mentioning Einstein's name whenever he had to refer to relativity theory in later years. His later articles suggest that he ignored not only Einstein's name but also Einstein's ideas;[53] that he clung to his own concept of 1905, regarding the relativity principle more as a conclusion from electrodynamics and confined to this subject rather than as an axiom of physics generally; that he viewed the Lorentz contraction as an independent hypothesis and not as a consequence derived from anything else—with behind it all still the ether, and hence a system at absolute rest.

Einstein and Poincaré met only once, at the first Solvay Congress in Brussels in 1911. The meeting, as Einstein reported to a friend, was not a success: "Poincaré was simply negative (toward the relativity theory) and with all his perceptiveness showed little understanding for the situation."[54] There was no occasion for a second meeting: Poincaré died in 1912 at the age of only fifty-eight.

Just as Poincaré kept silent about Einstein, so Einstein kept silent about Poincaré and about what he had read of Poincaré's work. This is significant, since Einstein owed more than just one suggestion to Poincaré, almost certainly including the definition of time. Did Einstein, while formulating his relativity theory, half repress Poincaré and half overlook him, and did he later feel so awkward about this that he repressed Poincaré altogether? Long after Poincaré's death, when Einstein for the first time mentioned his name, it was in a different context. In a lecture to the Prussian Academy of Sciences in Berlin, *Geometry and Experience*, he referred to "the acute-minded and profound Poincaré,"[55] though he was referring not to Poincaré's synchronization of clocks but to his conventional analysis of the relationship between physics and geometry. Not much later, in an interview published on the front page of the Paris daily *Figaro*, he spoke of his "great admiration for Poincaré."[56] Then followed three more decades of silence until Einstein, in old age, in a letter containing something like a thumbnail listing of authors who had influenced his development, mentioned Poincaré along with Hume and Mach.[57]

The following year there was a tempest in a teacup around the second volume of Sir Edward Whittaker's *History of the Theories of Aether and Electricity*—in many respects a masterpiece, but an oddity in its passages on relativity theory. Even the chapter heading "The Relativity Theory of Poincaré and Lorentz" indicated the direction of Whittaker's thinking; he declared that Einstein's contribution was marginal. Max Born, like Whittaker a professor at Edinburgh in Scotland, had tried to talk his colleague out of this strange whim, but in vain. In response to Born's warning of what was about to be published by a widely respected scientist, Einstein reacted irritably:

> Myself, I have always derived satisfaction from my efforts, but I don't think it sensible to defend my few results as my property, like some old skinflint defending the few coins he has laboriously scraped together. I'm not holding it against him. . . . After all, I don't have to read the stuff.[58]

Nevertheless, the episode left a mark on Einstein. Four weeks later he wrote to Bern, where a celebration was being prepared for the fiftieth anniversary of relativity theory: "I hope it will be ensured that the merits of H. A. Lorentz and H. Poincaré are also appropriately acknowledged."[59] After nearly half a century this was the first time that Einstein even mentioned Poincaré in connection with the special relativity theory.

Two weeks before his death, Einstein spoke with a young historian of science. At one point the conversation turned to the vanity which,

Einstein observed, was found "in so many scientists. You know," he told his visitor, "it has always hurt me that Galilei did not acknowledge the work of Kepler."[60]

After Einstein's death, Abraham Pais asked Einstein's secretary about a book he had lent Einstein. A few years earlier, he had asked Einstein what influence Poincaré's great treatise of 1906 on the dynamics of the electron had had on Einstein's own work. Einstein had never read it, and Pais, who had found an offprint of this not readily accessible article in an antiquarian bookstore, lent Einstein his precious copy, but it was never returned to him. Now the secretary was unable to find the article.[61] Poincaré's paper remained lost.

Poincaré and Einstein had passed like ships in the night, doing everything possible to avoid one another.

Einstein's relativity theory must have been a strange experience for Lorentz. For more than a dozen years the Dutch *praeceptor physicae* had struggled to adapt electrodynamics to the fact that movement relative to the ether was not observable. His efforts to "save" the relativity principle had worked, albeit at the cost of complicated arguments and a mountain of separate hypotheses. And now a fairly unknown man from Bern had simply turned the problem into a principle, and had actually succeeded with it. In 1906 Lorentz observed, with astonishment and some melancholy, that "Einstein simply postulates what we have deduced, with some difficulty and not altogether satisfactorily, from the fundamental equations of the electromagnetic field."[62] That, however, was a one-sided comment, because Einstein, on the other hand, had easily derived from his two principles much of what Lorentz had been forced to introduce as ad hoc hypotheses—from the Lorentz constant to the Lorentz force. Moreover, Einstein had established the validity of the principles and the kinematics based on them for physics as a whole—something Lorentz had not even attempted.

As it was not then possible to decide between the two theories by experiment, Lorentz tended to view the choice as a matter of taste. "Which of the two modes of thinking a person follows will probably be left to him"[63] is how he summed up his opinion in the Wolfskehl lectures he gave in Göttingen in the fall of 1910. Max Born, who had to edit these lectures for publication, thought this "absurd and reactionary."[64]

In a series of lectures in Haarlem, the Netherlands, in 1913, Lorentz was even more outspoken. He found "the older presentation more satisfactory, according to which the ether possesses a kind of substantiality, space and time are strictly separable from each other, and simultaneity can be defined without restrictions." Realizing that absolute simultaneity would imply infinite velocity, he also criticized "the bold assertion that super-light velocities are unobservable as a hypothetical restriction of what is accessible to us, one that cannot be accepted without reserve."[65]

Unlike Born, Einstein showed some understanding of Lorentz. When Lorentz's lectures went into print the following year, Einstein for the first time acted as a reviewer of a work concerning relativity theory. He made no mention of Lorentz's observation on the ether, time, or the velocity of light, but found everything else "clear and well explained." He added the recommendation: "No one with a serious interest in the subject should omit reading the little book."[66]

No one ever suggested that Lorentz's unwavering loyalty to the ether physics of the nineteenth century had anything to do with professional vanity, with obdurate pride in his own achievements, or with stubbornness. Lorentz was universally admired, not only as an authority and a wise guide through the problems of theoretical physics, but also as an integrated and harmonious personality. This is proved also by his numerous comments on relativity theory—precisely because his own view was different. And it is further confirmed by the relationship between Einstein and Lorentz.

Einstein had always regarded Lorentz as one of the giants on whose shoulders he himself stood, even if that giant was reluctant to see as far as he did. This intellectual admiration was soon matched by a happy personal relationship. When the two started to correspond in 1909, Einstein was enthusiastic even at a distance: "I admire that man more than anyone else, I might say I love him."[67] This boundless admiration was soon reciprocated, confirmed, and consolidated in personal contact. In the later phases of Einstein's life Lorentz will time and again appear as a kind of father figure, and as a splendid example of unclouded respect despite professional disagreement.

One of the major oddities in the customary accounts of the theory of relativity is the assertion that it came into being in close connection with the empirical findings of Michelson and Morley's ether-drift experiment. A typical example is an essay by Robert A. Millikan which introduced the special issue of *Review of Modern Physics* published on Einstein's seventieth birthday. Millikan states that the special theory of relativity may be "looked upon as starting essentially in a generalization from Michelson's experiment."[68] This may not have been the opinion of the man whose birthday was being celebrated, but it had been the opinion of physicists generally for some time. It had clearly been established with the first book on relativity theory, written by Max von Laue in 1911, in which Laue described the experiment as the "fundamental experiment of the relativity theory"[69] and it has since become a solid part of the folklore of physics.

It is argued, then, that the Michelson-Morley experiment led straight to the relativity principle. Yet a single glance at Einstein's 1905 paper shows that it contains no mention anywhere of this allegedly vital experiment. This raises the question of what Einstein actually knew of that experiment and what effect it had on his thinking—a question not only of biographical interest but of some importance for the genesis and justification of relativity theory.

In the 1950s the physicist Robert S. Shankland, who thought of himself as a scientific heir of Michelson, asked Einstein (who was then very old) when exactly he had first learned of Michelson's experiment. At first Einstein replied spontaneously that he had learned about it from H. A. Lorentz's writings, but only after 1905. "Otherwise I would have mentioned it in my paper."[70] This information could not have been correct, and two years later, following some reflection, he qualified his answer: "This is not so easy, I am not sure when I first heard of the Michelson experiment. I was not conscious that it had influenced me directly during the seven years that relativity had been my life. I guess I just took it for granted that it was true."[71]

However, Einstein had learned of that experiment during the first of those seven years, while he was still a student. In the summer vacation between his third and fourth years he reported to Mileva Marić that he had "read a very interesting paper"[72] by Wilhelm Wien, and this contains a list of thirteen experiments to prove the Earth's motion through the ether, including that of Michelson and Morley. In Lorentz's *Versuch* of 1895, which Einstein studied carefully several times, the experiment is described in detail and discussed in terms of its consequences. Since Lorentz was induced by Michelson and Morley's results to extend his electron theory by hypothesizing the contraction of dimensions, this is of decisive importance: a theory had reacted directly to an experimental result, and Einstein can scarcely have been unaware of this. But does this mean that the experiment was therefore important also for the development of Einstein's relativity theory?

The structure of Einstein's treatise of 1905 does not suggest that interpreting the ether-drift experiments in general or the Michelson-Morley experiment in particular had been of particular interest to him. In setting out the problem, he extensively outlined the structural asymmetry of the customary electrodynamic concept and followed this up with no more than a summary reference to "the failure of attempts to detect a motion of the Earth relative to the 'light medium' "—and that is all. On the basis of the two arguments, he then elevated the "principle of relativity" from an assumption to a presupposition. One of these "failed attempts" was Michelson's, but despite its great accuracy and its importance to Lorentzian theory, it was only one of many; and Einstein pointed out to Shankland that the aberration of starlight and the Fizeau experiment would have been a sufficient experimental basis for his arguments. Both of these were well known, like Faraday's induction experiment, whose interpretation had been Einstein's leitmotiv in his search for a comprehensive principle.

If the Michelson-Morley experiment played any role at all in Einstein's reflections, it was only indirectly, as part of the Lorentzian theory, because Einstein was of course aware that the contraction of dimensions must come out correctly in his own arguments. Indeed, some physicists found this step of Einstein's, from the problem to the principle, particularly attractive: in the artificial and contrived Lorentzian hypothesis, contraction had been invented solely for the interpretation of the Michelson experiment; but in relativity theory it followed effortlessly from Einstein's principles as a kinematic effect. Einstein must have heard of this, because at the beginning of 1908 he referred to the swift acceptance of his ideas and to the impression which his explanation of the theory had made on his colleagues: "If the Michelson-Morley experiment had not brought us into serious embarrassment, no one would have regarded the relativity theory as a (halfway) redemption."[73] This "redemption" evidently enhanced the famous "fundamental experiment of the relativity theory." Einstein himself occasionally paid tribute to it in systematic presentations,[74] though not in any of his reconstructions of his mental processes.[75]

Einstein's last word, like most of his statements on this topic, is not very precise, even though it was intended for publication:

> In my own development Michelson's result did not play any noteworthy part. In fact, I cannot even recall if I knew about it when I was writing my first treatise on the subject (1905). The explanation is that, for reasons of a general character, I had a firm idea of how this was compatible with our knowledge of electrodynamics. It is understandable therefore why the Michelson experiment did not play a decisive role in my personal struggle.[76]

This was certainly an honest formulation, but it also testifies to a failure of memory; Einstein was undoubtedly familiar with the experiment in his youth, even if he viewed it only as one element in the whole body of confirmations that no movement is demonstrable relative to the ether.

It is sometimes speculated whether relativity theory would have been discovered, or when it would have been discovered, if Einstein had not discovered it. Einstein himself, in old age, conveyed the impression that in his youth he had merely plucked a ripe fruit from the tree of knowledge: "There is no doubt," he wrote two months before his death, "that the special relativity theory, if we look at its development in retrospect, was ripe for discovery in 1905."[77] As early as 1906 he regarded it as "not improbable that Mach would have hit upon the relativity theory if, at the time when his mind was still youthfully fresh, the question of the constancy of the velocity of light had already engaged the attention of physicists."[78]

What enabled Einstein, rather than anyone else, to take this decisive step? If revolutionary achievements in science imply independence of the all-powerful traditions which often hold the leading figures in thrall, then Einstein's independence of thinking may, at least in part, have been due to his peripheral position at the Patent Office.

His success certainly suggests that working on the margin of scientific endeavor was not only no obstacle to him but perhaps a positive advantage. His "temporal monastery" in Bern was not intellectual isolation. It enabled him, through intensive reading, to absorb ongoing discussions, without being distracted by academic fashions or by career constraints from developing his own ideas. What is so astonishing about Einstein is not only the depth of the problems he addressed, but also the width of his interests. Only his combination of breadth and depth seems to explain his unique explosion of creativity in 1905, especially his discovery of relativity theory.

No one else saw the structural problems of electrodynamics in such close connection with radiation theory; and he alone, thanks to his "very revolutionary" paper of March 1905, on his "heuristic viewpoint" on light quanta, found it easy to dispense with the idea of a substantial carrier of electromagnetic waves. It was this liberating blow that enabled Einstein to bring together H. A. Lorentz's electrodynamic theory, with its concept of local time and its contraction of dimensions, the delicate status of relativity theory, and the prophetic insights of Henri Poincaré—a stroke of genius which turned the problems into a principle. The theory of relativity may have been ripe for discovery, but in 1905 only one man was able to discover it.

CHAPTER TWELVE

Expert II Class

THE NEWS THAT HISTORY had been made in physics in Bern in 1905 soon spread in professional circles, but it spread less rapidly in the city of Bern itself. Apart from his colleagues Besso and Sauter, Professor Gruner was probably the only person who realized that Einstein had accomplished quite exceptional things. In consequence, nothing much changed in his lifestyle. The only outward sign was that, for the good citizens, he was now Herr Doktor. There were advantages in this, as he later said when congratulating an acquaintance on his doctor's degree: "In my experience, it quite considerably facilitates relations with people."[1]

Thus only his doctorate, and not his other treatises, was referred to in the application which his chief, Friedrich Haller, addressed to the Swiss Federal Council, proposing Einstein's long-overdue promotion. He had "increasingly familiarized himself with technology, so that he now very successfully processes quite difficult technical patent applications and is one of the most highly respected experts of the Office."[2] On April 1, 1906, Einstein became an Expert II Class. His salary went up by 600 francs to 4,500 francs annually. This still left him in the lower range of officials in his grade. Although he is reported to have asked jocularly on payday what on earth he was to do with all that money,[3] he was not in fact satisfied with his salary and evidently tried on more than one occasion to get a better-paid job in the Post and Telegraph Directorate. He certainly later reported with some satisfaction that Federal Councillor Ludwig Forrer, his friend and patron, had been "very furious" when he discovered "that the Telegraph Directorate had not wanted me as an official a few years earlier."[4]

Einstein therefore remained a "good patent slave." He gave the Patent Office what was due to it, and the Patent Office for its part contributed something to his scientific output. Rudolf Ladenburg, reporting on a visit to Bern, recounted that Einstein had pulled out a drawer in his desk and announced that this was his department of theoretical physics. His duties at the office did not demand a lot of time, and so whenever he was free he would work on his scientific problems.[5] Needless to say, he did not go about announcing this publicly.

Not even his string quartet, with whom he met once a week during the winter for musicmaking, had any idea who their second violin really was—even though one of them was a physics teacher at the Freies Gymnasium in Bern. His fellow musicians remembered him as "an enthusiastic musician, a charming companion, and a modest person,"[6] but not as a new Copernicus. This was how Einstein liked it, and this was how things remained until he left Bern in the fall of 1909. Some people in the Patent Office, and also outside, found it almost unbelievable that the Expert II Class should have been offered a professorship at the University of Zurich.

During the seven years that Einstein spent in Bern he had lived in seven different apartments. We do not know the reason for this unsettled, un-Swiss way of life, but the fact that the Einsteins' first few apartments were rented furnished probably made moving easier. Even at the peak of his productivity—when the paper on the Brownian movement had just been completed—the Einsteins were on the move. They gave up their apartment at Kramgasse 49 in the old city center and on May 15, 1905, moved to Besenscheuerweg 28 in the Mattenhof district[7] on the outskirts. One advantage of this address was that Michele Besso lived nearby and he and Einstein were able to make the fifteen-minute walk to their office together. After all, the two had enormously important matters to discuss during the five or six weeks from the idea triggered by Besso to the completion of the paper on relativity toward the end of June 1905. The memorial plaque on the arcade pillar of Kramgasse 49, to the effect that Einstein created "his fundamental treatise on the relativity theory . . . in this house" should therefore be treated with indulgence.

"I have moved again,"[8] Einstein reported to his friend Solovine in the spring of 1906. As at the beginning of their married life, the Einsteins were once more in Kirchenfeld. They rented the upper floor of a small house in the typical local style, with a fine view of the Bernese Oberland mountains. This apartment, at Aegertenstrasse 53, probably the first with their own furniture, remained their home until they moved to Zurich.

Einstein greatly regretted the departure of "the good Solo": "Since you left I haven't been meeting anyone privately. And now the way-home conversations with Besso have come to an end too."[9] This complaint, perhaps a little exaggerated, was a reaction to the cliquishness of Bern society, and, no doubt, to Mileva's marked distrust of other people. But at home Einstein now had a small companion, his son Hans Albert, whose intelligence at age three fascinated his parents. The father remarked, proudly rather than critically, that the boy had "already grown into a rather fine impertinent lad,"[10] and the mother reported contentedly: "My husband often spends his free time at home just playing with the boy."[11]

Their income was probably adequate for a solid bourgeois lifestyle. Mileva's considerable dowry of 10,000 francs was regarded as a reserve and not touched; when they divorced, many years later, the full amount remained. If the Einsteins lived a little less grandly than some of his colleagues, this was because he had to support his mother, who was living with her sister Fanny and Fanny's husband, Rudolf Einstein, in Hechingen, in the Prussian enclave of Württemberg. Even so, there was enough money for vacation trips. In August 1905 the Einsteins visited their former fellow student Helene Savić in Belgrade—their visit may have been connected with inquiries about Lieserl—and subsequently they spent a week with Mileva's parents in Vojvodina. Over the following years they vacationed in the neighborhood of Bern, in resort villages, in the Simmental, the Valais, or the Bernese Oberland.

The incredible, even awesome, tempo with which Einstein completed four epoch-making papers between March and June 1905 could not last—fortunately, one might be inclined to say, since continuing to work at that intensity would have been bound to damage his health. But the reason was not physical exhaustion but exhaustion of subjects: "There is not always a ripe theme for musing over. At least not one that excites me."[12] Einstein had a sure instinct for choosing not only what problems to address but also what problems to pass over. "There would, of course, be the subject of the spectral lines, but I believe that there is no such thing as a simple connection between these phenomena and others already researched, so that the matter, for the moment, seems not too promising."[13] That was a shrewd judgment, as the internal structure of the atom was still unknown, the atomic nucleus had not yet been discovered, and it was not until 1913 that Niels Bohr would propose a quantum-theoretical model of the atom. In 1905, work on spectral lines could not have gone beyond an attempt at phenomenological interpretation, even for Einstein. The fact that he did not attempt it shows him to be a master of the art of the soluble, ever searching for a "connection between phenomena."

There was no shortage of soluble problems within the fields he himself had opened up. Thus in September 1905 he published the first version of the equivalence of mass and energy as the most spectacular consequence of the relativity principle. The next year would see a more general derivation of the formula $E = mc^2$ as well as a proposal for an experiment to decide between rival theories of the dynamics of the electron. The proposal was solidly buttressed by theoretical reflections, but the experimenters chose to go different ways. Einstein also generalized and deepened his theory of the Brownian movement and especially his "heuristic viewpoint" on light quanta; in November, eventually, another milestone was reached with his first quantum theory of solids. Mileva proudly reported that her husband now did not spend all his free time playing with the boy: "To his credit I have to say that this is by no means his only occupation outside his official work; the treatises written by him are piling up quite frighteningly."[14] Six publications—all of them important—were the rich harvest of 1906.

▪ ▪ ▪

Such an enormous scientific output cannot have been easy, even for a genius like Einstein. Even when his creative ideas had been developed, a lot of routine work remained to be seen to, and this was on top of his work at the Patent Office. Ideas had to be arranged in a form suitable for publication, the mathematics had to be tidied up, and a fair copy had to be written for submission to the editor of *Annalen*, to be passed on to the printer. Cautious authors would keep a copy, so as to have a reliable version available for checking proofs or in case the manuscript was lost in the mail. Einstein probably waived this precaution, relying on his notes and on his memory. Now and again—quite certainly with his dissertation on the determination of molecular weight—there might be queries or requests for changes from the editor. As a rule, however, the proofs would arrive from the printer after a month or more; this was an opportunity for correcting mistakes, checking the calculations, and making last-minute revisions. After another month a parcel would arrive with the offprints of the paper. Einstein built up a small reference library of his own offprints, in which he occasionally scribbled afterthoughts or corrected printing errors.[15] His manuscripts, drafts, and notes were usually thrown away once the offprints arrived.

These offprints were, and still are, the currency of mutual acknowledgment among scientists. An author would send one to a colleague to draw attention to himself and his work, or a reader of a journal might show interest and respect by requesting an offprint from an author. As we have seen, Einstein as early as 1905 was involved in this informal information network of physicists—sporadically at first and more regularly later. Some inquiries were addressed to "Herr Professor Einstein" at the University of Bern,[16] and the writers no doubt were astonished to find that this author was employed not at the university but at the Patent Office. Some recipients of his letters, especially if they were renowned professors, may also have been taken aback to find them written on graph paper with a ragged edge, carelessly torn from some copybook.

It was the custom then among physicists, once a lively correspondence had developed, to lend it a more personal note by exchanging photographs. Einstein fell into line with this practice,[17] and many a young colleague received a fine picture showing an elegantly clad gentleman in a smart check suit, resting his arm on a writing desk.

Although most of the early letters are lost, we may assume from later evidence that Einstein had always been an enthusiastic letter writer where scientific topics were concerned. Some idea of this is conveyed by his partially preserved correspondence with Wilhelm Wien, who in 1906, after the premature death of Paul Drude, became editor of *Annalen* and thus one of the most influential physicists in Germany. Throughout the summer of 1907, including his vacation in the Simmental, Einstein bombarded the "highly esteemed Herr Professor" with long letters and terse postcards—so much so that he asked Wien "not to think too badly of me because of my flooding you with letters."[18]

This correspondence was concerned with difficult questions concerning the interpretation of the velocity of light as a nonexceedable limit for signals. In the heat of battle Einstein occasionally tripped up, but even as a young man he never lost sleep over his mistakes: "On closer examination, unfortunately, everything I reported to you so precipitately in my last letter has turned out to be wrong,"[19] he once wrote to Wien. But in the end it was perfectly clear that the Maxwellian theory, which was compatible with the relativity principle, ruled out any transmission of a signal at a velocity exceeding that of light, although this did not give rise to a publication.

The best opportunity for meeting influential professors and maneuvering into position in the academic job market was provided by congresses. Einstein did not seize that opportunity, and perhaps even avoided it. Physicists then did not have conventions of their own but would meet in specialized groups within the framework of the annual general meeting of the Deutsche Gesellschaft der Naturforscher und Ärzte, the German Society of Scientists and Physicians. Its meeting in Stuttgart in 1906 would have offered Einstein a good chance to become acquainted with the leaders in his field, if only because of its comparative nearness to Bern. And he had no doubt seen the notice in

Physikalische Zeitschrift that Max Planck was going to speak there on *Kaufmann's Measurements . . . and Their Significance for the Dynamics of the Electron*, a subject that would have drawn attention to the author of *The Electrodynamics of Moving Bodies.* Nonetheless, *Relativtheorie* was discussed there in its author's absence.

The following year, when the meeting was held in Dresden, Einstein was again absent. In 1908, eventually, he did plan to go to Cologne, but this did not come about because he used his short leave for recuperation.[20] The two weeks' annual leave from the Patent Office to which he was entitled was certainly not very generous, but if Einstein had really wanted to attend he probably could have managed a few extra days. Probably he was not all that eager. The poorly paid university assistants' posts had little attraction for a well-paid civil servant,[21] and besides he may have believed that one day the mountain would come to Muhammad. And indeed, the following year, 1909, Einstein attended the annual meeting in Salzburg, now as a "guest of honor" entrusted with one of the keynote addresses.

We do not known whether Professor Gruner encouraged him to do so, or perhaps even Professor Kleiner of Zurich University, or whether it was on his own initiative that Einstein for a second time applied for *Habilitation* as a privatdozent. At any rate, on June 17, 1907, he submitted an application to the director of education of the canton of Bern.[22] Enclosed with his petition were his dissertation and doctor's diploma, a curriculum vitae of only nine lines, and "seventeen papers from the field of theoretical physics."

It is probable that he had discussed this step in advance with Paul Gruner, since, as with his first attempt, he wanted to become a privatdozent without writing a special *Habilitation* thesis; this procedure was possible, under the university's rules, "for other outstanding achievements." No doubt Gruner would have persuaded Aimé Forster, the (by then rather senile) full professor of physics, that the offprints submitted by Einstein far surpassed any normal *Habilitation* thesis, but once again, things did not work out.

Einstein's application was immediately circulated among the faculty members. By July 10 they had all read it. The fact that it was not put on the agenda before the impending summer vacation, which began on July 10, suggests that some of the professors were unhappy about the lack of a thesis. Not until October 28 was the application discussed. Only Gruner, who was no more than a "titular professor," supported a positive decision, "in view of the important scientific achievements of Herr Einstein, without demanding a special *Habilitation* thesis."[23] Professor Forster, the head of the department—"about whose incompetence," as Einstein later remembered, "stories were circulating among the younger people"[24]—recommended that the application be accepted "under the customary procedure." It was therefore decided, "after prolonged discussion, that the petition be refused until Herr Einstein has submitted a *Habilitation* thesis." Thus ended Einstein's second attempt to become a "great professor."

No doubt he again railed against the "pigsty,"[25] as he had after his first attempt four years earlier, but this time with more justification, and also in the certain knowledge that the universities would not be able to ignore him much longer. In old age, looking back on this episode, he remarked that "it is often the case that in small departments a few old fogies—as we ourselves are now—will stick together and run the show."[26]

By then the major scientific publishing houses had begun to notice Einstein. The first to approach him was the renowned firm of Teubner in Leipzig, whose proprietor assured him in September 1907 that "my presses will always be at your disposal in case you have any literary plans."[27] The following year the firm of S. Hirzel proposed that he should produce "a little monograph on the more recent advances in physics and chemistry," written "in an easy, not to say popular, style to make it accessible to the chemist as well as the physicist."[28] Although the idea appealed to him and he initially hoped to "undertake the task, even though I am seriously overloaded with work,"[29] he changed his mind two weeks later: "Unfortunately I am quite unable to write that book, because I am unable to find the time for it."[30]

When Eilhard Wiedemann suggested that he write a book on relativity theory, presumably for the publishing house of Vieweg in Braunschweig, Einstein declined after a lengthy period of consideration—this time not only for reasons of time but also because of the subject matter: "I cannot imagine how this topic could be made accessible to broad circles. Comprehension of the subject demands a certain schooling in abstract thought, which most people do not acquire because they have no need of it."[31]

Even in later years publishers found it difficult to extract a manuscript from Einstein. Despite the large extent of his publications, he wrote only two books: a popularized exposition of relativity theory[32] and a reworking of four lectures on the same subject at an academic level.[33] He never wrote a textbook, or even an authoritative monograph on a particular field of research. His interest was not in writing about what was already common knowledge, but in pursuing what he himself did not yet know.

In September 1907, while his *Habilitation* application was still pending at the University of Bern, Einstein had agreed to write a comprehensive article on the theory of relativity. What began as a commissioned job turned into a stroke of genius, perhaps his greatest. Johannes Stark, Einstein's senior by only five years but already a professor (albeit a provisional one) in Greifswald, had founded the *Jahrbuch für Radioaktivität und Elektronik* (*Yearbook of Radioactivity and Electronics*) in 1904. It was in this annual that Einstein's article was to appear. Einstein accepted "gladly" but asked Stark to help him with the literature, as he was unable "to inform myself on everything that has appeared on this subject, as the library is closed during my free time. Apart from my own papers, I am acquainted only with a paper by H. A. Lorentz (1904), one by E. Cohn, one by Mosengeil, and two by Planck. Other theoretical papers concerning the subject have not come to my notice."[34] Presumably Einstein was listing only offprints sent to him, because he must have read a good deal more. Nor can access to journals have been quite as difficult as Einstein made out. The list therefore is evidence not so much of the state of his knowledge as of his selective treatment of literature.

Einstein had only two months to write his *Jahrbuch* article.[35] After one month, he informed Stark that he had "so arranged the work that anyone could find his way with comparative ease into the relativity theory and its applications so far."[36] He had devoted much care "to the clarification of the assumptions used," and he was anxious, by means of "clarity and simplicity of the mathematical development," to make "the work more attractive."[37] In this, despite the pressure of time, he succeeded superbly.

The report, *On the Relativity Principle and the Conclusions Drawn from It*,[38] gave an excellent overview of the foundations and range of the principle for electrodynamics, mechanics, and thermodynamics. The advances achieved by the theory as formulated in this report were later most strikingly summed up by Einstein in his Nobel lecture:

> It reconciled mechanics and electrodynamics. It reduced the number of logically independent hypotheses of the last-named. It enforced a cognition-theoretical clarification of the basic concepts. It unified the impulse theorem and the energy theorem; it proved the essential unity of mass and energy.[39]

However, what would not fit into the relativity principle was the paradigm of all physics, Newton's theory of gravity. There were also a few other problems, predominantly having to do with cognition theory and aesthetics.

We do not know at what point between 1905 and 1907 Einstein first felt that his relativity theory as formulated in *On the Electrodynamics of Moving Bodies* could not be the final word. One problem probably gave rise to further thought fairly soon: the theory was limited to "inertial systems," that is, to referential systems in uniform nonaccelerated motion relative to one another. This restriction, as he later observed, "was really more difficult to tolerate than the privileged status of one single state of motion, as was the case in the theory of a resting luminiferous ether, because that theory at least proposed a real reason for that privileged status, namely the luminiferous ether."[40] With the discarding of the ether, the demand for a broadening of the theory seemed to Einstein the most natural thing in the world. But what he called his "need of generalization"[41] was not enough; there was another stumbling block: "It was only when I endeavored to present gravitation in the framework of this theory that I realized that the special theory of relativity was only the first step in a necessary development."[42]

Again, we do not know precisely when these endeavors began, or how intensive they were, but we do know when the breakthrough came—in October or November 1907, when the first half of his *Jahrbuch* article was ready and the second remained to be written.[43] On November 1 he was not yet sure if he would deal with gravitation at all; otherwise, he would surely have mentioned this to his editor. When the article was finished on December 1, the first four parts, essentially an overview, were followed by a fifth part which, on nine pages, contained entirely new material.

Under the heading "The Relativity Principle and Gravitation," Einstein makes a connection between the generalization to any referential system on the one hand and the relativistic treatment of gravity on the other. What is developed here, on the basis of a convergence of the two problem areas—a convergence as surprising as it is enigmatic—is not an axiomatic theory but only the outline of the beginning of a lengthy development which, eight years later, would result in the "general theory of relativity." But the outline already is bold, even revolutionary. If the relativity theory of 1905 was a revolution, then this revolution too "devours its children." The principle of the constancy of the velocity of light, only just established, is now "modified" in the sense that the velocity and direction of propagation of light are influenced by a gravitational field. While his colleagues were still struggling to assimilate the new ideas he had put forward in 1905, Einstein was already marching on. That was his destiny and his greatness.

Christmas Eve 1907 was not festive in the Einstein household. He was of Jewish origin, and her Serbian Orthodox Church did not observe the birth of Christ until January. So he used the break to write letters. "During October and November I was very busy with an article on the relativity principle, partly reporting and partly dealing with new matters," he informed Conrad Habicht. "Now I am concerned with another relativity-theory reflection on the law of gravitation, by which I hope to explain the still unexplained secular changes in the perihelion distance of Mercury."[44] He had thus focused on what, despite its minuteness, was a serious stumbling block in Newton's theory of gravitation. In a short postscript Einstein added: ". . . but so far it doesn't seem to work out." It would take eight laborious years before it did work out—before, on November 15, 1915, Einstein had the general theory of relativity in front of him.

PART III

THE NEW COPERNICUS

CHAPTER THIRTEEN

From "Bad Joke" to "Herr Professor"

"I MUST CONFESS TO YOU that I was amazed to read that you have to sit in an office for eight hours a day! But history is full of bad jokes."[1] This is how Johann Jakob Laub, Wilhelm Wien's collaborator in Würzburg, reacted to the news that the "Esteemed Herr Doktor," from whom, at the beginning of 1908, he had requested an offprint and for whose sake he was willing to come to Bern for three months, was to be found not at the university but at the Patent Office.

Much the same may have been felt by Arnold Sommerfeld, since 1906 *Ordinarius*, or full professor, of theoretical physics in Munich. At the beginning of January he had written a letter which had given Einstein exceptional pleasure: "No other physicist has yet approached me with such frankness and benevolence."[2] At the same time, Einstein felt it necessary to tone down Sommerfeld's compliments: "Because of my lucky idea of introducing the relativity principle into physics you (and others) now greatly overrate my scientific abilities, so much so that I feel quite shaken."

Those physicists who had already seen the light about his work would probably have felt "quite shaken" at the thought that Albert Einstein was still employed at the Patent Office. He had by then earned the kind of reputation that made the offer of a university chair only a matter of time.

Einstein himself had been thinking of a change in his career ever since the beginning of the year, and he had probably discussed the matter with Jakob Ehrat, who had come from Zurich over Christmas and who was himself thinking of becoming a Patent Office "expert."[3] To Marcel Grossmann he voiced his "sincere wish to be able to continue my private scientific activities under less unfavorable circumstances."[4] The efforts of the past three years—twenty-five publications, culminating in the tour de force of the *Jahrbuch* article, along with the growing volume of his scientific correspondence—would have represented not only an intellectual but also a physical achievement even if they had been his main occupation. How much more exhausting must all that have been on top of his work at the Patent Office.

Marcel Grossmann, Einstein's "lifesaver," had become a professor at the Zurich Polytechnic in 1907, after a career in the school service, and Einstein once more turned to him for advice. Einstein was aiming not at an academic post, but, with almost touching modesty, at a teaching post at the Technical College in Winterthur,[5] the institution where he had worked as a stand-in teacher for two months in the spring of 1901. "I now ask you: how does one go about this? Could I possibly call on somebody and verbally convince him of the great worth of my admirable person as a teacher and citizen? Wouldn't I make a bad impression on him (no Swiss-German dialect, my Semitic appearance, etc.)? Would there be any point in my stressing my scientific papers on that occasion?"[6]

We do not know if Einstein actually applied for a post in Winterthur, but he learned—probably from Grossmann—of a new vacancy at the gymnasium in Zurich. He first corresponded with the principal about the salary, found it acceptable, and on January 20 decided to apply: "With reference to the advertisement of a teaching post for mathematics and descriptive geometry, I hereby apply for that position. I would add that I would also be prepared to teach physics."[7] He enclosed his dissertation as well as "the rest of my scientific papers published hitherto." There were twenty-one applicants for the job; three of them were short-listed. Einstein was not among the three, and as the records contain no assessment of him, we must assume that his application was not even considered.

Meanwhile, though, Einstein had another iron in the fire—his *Habilitation.* He had abandoned his earlier opposition to the formalities and had now prepared the prescribed thesis. On February 11,

1908, he informed Professor Gruber of his change of mind: "My conversation with you in the City Library, as well as the advice of several friends, has now induced me to change my intention for the second time and after all try my luck with *Habilitation* at the University of Bern. To this end I have submitted a *Habilitation* thesis to the Dean."[8] And now the proverbially slow Bernese were suddenly in a hurry. Some of the professors were probably afraid that their earlier denial of a *Habilitation* to the only physicist in Bern who was known beyond the borders of Switzerland might cast a bad light on the department rather than on the candidate. The procedure was now speedily set in motion. The minutes of a faculty meeting on February 2 report that Einstein had submitted a thesis entitled *Conclusions from the Energy Distribution Theorem of Black Body Radiation, Concerning the Constitution of the Radiation.* "The thesis has been circulated among the faculty members. Herr Prof. Forster proposes in writing that the *Habilitation* thesis be accepted and Herr Einstein invited for a trial lecture. The faculty makes this proposal a resolution."[9] On Thursday, February 28, the trial lecture, *On the Limits of Validity of Classical Thermodynamics,* was given, along with a colloquium. The faculty thereupon unanimously recommended that the candidate be made a privatdozent for theoretical physics, and the following day the director of education of the canton of Bern issued the appropriate document. By the same mail as the document, Einstein's thesis was returned to him at Aegertenstrasse. He did not keep it. To judge by its title, it must have been a preliminary study for work to be published the following year.[10]

Even before the formal proceedings, Einstein had let it be understood by Professor Gruner—to whose lectures his own colloquia were to represent a supplement—that he was anxious that "the time that I have to spend on lecturing . . . will be optimally used, i.e. I would like to run a class adapted to the degree of knowledge and the interests of the students."[11] While the professor and the privatdozent very quickly agreed on the delimitation between the main lectures and Einstein's class—this conversation taking place at the tourist café Chalet Bovet—the "knowledge and interests" of the few physics students in Bern were quite another question. In his first class in the summer semester of

1908 Einstein had an audience of exactly three[12]—and these were not students but his loyal friends from the Patent Office, Michele Besso and Heinrich Schenk; and Lucien Chavan from the Postal and Telegraph Administration. On Tuesday and Saturday of each week, they had to get up early and climb up the Grosse Schanze, where Einstein would begin the class (on "molecular theory of heat") at seven in the morning so that he and his colleagues could start work at the Patent Office at eight. Chavan at least, as is shown by his meticulous notes, written in French,[13] did not miss a single class.

In the winter semester, Einstein moved his class to the evening, from six to seven o'clock, his topic now being "radiation theory." At this more civilized time, the three friends were joined by a genuine student, Max Stern from Lithuania; but he was not a physicist but a student of insurance mathematics with an interest in science. When, in the 1909 summer semester, the three friends no longer chose to pursue their education under Einstein, Max Stern remained his only student. Einstein thereupon canceled the class.

The circle of friends of the newly appointed privatdozent was soon joined by Johann Jakob Laub, who arrived in Bern in March 1908 and would henceforward pride himself on being Einstein's "first collaborator." Born in Austria, Laub had studied in Göttingen and taken his doctorate in Würzburg under Wien, who directed him toward relativity theory. By 1907 he had published two papers in this field in *Annalen* and thus belonged to what Planck called the "modest handful of representatives of the relativity principle" who had actually published their work. Einstein must have been pleased to hear from Laub, his junior by three years, how intensively his work was being discussed in Göttingen, Berlin, Munich, and Würzburg.

In Einstein's free time he and Laub concerned themselves with the problematical definition of the concept of force within the framework of relativity theory. They published a joint paper in 1908, *Electromagnetic Fundamental Equations for Moving Bodies*, but had to publish a correction to it before the year was out.[14] They next tried to solve the question of what forces act on bodies at rest in an electromagnetic field. This they did by reducing the problem to the movement of elementary charges, which they were able to calculate by their methods. Here, too, though, an amendment became necessary a year later.[15] "The way we bungled then," Einstein later said of their joint papers, "is incredible. *Si tacuissemus* . . . And yet a correction is something exalting. And a patched-up job is still better than one full of holes."[16] But in this case even the corrections were of no use; ten years later Einstein would write to a student in Zurich who had taken up similar problems in his thesis: "It has long been known that the values which I then derived with Laub are wrong."[17]

After Laub had left Bern in mid-May, he thanked his "dear friend Einstein" for the "great time I spent at your house. I am now back in Würzburg and thinking back with pleasure on our splendid discussions in Bern."[18] The discussions were continued by letter, and Laub, who soon joined Lenard in Heidelberg and traveled a good deal, became an important source of information for Einstein on the academic scene in Germany.

Their two joint papers are the first in which Einstein figured as a coauthor, and they were the last of his publications on problems of relativity for three years. He now turned to radiation theory and, to a lesser degree, to the development of an instrument for measuring very small electrical charges. In this context we will see him as a handyman, inventor, and experimenter.

Einstein had thought up the *Maschinchen*—the "little machine"—as he lovingly called it, in the course of his work in 1907, when he extended his analysis of molecular-kinetic fluctuation phenomena within thermodynamics to the fluctuations of minute amounts of electrical charge.[19] This electrostatic analogue to the Brownian movement should be observable in a condenser as a voltage fluctuation—but this would require an accuracy of within less than one-thousandth of a volt, whereas the best available electrometers were sensitive only to a few thousandths of a volt. An invention was called for.

To increase the accuracy of measurement Einstein devised an influence machine of variable condensers connected in series, to be charged at low voltage and high capacity and discharged at high voltage and low capacity into its successor in the condenser bank.[20] It so happened that Conrad Habicht's younger brother Paul had just started a small instrument-making business in Schaffhausen. Paul, as Einstein wrote to Conrad Habicht, would there "construct the little electrostatic machine in the shape I have now given him. I am very curious to see how much can be achieved—I am rather hopeful."[21] Einstein even tried to act as a patent expert in his own cause, but evidently without success: "I have dropped the patent, mainly because of the lack of interest shown by manufacturers."[22] This, however, did not lessen his enthusiasm. More important than a patent was the connection between the "little machine" and two of his great treatises. His paper on the Brownian movement had now yielded a method not only for an experimental demonstration of its electrostatic analogue, but also for measuring radioactivity and, in consequence, for testing the equivalence of mass and energy that followed from relativity theory.

Einstein's enthusiasm was such that he actually suggested to Sommerfeld in Munich that he should charge one of his assistants with this experimental task "which is close to my heart. It concerns a little electrostatic machine for measuring purposes. . . . If you, or he, would be interested, I would be glad to let you have a more detailed account of the business."[23] In his publication, which followed soon after, Einstein pointed out that "increasing the sensitivity of electrostatic measuring methods is of importance to the investigation of radioactivity" and expressed the hope "that some physicist may show an interest in this matter. I would be happy to pass on to him my further reflections on this subject."[24]

One such physicist soon reported to him from Fribourg, Switzerland, only some twenty miles away. This was Albert Gockel, professor of cosmic physics at Fribourg University, whose work evidently included measuring minute amounts of electricity in the atmosphere. No sooner had Laub left Bern than Einstein visited Gockel in his laboratory on May 16.[25] This was not to be his only trip to Fribourg, since Gockel intended to build a "little machine" himself, and the two men needed to discuss a lot of tricky questions, such as whether the contacts should be of mercury or gold.

Things were moving ahead in Bern too. "I have found a skillful mechanic here," Einstein wrote to Laub, "who is now building the little electrostatic machine according to my data. It's supposed to be ready in the next few days. I want to see for myself what it can do, and what truth there is in stories about the cussedness of contacts."[26] As there was no electrometer in the whole of Bern, Einstein built his own, together with a voltage battery. "You wouldn't be able to resist smiling," he wrote to Laub, "if you could see my home-botched glory. . . . At any rate, my results so far are encouraging."[27] A few weeks later he had a second little machine built for him, so that, as he wrote to Gockel, "I can now call a kind of 'laboratory' my own."[28] The experiments were conducted either in the physics rooms of the two gymnasium schools in Bern or at Einstein's home on Aegertenstrasse, where one or both Habichts would visit for working weekends.

Einstein had every right to be proud of having made possible "a previously unattained sensitivity in the measuring of electrical quantities" and to hope that soon "there would be no obstacle to an experimental verification of the validity limit of electrostatics as demanded by the molecular theory."[29] But he also had occasion to complain of the disadvantages of his peripheral existence as a researcher: "It is no small matter when one is so short of time and when one is dependent on one's own (narrow) purse."[30] Fortunately, the little machine was in good hands with Paul Habicht; soon it was being manufactured on a small industrial scale, and it would bring much joy to Professor Einstein.

There was also some tinkering going on in the Einstein household. One of Hans Albert's lasting memories is of his father building a little cable car out of matchboxes. "This was one of the nicest toys I had at the time and it worked," he said later. "Out of just a little string and match-boxes and so on, he could make the most beautiful things."[31] But there was not only play. It was probably due to his mother's teaching that the four-year-old scribbled his pet name, Buio,[32] on letters, such as those to Solovine. Hans Albert recalls that education at home was conceived as a "contrast program" to school. Thus Einstein tried to interest his young son in music, though he met with no response. If the boy misbehaved, Einstein discharged his fatherly duties in what was then the customary manner: "Every once in a while . . . he beat me up, just like anyone would do."[33]

At times Einstein's sister, Maja, would stay with them.[34] After graduating from the Aarau teacher training college, she had gone to Berlin in 1905 as one of the pioneers of women's university education and had studied Romance languages—albeit as a "guest student," since regular university enrollment was not granted to women in Prussia until 1908. Maja had come to Bern for the 1907 summer semester to work at the university on a doctoral thesis about an Old French manuscript. The following year she went to Paris for a few months for the sake of the libraries there, and shortly before Christmas 1909 she received her doctorate from Bern University. The Fräulein Doktor remained in Bern longer than her brother, in 1910 marrying Paul Winteler—one of the sons of Albert's "second parents"—who had studied law in Bern. With her husband, she moved to Lucerne in 1911.

By the summer of 1908 Einstein had earned a vacation. In August he went with his family to the Bernese Oberland—for one week to Mürren (which was not yet fashionable then) and for one week to Isenfluh. Toward the end of August Max Planck, too, was vacationing in the neighborhood but to Einstein's disappointment did not, as he had promised a year earlier, seek him out. He wrote to Einstein about the difficulty "of suggesting a definite place and time for a meeting, as the weather is playing too vital a part in all our plans. . . . Wouldn't it be more practical if we met at the *Naturforscher* meeting in Cologne? I will certainly be there. There would be more time and the right mood for scientific discussions then."[35]

In fact, Einstein had planned to go to Cologne and had made arrangements with colleagues, such as Stark, to meet at the event, "which, if at all possible, I will attend."[36] But (as we have seen) this did not come off: "To my great regret it was not possible for me to come to Cologne. It was absolutely necessary that I should use my short break from the Office for [relaxation]."[37] Had he gone to Cologne, he would have witnessed a magnificent piece of propaganda for the theory of relativity. Whether he would have approved is another matter.

The meeting of the mathematical-physical section opened in the afternoon of September 21, 1908, with rhetorical fanfares: "Gentlemen!

1. The young Albert Einstein, with his sister, Maja, Munich, 1894.

2. Cantonal school class photograph, Aarau, Switzerland, 1896 (Einstein is seated at left).

3. Einstein, Conrad Habicht, and Maurice Solovine comprised the "Akademie Olympia."

4. As a student at the Zurich Polytechnic, 1898.

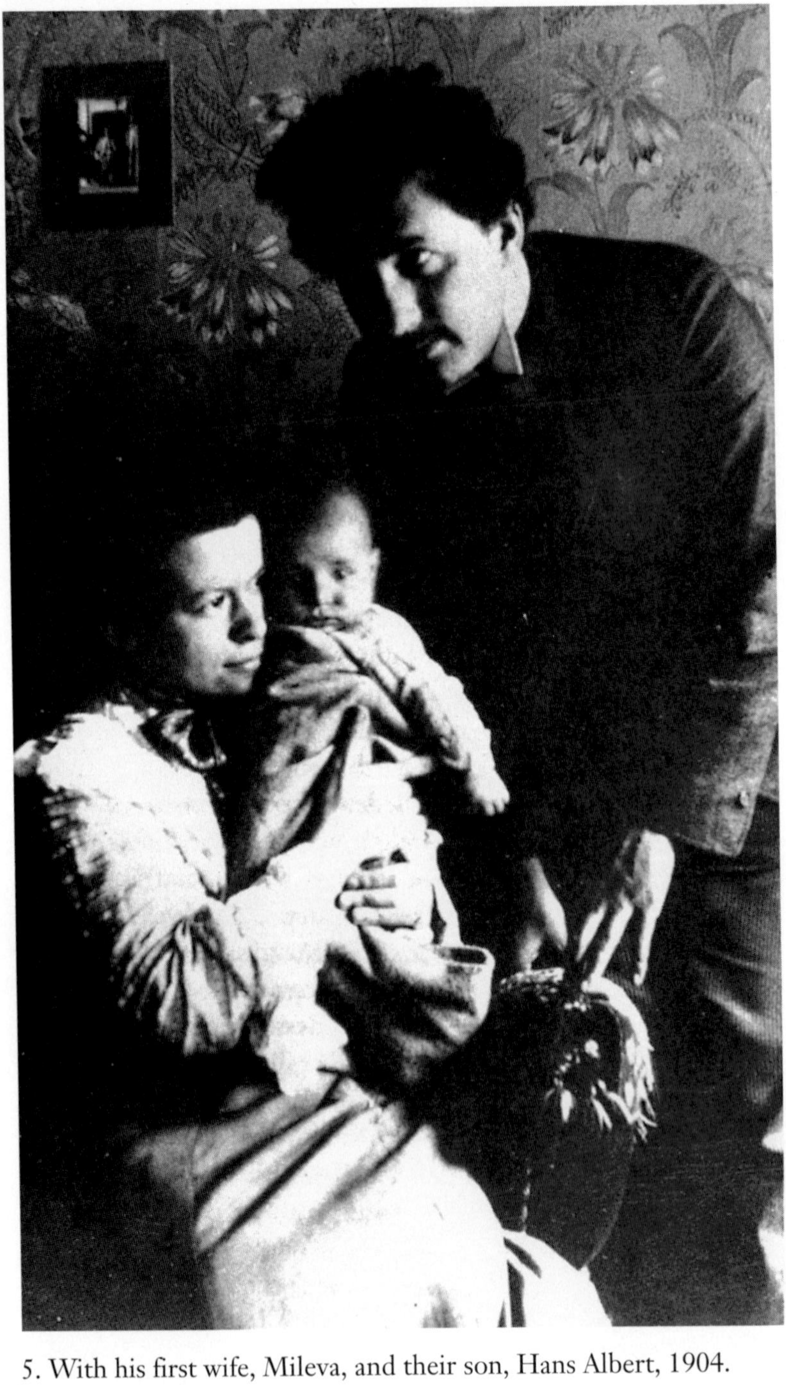

5. With his first wife, Mileva, and their son, Hans Albert, 1904.

6. At the Swiss Patent Office, Bern, about 1906.

7. First Solvay Conference, Brussels, 1911 (Einstein is standing, second from right).

8. Fifth Solvay Conference, Brussels, 1927 (Einstein is fifth from right, front row).

9. Einstein in 1912, the year he was appointed professor of theoretical physics at the Swiss Federal Technical University, Zurich.

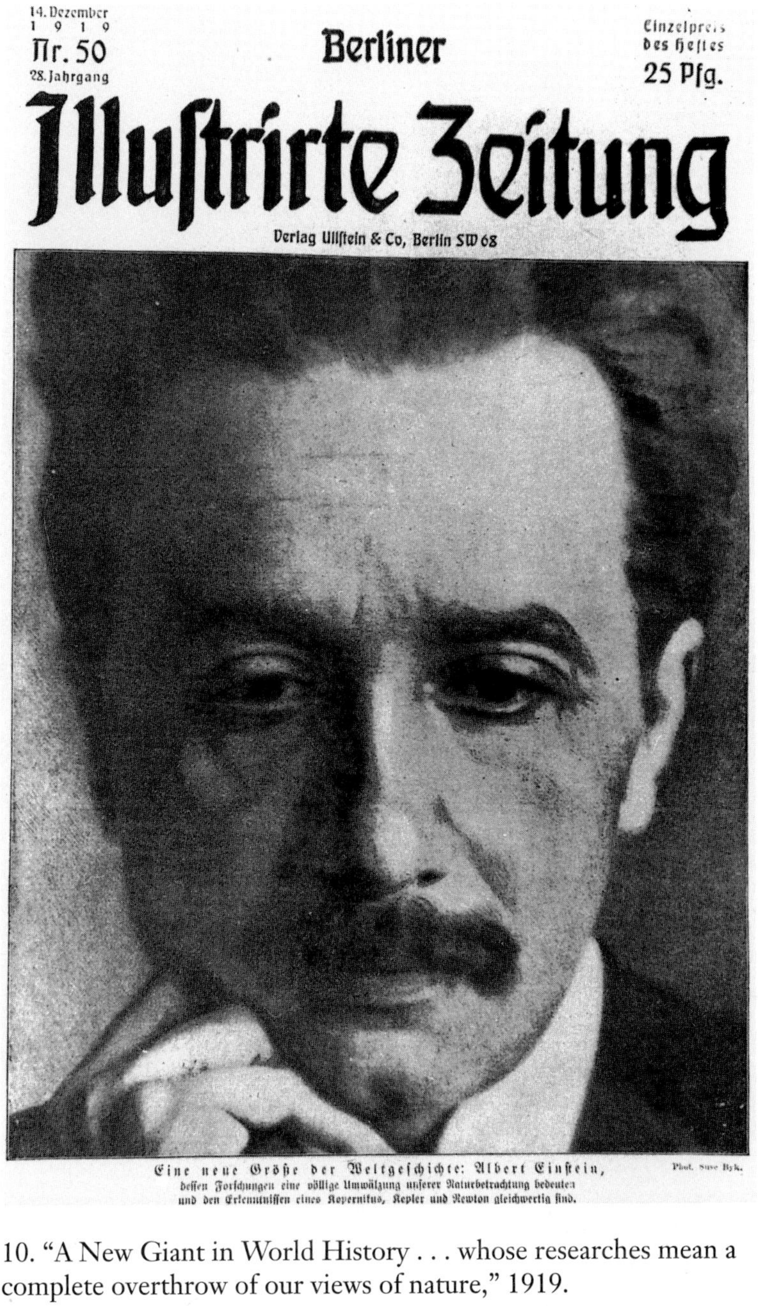
14. Dezember
1 9 1 9
Nr. 50
28. Jahrgang

Berliner

Einzelpreis
des Heftes
25 Pfg.

Illustrirte Zeitung

Verlag Ullstein & Co, Berlin SW 68

Eine neue Größe der Weltgeschichte: Albert Einstein, dessen Forschungen eine völlige Umwälzung unserer Naturbetrachtung bedeuten und den Erkenntnissen eines Kopernikus, Kepler und Newton gleichwertig sind.

Phot. Suse Byk.

10. "A New Giant in World History . . . whose researches mean a complete overthrow of our views of nature," 1919.

11. Lecturing at the Collège de France, Paris, 1922.

12. Einstein about 1920, a time of intense public interest in the theory of relativity.

13. With Hendrik Antoon Lorentz, who was Einstein's scientific father figure.

14. With his second wife, Elsa, in Berlin, 1921.

The concepts of space and time which I wish to present to you have sprung from an experimental physical soil. Therein lies their strength. Their tendency is a radical one. Henceforward space by itself and time by itself will totally decline into shadows, and only a kind of union of the two shall preserve independence."[38] The man who so bombastically prepared his audience for the epoch-making significance of his exposition was Hermann Minkowski, the famous mathematician whose lectures at the Zurich Polytechnic young Einstein had avoided but whose class on analytical mechanics he had appreciated as the only theoretical lecture there. In 1902 Minkowski had been invited to Göttingen to a chair specially created for him at the insistence of David Hilbert. There Minkowski subscribed to the belief of the two grand masters of Göttingen mathematics—Felix Klein and Hilbert—that the fundamentals of physics were really too difficult for physicists and should be handled by mathematicians.

Such a mind could not long overlook Einstein's paper *On the Electrodynamics of Moving Bodies*. Jointly with Hilbert, Minkowski organized a few seminars on new developments in electrodynamics in 1907–1908, which were followed by a lecture in Göttingen and a few publications of high quality. As for the paper of his erstwhile student, he observed to his assistant Max Born: "I really wouldn't have thought Einstein capable of that!"[39] Max Born even reported that Minkowski had told him of his "great shock when Einstein published his paper in which the equivalence of different local times of observers moving relative to each other was pronounced; for he had reached the same conclusions independently but did not publish them because he wished first to work out the mathematical structure in all its splendor."[40] There is no evidence at all of "same conclusions"; one can excuse the statement only by Minkowski's tendency to confuse physics with mathematics. That he should have discovered anything like the theory of relativity independent of and before Einstein is very doubtful.

But there could be no doubt about the glory of the mathematical structure which Minkowksi had found in and extracted from the theory of relativity and which he now presented to his audience in a flood of words. Minkowski presented not only a new and exceedingly elegant form of Einstein's theory, but simultaneously a highly stylized vocabulary which would earn relativity theory special notice among physicists and more generally. He spoke, for example of a multiplicity of "world points" that made up the "world," within which the "eternal life of the substantial point" formed a "world line." Einstein's principle of relativity was rededicated as the "postulate of the absolute world." For the first time in the context of relativity theory, he stated that "the relations under review only unfold their inner being of great simplicity in four dimensions," and the mystical thrill of the "fourth dimension" was heightened by the fact that in this concept, time figures as an imaginary coordinate.

Minkowski concluded his address with an apodictically formulated balance sheet: "The absolute validity of the world postulate is, as I would like to believe, the true core of an electromagnetic world picture, first hit by Lorentz, further carved out by Einstein, and now fully exposed."[41] Even if not all his listeners were able to follow what came between his programmatic introduction and his definitive conclusion, they probably gained the impression that here a new physics had been born, cerebrally, out of mathematics, without recourse to experiments and relying solely, as Minkowksi assured them, on "prestabilized harmony between pure mathematics and physics."

Only a few of Minkowski's listeners would have realized that relativity theory was here running a risk of being deprived of its physical foundations, but among those few was Jakob Laub. Having studied the theory of relativity at its source, he acquainted himself, on his return to Würzburg, with Minkowski's work and was astonished at the enthusiasm it was arousing, especially on the part of the mathematician Matthias Cantor. If Einstein's work were not available, he wrote to Bern, "we would find ourselves with Minkowski's transformation equations for time (as far as the physical interpretation is concerned) in the same situation, at best, as with Lorentz's 'local time.' "[42]

Minkowski's torrents of words and his concept of four-dimensional space-time did not, from a physical point of view, offer anything new, compared with Einstein's theory or indeed even with Lorentz's. Treating time as an imaginary magnitude is merely a mathematical device which illustrates certain analogies between the three spatial coordinates on the one hand and time on the other, and which makes it possible for the Lorentz transformations of relativity theory to be represented in a four-dimensional Euclidian space and to be described with the same mathematics developed for rotations in a three-dimensional space. Even Minkowski's dramatically emphasized "union" of space and time should be "relativized" inasmuch as the transformation equations do not allow for confusion between the two: a time coordinate always remains recognizable as such.

When, a few months later, Einstein read Minkowski's lecture in the February issue of *Physikalische Zeitschrift* he was not impressed by it, and he regarded the four-dimensional formulation as "superfluous erudition."[43] He is credibly reported to have said with a sigh: "Since the mathematicians pounced on the relativity theory I no longer understand it myself."[44] Later, he poked fun at the mystical frisson of the "fourth dimension" as a "sensation not unlike that of a ghost in the theater. And yet, no statement can be more banal than that our familiar world is a four-dimensional time-space continuum."[45]

If Minkowski's four-dimensional presentation had been only a matter of elegance, it could have been left to the tailors and shoemakers—a dictum of Boltzmann's often quoted by Einstein. But it combined elegance and usefulness, as Einstein was to discover for himself during his efforts, in 1912, to generalize his relativity theory. Now he no longer saw Minkowski's formulations as "superfluous erudition" but paid tribute to the "important idea, without which the . . . general theory of relativity might have remained stuck in its diapers."[46] Einstein was not able to express his gratitude to Minkowski in person; four months after the Cologne lecture, at the age of forty-four, Hermann Minkowski died of appendicitis.

After the Cologne meeting the question was being asked more and more how much longer the "bad joke" in Bern could continue. Einstein's former fellow student Kollros recalls how even well before that meeting—in April 1908—at the International Physicists' Congress in Rome, he was strolling through the gardens of the Villa d'Este with Lorentz and Minkowski: "Both of them acknowledged the great importance of the ideas introduced by the twenty-six-year-old scientist."[47] Minkowski's lecture in Cologne made it even clearer that Einstein belonged in a university rather than in a patent office. The difficulty was that there was then no vacancy for a physicist in his specialty.

In the first decade of the twentieth century theoretical physics had still not quite "come of age" in the academic world.[48] At most universities it was represented by "extraordinary"—that is, nontenured—professors, whose status was below that of a "full" professor of physics; they came under the full professor administratively and were paid only modestly. Few universities could afford, in addition to the full professor (who in that case was called professor of experimental physics), a second, equal full professor, the professor of theoretical physics. Berlin and Göttingen did have such an arrangement, though, as did Munich when Arnold Sommerfeld was appointed in 1906. The young privatdozent from Bern would have been an ideal choice for one of the nontenured professorships of theoretical physics, but there were barely two dozen such posts in the German-speaking countries, and all of these were filled. The only exception was the University of Zurich, but it had no actual vacancy; the establishment of a nontenured professorship remained to be wrested from the authorities. Only after involved arguments was the post created and Einstein invited to take it.

Professor Alfred Kleiner in Zurich had been trying for a number of years to lighten his teaching load by having a theoretical physicist appointed. He had met with opposition from the cantonal education authorities; but when Paul Gruner was invited to Bern in 1906 as a nontenured professor, Kleiner rekindled his hope that even the thrifty authorities in Zurich might now see the need for a second professor of physics.

What is more, Kleiner already had a candidate for the post—his former assistant Friedrich Adler, who in the meantime had found an interesting job at the German Museum in Munich. Kleiner persuaded Adler to return to Zurich and saw to it that Adler obtained his *Habilitation* right away, in December 1906.[49] The post was slow to be established, even though the Social Democrat members of the Education Council were eager to bring Adler to the university as its first "comrade." But when, in the winter semester of 1908–1909, Kleiner was elected rector of the university, it looked as if the new professorship would soon be authorized.

Friedrich Adler, a man of many talents, had always vacillated between becoming a physicist, a philosopher, or a politician. His father was urging him toward a professorship in physics, while his own inclinations were in the other two directions. Adler believed that if he was given the professorship he would be able to follow his philosophical interests rather than physics. It is likely that Kleiner gradually came to the conclusion that Adler might not turn out to be a fully committed physicist and would, therefore, not relieve him of as much of his workload as he had hoped. At any rate, in a tortuous conversation on June 19, 1908, recorded by Adler from memory,[50] the professor informed Adler that he would probably not be heading the list of candidates.

It is also possible that influential people may have suggested to Kleiner that he should consider another scientist, one whose publications had lately aroused a lot of attention. Adler named him in a letter written to his father that same day:

> I forgot to say who will most likely get the professorship—a man who on principle and from the point of view of the people involved, should certainly get it rather than myself, and if he gets it I will, apart from any awkwardness, be very pleased. This is a man by [the] name of Einstein, who was a student at the same time as I and with whom I attended several lectures. . . . For the people involved the situation is of course that, on the one hand, they have a bad conscience about the way they treated him in the past, and on the other it is felt to be a scandal, not only here but also in Germany, that a man like that should sit in the Patent Office. . . . Objectively therefore, if the business goes the way I expect, it is a fine thing that this man has asserted himself despite all difficulties, and it strengthens one's belief that one can do what one wants to do.[51]

A few days later Einstein's class on "molecular theory of heat" had a fourth person in the audience, his former supervisor Professor

Kleiner. Kleiner had come "to inspect the beast," as Einstein wrote to Laub: "That day I really did not lecture wonderfully—partly because I had not prepared myself well and partly because the situation of being inspected was getting on my nerves."[52] When Kleiner made some critical remarks about Einstein's teaching, the candidate agreed and added that "after all, they need not have invited me."[53] The professor, irritated, no longer had that intention anyway; and Friedrich Adler reported to his father that, according to Kleiner, Einstein was "a long way from being a teacher, that he holds monologues. The situation therefore has changed, and the Einstein business is closed."[54] When Einstein heard about it he reacted stoically: "The business with the professorship has fallen through. There are quite enough schoolmasters even without me."[55]

His stoicism, however, ended when he learned from Laub that Kleiner's impression of the "visitation" had reached Würzburg through the academic grapevine: "I now seriously reproached Kleiner in a letter for spreading unfavorable rumors about me, thereby making my difficult position a definitive one. Because such a rumor must kill any hope of getting into university teaching."[56] Kleiner meanwhile, as we know from Friedrich Adler, had come around to the view "that he must propose Einstein in the first place, because he could only make a proposal that would get through, and everyone was astonished that Einstein still had no position."[57] Kleiner therefore informed Einstein that he would gladly invite him to Zurich provided Einstein could first convince him that he had some talent as a teacher.

On Einstein's suggestion, therefore, a lecture was arranged at the Physikalische Gesellschaft (Physics Society) in Zurich, and in mid-February 1909 Einstein set out for this "exam." He was glad that his colleague Ehrat and Ehrat's mother could put him up at their home: "One is far less aware of one's life-and-death situation than if one has to blunder about among strangers."[58] This time the candidate met with approval. This is how Einstein summed it up for Laub: "I was really lucky. Totally against my usual habit I lectured well."[59] After his return to Bern, he wrote to Ehrat: "There is now a real prospect that we will many more times sit comfortably together, because the stern Kleiner, on whom I called on Friday, expressed himself very benevolently about the result of my 'exam' and hinted that certain things would probably follow soon."[60] Kleiner immediately requested a curriculum vitae from Einstein and inquired if he would be able to start the following semester. Einstein said that he would, "without having officially informed myself on this point. But I know that a former colleague left the Office one month after giving notice."[61] However, the appointment did not come through in time for the 1909 summer semester, because overcoming all kinds of opposition within the Zurich bureaucracy took some time.

The "stern Kleiner" immediately composed his assessment of Einstein. Needless to say, he emphasized that his candidate was "one of the most important theoretical physicists" of the day "since his treatise on the relativity principle [was] fairly universally acknowledged as that." More original was Kleiner's characterization of Einstein's publications, in which he saw "an extraordinary acuteness in the conception and pursuit of ideas, and a profundity aiming at the elemental. Also remarkable is the clarity and precision of his style; in many respects he has created a special language, which in a man of thirty is a clear sign of independence and maturity."[62] Only on Einstein's ability as a teacher was Kleiner reluctant to pass a final judgment, though he expressed the belief "that Dr. Einstein will prove his worth also as a teacher, because he is too intelligent and too conscientious not to be open to advice whenever necessary."

This recommendation was received by the faculty commission set up for appointment of the new *Extraordinarius*, the nontenured professor. The commission had nine names before it. Friedrich Adler was not even considered. The commission inclined toward Walter Ritz, a privatdozent in Göttingen, "because he is Swiss and, in the judgment of our colleague Kleiner, exhibits 'an exceptional talent, bordering on genius.' "[63] Ritz, however, had to be excluded from consideration because he was incurably ill with tuberculosis.[64] Thus everything now pointed to Einstein.

On March 4, 1909, a secret ballot among the full professors of Science Section II of the Department of Philosophy produced ten votes in favor of Einstein and one abstention. The result was immediately passed on for confirmation to the director of education of the canton of Zurich, along with some remarks, evidently considered useful, on Einstein's Jewish origin. In these, the professors revealed the same discreet anti-Semitism,[65] then typical of academic circles, which they were evidently trying to counteract. Kleiner, in his judgment, had emphasized that "about the personal character of Dr. Einstein nothing but the best reports are made by all who know him."[66] He himself had known Einstein socially for six years and was "unhesitatingly prepared to have him as a colleague in my immediate proximity." As a kind of supplement to this character reference, the dean, Professor Stoll, an anthropologist, had this to say:

> The above remarks by our colleague Kleiner, based as they are on many years of personal contact, were the more valuable to the commission, and indeed to the department as a whole, as Herr Dr. Einstein is an Israelite, and as the Israelites are credited among scholars with a variety of disagreeable character traits, such as importunateness, impertinence, a shopkeeper's mind in their understanding of their academic position, etc., and in numerous cases with some justification. On the other hand, it may be said that among the Israelites, too, there are men without even a trace of these unpleasant characteristics and that it would therefore not be appropriate to disqualify a man merely because he happens to be a Jew. After all, even among non-Jewish scientists there are occasionally people who, with regard to a mercantile understanding of their academic profession, display attitudes which one is otherwise accustomed to regard as specifically "Jewish."
>
> Neither the commission, nor the department as a whole, therefore thought it compatible with its dignity to write "anti-Semitism" as a principle on its banner, and the information which our colleague Herr Kleiner was able to furnish on Herr Dr. Einstein has put our minds completely at rest.[67]

Despite the extremely carefully formulated proposal, there was massive opposition to Einstein in the Directorate of Education—not, however, because of anti-Semitism but on political grounds. The Social Democrats, who held the Directorate of Education in the cantonal administration, were bitterly disappointed that anyone other than their comrade Friedrich Adler should have been proposed by the university. Adler, for his part, was in the awkward position of being favored by his political friends in the administration, but not by the professor or the faculty. His interests—which (as we have seen) had oscillated for some time between mechanics, Mach, and Marx—were gradually turning toward Marx; and he decided to put an end to the affair, which had been going on over a year. He did so with a grand gesture, by telling everyone that Albert Einstein from Bern was the better physicist and therefore should get the professorship.

There remained some problems with salary. In line with their custom, the Zurich authorities wanted to give the nontenured professor a salary significantly lower than that of a gymnasium teacher. This would have been about half of what Einstein had been receiving at the Patent Office. Einstein remained inflexible on this point, and he got his way: "My pay is roughly the same as at the Office. Initially they wanted to give me a lot less, but in that case I would have bowed out."[68] On the other hand, he had to undertake to teach six to eight hours a week,[69] although nontenured professors, according to the Zurich education law, had to teach only four to six hours.

On May 7, 1909, by which time the summer semester was in full swing, Einstein was appointed by the Governmental Council of the Canton of Zurich, for the period of six years customary for nontenured professors, with a salary of 4,500 francs, plus "listener" and "examination" fees due under the regulations. He was to assume his position at the beginning of the winter semester on October 15, 1909. "So now I am an official of the guild of whores"[70] was his bitter summing-up of the prolonged and annoying affair. If he had known of all the backstage details, his comment would have been even more vehement.

Before Einstein took up his professorship, he received an honorary degree. This, too, did not come to pass without incident, but it was of an innocent and comical nature, as he himself recalled four decades later: "One day I received a large envelope at the Patent Office, containing an elegant sheet of paper with some words in picturesque print (I even believe in Latin) which seemed to me impersonal and of little interest, and therefore landed at once in the official wastepaper basket."[71] Only later did he learn that this was an invitation to a celebration on July 8 of the 350th anniversary of Calvin's founding of Geneva University and that, on this festive occasion, he was to be awarded an honorary doctorate. When there was no response from Bern, the Genevans got their fellow citizen Louis Chavan to persuade Einstein to travel to Geneva, without Einstein's having any clear idea of what awaited him there:

> So I left on the proper day and already that evening met a few Zurich professors in the restaurant of the hostelry where we were accommodated. . . . Everyone explained in what capacity they had come. When I kept silent, they put the question to me, and I had to confess that I had not the faintest idea. But the others were informed and briefed me. The following day I was to march in the procession, and I only had a straw hat and an informal suit with me. My proposal that I would dodge it was firmly rejected, and the ceremony accordingly took a droll course as far as my participation was concerned.

As it was raining heavily, the festive procession through the narrow streets of the old city to the Cathedral of St. Pierre struck one reporter as "rather too quiet and like a funeral cortege,"[72] but there was no mention of a man in a straw hat among all the dignitaries in their gowns, uniforms, and chains of office, who had arrived from the four corners of the earth. During the subsequent ceremony at Victoria Hall, the richly decorated Geneva concert hall behind the Place Neuve, there was a kind of academic mass baptism, during which no fewer than 110 honorary doctorates were handed out. Among those thus honored were some of the great figures of science, like Marie Curie and Wilhelm Ostwald, but also a certain Ernst Zahn, a dialect poet and station restaurateur at Göschenen—and of course Albert Einstein. He probably owed his first doctorate (honoris causa) to Charles Eugène Guye, professor of physics in Geneva, who had investigated the velocity-dependence of beta rays and in that context had no doubt come across the theory of relativity.

Einstein described the conclusion of the day as follows:

> The celebration ended with the most opulent festive repast that I have attended in my entire life. I said to a Genevan patrician who

> sat next to me: "Do you know what Calvin would have done if he were still here?" When he shook his head and asked for my opinion, I said: "He would have built a huge stake and burnt us all for our sinful gluttony." The man did not say another word and thus ends my recollection of that memorable celebration.

Immediately before this short trip to Geneva, on July 6, 1909, Einstein had handed in his notice at the Patent Office, effective from October 15. Haller placed on record that the Expert II Class had "performed highly valued services. His departure is a loss to the Office. However, Herr Einstein feels that teaching and scientific research are his real profession, and for that reason the Director of the Office made no attempt to bind him to the Office by better financial arrangements."[73]

As for his "real profession," he must have been pleased that about the middle of March a Japanese by the name of Ayao Kuwaki had come to Bern to see him—not, admittedly, from Tokyo but from Berlin, where since 1907 he had been studying under the most famous physicists. He did not want to return to Japan without having met Einstein. The meeting must have been a success, because Kuwaki's journey was taking him through Paris and Einstein commended him to the care of Solovine "in the belief that you are sure to enjoy meeting him."[74] Kuwaki, a future professor at the University of Fukuoka, was Einstein's first contact with Japan, whose physicists would soon make important contributions to relativity theory.

Although Einstein owed his professorship, his honorary degree, and his growing reputation to his relativity theory, he now, without any regrets, left it to other scientists while he himself devoted all his energies to radiation theory—along with his by no means merely playful work with his "little machine." "I am ceaselessly concerned with the constitution of radiation,"[75] he wrote to Laub, who had meanwhile joined Philipp Lenard in Heidelberg. "This quantum question is so enormously important and difficult that everybody should work on it."

Einstein was not alone in this view. In his lecture at the International Congress of Physicists in Rome in April 1908, Lorentz had first suggested that it would not be possible to integrate Planck's radiation formula into the Maxwellian field theory, and that it was therefore "more revolutionary" than Planck would wish to admit. Although for a number of years to come Planck was reluctant to recognize his formula as the beginning of the end of classical physics, and particularly reluctant to follow Einstein's views, his conviction began to crumble in 1908. Nothing, or not much, of this appeared in Planck's publications, but it was reflected in his extensive correspondence with Lorentz and with Wien.

From his position at the Patent Office, Einstein had taken part in this debate in a survey paper, *On the Present State of the Radiation Problem*. As soon as he received his offprints, he sent one to Hendrik A. Lorentz in Leyden, as "the modest result of several years' reflections. I have not succeeded in penetrating to a real understanding of the matter."[76] Unlike Planck, whom Einstein, despite his great respect, criticized for "finding it difficult to enter into the arguments of others" and who "made totally wrong objections to [his] last radiation paper,"[77] Lorentz must have replied most sympathetically. Not only was this the start of an "exceedingly interesting correspondence,"[78] but from then on, albeit at first only in writing, Lorentz became Einstein's scientific father figure.[79]

All through the summer, Einstein worked on his radiation theory. "You can hardly imagine," he complained to Johannes Stark, "what pains I have taken to think up a satisfying mathematical execution of the quantum theory. So far, however, I have not been successful with it."[80] He would also have liked to produce something definitive, as he had to give his first major lecture in September. But even without a breakthrough in quantum theory, his appearance at the Naturforscher convention in Salzburg was a memorable event, both for Einstein's reputation as the most important physicist of the younger generation and for the history of physics.

Only a handful of his younger colleagues had visited Einstein in Bern. Now every physicist, including the top names, had an opportunity in Salzburg to "take a closer look at the beast." Einstein in turn was able to meet his correspondents face to face: the universally respected Planck and Sommerfeld, for instance, as well as some younger scientists.[81] And he must have enjoyed the deliberately crowd-pleasing opening address, in which Henry Siedentopf, of the Carl Zeiss Works in Jena, was able, with his newly developed ultramicroscopic methods, to make the Brownian movement impressively visible by projecting it on a wall. At the center of interest, however, were "problems of radioactivity on the one hand and the relativity principle on the other."[82]

Einstein could very easily have allowed himself to be lionized as the founder of the relativity principle and, in accordance with academic custom, could have shown his thanks by a comprehensive lecture on the subject. Indeed, Planck had probably invited him with just that intention.[83] But that was not Einstein's style, either in Salzburg or on similar occasions in the future. In the words of Max Born (whose own lecture was *On the Dynamics of the Electron in the Kinematics of the Relativity Principle*) Einstein now left relativity "to lesser prophets."[84]

Einstein himself chose instead the theme on which, in his belief, "everyone should work," *The Nature and Constitution of Radiation*.[85] The assembled crème de la crème of scholarship, who had come to hear him on the afternoon of September 21, was assured

> that we are . . . standing at the beginning of a not yet assessable but undoubtedly most significant development. What I am about to present is largely my personal opinion, or the result of reflections, which have not yet been adequately verified by others. If I nevertheless present them here, this is due not to an excessive confidence in my own views, but to my hope that one or the other among you may be induced to concern himself with these problems.[86]

These words reveal a lot: his typical modesty, combined with a decent reference to his scientific existence on the periphery, outside the universities, and to the widespread disregard of his work on radiation and quantum theory—but also an attempt to point his colleagues in appropriate directions.

Right at the beginning, he set out his conviction—which still met with considerable opposition—that light could no longer be understood and described solely as a wave phenomenon, but that, at the same time, something like a granular structure must be ascribed to it: "It cannot be denied that there exists a large group of radiation-related facts which show that light possesses certain fundamental properties which can much more easily be understood from the standpoint of Newtonian emission theory than from the standpoint of wave theory." For the future, he ventured to predict that "the next phase of development in theoretical physics will bring us a theory of light that will be susceptible to being understood as a kind of fusion of the wave theory and the emission theory of light." This may be seen as the first announcement of the combination of light as wave and light as particle, an interpretation which would later, in fully developed quantum mechanics, be called "complementarity." He saw the aim and purpose of his lecture as the argument "that a far-reaching change in our concepts of the nature and constitution of light is indispensable."[87]

Of course, Einstein also referred to relativity theory, but only to the extent that the abandonment of a pure wave theory of light seemed inevitable. The abolition of the ether in relativity theory had already "changed ideas on the nature of light in so far as it does not understand light as a sequence of states of a hypothetical medium, but instead as something existing independently, just like matter."[88] Einstein derived a second argument from the equivalence of energy and mass, whose relationship he once more briefly developed for his listeners, in order to attach the conclusion that this "something existing independently like matter" shares with a "particle theory of light the characteristic of transferring inert mass from the emitting to the absorbing body."[89]

After these preliminaries, he turned to radiation theory proper and demonstrated to his audience that a direction has to be ascribed not only to the absorption of radiation but also to its emission—in direct contradiction of the Maxwellian theory that radiation is emitted as a spherical wave. Einstein demonstrated, moreover, not only that the concept of energy quanta made it possible to derive Planck's radiation formula, but that from the validity of that formula necessarily followed a quantum structure of radiation, and that, in consequence, the Maxwellian equations could no longer be regarded as strictly correct.

Hardly anyone, however, was prepared to follow Einstein this far.

In the discussion which followed, his only supporter was Johannes Stark. Planck, representing the majority view, was reluctant—though with the greatest respect for Einstein's achievement—"to assume the light waves themselves to be atomistically constituted, and hence to give up the Maxwellian equations. This seems to me a step which, in my opinion, is not yet necessary."[90]

Even though Einstein was unable to convince either the authorities or his younger colleagues by his bold outline of a future radiation theory, his first appearance before a major audience of physicists was nevertheless a complete success. Max Born felt that in Salzburg "Einstein's achievement received its seal before the assembled world of scientists."[91] And a few decades later, it would become clear that Einstein's Salzburg lecture, in Wolfgang Pauli's words, could be "seen as one of the turning points in the evolution of theoretical physics."[92]

CHAPTER FOURTEEN

Professor in Zurich

IN MID-OCTOBER, in time for the start of the semester, Albert Einstein arrived in Zurich, with Mileva and their son Hans Albert. In the excitement of the move he had overlooked the police and army formalities associated with removal to a different canton, and he therefore mailed his "service book" and residence permit to Lucien Chavan with a request to notify the Bern police and the district army headquarters of his change of residence. "Things have started moving here,"[1] he wrote to his friend in Bern. "I like my new position a lot. But it's exhausting work."

The Einsteins moved to the middle floor of a three-story building at Moussonstrasse 12 on the slope of the Zürichberg, the hill overlooking the city. It was in the immediate neighborhood of the Physical Institute of the Polytechnic and also close to the university's institute on Rämistrasse 69, which Einstein could reach by strolling a few hundred yards down the Gloriastrasse. On moving in, they found to their surprise that the Adlers had an apartment in the same building. "We are on very good terms with Einstein, who lives above us, and, as it happens, we're closer to them than any of the other academics. They run a . . . Bohemian household [like our own],"[2] Friedrich Adler reported to his father. Hans Albert and the Adlers' daughter, Assinka, became friends and would play in the garden or in the street with a crowd of eight to ten children of their own age. When the noise got too much for the parents, they could withdraw to an attic for undisturbed discussions. "The more I talk to Einstein," Friedrich Adler wrote in the same letter, "the more I realize that my favorable opinion of him was justified. Among today's physicists his is not only one of the clearest, but also one of the most independent minds, and we find ourselves in agreement on questions which the majority of other physicists would not even understand."

Following his sabbatical leave during the summer semester, Adler was standing in for the *Ordinarius*, the head of the department, who in turn was substituting for the *Rektor*, the principal. In consequence, Adler gave a number of lucrative freshman lectures, as well as the laboratory workshops that went with them, so that the two neighbors were now also colleagues. Einstein gave a four-hour lecture, "Introduction to Mechanics," attended by seventeen students; a two-hour class, "Thermodynamics," attended by nineteen students; and a one-hour "Physical Seminar" on topical problems of research, attended by twelve students.

"My new profession is very much to my liking," he reported to Jakob Laub. "I am on very close terms with my students and hope I'll be able to give some ideas to some of them."[3] But the new professor also discovered that "*very* great demands are being made on me. I am taking my lectures very seriously, which means that I have to devote a lot of time to preparation. Six hours a week plus one evening seminar may not sound too bad, but it is a lot."[4] On a postcard to his friend Besso in Bern he observed that "my really free time is less than in Bern. But one learns a lot doing it."[5] To Sommerfeld, too, he said regretfully that "my new post makes greater demands on me than I had expected."[6] He added, "It's due to my poor memory, as well as to the fact that until now I had concerned myself with my subject merely as an amateur." This may have been taken as a kind of fishing for compliments, but it probably meant no more than that in the past he had practiced physics as a hobby, much like playing the violin. Einstein gave his official inaugural lecture—his first but by no means his last—on December 11, on *The Role of Atomic Theory in the New Physics.*

At the lectern Einstein really was a beginner. The two classes he had taught in Bern as a privatdozent to a handful of his friends could hardly be considered training for the demands of a major university course. He was well aware of that, and for this reason did not want to write a book about relativity theory: "As I have little teaching experience so far, it would be downright irresponsible to undertake further obligations until I've become more familiar with my new profession."[7] As for the time-consuming preparations mentioned by Einstein in his letters, his students certainly saw no evidence of these: Hans Tanner later recalled that "the entire manuscript he carried with him consisted of a scrap of paper the size of a visiting card, on which he had outlined the ground he intended to cover with us."[8] Tanner was the only student to take his doctor's degree under Einstein, and a regular attender of all his lectures at Zurich University.

Actually, Tanner regarded the scanty notes as an advantage, because "Einstein had to develop everything out of himself, so that we gained a direct insight into his working technique. . . . We were thus able to witness the often curious paths along which a scientific result is sometimes reached." This kind of participation in the creative process of science may not always have been easy for students accustomed to pedagogically structured, methodical teaching; but any problems of comprehension were mitigated by the fact that Professor Einstein had what in those days was a totally unprofessorial and indeed informal relationship with his students—just as he had promised his own revered teacher, Sommerfeld. Thus Einstein encouraged his students to ask questions at any time during his lecture, whenever something was unclear to them, and this academic method proved successful: "It was not long," Tanner reports, "before we abandoned all shyness in case we asked a silly question."

In the audience at the thermodynamics lecture, which was intended for advanced students, was also Adolf Fisch, who had graduated from the cantonal school in Aarau along with Einstein, had subsequently studied at the "Poly," and now was a teacher in Winterthur. Fisch recalls that "Einstein took great trouble to offer the students something of substance and something new. He kept asking if he was being understood. During the breaks he would be surrounded by men and women students anxious to ask questions, and he would patiently and kindly try to answer them."[9] The atmosphere was particularly informal after his evening seminar. Right from the first semester, it became his custom to proceed to the Terrasse café on the Bellevue, where the

Limmat leaves Lake Zurich, and continue discussions there until closing time. Arnold Sommerfeld would have been pleased with this young professor.

Einstein's relations with his colleagues were also entirely amicable. He must have been relieved to find that Alfred Kleiner, formerly his doctoral supervisor and now, as *Ordinarius*, his superior, was putting no difficulties in his way. "Kleiner is odd but tolerable,"[10] he wrote to Besso; and to Laub he even described Kleiner as "a very nice person. He's treating me like a friend and is not holding anything against me."[11] Two months later he wrote to Laub that although the head of the institute was "not a superb physicist, he is a splendid person for whom I have a great liking."[12] Having observed the scientific elite assembled in Salzburg, Einstein evidently took a tolerant view of Kleiner's professional mediocrity: "It seems that scientific reputation and personal qualities do not always go hand in hand. To me a harmonious person is worth more than the most sophisticated formula-basher or system inventor."[13] Nevertheless, his relations with Kleiner had a professional character; his friends among his colleagues came from elsewhere.

There was a joyful reunion with Marcel Grossmann, Einstein's "lifesaver" from his student days and now professor of geometry at the Polytechnic. He also developed a friendship with another mathematician at the "Poly," with whom his relations had been rather cool in earlier years, Adolf Hurwitz. As a student, Einstein had cut Hurwitz's seminars but had tried, unsuccessfully, to become his assistant. Now they were brought together by their love of music—chamber music was played on Sundays at Hurwitz's home. And Aurel Stodola, professor of mechanical engineering, an authority on steam and gas turbines—and a native of Bohemia—actually attended the lectures of the new *Extraordinarius.* Stodola had briefly met Einstein while the latter was a student at the Polytechnic, and now their acquaintance ripened into genuine friendship. When Stodola retired in 1929, Einstein dedicated to him an extensive article in a *Festschrift* for the occasion, as well as a piece in a Zurich daily paper. In the newspaper article, Einstein recalled how "to his delight and his uncomprehending alarm Stodola's splendid figure appeared in the auditorium" to attend the new professor's lectures on developments in theoretical physics, "partly for the sake of pure knowledge and partly with a view to utilizing what he had heard. When the class was over, Stodola, always readily spotting the essential point, would ask profound questions which often contained justified criticism in a refreshing form."[14]

Einstein's closest friend, however, was a medical man, Heinrich Zangger. Zangger, his senior by five years, had gained international renown as director of the Forensic Medicine Institute at the university and as one of the pioneers of "disaster medicine," and his views could not easily be overlooked in Swiss political and academic circles. Zangger had met Einstein in 1905, when he had been wrestling with some unfamiliar mathematical problems and Aurel Stodola had suggested that he consult Einstein in Bern—who in fact had been able to help him. As dean of the medical faculty, Zangger had supported Einstein's appointment in Zurich, and after his arrival the two men became friends. Einstein later recorded that "his range of interests was virtually unlimited";[15] he was a man with whom physical problems could also be discussed and who provided an impetus for at least one of Einstein's publications.[16] Einstein, moreover, credited him with "sound judgment also with regard to persons and things on which his professional knowledge was really much too meager." This proved a considerable advantage, as Zangger was probably Einstein's most committed champion in dealings with the Swiss authorities, both in Zurich and with the federal government in Bern. Later, after Einstein and Mileva had parted, Zangger was a patient mediator between them whenever they were unable to resolve their problems themselves.

With the exception of a few physicists and mathematicians, and perhaps Heinrich Zangger, hardly anyone would then have been aware that the newly appointed professor was one of the giants in his field. To most of his colleagues, he probably appeared as he had while still a student in Zurich—a rather awkward eccentric with a sharp tongue, who "with his somewhat shabby clothes, his too short trousers, and his steel watch-chain"[17] did not fit the accepted image of a Swiss professor. This view was prevalent not only among the faculty, but also among the students and the assistants.

This changed dramatically when, during the spring break in March, Einstein was visited by Walther Nernst, an unchallenged scientific authority, and moreover wealthy and popular as a result of his invention of the "Nernst lamp"—he had sold the patent to Emil Rathenau of the AEG for a million gold-marks. George Hevesy, then working as an assistant at the Physical Chemistry Institute of the Polytechnic, recalls that it was Nernst's visit which made "Einstein famous in his circle. Einstein had come to Zurich as an unknown. Then Nernst arrived, and people in Zurich were saying: 'That Einstein must be a clever fellow if the great Nernst comes all the way from Berlin to Zurich to talk to him.' "[18]

Nernst—the first physicist to visit Einstein in Zurich—was interested in him because he had taken seriously Einstein's paper of 1906 on the quantum-theoretical interpretation of the specific heat of solids. Nernst's theorem on the behavior of thermodynamic magnitudes approaching absolute zero was referred to by its inventor simply as "my theorem" but would soon be called the third law of thermodynamics by his colleagues, and it could be put on a new footing by Einstein's reflections. At his Physical Chemistry Institute of Berlin University, Nernst had set in motion an extensive program for the experimental investigation of these relationships in which Heinrich Rubens, professor of experimental physics, participated both with teams within the university and at the Reich Physical-Technical Institute in Charlottenburg. When, in 1909, Nernst came to believe that Einstein's quantum theory of solids best fit both the measured values and his theorem, Einstein was absolutely delighted. Henceforward he had—along with Planck, who especially liked his relativity theory—another champion in Nernst. Nernst admittedly did not understand much of relativity theory but (unlike Planck) was ready to follow Einstein's quantum ideas, even though only pragmatically and in his own specialized field, not in radiation theory generally.

Einstein was enormously pleased with the visit. No sooner had Nernst left than he informed Jakob Laub: "My predictions with regard to specific heat seems to be brilliantly confirmed. Nernst, who just visited me, and Rubens are busily engaged on their experimental verification, so that we will soon be enlightened about them."[19] The data for diamond had given a good agreement, but further investigations of other materials soon showed that this had been an exception. However, over the next few years Nernst's concern with this problem moved quantum theory into the center of scientific interest and thereby greatly helped this "revolutionary" physics to gain acceptance. That in this field, too, Einstein had laid the foundation[20] further enhanced his reputation.

Famous or not, Einstein found that teaching exacted a price; and in the summer semester of 1910, in addition to his lectures and seminars, he was put in charge of "Daily Practical Work for Advanced Students," who numbered twelve. For a theoretician, this appears to have been a nerve-racking duty: he admitted to Hans Tanner that he scarcely dared to "pick up a piece of apparatus for fear it might blow up."[21] To Sommerfeld he complained: "My anxieties about the laboratory were largely justified."[22] On the other hand, he now had an assistant to relieve him of some of his workload. This was Ludwig Hopf, a native of Nuremberg, five years younger than Einstein, who had just taken his Ph.D. under Sommerfeld in Munich. Einstein and Hopf had met at the Naturforscher convention in Salzburg and found themselves in tune not only in physics but also musically. Hopf was a competent pianist, and playing duets with him was a form of relaxation for Einstein. And Einstein's musings at the time were indeed tiring, concerned as they were exclusively with quantum theory and its almost insoluble puzzles.

"The quantum theory is a certainty for me,"[23] Einstein declared jubilantly after Nernst's visit. But that was true only of solids and their specific heat, and then only with the qualification that the theory was "still rather unsatisfactory" because "it presupposes the invalidity of our mechanics, and all attempts to adapt molecular mechanics to the imperious demands of experience have been unsuccessful."[24] Matters were even worse regarding the quantum theory of radiation, whose problems had concerned Einstein ever since Planck published his radiation formula at the turn of the century. "In the matter of light quanta I have not yet arrived at a solution, though I have discovered some significant things,"[25] he wrote to Jakob Laub on the last day of 1909. The next sentence sounded like a New Year's resolution: "I'll see if I can't hatch this favorite egg of mine after all." Ten weeks later came a similar message: "With regard to the quanta I have found some interesting things, but nothing that's ready yet."[26] One of the "interesting things," described by Einstein as the "core of the whole question," was, as first mooted in 1905 and turned over one way and another in 1909, the compatibility of a corpuscular and, simultaneously, a wave character for radiation. In this context Einstein in a letter to Sommerfeld once again referred to the "Almighty" and his sophistication: "Can the energy quanta on the one hand and Huygens's principle on the other be combined? Appearances are against it, but the Almighty—it seems—managed the trick."[27]

Einstein was obviously delighted that Sommerfeld wanted to visit him in Zurich at the end of the semester, but he did not feel like encouraging him: "Because I haven't been able to produce anything halfway complete on the problem of the quanta."[28] That did not put Sommerfeld off. At his institute in Munich his colleagues were surprised that their professor should be in such urgent need of recuperation even before the end of the semester that he had to travel to Zurich for a week.[29] The nature of that recuperation is revealed in a letter from Einstein to Jakob Laub: "Sommerfeld was with me for a week to discuss the light problem and some points of relativity. His presence was a real pleasure to me."[30] Einstein had now gained a respected ally in the matter of quanta. Quite unlike Planck, Sommerfeld had "to a great extent associated himself with my view on the application of statistics."

But despite their combined efforts, the week's discussions produced nothing that even came near to a breakthrough: "I haven't got any further with the constitution of light. There is something very fundamental hidden behind it."[31] This assumption was to be confirmed over the next fifteen years. Meanwhile, however, Einstein succeeded, jointly with Ludwig Hopf, in producing two papers[32] which supported his thesis that a quantum theory of radiation would call for a radical abandonment of classical physics. "The result," Wolfgang Pauli summed up, "was disappointing for all those who were still vainly hoping that

Planck's formula might be derived merely by a change in the statistical assumptions rather than through a fundamental break with the classical ideas concerning elementary microphenomena."[33]

At one point in the autumn Einstein believed he could see the light at the end of the tunnel: "At this moment I am very hopeful of solving the radiation problem, moreover without any light quanta. I am very curious how the business will turn out."[34] In this not merely revolutionary but downright reckless approach he even toyed with the idea of abandoning well-tested and sacred principles of physics: "One would have to give up the energy principle in its present form." But a week later Einstein characterized his strenuous efforts by invoking the Almighty's adversary: "Again nothing has come of the solution to the radiation problem. The Devil merely played a poor joke on me."[35] It is most regrettable that Einstein did not keep his notes: it would have been fascinating to glance over his shoulder as he designed a physics with a modified or even abandoned energy principle.

Although he owed his fame and his professorship to his relativity theory, and although he occasionally pondered it in Zurich,[36] Einstein did not publish anything in that field, nor did he deal with it in his lectures or seminars. His only lecture on the subject was not at the university but, at the beginning of May, to the Naturforschende Gesellschaft at the Guild House Zum Rüden on the Limmat embankment. On that occasion he is reported to have covered a small blackboard with clocks, to illustrate the concept of simultaneity, and, after an exhausting lecture, to have asked: "What is the time, actually? I don't have a watch."[37]

Einstein soon discovered that relativity theory was arousing interest even among circles unconnected with physics. This was partly due to Ludwig Hopf, who was fascinated not only by physics but also by psychoanalysis, or "depth psychology." Immediately after his arrival in Zurich, Hopf had called on the psychiatrist Carl Gustav Jung and had introduced Einstein to him. Einstein was Jung's dinner guest on several occasions, and there also met Eugen Bleuler, the director of the internationally famous Burghölzli psychiatric institution, as well as a number of other medical men interested in relativity. "He tried, with more or less success, to teach us the fundamental arguments," Jung recalled, "but as non-mathematicians we psychiatrists found it difficult to follow his argument."[38] Einstein, for his part, seems to have had difficulties with the psychiatrists, and in consequence the conversations were not continued for long. Jung's impression of his talks with Einstein was that one could "hardly imagine a greater contrast between the mathematical and the psychological mentality. The one is quantitative in the extreme, while the other is qualitative in the extreme."[39] Evidently, Jung not only made the common mistake of regarding mathematics as the essential aspect of physics but also failed to perceive the intuitive content of creative natural science.

Needless to say, Professor Einstein continued to care about his beloved "little machine." Paul Habicht, the instrument manufacturer, came over from Schaffhausen several times and, to Einstein's delight, "got the little machine to function all right."[40] His brother Conrad, Einstein's friend from the happy days of the Akademie Olympia and by then a teacher of mathematics in Schier, canton Grisons, also traveled to Zurich. They both stayed with the Einsteins, and the three of them tinkered and experimented at the university laboratory. For the spring break Einstein "cordially invited [them] to make the final experiments with the little machine and then button up the business. It's got to be finished at long last before someone else gets in before you."[41] The three worked hard and successfully. The Habicht brothers applied for, and obtained, a patent for the "little machine"; and when the paper was published,[42] Einstein, the initiator, renounced any authorship and contented himself—apart from a citation of his original publication and the obligatory acknowledgment—with a notation that the experiments had been conducted "jointly with A. Einstein at the Zurich University laboratory." The new machine came up to expectations: potentials of less than one-thousandth of a volt could be measured with it, so that "a single radioactive elementary process . . . could readily be demonstrated with the instrument."[43]

Einstein followed the further career of the "little machine" with intense interest. Paul Habicht presented the perfected invention at the Berlin Physical Society and scored a huge success. "The fellows nearly stood on their heads,"[44] Einstein reported to Besso. "I'm tremendously pleased. Habicht already has quite a few orders."[45] However, Einstein was mistaken in regarding the machine's future as secure. For one thing, its accuracy left something to be desired, and for another it was, within a few years, rendered obsolete by the new electronic amplification technique. When Paul Habicht died in 1948, Einstein in his letter of condolence to Conrad included a "recollection of the old days, when together with your brother I worked on the little influence machine for the measurement of small voltages. That was fun, even though nothing useful came of it."[46]

Compared with his reflections about the intricacies of quanta, his work with the "little machine" must have been something like light relief for Einstein. Much the same was probably true of a paper on fluctuation problems in classical physics. This was once more concerned with the reality of molecules and the childlike—but by no means silly—question of why the sky is blue. His starting point was the "Tyndall effect," first described in 1869, in which a bluish "opalescent" tint appears in a gas or a liquid when a beam of light of whatever color combination is sent through the medium. It seemed plausible that this "opalescence" must be due to scattered light, but it was not clear by what the light of the primary beam was scattered. Tyndall initially assumed that the scattering was caused by minute contaminations in the air, but Lord Rayleigh subsequently demonstrated mathematically that the light was being scattered by the irregularly distributed molecules of the air itself. This view was accepted until 1908, when Marian von Smoluchowski[47] showed that the scattering of the light was caused by density fluctuations in the gas or liquid, provided these fluctuations extended over minute volumes within the range of one wavelength of the light.

Einstein, who greatly esteemed Smoluchowski as one of the "most subtle of contemporary theoreticians,"[48] intended to develop this idea, and particularly to obtain an exact formula for the intensity of the laterally scattered light: "I am at present writing a paper on the opalescence of gases and liquids,"[49] he informed his friend Laub in the summer of 1910. "Quantitative implementation of Smoluchowski's theory. I've finished with the basic part. It is an entirely strict theory." What Einstein dispatched to *Annalen* in October[50] was a mathematically rather complicated derivation of a formula explaining why the light from the sky opalesces blue during the day and reddish in the morning and evening. Einstein saw the "main result" of his investigation as the fact that his formula "permits an exact determination of the constant *N*, i.e. the absolute size of the molecules."[51] Thus he once more suggested a method for the experimental determination of the Avogadro number. That he made use of the blue of the sky to convince the last doubters of an atomic view of matter was entirely in line with his endeavor, formulated ten years previously, "to recognize the unity of a complex of phenomena which appear as totally separate things to [sense] perception."[52]

Smoluchowski was delighted with Einstein's "ingenious calculation"[53] and regarded his paper on opalescence as a "substantial step forward" in science. His own experiments for verifying Einstein's formula proved exceedingly difficult, however, because what is magnificently visible in the vast volume of the atmosphere is not easily imitated in a small laboratory setup. Eventually, despite the difficult working conditions at the University of Cracow in Poland during World War I—he had become a full professor there in 1913—Smoluchowski achieved a fine demonstration of the phenomenon justifying quantitative statements: "Improvised photometric measurements yielded results which . . . satisfactorily agree with the theoretical formula."[54] When Smoluchowski died in 1917, at the age of only forty-five, Einstein in an obituary mourned "not only the brilliant researcher but also a fine, sensitive, and benign person."[55]

The "strict theory" of opalescence was Einstein's last paper on classical mechanics and simultaneously his last major publication as *Extraordinarius* at the University of Zurich.

Ever since April 1910 Einstein had been, in a sense, "on call" in Zurich. After less than six months it became clear that, as had been expected, the post of an "extraordinary" professor could only be a halfway station for him, pending an offer of a regular chair. This offer came from the German University in Prague, and if the appointment procedure[56] had not again been so protracted Einstein might have left Zurich after two semesters there instead of after three.

The—initially informal—inquiry as to whether Einstein would be prepared to move to Prague must have come in March 1910, because on March 30 Friedrich Adler wrote to his father that Einstein had been asked "if he would accept a post at another university."[57] On April 29 Einstein informed his mother, who was then staying with her sister and brother-in-law in Berlin, of "some rather interesting news. I will probably be invited to a great university as a full professor, with a considerably bigger salary than I'm getting now. I'm not yet allowed to say where it will be."[58] But he did tell Friedrich Adler, because Adler reported to his father the same day that it was the German University in Prague, and that Einstein headed the list of names for the post of professor of theoretical physics.[59]

In Prague the academic procedures had begun as early as January, as Hofrat—an honorific in the Austrian monarchy—Ferdinand Lippich, the incumbent of a chair, intended to retire in the summer semester of 1910. The department had therefore set up a small commission, which included the mathematician Georg Pick and the experimental physicist Anton Lampa, whom Einstein had met in Salzburg at the Naturforscher convention. On April 21 the department approved the recommendations of that commission to the effect that the chair of mathematical physics should become a chair of theoretical physics, with the cabinet for mathematical physics being simultaneously converted into an institute of theoretical physics. It also approved the appointment proposal, with Einstein heading the list, and immediately passed it on to the Ministry of Education in Vienna.[60] In order to lend greater weight to its recommendation to the imperial and royal authorities, the department had asked Max Planck for an expert opinion. It is probable that Planck described Einstein as one of the most important physicists and the inspired inventor of relativity theory, though with regard to quantum theory, he would point out that he could not judge yet whether Einstein was always right.[61] Fortunately, a book by Max Planck had been published just then, in the spring of 1910, with a euphoric statement naming Einstein's relativity theory in the same breath as Copernicus; this was bound to impress the ministers and indeed Emperor Francis Joseph. This is what Planck had written on the theory of relativity and the resulting revision of the concept of time:

> In boldness it probably surpasses anything so far achieved in speculative natural science, and indeed in philosophical cognition theory; non-Euclidian geometry is child's play in comparison. And yet the relativity principle, in contrast to non-Euclidian geometry, which so far has been seriously considered only for pure mathematics, has every right to claim real physical meaning. This principle has brought about a revolution in our physical picture of the world, which, in extent and depth, can only be compared to that produced by the introduction of the Copernican world system.[62]

It is probable that toward the end of April Einstein had been informed by Anton Lampa about the main aspects of the developments in Prague. Then, for a while, there was no news either from Prague or from Vienna, and after that the news was not good. In July he wrote to Sommerfeld: "I won't get to Prague. The ministry—as I hear from Prague—has made difficulties."[63] To Laub he was more outspoken: "I was proposed only by the department; the ministry, however, has not accepted the proposal because of my Semitic origin."[64] It is impossible now to establish the precise role played by anti-Semitism, but we do know that the ministry wished to appoint not the foreigner listed in first place, but the second candidate, Gustav Jaumann, professor at the Technical College in Brünn (now Brno) who was Austrian. Such disregard of proposals from universities was by no means unusual in the Austrian monarchy. The previous year the department had favored Johannes Stark, but the appointment had gone to Anton Lampa,[65] an Austrian, who was now championing the foreigner Einstein, though unsuccessfully. The "most humble submission of the most obedient Minister of Education and Instruction, Karl Count Stürgkh," to His Imperial and Royal Majesty Francis Joseph contained the statement: "Although the Collegium of Professors attaches special importance to the appointment of Professor Dr. Einstein, listed in first place, in view of his brilliant achievements in the area of modern theoretical physics,

I yet believe that negotiations should first be initiated with Professor Jaumann in Brünn, listed in second place."[66] Here ends the first part of the appointment procedure.

Although Einstein would have to wait for some time for his full professorship and the "big salary," the offer from Prague meanwhile was benefiting him in Zurich. His students, on the initiative of Hans Tanner, had addressed a petition, signed with fifteen names, to the "Honorable Directorate of Education of the Canton of Zurich" requesting it "to do its utmost to preserve this outstanding scientist and teacher for our university."[67] They pointed out that "in an admirable manner he succeeds in presenting the difficult problems of theoretical physics so clearly and comprehensibly that it is a great pleasure for us to follow his lectures; moreover he manages to establish such perfect rapport with his listeners that we are convinced that such teaching would prove of great benefit to our university."[68]

The Directorate of Education shared that opinion and within three weeks submitted a proposal to the Governmental Council, even though the refusal from Vienna had by then become known in Zurich. "It appears," the protocol of the Governmental Council states, "that the threatening danger of Herr Professor Einstein's departure has for the moment disappeared owing to a negative attitude by the supreme state authorities of Austria. The Educational Council nevertheless feels that Herr Professor Einstein should again be tied to our university by some kind of improvement in his position."[69] It was decided, therefore, to raise Einstein's annual salary "in the event of his further staying at the University of Zurich on October 15, 1910, from 4,500 francs to 5,500 francs."[70] The raise was well timed, for two weeks after this decision Einstein became a father for the third time. His second son, Eduard, was born on July 28.

Although both Einstein and Mileva were fond of Zurich, he was acutely aware that as an *Extraordinarius* he was not an equal member of the faculty. And although his financial situation would be less strained in the future, he was clearly determined to take the next opportunity of a more radical change in his position. This opportunity came with a surprise turn of events in Prague.

▪ ▪ ▪

Sometime in the summer of 1910 Gustav Jaumann must have heard that while the ministry in Vienna wished to appoint him to the post in Prague, the faculty had listed him only in second place. To this he is reported to have angrily declared that he "would have nothing to do with a university that was chasing after modernity while being blind to real merit."[71] Jaumann confronted the ministry with exorbitant salary demands and thus caused the breakdown of negotiations.[72] In consequence, Count Stürgk, the minister, had no choice but to come back to the foreigner, Einstein.

On September 20 Einstein received an invitation for a discussion of appointment terms, and on September 24 he traveled to Vienna.[73] His salary was agreed to at 8,672 Austrian crowns,[74] at the official rate of exchange a little over 9,000 Swiss francs and thus a handsome increase over his pay in Zurich. The fact that as an Austrian professor he would also have to become a subject of His Imperial and Royal Majesty did not bother him unduly, as the authorities were prepared to overlook his retention of Swiss citizenship.

More difficult, evidently, was the problem of religion. In Switzerland Einstein had always described himself as "without religious denomination" on official questionnaires, but this was not acceptable in Francis Joseph's empire. In the view of the old monarchy it was inconceivable that a person without religious denomination could swear a proper oath of allegiance. When the problem was explained to Einstein, he simply declared that he was a Jew, whereupon "Mosaic" was entered on the form.[75] Einstein scarcely saw this concession to Austrian bureaucracy as a return to the religion of his forebears. He was simply prepared, in return for a full professorship, to render unto Caesar that which was Caesar's, without feeling particularly disturbed about it. When his friend Paul Ehrenfest, whom he wished to propose as his successor in Prague in 1912, declined to make this concession, Einstein showed no understanding: "It worries me that you have the spleen of being without religious denomination; why not drop it for the sake of your children? Besides, once you are a professor here you can return to this curious hobby."[76]

Although all obstacles had now been removed, the appointment was some time in coming. In December Einstein wrote: "It's by no means certain that I will get away from Zurich. True, it's been stated in various places that I've been appointed to Prague, and indeed I was promised appointment. But so far no appointment has come."[77] Not until December 16 did the minister, Count Stürgkh, submit the proposal to the emperor, who approved it only on January 6, 1911.[78] On January 20 Einstein applied for release at the end of the current semester from his duties in Zurich. On February 10 the Governmental Council met his request, regretting "that the university is losing the outstanding scholar and that the cantonal authorities had not been given a chance to try to keep him for our university in the future."[79]

It is not quite clear why Einstein did not try to obtain in Zurich what was being offered him in Prague. After all, Prague was not a "great university," as he had said to his mother, and certainly not a center of research in physics. In fact, "there was no doubt that at this time the German University in Prague was suffering from a loss of importance among German universities."[80] Einstein possibly believed that there was nothing more to be gotten out of Zurich, and perhaps he was also reacting to certain "conflicts" within the department, referred to in a letter from Kleiner to a colleague, to the effect that "Einstein knows that he cannot expect any personal engagement on the part of representatives of the department."[81] At any rate, it was settled that Einstein would move to Prague as a full professor in the spring.

At the beginning of his last semester in Zurich, while he was waiting for confirmation from Vienna, he had a pleasant surprise. In early November he had received a letter from Emil Fischer, the famous professor of chemistry in Berlin: "Your great theoretical papers in the field of thermodynamics," he wrote, "have caused a sensation in the world of science, and in our circle there is frequent talk of them, especially since Herr Nernst has occupied himself with the experimental verification of your conclusions concerning the law of Dulong and Petit."[82] But the real surprise was the news that Fischer was acting as a go-between for a man in the chemicals industry, who was very pleased "that German scientists like yourself, Herr Planck, and Herr Nernst have taken over the leadership in this field, and he feels that it is the duty of wealthy people in Germany to promote that brilliant work a little by material support." This support proved to be exceedingly generous—altogether 15,000 marks in three annual installments of 5,000 marks each, "the first at once, and the other two in 1911 and 1912." The recipient of the donation was to use it as he himself saw fit; it would not be tied to any obligation or restriction.

Einstein, who had given up German nationality fifteen years previously, did not object to being included among "German scientists"; he conveyed his sincere thanks, first to Emil Fischer, whose praise had "greatly honored and even more greatly embarrassed" him, considering that "every day I am keenly aware of how impotently I am facing the urgent problems of our science,"[83] and also to the man "who is prepared to donate such considerable sums" and whom he wished to assure "that I will apply the sums entrusted to me as conscientiously as possible."

The generous donor was Dr. Franz Oppenheim, a cofounder and shareholder of the Aktiengesellschaft für Anilinfabrikation, better known as Agfa, who lived in a princely mansion in Berlin-Wannsee and could rightly regard himself as one of the "wealthy people." He served as treasurer to the German Chemical Society and was one of its greatest benefactors.[84] He was "very pleased"[85] to learn of Einstein's readiness to accept his anonymous offer. He requested his go-between Emil Fischer "not to disclose my name, because I believe this way would be more agreeable to Professor Einstein." Einstein was delighted that "the capital promised would indeed greatly facilitate his scientific work."[86] There is no information on how the money was used.

In January 1911 Einstein received not only the confirmation of his appointment in Prague, but also an "exceedingly cordial" invitation from Hendrik Antoon Lorentz for a lecture in Leyden. "You can hardly imagine," Einstein wrote to his scientific father figure, "how much I am looking forward to making your personal acquaintance."[87] At the same time he thought it a "curious undertaking to bring theoretical physics to Leyden. And yet, I am not gripped by any sense of timidity, as I am convinced that with you and those around you I will encounter a friendly attitude and not severe criticism."

As Mileva's mother happened to be visiting them in Zurich and looking after the two children, the Einsteins could travel together. They boarded their train on Wednesday, February 8, and from Basel, where it stopped, sent a postcard to Friedrich Adler: "If the house burns down or anything else amusing happens, please cable us c/o Prof. H. A. Lorentz, Leyden."[88] The following day they arrived in Leyden, where the Lorentzes put them up. The lecture was on Friday, followed in the evening by long conversations with Lorentz; these continued the next day, first in a small circle with brilliant colleagues such as Heike Kamerlingh Onnes and Willem Hendrik Keesom, and in the evening once more just with Lorentz. On Sunday the Einsteins left for Antwerp, where, before returning home, they visited the "favorite uncle" Caesar Koch.

No sooner was he back in Zurich than Einstein thanked Lorentz: "Now I am back here in my study, filled with the most beautiful memories of the wonderful days which I was permitted to spend in your proximity. . . . There is so much kindness and human warmth radiating from you that throughout my stay it was not even possible for the tormenting conviction to develop that I am an undeserving recipient of all that kindness and honor."[89] The image that Einstein had of Lorentz was even brighter after his visit, and when Einstein later declined an appointment to the University of Utrecht in the Netherlands, he apologized to Lorentz "with a heavy heart, like one who has done a wrong to his father."[90] Einstein's profound respect for Lorentz continued after Lorentz's death. At his graveside Einstein pointed out that "his unfailing kindness and generosity, his sense of justice, combined with an intuitive insight into people and conditions, made him a leader wherever he found himself. Everyone followed him gladly, because they felt that he never wanted to dominate them, but always only to serve them."[91] And even in his old age, Einstein recorded: "To me, personally, he meant more than all the others I encountered during my life."[92]

■ ■ ■

Before his move to Prague, Einstein wished to tidy up his desk drawers. A few lesser papers were completed and sent off to *Annalen*, along with a whole bunch of corrections, including the big mistake in his doctoral thesis that Ludwig Hopf—who was now to accompany him to Prague as his assistant—had checked and put right. Then he had to look after Hans Tanner, his doctoral student, who had not been able yet to complete his thesis on a subject related to the kinetic theory of gases. Einstein helped him obtain an assistant's post at the University of Basel.[93] Tanner was not only Einstein's first but also his last doctoral student. As for his own extraordinary professorship, he had hoped "that Adler would succeed me."[94] Adler, however, since April 1910 editor-in-chief of the Social Democrat paper *Volksrecht*, had offered no more than a one-hour "Cognition-Theoretical Introduction to Physics" for the winter semester, and had decided by then to make politics his life's work.

The journey to Prague was broken in Munich, where Einstein met with Sommerfeld. On that occasion he made the acquaintance of Peter Debye, a young Dutchman, who was Sommerfeld's assistant and became Einstein's successor in Zurich.

On April 1 the Einsteins arrived in Prague. He had promised his friends in Zurich to let them know as soon as he was ready to consider a position at another university, so that Zurich could make an attractive offer to bring him back.

CHAPTER FIFTEEN

Full Professor in Prague—But Not for Long

"Here I have a magnificent institute, where work is very pleasant."[1] Immediately on his arrival in Prague, Einstein revealed what mattered to him most. An apartment for the family was soon found, on the left bank of the river, in the district of Smíchov, where a few streets of new middle-class apartment blocks had just been built. In one of these buildings[2] the Einsteins rented a spacious apartment on the mezzanine. For a while Mileva's mother lived there with them, as well as—now part of their bourgeois lifestyle—a maid, Fanny, who was a young mother herself and who, after unsuccessful attempts to find a suitable "shelter," was allowed by the Einsteins to keep her child at their home.[3] And there was still a guest room, which was being offered to any friends from Switzerland, because "Prague is a beautiful city for sightseeing."[4]

Einstein was now living in an ancient city full of history and mystery. From the huge Hradčany Castle the kings of Bohemia had once ruled the land, and three emperors had reigned over the German empire. But Einstein found Prague "less cozy" and complained about "the Czech language, bedbugs, terrible water, etc."[5] One wonders about his anxieties before his journey to Prague, since after his arrival he conceded that "the Czechs, incidentally, are much more harmless than one thinks"[6]—though, "animosity between Germans and Czechs seems considerable."[7] The Germans, too, among whom he was automatically reckoned and with whom he presumably mixed exclusively, remained strangers to him: "These are not really people with natural sentiments; they are cold and an odd mixture of snobbery and servility, without any kind of benevolence toward their fellow humans. Ostentatious luxury and side by side with it creeping destitution in the street. Poverty of ideas without faith."[8] It did not look as if he would ever feel at home in Prague or as if he would remain there longer than necessary. Even a few weeks later his reports still read as if he had been shipwrecked on some inhospitable shore beyond civilization: "Life here is not as pleasant as in Switzerland, quite apart from the fact that we are strangers here. There is no water here that can be drunk without boiling it first. The population for the most part does not know German and behaves in a hostile manner toward Germans."[9]

Einstein thus found himself unexpectedly confronted with the conflict between the nationalities that would eventually cause the breakup of the Habsburg empire. Some ninety percent of the population of Prague were of Czech nationality; they resisted the political, economic, and social domination of the ten percent of German-speakers in a manner that must have been perceived as "hostile" not only by Einstein. Just as the Czechs refused to speak German, so—and even more—the Germans rejected the Czech language. In this attitude, Einstein followed them, surely not because he was overbearing but simply for convenience. He made no effort to learn the language of the majority and had virtually no contact with Czechs.

In a certain sense Einstein's chair was itself a consequence of this nationalities conflict. The Universitas Carolina Pragensis had been founded in 1348 by the emperor and king of Bohemia Charles IV, as the first university in the Holy Roman Empire of the German Nation. It was modeled on the ancient universities of Bologna and Paris and given the same privileges. In the nineteenth century, after more than five hundred years of colorful history during which religious squabbles had occasionally disrupted academic pursuits, the national and linguistic conflict became so violent that by a decree of the emperor the ancient institution was divided into a Czech and a German university in 1882. The first rector, or principal, of the German University had been Ernst Mach, whom Einstein revered and who, though long retired, was then still living and working in Vienna. In addition to the two universities there were also two technical universities, likewise divided by language and nationality. With four universities one might have expected a lively intellectual atmosphere in Prague, but jealousy and discord between the institutions prevented any productive contact and resulted in quadruple mediocrity.

If Einstein had really needed stimulating scientific contacts, he would have been lost in Prague. "Einstein was totally lonely in Prague, even though there were four universities,"[10] is the recollection of Otto Stern, who came to Prague in the spring of 1912 to study under Einstein. "At none of them was there anyone which whom Einstein could discuss the matters which really interested him." Einstein's efficient assistant Ludwig Hopf, who had come to Prague with him, left after a few months because of the poor pay and went to the Technical University in Aachen, where he later became professor of hydro- and aerodynamics. As a successor to Hopf, Einstein's colleague Lampa recommended an advanced mathematics student named Emil Nohel. However, Nohel does not seem to have been much of a replacement for Hopf. Einstein never published anything with Hopf and never mentioned Hopf in his letters. Nor is anything known about any contributions Nohel might have made to science. He may have been a good assistant and a pleasant discussion partner, but not in the areas which really interested Einstein. Einstein, though, seemed content with his "beautiful institute with its rich library" and the chance of "indulging, without much interruption, in my scientific musings."[11]

While the professors of the humanities worked in the splendid buildings of the Carolinum and Clementinum in the old city, a neoclassical building had been erected for the scientific disciplines on the edge of the new city.[12] There, on the third floor, Einstein had a large, light room as his study, with four windows looking out on an idyllically neglected garden surrounded by stone walls. His apartment had probably been chosen with an eye to its not excessive distance from his institute. Besides, the walk was delightful—crossing the river by the Palacký Bridge, passing on the right the famous gothic Emmaus Abbey (which was one year older than the university and where visitors were shown a room in which Johannes Kepler had lived and worked three centuries earlier), and on past palaces and churches to the institute.

"I'm greatly enjoying my position and my institute here,"[13] Einstein wrote to Besso, but soon a few drops of bitterness entered his correspondence. The philosophy department at Prague (unlike the one at Zurich) was not divided into one section for the humanities and another for mathematics and science; as a result, all problems had to be discussed by some forty full professors at faculty meetings. The power struggles and intrigues among the professors amused rather than irritated Einstein. "Attending those meetings," he told his successor, Philipp Frank, "often saved me a visit to the theater."[14] What did irritate him was the tiresome bureaucratic paperwork, which flourished in the Austrian monarchy even more than anywhere else. "No end to form-filling for the most trivial rubbish,"[15] he complained to Grossmann, and to his fatherly friend Alfred Stern he exploded: "This ink-shitting has no end!"[16]

His lectures too—initially again "Mechanics" and "Thermodynamics"—soon disappointed him: the students, it seemed to him, were "less interested and less hard-working than in Switzerland"[17]; they were "not keen on the subject."[18] After a year of fruitless teaching Einstein was longing for intelligent students, "but the lack of interest among the students for my fine subject is depressing. In my seminar I have one massive male and otherwise only two more or less useful women students whom, in the circumstances, I should be glad to have."[19] There were also other shortcomings: "The Institute of Experimental Physics is pitiful, and there is no real drive in the whole business."[20] After six months he was pleased to see the first chance of another move beginning to emerge, because "it is a sure thing that I will leave this semi-barbaric Prague with an easy mind."[21]

Yet on his arrival in Prague—unlike Zurich—Einstein had been instantly hailed as a celebrity. Newspapers such as *Prager Tageblatt* announced his appointment, emphasizing his achievements in "relativity theory, electron theory, thermodynamics, and the theoretical determination of the size of molecules."[22] As a result, the public lecture he gave on May 24 at the monthly meeting of the Bohemian section of the Society of German Scientists and Physicians[23] turned into a social event. "The whole of Prague's intelligentsia" had assembled, filling "the largest auditorium that could be found in any of the scientific institutes,"[24] recalled a mathematician who had come over from the technical university to hear Einstein. "Einstein had an exceedingly modest manner. With this he conquered all hearts. . . . He spoke vivaciously and clearly, and in no way stiltedly, but quite naturally and at times with refreshing humor. Many a listener will have been surprised that the theory of relativity is something so simple."

Such an ornament to academic life could not long remain lonely in Prague's sparkling intellectual community. He was soon a welcome guest in the salon of Bertha Fanta, an intellectual who was interested in music and who owned the Unicorn pharmacy in the old town square, in a medieval building bearing the same name. There a number of young Jews, who had previously met at the Café Louvre, now held their philosophical and literary discussions. Among the more prominent members of that group were the writer Max Brod and (albeit rarely) his shy friend Franz Kafka, as well as the philosopher and Zionist Hugo Bergmann, Bertha Fanta's son-in-law.

Einstein, it seems, remained equally unaffected by the views of the Fanta circle and by the Jewish atmosphere of Prague—the old Jewish quarter with its several synagogues, and the ancient Jewish cemetery dating from the fifteenth century, with its twenty thousand tombstones piled on top of and between one another. Hugo Bergmann, a philosophy graduate and librarian in the Clementinum, used to attend Einstein's seminars as a guest and often accompanied him home from the institute to the Einsteins' apartment on the other side of the river. Yet Bergmann could not recall ever discussing Jewish topics with Einstein during 1911–1912.[25]

Max Brod was then writing a historical novel about the last year in the life of Tycho Brahe, the great Danish astronomer, who had determined the movements of the planets with an unprecedented accuracy and who had been brought to Prague by Emperor Rudolf II as his court astronomer. Johannes Kepler, initially Tycho Brahe's assistant and later his successor as court astronomer, derived from Tycho's measurements the laws of planetary motion which to this day bear his name. In Tycho, Max Brod endeavored to portray an elderly scholar and God-seeker, obstinately clinging to his "system" and ultimately proving a failure in both his science and his life. Deliberately deviating from historical fact, Brod portrayed Kepler as the young genius favored by fortune. Brod believed that he had found in Einstein a model for the character of Kepler. Einstein, he said,

> time and again filled me with amazement, and indeed enthusiasm, as I watched the ease with which he would, in discussion, experimentally change his point of view, at times tentatively adopting the opposite view and viewing the whole problem from a new and totally changed angle. He seemed to take a downright delight in exploring all possibilities of the scientific treatment of a subject with unflagging boldness: he would never tie himself down, with virtuosity and jocularity he never avoided any multiplicity and yet always retained a sure and creative grip.[26]

It was indeed on this fine and accurate portrait of Einstein that Brod modeled his Kepler, a man "who knows no other conscience than the requirements of his work."[27] And when Max Brod's *Tycho Brahe's Way to God* appeared in 1915, the model was instantly recognized. "This Kepler," Walther Nernst is reported to have said to Einstein, "that's you."[28]

Soon afterward Einstein borrowed Brod's book from a colleague and "read it with great interest." He accepted Brod's portrait or at least did not contradict it: "It's written in an interesting way, by a man who knows the cliffs of the human soul."[29] But his remarks about the author and his Prague circle were rather more reserved: "Incidentally, I believe I met the man in Prague. He seemed to belong to a philosophically and zionistically infested circle there, loosely grouped around the university philosophers, a little band of almost medievally . . . unworldly people." Evidently Einstein, on his occasional visits, had left more of a mark on the Fanta circle than its members had left on him.

The park below the windows of Einstein's study at the institute belonged to what was then called a lunatic asylum. Einstein would take his visitors to the window and, pointing to the inmates strolling under the ancient trees, observe: "Here you see that portion of lunatics who do not concern themselves with quantum theory."[30] To avoid going out of his own mind, he had now resigned himself to the riddles of quantum theory. "I no longer ask myself," he wrote to Besso, whether those quanta really exist. Nor do I try any longer to construe them, because I realize that my brain is unable to penetrate to them."[31] This did not, of course, mean that he abandoned work on quantum theory altogether. Rather, it represented an adoption of his well-tested "heuristic viewpoint" without any attempt to comprehend the nature of quanta: "But I am exploring the consequences as carefully as possible in order to learn something about the range of applicability of that concept."[32] This was a sad retreat to the pragmatic—to what was soluble—even though "the theory of specific heat has achieved veritable triumphs with Nernst discovering with his experiments that everything is roughly as I predicted."[33]

Nevertheless, Einstein had to continue to concern himself with quanta, if only because in mid-June 1911 he had received an invitation from the Belgian industrialist Ernest Solvay to a conference scheduled for the fall, when a select circle of researchers would review the present state of radiation and quantum theory. "I gladly accept the invitation to the meeting in Brussels and will be pleased to write the paper assigned to me," Einstein wrote to Nernst. "The whole enterprise is very much to my liking and I have little doubt that you yourself are its soul."[34] He was entirely correct in this assumption.

Walther Nernst and the philanthropic industrialist Ernest Solvay had conceived the idea of a scientific summit as early as the spring of 1910. Solvay, who thanks to his method of soda manufacture had become the owner of a vast chemical concern—and fantastically rich—had not only philanthropic but also scientific inclinations. These included a bizarre theory of gravitation and matter,[35] which he disseminated with an amateur's unconcern. Because of his wealth, it encountered little opposition. But no one wanted to listen to it either, and Solvay had therefore inquired of Nernst how he could bring his ideas to the notice of such leading physicists as Lorentz, Planck, Poincaré, and Einstein. Nernst seized the opportunity and suggested a scientific summit at which the kinetic theory of matter and radiation theory could be discussed.[36]

It is worth noting that Planck "accepted with the fullest conviction" the idea of a meeting of top physicists, while "not concealing serious misgivings about its feasibility."[37] He suggested to Nernst that the conference should be postponed for a year, because too few physicists regarded the theory as being in need of reform and many of the intended participants would probably not turn up. "From the whole list of those envisaged by you, I believe that, apart from us, only Einstein, Lorentz, W. Wien, and Larmor are seriously interested in the matter."[38] Planck was wrong. When the invitations—on Nernst's suggestion, personally signed by Solvay—went out in June 1911, all of them were accepted in terms as enthusiastic as Einstein's.

The Conseil Solvay was scheduled for the week from October 29 to November 3 in Brussels. Eighteen participants were expected; twelve of them were asked to give papers, which were to be available for study even before the conference, to make the discussions the more productive. Nernst had asked Einstein to give the concluding paper, *On the Present State of the Problem of Specific Heat.*

Einstein had not intended to take any leave during the summer, "because I urgently need the vacations for work."[39] After a break of three years he had resumed his reflections on gravitation within relativity theory and had established that the effect of gravity on the propagation of light should be observable. This advance over his ideas of 1907 galvanized him and became the starting point of his general theory of relativity, whose development would absorb all his efforts over the next few years. In this situation, he may have found writing his lecture an irritating distraction, especially as he did not even have time to answer letters: "I am plagued by my drivel for the Brussels congress."[40] At the end of September he set out for three weeks of traveling—first to the Naturforscher convention in Karlsruhe, where he made the acquaintance of Fritz Haber; then to Zurich, where he gave eight lectures at a vacation course at the Polytechnic,[41] which he had promised to do even before his departure for Prague. On October 28 he boarded a night train in Prague and traveled via Cologne to Brussels, arriving at six in the evening. He was just in time for the festive reception which marked the opening of the event, which he had by then dubbed a "witches' Sabbath."[42]

For the conference, Ernest Solvay had booked the Grand Hotel Metropole in the center of Brussels. His guests were accommodated like kings, and, in addition to having their travel expenses reimbursed, each received 1,000 Belgian francs. Far more important to Einstein, though, was the fact that he was accepted, as a matter of course, into the circle of Europe's leading physicists. He had already met the German participants—Nernst, Planck, Sommerfeld, Rubens, Wien, and Warburg—and also the two Dutchmen, Lorentz and Kamerlingh Onnes. But now for the first time he met the British scientists James H. Jeans and Ernest Rutherford, as well as the physicists arriving from Paris: Henri Poincaré, Paul Langevin, Jean Perrin, and Marie Curie. In a list of participants compiled by nationality, he was included in the Austrian contingent, along with Professor Friedrich Hasenöhrl from Vienna. The function of scientific secretaries of the conference was performed by Frederick A. Lindemann from England—later, as Viscount Cherwell, Winston Churchill's scientific adviser, but then a successful coworker of Nernst's in Berlin—and by the Frenchman Maurice de Broglie.

The event was chaired by Hendrik A. Lorentz, who was universally respected and moreover came from a small—as it were, neutral—country. "H. A. Lorentz presided with incomparable tact and incredible virtuosity,"[43] Einstein wrote enthusiastically of his father figure. "He speaks all three languages equally well and has a uniquely acute scientific mind and delicate tact. A living work of art."[44] On the one hand, Einstein had the satisfaction of hearing his earlier work—except his work on purely experimental topics—mentioned and commended in most of the papers; but on the other hand he complained afterward that the congress "didn't get me much further, because I didn't hear anything I didn't know already."[45]

In his own paper, no doubt deliberately scheduled by Nernst as the conclusion, Einstein disregarded the limits of the subject he had been assigned, widening it to an examination of the "supremely important but unfortunately still essentially unresolved question: how should mechanics be reshaped in order that it may satisfy both the radiation formula and the thermal properties of matter?"[46] Among the participants, however, there was no unanimity, either before or after the congress, on the need for a new quantum mechanics. "Poincaré was simply generally negative, and for all his sharp mind showed little understanding for the situation. Planck is obsessed with some undoubtedly wrong preconceptions . . . , but no one really knows anything. The whole business would have been a joy for diabolical Jesuit fathers."[47] For Besso, he summed up the state of physics and the events in Brussels: "The congress there altogether resembled a lamentation on the ruins of Jerusalem. Nothing positive came out of it."[48] His personal profit consisted in the fact that he had been acknowledged as a leading mind by the elite of his discipline and that he had made useful contacts. "I spent a lot of time with Jean Perrin, Paul Langevin, and Madame Marie Curie and am quite charmed by these people. The last-named even promised to visit us with her daughters."[49]

Einstein impressed all the other participants in almost the same way that Lorentz had impressed him, except perhaps for his tact. He argued for his ideas with complete assurance and—as is shown by the conference proceedings—participated in the discussions with refreshing directness. Immediately after the congress Frederick A. Lindemann wrote to his father: "I got on very well with Einstein, who made the most impression on me except perhaps for Lorentz."[50] And later Lindemann recalled that for his colleague Maurice de Broglie "Einstein and Poincaré were in a class of their own."[51]

The congress received a good deal of publicity in the press, but not so much for its scientific discussions as for the "Curie-Langevin affair" which was revealed by the Paris paper *Le Journal* on November 4, when Curie and Langevin were still Solvay's guests in Brussels. In a dramatic article, Marie Curie was accused of trying to entice Langevin away from his wife and four children, which—the paper suggested—was why the two had eloped to Brussels. Ernest Solvay was not amused. Einstein commented on the business soberly: "That Langevin wants a divorce has been known for a long time. If he loves Madame Curie and she loves him they have no need to elope, because they have plenty of opportunity in Paris of seeing each other. But I don't have the impression at all that there is anything special going on between them, but have found all three [author's note: Einstein here includes Perrin] harmlessly relaxed company. Nor do I believe that Madame Curie is out to dominate, or out to do anything else of the sort. She is a simple honest woman, almost buried under her duties and obligations. She has a sparkling intelligence but, despite her passionate nature, is not attractive enough to be a danger to anybody."[52] On this last point, however, Einstein was wrong.

Yet despite this distasteful hullabaloo in the press and its not exactly outstanding scientific yield, the congress was a complete success. Ernest Solvay endowed permanent institutes of physics, chemistry, and sociology and announced that the Conseil Solvay would become a biennial event. In his final address Solvay expressed the hope that in 1913, at the second conference, he would be able "to present for discussion my gravito-materialistic theory, just as you will do yours." Despite this threat, the Solvay conferences became established as physicists' summits. If Einstein was unable to attend—as happened twice in the 1920s—some of his colleagues felt that it was like holding a Council of the Church without the pope.

Einstein made use of his stay in Brussels to pay a brief visit to his uncle Caesar Koch, and on his return trip he made a detour via Utrecht, in the Netherlands, for the purpose of career negotiations.

Einstein was now, at thirty-two, a famous man. Even though, by the standards of his day, thirty-two may have been on the young side for a full professorship, his reputation was such that he would not have to remain much longer in what he called "semi-barbaric" Prague. The first inquiry had come on August 20, 1911, more than two months before the Solvay conference, from the University of Utrecht. Professor Willem Julius, a pioneer in solar corona research, had, informally at first, offered Einstein a newly vacated chair. Einstein declined, but as Julius was unwilling to admit defeat, a complicated correspondence developed between them,[53] with Einstein keeping all his options open. Although the visit to Utrecht was very enjoyable, on November 15 Einstein sent a definitive refusal from Prague. Utrecht had probably been only a means of accelerating his return to Zurich.

To his friend Zangger, Einstein had described the offer from Utrecht as certainly worth considering: "The salary is 6,000 Dutch guilders. I must confess that I find the thing attractive. Kamerlingh Onnes and Lorentz in my immediate proximity. . . . Do you think that matters could be decided so quickly in Zurich?"[54] While he was teaching a vacation course at the Zurich Polytechnic, Einstein no doubt discussed the chances of a full professorship there with Grossmann and Zangger and judged them favorable. The Polytechnic had just then been upgraded to the Eidgenössische Technische Hochschule, the Swiss Federal Technical University, with all academic privileges, such as awarding doctorates. As part of the reorganization, basic research was to be intensified. The former Section VIA, in which Einstein had been a student, now was an independent Section VIII, and it so happened that Marcel Grossmann was its chairman.

Zangger now mobilized his political connections and, while Einstein was still in Zurich, wrote to Federal Councillor Ludwig Forrer. Although Forrer was in charge of the Postal and Railway Department and therefore not responsible for university appointments, he promised to speak to his opposite number in the Department of the Interior in support of Einstein. To reinforce his argument, Zangger had explained to Forrer that the Technical University urgently needed a theoretical physicist as a full professor, and that Einstein's appointment would be all the easier because he would not need a laboratory or an assistant. More important still, Zangger dismissed any suspicion that, while Einstein might be famous, he had little interest or skill in teaching. "On this point I may be permitted a personal judgment," Zangger wrote to Forrer, "because, while Einstein was in Zurich, I used to hear him lecture for several hours each week."[55] Zangger's judgment amounted to a fair portrait of Einstein as a teacher:

> He is not a good teacher for mentally lazy gentlemen who merely want to fill a copybook with their notes and then learn it by heart for an exam, he is not a smooth talker, but anyone wishing to learn honestly how, deep down, to construct his physical ideas, carefully to examine all premises, see the pitfalls and problems,

> review the reliability of his reflections, will find Einstein a first-class teacher, because all that emerges impressively in his lecture which compels intellectual participation and unrolls the whole extent of the problem.

When Einstein returned from the Conseil Solvay and found no news from Zurich, he became impatient. Career negotiations and the intrigues and the waiting game associated with them were not to his taste: "It is hideous always to be faced with decisions like Heine's donkey between two bundles of hay; one is better off when one is pushed one way or the other by fate."[56] He wanted to be rid of these matters in his own way: "I have just turned down Utrecht," he informed Zangger, "and the dear Zurichers can also get stuffed—except yourself."[57]

No sooner had he mailed this intemperate message than a letter arrived from Grossmann in Zurich, evidently reporting favorable progress in the negotiations. To this Einstein reacted quite differently: "Certainly I am inclined, in principle, to accept a professorship for theoretical physics at your Polytechnic. I am exceedingly pleased at the idea of going back to Zurich. This prospect induced me a few days ago not to accept an invitation from the University of Utrecht."[58] He agreed with Grossmann that "the students of Section VIII at the Poly are still offered too little, or too little of modern matter" and he looked forward "to contributing to the filling of that gap." He had also given some thought to his assumption of the post and expressed the wish "that this might be scheduled for the beginning of the academic year, because that would be the best in the interest of teaching—that is, therefore, next fall." The business did not seem particularly urgent. But this was soon to change.

When Lorentz wrote to inquire "to what extent is it now definitive that you are going to Zurich,"[59] Einstein feared that, in the event of a renewed invitation from Lorentz, he might be unable to decline. He immediately informed Grossmann of the danger: "Unless I am officially tied down to Zurich I would find it difficult, as you can well imagine, to say no to him. I am asking you therefore to see to it *prestissimo* that the negotiations are initiated.—I sincerely apologize for putting you under pressure in this way, but I am in such a delicate situation that I know of no other way out."[60]

As soon as he received the letter, the faithful Grossmann "immediately put his skates on" and informed Rudolf Gnehm, the president of the Swiss Education Council, who lived in Zurich, that "it would be advisable to step up the tempo."[61] Zangger increased the pressure by pointing out to Gnehm that Vienna also had lately been casting a line out; he did not wish "to have it on his conscience not to have spotted the danger in time."[62] Gnehm would have actually been prepared to travel to Prague to negotiate with Einstein, but business accumulating at the end of the year made the trip impossible. After making arrangements by telegram, Einstein therefore went to Switzerland shortly before Christmas and then, accompanied by the globe-trotter Zangger, to Bern. Although on his return journey he did not have a contract in his pocket, he had been assured that it was just a matter of formalities.

That these formalities were speedily executed was largely due to groundwork by Pierre Weiss, since 1902 the successor of the Professor Pernet who had once reprimanded Einstein for "lack of application." Weiss, an excellent physicist, had been put in charge of the appointment procedure, in place of the elderly and ailing Professor Weber, whose memory of his student Einstein might in any case have been somewhat clouded. Pierre Weiss had shown the foresight of asking Marie Curie and Henri Poincaré for expert opinions; their popularity, he rightly judged, would impress the politicians in Bern. Marie Curie, whose area of research was rather remote from Einstein's interests and who could not have understood very much about relativity theory, nevertheless gave him a fine testimony:

> I have greatly admired the papers which M. Einstein published on questions of modern theoretical physics. I believe moreover that the mathematical physicists are agreed that those papers are of the highest order. In Brussels, where I attended a scientific conference in which M. Einstein also participated, I was able to admire the clarity of his intellect, the breadth of his information, and the profundity of his knowledge. Considering that M. Ein-

> stein is still very young, one is justified in placing great hopes in him and in regarding him as one of the leading theoreticians of the future. I believe that a scientific institution which grants M. Einstein the work facilities desired by him by creating the kind of Chair he deserves, will not only partake of a great honor but also with certainty render a great service to science.

Henri Poincaré, who had not greatly impressed Einstein in Brussels, likewise commended his young colleague as "one of the most original minds I ever came across" and was just as generous in his praise as Madame Curie. This makes his reservations all the more interesting: "I dare not say that his predictions will stand up to experimental verification, provided that becomes feasible one day. Indeed, as he has explored all kinds of roads, one must, on the contrary, expect many of his efforts to end in a blind alley. At the same time, however, one may hope that one of the paths chosen by him proves the right one, and that is enough."[63] And if all this was not enough, the proposal to the Swiss Education Council contained a wealth of quotations from the contemporary literature, such as Max Planck's comparison of relativity theory to the introduction of the Copernican world system (with which the Prague department had already impressed the ministry in Vienna) and Sommerfeld's pronouncement at the Naturforscher convention in Karlsruhe "that the relativity principle scarcely belongs among topical questions any longer. Though only six years old—Einstein's paper was published in 1905—it seems already to have become part of the solid stock-in-trade of physics."[64]

Such an exceptional person would be entitled to having his teaching duties defined in a somewhat unusual way. "The very fact that this Chair does not entail any obligations of giving general physics lectures for a big student audience or of supervising laboratory exercises is, in a particular measure, in line with Einstein's nature."[65] The new professor would teach only senior students, which would give him "a small group of motivated students with appropriate basic training, who would especially benefit from a personality such as Einstein."

One problem yet to be resolved was the fact that Einstein had meanwhile become "expensive" and would not want to lose out financially as a result of a move to Zurich. In Prague he was already getting just over 10,000 Swiss francs per annum, the maximum salary in Zurich. He was therefore granted an "annual supplement of 1,000 francs" from a special fund at the disposal of the Swiss government, "along with a share of school fees and attendance charges in accordance with the provisions of the Rules."[66] At the next session of the Education Council, on January 22, 1912, note was taken of Rudolf Gnehm's exhaustive report, and a recommendation was addressed to the Federal Council to the effect that Albert Einstein be "elected" professor of physics at the Swiss Technical University for ten years, the customary period for full professors in Switzerland. The Federal Council complied within a week. On January 30 Zangger reported to Prague that the election had been held, and on February 2 Einstein thanked the Highly Esteemed Herr Federal President: "It gives me great pleasure to be allowed to work as a teacher in Switzerland soon again."[67] We have a postcard, dated the same day, to his fatherly friend Alfred Stern: "Two days ago (Hallelujah!) I was appointed to the Polytechnic in Zurich and have already given notice here of my Royal-Imperial departure. Great joy about it among us all and the two bear cubs."[68]

Einstein was not only glad to be back in Zurich on favorable terms; he was also relieved to be spared further decisions and to be able to decline any further offers. Emil Warburg offered him a position in his Physical-Technical Reich Institute in Berlin, and an unusually well endowed offer also came from Vienna. "But I declined considering anything else at all until I settled down in Zurich," he wrote to Zangger. "It would have been very ignoble to 'sell myself' surreptitiously in this way."[69] He similarly refused an invitation from Columbia University in New York[70] to follow in the footsteps of Planck and Lorentz and give a series of lectures as a visiting professor in the fall of 1912 or the spring of 1913. Physics was more important to him than even an attractive and well-paid excursion across the Atlantic: "Unfortunately I am so overburdened with work that I cannot consider such an undertaking."[71] There was also another reason, which he revealed to his friend Zangger, though not to the Americans: "I don't much care for this public lecture business."[72]

There was a sequel to the offer from the Netherlands. Sometime in February, Einstein learned that Lorentz's inquiry in December, about whether Einstein's move to Zurich now was a certainty, had to do not with the professorship at Utrecht but with his own successor. Lorentz, though still only fifty-eight, had decided to give up his chair in order to devote himself to research and his numerous international obligations. When Einstein realized that his father figure had chosen him as a successor, he wrote to Zangger: "A good thing I had committed myself to Zurich by then, because otherwise I would have absolutely had to go there."[73] The prestigious chair in Leyden, instead of going to Einstein, now—to the surprise of the scientific world—went to Paul Ehrenfest.

If Lorentz was Einstein's father figure, Ehrenfest was Einstein's great friend. Their friendship began in February 1912 when Ehrenfest visited Einstein in Prague. "Within a few hours we were friends," Einstein recalled in his obituary for Ehrenfest, "as if made for one another by our strivings and longings. We remained linked in sincere friendship until he departed this life."[74]

In his diary, Ehrenfest recorded his own impressions of that first meeting. On February 23, he noted:

> At last arrival in Prague—gray. Get off. Walk to the exit. Einstein (cigar in mouth) there, with wife. Straight to a café. . . . Talk about Vienna, Zurich, Prague. . . . On the way to the institute first argument about everything. Rain in the street—mud—all the time discussion. Institute: lecture hall—up the stairs into Theoretical Physics. . . . Continued arguing with Einstein.[75]

Einstein at last had a partner for the productive dialogue he had so missed in Prague, and as they both not only wielded but appreciated a sharp tongue, what Ehrenfest called "argument" was probably the best thing that could have happened to them. Ehrenfest stayed with the Einsteins; they went to a concert in the evening and returned home toward midnight. "Tea. From 12–2.30. Argued with Einstein. Very late to bed." The following morning the same situation at the institute: "We start arguing at once. . . . Later Einstein tells me about his gravitation paper."

The next day was Sunday, "so we played the piano. Brahms violin sonata (boy sings!)—Yes, we'll be friends.—Was terribly pleased." About a walk in the afternoon, with Einstein pushing the baby carriage and taking huge strides, Ehrenfest noted that "Einstein often uses a strong word in the boy's hearing"; he also noticed a hole in Einstein's coat. Thus it continued for four days: Einstein family idyll and "argument" late into the night. On Tuesday evening Ehrenfest gave a lecture on radiation theory to the Mathematical Society, for which Einstein had drummed up all available colleagues, and on Thursday he left: "With Einstein to cake shop (chocolate), then along the Vltava. . . . Einstein complains about Goethe. Sightseeing in the city. To the station—let's go—entrance. Einstein helps carry. Along the train. To the last coach. Einstein very kind and cordial.—Parting. What will the future bring?"

At Easter Einstein paid his first visit to Berlin "in order to talk shop with various people. I have discussions there with Nernst, Planck, Rubens, Warburg, Haber, and an astronomer."[76] These were more or less the Berlin delegation to the Solvay conference in Brussels a few months earlier. There was also the famous chemist Fritz Haber, director of the newly founded Kaiser Wilhelm Institute for Physical Chemistry and Electrochemistry, as well as the assistant at the university observatory in Babelsberg near Berlin, Erwin Freundlich—who was most anxious to test by astronomical observations Einstein's prediction of the deflection of light in a gravitational field. It seems probable that the pillars of Berlin's scientific establishment were also cautiously exploring the possibility of luring their visitor to Berlin on a long-term basis. As early as 1909 Walther Nernst had put out some feelers with a view to forestalling the Austrians by making Einstein an offer. This emerges from a rather cryptic postcard: "I've made inquiries re Einstein, but have no news yet. Besides, the matter is probably independent of whether E. is a little better off now."[77] Shortly before his visit to Berlin, Einstein had declined an invitation by Warburg to the Physical-Technical Reich Institute; but Nernst evidently believed that, in view of the many possibilities in Berlin, it should be possible to put together an offer that Einstein would find irresistible. And in fact, almost exactly two years later, Einstein did arrive in Berlin, complete with moving van and family.

On his visit to Berlin Einstein also met his cousin Elsa again, whom he had not seen since his childhood in Munich. Her name now was Löwenthal; she was divorced and lived with her two small daughters on the top floor of a comfortable apartment block in Berlin's Bayrisches Viertel. The floor below was occupied by her parents: Einstein's "rich uncle" Rudolf from Hechingen, who had loaned money to his father and later "pestered" him,[78] and his aunt Fanny, his mother's sister. Einstein's mother, Pauline, had come to Berlin with her relatives from Hechingen in 1910, but in 1911 she had accepted a post as a housekeeper in Württemberg. Einstein seems to have felt comfortable in this family circle, the more so as a permanent source of contention had been removed. Einstein by then had accepted his mother's—and probably the other Einsteins'—view that his marriage to Mileva had been a great mistake, if not a disaster, and the prospect of a separation and eventually a divorce was beginning to emerge.[79]

The Einsteins' son Hans Albert, while never entirely clear about the causes, later attributed the estrangement between his parents to the fact that his father believed the family would take up too much of his time; he felt he had a duty to concentrate solely on his work.[80] It is quite true that Einstein's letters at that time sometimes seem breathless, and hectic: he was "working furiously on the gravitation problem,"[81] he complained that he "really lacked the least bit of time,"[82] and he "worked like a madman."[83] But then that was the only lifestyle he liked, and Mileva probably bore it patiently.

Einstein himself, in later years, listed other difficulties as well. Mileva had been "very cold and mistrustful toward all people who came close to me in one way or another."[84] This was true, with few exceptions, even of Einstein's male colleagues and friends; when it came to his female acquaintanceships, though they were entirely within the customary bounds of social contact, Mileva would act like a woman possessed. Thus, on one occasion in Zurich, when he exchanged youthful recollections with Annelie, the daughter of the *hôtelier* of the Paradies in Mettmenstetten—and by then a married woman—Einstein had great difficulty avoiding a public scandal provoked by his wife.[85] Her depressed mood seemed to get worse, and her isolation in Prague, where she did not feel part of either nationality and longed to return to Switzerland, did not improve matters. Finally there was the "relationship bordering on hostility" between Mileva and Einstein's mother, "which of course also contributed to the divorce from my first wife."[86]

It was probably then that he arrived at the conviction which, with his sense of fairness, he was later to describe as follows: "The marriage, entered by me from a sense of duty and against the passionate opposition of my parents, was unhappy—for which Mileva by no means bore the main responsibility. With an inner resistance I had undertaken something that simply exceeded my strength."[87] However, this realization was also helped along by another factor—his cousin Elsa. No sooner was he back in Prague again than he received a letter from her, addressed to the institute rather than the apartment, to which he happily replied: "I have grown so fond of you during those few days that I can hardly tell you how." He promised to come to Berlin again before the end of the semester, and he basked in his memories: "I feel quite blissful when I think of our excursion to Wannsee. I'd give a lot to be able to repeat it." Because of Elsa he would now have preferred to go to Berlin rather than to Zurich, but "the prospects of my being called to Berlin unfortunately are rather slight, as I am bound to tell myself on sober consideration." In conclusion he assured his cousin "with every conviction that I regard myself as a fully adequate male. Perhaps there'll be an opportunity to prove it to you."[88]

A week later he had come to the conclusion that he could not, after all, escape the marital yoke: "I suffer greatly from the fact that it is denied to me to love, really to love a woman whom at least I get to look at! I suffer even more than you, because you suffer only from what you *don't* have. Nevertheless I yield to the inevitable in order to prevent something even worse."[89] Two weeks later Einstein once more said good-bye—"I'm writing to you for the last time and return to the unavoidable"[90]—but not completely, because he promised to notify Elsa of an address where she could write to him in Zurich. Two years later Einstein took up a prestigious position in Berlin, and his separation from Mileva became inevitable. After three more years Einstein's address was the same as that of his cousin: Berlin-Wilmersdorf, Haberlandstrasse 5.

But what were marital crises and love affairs compared with his "scientific endeavors"? They certainly did not impair the joy he derived from his work. During his last semester in Prague, in the spring of 1912, he found an almost adequate discussion partner: Otto Stern, age twenty-four, who had just received his doctorate in physical chemistry in Breslau and wished to learn more about thermodynamics, had chosen Einstein as its most up-to-date representative.

When Stern came to Prague, he expected to find a scholarly gentleman with a big beard. He found no one who looked at all professional in the institute, until eventually "there, sitting at a desk, was a man without jacket, without tie, in the kind of shirt Italian road-workers wear, with a large triangular tear in the back of it. It turned out that this man was Einstein. He was terribly nice."[91] Young Otto Stern benefited from Einstein's isolation in Prague and learned "a lot about the problems of the quantum theory, simply because Einstein had only me to talk to—he had no one else. That, of course, was a great stroke of luck for me."[92]

Unless Otto Stern's memory deceived him, Einstein was then racking his brains over the law of radioactive decay, and in this context he was entertaining ideas which amounted to a quantum theory of emission and absorption of radiation. Nothing of this matured into publication; however, Einstein did publish a paper[93] which linked up with the final paragraph of his "revolutionary" treatise of 1905 on light quanta. This dealt with the effect of light on chemical reactions and offered a derivation of the law of photochemical equivalence from thermodynamic principles, entirely in line with his idea of thermodynamics as an eternally valid "theory of principle." However, the quantum hypothesis was not now a premise of the law of photochemical equivalence (as it had been in 1905) but resulted, almost as a spinoff, from the thermodynamic exposition. The paper drew a foolish attack from Johannes Stark, who felt he had to defend his intellectual property.[94] Einstein reacted calmly,[95] the more so as he was able to show that Stark had missed the point of his argument, and had claimed as his own what had already been contained in Einstein's publication of 1905. The paper, moreover, was immortalized when a physical unit was named after its author: the (frequency-dependent) energy of one mole of light quanta is called one Einstein.

Einstein's departure from Prague was not without upheavals—another consequence of his growing reputation. On May 21, Emperor Francis Joseph had agreed that "Dr. Albert Einstein should leave his teaching post as of the end of September."[96] Two days later the announcement appeared in *Prager Tageblatt*, and, as Professor Einstein was now a celebrity, the news was accompanied by all kinds of speculation about the reason for his short stay in Prague. The decline of research in Austria-Hungary was compared with the situation in Germany, despite the fact that Einstein was going to Switzerland. There were also suggestions "that because of his Jewish origins he had been treated badly by the education authorities in Vienna and therefore did not wish to stay in Austria."[97] In spite of his dislike of Prague, Einstein found this gossip annoying. He wrote a letter to the ministry in Vienna which must have been extremely cordial; the official in charge was still raving about it to Einstein's successor, Philipp Frank: "From Herr Einstein I received such a wonderful letter as an official does not usually get from a professor employed by us. I very often have to think of that letter."[98] The letter has been lost, but it was undoubtedly similar to a statement which he published in the Prague *Neue Freie Presse* and which was reprinted by the more widely read *Prager Tageblatt*:

> I have to emphasize that I have had no cause for dissatisfaction in Prague. On my appointment the ministry was most obliging toward me, and during my activity in Prague I likewise had no difficulties with the education authority. My decision to leave Prague is due solely to the fact that when I left Zurich I promised that I would be pleased to return there under acceptable conditions.[99]

He added that the "national conditions in Prague . . . never affected or disturbed" him. "Likewise I never experienced anything or noticed anything of any religious prejudice, such as has been suggested." Even so, he could hardly wait for the semester to end. For months he had rejoiced: "At the end of July comes the haulage man,"[100] and so it was. On July 25 the Einsteins left Prague.[101]

By far the most important aspect of Einstein's sixteen-month stay in Prague was that, on the quiet premises of the Institute of Theoretical Physics, he had found the necessary leisure to resume his thoughts on gravitation within the relativity theory and to develop them further. Later, in his preface to the Czech edition of his popular little book on the relativity theory, he acknowledged this gratefully.[102] His return to Zurich, then, was not only the fulfillment of a long-cherished wish but also a lucky event for science as it traveled the long and difficult road to the general theory of relativity. This road had begun at the Patent Office in Bern in the fall of 1907. Let us, therefore, return once more to that "temporal monastery."

CHAPTER SIXTEEN

Toward the General Theory of Relativity

We know fairly accurately when Einstein took the first step along the road to the general theory of relativity. It must have been toward the end of October and the beginning of November 1907, while he was working on the *Jahrbuch* article—his article for the *Yearbook of Radioactivity and Electronics.* "While I was writing this, I came to realize that all the natural laws except the law of gravity could be discussed within the framework of the special theory of relativity. I wanted to find out the reason for this, but could not attain my goal easily."[1] Einstein recalled the "breakthrough" quite clearly: "I was sitting on a chair in my patent office in Bern. Suddenly a thought struck me: If a man falls freely, he would not feel his weight. I was taken aback. The simple thought experiment made a deep impression on me. It was what led me to the theory of gravity."[2]

This account is surprisingly reminiscent of the original beginning of gravitation theory. It had come to the twenty-three-year-old Isaac Newton, according to his own recollection, as a result of "the fall of an apple," when, in 1666, he was sitting in the garden in his native Woolsthorpe "in a contemplative mood."[3] Newton then had the inspiration that the force attracting the apple to the ground must be the same as the force keeping the moon in its orbit, provided that force diminished with the square of the distance. Newton's development of this inspiration on the basis of Galileo's mechanics and Kepler's laws of planetary motion eventually culminated, in 1687, in his *Principia Mathematica*—the monumental paradigm of the exact sciences and for more than two centuries the model for all physics.

Here, then, 241 years later, was another inspiration, again evoked by free fall. But now the free fall is imaginary, and the falling object is a human being—moreover, a physicist who is still determined on experimenting, despite his unenviable position. If this free-falling physicist for his part dropped a few other objects, he would discover that, relative to him, these were in a state of rest or uniform motion. It has been known ever since Galileo that all bodies fall with the same velocity; therefore, the physicist would be entitled to regard his state as being at rest, a state in which no gravitational field is perceptible.

The problem of weightlessness has meanwhile been demonstrated to a global public by the bizarre antics of astronauts on television. To Einstein, turning this thought experiment one way and then another was sufficient to make it the starting point for a generalization of relativity theory to apply to any kind of coordinate system and, at the same time, to become a theory of gravitation. Even though this eventually led to breathtaking heights of abstraction, the origin was of typically Einsteinian simplicity: "For an observer," he later wrote in the draft of an article which, because of its length, was never published, "in free fall from the roof of his house there exists no gravitational field—at least in his immediate proximity."[4] When, after long years of quest, torturing uncertainty, and hard work, the theory was finally ready, he saw that point of origin as "the luckiest idea of my life."[5]

Soon after the publication of Einstein's relativity theory, a few theoreticians such as Henri Poincaré[6] and Hermann Minkowski[7] took up the problem of integrating Newton's theory of gravitation into the new theory of motion. At first glance this did not seem too difficult, but it soon proved full of pitfalls. Complications arose because in Newton's theory gravitation propagated through space instantaneously—that is, without any time delay—whereas in relativity theory the velocity of light represented an upper limit which could not be exceeded. The tentative relativistic gravitation theories all involved sacrificing such pillars of theory as the axiom of the equality of action and reaction or the principle of the equivalence of mass and inertia.

Einstein too had gone down that road, without knowing the work of his famous colleagues;[8] but unlike them he published nothing in this field, because "these investigations led to a result which made me exceedingly suspicious."[9] Whichever way he tackled the problem, his result "did not agree with the old experience that all bodies are subject to the same acceleration in a gravitational field."[10]

The "old experience" was Galileo's argumentation. In Newtonian mechanics this resulted in the identity of heavy mass and inertial mass. "Inertial mass," or simply "inertia," is the term for the resistance offered by a body to the application of a force; "heavy mass," or "gravitational mass," or simply "mass," is the quantity involved in the gravitational attraction of two bodies. That these two masses, defined quite differently, are identical[11] was not only extensively discussed by Newton, but also seemed to him worth experimental testing. Critical minds, such as Heinrich Hertz and Ernst Mach, had profoundly analyzed that identity, as Einstein was aware; and the Hungarian Baron Roland Eötvös had devoted decades of his life to experiments for an accurate determination of that identity—something Einstein did not know at the time. For most physicists, however, this identity had become so much a matter of course that they scarcely saw it as a problem. When Einstein later observed that past mechanics had " 'taken note' of this important theorem but not 'interpreted it,' "[12] he was, if anything, overly complimentary to the official tradition of physics. In fact, the theorem was not even "taken note" of; it was simply accepted tacitly. To Einstein, the theorem provided an opportunity for discarding all previous attempts to fit gravitation into relativity theory and for giving the problem a new twist by linking gravitation and coordinate systems accelerated relative to each other. This he accomplished in the fifth and final part of the *Jahrbuch* article.

"Is it conceivable that the principle of relativity also applies to systems that are accelerated relative to each other?"[13] This was Einstein's initial question, one that "will occur to anybody who has been following the applications of the principle of relativity so far." He converted his thought experiment—the observer falling from the roof of a house—into the abstract form suitable for a physical journal and so elucidated why "in the discussion that follows [he would] therefore assume the complete physical equivalence of gravitational field and a corresponding acceleration of the reference system." Five years later he would raise this assumption to the "equivalence principle." He immediately points to the "heuristic value of this assumption" in that "it permits the replacement of a homogeneous gravitational field by a uniformly accelerated reference system, the latter case being to some extent accessible to theoretical treatment."

Hidden behind the qualification "to some extent" is the fact that what will be offered is not so much a theory as a "theoretical treatment"—acute reflections, at times bordering on tricks, as well as mathematical approximation arguments. This was due, in the first place, to the circumstance that in relativity theory the definitions of time and space are valid only for nonaccelerated systems of reference and, strictly speaking, cannot be transferred to accelerated systems. Einstein helps himself by juggling with three systems of reference—two nonaccelerated and one accelerated, with their zero points coinciding at a certain moment in time. By means of this ingenious trick he can now transfer the methods of his relativity theory to the accelerated system, albeit only by involved arguments and with one important limitation: "The principle of constancy of the velocity of light can be used here too to define simultaneity if one restricts oneself to very short light paths."[14] By varying this procedure and returning from the accelerated system to the equivalent gravitational field, Einstein manages to deduce astounding consequences.

First he demonstrates that a gravitational field affects clocks, in such a way "that the process occurring in the clock, and more generally, any physical process, proceeds faster the greater the gravitational potential at the position of the process taking place."[15] The difference is minute, and its confirmation would require vast differences of heavy mass. These cannot be found on Earth but are found in the planetary system: thus a clock on the sun would have to run faster than one on the Earth. Einstein leaves the details of the calculation to the reader but presents a conclusion that is verifiable at least in principle: "There exist 'clocks' that are present at locations of different gravitational potentials and whose rates can be controlled with great precision; these are the producers of spectral lines. It can be concluded from the aforesaid that the wave length of light coming from the sun's surface, which originates from such a producer, is larger by about one part in two millions than that of light produced by the same substance on earth."[16]

Einstein does not waste any time considering whether this effect—later to be described as a "red shift" in the gravitational field—is observable. He proceeds immediately to the next aspect, the effect of gravity on electromagnetic processes, which amounts to a treatment of the Maxwellian equations in the gravitational field. Using the methods tested in his first example, he demonstrates that the Maxwellian equations have the same form in an accelerated system as in a nonaccelerated system, although the velocity of light has to be replaced by an expression containing the gravitational potential. "From this it follows that those light rays . . . are bent by the gravitational field." With regard to experimental verification of this prediction, Einstein evidently thinks only of observations on Earth and therefore comes to a melancholy conclusion: "Unfortunately, the effect of the terrestrial gravitational field is so small according to our theory . . . that there is no prospect of a comparison of the results of the theory with experience."[17]

On the final page, Einstein briefly discusses conservation of energy in the accelerated system and gets "a very remarkable result": any energy, including electromagnetic energy, must, in the gravitational field, have a potential energy ascribed to it, which is equal to that of a mass of magnitude E/c^2. Hence the formula $E = mc^2$ is valid "not only for the *inertial* but also for the *gravitational* mass,"[18] a fine rounding off of the concept of the equivalence of mass and energy, whose universal validity is now ensured—and with it an abrupt end to the article.

This was probably the first though certainly not the last time that Einstein worked under extreme time pressure. Papers for journals can be published at any time in an ongoing series of issues; but Johannes Stark had ordered Einstein's contribution for his *Jahrbuch*, his annual, for December 1, 1907.[19] On November 1 Einstein confirmed that he had finished the first part; "I am now busy working on the second part in my scant free time. . . . I certainly hope to be able to send you the manuscript at the end of this month."[20] The manuscript did in fact reach Stark in Greifswald on December 4, but the final nine pages on "Relativity Principle and Gravitation" testify to Einstein's feverish wrestling with new problems under the threat of a deadline. The presentation tends to be sketchy, and the physics is not always argued to the last consequence. It is in the nature of the subject that Einstein did not have any strict methods available to him, but some approximations are awkward and more involved than necessary, and the limits of their validity are left unclear. These pages therefore lack the aura of perfection that distinguished his relativity theory of 1905 and his other major treatises.

Actually, however, the reason for this was not so much pressure of time as the fact that Einstein could not here present something complete, but only the outlines of a future theory, whose final shape he could not even surmise. All the more admirable, then, is his boldness in modifying the principles of his relativity theory—which was scarcely two years old—in order to comply with the equivalence of mass and inertia. This is a triumph of his instinct as a physicist and a crowning completion to his burst of creativity at the Patent Office in Bern, along with a lasting realization: that "a sensible theory of gravitation can only be expected from an extension of the relativity principle."[21]

Einstein's colleagues, however, do not seem to have taken notice of this bold achievement. Max Planck alone asked for a few elucidations, in a letter;[22] otherwise no reactions to that part of the article are found either in the published literature or in Einstein's correspondence. His colleagues evidently were still so fully occupied with comprehending the constancy of the velocity of light, and the principle of relativity, that no interest could be aroused by a discussion of new and seemingly exotic ideas.

For some time after the submission of his article, Einstein was still racking his brain over gravitation theory. On Christmas Eve 1907 he wrote to Conrad Habicht that he was "busy on a relativity-theory examination of the law of gravitation, by which I hope to be able to explain the hitherto unexplained secular changes in the perihelium distance of Mercury."[23] This was the only flaw which astronomers had established in the Newtonian world picture, and Einstein now focused on it. It was not until eight years later that he succeeded in solving this problem, with the armamentarium of the (by then complete) theory, and at the same time put it to the first test of its validity.

For over three years this remark on Mercury's perihelium was to be Einstein's last reference to gravitation theory. He published nothing on the subject, and while his letters show an occasional interest in accelerated systems of reference, there is no mention in them of gravitation. This may be partly due to his intensive occupation with radiation theory, though Einstein had never found it difficult to reflect on more than one topic at the same time. Later in life he had this to say about the years from 1909 to 1912: "All the time that I was teaching theoretical physics at Zurich and Prague universities, I ceaselessly pondered about the problem."[24] But the actual content of his pondering remained his secret. Nor is any light shed on it by his next paper, which was received by *Annalen* on June 21, 1911.

In May 1911, shortly after his arrival in Prague, Einstein had completed a voluminous treatise on molecular movements in solids. His second publication during this time in Prague was *On the Influence of Gravity on the Propagation of Light.*

As he explained at the beginning of the paper, he was reverting to a subject of 1907

> because my former treatment of the subject then does not satisfy me, but, more importantly, I have now come to realize that one of the most important consequences of that analysis is accessible to experimental test. In particular, it turns out that, according to the theory I am going to set forth, rays passing near the sun experience a deflection by its gravitational field, so that a fixed star appearing near the sun displays an apparent increase of its angular distance from the latter, which amounts to almost one second of arc.[25]

This is a spectacular prediction, but Einstein's problems are the same as in 1907, and his methods are almost the same. The principal formulas of the two papers are therefore identical, though this does not mean that Einstein copied himself. Some arguments, such as that for the red shift, are new, and some are clearer and more detailed. He hints at the need for a revision of the fundamentals by calling the theory of 1905 the "ordinary relativity theory" and by making it clear that this will probably not be the last word: "The principle of the constancy of the velocity of light does not hold in this theory in the formulation in which it is normally used as the basis of the ordinary theory of relativity."[26]

His treatment of the problem goes beyond his 1907 paper in that the velocity of light in a nonhomogeneous gravitational field is now a function of the location. In the final section Einstein derives the deflection of a ray of light in any gravitational field, using an analogy with propagation of light in a medium having a variable refractive index. For the special case of a ray passing close to the sun he obtains 0.83 second of arc. This is a minute angle, which would be subtended, for instance, by a quarter—a twenty-five-cent coin—at a distance of three miles. Nevertheless, this small deflection could be demonstrated if, during an eclipse of the sun, the area of the sky behind the sun was photographed and compared with customary nocturnal photographs. "It is greatly to be desired," Einstein concludes his paper, "that astronomers take up the question broached here, even if the considerations here presented may appear insufficiently substantiated or even adventurous."[27]

The possibility of an astronomical verification of the bending of light galvanized Einstein. Even before his paper was published in September, he had confronted the Prague astronomers with his problem and in fact had met with interest on the part of Leo Wenzel Pollak, a "demonstrator" at the Institute of Cosmic Physics of the German University. Pollak approached Erwin Freundlich, an assistant at the Royal Prussian Observatory in Berlin, and Freundlich henceforth made the astronomic testing of Einstein's prediction his life's task. Einstein wrote to Freundlich that he was "extremely pleased that you are taking up this interesting question. I am well aware that its verification by experience is no easy matter, because refraction in the solar atmosphere may interfere. One thing, however, can be stated with certainty: If there is no such deflection, then the assumptions of the theory are incorrect. It should be remembered, of course, that these assumptions, though they seem to suggest themselves, are rather bold."[28]

The next solar eclipse suitable for astronomical purposes was expected in southern Russia in September 1914. This was a hard test of Einstein's patience; he did not want to wait so long. He therefore welcomed Freundlich's proposal that plates taken by the Hamburg Observatory on earlier occasions might be utilized: "I am extremely glad that you are taking up the question of the bending of light with such zeal and I am curious to know what the examination of the available photographic plates will yield."[29] It yielded nothing: the plates proved unsuitable. The alternative option of using not the sun but Jupiter, the most massive planet, as the center of deflection had been discussed by Einstein in his paper—but the effect would be a hundred times less than with the sun, and hence not observable. "If only we had a substantially bigger planet than Jupiter," Einstein lamented. "But nature has not troubled to make the discovery of its laws easy for us."[30] There was therefore no alternative but to wait for the eclipse, and to develop the theory further. The location-dependent velocity of light provided Einstein with a good starting point.

The extent of the validity of the equivalence principle for uniformly accelerated systems and constant gravitational fields had probably been exhausted by Einstein's approximation methods; no further advance seemed possible along those lines. But he saw an opportunity for a new beginning. Even before his paper on the deflection of light appeared, he wrote to Jakob Laub: "The relativity-theory treatment of gravitation causes serious difficulties. I think it probable that the principle of the constancy of the velocity of light in its familiar form is valid only for spaces with constant gravitational potential."[31] That the velocity of light in any gravitational field should vary from one location to another—though a natural extension of the situation previously analyzed, when the velocity of light was constant but also dependent on the gravitational potential—was nevertheless an adventurous idea, because time could no longer be defined as it had been in the "ordinary theory of relativity."

Over the next few months Einstein was exceedingly busy with journeys to the Naturforscher convention in Karlsruhe, lectures in Zurich, the "witches' Sabbath" in Brussels, and career negotiations. Around Christmas he reported to Ludwig Hopf that he was "working like a cart-horse, even if the cart does not always move a lot. But some things have after all come off. . . . I have now derived the theory of gravitation for a static field with great strictness. The matter is very beautiful and astonishingly simple."[32] In February and March 1912 he sent two papers[33] to *Annalen*, developing the idea of a variable velocity of light in the event of a static gravitational field constant over time. As the velocity of light is location-dependent, Einstein made the experiment of replacing the gravitational potential in the classical theory with the velocity of light. As a result, some of the equations look as if they had been taken straight from Newtonian theory, with only the perplexing difference that the location-dependent velocity of light takes the place of the gravitational potential. It could therefore be argued, in a sense, that in those papers it is time that is "bent" whereas space continues to be "flat"—that is, Euclidian.

These papers are by no means so "very beautiful and astonishingly simple" as Einstein had made out to Hopf. They are complicated, uncertain gropings for a transition from the "old relativity theory"—as he now called the theory that had a constant velocity of light—to a new theory. And although he believed himself to be on the right road, he had no clear idea where it would take him. After mailing the second paper he at last found time to write to his friend Besso again: "I have lately been working furiously on the gravitation problem. At the present moment I have finished with statics. Of the dynamic field I don't know anything yet, that's to come next. . . . Each step is damnably difficult, and what I have derived so far is bound to be the simplest part."[34] But he had learned a lot about the relationship between light and gravity and now had some ideas of how to proceed: the new dynamic theory would, in the language of mathematicians, be "nonlinear," and for any accelerated systems not only time but also space would be "bent." This would require a new set of mathematical instruments.

■ ■ ■

As a student Einstein had not been particularly fond of mathematics and had shown interest in it only to the extent necessary. Nothing about this attitude had changed by the time he went to Prague. At the Solvay conference in the fall of 1911 he had surprised his colleagues by his statement that he had only a slight knowledge of mathematics.[35] He was not fishing for compliments: in mathematics Einstein could not have been compared to people like Henri Poincaré or Arnold Sommerfeld. This does not, of course, mean that in the conventional sense he was a poor mathematician. Indeed some of his papers, such as the one about "critical opalescence," are mathematically highly involved. But ultimately, mathematics to Einstein was just a tool.

That was not how the mathematicians saw it. Thus Hermann Minkowski told his students in Göttingen: "Einstein's presentation of his subtle theory is mathematically cumbersome—I am allowed to say so because he learned his mathematics from me in Zurich."[36] Einstein returned this backhanded compliment by (as we have seen) initially dismissing Minkowski's four-dimensional formulation as "superfluous erudition." To Otto Stern he expressed his mistrust of sophisticated mathematics in stronger terms: "You know, once you start calculating you shit yourself up before you know it."[37]

In Prague, however, he realized that his mathematical knowledge needed broadening. Quite often he would run down the stairs to the Mathematical Institute to confront Professor Georg Pick with his problems and to get Pick to recommend books to him.[38] The fact that he eagerly studied a book on determinants was probably due to his concern with Minkowski's presentation of relativity theory and the mathematics that went with it. It seems probable that he eventually familiarized himself with that "erudition" through two didactically outstanding papers written by Arnold Sommerfeld for *Annalen*, intended to meet the needs of physicists. And, of course, he had read the first book on relativity theory, published in 1911. This was the book he himself should have written for Eilhard Wiedemann's series *Die Wissenschaft* (*Science*), but which, in 1908, he had declined. Instead, Max von Laue wrote it, offering a sovereign overview of the subject based on a four-dimensional presentation.[39] Einstein was full of praise for the book and its author: "it is a little masterpiece, and some of it is his own intellectual property."[40]

Einstein by then suspected that Minkowski's (pseudo-)Euclidian geometry based on right angles and straight lines could not be the world of his own future theory. After all, time was somehow "bent" in the gravitational field, and spatial relations were bound to become more complex for systems of reference accelerated relative to one another. In his February paper, using the simple example of a uniformly rotating system of reference, he had already hinted that the ancient wisdom describing the ratio of any circle's circumference and diameter by the number π would no longer be valid in a relativistic world:[41] the circumference of the circle would be shortened because of the Lorentz contraction and its ratio to the unshortened diameter could therefore no longer be π.

But if the two-thousand-year-old Euclidian theory—which in his boyhood had so excited him in his "sacred little geometry book"—was no longer to be valid, what then? To give up geometry altogether was impossible. How Einstein found a way out of this dilemma he explained to his audience in Kyoto on December 14, 1922:

> Describing the physical laws without reference to geometry is similar to describing our thought without words. We need words in order to express ourselves. What should we look for to describe our problem? This problem was unsolved until 1912, when I hit upon the idea that the surface theory of Karl Friedrich Gauss might be the key to this mystery. I found that Gauss's surface coordinates were very meaningful for the understanding of this problem. . . . I happened to remember the lecture on geometry by Carl Friedrich Geiser who discussed the Gauss theory. I found that the foundations of geometry had deep physical meaning in this problem.[42]

The surface theory developed by Gauss in Göttingen in 1828 can be described as an attempt to draw conclusions about the shape of a curved surface by measuring small distances on it. To that end Gauss had introduced generalized coordinates which were no longer based on axes intersecting at right angles. By extending this method to four dimensions, Einstein found in these coordinates the language for which he had been desperately searching, in which curved tracks of light and the paths of gravitating masses might be described. But there was a price to pay for the new freedom which Einstein hoped to gain by generalizing the flexible Gaussian coordinates. The simple definition of distance in "plane" Euclidian geometry, which is valid also in Minkowski's four-dimensional world of relativity theory, had to be replaced by a complicated expression in which a so-called "metric tensor" with altogether ten components played a central part.

This tensor now moved into the foreground of Einstein's reflections, because it was able to meet several of the demands of his new theory. It made it possible to describe any accelerated system of reference; it governed the behavior of measuring rods and clocks and hence the "curved" four-dimensional space-time; and it simultaneously described the gravitational field and the paths of gravitating masses within the curved space-time. Thus a single mathematical function of Newtonian theory, the "scalar potential," had turned into a construct of ten components which, moreover, were bound to be linked with one another in a complicated way. Einstein's problem of deriving the field equations and the laws of motion from this "metric tensor" was not unlike Baron Münchhausen's attempt to pull himself out of a swamp by his own pigtail. Through the "metric tensor" the distribution of matter determines the curvature of space and hence the paths of material bodies, whose movement in turn changes the curvature of space—a close relationship called "nonlinear" by mathematicians and as a rule exceedingly difficult to handle.

While Einstein was in Prague, his development of the theory got stuck in these mathematical problems, in particular the derivation of the laws of motion. Writing to Ludwig Hopf six weeks before his departure from Prague about the work he had done until then on the statics of gravitation, Einstein was "very confident of the results. But the generalization will be difficult."[43] Two months later his report—by then from Zurich—was ecstatic: "Things are going brilliantly with gravity. Unless everything deceives me, I have now found the most general equations."[44]

▪ ▪ ▪

The Einsteins had left Prague on July 25. They would have arrived in Zurich a few days later and would have been busy looking for an apartment and settling in. At the "Residents' Control" their registration was not recorded until August 10, because Einstein had other things on his mind. It seems that immediately on arrival he burst into the home of his old friend Marcel Grossmann: "Grossmann, you've got to help me, or I'll go crazy."[45] Einstein's hopes were not disappointed; his memory of his friend's reaction remained indelible: "Instantly he was all afire, even though, as a true mathematician, he had a somewhat skeptical attitude to physics."[46]

But Grossmann in fact was being appealed to only as a mathematician. Einstein had thought the gravitation problem through to a point where all questions had been reduced to purely mathematical problems. Did differential equations exist for the components of the metric tensor, which, for any nonlinear coordinate transformations, retained their form—which, in other words, were "invariant" with regard to these transformations? Such differential equations should be of the second degree because, in the limiting case of low velocities and small densities of material, only thus could an effortless link-up be established with Newtonian theory.

Grossmann, though not really an expert on differential geometry, was able to help and thus once again proved a "lifesaver" for Einstein. He probably referred Einstein to the geometry developed by Bernhard Riemann,[47] but warned him that this was exceedingly difficult because Riemann's equations were nonlinear. No doubt Einstein excitedly tested these new possibilities, probably still amid the packing cases, and soon he was enthusiastic about his results. Nevertheless, his announcement to Ludwig Hopf on August 16 that things were "going brilliantly," and that he had found "the most general equations," was a little premature. Another nine months of hard work were to pass before publication.

Henceforth Einstein was not interested in anything else. When Arnold Sommerfeld wished to invite him to Göttingen for a series of lectures on the quantum riddles, Einstein assured him "that I have nothing new to say on the quantum subject that would merit interest." A few lines further on he explained his reason: "I am now working exclusively on the gravity problem and I believe that, with the help of a mathematician friend here, I will overcome all difficulties. But one thing is certain—that never in my life have I tormented myself anything like this, and that I have become imbued with a great respect for mathematics, the more subtle parts of which I had previously regarded as sheer luxury! Compared to this problem the original relativity theory is child's play."[48] Sommerfeld forwarded this letter to David Hilbert, accompanied by the regretful note that "Einstein is evidently stuck so deep into gravity that he is deaf to anything else."[49]

The mathematician friend, of course, was Marcel Grossmann. According to Einstein's recollection, Grossmann was most willing to cooperate, "though with the proviso that he would bear no responsibility at all for any assertions or interpretations of a physical character."[50]

For Einstein there followed months of most intense work and concentration. To Ehrenfest he mentioned the "positively superhuman efforts which I devoted to the gravity problem,"[51] and to Ernst Mach he wrote that the work, "after infinite pains and tormenting doubts, has now at last been completed."[52] Not until May 1913 was the paper ready for publication. "Deep down I am now convinced that I have hit on the right solution, but also that a murmur of outrage will run through my colleagues when the paper gets published, which will be the case in a few weeks."[53] In a departure from the usual practice, it was printed first as a separate brochure and only some months later published in a journal.[54]

It had been Grossmann's request that, in their collaboration, the physics would be left to Einstein, while he would confine himself to the mathematics. The second part of the paper, therefore, headed "Mathematical Part," was signed by Grossmann alone; the first, the "Physical Part," was signed by Einstein. The overall title, *Draft of a Generalized Theory of Relativity and of a Theory of Gravitation*, suggests certain misgivings which Einstein, despite his occasional enthusiasm, was never entirely able to dispel. In fact, the complete theory, as developed a good two years later, was missed by no more than a whisker, though some interpreters, with the benefit of acquaintance with the finished theory, accused the two authors of real blunders—one mathematical error by Grossmann and quite a few physical errors by Einstein.

If these were indeed blunders, then Einstein had good reason for committing them. The most serious was the plausible demand that, in the limiting case of weak static gravitational fields, a first approximation must lead to Newton's theory. As his notebooks prove,[55] he certainly found the "correct" equations but again rejected them because they did not permit that approximation. In order to make some progress Einstein sacrificed the "general covariance" of the field equations, i.e., the demand that these equations should retain their form under any coordinate transformations. This can have only been an act of desperation, because it put in question the fundamentals of the theory. As if that were not enough, he immediately proved, with an acute argument, that the field equations could not be covariant anyway—thereby demonstrating that he had not yet understood all the subtleties of absolute differential calculus. Later Einstein observed: "These were errors of thought which cost me two years of excessively hard work."[56] For the time being he confined himself to linear transformations and so fought his way through the labyrinth of mathematics and physical principles. The hard results of this "draft" theory essentially link up with his Prague papers: In the limiting case of weak fields, as found in the planetary system, space is "flat" again and only the path of light is "bent." Deflection in the sun's gravitational field is therefore found to be 0.83 second of arc.

Max Born wrote an enthusiastic review of the new theory. Even if experiment did not bear it out, even if the predicted deflection of light proved to be nonexistent, he believed that Einstein's "bold theory could not be denied admiration. The colossal strength of abstraction and generalization which its author here demonstrates cannot fail to leave its mark on the reader who is not scared off by the complexity of the formulas."[57]

By the time Einstein read this, he already knew better. For all his creative enthusiasm, he still was his own best and severest critic. Shortly after the publication of his "draft" theory he wrote to Lorentz:

> I am happy that you are taking up our treatise with such warmth. But unfortunately the business still has so many major hitches that my confidence in the reliability of the theory fluctuates. The draft is satisfactory in so far as the effects of the gravitational field on physical processes are concerned, because the absolute differential calculus here permits the creation of equations which are covariant with regard to any substitutions. The gravitational field, as it were, appears as the skeleton on which everything hangs. *But the gravitational equations themselves unfortunately do not have the property of general covariance.* Certain alone is covariance with regard to *linear* transformations. But then the whole confidence in the theory is based on the conviction that acceleration of a system of reference is equivalent to a gravitational field. Unless therefore all the equation systems of the theory . . . permit, in addition to linear, also other transformations, the theory refutes its own point of origin and therefore hangs in the air.[58]

Filled with these misgivings, Einstein still had to face critical discussion, and this he did without hesitation. His first lecture,[59] jointly with Grossmann, was at the annual convention of the Swiss Naturforschende Gesellschaft in Frauenfeld on September 9, 1913. This was a kind of trial run for the annual meeting of the Society of German Scientists and Physicians in Vienna.

Few of his colleagues would have read Einstein's more recent papers, let alone studied the mathematically demanding "draft," but many would have read Max Abraham's invectives against Einstein in *Annalen* or heard rumors that Einstein had thought up a totally incomprehensible new theory. The great lecture hall of the Physical Institute in Vienna was therefore filled to capacity when, on September 23 at nine in the morning, Einstein presented his survey *On the Present State of the Gravitation Problem.*[60] He readily admitted that there were still "major hitches" and conceded that his "draft" was "more in the nature of a scientific credo than a secure foundation."[61] For the benefit and entertainment of the many scientists who were not physicists he offered some amusing illustrations, such as that of two physicists "awakening from drugged sleep and noticing that they are in a closed box with opaque walls, but equipped with all instruments,"[62] and who, despite all their efforts, are unable to discover whether their box is resting on a celestial body such as the Earth or whether some external force imparts to it a uniformly accelerated motion. With great fairness he first presented the theory advanced by Gunnar Nordström in Helsinki and then his own, modestly suggesting that he regarded his own as "more natural": "Admittedly it leads to equations of considerable complexity, but the equations looked for all arise out of surprisingly few hypotheses and they satisfy the demands of the relativity of inertia."[63]

The organizers regretted that "many of the better known German names [were] absent."[64] Not only were Planck and Sommerfeld missing, but Abraham and Nordström had not arrived either. But Professor Gustav Mie from Greifswald was there, and in the course of the "very lively discussion"[65] got very angry because the speaker had not mentioned his theory at all. Einstein explained that he had "not spoken of Herr Mie's theory because in it the equivalence of mass and inertia is not strictly observed. It would have been illogical of me to proceed from certain postulates and then not to adhere to them."[66] Mie retorted with a threat that he would "shortly publish a treatise in which it will be shown that the Einsteinian theory likewise does not absolutely meet the demand of the equivalence of inertia and mass." Einstein suggested that discussion of this question "might with advantage be postponed until Herr Mie has published his misgivings on that point." In fact, Mie wrote a violently polemic paper, but Einstein managed to see it positively: "I am delighted that my colleagues are concerning themselves with my theory, even though, for the moment, only with the intention of killing it dead."[67] He waited in vain for approval. "With regard to my gravitation paper physical mankind is keeping rather passive,"[68] he observed a few months later.

Einstein's only ally, and ultimately the only judge of his theory, was nature itself. He had concluded his lecture in Vienna with his conviction that the verdict would come from "photographs of stars appearing close to the Sun during solar eclipses. I hope that the eclipse of 1914 will bring about that important decision."[69] From personal remarks, however, it is clear that what he was expecting was not so much a decision as a confirmation of his theory: "In myself I am fairly convinced that the light rays do in fact undergo bending."[70]

Full of impatience, Einstein bombarded the Berlin astronomer Erwin Freundlich with letters, thanking him profusely for his interest, urging him to hurry, and explaining to him the significance of the problem: "Nothing more can be done along theoretical lines. In this matter it is only you, the astronomers, that can next year perform a downright invaluable service to theoretical physics. We will obtain reliable information on whether it is correct to generalize the relativity theory further, or whether we have to come to a halt at the first step."[71]

In order to shorten the period of uncertainty—a period of at least a year—Freundlich came up with an idea to which Einstein instantly reacted with enthusiasm: "I am extremely interested in your plan to observe stars close to the Sun during the day."[72] When in September 1913 Freundlich, on his honeymoon, passed through Zurich with his young wife, this idea no doubt was his and Einstein's principal topic of conversation. Soon, however, Einstein was told by the Zurich astronomer Professor Maurer that Freundlich's plan would hardly prove feasible. Still, Einstein did not give up. In October he approached George Hale, the founder and director of the Mount Wilson Observatory in California, an observatory equipped with unique instruments, with the question, "Using your greatest magnification, to what proximity to the Sun can bright fixed stars be seen in daytime (without solar eclipse)?"[73] To emphasize that this inquiry came from a responsible person, Professor Maurer added a few friendly words, as well as the official seal of the Eidgenössische Technische Hochschule (Swiss Technical University). As expected, Hale saw "no possibility of observing the effect in sunlight" and referred to eclipses.

Einstein thus had no choice but to wait for the next solar eclipse and to enable Freundlich to mount an expedition to southern Russia. He wrote to Planck, who had "involved himself in the matter with real seriousness,"[74] but he did not expect a commitment on the part of the

Prussian Academy of Sciences. He reassured Freundlich: "If the Academy won't play ball, then we'll get that little bit of Mammon from private quarters. . . . If everything fails, I'll pay for the thing out of my own slight savings, at least the first 2,000 marks. After mature consideration you therefore go ahead and order the plates, and don't let time slip by because of the money question."[75] To anticipate: Einstein did not have to touch his savings, or even the 15,000 marks he had received through Emil Fischer from Franz Oppenheim (whose identity was still unknown to Einstein). Eventually, thanks again to Emil Fischer, the bill was paid by the firm of Krupp in Essen.

While he was waiting for the solar eclipse, Einstein, as he wrote to Freundlich, had once more "thought the theory through from all sides" and could "say nothing else than that I have every confidence in the business."[76] Simultaneously, he was trying hard to develop his theory further, so it would no longer "hang in the air," as he had said to Lorentz. In a further publication with Grossmann[77] he managed, at least for a few important special cases, to free himself of the "draft" theory's limitations to linear transformations. The new techniques were then successfully tested in a joint paper with Adriaan Daniel Fokker, who was Dutch and, having taken his doctorate under Lorentz, had come to Zurich for one semester. At the beginning of March, Einstein once again asked Besso not to "blame me too much for my unheard-of silence, I have been working like a madman, and what's best about it, with great success."[78]

Proceeding from certain coordinate conditions in his paper with Grossmann, he and Fokker now succeeded[79] in finally treating gravity in a form in which the demand for covariance was strictly observed. He imparted the news to his friend Zangger as an "unheard-of success": "We have successfully proved that the gravitational equations are valid for any moving systems of reference, that therefore the hypothesis of the equivalence of acceleration and gravitational field is absolutely correct, in the widest sense. Now the harmony of the reciprocal relations in the theory is such that I no longer have the least doubts of its correctness."[80] In this ecstatic mood he was even prepared to dispense with verification by experience. Besso must have wondered what had become of his friend, who had formerly been so empirically minded, when he read: "I am now completely satisfied and no longer doubt the correctness of the whole system, no matter whether the solar eclipse observation succeeds or not."[81] This confidence in his own vision was something new, and remarkable even in Einstein. At the time he wrote those words he had absolutely no experimental data, and even in the area of theory he had nothing but his faith in the analysis of profound and simple postulates. In retrospect we know that important steps still lay ahead. We also know that he was on the right track, that he was close to his goal, but that he had not yet arrived.

Perhaps he himself suspected that something essential was still lacking. Only thus can we understand the simile by which he explained his position to his friend Zangger: "Nature is showing us only the tail of the lion. But I have no doubt that the lion belongs to it, even though, because of its colossal dimensions, it cannot directly reveal itself to the beholder. We see it only like a louse sitting on it."[82] It took another eighteen months before, in Berlin, he was to see the lion as a whole.

During the three semesters that he spent at the Swiss Federal Technical University in Zurich, nothing was more important to Einstein than the theory of gravitation. But of course there were other matters, which will now have to be covered.

CHAPTER SEVENTEEN

From Zurich to Berlin

EINSTEIN'S ARRIVAL IN ZURICH at the end of July 1912 was like a homecoming. The Einsteins again lived on the sunny slope of the Zürichberg, but this time a little higher up, near the church at Fluntern. The apartment was in a newly built two-family house; it had six rooms, central heating, and all modern conveniences.[1] Two years previously the annual rent, 2,600 francs, would have eaten up more than half of Einstein's salary. Now, as a full professor at the top salary grade, he no longer had to live carefully. The detour via Prague had paid off.

In the Physical Institute of the Polytechnic he had a spacious, sunny office above the south entrance, with a panoramic view of the city all the way to the Uetliberg. While still in Prague, Einstein had commented, not too tactfully, on the changes that had taken place in Zurich: "Grouchy Weber has died, so that things will be quite agreeable also personally."[2] Pierre Weiss, a native of Alsace, was now director of the institute, but not the superior of the professor of theoretical physics; also, Einstein got on well with him. Apart from Marcel Grossmann, who was on the board, Louis Kolross, another former fellow student of Einstein's, was now professor of descriptive geometry—so that of the five graduates of the class of 1905, three had returned as professors. Of the older generation, whose lectures Einstein had either suffered or cut, the mathematicians Geiser and Hurwitz were still in their posts. Soon, just as two years before, there were regular chamber music meetings at Hurwitz's home.

At the university, Kleiner was still the *Ordinarius.* To Einstein's regret, Peter Debye, his own successor as "extraordinary professor," had left Zurich again and accepted the post in Utrecht which Einstein had declined. Instead, Max von Laue joined the university in October. In Einstein's opinion this was a most welcome appointment, "not only for the university, but also because it will give me great pleasure to be personally in contact with Laue, whom I esteem very highly both as a colleague in my field and as a person."[3] In time for his appointment proposal, Laue, then still at Sommerfeld's institute in Munich, had taken his first photographs of the diffraction of X-rays by simple crystals. When Laue had sent one such photograph to Prague, Einstein had immediately congratulated him "on your wonderful success. Your experiment is one of the finest things to have happened in physics."[4] That atoms and molecules could now be seen, not just indirectly as in the Brownian movement but almost directly, was an overwhelming experience for Einstein: "It is the most wonderful thing I have ever seen,"[5] he wrote to Hopf. "Diffraction on individual molecules, whose arrangement is thus made visible. The photograph is much sharper than one would expect, given thermal agitation." Simultaneously it provided definitive proof that X-rays were electromagnetic waves, like light—but Einstein, of course, had been assuming that for many years anyway.

Two years later Laue received the Nobel Prize for this achievement. With Einstein following in 1922 and Peter Debye in 1936, this was a remarkable balance sheet for Zurich's appointment policy. All three physicists who had begun their academic career as "extraordinary" professors between 1909 and 1912 eventually received this highest of scientific accolades.[6] Alfred Kleiner may not have been an outstanding scientist, but he evidently had an eye for talent.

Einstein resumed teaching with a three-hour lecture on analytical mechanics and a two-hour lecture on thermodynamics, as well as a seminar. According to the recollections of Otto Stern, who had come to Zurich with, and because of, Einstein to obtain *Habilitation* in physical chemistry, these lectures were "always very fine, but not for beginners."[7] This was entirely in line with the intentions of the Educational Council; advanced students were to be offered something special. But because Einstein never properly prepared himself, even advanced listeners often had difficulty in following him. "But then this was Einstein, and even when he messed about it was always very interesting. He always did so in a most sophisticated, and above all a very physical, way."[8]

Einstein was more in his element in his weekly colloquium, for which Max von Laue and a few students came over from the university. As it had been two years previously, the colloquium would usually be continued in a restaurant or at the Café Terrasse. The principal topics there would be radiation and quantum theory, the existence of so-called zero-point energy, and the theoretical interpretation of magnetic phenomena.

There was another reason why Einstein became the center of communications in the institute—a reason that had nothing to do with physics. Pierre Weiss had decreed a no-smoking rule for the entire building and made sure it was strictly observed. Needless to say, this could not apply to his colleague Einstein, with the result that his office became a kind of smokers' corner. Anyone wanting a smoke would come to Einstein, and people wishing to see Einstein had only to say that they wanted a quick smoke.

As far as his great passion, the theory of gravitation, was concerned, Marcel Grossmann was probably the only person in Zurich he could discuss it with, and then only in its mathematical aspects. Laue, who had been one of the first champions of relativity theory and who had made some valuable contributions to it, recalled "much argument with Einstein on this subject,"[9] but Laue probably disappointed Einstein by not showing much interest in his endeavors to generalize it. "Laue is not accessible to fundamental reflections"[10] was Einstein's summing up of their three semesters as close colleagues.

Although he was not at all enthusiastic about public lectures, Einstein had accepted an invitation to the annual meeting of the Société Française de Physique in Paris at the end of March. He traveled with Mileva, while the children remained in Zurich in the care of Mileva's mother. He must have looked forward not only to the change of scene, but also to seeing his French colleagues, as well as his "good Solo"—

Solovine—who was scraping a living in France as a writer and with whom he intended to "loaf about" Paris. "If only there wasn't that damned lecture which—horrible thought—I am to give in French."[11] But he made the lecture, given on March 27, 1913, easy for himself, and not only because of the language problem. He had nothing new to report on quantum theory, and he did not wish to speak about the theory of gravitation, whose "draft" was then in its final stage. He therefore chose for this subject his last attractive paper from his days in Prague, on the law of photochemical equivalence.[12] This was probably not what his audience had expected.

In Paris, Paul Langevin did Einstein the favor of listening with interest when Einstein talked about the hard time he was having with the theory of gravitation. It is likely that Einstein also discussed with Langevin and Marie Curie a subtle argument[13] according to which the hypothesis of equivalence—the "essential identity of mass and inertia"—is supported, in a surprising way, by the mass changes in radioactive decay. Back in Zurich, Einstein thanked Marie Curie for "the hours you devoted to us"[14] and again reminded her of the little "mountain trip" they had both planned for after the end of the semester.

At the beginning of August, Marie Curie arrived in Zurich with her two daughters and their governess. Einstein set off with his guests and his son Hans Albert for a hike in the Engadine. Although Einstein had some difficulty with his French and Marie Curie understood hardly any German, there do not seem to have been any difficulties with communication. In her biography of her mother, Eve Curie described this hike: "In the vanguard gambolled the young ones, who were enormously amused by the journey. . . . Sometimes they caught words on the fly which seemed to them rather singular. Einstein, preoccupied, passed alongside the crevasses and toiled up the steep rocks without noticing them. Stopping suddenly, and seizing Marie's arm, he would exclaim: 'You understand, what I need to know is exactly what happens in a lift when it falls into emptiness.' Such a touching preoccupation made the younger generation roar with laughter."[15] That summer Eve Curie, like Hans Albert Einstein, was nine.

■ ■ ■

Before the end of the summer semester, Paul and Tatyana Ehrenfest had come to Zurich from Leyden for a couple of weeks. They were staying at a *pension*, but Ehrenfest spent as much time as possible with Einstein. Only once did he record in his diary: "A day without Einstein."[16] For the two physicists, this was an opportunity to consolidate the friendship begun in Prague. They again made music together, roamed the hills around Zurich, and discussed physics with their characteristic sharp-tongued passion. Ehrenfest, a temperamental person, proved a stimulating addition to Einstein's and Laue's already lively colloquium. Even in old age, Max von Laue recalled "Einstein and Ehrenfest striding in front of a large swarm of physicists, climbing the Zürichberg, and Ehrenfest there erupting into triumphant shouting: 'I understand it now!' "[17]

Just then the new atomic model, proposed that spring by Niels Bohr, was being fiercely discussed. Einstein no doubt would have learned about it from Ehrenfest. While their colleagues were excitedly debating whether Bohr's model could make any physical sense at all—Otto Stern and Max von Laue were actually horrified by it—Einstein surprised Stern with the remark: "You know, I've been thinking along much the same lines myself for some time."[18] But the theory of gravitation preoccupied him so much that he was unable to take an active part in this other problem. Two years previously he had, in a sense, taken his leave of quanta, so that he was only an observer during this exciting phase of the birth of atomic physics.

He showed similar reserve when, toward the end of October, he traveled with Max von Laue and Pierre Weiss to Brussels for the Second Solvay Conference. He had gratefully accepted the invitation, but had urgently asked not to have to give a paper, as he was overwhelmed with work.[19] On the general subject, the structure of matter, he had nothing new to communicate; nor did he learn anything new because, rather strangely, the Rutherford-Bohr model of the atom was excluded from the discussion. The group photograph, however, reflected Einstein's enhanced reputation: while in 1911 he had been placed at the end of the line, he now stood at the center. Moreover, as his colleagues realized, his outstanding position was soon to be enhanced by a move to Berlin, to the Prussian Academy.

▪ ▪ ▪

As he had done at the University of Zurich and later in Prague, Einstein now, as professor at the Polytechnic, began to think about a change even before the end of his second semester. On July 12, 1913, Max Planck and Walther Nernst had come to Zurich to make Einstein an attractive offer to move to Berlin. Nernst had first thought of winning Einstein for Berlin in 1910, and in January 1913 Haber had discussed with the Ministry of Education "whether a post might be created for this extraordinary man at the Institute under my direction."[20] The scientific establishment of Prussia was clearly pursuing several objectives: optimum working conditions for the most productive researcher in their field; a fascinating addition to the already impressive circle of scientists in Berlin; and, finally, the greater glory of Prussia and the Prussian Academy. That this man Einstein would hardly be a model subject of His Majesty the Kaiser—considering that Einstein was of Jewish origin, had no military service, and had actually been released from German citizenship on his own initiative, even if it was only the citizenship of Württemberg—all this would be graciously overlooked. When it was a matter of acquiring an intellectual elite, the Prussians could be liberal, and anyone who wanted the "new Copernicus" had better ask no questions about his lieutenant's warrant.

The best opportunity to make Einstein an irresistible offer arose through the death of Jacobus van't Hoff, who had held a salaried post at the Prussian Academy since 1886 and who had won the first Nobel Prize for chemistry. There were only two such posts, because the academy was not so much a research institution as a club of outstanding scholars. Membership was regarded as a high distinction and, with Prussian thriftiness, carried a honorarium of 900 marks per annum. When van't Hoff died, the academicians were bound to offer the post to the first Nobel laureate for physics, Wilhelm Conrad Röntgen. But at sixty-seven, Röntgen thought he was too old to move; he would rather stay in Munich and therefore declined with thanks. This made it possible for the post to be offered to Einstein.

The offer—at Franz Haber's initiative, arranged in advance with the bureaucracy—proposed that Einstein, in addition to the honorarium, would receive the maximum salary for professors in Prussia:

12,000 marks a year. Half of this salary was to be met from the academy's coffers and the other half by a donation from the industrialist Leopold Koppel.[21] The position included a professorship at the Friedrich-Wilhelm University with all academic privileges but without any duties. Einstein would be able, just as he pleased, to give lectures and hold seminars—or not. As if that were not enough, Einstein would be offered the directorship of an Institute of Theoretical Physics to be established within the framework of the Kaiser Wilhelm Society. This, as it soon turned out, was more than Einstein desired or thought sensible, but in the meantime his only duties would be to live in Berlin and not miss too many of the meetings of the Prussian Academy.

On June 12, 1913, Max Planck submitted an official proposal for Albert Einstein to be elected a regular member of the academy at a personal salary of 12,000 marks. The proposal was obviously formulated by Planck, but was also signed by Walther Nernst, Heinrich Rubens, and Emil Warburg. With Prussian bombast, Planck had strung together so many commendations that his single reservation could not detract from them: he noted that Einstein had occasionally gone overboard in his speculations, such as his light quanta hypothesis; however, one should not hold this against him "because, without sometimes taking a risk, no real innovation can be introduced even in the most exact natural science."[22] Einstein's theory of gravitation, Planck suggested, would have to be judged in the future—even though, as we will see, a year later he would use his *laudatio* on welcoming Einstein to the academy to record his misgivings about any generalization of the relativity principle.

The proposal made its way through the committees and on July 3 was passed on to the Physical-Mathematical Class of the Prussian Academy for a secret ballot. Following the traditional ritual this was performed with black and white balls. The result was twenty-one white balls and only a single black ball. A few subtle academicians had been urging that Einstein's personal salary should be prorated or scrapped altogether in the event that he received any other official income.[23] Others voiced "fundamental misgivings about the participation of a private person in the appointment of a new member of the Academy,"[24] arguing that it would be "more dignified" not only for

Einstein but also "for the Academy if the Class made the full sum of 12,000 Mark available to the appointee and if Herr Kommerzienrat Koppel instead transmitted his contribution of 6,000 Mark to the Academy in the form of a donation."[25]

The way was now clear for Max Planck and Walther Nernst to travel to Zurich to see Einstein, who of course knew that Berlin was interested in him but was taken by surprise by the details of the offer. Planck and Nernst were anxious to return to Berlin with an acceptance but were tactful enough to allow Einstein a day to consider, as he requested. The two emissaries made an excursion to the Rigi mountain by Lake Zug, and arranged with Einstein that in the event of his refusal he would meet them at the station on their return with a bunch of white flowers. In the event of his acceptance, he was to carry a bunch of red flowers. Einstein carried red flowers.[26]

On July 24, 1913, the plenary meeting of the academy approved the arrangements and, through the official channels, by way of the Ministry of Education, submitted them to His Majesty Wilhelm II, Kaiser of Germany and King of Prussia. Even though these were mere formalities, they took time. "Einstein will move to Berlin at Easter,"[27] Nernst wrote to Lindemann in August. "Planck and I recently visited him in Zurich and the Academy has already elected him; we have great hopes of this."

It seems that Einstein kept the Berlin offer, and his acceptance, a secret for the time being; otherwise Ehrenfest, who was in Zurich shortly afterward, would have recorded it in his diary. Only to his cousin Elsa did Einstein enthusiastically report the joyous news, but he asked her to keep it quiet: "Don't tell a soul about it. It still needs a resolution by the Plenary Meeting of the Academy, and it would look bad if anything leaked out before then."[28] His next mention of this is in a letter to Jakob Laub of July 22; by then Laub was a professor in Argentina, a long way off: "The thing is that at Easter I am going to Berlin as an Academy person without any obligations, rather like a living mummy. I'm already looking forward to this difficult profession!"[29] This does not sound as if he was convinced that he had made the right decision.

It is not entirely clear why Einstein accepted the invitation to Berlin. He had never had a longing for Germany—rather the opposite. The Swiss republicans' deep-rooted mistrust of the "great canton in the north" with its saber-rattling and absurd striving for a "place in the sun" had been instilled in him, by "Papa" Winteler, while he was still a schoolboy in Aarau. He hardly knew Berlin, having been there only once—in March 1912—but he must have realized that as a "native" Swabian he was closer in mentality to the southern Germans, with their enjoyment of life, than to the Prussians' stiff demeanor and glum ethic of duty. He had run away from a Bavarian gymnasium, and now, far worse, his two sons would have to attend Prussian schools.

Nevertheless, his first recognition as a scientist had come from Germans—Max Planck, Walther Nernst, Wilhelm Wien, and Arnold Sommerfeld—and this he had not forgotten. When he was asked by the editors of the newly founded journal *Die Naturwissenschaften* (*The Natural Sciences*) to write a tribute to Max Planck on the occasion of Planck's election as rector of the University of Berlin for the winter semester of 1913, Einstein seized on the opportunity to voice his gratitude for Planck's support of relativity theory: "The notice which this theory encountered so quickly among my fellow scientists is largely due to the determination and warmth with which he championed this theory."[30] But when Einstein had irrevocably accepted the invitation of the Prussian Academy, he wrote to Besso that he saw, "not without certain misgivings, the Berlin adventure drawing nearer,"[31] and he believed that Planck's and Laue's lack of interest in the generalization of relativity theory reflected a limitation typical of the German national character: "The free, uninhibited gaze is altogether something the (adult) German does not have (blinkers!)."[32] Why then did he go to Berlin?

The higher income can hardly have been the reason. He did not spend much on himself, and as an *Ordinarius* in Zurich he was certainly able to ensure that his family would live comfortably, without any financial restrictions. The fact that his earnings in Berlin would be greater by one-third was certainly not unwelcome, but it was not likely to influence his decision. Besides, the salary was by no means exorbitant for an academic in Prussia. Although the top salary of 12,000 marks was granted to only a few professors, many of those on a lower basic salary would vastly exceed that figure through lecture and examination fees.[33] Einstein's salary, in consequence, was more or less in line with the average earnings of a full professor. For instance, the directors of the Kaiser Wilhelm Institute, an institution devoted solely to research, had a salary of 15,000 marks, designed to compensate them for the loss of the usual additional earnings.[34] For professors of medicine, with a flourishing private practice, such an income would not be particularly impressive, nor would it for an economic wizard such as Walther Nernst, who was both mocked and admired as an "honorary *Kommerzienrat.*"

More important, probably, than his salary was Einstein's unfortunate experience that his teaching had been distracting him from his gravitation theory. In a letter to Laub he had hinted that the two activities had been too much for him: "I am so run down from my work during the past year that I really deserve something."[35] When news of his move became known in August, probably through academic gossip originating in Berlin, Einstein gave his workload—his lectures—as the main reason. Thus he wrote to Lorentz, who had congratulated him on being elected to the academy: "I couldn't resist the temptation of a post in which I would be free from all obligations and be able to indulge wholly in my musings."[36] When Ehrenfest, not without a hint of irony, congratulated him on his impending "Berlinization," Einstein's justification was similar: "I accepted this odd sinecure because lecturing in class was getting oddly on my nerves, and there I won't have to do any lecturing."[37] This was curious reasoning, perhaps, for a man who throughout seven years of six-day weeks at the Patent Office had found it possible to "indulge in his musings."

Certainly, Einstein was not an inspired teacher, but at times, when his theoretical reflections were not getting him anywhere, he had found teaching an honest way to earn his living. "In my own research it often happens that I pursue and explore some thought, only to realize in the end that I have been wandering in a maze," he had once told Hugo Bergmann in Prague, on the way home from the Physical Institute. "All those weeks of hunting a phantom would have been lost, had I not been giving lectures and thus done something useful all that time."[38] Still, the chance of doing without that reassurance and devoting himself solely to speculation must have seemed extremely attractive to him, even though it was offered him by the Prussian Academy in Berlin.

His friends and colleagues in Zurich seem to have reacted to Einstein's readiness for his "Berlin adventure" partly with astonishment and partly with disappointment, but also with a conviction that if anyone deserved such a position—without obligations or worries—it was Einstein. When he told Aurel Stodola of his call to the Prussian Academy, Stodola was overcome: "a tremor ran through me and a tear of joy welled up in my eye because ideal justice had been meted out to somebody on this earth."[39]

And there was another reason, which he did not then disclose to anyone. Two years later, he would casually mention it in a letter to Zangger, and then it seemed the main reason: "I live a very withdrawn life but not a lonely one, thanks to the care of a female cousin who actually drew me to Berlin in the first place."[40] We do not know when it was that Einstein gave his cousin his address at the Physical Institute of the Zurich Polytechnic. But we do know that Elsa Löwenthal wrote to him on his thirty-fourth birthday, March 14, 1913. He was delighted "that you thought of me,"[41] but kept this letter and the next few letters free of amorous effusions. After all, he had decided to submit to the inevitable. This changed after the visit from Nernst and Planck: "Next year, if not before, I am coming to Berlin for good," he wrote jubilantly. "It is a colossal honor that's being bestowed on me. . . . I am already looking forward to the wonderful time we will be spending together."[42] Thus it continued at short intervals. Elsa was no less happy at this turn of events; she even, on her own initiative, called on Fritz Haber, who was pulling strings for Einstein at the academy, in order to speed up the procedure. "The boldness with which you bearded Haber is typical Elsa. Had you told anyone about this, or did you settle it alone in your black soul? I wish I had been present."[43]

After the Naturforscher convention Einstein visited his cousin. Back in Zurich he wrote to her that he was "no longer the same person as before. I now have someone of whom I can think with unclouded pleasure and for whom I can live."[44] As once with Mileva, he looked forward to a joint "small gypsy ménage."[45] Elsa's literary reading inspired Einstein to theatrical flights of fancy in his youthful "anti-Philistine" vein: "Traveling people, both of us, destined to high-wire dancing from among the swarm of Philistines, though—thankfully—not on a real wire, but only on the sunlit heights of human insanity! I want a splendid little dance!"[46] At the same time he steadfastly resisted Elsa's advice about a healthy way of life and commended his own lack of consideration for himself: "I have firmly resolved to bite the dust, when my time comes, with the minimum of medical assistance, and until then to sin cheerfully just as my dastardly soul prompts me: smoke like a chimney, work like a beaver, eat without thought or choice, walk *only* in really agreeable company, in other words rarely, sleep irregularly, etc."[47] A few years later Elsa had to bear the consequences of that lifestyle when she had to nurse him in Berlin and cook a special diet for him.

When, during Einstein's visit to Berlin toward the end of September 1913, the two cousins had become very close, Elsa thought it appropriate to give him a few implements for his personal hygiene. It was an attempt to tame a recalcitrant. While he used the hairbrush, he had "retired the toothbrush on genuinely scientific considerations: pigs' bristles can drill through diamond, so how should my teeth stand up to them."[48] When Elsa evidently repeated her admonitions about the purpose and usefulness of hygiene, he eventually burst out: "But if I begin to groom my body then I'm no longer myself. . . . If I am so unappetizing to you you'd better find a friend more acceptable to feminine tastes. But I'm retaining my indolence, which also has the advantage that many a 'jerk' leaves me alone who might otherwise call on me.—With a vigorous curse, therefore, and a blown kiss from a hygienic distance, your honestly filthy Albert."[49] In his next letter he thanked Elsa for not holding his outburst against him with a "blown kiss from a hygienic distance from the incorrigible filthy fellow. We are agreed that we will light up each other's lives amidst this affliction."[50]

The affliction, of course, was Einstein's long since wrecked marriage, from which, under the law then in force, there was no escape against Mileva's will. "Do you think it's that easy to get a divorce," he complained to Elsa, "if one has no evidence of the other party's guilt?" Although he suspected an adulterous relationship (as it was then called) on his wife's part, he had no evidence that would stand up in a court, and perhaps was not even convinced of it himself. Instead "I treat my wife like an employee, except that I can't give her notice. I have my own room and avoid being alone with her."[51] The coming move thus suited him as far as his domestic warfare was concerned: "My wife moans to me ceaselessly about Berlin and her fear of her family. She feels persecuted and is afraid; the end of March, she says, was her last peaceful minute. Well, there's some truth in it."[52]

Nevertheless, Mileva went to Berlin on her own shortly after Christmas, in order to find an apartment with the help of Fritz Haber, or presumably his wife, Clara Immerwahr, a doctor of chemistry and, like Mileva, a pioneer of women's university education. Haber's institute, where a room was reserved for Einstein, was situated, like most of the institutes of the Kaiser Wilhelm Society founded in 1911, in Dahlem, a rural suburb to the southwest of Berlin, originally an imperial domain. Some ten minutes' walk from Haber's institute Mileva rented an apartment in a house that seemed rather too big for a neighborhood of villas.[53] Her fears were to come true: she did not long remain in that house.

Toward the end of November the official letter came from the Prussian Academy, informing Einstein that his election had "been confirmed as of November 12 by All-Highest Decree of His Majesty the Kaiser and King."[54] He allowed himself nearly two weeks before he acknowledged his election in the customary modest phraseology: "When I think that every working day reveals to me the weakness of my thinking, I cannot but accept the high distinction bestowed on me with a certain diffidence."[55] To Otto Stern he reduced his acceptance to his own denominator: the Berliners, he said, seemed to him "like people anxious to acquire a rare postage stamp."[56]

One of the major oddities of Einstein's election and the appointment procedure was that the question of his citizenship was never mentioned by the normally punctilious Prussian authorities. To Nernst and Haber, Einstein had expressed a wish to remain a Swiss citizen, and evidently he never heard anything to the contrary. He therefore left for Berlin with the firm conviction that he continued to be a Swiss citizen and nothing else, and the Prussians for their part did nothing to claim him as a subject of the Kaiser. Not until many years later—when, in connection with his Nobel Prize, some diplomatic complications arose between Germany and Switzerland—was it officially determined[57] that Einstein was a "Reich German" and had been since assuming his post in the academy in 1914. Before then, neither he nor the German authorities had been aware of it.

In a lecture to the Zurich Physical Society on February 9, Einstein balanced the scientific books of his three semesters at the Polytechnic; the bottom line was the theory of gravitation.[58] There was a grand farewell supper at the Kronenhalle, and on the way home Einstein said to his colleague Kollros: "The Berliners are speculating with me as with a prize-winning laying hen; but I don't know if I can still lay eggs."[59] Unlike his departure for Prague three years earlier, this farewell included no talk of a return. The Swiss had no position to offer like the post in Berlin. The parting seemed final.

When their furniture was loaded onto the moving van, Mileva and the two boys went to Locarno for three weeks. Three-year-old Eduard had been ill throughout the winter and needed to recover in the mild climate of the Ticino. Einstein left Zurich on March 21. He first went to Leyden, together with Fokker, whom he would have liked to take to Berlin as his assistant, though that was not feasible. In Leyden Einstein lectured on his recently completed papers on the theory of gravitation and discussed with Ehrenfest the latest developments in quantum theory. He and Ehrenfest jointly visited Lorentz in Haarlem. Einstein thanked Ehrenfest for "an unforgettable week"[60] in the Netherlands before arriving in Berlin at the beginning of April, moving with Mileva and his two sons into their Dahlem apartment, and starting out on his new life as a "living mummy."

"Things are very nice here in Berlin,"[61] Einstein wrote to Ehrenfest, describing his first impressions. "A pretty room and an interesting colleague in Haber. Otherwise I have not yet seen any physicists." A few weeks later he reported to Zurich: "Settling in successfully against expectation; except that a certain drill with regard to clothes, etc., that I have to submit to on the orders of a few uncles so as not to be counted among the rejects of local humanity, slightly upsets my peace of mind."[62] The "uncles" presumably were his fellow academicians, whose acquaintance he had made at the weekly meetings.

The Prussian Academy of Sciences had its premises in the center of Berlin, in the western wing of the impressive Prussian State Library on Unter den Linden—easy for Einstein to reach by the Wannsee Railway from Lichterfelde. Every week the fifty or so "ordinary members" met either in plenary session or in one of the two "classes," "philosophical-historical" or "physical-mathematical." The business of the two classes was managed by two "permanent secretaries," who, in a system of rotation, also participated in the management of the academy.

Einstein was not exactly impressed by the meetings—which was hardly surprising for a man of thirty-five in a circle of elderly gentlemen who were addressed as "*Geheimrat*" (Privy Councillor) or "Your Excellency." The academy reminded him "in its character of any university department. It seems that most members confine themselves to displaying a peacock-like grandeur in writing; otherwise they are quite human."[63] No doubt he was amused by the pedantic observance of formalities and by the careful performance of traditional customs. Thus every member had to give a lecture in plenary session on a date laid down in advance by the so-called "reading card." But as, for instance, an Asian scholar was unlikely to show interest in the expositions of a chemist, or even be able to follow them, it was inevitable that some of the audience would drowse or even nod off. Any suggestion that such obsolete practices be changed had invariably been rejected. Einstein learned to treat the academy with the indulgence to which it was entitled as an institution that had been founded by Leibniz and reformed by Frederick the Great on the model of the Académie Française, and in this form had survived for two centuries. He became an obedient academician, was rarely absent from its meetings, and above all observed the rule that the findings of his research had to be published in the *Sitzungsberichte—Proceedings*—which greatly enhanced the value of that publication.

"I haven't got round to music-making yet because there's always such a lot of other things happening," Einstein commented on the effects of metropolitan life, to which the string quartet planned with the astronomer Freundlich had, for the time being, fallen victim. As for the self-assurance of the Berliners and their proverbial arrogance, he had an explanation that showed him as not too sympathetic: "I now understand the smugness of the Berliners. There is so much happening outside that they are not painfully aware of their inner vacuity as they would be in a quieter spot."[64] But soon, in a letter to Ehrenfest, he raved about the "enormously stimulating"[65] atmosphere, which alone made a trip to Berlin worthwhile. The university, with Max Planck, Heinrich Rubens, and Walther Nernst; the Physical-Technical Reich Institute, under the directorship of the brilliant experimenter Emil Warburg; and the two Kaiser Wilhelm Institutes in Dahlem, under Fritz Haber and Emil Fisher—these represented a concentration and quality of research that could not be matched anywhere else. Many highly gifted young people, such as Lise Meitner, Otto Hahn, James Franck, and Gustav Hertz, had come to Berlin for that reason and had there established their international renown.

All this was something Paul and Tatyana Ehrenfest wished to experience in person, and so they visited the Einsteins in their new apartment for a week at the end of May. They discussed physics, went sightseeing in the city, and took the children to the zoo. Ehrenfest found the company of ten-year-old Hans Albert entertaining, but he wondered about Mileva's gloomy mood.

No doubt Ehrenfest guessed the reason for Mileva's gloom, because Einstein, a few days after arriving in Berlin, had written to him about his "genuine pleasure in my relations here, especially in a female cousin of my own age, with whom I am linked in old friendship. Mainly for this reason I can very well tolerate the big city, odious as it is to me otherwise."[66] Mileva had no contact with those relatives and probably did not seek any—least of all with the cousin. This must have led to conflicts which rendered any further life together impossible. It was probably in June 1914 that Michele Besso came to Berlin to take his friend's helpless wife and the children back to Switzerland.[67] When Einstein accompanied Mileva, his sons, and his friend Besso to the railway station, Fritz Haber went along with him. "In tears he returned from the station,"[68] Haber said.

For Einstein that parting was definite. Mileva, on the other hand, wavered for years between hope and despair. For a long time she was unable to accept the separation and persisted in an attitude which reminded Einstein of "the classical image of Medea."[69] He suffered from the "gloomy veil" which hung over his relationship with his two sons, to whom he had clung with paternal love and tenderness. Hans Albert in particular experienced the years of separation as the "worst time, because then nobody knew what the future would bring."[70] Not until the divorce five years later did the "gloomy veil" lift sufficiently for Einstein to travel with his sons, visit them in Zurich, or invite them to Berlin.

The theory of relativity engaged Einstein's attention during his first few weeks in Berlin, but in an unfamiliar version—a presentation for the common reader. By then he and his theory had attracted so much attention also outside scientific circles that the editor of *Vossische Zeitung*, the most highbrow of the Berlin dailies, asked him for a contribution. "I am happy to comply with this request," Einstein began his first newspaper article. "Although a deeper insight into the relativity theory cannot be achieved without considerable effort, it may yet be interesting even for the outsider to learn something about the method and results of this branch of theoretical research."[71] His article, *About the Relativity Principle*, appeared in the Sunday supplement of April 26; it extended over two long columns, was engagingly written, and dealt predominantly with the special theory of relativity. Shortly before the conclusion he briefly referred to the generalization, pointing out that on this development "the opinions, even of those physicists who value the relativity principle, are still divided." However, "the future will show whether this generalized theory of relativity, which is very satisfying in terms of cognition theory, corresponds to reality."

Einstein's cautious comment about divided opinions was something of an understatement. Actually, agreement and support came only from Max Born, a privatdozent in Göttingen, and from the astronomer Erwin Freundlich—both young hopefuls rather than authorities in the world of science. The extent to which Einstein was still isolated with regard to his general theory of relativity became obvious at his solemn induction into the Prussian Academy. This was at a public meeting on July 2, 1914, the traditional "Leibniz Day" commemorating its founder. In his inaugural address Einstein outlined his concept of the method of theoretical physics, whose finest task he saw as discovering nature's general principles.[72] He described the difficulties of a theory of heat radiation as an illusion produced by a complex of facts "for the treatment of which the principles are lacking," contrasting it with the theory of relativity as the paradigm of a theory based on "principles." Having highlighted the successes of the "ordinary" theory, he addressed the necessity and purpose of its generalization:

> On the other hand, this theory is not fully satisfying from a theoretical point of view because the relativity principle formulated a moment ago prefers uniform motion. If it is true that, from a physical point of view, uniform motion cannot be ascribed an absolute meaning, then the question arises whether this statement should not also be extended to non-uniform motion. It emerges that one arrives at an entirely definite extension of the relativity theory if one bases oneself on a relativity principle in this extended meaning. This leads one to a general theory of gravitation, one which embraces dynamics. For the moment, however, we lack the data by which we could test the justification for introducing the postulated principle.

Max Planck, as secretary of the physical-mathematical class, welcomed the new member as a man whose "real love is the discipline in which a personality most freely unfolds, in which imagination flowers most richly, and in which a scientist can most easily abandon himself to the comfortable feeling that he cannot easily be replaced by another. Admittedly, he also, more than anywhere else, runs the risk of occasionally losing himself in dark regions and unexpectedly encountering hard contradiction."[73] In the past it had been Einstein's interpretation of light quanta with which Planck took issue; now he could "not resist the temptation to record my objection" to the generalization of the relativity principle as well.

This retort by Planck at a public meeting of the academy must have made Einstein realize that, with regard to the generalization of relativity theory, he stood practically alone. This, however, did not detract from Einstein's and Planck's mutual esteem. "Not without some pride," Planck pointed out that in physics, "more easily than in other sciences, the sharpest factual differences can be conducted with personal high esteem and in cordial friendship." Nevertheless, the persistence with which Planck emphasized his dissent on a festive occasion is surprising, especially as observations of the solar eclipse in southern Russia on August 21 were expected to decide the dispute. That this did not happen was due to a catastrophe: World War I.

PART IV

THE NOISE OF WAR AND THE SIZE OF THE UNIVERSE

CHAPTER EIGHTEEN

"In a Madhouse": A Pacifist in Prussia

"UNBELIEVABLE WHAT EUROPE has unleashed in its folly"[1] was Einstein's comment on the outbreak of World War I, in a letter to his friend Ehrenfest in the Netherlands, which had remained neutral. To Einstein, as to the British foreign secretary, "the lamps had gone out all over Europe" at the beginning of August 1914. He was bewildered by the enthusiasm for war that he saw in all classes of society—among the Social Democrat workers and their party just as much as his fellow professors. Amid the patriotic roar of mobilization and the jubilation over the first victories, he felt lonely and a stranger:

> At such a time as this one realizes what a sorry species of animal one belongs to. I doze along quietly with my musings and only experience a mixture of pity and revulsion.[2]

The universal enthusiasm for what he saw as a disaster made him realize what it was that divided him from the current political opinions of most people—his pacifism and his internationalism.

Einstein's pacifism was an instinctive, fundamental attitude, not so much "considered" as "congenital." Revulsion against all force and all militarism needed no justification; to him this was as natural as the air he breathed or the pleasure he took in thinking. This trait extended to a rejection of anything competitive—not only to weapons and lethal intent, but similarly to sports and even to such a seemingly peaceful activity as chess. He neither played nor liked chess, because he had always felt repelled by the power struggle and competitive spirit even in this intelligent game.[3]

He had arrived in Berlin not without some misgivings, and when he discovered that a corner turret of the Kaiser Wilhelm Institute of Chemistry had a roof in the shape of, and in honor of, the traditional Prussian *Pickelhaube* helmet, he probably laughed out loud. But now, with the outbreak of war, the Prussian nature of his colleagues emerged in terrible earnest. Max Planck, otherwise so sensible, in his capacity of rector sent his students into battle with a sermon on the "just war": "Its unprecedented patience exhausted, Germany has drawn its sword against the breeding-ground of insidious perfidy."[4] The mercurial Walther Nernst—who, like Planck, was exempt from military service because he wore eyeglasses—did not wish to be outdone by his students and, being an enthusiastic motorist, enrolled at age fifty in the Voluntary Drivers' Corps. Under the supervision of his wife, he promptly practiced a few of the military basics, such as correct marching and saluting, in front of his house. With his private car, supplied with a stock of rubber plugs in case the enemy punctured his gas tank, he arrived at Second Army headquarters as a general staff courier, just in time for the rapid advance on Paris; he and the army approached close enough for him to see the glow of the lights at night.[5]

That was all the use Germany's military leaders had for the country's best scientists. But then, in line with the Kaiser's pronouncement, they believed the war would be won speedily and the troops would be back home for Christmas.

However, the rapid advance was followed by the Battle of the Marne and an equally rapid retreat. By the end of September the armies had dug in for what would be four years of seemingly endless trench warfare on the western front. Thus the professors' "paper war," their speeches and writings—which during the first few weeks had been more of a decorative accompaniment to the din of battle—took on a new quality. The task was now the moral rearmament of the German population and the championship of the German cause abroad, a cause that had suffered severely as a result of Germany's attack on neutral Belgium.

The first and most unfortunate product of those endeavors was an

Appeal to the Cultured World, drafted by a not exactly outstanding writer, Ludwig Fulda, and signed by ninety-three intellectual giants of the German empire. On October 4 it was published in the leading German dailies[6] and simultaneously translated into ten languages throughout the world. With this appeal the representatives of German scholarship and culture wished to protest "the lies and defamations with which our enemies are trying to besmirch Germany's pure cause in the hard life-and-death struggle forced upon it." There were six points, each introduced by the invocation "It is not true. . . . " The writers denied "that Germany bears the responsibility for this war"—which, as propaganda, was perhaps arguable. But they also denied "that we criminally violated the neutrality of Belgium"—a point arguable only in the sense that the invasion had been undertaken not so much "criminally" as in accordance with the plans of the German general staff. They further disputed "that the life or property of even a single Belgian citizen was touched by our soldiers"—which was an outright lie after the brutal passage through Louvain and the cruel destruction of that city. The appeal declared that any "undisciplined brutality" was alien to the Germans' conduct of the war, whereas the French and British were accused of "having allied themselves with Russians and Serbs and presenting to the world the shameful spectacle of Mongols and Negroes being driven against the white race." In concluding this patriotic effusion, they assured the world "that we shall fight this struggle through to the end as a cultured nation to whom the legacy of Goethe, Beethoven, and Kant is just as sacred as its hearths and plots of land."

If Einstein read this manifesto in *Vossische Zeitung* on October 4, he must have rubbed his eyes in disbelief, for a number of reasons. The apodictic assertion that it was not true "that the life or property of even a single Belgian citizen was touched" was immediately refuted by a dramatic account of the German assault on Antwerp filling the entire front page of the same daily. To Einstein this would have been cause for personal concern, since his favorite uncle, Caesar Koch, lived in Antwerp. But he would also have been shocked by the ninety-three signatories. Along with famous names from the world of culture—such as the painter Max Liebermann, the poet Gerhart

Hauptmann, and the producer Max Reinhardt—were the names of fifteen scientists. That these included such conservative nationalists as Philipp Lenard and Wilhelm Wien was not surprising, but the signatures of Planck, Nernst, Haber, and Fischer would have pained him. From Einstein, as a Swiss citizen, no signature would have been expected anyway, so he was able to avoid conflicts. However, as a reaction to the chauvinistic blindness of his colleagues and out of concern that the "appeal" might disrupt scientific cooperation beyond the duration of the war, he decided to collaborate on an alternative document—his first political activity.

The initiator of the *Appeal to Europeans* was the physician and physiologist Georg Friedrich Nicolai, "supernumerary professor" at Berlin University and a renowned heart specialist, consulted even by imperial highnesses. Out of respect for what he understood by German culture, he felt it his duty to protest the *Appeal to the Cultured World.* Nicolai drafted a countermanifesto and submitted it to his "like-minded friend Albert Einstein."[7] The fruit of their joint efforts was a clear distancing from the *Appeal to the Cultured World* and its signatories: "Such an attitude cannot be excused by passion, it is unworthy of what until now the whole world has understood by the term culture, and it would be a disaster if it were to become the common property of educated people."[8]

Nicolai and Einstein warned that this war would "leave scarcely a victor but probably only vanquished" and urged that "educated men of all states" should at least ensure that "the conditions of the peace did not become the source of future wars" and that "an organic entity be created from Europe."[9] These were restrained phrases, appeals to educated people of goodwill, avoiding any attribution of guilt or any analysis of the causes of the war. This countermanifesto was at least a recommendation to avoid annexations and to create a durable system of peace for Europe. In a war-intoxicated climate, it was certainly radical and probably at the very limit of what the military censorship could be expected to pass.

It seems likely that the manifesto was deliberately expressed in general, imprecise terms in order to attract broad agreement. But if so, that expectation was disappointed. From mid-October onward the paper circulated among members of the staff at the university in Berlin, generating, as Nicolai recalled, "a lot of friendly agreement" but no signatures. The only ones to support the *Appeal to Europeans* were the aged astronomer Wilhelm Förster, who had been one of the ninety-three signers of the original "appeal" but by then probably regretted it; and an otherwise unknown Dr. Otto Beck. Any public initiative, under these circumstances, seemed pointless. "Deeply distressed," Nicolai summed up, "we realized our isolation."[10] The *Appeal to Europeans* first appeared in print in Zurich in 1916 as part of Nicolai's book *The Biology of War*. This book—despite its title, which sounds like a vulgarization of Darwin—was a very lucid refutation of the current political view of war as not only normal and unavoidable but actually desirable.

In Germany the countermanifesto was of importance only to a small handful of pacifists and was barely noticed otherwise—though it was enough to ruin Nicolai's academic career. For Einstein, its consequence was that he acquired a reputation as a political oddball—but this was viewed with indulgence by his colleagues and the authorities because he was a foreigner. And he made the most of this situation. Among the more prominent professors at Berlin, he was apparently the only one to experience revulsion against the war, and to show it openly, though within the bounds of conventional civility. In his letters to friends in neutral countries, he was more outspoken. He wrote to Ehrenfest toward the end of 1914:

> The international catastrophe weighs heavily on me as an internationalist person. It is hard to understand, as one lives through this "great epoch," that one belongs to this crazy degenerate species that claims to possess freedom of will. If only somewhere there were an island for the benign and prudent! There I too would be a fervent patriot.[11]

In spite of his decidedly pacifist views, then, Einstein did not become an isolated outsider in Berlin. Perhaps for the first time, he exhibited his remarkable gift for dividing his life into separate compartments, with the result that his political beliefs put no serious strain on his scientific work or his personal relations. He could not and would not accept the unconditional patriotism of the professors who saw themselves, in the words of the great physiologist Emil Du Bois-Reymond, as the "intellectual Life Guards of the Hohenzollerns." In fact, he despised such patriotism, and his colleagues were aware of this—or, at least, some of them must have suspected it. Nevertheless, the scientists, for whose sake he had come to Berlin, remained on friendly terms with him, and he reciprocated their sympathy and affection beyond the purely scientific sphere.

As far as the scientific scene was concerned, Einstein was as integrated into it as if he were a loyal Prussian subject. The patriotic hullabaloo within the German Physical Society might have induced him to resign, but he was elected a member of its board in 1914 and beginning in May 1916 actually served as its chairman for the next two years.[12] As a Swiss republican, he had nothing but scorn and mockery for the chauvinistic grandiosity of the Hohenzollerns, and yet he allowed himself to be appointed to the board of trustees of the Physical-Technical Reich Institute by his majesty the Kaiser's "most graciously" granted "all-highest decree."[13] Despite his "emotional socialism," he had come to terms with the fact that his salary at the Academy was indirectly paid by a donation from the banker and industrialist Leopold Koppel, and he found nothing to object to when Koppel created the Kaiser Wilhelm Foundation for Military Engineering Sciences. As director of the Kaiser Wilhelm Institute of Physics—which because of the war did not begin its work until 1917, and then on a modest scale—he had to harmonize his management with "commercial counsellors" and "privy councillors" such as Leopold Koppel and Wilhelm von Siemens, representatives of the very strata of society which, on other occasions, he would hold responsible for the pernicious Wilhelminian ideology.

Other men might have tried to find a way out of such a dichotomy. But Einstein bore these contradictions smilingly or simply ignored them. He was helped in this by his earthy humor. "I am beginning to feel well amidst this present mad turmoil by deliberately detaching myself from all the things which concern the mad general public," is how he described his state of mind to Zangger at the beginning of

1915. "Why should one not, as a servant in a madhouse, be able to live cheerfully? After all, one respects all those lunatics as the people for whom the building exists in which one lives. Up to a point one can choose the institution—but the difference between them is less than one might have thought in younger years."[14]

If ever there was a cheerful opponent of the war, it was Einstein, as portrayed in a diary by the French writer Romain Rolland, after a meeting in Switzerland in 1915 which included exceedingly serious conversations about the tragedy of the war:

> Einstein is still young, not very tall, with a broad and long face, a full mane of slightly curly and dry deep-black graying hair, rising above a high forehead, with a fleshy and arrogant nose, small mouth, thick lips, a closely trimmed little mustache, full cheeks, and a round chin. He speaks French rather laboriously and mixes in some German. He is very vivacious and serene; he cannot help giving the most serious thoughts a jocular form. . . . Einstein is incredibly free in his judgments on Germany, where he lives. No German enjoys that freedom. Anyone else but he would have suffered from a sense of isolation in his thinking during this frightful year. Not he. He laughs. . . . I ask him if he voices his views before his German friends and if he argues with them. He says no. He confines himself to asking a lot of question—as Socrates did—in order to upset their peace of mind. He adds: "People don't like that very much."[15]

Immediately after his first endorsement of a political declaration, Einstein for the first time joined a political organization. This was the Bund Neues Vaterland (New Fatherland League), an association of men of the most varied backgrounds. Along with an early peace, without annexations, its aim was an organization of European countries that would make future wars impossible. These objectives commanded Einstein's full support, and so he was one of the founders when the group was established on November 16, 1914. Its membership ranged from the influential banker Hugo Simon, who after the war became the Prussian minister of finance, to the young left-wing radical journalist Ernst Reuter, who acted as a kind of secretary and who much later became well known as a Social Democrat and governing mayor of West Berlin.

The New Fatherland League was not so much a party as a political club. Its members met every Monday evening, discussed the situation, drafted leaflets, tried their hand at smuggling pacifist literature, and planned schemes to aid sympathizers in difficulties, often very close to or even beyond the bounds of legality. A young Swiss woman, who had attended one of Einstein's classes at the Zurich Polytechnic, now came to know him as a man "who gave support and help wherever, however, and to whomever he could."[16] He did not shrink from risky undertakings: the young woman noticed that "Einstein took an almost impish pleasure in pulling a fast one on the military authorities. Each time we succeeded in smuggling letters or books into a prison, he would laugh aloud in amusement."

As an "internationalist person" he was particularly anxious to keep scientific contacts from being impaired by war psychosis and to preserve them beyond the war. "Are future centuries really to say of our Europe that three centuries of intensive cultural work have not led it further forward than from religious madness to national madness?"[17] he wrote to Romain Rolland in the spring of 1915. What depressed him most was the failings of his colleagues: "Even the scholars of the different countries act as if eight months ago they had their cerebrum amputated." Clearly, he was not unaware that the "paper war" was being fought on both sides of the front, and that French professors in particular were scarcely second to their German counterparts in chauvinistic declarations. That, however, was no consolation. "I would love to do something," he lamented to Ehrenfest, "to keep our colleagues from their different fatherlands together. Isn't that handful of intensively thinking people [the] only 'fatherland' for which the likes of us have any serious affection? Are these people, too, to hold opinions which are a function of location?"[18]

At home in Berlin, Einstein tried, somewhat in the manner of Socrates, to lead his colleagues toward a realization that they would have done better not to sign the *Appeal to the Cultured World*. In this he actually achieved a measure of success. He reported to Lorentz that the "appeal" was "being regretted by all calmly thinking people here.

The signatures were appended carelessly, sometimes without previous reading of the text. That was the case, for instance, with Planck and Fischer, who have stood up very courageously for the maintenance of international ties."[19] Lorentz had expressed the hope that the German scientists would officially distance themselves from the "appeal," but on this point Einstein had to disillusion him:

> I will hardly be able to bring about a formal revocation of the notorious manifesto, even though the realization has gained ground that this was an unfortunate and ill-considered step. I consider it more important still that men of good will should go forward together on those things which are of real significance for the future. The repentant are not primarily identified by recantation.[20]

Einstein thus called for understanding of even his German colleagues—even for understanding of their follies, provided these were not too bad. Meanwhile he found that the colleagues close to him were altogether "strictly internationally minded as scientists," whereas the historians and philologists were "mostly chauvinistic hotheads."[21]

Given Einstein's implacable remarks on the war it is surprising that he remained silent on an aspect of the war which he witnessed at close quarters—the alliance between the natural sciences and the military. This relationship has a tradition of over two thousand years, if one can believe the stories of the terrifying stone slings and burning mirrors which Archimedes is said to have built for the defenders of Syracuse against the besieging Romans in 215 B.C. In fact, war was a stimulus in the development of modern physics: the study of parabolic trajectories by Niccolò Tartaglia and Galileo Galilei stemmed from a need to calculate the trajectories of cannonballs. The tradition was continued and consolidated by the development of modern science, and during World War I it reached a new intensity and hence also took on a new moral quality.[22]

That scientists should wish to make a contribution to victory was considered by both sides a natural expression of patriotism. Of the participants in the Solvay Conferences, for instance, Einstein's friend Paul Langevin was working on methods of U-boat spotting; and Nernst's favorite disciple, Frederick A. Lindemann, was engaged in his native Britain at the Royal Aircraft Establishment in Farnborough as an aircraft designer and death-defying test pilot. But while these activities might have been regarded as acceptable, even for an internationally minded scientist, that could no longer be said of some of the military research in Berlin, especially in chemical warfare.

At first the generals in Germany showed little inclination to allow ivory-tower civilians to interfere in their business. But when the speedy advances were followed by equally speedy retreats, and when the troops dug in for trench warfare, they had to prepare for a prolonged war—and at that stage scientists such as Walther Nernst could be employed more usefully than as volunteer drivers.

Among the Berlin professors who placed their abilities at the service of the generals, the most zealous was Einstein's "good colleague" Fritz Haber. Like Einstein, Haber was of Jewish origin, but he had had himself baptized and, in his overcompensating eagerness to assimilate, was, even in appearance—he wore a monocle and had dueling scars—something of a caricature of a Prussian. True to his motto, "In peacetime for humanity, in war for the fatherland," he had, as soon as war broke out, reorganized his Kaiser Wilhelm Institute of Physical Chemistry to deal with war-related tasks.

Haber would in any case have earned his country's gratitude, because, even before the war and with wholly peaceful intentions, while still a professor at the Technical College in Karlsruhe, he had achieved something of greater importance to the German war effort than any planning by the general staff. On the basis of Nernst's work on the kinetics of reactions—and at times in unfriendly rivalry with Nernst—Haber had developed the high-pressure synthesis of ammonia from atmospheric nitrogen into a process suitable for mass production, and along with Carl Bosch of the Badische Anilin- und Sodafabriken BASF in Ludwigshafen, had made it into a large-scale industrial technology. When Haber began his work on ammonia synthesis in 1907, his aim was to manufacture fertilizers, because ever since the turn of the century, supplies of natural saltpeter from Chile had been threatened with exhaustion. In war, however, nitrate has another use—as a component of explosives. It soon emerged that the general staff had budgeted for only a short war. As the British navy seriously disrupted supplies from Chile, the generals would have been forced to cease operations sometime in 1915. That they were able to continue firing their weapons was something they owed to the Haber-Bosch process and to the circumstance that the first plant for ammonia synthesis had gone into operation toward the end of 1913. On this model, huge factories were built during the war. Thus German researchers had scored a victory over the Royal Navy, at least in the nitrate sector, a victory more important than many a naval battle, but also one which prolonged the misery of war.

When Walther Nernst returned from the front to his institute at the beginning of the semester to concern himself with explosives, he was asked by the War Ministry if he could devise a means of driving the enemy out of his trenches, in order that Germany might recapture the initiative through a war of movement. Nernst found the task attractive, as it was in line with his view that in a modern scientifically waged war, what mattered was not killing the enemy but merely rendering him incapable of combat. He therefore experimented first with tear gas and then with irritants designed to paralyze without causing permanent damage. The generals were not impressed with Nernst's humane objectives; they wanted speedier, more effective solutions. For that, Fritz Haber was the right man.

Haber had earned the respect of the generals as the intellectual head of the "saltpeter promise"—the commitment by the chemical industry to make up for the poor planning of the general staff with regard to raw materials. In addition, he had speedily developed antifreeze additives for gas, diesel, and lubricants, ensuring that motorized operations could continue even in the Russian winter. And now he was working on explosives and on gas warfare. The irritants proposed by Nernst were rejected as ineffective. Haber preferred chlorine, for a number of reasons: it could be produced easily in large quantities by the chemical industry; it was relatively easy to transport in liquid form in pressurized containers; and, being heavier than air, it would creep through the trenches like a thick fog, doing its deadly work.

In January 1915 laboratory tests in Dahlem were completed. Small-scale field tests in Russia followed, and on April 22, 1915, when sufficient quantities of chlorine had been produced, 160 tons of it, under Haber's supervision and guidance, were blown over the French trenches along seven kilometers of frontline at Ypres in Belgium. Five thousand men died, and ten thousand suffered severe damage to their health as a result of corrosion of their air passages. Fritz Haber had inaugurated chemical warfare. The generals were impressed, but so much taken by surprise by the effectiveness of the new weapon that they had made no preparations for a breakthrough at the gap the gas had torn in the front. By the time the next attack was launched, gas masks had been developed. Thus while gas attacks made the war more brutal, they were unable to influence its course in any decisive way.

The attack at Ypres was a breach of the Hague Convention of 1907, which prohibited asphyxiating gases in warfare. Germany had signed the convention, but this breach of the law does not appear to have worried Haber—who was reported to have shed tears of joy over his promotion to captain. He became director of a department for chemical warfare in the War Ministry and enlisted many young scientists as collaborators, including Otto Hahn. New and ever more deadly toxic gases were being developed at Haber's institute in Dahlem, based on mustard gas and phosgene. This was not without incident. On December 17, 1914, there was a terrible explosion in the laboratory, killing one of Haber's closest and most efficient collaborators, Professor Otto Sackur. *Physikalische Zeitschrift* carried an obituary of this young, highly gifted scientist, in the same style as its obituaries for scientists killed in action: "he too died for his fatherland."[23]

All this was happening under Einstein's eyes. Nernst and Haber were his friends, and his own office was in Haber's institute. Whether he excluded these two pioneers of chemical warfare from those scientists whom he described as "internationally minded" we do not know. Einstein could have been biased by solidarity with his colleagues and might have regarded scientists as people whose intelligence would induce him to overlook what in anyone else would have drawn his biting criticism. Perhaps he did not wish to jeopardize his friendship with the two men; or perhaps poison gas seemed to him only a marginal aspect of a ghastly war. Still, it certainly remains a riddle why he should have found "chauvinistic hotheads" primarily in the humanities. Even disregarding morality and the usages of war, Nernst's and Haber's activities did at least as much damage to the reputation of German scholarship in the enemy countries and in the neutral countries as did the warmongering speeches and writings of historians and philologists.

Admittedly, we do not know how often Einstein would still have visited his official study in Haber's Kaiser Wilhelm Institute, now reorganized into a war laboratory, at Faradayweg 8 in Dahlem. After all, he was now able to work at home uninterrupted. "I live all on my own in my big apartment in unmitigated cosy comfort,"[24] he wrote to Ehrenfest. At first, Mileva and their two sons had been staying at a *pension* on Bahnhofstrasse in Zurich; as soon as she found an apartment on the Zürichberg, toward the end of 1914, he arranged for most of the furniture to be freighted to Switzerland, and he himself moved into a smaller apartment,[25] situated more centrally, near the Kurfürstendamm.

There he seems to have led a bachelor existence with a minimum of comfort. A student from Switzerland who came to ask Einstein for three of his papers found him in stocking feet: "He spent a long time searching for one [of the papers,] wondered where it was, complained about his untidiness and forgetfulness, and failed to find it. We moved from room to room and stood helpless in front of the shelves. He has a rather bleak apartment and seems to live there on his own without a housekeeper."[26] Actually, this lifestyle was entirely in line with Einstein's needs. After ten years of marriage, he was tired of family life, though for his sons he felt something like longing. "I am very content with the separation, even though I only rarely have news of my boys," he wrote to Besso. "The quiet and the peace of mind are doing me a lot of good, and so is the extremely agreeable, truly pleasant relationship with my cousin; its durability is guaranteed by the avoidance of matrimony."[27] Elsa Löwenthal's apartment in the Bayrisches Viertel was only a fifteen-minute walk away.

■ ■ ■

The onset of war had a very direct effect on Einstein's scientific interests, entirely peaceful though these were. "My good astronomer Freundlich," he wrote in the third week of the war, "instead of experiencing a solar eclipse in Russia, will now be experiencing captivity there. I'm worried about him."[28] Erwin Freundlich, who ever since 1911 had linked his whole career to the astronomical verification of Einstein's theory of gravitation, had succeeded in obtaining the (perhaps grudging) agreement of the director of the Royal Observatory at Potsdam-Babelsberg, Privy Government Councillor Hermann Struve, to mount an expedition to Russia for the solar eclipse there on August 21. The purpose was to take photographs which, according to Einstein's prediction, would show the deflection of light in a gravitational field.[29] The total cost was to be 5,000 marks. The Prussia Academy had made 2,000 marks available for the conversion of instruments and the purchase of photographic plates, and the Krupp Foundation had contributed 3,000 marks for travel and freight costs.

On July 19, 1914, Freundlich left Berlin with his colleague Walther Zurheilen and a technician from the firm of Carl Zeiss. After a week's travel they arrived at Feodosiya in the Crimea, where they enjoyed the hospitality of the well-equipped expedition of the Argentine observatory of Córdoba and were able to use its instruments. All together, Freundlich had prepared four astrographic cameras, to ensure that at least a few useful pictures would be obtained during the optimal observation period of barely two minutes. The leader of the Argentine expedition, the outstanding astronomer Charles Perrine, had considerable experience in the photographic observation of solar eclipses. Actually, Perrine's original intention had been to photograph a hypothesized minute planet near the sun, called Vulcan, whose existence had been postulated by Jean LeVerrier as early as 1859 to explain irregularities in the orbit of Mercury.[30] (Einstein, as will be shown, was soon to solve this problem in a totally different and—to the astronomers—unexpected way.) At Freundlich's request, Perrine, at the last solar eclipse in Brazil, in 1912, had extended his research to include the deflection of light, but clouds and torrential rain had brought the wrong kind of darkness: "We . . . suffered a total eclipse instead of observing one,"[31] Perrine recorded.

This time it was the war that interfered with science, from the outset. Preparations had just begun when Freundlich and his German colleagues overnight became enemy aliens. They were interned in camps, and their equipment was confiscated. The Argentinians were unable to arrange for adequate replacements in time for August 21, especially as Perrine's arrival was delayed by wartime conditions. As it turned out, it made no difference in the end, because once again clouds appeared at the height of the eclipse. Equally unsuccessful were the endeavors of an American expedition from Lick Observatory in California, under William Wallace Campbell,[32] which had taken up a position south of Kiev. Although the Americans, as neutrals, were not interned, the sky over Campbell's instruments also turned dark at the decisive moment. "Total failure," Mrs. Campbell recorded in her diary. "Thick gray cloud at eclipse time and lovely clear sunshine afterwards."[33] The Americans were able to leave Russia unmolested, but they did not get their instruments back until four years later, with the result that any repetition of the attempt elsewhere in the world was ruled out during the intervening period.

Einstein need not have worried about his "good astronomer" Freundlich, however. Although Freundlich initially had a kind of prisoner-of-war status, he, along with other Germans, was exchanged a few weeks later for Russian officers who had been taken prisoner. By the end of September he was back in Berlin.

Einstein, meanwhile, had to fall back on his own profound, unshakable faith in his theory of gravitation and go on waiting for it to be confirmed by astronomical observation. He used the intervening time for a comprehensive account of what had been achieved in his theory up to that point.

It was at the plenary meeting of the Prussian Academy on October 19, 1914, and ten days later at the meeting of the physical-mathematical class that Einstein gave his first lectures; these resulted in a contribution of over fifty pages to *Proceedings*, intended as a comprehensive overview of the theory. Beginning on a relaxed note, he observed that in the papers he had written so far, some with Marcel Grossmann, on the generalization of relativity theory, "as a heuristic aid in those investigations, a motley mixture of physical and mathematical requirements were used, so that it is not easy, on the strength of those papers, to survey and characterize the theory from a formal physical viewpoint."[34] It was mainly this gap that Einstein hoped to close with his long paper, which was designed simultaneously as a kind of textbook for interested colleagues, even though its title—*The Formal Foundation of the General Theory of Relativity*—shows that he must have been fairly certain about having now cast the theory in its definitive form.

He had by then sufficiently emancipated himself from Grossmann's mathematical instruction and advice that, to demonstrate his mathematical independence, he was able to present and promise a kind of "tensor calculus for physicists—simple derivations for the fundamental laws of absolute differential calculus, which are probably new, and which, it is hoped, will enable the reader fully to comprehend the theory without reading any other, purely mathematical, treatises."[35] On this basis he offered a few new, elegant conclusions as well as a new derivation of the gravitational equations; these were still the curious hybrid noncovariant equations of the earlier "draft" paper, but Einstein had every reason to be satisfied with the results. Above all, his new derivations reveal that what he had created was not only a theory of gravitation but a generalized theory of relativity as a structural framework for any physics. Newtonian theory as an approximation, the deflection of light, and the red shift appeared in the already customary form; but on the crucial point, the general covariance of the equations, he had made no progress.

Einstein openly discussed some difficulties with his theory and did not shrink from admitting that many important questions would still have to be left open or would not be susceptible to the requisite proof. Einstein's tone was unusual for the venerable *Proceedings of the Academy*—all the more so when he conceded that "after all these admissions [he] could not help observing a compassionate smile on the reader's face."[36]

After this amiable opening, Einstein first explained to the reader that Euclidian geometry, not as it was practiced in the rarefied air of mathematical axiomatics but rather as it referred to the real world, was also a physical science, whose theorems might or might not be correct in this context. He even hinted that his restriction of the coordinate systems of weak fields and small distances to Minkowski's pseudo-Euclidian geometry was "by no means the simplest or the most obvious" possibility.[37] Another year was to pass before he returned to the "simplest and most obvious" possibility and before he actually completed his theory.

His German colleagues continued to view the theory with skepticism or to ignore it. If Einstein had come to Berlin mainly in the hope of a fruitful exchange of ideas on relativity theory, his disappointment could not have been greater. As it was, he was merely surprised and remarked that "in future people will be astonished that the idea of general relativity [was] encountering so much resistance."[38] At any rate, the long article in *Proceedings* gave rise to some comment from competent colleagues abroad. Ehrenfest and Lorentz were joined by two Italians: Paolo Straneo in Rome, a fellow student of Michele Besso's; and the mathematician Tullio Levi-Civita in Padua, the author of those classic publications on differential calculus which Einstein and Grossmann had studied together in Zurich.

Levi-Civita's interest in relativity theory gave Einstein particular pleasure: "You can well imagine," Einstein wrote in response to his first letter from Padua, "how rarely anyone thoroughly concerns himself with this matter if he confronts it independently and critically."[39] Levi-Civita drew Einstein's attention to a few unclear points and even to some mathematical mistakes. Einstein was grateful, and delighted with the Italian elegance with which it had been done: "You do it very nicely in your letters; first you obligingly caress me, so I don't make a sullen face when I read the new objections."[40] However, the exchange of ideas was confined to mathematical subtleties which did not make it necessary for Einstein to attempt to rewrite his theory. He waited, and during the first winter of the war he took on the unfamiliar role of experimenter, which he had not performed since his work with his "little machine." It was typical of Einstein that even his experimental activity concerned fundamental questions of physics.

▪ ▪ ▪

"In my old age I'm developing a passion for experimenting,"[41] Einstein wrote to his friend Besso after observing for the first time what would come to be called the "Einstein–de Haas effect." He conducted this experiment as Emil Warburg's guest at the Physical-Technical Reich Institute, collaborating with the Dutchman Wander Johannes de Haas, the son-in-law of Hendrick A. Lorentz. De Haas was a year older than Einstein, had first practiced as a notary, and had developed his interest in physics relatively late. He had taken his doctor's degree in Leyden in 1912 and had since been extending his qualifications in Berlin, where in 1914 he had found a position as assistant at the Reich Institute.

The idea of the experiment came from Einstein and went back to experiments he had performed in the physics room of the municipal gymnasium in Zurich.[42] He probably took it up again because it could throw light on the most varied aspects of physics: the origin of magnetism, the zero-point energy of quantum theory, and the nonradiating orbits of electrons in the model of the atom proposed by Niels Bohr two years previously. Actually, this experiment illustrates that even a physicist of Einstein's status may find himself obtaining the results he expects from his theoretical arguments—results that need not always be correct.

The underlying idea of the experiment was nearly a hundred years old: André Marie Ampère had suggested that magnetism might be related to moving electrical charges and, more particularly, that the permanent magnetism of iron might be explained by hidden electric currents within a magnet. Einstein developed this into the hypothesis that the permanent magnetic moments of the atoms are the result of circular electric currents produced by electrons orbiting the atomic nucleus along closed paths. Thus the electrons give rise not only to a magnetic moment, but also to angular momentum, or torque. According to his hypothesis Einstein was able to determine the ratio of magnetic moment to angular momentum, and to demonstrate that this was independent of the radius of the orbits and of the velocity of the electrons. It has since become customary to write the formula in a way that leads to the determination of a single magnitude, the "gyromagnetic ratio" *g*. If Einstein's hypothesis was correct, the result would have to be $g = 1$.

To test these ideas Einstein and de Haas suspended an iron rod from a fine thread inside a coil. When the current was switched on in the coil, the magnetic moments of the iron atoms in the magnetic field should orient themselves in a definite direction, and the same should be true of the angular momenta. But as angular momenta are subject to a conservation law, the iron rod should compensate the orientation in the magnetic field by a torsion movement. If the direction of the magnetic field was rapidly changed, the iron rod should rotate like a spinning top, and that rotation could then be observed. The experimental idea sounds relatively straightforward, and in fact it has been reported that, while still in Bern, Einstein along with two teachers had discovered that "by suddenly changing the direction of the electrons a rotation can be produced as a reaction"[43] However, the quantitative determination of *g* was rather complicated and required some tricks at the limits of what was then possible.

Allowance had to be made for countless sources of error, among which compensating for the terrestrial magnetic field by using large coils was one of the simpler problems. First of all, the iron rod surprised the experimenters by performing "the most adventurous movements."[44] By suspending it from a thin, rigid glass thread and by applying a sophisticated resonance method, they succeeded in eliminating several sources of interference and in demonstrating both qualitatively and quantitatively that angular momentum was produced by magnetization. For the gyromagnetic ratio they found the value that had been theoretically assumed: $g = 1$ (more accurately, 1.02). Given the pitfalls of the experiment, this was almost too good. "Even though the good agreement may be due to chance," Einstein and de Haas wrote when they published their results, "as we must assume about 10 percent of uncertainty in our determination, we have certainly proved that the consequence of the theory of orbiting electrons, set out at the beginning, has been confirmed also quantitatively, at least approximately, by the experiment."[45]

On February 19, 1915, Einstein gave a report on this experiment and its results to the Physical Society. De Haas returned to the Netherlands in April, and the partners henceforth worked separately on improving their apparatus—Einstein at the Reich Institute and de Haas in Leyden—without any substantial changes in the findings.[46] Throughout the summer and fall of 1915 Einstein kept returning to the laboratory, to magnets, coils, and mirrors. On February 25, 1916, just after the dramatic weeks of completing his general theory of relativity, he presented to the Physical Society a simplified variant,[47] designed to make this splendid effect known to his colleagues.

Einstein was invariably full of enthusiasm. "A wonderful experiment," he had raved to Besso even before the first results had been published. "A pity you can't see it. And how insidious Nature is when one is trying to get at it experimentally."[48] He was right about "insidious Nature," though, because not much survived of his conclusions. As for "zero-point energy," that concept had been introduced into the debate by Max Planck in 1911 to explain certain anomalies of the specific heat of solids at low temperatures, but Planck had created more problems with this than he had solved. Einstein had tried his hand at it with doubtful success, and had accurately described the position in 1915: "No theoretician now utters the words 'zero-point energy' without a half-embarrassed and half-ironical smile on his face."[49] Magnetism persists at very low temperatures, even close to absolute zero; thus Einstein concluded that the movement of the orbiting electrons also continued even at low temperatures, which was correct. But he concluded further that his experiment would therefore prove the existence of zero-point energy—and that was a rash conclusion. A dozen years later, zero-point energy would be correctly interpreted as a quantum-mechanical effect; it had nothing to do with the constancy of magnetic moments at low temperatures.

Moreover, Nature was even more insidious than Einstein had imagined when it came to measuring the gyromagnetic ratio. To cut a long and involved story[50] short: because the effect was of such importance to fundamental problems of physics, the experiment was repeated by other physicists over the next few years, including the reliable Emil Beck at the Zurich Polytechnic and researchers in America and Sweden. Instead of the value obtained by Einstein and de Haas, $g = 1$, they all obtained almost exactly twice that: $g = 2$. At the time, there was no theoretical explanation; not until the end of the 1920s was the problem solved by quantum mechanics, electron spin, and Werner Heisenberg's theory of ferromagnetism, based on electron spin. Einstein's simple hypothesis, in consequence, became obsolete; ferromagnetism was understood as a pure spin effect, with Ampère's molecular currents contributing virtually nothing to the phenomenon. Even more important for the Einstein–de Haas effect, quantum-mechanical theory postulated a gyromagnetic ratio of $g = 2$—that is, exactly the result of all the experiments except those of Einstein and de Haas.

Einstein was reluctant to concede defeat. Even after $g = 2$ had been universally accepted, he suggested that the magnetic rotation effect be carefully examined "because there is still no certainty on the numerical factor."[51] That was a little too obstinate. Famous as Einstein was for unhesitatingly submitting corrections to his own papers, he failed to do so in this instance.

For the summer of 1915, Einstein had planned a trip to Switzerland. "I greatly look forward to breathing some Zurich air again," he wrote to Zangger, "although my life here is ideally beautiful except for the things which really don't concern me in the least."[52] In fact, those "things" did concern him, much more than he admitted to himself. No doubt with an eye to his trip he had, in March, written to Romain Rolland, the French writer living in Switzerland, who was watching his life's work—mediation between French and German culture—being wrecked by the war. Einstein had learned of Rolland's determined pacifism through the New Fatherland League and about

> how courageously you have engaged your existence and your person for the removal of the fatal misunderstanding between the French and the German nations. I feel impelled to assure you of my unbounded admiration and respect. . . . I place my weak

> forces at your disposal in case you think that, either through my place of residence or my connections with German and foreign representatives of the exact sciences, I can serve you as an instrument.[53]

Rolland, who before this letter may not have known of Einstein, replied immediately: "Your generous letter profoundly moved me."[54] At Einstein's suggestion they kept in touch through Heinrich Zangger in Zurich until such time as Einstein would come to Switzerland.

Meanwhile, Einstein accepted an invitation to go to Göttingen for a week, where, at the request of David Hilbert and Felix Klein, he presented his generalized relativity theory in six two-hour lectures. Hilbert—who, after Poincaré's death, was unquestionably the world's foremost mathematician—had since the winter semester of 1914–1915 devoted a seminar to the fundamentals of physics and in this context had dealt in particular with the theories of Gustav Mie and Albert Einstein. Einstein was delighted by the interest of the mathematicians at Göttingen and, as he reported to Sommerfeld, had "the great pleasure to be understood in every detail. I am quite enthusiastic about Hilbert. A great man!"[55] This assessment was also based on Hilbert's condemnation of the war as an appalling folly and on the fact that Hilbert (unlike Felix Klein) had refused to sign the *Appeal to the Cultured World.* Of course Einstein did not mention this to Sommerfeld, since he knew that Sommerfeld would approve of even the most stupid patriotic appeals. (Einstein, though, did not know that Sommerfeld vehemently protested the publication of a paper by Lorentz in *Naturwissenschaften;* or that Sommerfeld believed that Lorentz's son-in-law de Haas "had better be removed from the Reich Institute."[56])

The main reason for planning a trip to Switzerland, of course, was Einstein's wish to see his sons. However, there were difficulties. "The fact is my boy wrote me a very brusque card, resolutely rejecting a tour with me,"[57] he wrote to Zangger, explaining the postponement of his visit. "I wouldn't even be able to see my children as my wife is traveling somewhere, or perhaps has traveled already." That was why, together with his cousin Elsa, her daughters, and presumably also her parents, he went to the Baltic—then the vacation venue of well-situated Berliners—to a small place called Sellin on the island of Rügen. He was "enormously looking forward to making the acquaintance of the sea" and was generally more than content with his personal situation.

The trip to Switzerland materialized after all, however, at the beginning of September. Mileva may still have hoped that Einstein would return to his family, but he wanted only to see his sons and went for a number of walks and hikes with eleven-year-old Hans Albert. On September 16, he and Zangger went to Vevey on Lake Geneva and there met Romain Rolland. Rolland recorded in his diary their lengthy conversations on the terrace of the Hotel Mooser; more dramatically than any letters written by Einstein from Berlin or any statements made by him in Germany, Rolland's observations reveal what Einstein thought of the war and the country of his birth. Rolland wrote:

> What I hear from him is not exactly encouraging, for it shows the impossibility of arriving at a lasting peace with Germany without first totally crushing it. Einstein says the situation looks to him far less favorable than a few months back. The victories over Russia have reawakened German arrogance and appetite. The word "greedy" seems to Einstein best to characterize Germany. Everywhere the will to power is spreading, the admiring belief in force, and a firm resolution for conquests and annexations. The government is much more moderate than the people. It wanted to [withdraw from] Belgium, but could not do so; the officers threatened rebellion. The big bankers, the industrialists, the business companies are all-powerful; they demand compensation for the sacrifices they have made; the Kaiser is but a tool in their hands and in the hands of the officers. . . . As for the intellectuals in the universities, Einstein divides these into two very distinct classes—the mathematicians, the physicists, the men of the exact sciences, who are tolerant, and the historians and philologists who talk like lunatics from sheer national passion. The bulk of the nation is astonishingly docile, "tamed" (Einstein is fond of this word of Carl Spitteler). Einstein primarily blames education, directed as this is totally toward national pride and blind submission to the state. He does not believe that any racial features are responsible, as the French Huguenots, who came there as refu-

gees two hundred years ago, have acquired the same characteristics. The socialists are the only (up to a point) independent element, but it is a minority that rallies around Bernstein. The New Fatherland League makes very slow progress and has only a very small following.

Einstein does not expect any renewal of Germany out of itself; it lacks the energy for it, and the boldness for initiative. He hopes for a victory of the Allies, which would smash the power of Prussia and the dynasty. . . . Einstein and Zangger dream of a divided Germany—on the one side Southern Germany and Austria, on the other side Prussia. But such a radical defeat of the Reich is more than doubtful. In Germany everyone is convinced of victory; it is officially expected that the war will go on for at least another six months. Nevertheless Einstein believes that the very well informed people realize that the situation continues to be serious and that it will become more serious if the war goes on for much longer. The worst shortages will probably not be foodstuffs, but chemical products vital for the war. It is a fact that the truly admirable inventiveness of the German scientists offers new compounds as substitutes for products in short supply. Einstein says one cannot imagine the organizational skill that has emerged and that includes all capable brains. All the university professors in the natural sciences have undertaken military orders or tasks. Einstein alone has refused to cooperate.

We speak of the deliberate blindness and the lack of psychology in the Germans. Laughing aloud, Einstein relates that after every meeting of the Berlin University Senate all the professors meet in a restaurant and that *invariably* the conversation begins with the question: "Why are we hated in the world?" Then a discussion follows, in which everyone supplies his own answer, yet they all most carefully steer clear of the truth.[58]

When Einstein left by train for Bern the following morning, Rolland saw him to the station. As a mild pacifist and an admirer of German culture, Rolland was not always happy with, and was sometimes even irritated by, Einstein's judgments and prejudices concerning Germany. "Observing Einstein," he noted in his diary, in conclusion, "one notices that, just as the very small number of spirits who have kept free from a slave mentality, he tends, out of a reaction to that slave mentality, to see the worst side of his nation and to condemn it almost as severely as its enemies are doing."[59]

Back in Berlin, Einstein was confronted with the task of expressing his ideas on the Germans and on the war in a form that would not be too shocking to the censor or the public. The Berliner Goethebund (Berlin Goethe League), a peaceable but by no means pacifist cultural organization, had asked him for a contribution for a planned *Patriotic Album*. In October 1915, under the heading *My Opinion of the War*,[60] Einstein set out in three pages what he thought would be just about acceptable for the Goethebund and the German public. Arguing psychologically, he saw the roots of war "in a biologically determined aggressive feature of the male creature." To search for the objectives or the causes of war he regarded "as rather irrelevant; they can always be found when passion needs them." He consistently rejected any kind of war and urged the creation of a political order along the lines championed by the New Fatherland League:

> The best minds from all epochs are agreed that war is one of the worst enemies of human development, that everything should be done to prevent it. Despite the unspeakably sad circumstances of our present times I remain convinced that a political organization in Europe would be attainable in the not too distant future, which would rule out European wars in the same way as the German Reich now rules out a war between Bavaria and Württemberg. No friend of spiritual evolution should fail to support this important objective.

That much the Goethebund was prepared to publish—but not Einstein's biting mockery of patriotism as a shrine in the emotional makeup of the citizen, "containing the moral requisites of bestial hatred and mass murder, which he then obediently takes out in order to apply them." Einstein recommended that this patriotic altar be replaced by a piano or a bookcase. Another passage not considered fit to print expressed his concept of internationalism: "How close a person or a human organization is to me depends solely on how I judge their intentions and abilities. The state, to which I belong as a citizen, plays not the slightest role in my emotional life; I regard a person's relations with the state as a business matter, rather like one's relations with a life assurance company."

In defending his critique, Einstein argued that "the glorification of war in peacetime, as well as all other emotional and ideological complexes which prepare for war in peacetime," should "be vigorously opposed by all genuine friends of human progress. This, in my opinion, includes everything that goes by the name of 'patriotism.' "[61] Nevertheless, he eventually agreed to the publication of the abridged version; and even in its abridged form his essay was one of the most courageous statements published in Germany against the war.

The article must have been very important to Einstein, for it was written during the "most exciting, most strenuous, but also the most successful period of my life."[62] His defense of the two paragraphs which the officers of the Goethebund intended to excise was written on November 11, 1915, the same day on which he had scheduled the second of his four lectures marking the conclusion of his work on generalizing the theory of relativity.

CHAPTER NINETEEN

"The Greatest Satisfaction of My Life": The Completion of the General Theory of Relativity

"IT'S A STRANGE THING about scientific endeavor: often there is nothing more important than realizing where it is not desirable to spend time and effort. On the other hand, one should not pursue objectives that are easily achieved. It is necessary to develop an instinct for what, with an all-out effort, is just about attainable."[1] This is how Albert Einstein, in a letter to a former student of his at the Zurich Polytechnic, described the difficulties of a physicist confronted with the diversity of nature. His attitude toward his own objectives and subjects is illustrated by his comments on his activities in the spring of 1915. Of the experiment on the gyromagnetic effect, which he had once described so enthusiastically, he now said that it "could have been done by any fool." This apparently derogatory remark was intended not to belittle his work as an experimenter but rather to serve as a yardstick for assessing his principal efforts: "But the general theory of relativity is of a different kind. To have now really reached that objective is the greatest satisfaction of my life, even though none of my colleagues in the field has as yet been able to realize the depth and the necessity of this road."[2]

Einstein wrote this in May 1915, and his obvious confidence was based on his presentation of the theory in the comprehensive article published six months earlier in the *Proceedings* of the Academy: *The Formal Foundations of the General Theory of Relativity.* All the evidence of the next few months serves to show that, throughout the summer, he believed that in this article he had accomplished something like a definitive version of the theory. By the end of June, the mathematicians at Göttingen had "understood him in every detail"; and two weeks later he already had "the intention to write a special little book as an introduction to the relativity theory, its treatment aiming from the outset at a general theory of relativity."[3] At the end of August, he was rather proud that he had "completely convinced"[4] Felix Klein and David Hilbert in Göttingen.

He was, moreover, confirmed in his belief by his "good astronomer" Erwin Freundlich. Because of wartime conditions—and clouds—it had so far been impossible to prove that light would be deflected by the gravitational field of the sun. But Freundlich, immediately on his return to Berlin, applied himself to an alternative—albeit experimentally more difficult—test of the theory through the red shift of spectral lines in a gravitational field. As early as 1907, when he first reflected on the theory of gravitation, Einstein had worked out a value of two millionths for the shift, and that value was found even in the more sophisticated versions of the theory. Although this effect was measurable, it would be slight, and it would also be blurred by incalculable disturbances from the turbulent conditions in the outer layers of the sun. Instead of sunlight, therefore, Freundlich examined the spectra of binary stars of greatly varying magnitude, and after months of hard work he discovered that "undoubtedly phenomena exist to support the existence of a gravitational shift."[5]

Einstein was pleased with this news and in February 1915 wrote to Besso that there was even "an approximative quantitative confirmation which agrees satisfactorily."[6] By May, he had persuaded himself that the theory was already "brilliantly confirmed."[7] However, his jubilation was premature: Freundlich had made some gross errors in estimating the masses of the fixed stars. A few astronomers gave the presumptuous assistant—for that was all Freundlich was at the time—a good dressing-down, but Einstein remained loyal to him. In the end, it was not Freundlich's errors that made Einstein doubt his theory.

It was probably after his return from his trip to Switzerland that Einstein had to get used to the idea that his formulas for gravitation could not be correct. At the beginning of October he realized "that my previous argument was deceptive."[8] This gave him "some bad moments,"[9] especially when he discovered "that my past field equations for gravitation were totally untenable."[10] When, after seven weeks of feverish work, all difficulties were overcome, he outlined the causes of his growing misgivings to Sommerfeld: First, he had proved that even in the simple case of a uniformly rotating system of reference, the gravitational field does not agree with the field equations. Second, his calculations of Mercury's perihelion had yielded a result that was too small.

The first argument, the violation of rotational symmetry, must have hit him like a thunderbolt. Up to his great overview article in the autumn of 1914, he had regarded this invariance between rotations as a matter of course, without bothering about proof. When he recalculated in the autumn of 1915, the problem immediately emerged: the equations could not be correct.[11]

The second argument was even stranger. He had been doing calculations on the perihelion of Mercury as early as 1907,[12] a task at which the best astronomers had been failing since 1859, when Urban Le Verrier discovered this flaw in the otherwise perfect machinery of Newtonian celestial mechanics. Mercury, the planet closest to the sun, moves along an orbit that is not a closed ellipse but a kind of petal pattern characterized by a slow precession at the point nearest to the sun. The deviation from a perfect ellipse is very slight—less than one minute of arc in a century—but despite its tiny size this rotation of the perihelion was a major problem, for which no satisfactory solution could be found within Newtonian mechanics.

From the very first phase of his relativistic considerations of gravitation, Einstein had hoped he would be able to solve that riddle. His first reference to his early calculations comes in a letter to Sommerfeld, dated November 28, 1915, in which he confesses that, instead of the astronomically observed value of forty-five seconds of arc, he had in the past obtained only eighteen seconds of arc. But he was making this admission at a time when his theory, in a new and definitive version, had already passed the test. Equipped with this new theory, he had once more returned to the old problem in May 1913, at times jointly with Besso, though success continued to elude him.

October 1915 seems to have been a time of hectic work. For the Prussian Academy's plenary session on Thursday, November 4, he had announced a lecture, *On the General Theory of Relativity*, which he continued a week later, at the meeting of his specialized class. With unconcerned openness, which must have been something like a shock to his academic colleagues, he began with a résumé of his mistakes and false turns, in a manner probably unprecedented in the academy:

> Over the past few years I have endeavored to establish a general theory of relativity on the assumption of the relativity even of non-uniform motion. I actually believed I had discovered the only law of gravitation which was in line with a logically conceived general postulate of relativity, and I tried to demonstrate the necessity of just that solution in a paper published in these Proceedings last year.—A renewed critical examination has now shown me that this necessity absolutely cannot be proved along the lines then chosen. That this nevertheless appears to be possible is due to an error.[13]

Einstein sketched out a serious mathematical objection proving that all his efforts to chose special conditions of coordinates had been futile. He continued: "For that reason I completely lost faith in the field equations presented by me and sought for a way that would limit the possibilities in a natural manner. I therefore returned to the postulate of a more general covariance of the field equations, which, three years ago when I was working with my friend Grossmann, I had abandoned with a heavy heart. In point of fact, we had already got quite close to the solution of the problem then given." In Zurich, he had entered the almost correct equations in his notebook[14] (which is still extant). But he then rejected them, believing—erroneously—that they did not lead to the Newtonian approximation, and also because of some stringent considerations of causality and unambiguity. Now, having rid himself of these errors, he had returned to the general covariant theory rejected three years earlier and, applying Riemann's curvature tensor, had derived equations that in approximation led to Newton's kinetic equations, which were invariant with regard to all coordinate transformations, and which had come out so effortlessly and logically that they could not be other than correct: "Hardly anyone who has truly understood it will be able to escape the charm of this theory."[15]

On November 18 Einstein again appeared at the Prussian Academy's plenary meeting and presented "an important confirmation of this most radical relativity theory. It actually turns out that the secular rotation of the orbit of Mercury in the direction of the motion of the orbit, first discovered by Le Verrier, amounting to about 45 seconds of arc in a century, is qualitatively and quantitatively explained without any special hypothesis having to be postulated for it."[16] At this stage Einstein was not yet able to offer exact, complete solutions of the field equations, but that was not really necessary: the first—by now "classical"—problems could be tackled by approximation methods.

At this point, Einstein made the strange discovery that although the first approximation yielded the Newtonian kinetic equations, the ten components of the metric tensor could not be reduced to the Newtonian gravitation potential. Specifically, this meant that a curvature of space is found even in weak fields. This has no consequences for the red shift in the gravitational field, but it does have consequences for the deflection of light passing the sun. Thus the amount first obtained by Einstein in Prague in 1911—0.85 seconds of arc—was now doubled[17] because of the curvature of space by the sun's mass: Einstein predicted a deflection of light of 1.7 seconds of arc. This was exciting enough, but it was nothing compared with the second approximation, which produced a perihelion precession for Mercury stunningly close to the observed values.

"The result of Mercury's perihelion movement fills me with great satisfaction,"[18] Einstein wrote to Sommerfeld. "How helpful I now find the pedantic accuracy of astronomy that I often made quiet fun of in the past!" His equations had thus passed their trial by fire, and nature itself had confirmed that his eight-years odyssey through the abysses of gravitation theory had ended in a wonderful success. Einstein was euphoric. "Imagine my joy over the practicability of general covariance and over the result that the equations correctly yield the perihelion movement of Mercury,"[19] Einstein wrote to Ehrenfest, describing what was perhaps the most profound experience of his life. "For some days I was beyond myself with excitement." Later he told his Dutch friends[20] that the discovery had set off palpitations, and that he had the impression that something had burst within him.

Today, Einstein's theory of gravitation wears a halo as "probably the greatest scientific discovery ever made";[21] otherwise, one might be tempted to present it as a comedy of errors, though of course on the very highest plane. Not surprisingly, a few flaws and even mistakes slipped into Einstein's reports to the academy in November 1915. "The final errors in this struggle," Einstein confessed to Sommerfeld, "have now unfortunately been immortalized by me in the Academy papers which I will soon be able to send you."[22] There was something not quite right even in Einstein's field equations. It must, then, have been a load off his mind when he discovered that the necessary modifications of the approximation procedure of November 18 had no bearing on the conclusions, so that the important results on the deflection of light and on Mercury's perihelion remained correct. Undaunted, he faced the Physical-Mathematical Class the following Thursday, November 25, and in a mere three compact pages of typescript presented the correct justification of what were now definitively the correct "field equations of gravitation." The task he had embarked on eight years earlier was accomplished. "With this, finally, the general theory of relativity is completed as a logical edifice,"[23] he wrote in *Proceedings of the Academy.* "The postulate of relativity in its most general form, which makes the space-time coordinates physically irrelevant parameters, leads with cogent necessity to a very definite theory of gravitation, one which explains Mercury's perihelion movement."

As soon as Einstein received the offprints of his academy lectures, he sent them to his colleagues and friends, with covering letters reflecting his happiness. "The theory is of incomparable beauty,"[24] he exulted to Zangger. He vigorously urged Sommerfeld to study the equations: "Make sure you have a good look at them; they are the most valuable discovery of my life."[25] To Besso he wrote: "My boldest dreams have now come true,"[26] ending with regards from "your content, but rather bushed Albert." A little later, once more to Besso: "Do read the papers! They bring the final release from misery."[27] Despite his euphoria, it was not long before his humor burst through again: "It's a cinch with that Einstein,"[28] he said, mocking his own enthusiasm for the earlier stages of the theory. "Every year he revokes what he wrote the year before." This time there was no need to revoke anything—and he knew it.

During the decisive phase Einstein even had a congenial colleague, though this caused him more annoyance than joy, as it seemed to threaten his primacy. "Only one colleague truly understood it, and he now tries skillfully to appropriate it,"[29] he complained to Zangger about what he evidently regarded as an attempt at plagiarism. This colleague was none other than David Hilbert, with whom, as recently as the summer, Einstein had been "absolutely delighted." What must have irritated Einstein was that Hilbert had published the correct field equations first—a few days before Einstein.

Einstein presented his equations in Berlin on November 25, 1915, but six days earlier, on November 20, Hilbert—in a paper submitted to the Royal Society of Sciences in Göttingen[30]—had derived the identical field equations for which Einstein had been searching such a long time.[31] How had this happened?

David Hilbert had concerned himself intensively with physics for a number of years; had read everything about electrons, matter, and fields; and in this context had invited Einstein to Göttingen toward the end of June 1915 to lecture on relativity theory. Einstein had stayed at the Hilberts' home, and one must assume that the week he and Hilbert spent together would have consisted of dawn-to-dusk discussions of physics. They continued their debate in writing, although Felix Klein records that "they talked past one another, as happens not infrequently between simultaneously producing mathematicians."[32] Hilbert was in fact aiming at greater things than Einstein: at a theory of the entire physical world, of matter and fields, of the universe and electrons—and in a strictly axiomatic structure.

In November, when Einstein was totally absorbed in his theory of gravitation, he essentially corresponded only with Hilbert, sending Hilbert his publications and, on November 18, thanking him for a draft of his treatise. Einstein must have received that treatise immediately before writing this letter. Could Einstein, casting his eye over Hilbert's paper, have discovered the term which was still lacking in his own equations, and thus "appropriated" Hilbert? This is not really probable: Hilbert's treatise was exceedingly involved, or indeed confused—according to Felix Klein, it was the kind of work "that no one understands unless he has already mastered the whole subject."[33] It cannot be entirely ruled out that Hilbert's treatise made Einstein aware of some weaknesses in his own equations. Nevertheless, his eventual derivation of the equations was a logical development of his earlier arguments—in which, despite all the mathematics, physical principles invariably predominated. His approach was thus quite different from Hilbert's, and Einstein's achievements can, therefore, surely be regarded as authentic.

For a few weeks relations between Einstein and Hilbert were clouded; at least, we know that Einstein was convinced that his Göttingen lectures and some of his other thoughts had—perhaps inadvertently—been plagiarized by Hilbert. It may well be, though, that he was somewhat mollified when he saw the printed version of Hilbert's treatise, since Hilbert, in the very first sentence, paid tribute to "the gigantic problems raised by Einstein and the brilliant methods developed by him for their solution,"[34] which represented the prerequisites of a new approach to the fundamentals of physics. Thirty years later, Einstein told his assistant Ernst Straus, who in turn after another thirty years told Abraham Pais, that "Hilbert had sent him a written apology, informing him that he had 'quite forgotten that lecture.' "[35] If that is what happened, then it must have satisfied Einstein, for just before Christmas he wrote to Hilbert: "There has been between us something like bad feeling, the cause of which I don't wish to analyze further. I struggled against a resulting sense of bitterness, and I did so with complete success. I once more think of you in unclouded friendship, and would ask you to try to do likewise toward me. It is, objectively speaking, a pity if two fellows who have worked their way out of this shabby world cannot find pleasure in one another."[36] The reconciliation worked so well that no one else seems to have noticed any friction, and a legend later arose that there had never been anything but friendly feelings between Einstein and Hilbert.[37] Hilbert, like all his other colleagues, acknowledged Einstein as the sole creator of relativity theory.

When Rudolf Jakob Humm, a Swiss studying mathematics in Göttingen, visited Einstein in Berlin in May 1917, he mentioned Hilbert's attempt to derive quantum theory from the theory of gravitation. Einstein made "a roguish face: that would probably not be possible, even though the theory of gravitation was the more general. The relativity idea simply could not yield more than gravitation. . . . The idea of constructing a world out of one's fantasy was fine and might produce something . . . but he felt reservations about all such attempts to construct the world from imagination. It would be a piece of excessive boldness to form a finished picture of the world already, seeing that there are still so many things which we cannot even surmise."[38] A few years later Einstein himself had taken over Hilbert's goals, though not his methods, and had set out on his own track toward a unified field theory.

With the general theory of relativity now completed, there was need for a comprehensive presentation of what had been safely established, the more so as Einstein's progress, in Felix Klein's account of the development of his ideas on gravitation, "as befits a genius, [had been] irrational—a mixture of philosophical requirements with a powerful physical instinct and a progressive penetration into the preparatory studies of the mathematicians."[39] This was reflected in Einstein's publications. They were, for the most part, provisional workshop reports, at times building on one another and at times correcting or revoking something published earlier.

Einstein was well aware of this and was thinking of writing a comprehensive presentation. However, he found it difficult to start, "as in all things that are not driven by an intense desire. But unless I do it, the theory will not be understood, straightforward as it basically is."[40] To Lorentz, who had understood it all and had congratulated him on his "brilliant result," Einstein wrote: "The series of my papers on gravitation is a chain of errors which nevertheless gradually led me nearer to my goal. That is why the fundamental formulas, at long last, are good, though the derivations are atrocious. This flaw must first be eliminated."[41] By way of timid hints he even tried to persuade Lorentz, whom he rightly regarded as a master of presentation, to undertake this task, saying that he himself "had unfortunately been denied by nature the gift of communication, with the result that what I write may be correct but is highly indigestible."[42] Lorentz evidently did not rise to this bait, so that Einstein had no choice but to present an overview of his theory himself. Actually, in the end the work proceeded quite speedily. The treatise, fifty pages long, was completed in March 1916 and published in *Annalen der Physik*.[43] The publisher of the journal, Johann Ambrosius Barth, printed it as a separate brochure—Einstein's first book.

From then on, the terminology was laid down: Einstein's theory of 1905 was no longer the "ordinary" or "usual" but the "special" theory of relativity, in contradistinction to the "general" theory of relativity—even though this name unfortunately obscured its close relationship to gravitation. This treatise of 1916—it had the title *Foundations of the General Theory of Relativity*—was for a long time the authoritative presentation of the new theory. It was included in an anthology on the relativity principle and translated into foreign languages.

As early as February 1916, Max Born had published a comprehensive survey[44] for the benefit of the many physicists who, being experimental scientists, had difficulties with abstract theories presented in the unfamiliar garb of absolute differential calculus. Einstein was grateful "to have been so completely understood by one of my best colleagues. But quite apart from its objective contents, I am happy to note the sense of joyful sympathy that shines from your work."[45] Only a few months later, there appeared a booklet by Erwin Freundlich[46] with similarly reduced mathematics. In a brief foreword Einstein commended it for "making the basic ideas of the theory accessible to anyone who has some kind of acquaintance with the reasoning methods of the exact sciences."

Perhaps it was these models, especially the successful article by Born, that induced Einstein to attempt a presentation with as little mathematics as possible. By the end of 1916 the new manuscript was finished, and at the beginning of the following year it was published as a little book by Vieweg in Braunschweig, under the title *On the Special and the General Theory of Relativity, Generally Comprehensible.* It went into numerous editions and was widely translated.

Thus all the conditions had been created for the dissemination of the theory among a readership educated in mathematics or physics. But there was still no popularization for the "interested layman." Einstein readily admitted that his little book might as well be called "generally incomprehensible," and (no doubt with his often observed "roguish mien") he approvingly passed on Max Planck's comment: "Einstein believes his books will become more readily intelligible if every now and again he drops in the words 'Dear reader.' "[47]

It may be useful, therefore, if, in our own words, we once more outline what is difficult to understand in the theory and try to explain why Einstein described it as his "most beautiful discovery."

To do justice to Einstein's achievement, one has to proceed from the starting point provided by Newtonian physics, inherited by Einstein as a paradigm that had proved its worth throughout two centuries. This was an image of the world as a vast machine functioning according to simple laws which could be formulated, in essence, by two pairs of concepts: time and space; bodies and forces. Space, understood by Newton as "absolute" space, was a kind of container within which bodies moved—in other words, changed their location over time. Time, in the same way, was understood as "absolute" time. In the absence of forces, bodies moved uniformly and rectilinearly; if forces acted on them, they moved along paths which, in principle, could be calculated.

This theoretical framework had stood the test of time in the most varied fields, from the well-nigh miraculous accuracy of celestial mechanics to the propagation of sound in gases. Its success was so overwhelming that later generations might overlook a few conceptual pitfalls, though Newton himself had been aware of these. One flaw was the attraction between bodies—gravity, which was supposed to propagate through space at infinite velocity; or, more accurately, was supposed to be independent of time. Another flaw was an asymmetry between bodies and space. "Absolute" space acted on bodies in the form of inertial resistance to acceleration—resulting, for instance, in a flattening of the Earth as it rotated "in space"—but bodies did not act on space. Such an asymmetry is not a priori unreasonable, but to such questioning minds as Ernst Mach and Heinrich Hertz it was unsatisfactory in terms of cognition theory; and to Einstein "absolute" space seemed like a stage ghost. It ran counter to his instincts as a physicist.

Despite its profound modification of the concept of time, the special theory of relativity did not so much change these problems as exacerbate them: the Newtonian theory of gravitation, because of its instantaneous propagation of effect, could be integrated into the special relativity theory, as the framework of all physics, only at the price of questionable hypotheses. Einstein went his own way; his starting points did not at first glance seem to have anything to do with each other, but in his view they were fundamentally linked.

For one thing, he extended the relativity principle to any accelerated system of reference. To him this step seemed obvious, but after centuries in which physics had been tied to inertial systems it was actually such a bold step that Max Planck (for one) for a long time refused to follow it. Next, Einstein regarded the identity of inertia and mass—"hitherto only registered but not interpreted"—as having a profound significance in physics, far beyond its ability to explain an always equally accelerated free fall. Imagining a man falling from his roof and not experiencing a gravitational field, Einstein realized that, at least in certain situations, acceleration and gravitation could not be distinguished. This principle of equivalence suggests that a generalized theory of relativity must, at the same time, be a theory of gravitation, and conversely that a satisfactory theory of gravitation can be formulated only within the framework of a generalized relativity postulate.

That meant abandoning the preferential systems of reference in classical physics—including the special relativity theory—the inertial systems in which the laws of physics emerge in their simplest forms as peculiar consequences of absolute space and absolute time. The fundamental character of the general theory of relativity is expressed in the fact that the coordinates of space and time have no physical meaning. Within the mathematical framework of "absolute differential calculus"

Einstein was able to unite what, according to classical tradition, should be strictly distinguished: space and gravity; geometry and gravitation. Masses no longer interact through a law of forces; instead, they bend or curve space. Geometry, in a generalization of the Pythagorean theorem, is presented by a flexible "metric tensor," which is determined by the distribution of masses and, at the same time, determines the geodesics—the "shortest paths" of the motion of these masses. Thus space is no longer a kind of container for masses, but rather an object with its own inner dynamics. The properties of space and the movements of masses are described by field equations in which the "metric tensor," "Riemann's tensor of curvature," and the energy impulse tensor are linked with each other. This is achieved within "absolute differential calculus" in a manner free from any arbitrariness and almost matter-of-fact.

Thus for the first time a physical theory had been created that was free from grafted-on external concepts, such as "absolute space" in Newtonian theory. The concepts of the general theory of relativity and their meanings arise effortlessly within the theory itself. Einstein had attained a new level in the contemplation of nature, one that has been regarded as a model ever since but has never been repeated.

Some of Einstein's admirers were tempted to see the general theory of relativity as a triumph of speculation over empiricism. That kind of misunderstanding made Einstein "downright angry," even if his friend Besso was involved: "This development teaches us something entirely different, indeed almost the opposite, namely that a theory, in order to merit confidence, must be based on generalizable facts."[48] For the general theory of relativity the "equivalence of inertia and mass" was such a fact—even though he did not learn of Baron Eötvös's precision experiments until 1912. To Einstein, then, facts were not only the starting point of his theory but also the keystone for any test of it. We have seen how elated he was by his correct derivation of the precession of the perihelion of Mercury. Now he would have liked to force people to undertake the other two verifications of the theory.

In his great article for *Annalen* he had calculated the deflection of light not only for the sun (1.7 seconds of arc) but also for Jupiter: 0.02 second of arc. There were good reasons for the second calculation, minute though the result was. Since the war ruled out any further expedition to observe a solar eclipse in the near future, he was anxious for the deflection of light to be tested using Jupiter, a method he and Freundlich had first considered as early as 1912 but had then dismissed as impracticable because the effect was so slight.

Freundlich now reverted to this method as a makeshift solution and no longer found it to be hopeless. As soon as Einstein had finished his series of papers on relativity theory for the Prussian Academy, therefore, he called on the highest official in the Ministry of Education, describing to him the verifiable results of relativity theory. On the deflection of light, Einstein wrote: "This consequence is not only the most interesting and surprising of them all, but undoubtedly also the most characteristic of the theory. Yet this consequence has not so far been put to the test. . . . Through a careful study of the available means of observation Herr Freundlich has now come to the conclusion that the light-bending effect might be proved also by the planet Jupiter, albeit only with the most sophisticated photographic measurements and a multitude of observations."[49] This was not really intended for the instruction of the official; it was support for Einstein's request "that Herr Freundlich be freed for a few years from his regular duties of determining the position of stars, in order that he may devote himself without interference to the task here outlined."

Freundlich's boss, Privy Councillor Struve, director of the Babelsberg Observatory, regarded Einstein's attempt to influence the observatory's research program through personal intervention at the ministry as regrettably tactless; and in any case he would have liked to kick Freundlich out because of his persistent neglect of routine duties. Therefore, Struve rejected the Jupiter project not just firmly but arrogantly: "Even a 'multitude of the most sophisticated measurements' by expert observers, let alone by those who do not come under this heading, will not yield any useful result and merely cause a needless expenditure of time and effort."[50] Einstein was unwilling to accept this and complained: "Only the intrigues of miserable people prevent this final important test of the theory from being carried out."[51] But it was no use; observations of Jupiter were not even begun. Einstein had to wait for the next solar eclipse, and this was not due until June 8, 1918, in Finland.

The red shift of spectral lines in a gravitational field was likewise a source of continuing controversy among astronomers. Freundlich believed that he had proved this effect, but he had made some mistakes in his publications, for which he was being fiercely attacked, especially by the astronomer Hugo Ritter von Seeliger, president of the Bavarian Academy, in Munich. It was, of course, clear to everyone concerned that although Freundlich was being whipped, the real target was Einstein. Einstein stood loyally by Freundlich, and also stood by Freundlich's questionable results, which to him seemed to be "sufficiently reliable confirmation, at least qualitatively, of the shift of the spectral lines."[52] Struve, on the other hand—quite correctly—regarded the phenomenon "as not proved by the totally superficial investigations made so far."[53]

Seeliger and Struve, however, went even further. Although they did not dispute Einstein's calculations for the perihelion of Mercury, they denied that this was a verification of relativity theory, arguing that the movement could also be explained within Newtonian gravitation theory. They did manage to produce such an explanation, but only at the cost of such improbable assumptions that, to Einstein, refuting them was a case of "fighting battles that were already won."[54] None of this improved his relations with the leading German astronomers.

Einstein's annoyance with the astronomers was more than compensated for by his success among theoretical physicists. Even the stubborn skeptic Max Planck, as Einstein reported, was beginning "to take the matter more seriously, even though he is still resisting a little. But he is a splendid person."[55] Soon Planck gave up his opposition entirely and indeed came to regard the general theory of relativity as the realization of his demand for absolute structures. "If at this moment the theory has still many opponents," Einstein wrote to Hermann Weyl, the mathematician who had moved from Göttingen to Zurich, "I am consoled by the following circumstance: the otherwise determined intellectual strength of the adherents greatly exceeds that of the opponents. That is an objective testimony to the natural and reasonable character of the theory."[56]

Einstein could also have found consolation in the first successful applications of the theory. These came from Karl Schwarzschild, the highly intelligent director of the Astrophysical Observatory on Potsdam's Telegraphenberg (Telegraph Hill). Schwarzschild would have been an altogether congenial partner in astronomical and astrophysical matters; what prevented Einstein from collaborating with him was the fact that Schwarzschild had volunteered for army service as early as 1914 and was now calculating artillery trajectories on the Russian front. However, this does not seem to have taken up all his time, for as soon as he had received Einstein's reports to the Prussian Academy he managed to provide the first complete solution of the field equations in the case of a large mass with a spherical gravitational field. The Mercury problem, calculated by Einstein by means of approximation methods, was thus accurately solved.[57] On behalf of the absent Schwarzschild, Einstein submitted his calculations to the academy on January 13, 1916. Schwarzschild's precise calculations made no difference to the result found by Einstein's approximations.

A month later Einstein presented to the academy another paper that Schwarzschild had sent to him from the Russian front. In this, Schwarzschild, using the general theory of relativity as an instrument, had investigated a spherical accumulation of matter, such as is represented by the sun and other stars.[58] He regarded a star as a liquid sphere, which is a crude oversimplification of the real state of affairs, but thanks to this simplification he was able to solve the problem exactly and draw some astonishing conclusions from his calculations. If the mass of a star is concentrated within a sphere of a definite radius—later to be called the "Schwarzschild radius"—then the curvature of space becomes too great for light to escape from the star. Hence anything occurring within the Schwarzschild radius remains an eternal secret for an outside observer; equally, from a position within the Schwarzschild radius there is no possibility of making any contact with the external world. This bizarre situation would arise if the mass of the sun were compressed into a sphere with a radius of three kilometers; for the mass of the Earth, the Schwarzschild radius is less than one centimeter. Such densities of material are so extreme that for a long time theoreticians of relativity, Einstein included, were unable to do much with these findings. Only considerably later, with new knowledge provided by nuclear physics about stellar dynamics in a gravitational collapse, was it realized that Schwarzschild had in fact calculated the first simple model of what, almost exactly half a century later, came to be known as a "black hole."

In March 1916 Schwarzschild returned to Berlin, seriously ill. He was suffering from a rare skin disease for which there was then no cure. He spent two months in a hospital and died on May 11, 1916. On June 8, Einstein gave a memorial address at the Prussian Academy for his colleague, who had lived only to the age of forty-two.

Needless to say, Einstein did not just sit with his hands in his lap after his "most beautiful discovery," the field equations. Along with some formal and physical improvements he developed some remarkable aspects which opened up entire areas of research that have kept physicists busy to this day and will probably continue to engage them in the next millennium. The consequences are irrelevant for terrestrial conditions; they concern problems of space—cosmology and gravitational waves.

The first of these interesting applications was ready in June 1916.[59] Through this application, Einstein demonstrated—though this had been an implicit starting point of the theory—that gravitational fields, in line with the general theory of relativity, propagate at the velocity of light. In much the same way that Heinrich Hertz had deduced from Maxwell's equations that electromagnetic waves were created by the accelerated movement of electrical charges, so Einstein from the field equations proved the existence and the principal properties of gravitational waves, which are triggered by masses moving relatively to one another, though admittedly only on paper and not in experiment.

Soon he was no longer satisfied with his paper: he thought his "presentation of the subject not sufficiently transparent and moreover marred by a regrettable calculating error."[60] He therefore took up the subject again in February 1918, corrected the mistake, and arrived at his so-called "quadrupole formula" for the energy radiation of a mechanical system—a formula discussed to this day.

Unlike electromagnetic waves, which are easy to generate and whose existence is easy to prove (in fact, they are the basis of entire industries), gravitational waves were expected from the outset to be barely susceptible to verification, because of their extremely slight effects on mechanical systems. For a while they were even regarded as a mathematical artifact, a mental construct not matched by anything in physical reality. Not until the 1960s had confidence in these theoretical constructs become sufficiently consolidated and experimental skills sufficiently developed for physicists to consider seriously the construction of a physical apparatus to react to gravitational waves caused by cosmic catastrophes such as supernovas. And even now, after about a dozen teams of researchers have concerned themselves with developing and building the necessary antennas, all efforts have been in vain. Although the most modern antennas would probably respond to a supernova within our Milky Way, such occurrences are not exactly frequent: the last one was in 1604. Supernovas are observed in distant galaxies at a rate of about one each month, but for these the sensitivity of the antennas is still much too low. A great opportunity was missed in February 1987, when supernova 1987A flared up in the Magellanic Cloud, close to our own galaxy. Its gravitational waves might just have been verified, but all three detectors, though operational in principle, happened to be switched off at the time of this unexpected event.

However, the experimental relativists received another unexpected present from the skies. This was a "binary pulsar," a rotating system consisting of a pulsar and another heavenly body.[61] The time signature of pulsar radiation represents an excellent clock, and extremely accurate observation of this heavenly timepiece made it possible to confirm a change in the pulsar's speed of revolution of about one ten-millionth of a second annually. This result in 1978 agreed accurately with the binary pulsar's loss of energy through the emission of gravitational waves according to Einstein's of 1918 formula.

Thus it took sixty years to confirm the existence of gravitational waves at least in the cosmic laboratory of a binary pulsar, and it may well take a comparable time span before gravitational waves are directly proved on Earth.

The next range of problems, which Einstein addressed in the spring of 1916, was to provide answers to two questions—about the structure of the universe and the origin of inertia. He described his endeavors to Besso: "In gravitation I am now looking for the limiting conditions in the infinite. Surely it is interesting to reflect to what extent there exists a finite world, i.e., a world of naturally measured finite extension, in which all inertia is really relative."[62]

To Einstein, in a philosophically satisfactory physics there could be no inertia with regard to Newton's "absolute space," but only inertia as a result of the interaction of masses. This belief, inspired by Ernst Mach's critical analysis of the fundamental concepts of mechanics, represented—along with the equivalence principle and the generalized relativity postulate—the third leitmotiv of the general theory of relativity. In his *Outline of a General Theory of Relativity* of the spring of 1913, Einstein had already claimed that it was formulated in line with "Mach's bold idea that inertia has its origin in an interaction of the mass point observed with all other mass points."[63] Ultimately this means nothing less than that the inertia of a given body derives from its interaction with all the masses in the universe, so that, say, a bowler overcoming the inertia of a bowling ball is solidly linked, in a physical sense, to the entire universe.

Easy as it was to formulate this "Mach principle," its actual position in the finished theory was exceedingly difficult, the more so as nothing much was known then about the distribution of masses in the universe beyond the fact that all attempts to understand it in terms of the Newtonian theory of gravitation had failed, in the face of internal contradictions. When, after more than six months of reflection, Einstein was ready to submit his findings to the Prussian Academy in February 1917, he announced to Ehrenfest that he had "once again committed something in the theory of gravitation that threatens to get me interned in a lunatic asylum. I hope you don't have one in Leyden, so that I may visit you again in safety."[64]

Einstein had every reason to be concerned about the reaction of his colleagues, especially as he had wriggled out of the problems of infinite space by a bold device that was not far from a retraction. "The fact is, I have come to the conclusion,"[65] he informed his astonished colleagues, "that the field equations of gravitation hitherto presented by me are in need of a slight modification, in order, on the basis of the general theory of relativity, to avoid those fundamental difficulties which we . . . have shown to apply to the Newtonian theory." By way of explanation he then guided "the reader along the somewhat indirect and rough path covered by myself, because only thus can I hope that he will show interest in the final result."[66]

After an extensive discussion of the contradictions in the Newtonian picture of the universe and the impossibility of reconciling those contradictions within relativity theory for an infinite universe, the frustrated reader is gently led toward an alternative. This is a spatially closed world with an average uniform distribution of matter, which, moreover, is to be static—i.e., not subject to changes over time—and, more especially, has no beginning and no end in time. For such a structure the field equations offered no solutions; indeed, if they were the only possible equations "we would have to conclude that the relativity theory does not permit the hypothesis of a closed world."[67] As a way out, Einstein added to his field equations a "cosmological term" which left the general relativity postulates intact and could be chosen so that results "on the small scale," such as the perihelion of Mercury, would not be affected by it, while "on the big scale" a satisfactory solution would be obtained for the universe.

According to this model, the universe is unlimited, though finite in volume and in all dimensions—like the two-dimensional surface of a balloon in a three-dimensional system, which has no boundaries but nevertheless has a clearly defined area. The globally uniform distribution of matter ensures the curvature of space on the large scale, while local density fluctuations, such as the sun or some other star, effect an additional curvature within which, in a manner of speaking, the planets are trapped. In such a universe a beam of light must, of course, return to itself. In other words, a star should be visible not only in one direction but also in the exactly opposite direction, if it were not that the dimensions of the universe are too great and that starlight is too dim.

Needless to say, there was little prospect of verifying this effect—a "double presence" of distant stars—and hence Einstein's model of the universe, by astronomical observation. With the best telescopes then available, astronomers could see only some ten thousand light-years, whereas Einstein's calculations, based on accepted data on the distribution of stars, assumed that the universe had a radius a thousand times as large—i.e., ten million light-years. However, Einstein did not consider the actual figures of any importance compared with the general structure of his model, and therefore did not include them in his paper. Nowadays, when distances in the universe are assumed to be more than one million times greater still, Einstein's figures may seem minute. But well into the 1920s, the Milky Way was virtually identified with the universe—only when the Andromeda Nebula was identified as an independent galaxy did our image of the universe expand to become a space occupied by billions of galaxies. It is all the more admirable that, believing himself to inhabit such a relatively small cosmos, Einstein yet pointed the way toward a science of the universe.

Einstein was fully aware of the arbitrary way he had established his cosmological model. He himself described the " 'cosmological term' as an extension of the field equations not justified by what we actually knew about gravitation." He had felt obliged to resort to that device only because he envisaged a static model with a mean density of matter and because, without the "cosmological term," the field equations provided solutions only for a contracting universe or an expanding universe. At the conclusion of his paper he therefore specifically stated that no essential aspect of his cosmological reflections was affected by it, and that "a positive curvature of space from the matter present within it results also if that additional term is not introduced."

If only he had left it out, relied on his original solutions, predicted an expanding universe, and sat back to await the astronomers' discoveries! A dozen years later, when Edwin Hubble with his one-hundred-inch reflecting telescope at Mount Wilson in California discovered the escape velocity of galaxies and linked it with the expansion of the universe, Einstein would have been considered a prophet. As it was, he had to withdraw his "emergency"[68] modification of the field equations; later he called the introduction of the cosmological term "the biggest blunder he ever made in his life."[69]

Despite the "indirect and rough" detour by way of the cosmological term, Einstein's vision of a boundless but finite universe survived. At a single stroke it removed the Newtonian difficulty about matter disappearing in infinity; it provided a physical basis for the centuries-old riddle of inertia; and, by returning to the original field equations, it cleared the path to a dynamic theory of the universe as soon as this was suggested by astronomical observations. Even though Einstein's first cosmological model was not entirely correct, these "cosmological observations"—as he called them in the title of his 1917 paper to the academy—are a good illustration of "how a wrong solution of a fundamental problem can be incomparably more significant than the correct solution of an insignificant and uninteresting problem."[70] To that extent the "observations" are not only exciting applications of relativity theory, but fascinating ideas on the structure of the universe generally—once more comparable to the Copernican revolution.

As if the year 1916 had not yielded enough in terms of relativity theory, Einstein once more turned his attention to quantum theory, presumably because he believed that the general theory of relativity could lead to a deeper understanding of the microcosm of the atom. He had abandoned quantum problems six years previously; and he had enthusiastically welcomed the successes achieved in the meantime by Bohr's model of the atom, with its electrons orbiting on stationary paths. Einstein was "charmed" by Bohr and Sommerfeld's theory of spectral lines, which interpreted the emission and absorption of radiation as an electron crossing over from one stationary orbit to another. He saw this as a "revelation"[71] and was curious to know "what screws the Almighty applies to it."[72] Naturally, though, major problems remained, such as how the electrons managed to revolve about the nucleus without emitting electromagnetic waves as demanded by Maxwellian theory.

Einstein had come across the problem of the stability of atoms when he was analyzing gravitational waves, which were supposed to be emitted not only by the huge mass of the stars but also by electrons. This would clearly undermine the virtually eternal life of stable atoms. Although this was not a measurable effect, and probably never would be, "the atoms, because of the intra-atomic electron movements, would have to emit not only electromagnetic but also gravitational energy, albeit in minute amounts. As this is presumably not happening in nature, it seems that the quantum theory will have to modify not only Maxwellian electrodynamics but also the theory of gravitation."[73]

At that time Einstein could do no more than draw attention to the problem of uniting quantum theory and gravitation theory. Whether this problem would one day be solved by the quantum theory of black holes, only the future would show. Meanwhile, however, Einstein contributed a novel link between Planck's radiation formula and the new atomic theory. "A splendid idea has dawned on me concerning absorption and emission of radiation" is how he announced his new treatise to Besso. "An amazingly simple derivation of Planck's formula, I am tempted to say *the* derivation. Everything quite quantic."[74]

Einstein commended the "unparalleled boldness"[75] of Planck's derivation of 1900, meaning not only the problem itself but also the fact that it was based on assumptions that were not entirely free of contradictions. Einstein now succeeded in the first of two papers in eliminating that flaw. More interesting than the derivation itself was the general character of his methods. Einstein proceeded from Niels Bohr's basic—and by then well tested—assumption that the electrons within an atom occupy a number of discrete energy states, and are able, through emission or absorption of radiation, to pass from one of those states to another. Added to this was an assumption of thermodynamic equilibrium between radiation field and atom, as well as a consideration of the "classical" limiting case at high temperatures—and there was Planck's formula. This brief argument, on a mere two pages, also covers emission stimulated by the radiation field; thus the formulas already, by implication, contain the theory of the laser, though it was to take nearly half a century to be realized. More important to Einstein was "the simplicity of the hypotheses" and "the general way in which the argument effortlessly develops." This led him to "think it very probable that this will become the fundamental line of future theoretical presentation."[76]

This he immediately demonstrated in a second paper,[77] in which he showed that light is emitted not as a spherical wave but as spatially oriented "needle radiation," with the molecule suffering a retro-thrust of a definite magnitude, which must therefore likewise be ascribed to the emitted radiation. This tends toward a final departure from the Maxwellian equations and toward a confirmation of Einstein's "heuristic viewpoint" of 1905, according to which light is to be understood as a beam of particles.

Basically, Einstein in this paper derived all the characteristics for the definition of the photon—the quantum of radiation or the particle of light—even though he kept his terminology as conventional as possible. He was only too well aware that he was still alone in his quantum hypothesis of radiation. Six years later the formulas developed by Einstein were experimentally confirmed by the American physicist Arthur Compton, and soon afterward the whole scientific community was talking about the "photon." But as early as the summer of 1916, when he wrote this paper, Einstein pointed out that the "elementary processes make the establishment of a truly quantum-based theory of radiation appear almost inevitable."[78] To Besso he wrote: "With this, the light quanta are as good as certain."[79]

Despite their outstanding importance for quantum physics, these two papers may at first glance seem almost recreational, after Einstein's exhausting work on the general theory of relativity. They may have been just that—yet even Einstein did not often succeed in presenting scientific milestones with such lightness and elegance. What made them true strokes of genius, moreover, was Einstein's mention of what he then regarded as a "weakness of the theory": the fact that it "leaves the time and direction of the elementary process to 'chance.' "[80]

This is one of the rare times when Einstein resorts to quotation marks. On the one hand it is quite obvious in his theory that the time and direction of the emission of an individual light quantum are left to statistical chance, yet on the other hand he is reluctant to believe in "chance" in natural processes and therefore regards it as a "weakness of the theory." He probably suspected that the random nature of elementary processes would occupy a special place in quantum theory. Ten years later, with the development of quantum mechanics, opinions would be divided on just this issue. Whereas most physicists would no longer consider "chance" a questionable principle in quantum theory, Einstein from the outset had a feeling "that the real joke presented to us here by the eternal riddle-setter has absolutely not yet been understood."[81] A sense of unease about chance remained with him to the end of his life.

Along with his *annus mirabilis* of 1905, the period from November 1915 to February 1917—a little more than a year—was undoubtedly the most fruitful period in Einstein's work. During these fifteen months he produced fifteen scientific treatises, including the two significant contributions to quantum theory and, above all, the brilliant culmination of the general theory of relativity and the foundations of a scientific cosmology in the light of newly discovered possibilities.

He produced these achievements in what for him was a difficult time, amid family crises and the noise of war. Even though Einstein had a remarkable gift for insulating his work from such outside influences, he did not, in the end, remain unaffected by them.

CHAPTER TWENTY

Wartime in Berlin

"AT THIS DREADFUL TIME," Einstein wrote to one of his former students at the Zurich Polytechnic, "be happy and proud to be Swiss."[1] Quite apart from his own pride, his Swiss nationality and the little red Swiss Confederation passport had acquired an importance for him which he could scarcely have expected when he applied for citizenship in Zurich in 1900. His status as a neutral alien protected him from conflicts with colleagues and with the military authorities at Berlin and enabled him to make occasional trips to neutral countries.

His visit to Switzerland in September 1915 had confirmed his wish to return there again at the turn of the year. However, as the frontier was temporarily closed even to holders of valid passports, and as the journey was difficult and he himself weary and overworked, he postponed his visit to the spring: "I am greatly looking forward to the Swiss air and to being relieved of my muzzle."[2] He was looking forward also to seeing his two sons again, as well as his friends Zangger and Besso—Besso having just settled in Zurich.

At Easter 1916 his wish was fulfilled, but this visit to Zurich ended in disaster. A marriage between Einstein and his cousin Elsa was now no longer ruled out, and as a prerequisite Einstein hoped to formalize his anyway irreparable breach with Mileva by a divorce. For Mileva, who had never reconciled herself to the separation and who was probably secretly hoping that her husband would return to her and his sons, agreement to a divorce was out of the question. The scenes of a bourgeois marriage in dissolution had disastrous consequences, as emerges from Einstein's correspondence with Zangger and Besso: he took an "irrevocable resolution" not to see his wife again,[3] and he had to accept that his sons no longer wanted to be together with him and that Hans Albert, then twelve, would no longer even answer his letters. Moreover, Mileva became seriously ill, so that Zangger and Besso had to cope with the most pressing problems and look after the boys. In this they were vigorously supported by the attorney Emil Zürcher, who lived in the same building as Mileva, at Glockenstrasse 59, not far from the Physical Institute of the Polytechnic.

These friends did not conceal from Einstein that they regarded Mileva's breakdown as a result of his selfish endeavors to free himself from her, nor the fact that they had no sympathy with his attitude against the marriage. Defending himself in rather dramatic terms, Einstein explained to them that he simply could not live with Mileva; he felt unable to empathize with her problems. Also, he felt sure that she did not lack for anything: "She has a carefree life, she has her two splendid boys, she lives in a magnificent neighborhood, she is the mistress of her time, and she stands in the halo of innocence abandoned."[4] Only when Zangger, the medical man, expressed a fear that Mileva might not recover from her illness did Einstein he react with remorse: "Without your, Zürcher's and Besso's help I would be losing my head in this sad situation. I am very sorry for the woman, and I also believe that what she has been through with me and because of me is at least partially responsible for her serious illness."[5] When Mileva did improve after a few weeks, Einstein was visibly relieved and had Besso tell her that he "would no longer bother her about the divorce. The battle with my family has been fought through. I have learned to withstand tears."[6]

One does not get the impression that this promise cost him much pain. After all, it ensured that his beautiful relationship with his cousin would not be jeopardized by matrimony or the thought of matrimony. Still, he could not face going to Zurich again, and so he decided to implement another long-standing plan—to visit Leyden.

Einstein had always felt that Lorentz and Ehrenfest were the men who understood him best, both as physicists and in other respects. He had always sent the galleys of his papers to Leyden, almost before the ink was dry, but correspondence was a poor substitute for actual conversation. By the time he had completed his general theory of relativity, at the beginning of 1916, Einstein would have liked to drop everything and travel to the Netherlands, but because of wartime conditions the trip had to be postponed until the summer. Then, the authorities still made so many difficulties that he almost gave up hope: "I won't be going to Holland because traveling is being made hideously difficult. I'm thinking of waiting until we have peace."[7] Nevertheless, the following day found him calling at the Foreign Ministry, where he was informed that an official invitation from the University of Leyden might be helpful. This was instantly provided by Lorentz, but Einstein still had to get the original of his certificate of residence in Zurich[8] and overcome some further bureaucratic obstacles before, on September 27, he eventually left Berlin.

The small Dutch university town of Leyden must have seemed to him an oasis of peace, humanity, and happiness. He stayed with the Ehrenfests on Rozenstraat, enthusiastically welcomed not only by Paul and Tatyana but also by their children. He again made music with his friend, just as on their first meeting in Prague. On the second day of his stay he drove with his host to Haarlem, less than ten miles away, to see Lorentz. Einstein knew that Lorentz had understood everything about the general theory of relativity, and he was now waiting for it to be blessed by his father figure.

After dinner Lorentz invited his guests to his study, passed cigars around, and with great didactic mastery recapitulated the difficulties that Einstein had to overcome to arrive at the finished form of his theory. As long as twelve years later, Ehrenfest would recall that during Lorentz's summary, Einstein drew on his cigar less and less often and that, when Lorentz concluded with a subtly formulated question, the cigar had gone out altogether. Einstein was bending over sheets of paper covered with Lorentz's formulas, absently twisting between his fingers a lock of his full mane. "Lorentz was smiling at Einstein, as he sat there deep in thought, much as a father would regard his beloved son—confident that he would crack the nut offered him, but tense to discover just how. It took a while, but then Einstein joyously raised his head—he 'had it.' A little more to and fro, a slight misunderstanding, followed by clarification and complete mutual understanding, and then the two men's eyes shone at the riches of the new theory."[9]

In addition to Lorentz and Ehrenfest, the astronomer Willem de Sitter, also at Leyden, had immediately shown interest in the general theory of relativity and had accurately calculated its consequences for the motions of the planets. Einstein had given his friends in Leyden offprints of the authoritative *Foundations of the General Theory of Relativity* of the spring of 1916, and de Sitter had passed one copy on to Arthur Stanley Eddington, the secretary of the Royal Astronomical Society in England, where, because of the war, German publications were no longer available. Eddington instantly recognized the outstanding importance of Einstein's treatise; hurriedly embarked on a study of "absolute differential calculus," a discipline unfamiliar to him; and asked de Sitter to present the new theory, and especially its astronomical implications, for the benefit of his British colleagues in the *Monthly Notices* of the Royal Astronomical Society. De Sitter now had an opportunity to inform himself at the very source—Einstein in person—though neither of them could have expected that in his third contribution, published in 1917, de Sitter would far exceed the original intention by presenting an alternative to Einstein's cosmological model. Nor could they have foreseen that three years later the offprint of *Foundations* would give rise to spectacular consequences.

The scientific yield of the trip to Leyden was thus exceedingly encouraging: "General relativity has really come to life there,"[10] Einstein reported to Besso. "Not only Lorentz and the astronomer de Sitter are working independently on the theory, but also a number of younger colleagues. In England, too, the theory has struck root."

Back in Berlin, the two weeks in Leyden seemed like a "beautiful dream."[11] In emotional terms, Einstein thanked Ehrenfest for his hospitality: "The journey to Holland has done me an indescribable lot of good, physically and mentally. One can only bear solitude up to a certain degree."[12] A week later he reported to Besso on his "wonderful days" in the Netherlands: "I spent unforgettable hours with Ehrenfest and especially with Lorentz, not only stimulating ones but also refreshing ones. Altogether I feel that I am incomparably closer to those people."[13]

After enduring the wartime atmosphere in Berlin for another month, Einstein wrote to Lorentz: "I still feel the benefit of my refreshing trip to Holland. It was not only the meeting with those highly esteemed men with similar interests that made this experience so liberating, but more particularly our concordant views on extrascientific matters."[14] He would find no such agreement at his place of work. By the second year of the war, German militarism had penetrated into all areas of the state and society. The New Fatherland League had been banned in February 1916, as no discussion of war aims, let alone the possibility of a negotiated peace, was now tolerated. The only goal was victory with extensive territorial annexations—and what depressed Einstein most was that this was not only the policy of the government and the generals, but undoubtedly the attitude of the entire nation, including his colleagues among the professors and academicians.

Most physicists were now part of the war effort: the younger ones, as junior officers, were in the meteorological service, the artillery, or the chemical warfare industry; the older ones were in their laboratories, perfecting killing machines. Often they even engaged in foolish "psychological warfare." Thus the Nobel laureate Wilhelm Wien drafted a circular against *Engländerei*, "anglomania": in the future, the perfidious British were to be quoted in footnotes only when this was indispensable. Even Einstein's esteemed colleague and correspondent Sommerfeld was "naturally pleased"[15] to sign it.

Nevertheless, in May 1916 Einstein had allowed himself to be elected president of the German Physical Society for a period of two years. Whether he hoped to ensure that no chauvinistic tirades would be heard from the representative of this learned institution at least, or whether some other consideration induced him to accept the office, must remain an open question. For an "internationalist person" and a declared pacifist it certainly was an equivocal position, as indeed he hinted in a letter to his friend Zangger: "Admittedly, things are fine for me here and I float on the very 'top,' but on my own and rather like a drop of oil on water, isolated by my attitude and concept of life."[16] At best, he might hope to have a mitigating effect, and that only in conversation with people who, in view of his outstanding scientific achievements, were prepared to overlook the fads of a man without a fatherland. If he thought his post would enable him to do something for peace, then his expectations were disappointed.

The intellectual confusion created by the war soon seemed to Einstein to be virtually psychopathological. "When I talk to people," he wrote to Besso, describing the atmosphere in Berlin, "I can sense the pathological in their state of mind. I am reminded of the period of the witch trials and other religious perversions. The most responsible people, the most unselfish ones in private life, are often the most solid pillars of dogged stubbornness. Social attitudes have been dangerously derailed. I could not envision the people if I didn't see them before me."[17] He expressed much the same views in other letters, provided they were not addressed to his German colleagues: "I am convinced that this is some kind of mental epidemic," he said, summing up his diagnosis for Lorentz. "Only quite exceptionally independent characters can escape the pressure of prevailing opinion. There doesn't seem to be one of those in the Academy."[18]

The dubious role of science and technology in the war effort was being acted out before Einstein's eyes, though he mentioned it in his letters only occasionally. Thus, to Ehrenfest: "Our Jehovah no longer has any need to let pitch and sulfur rain down; he has become mechanized and has provided for an automatic operation."[19] And, more sadly, to Zangger: "Our entire much-praised technological progress, and civilization generally, could be compared to an axe in the hand of a pathological criminal."[20] However, Einstein's attitude did not prevent him from helping, in a small way, to fashion that ax—as a designer of aircraft wings.

It was probably Ludwig Hopf, his former assistant at Zurich University, who had interested Einstein in flying. Hopf had been posted from the Technical University in Aachen to the aircraft design center at Berlin-Adlershof, where, as an expert in hydrodynamics, he was now to study the related problems of aerodynamics. This had become an urgent task, since the British and the French had already developed alarmingly efficient air forces.

A particular problem was the shape of the wings—traditionally not unlike an ironing board. This shape had caused numerous crashes when planes were banking. Although we know nothing of Einstein's beginnings as a theoretician of aircraft wings, we do know that in the summer of 1916 he published a short paper, *Elementary Theory of Waves in Water and of Flying.*[21] Its answer to the question of why aircraft fly might still be appropriate in a high school textbook. In the very first sentence of the article, Einstein referred to the ideal of all aircraft designers, "birds gliding through the air,"[22] and then, on the basis of certain theoretical considerations, he proposed a hunchbacked profile for wings. However, Einstein's performance designing aircraft wings was less than totally successful; this was an episode about which he kept silent in later life.

He next tried his skill in another field of military technology, the gyrocompass. This clever instrument, developed just in time for the war at sea, especially the U-boat war, owed its invention to a young man named Hermann Anschütz-Kaempfe.[23] Inspired perhaps by Jules Verne, Anschütz spent some of his inherited wealth exploring the North Pole by submarine. This plan involved a tricky navigational problem because, for one thing, a magnetic compass is not reliable in the polar regions and, for another, encapsulation in a metal shell, such as a submarine, screens a magnetic compass from the Earth's magnetic field.

Navigation was not Anschütz's field (he had first studied medicine and then the history of art, writing his thesis on the Venetian painters of the sixteenth century); but he realized that a rapidly rotating top could be a useful alternative to a traditional compass. Its axis would, because of purely mechanical effects, arrange itself parallel to the Earth's axis; thus even in the belly of a submarine, it would maintain the correct orientation. When estimates from the Krupp Germania shipyards in Kiel for a submarine for polar waters turned out to be alarmingly high, Anschütz-Kaempfe concentrated on the gyrocompass. He obtained his first Reich patent in 1904 and greatly impressed the Kaiser's admirals with his invention. In 1905 he established a firm in Kiel, which henceforth supplied the German navy with its gyrocompasses.

The basic idea, of course, had not come from Anschütz. It was Jean-Bernard Foucault who in 1815, first in his own cellar and then—more spectacularly—in the Panthéon in Paris, demonstrated the rotation of the Earth with a pendulum and a year later with a gyroscope. Although there were great technical difficulties in constructing a reliable gyrocompass on the basis of Foucault's ideas, a number of enterprises, especially Elmer A. Sperry's gyroscope company in America and some smaller firms in Europe, were competing in this field. Naturally enough, this led to endless litigation over patents.

It was in one such patent dispute, between Anschütz and the Sperry Gyroscope Company, that Einstein, in November 1914, was appointed as an independent expert by the district court in Berlin.[24] It seems likely that Einstein welcomed this appointment, not only because the court offered a fee of 1,000 marks but also because it reminded him of his days at the Patent Office.

The Anschütz firm was accusing the Sperry company of infringing two of its patents. The attorneys for Sperry based their defense on the fact that one of the original Anschütz patents had itself been an infringement of a patent granted to a Dutchman named Van den Boos in 1885 and hence was null and void. In this involved situation Einstein initially seems to have been confused. He stated that a functioning gyrocompass could never have been constructed on the basis of the patent granted to Van den Boos; but the court was not satisfied with his oral statement and demanded a written, "evidential" expert opinion. This too was insufficient, and so on August 10, 1915, Einstein examined the Sperry compass experimentally at the German naval establishment in Kiel. In a supplementary statement, Einstein unambiguously declared that, first, the Anschütz patent was valid in law and, second, Sperry had infringed another Anschütz patent. Anschütz won the case.[25]

Hardly anything is known about any further relations between Einstein and Anschütz during World War I. However, during the final year of the war, in July 1918, Einstein did Anschütz a service by giving a "private expert opinion." After the war, Anschütz and Einstein developed very close contacts, and they collaborated intensively on the development of a fundamentally improved gyrocompass. In the 1930s virtually every navy in the world, except the British and the American, was equipped with gyrocompasses by the Anschütz firm, and the construction of these gyrocompasses also involved a patent of Albert Einstein's.

Aircraft wings and gyrocompasses would have been entirely appropriate areas of activity for any physicist claiming to be a patriot; but for Einstein, a resolute pacifist, they were certainly rather strange pursuits, the more so in light of his comment that the "much-praised technological progress . . . could be compared to an axe in the hand of a pathological criminal." Moreover, Einstein had described as particularly despicable cases of mental confusion those Germans who "are impatiently awaiting the proclamation of unrestricted U-boat warfare,"[26] while he himself, at the same time, was providing expert opinions for the inventor and manufacturer of the instruments that made submarine warfare possible. Even if these activities can be seen as a kind of technological playfulness by a great physicist—much like his "little machine"—it seems odd that, as a convinced pacifist, Einstein was evidently indifferent to the military implications of such work.

Soon after his return from Leyden, Einstein had news of his old friend Friedrich Adler: the newspapers reported on November 26, 1916, that Adler had shot and killed the Austrian premier Count Karl Stürgkh. This was the same Count Stürgkh who, as minister of education, had made difficulties about appointing the Jewish foreigner Einstein to a professorship in Prague. Since 1912 Stürgkh had been the head of government; he had instituted a strict, autocratic military regime and had dissolved the parliament. Adler, whom Einstein would have liked as his successor in Zurich, had definitely abandoned a university career and in 1912 had returned to his native Austria in order to be active there as a Social Democrat politician, writer, and journalist. For October 20, 1916, he had organized a public demonstration calling for the parliament to be reconvened. This was brusquely banned by Count Stürgkh. At about two P.M. the following day, Adler walked into the famous Hotel Meissl & Schadn, where Count Stürgkh was in the habit of eating lunch, stepped up to his table, shouted "Down with absolutism! We want peace!" and killed him with three shots in the head. Adler then allowed himself to be arrested without resisting.[27]

Although Einstein too wanted peace, he was unable to understand Adler's act, especially as the censored press had reported only the assassination and not its background. Still, he was determined to help his friend as best he could: "Your and your much revered husband's misfortune has deeply touched me, more than anything that I have experienced in this hard world," he wrote to Adler's wife, Katya. "He is one of the most excellent and purest men I have known. I cannot judge his deed, as I am unable to form a picture of its motives. I do not believe he would act rashly, he is much too conscientious for that. If you think that I can do anything for him or for yourself, please think of me and write me."[28]

Once Einstein was able to write to Adler in prison, he made a "strange request: when your case comes before the court I would like to be called as a witness; you go and demand that. Do not think that it would be pointless."[29] As soon as the trial was scheduled, before a special court, Einstein urgently appealed to his friends in Zurich to do something for their former colleague: "My feelings for him have revived so much that I would like to do something for him," he wrote to Zangger. "Here no one knows him, but surely in Zurich, the place of his former work, there must be sympathies for him. I ask you now in my name to initiate a speedy action in the Zurich Physical Society, so that a petition for mercy can be lodged with the competent authority."[30] To Besso he proposed specifically that the petition should stress that Adler had "shown himself as an unselfish, quiet, hard-working, kind-hearted, conscientious person who enjoyed universal respect. That it is therefore our profound wish for a word of intercession to be spoken in his favor."[31]

The physicists in Zurich may have regarded Einstein's endeavors to save an assassin from the death sentence as absurd, but after tough arguments they complied with his request. "It took a hot evening,"[32] Besso reported to Einstein in Berlin. The approved text more or less followed Einstein's suggestions, but carefully confined itself to a character testimonial, to the effect that in Zurich Adler "had shown himself to be an utterly conscientious, scientifically highly gifted, and philosophically thinking person . . . [who had] for many years loyally lived for his science."[33]

In prison, Adler returned to science, putting many physicists, including Einstein, in an embarrassing situation. "I woke up on Saturday with the solution to a little problem," Adler reported to Katya on February 17, "that had been troubling me for some years, namely Foucault's pendulum experiment."[34] The solution Adler had dreamed up led to a treatise designed to refute relativity theory. Einstein received the manuscript toward the end of April and remarked in astonishment that Adler "with the conviction of a prophet is setting out worthless subtleties, so much so that I find myself at a loss on what to say to him. I keep racking my brain about it. He is riding Mach's old nag to exhaustion."[35]

The situation grew even more embarrassing when Friedrich Adler's father, Victor Adler, saw the manuscript as a ray of hope for his son: it was to be used for having Friedrich declared insane. The court invited the expert opinion of psychiatrists and physicists, who for their part found themselves in an awkward situation, since Friedrich Adler insisted that he had shot Count Stürgkh while of sound mind, responsibly, and out of political necessity. Adler would sooner have faced a firing squad than let himself be saved by psychiatric expertise; besides, he was convinced that his treatise was a scientific achievement, and any use of it as psychiatric evidence would have deeply hurt him. In fact, there was nothing insane about his treatise; it was simply wrong.[36]

If only to ensure that his trial would have a political effect, Friedrich Adler insisted that he bore full responsibility for his action; his attempted refutation of relativity theory was therefore irrelevant. He was also able to dispense with Einstein's testimony because a hint had been received that—out of consideration for his father, the leader of the Austrian Social Democrats, who was respected even by the imperial House of Habsburg—his life would be spared. Through contacts in Vienna, Einstein had been aware, even before the trial, that "the general opinion is that no serious danger threatens him."[37] Sentence was passed on May 19, 1917. Adler had the satisfaction of seeing his impassioned closing speech against the war and for democracy published in full in the Austrian socialist press, and, in excerpts, even in the German papers. He received a death sentence, but he knew that this was not to be taken seriously.

In order to paint a favorable picture of his friend for the public and for the appeals court in Vienna, Einstein gave an interview to an editor of *Vossische Zeitung*, an excerpt from which was published a few days after the death sentence had been passed.[38] In it, he recounted the story of Adler's withdrawal from the extraordinary professorship in Zurich in Einstein's favor and, against his better judgment, described Adler as a reliable physicist striving for intellectual clarity. No mention was made of political matters or of the assassination, but these were perhaps hinted at when Einstein observed that "the objectivity reflected in his scientific work also governed his actions." That was courageous on Einstein's part, and also on the part of the newspaper.

As had been expected by those in the know, the death sentence was reduced by the appeals court to eighteen years of "fortress detention," with the unspoken assumption that there would be an amnesty after the war. This is exactly what happened.

During this imprisonment, Adler developed his earlier manuscript into a veritable book. Einstein corresponded with him and tried to point out to him the errors in his argument: "It's a pity we cannot endlessly discuss these matters," he wrote, asking Adler whether, if he came to Vienna, he would be allowed to visit him in prison. "I wonder which of us would first irritate the other. Who can tell?"[39] In 1920, when Adler was a free man and published his book under a long and pompous title,[40] Einstein probably did not read it. But in his old age he wrote: "It is to the eternal credit of the Austrians that they did not put Friedrich Adler to death after his assassination of Stürgkh."[41]

Einstein was barely thirty-eight when he had to worry seriously about what he was to describe as his "creaky corpse."[42] At the beginning of 1917 he fell gravely ill. On February 6 he had presented his *Cosmological Observations* to the Prussian Academy; a week later he had to give up a planned trip to the Netherlands. He explained his cancellation to Ehrenfest[43] as being due to a need to cure a liver complaint by a strict diet and rest. In March his doctor informed him that he had gallstones: "Taking the waters, strict diet. . . . I feel much better now, no more pains, and I look better."[44] But this was only the beginning of a string of illnesses continuing, with varying intensity, for the next four years.

Intensive work, poor food, and the irregular lifestyle of a bachelor were probably all factors undermining his health. Perhaps there was also a recurrence of problems which had affected him in his student days and the following two years. "Albert, unfortunately, now suffers frequently from his famous complaint,"[45] Mileva had then reported about his chronic digestive upsets, probably caused by his erratic eating habits.

Now it was no longer a student's careless life, but acute hardship that made nutrition, let alone a diet, a problem for him in wartime Berlin. A disastrous harvest in 1916, which had robbed Prussia even of its potatoes, was followed by the notorious "turnip winter," with only "ersatz bread" and other unusual foodstuffs available to appease hunger. Einstein was able to keep to his special diet only with packages of food sent by his relatives in southern Germany and, especially, by his friend Zangger in Switzerland—and, of course, with Elsa's cooking. Einstein's physician was Dr. Otto Juliusburger, who was known as an activist against alcohol. Both he and Zangger advised a cure at Tarasp in the Engadine. "But I find it difficult to raise the necessary superstition,"[46] Einstein wrote to Besso. While he agreed with his doctor that mankind was being stultified by hard liquor, he did not share Juliusburger's belief in the benefits of a spa. When the doctor insisted, his patient still could "not accept that I should have my vacations messed up in this way. . . . On the other hand, I undertake, which is incredible enough, to do, not to do, to drink, etc. anything else, in short to act medically loyally and obediently."[47] In the end, the patient had his way, and Einstein gradually recovered without going to a spa—though only temporarily.

What worried Einstein more than his own health was bad news from Zurich. Mileva was not recovering. Experts who were consulted first suspected tuberculosis of the brain but eventually agreed on scrofulosis, a disease of the lymphatic glands, which was then believed to be connected with hereditary tuberculosis. This diagnosis was also alarming with regard to the Einsteins' son Eduard, who had just reached school age and whose frequent sicknesses were now seen as inherited from his mother. The prognosis, conveyed from Zurich by Zangger, was not encouraging: "I am depressed by the condition of my young one. It is ruled out that he will grow to full manhood. Who knows if it wouldn't be better if he could depart before coming to know life properly."[48]

According to modern medical opinion, this anxiety was totally unfounded. The views then held about the link between scrofulosis and tuberculosis, and their hereditary character, have since been abandoned. But Einstein had to go by what was the best medical opinion of the leading specialists in Zurich, and therefore he took Eduard's prognosis seriously. He poured out his heart to Besso: "It is my fault he is here, and I blame myself, for the first time in my life. Anything else I didn't bother about too much, or feel responsible for."[49] He added: "I didn't know about the nature of scrofulosis, I didn't know that this is tuberculosis with the risk that children may inherit it. Indeed, to be quite frank, I knew nothing about scrofulosis and attached no particular importance to the glandular swelling that my wife exhibited. Now the misfortune has come, as it was bound to."[50]

We do not know whether Einstein ever learned that his fears and self-accusations were based on a misconception. But at any rate, he was happy that this cup passed by his son—and himself. Eduard grew up to be a delicate but essentially healthy young man. At the age of twenty misfortune struck him in a different way: he developed schizophrenia.

After his own convalescence, Einstein felt not only physically but also intellectually exhausted: "Scientific life is more or less dormant, and nothing is moving in my head,"[51] he complained to Zangger. As far as his health permitted, though, he presided over the meetings of the German Physical Society; on Wednesday he attended the traditional colloquium at the Physical Institute on the *Reichstagsufer;* on Thursdays he attended the meetings of the Prussian Academy; and he even resumed his two-hour lectures on relativity theory at the university.

One of the attractions of Berlin had been that he could please himself about lecturing. In the first year of his stay he had kept away from the lectern, but in the summer semester of 1915 he gave a two-hour class on relativity theory, followed in the winter semester of 1915–1916 by a class on "Statistical Mechanics and the Boltzmann Principle."[52] In the midst of his work on the general theory of relativity, this self-imposed obligation must have seemed a mistake to him, so he kept away from the university for the next two semesters. In the summer semester of 1917, however, he resumed his two-hour classes, alternately on relativity theory and statistical mechanics.

There were not many students, since most young men were in the armed forces. Only a few men, who either were unfit for active service or had been wounded, were scattered about the all but empty lecture room; and there would have been still fewer students if the Prussian authorities had not, in 1908, allowed women to matriculate. There was also mourning among Einstein's close colleagues. Nernst had lost both his sons in the war; Planck's elder son had been killed in action and his younger was a prisoner of war. Thus there was more occasion for condolences than for scientific exchanges; and even now, in the third year of the war, no end to the butchery was in sight. "Old Jehovah is still alive," Einstein wrote to Ehrenfest. "Unfortunately he strikes also the innocent. The guilty he strikes with such terrible blindness that they do not feel guilty. But from where does he derive the right to punish and to shatter? Perhaps also from brute force? I have become a lot more tolerant, without basically changing my opinions in the slightest."[53]

Max Born, Einstein's junior by three years, had become a close friend. Toward the end of 1914, Born had been appointed to an extraordinary professorship which the University of Berlin had established to relieve Max Planck. Although he was called up by the army, he nevertheless came to Berlin, to the artillery testing commission. There, in early 1916, he wrote for *Physikalische Zeitschrift* the report (described above, in Chapter 19) on the general theory of relativity. This was the paper which Einstein had "read with the happy sensation of having been completely understood and acknowledged by one of my best colleagues." As Born played the piano, their relationship soon embraced music as well. Born's wife, Hedwig, records that on his first visit to their apartment, Einstein "shook her hand and said: 'I hear you've just had a young one!' Thereupon he put down his violin, stripped off his little [wrist] rolls . . . and flung them into some corner. Then they played Haydn, of whom he was then particularly fond."[54]

When Einstein complained of isolation in Berlin, this may certainly have applied to his views—but he cannot have lived in quite the seclusion his letters suggested. His fame was gradually spreading beyond scientific circles, even though there was as yet nothing like the excitement over relativity in the 1920s. Educated people had begun to suspect that a genius was living among them, and they sought his acquaintance. Elsa Einstein—she had reverted to her maiden name—was delighted when strangers commented on her famous name. Indeed, one such comment led to a lifelong friendship between Einstein and the physician Hans Mühsam.[55]

A "Literary Society" of writers who were not exactly famous invited Einstein to participate in their weekly meetings at the Hotel Bristol. There he would drink tea and smoke cigars until late at night. To one literary man, who wished to know what "potential, invariant, contravariant, energy tensor," and other terms meant, he gave the memorable answer: "Those are technical terms!"[56] Philosophers, too, were beginning to show interest in the theory of relativity and to write books on the more general aspects of the new physics. The psychologist Max Wertheimer, a *Dozent* at the university and one of the founders of gestalt psychology, sought out the famous physicist to discover from him the secrets of "productive thinking." Many years later he recalled the "wonderful days, beginning in 1916, when for hours and hours I was fortunate enough to sit with Einstein, alone in his study, and hear from him the story of the dramatic developments which culminated in the theory of relativity."[57]

Einstein's contacts even reached to the center of power—not, of course, the landed aristocrats in the archconservative Herrenclub, to which Haber had an entree, but to such people as Walther Rathenau, the chairman of the board of trustees of the huge electrical-engineering concern AEG. Rathenau had set up a raw-materials division at the War Ministry as soon as the "short, swift" war that had been planned turned into a war with no end in sight. From Rathenau and Haber,

Einstein knew that Germany's military situation was hopeless, even though the fronts were deep in enemy territory. The blockade of the sea routes—whose effect the Germans felt in their cooking pots—was bound, sooner or later, to lead to the country's collapse.

In the spring of 1917, when Einstein had sufficiently recovered from his gallstones, he was anxious to travel to Switzerland. He had to wait until the beginning of July: "I can't manage it earlier because of my class, the Phys. Society, and also because of a toe which I was clever enough to break."[58] He left Berlin toward the end of June; visited his mother in southern Germany, where she was living with her brother Jakob; and at the beginning of July arrived in Zurich. He did not stay in Zurich long, "because that's too tiring for my creaky corpse."[59] For the same reason, he did not arrange to meet Romain Rolland again. But he went to Arosa with his elder son and completed his recovery at the house of his sister, Maja, and her husband, Paul Winteler, in Lucerne.

At the beginning of September he returned to Germany via Schaffhausen. After protracted formalities at the frontier, with painstaking inspections, he put up first at Benzingen, a small village near Sigmaringen. Back in Berlin, he did not return to his bachelor flat on Wittelsbacher Strasse, but moved into an apartment at Haberlandstrasse 5, next door to Elsa's. This new apartment had fallen vacant while he was away, and Elsa, as he learned during his short stay at Benzingen, had acted promptly. "My address now is Haberlandstr. 5," he informed Besso. "The removal is said to have been completed already."[60] Elsa was now able to organize his convalescence more thoroughly. He had a slight relapse in Berlin, but "the grub is good and I rest a lot."[61]

New duties awaited Einstein after his return. On October 1, 1917, he became the director of an institute. The original offer made to him in 1913 had, along with the academy post and the professorship, included the directorship of a "Kaiser Wilhelm Institute of Physics," yet to be founded. As this involved the Prussian Ministry of Finance, establishment of the institute was delayed and eventually postponed until victory. Einstein never complained. He needed no institute. He worked without an assistant and without a secretary in his spartan apartment, writing everything by hand, and being content to have no administrative duties.

However, private money was available, primarily from the pocket of the Kommerzienrat Leopold Koppel; according to the rules this had to be used, and so the institute had to be established. A board of trustees was appointed under the chairmanship of the industrialist and privy councillor Wilhelm von Siemens, as well as a board of management chaired by Einstein and including Haber, Nernst, Planck, Rubens, and Warburg as advisers. And that was the institute: so long as the war continued there could be no question of a building with staff, a library, or even a janitor. In practice the institute was located in Einstein's apartment and confined itself to the role of a funding institution, to which university scientists could apply for research grants. Einstein was given an annual honorarium of 5,000 marks;[62] there is no evidence that this was deducted from his academy salary.

All kinds of trivialities, such as the establishment of an institute petty cash box, had to be settled in writing, using pompous forms of address—for lack of a better title Einstein was invariably addressed as "Your Honor," which would have pleased his mother. Einstein requested, and for the first time was assigned, a secretary for three half-days a week at a wage of 50 marks per month.[63] This job was taken by Elsa's twenty-year-old daughter, Ilse. Initially Ilse wrote by hand; not until 1919, by which time the institute's business had greatly increased,[64] was a typewriter purchased as its first investment. This family enterprise was the smallest and most unusual of all the Kaiser Wilhelm Institutes, and it remained so while Einstein was in Berlin, probably because he himself did not want more.

Nevertheless, he immediately used his new position to further his scientific interests; he was now able, at last, to free the astronomer Erwin Freundlich from slavery at the Babelsberg Observatory. An adequate stipend from the institute enabled Freundlich to direct all his efforts to the astronomical consequences of relativity theory. Before long this arrangement was converted, as a special case, to Freundlich's employment as a scientific worker at the institute.[65]

However, Einstein had yet to learn to adopt administrative usage. Typical of his concept of research subsidies was his intention to approve an application by Peter Debye on the spot because he believed "that we cannot possibly employ our money better. . . . Waiting, in my opinion, is out of the question, because we only have one Debye."[66] Planck had to instruct the new director in proper administrative procedure—but even so Einstein arranged that Debye's application, submitted at the beginning of July, was approved by the board on July 27. This practice resulted in the creation of files which now have the particular charm of containing the handwritten notes and signatures of four eventual Nobel Prize winners, sometimes on the same sheet of paper. There were even occasions, as in Peter Debye's case, when the applicant himself eventually received a Nobel Prize.

Out of consideration for his health, Einstein still had to limit his duties at the institute, and his other work. Things seemed to be improving: "My health is quite fair now,"[67] he reported to Zangger at the beginning of December. "I have gained four pounds since the summer, thanks to Elsa's good care. She cooks everything for me herself, as this has proved necessary." But before the end of 1917 there were renewed acute attacks. A specialist in gastroenteritic and liver complaints, using the new X-ray technique, diagnosed a duodenal ulcer as the main cause of the persistent problem. Einstein had to keep to his bed for many weeks.

It was probably not just his sickness that made Einstein feel burned out. "Anything really new is invented only in one's youth,"[68] he wrote to Zangger. "Later one becomes more experienced, more famous—and more stupid." To Besso he reported from his sickbed that, as for scientific work, he was engaged "only on trifles. Instead I read and study a lot, which is not to be sniffed at either."[69] At any rate, it was during those weeks in bed that, in his second paper on gravitational waves, he derived his quadrupole formula, which he presented to the academy on February 14—anything but a trifle.

After that, there really was a prolonged pause in high-level publications, but not necessarily in work. "The K. W. Institute involves a rather large correspondence; otherwise, too, correspondence is steadily increasing."[70] As president of the German Physical Society—which he was until May 1918—Einstein was in charge of the preparations for the celebration of Max Planck's sixtieth birthday on April 26. Its success meant a lot to him, "because I am very fond of Planck and he is sure to be pleased when he sees how fond we all are of him and how we revere his life's work."[71] Einstein gave an address on "The Motivation of Research,"[72] which turned out exceptionally well and said a great deal about what, despite all their political differences, united the two men and about what, despite war and isolation, was keeping Einstein in Berlin.

"The general th. of rel. is meeting with downright enthusiastic acceptance by my colleagues in the field,"[73] Einstein wrote to his friend Zangger, expressing his ambivalence about his situation. "How is it possible that this culture-loving period is so appallingly amoral?" To Lorentz he wrote, at the New Year: "I am continually most depressed about the boundless sadness which we have to live through. Even my customary escape into physics does not always help."[74] From Planck at the New Year, along with best wishes for a thorough improvement of his health, he received a little admonishment: "I hope that in this year your sympathies will also grow for the German side, which is time and again ready for peace, even though we are militarily better off than ever before."[75] This did not impair Einstein's affection for Planck. He esteemed Planck as a physicist and a noble character—the rest he had learned to live with.

Einstein had given up his pacifist activities because he saw no more purpose in them. When Georg Nicolai wanted to mount another action—"Nothing is more difficult than to say no to Nicolai"[76]—he simply tried to disregard Einstein's refusal.

Nevertheless, the military authorities had included Einstein on a list of thirty-one "renowned pacifists"[77] resident in Berlin and surroundings, who, because of their "endeavors of international character," could be granted permission to travel abroad only after authorization by the High Command. It is probable that Einstein was unaware of this, because after the date of that ruling (January 1918), he had not been planning any foreign trips anyway.

Einstein's reluctance to engage in any activities like those he had undertaken at the beginning of the war was due to a recently acquired conviction that, if things were to get better, a defeat of Germany would be preferable to a negotiated peace. "This country has developed a religion of power through the success of its arms in 1870 and through its successes in trade and industry,"[78] he wrote to Romain Rolland:

> This religion dominates nearly all educated people; it has almost totally replaced the ideals of Goethe's and Schiller's time. . . .
>
> I am firmly convinced that this delusion of minds can be checked only by the harshness of reality. The people must be shown that it is necessary to show consideration for non-Germans as persons of equal worth, that it is necessary to earn the trust of foreign countries in order to live, that with brute force and perfidy one does not reach the goals one has set oneself.

He wrote this in August 1917, during his vacation in Switzerland—admittedly without any danger that the German censors would read it, but still with the risk that these views would be publicized by Rolland. Also, if the High Command in Berlin had learned of these sentiments in some roundabout way, the neutral foreigner Einstein would probably have been expelled, and his revered colleague Planck might have viewed this as entirely justified.

In April 1918, however, Einstein tried once more to make the voice of reason heard. He drafted a circular in the style of a manifesto to his German colleagues, presumably after consultation with and encouragement from the theologian and philosopher Ernst Troeltsch, who had come to Berlin in 1915. The Germans had just concluded the "victorious peace" of Brest-Litovsk with the Russians and had exacted massive annexations and reparations; they now believed that the summer would bring the decisive battle and victory on the western front. The public mood was once more patriotic fervor; few people realized that the situation was hopeless and defeat was only a matter of time. In his appeal, Einstein was concerned not with avoiding defeat for Germany by negotiating for peace, but with preparing to revive international scientific contacts in peacetime. In his long letter to his "highly esteemed colleagues," he said:

> At countless times during these tragic years of universal national delusion, men of science and the arts have issued public statements which already have done immeasurable damage to the sense of solidarity, which so promisingly developed before the war, among those who devote themselves to higher and freer objectives. The clamor of narrow-minded priests and servants of an empty power principle is so loud, and public opinion is so misguided by the deliberate muzzling of the entire media that, in a sense of hopeless confusion, those of better intentions dare not raise their voices.
>
> This serious situation faces those who, by their successful intellectual achievements, have attained superior recognition from the intellectual workers of the whole civilized world with a task which they must not shirk—they should make a public avowal that may serve as support and consolation to those who, in their isolation, have not yet lost faith in moral development.[79]

Einstein was also planning a book with contributions from renowned scholars. To stress its international character, men from the belligerent states as well as from neutral countries were to be approached. He had already interested the publishing firm of Rascher in Zurich in his project. As no one was to be invited "who has blemished his name by chauvinistic declarations of any kind,"[80] Einstein could send his appeal to only a handful of German professors. Apart from Troeltsch, these were the economist Lujo Brentano in Munich, the lawyer Walther Schücking in Marburg, the sociologist Alfred Weber in Heidelberg (but not Max Weber), and David Hilbert in Göttingen. To Hilbert, Einstein added a handwritten postscript: "Among mathematicians and physicists you are the only one I can write to in this matter." Hilbert reacted positively—"Needless to say, the ideas in your letter arouse my sympathies"[81]—but he first wanted to consult with friends. The result was that, while he still approved of Einstein's ideas, he did not approve of the proposal for specific action. "Such declarations," he wrote, summing up the result of his inquiries in the Göttingen university milieu, "would be tantamount to self-denunciations, giving rise to great joy among all our enemies in the faculties. Not even your name would offer us protection; after all, the very word 'international' acts on our colleagues, when they are *in corpore*, like a red rag." Hilbert proposed to wait until after the war and then to draft an appeal "in which every word must be unassailable and be like a cudgel blow, a statement on what science is and what obligations follow from it."[82]

After a month Einstein was compelled to admit that he "lacked the basis of the enterprise intended by me, because the very people whom I would have regarded as important do not predominantly feel international."[83] He had done what was within his powers, albeit with the melancholy heroism of a vain gesture. He had acquired a place of honor among the handful of pacifists and internationally minded individuals. Now he could do no more than wait for the end of the war—and regain his health.

CHAPTER TWENTY-ONE

Postwar Chaos and Revolution

THE YEAR 1918 BEGAN BADLY. Einstein's duodenal ulcer was a life-threatening condition, even though he tried to look on its bright side: "Since I took to my bed I've been quite well,"[1] he reported to Besso at the beginning of January. "I'll gladly stay in bed for another 4 weeks or longer, the more so as this is preferable also in view of the inadequate heating." The special food he needed for his strict diet continued to be sent by his relatives in southern Germany and by friends in Switzerland.

However, six weeks in bed did not result in any marked improvement. Although from his sickbed he had completed his second paper on gravitational waves, he was still dejected. "Somehow I can't manage the jerk upwards."[2] Gradually, he was allowed to get up, but he had to avoid any physical strain. "The other day I had a nasty attack, evidently brought on by my playing the violin for an hour or so."[3] In May his liver played up again: he had jaundice. Not until the summer did he regain his strength: "My health is decidedly better now than last year. Elsa indefatigably cooks my chicken feed, and I keep quiet, and am mostly on my balcony. Everyone's saying that I've never looked so healthy."[4] And even then the impression was deceptive. Einstein complained that he was "still dependent on care, as on the slightest occasion my belly revolts. And why should it have less right to do so than the brain?"[5]

Professor Rudolf Ehrmann had been brought in to supervise Einstein's health. He was a stomach specialist and the medical director of the

Neukölln Hospital, and he was the same age as Einstein. Einstein remained not only his patient but also his friend for the rest of his life. After the Nazis' seizure of power, Ehrmann had to leave Germany and took up practice in New York; from there, he would go to see Einstein in Princeton whenever the medical situation called for it.

Ehrmann would soon discover that his patient "loves medical men but not medicines."[6] Not only did Einstein lack confidence in spas; he was suspicious [even of] the "new medical wizardry of X-rays."[7] Einstein once observed that "the only diagnoses in which I have confidence are those *post mortem*—otherwise none."[8]

To recuperate, Einstein and his "small harem"[9]—Elsa and her two daughters—toward the end of June went to Ahrenshoop on the Baltic for eight weeks. This was a quiet fishing village on a narrow spit of land near Rostock. He thought it "wonderful, no telephone, no obligations, absolute tranquillity. . . . I lie on the beach like a crocodile, get baked by the sun, never read a newspaper, and don't give a damn about the so-called world."[10] The eight weeks did him good; there were none of those terrible attacks. "Admittedly I must absolutely avoid long walks or sudden movements, and more or less stick to a diet."[11] Lazing on the beach and being able to spend the whole day barefoot gave him such pleasure that he jocularly suggested to Born: "If only we could introduce this last lovely custom (voluntarily) in Berlin!"[12] Not until he was in America could he indulge this practice—and then he became instantly famous as the "man without socks."

But even in Ahrenshoop Einstein did not quite escape all his problems. The trickiest came with a letter from Zurich, toward the end of his vacation. It was from Edgar Meyer, Alfred Kleiner's successor as professor of physics at the university. In response to Heinrich Zangger's efforts, the university and the Polytechnic had joined forces to offer Einstein a very special joint professorship, tailored entirely to Einstein's personal needs. This might have been an occasion for joy, but it actually brought him "painful embarrassment."[13]

Einstein was touched by this offer—so touched that he called Zurich his "native city." But he could not refrain from recalling how happy he would have been "eighteen years ago with a lousy assistant's post"; that wound still smarted. But again, it went without saying that political and general conditions in Switzerland were incomparably more attractive than those in Germany. Besides, there would be a collapse of the political system in Germany after the war, which meant that the Swiss franc was more solid a currency than the mark. Still, Einstein did not wish to leave Berlin, and he could not be in two places at once. "What's to be done? Difficult days of pondering lie behind me. Proof: I dreamed I had cut my throat with my razor."[14]

Einstein urged Besso to visualize what was keeping him in Berlin. "If you could see the beautiful relations which have developed between my closest colleagues and myself (especially Planck!), and how everyone here has done his best for me, and is doing his best, and if moreover you remember that my papers have become effective only through the understanding they have encountered here, then you will understand that I cannot make up my mind to turn my back on this place."[15] Much of this seems exaggerated—not his good relations with Planck and his other colleagues, but Berlin as a place where he found special understanding and could be especially effective. For one thing, he had used rather different terms in the past, especially about the intransigent astronomers. Moreover, since Schwarzschild's death no one in Berlin except himself was working on the general theory of relativity. And was there not also Leyden, Göttingen, or Munich? And had not Hermann Weyl meantime become a highly productive relativist in Zurich? And after the war would Einstein not have found it easier from Switzerland to resume contacts in the "enemy countries," with English, French, and Italian scientists?

Of course, there were other considerations as well. There would be "great difficulties if I pitched my tent in Zurich again, tempting though it is to be near my children. My past experience with visits to Switzerland does not greatly encourage me in this respect. Here everybody approaches me only to within a certain distance, so that life unrolls almost without friction."[16] Einstein, in short, was torn between loyalty to what he called his "native city" and the scientific climate in Berlin—and between Elsa and Mileva. On both counts, Berlin seemed preferable.

But because he did not wish to disappoint Zurich, and because "saying no would have been unspeakably painful, I did something that I normally detest—I resorted to a compromise."[17] He therefore proposed to Edgar Meyer that he would come to Zurich twice a year, each time for four to six weeks with about twelve lectures. This was to be done not within the framework of a firm appointment, but as a lectureship contract. He did not wish to make any money out of it, but would ask only for reimbursement of travel and living expenses. When, after some argument between the university and the Zurich education council, this arrangement was approved, he informed the council, in response to a query, that he would be glad "to give his lectures free of charge, to show his gratitude to his fellow-citizens."[18] It was only because of difficult wartime conditions that he was asking for reimbursement of his travel and living expenses, to a sum of 1,200 francs. He was glad that he would be able "once more to participate in the intellectual life of the University of Zurich."

There were other developments in Zurich as well. After no doubt interminable conversations with Besso and Zangger, Mileva had evidently realized that her husband was not going to return to the family nest, and she had agreed to a divorce. This relieved the tension. Einstein was once more receiving letters from his sons, and "Mileva, too, writes in a friendly manner."[19] The attorney Emil Zürcher acted as an honest broker between the parties and helped to settle the financial aspects, which were far from simple.

When Einstein had let his family return to Zurich in the summer of 1914, he had every reason to assume that his salary at the Prussian Academy in Berlin would make it easy to provide adequately for them in Switzerland. Each year he transferred to them 7,000 marks, more than half his income. If necessary, he sent even more; it was necessary quite often. "It is fortunate that I got the post here,"[20] he wrote to Besso toward the end of 1916, "otherwise I'd be totally ruined financially." For himself he had enough, though, and he had even saved a little, "as I've been living exceedingly thriftily."

In spite of his thriftiness, however, the financial situation was now getting increasingly difficult. On the one hand, Mileva's and Eduard's frequent sicknesses and their stays in sanatoriums had entailed major expenses; and on the other hand the mark's value against the Swiss franc had declined in the course of the war by more than one third. Whereas before the war 100 francs had been worth only 83 marks, by 1916, 100 francs were worth 106 marks, and in 1917 even 135 marks.[21] In 1917, in consequence, Einstein had been obliged to send 12,000 marks to Switzerland—his entire salary from the academy. He therefore had to call for greater economy in Zurich: in the future, he would send 6,000 francs, or 8,500 marks at the prevailing rate of exchange, and no more.[22] This was still substantially more than his salary during his early years as a professor, and should have been adequate for a comfortable lifestyle.

Discussions about financial support for Mileva and their sons after divorce were made easier by a curious idea—we can no longer tell who first brought it up. This concerned the Nobel Prize. The awards had been suspended in 1916 because of the war, but once they were resumed it would only be a matter of time before Einstein received one. The prize money was large, and the Swedish crown was as solid a currency as the Swiss franc; thus it would be possible to live quite comfortably on the interest alone. It was therefore agreed, in the summer of 1918, that Mileva was to receive the principal of the (as yet hypothetical) prize, but she would not be able to dispose of it "without the agreement of Professor Einstein."[23]

In the winter semester of 1918–1919 Einstein held his by then almost traditional two-hour class on relativity theory. He was still concerned with its basics when, on September 9, he recorded in his lecture diary that the class was "canceled because of Revolution."[24] The German Reich was falling apart, and Einstein was delighted. On October 3, the government had addressed a request for an armistice and peace to Woodrow Wilson, the president of the United States, who responded with a demand for Germany's democratization. The Germans, who until then had seen themselves as the imminent victors, were stunned. The army was exhausted. On November 4, the sailors of the German navy mutinied in Kiel. The revolt spread like wildfire. On November 9 it reached Berlin and became a revolution. Workers' and soldiers' councils proclaimed a general strike. The people poured into the streets, advanced on the Reichstag, and demanded the immediate termination of the war. On the same day when Einstein's class was canceled in the morning, the Republic was proclaimed in the afternoon. That night the Kaiser abdicated and went into exile in the Netherlands.

All this had a rejuvenating effect on Einstein. With a sense of history he sent out postcards all over the world, declaring—as Goethe had declared during the shelling of Valmy—that he had witnessed these great events. "A great thing has happened," he wrote to his sister and brother-in-law in Lucerne two days after the revolution. "To think that I have lived to see it! No bankruptcy is too great that one wouldn't gladly accept it for the sake of this compensation. Here, militarism and the Privy Councillor nonsense have been thoroughly liquidated."[25] Einstein's mother, who was now living in Zurich with her brother Jakob Koch, also had news from him the same day: "Don't worry. So far everything has gone smoothly, even impressively. . . . Now I am feeling really well here. The bankruptcy has worked wonders. Among the academics I am a kind of *Obersozi* [top-rank Red]."[26] Pauline Einstein, the daughter of a merchant by royal appointment to the Württemberg court, was, once again, amazed by her son.

Einstein was hoping that the religion of military might would be gone forever; that the nobility, the landed gentry, the officers' and officials' cliques would be stripped of their power; and that Germany would be on the road to democracy. The meetings of the Prussian Academy amused him more than before; for his colleagues on the privy council, whose world had collapsed, he had more scorn than pity: "The dear old people are, for the most part, disoriented and dizzy. They perceive the new era as a kind of sad carnival and mourn the old order, whose disappearance means such a liberation to the likes of us. . . . I'm enjoying the reputation of an unblemished Red. In consequence, yesterday's heroes approach me with wagging tails, in the belief that I can arrest their fall into the void. Funny world."[27]

On the model of workers' and soldiers' councils, a students' council had formed in Berlin; its first revolutionary act was to depose the rector of the university, along with the deans, and then to have them locked up.

It was thought that the "top-rank Red" Einstein might have some influence with the revolutionaries, and he was ready to try and get the imprisoned professors released.[28] Accompanied by Max Born and the psychologist Max Wertheimer, whose support he had asked for by telephone, and with the manuscript of a little speech in his coat pocket, he rode by streetcar to the Reichstag building. The armed revolutionaries did not quite know what to make of the three professors; but when someone recognized Einstein, that was as good as a pass.

The students' council was in session in one of the conference rooms of the Reichstag, discussing new statutes for the university. Einstein and his companions had to wait, while listening to the arguments. Eventually the famous man was asked his opinion. He was not at all in agreement with the innovations just decided on, and he extemporized a little plea for the preservation of academic freedom, which he described as "the most valuable aspect of the institution of German universities." "Your new statutes seem to abolish all that and to replace it by precise regulations. I should be sorry to see the ancient freedom come to an end."[29] There was an embarrassed silence, followed by the explanation that releasing the rector was not in the power of the students' council, but was a matter for the new government.

So the three went to the Reich Chancellery on Wilhelmstrasse. There, too, revolutionary chaos reigned, but Einstein was instantly recognized and admitted to see Friedrich Ebert, who had just been proclaimed president of the Reich. Although Ebert had more urgent things to worry about than the release of a few detained professors, he issued a written instruction to the responsible minister and "the business was settled in no time."[30]

There had probably been no opportunity for Einstein to make the little speech he had prepared. According to the extant manuscript, he would have introduced himself to the "comrades" as an "old democrat who had no need to relearn anything." This would have been followed by a short lesson in democracy, which included the warning that "the old class tyranny of the Right should not be replaced by a class tyranny of the Left. Do not allow feelings of revenge to lead you to the fatal belief that a provisional dictatorship of the proletariat is necessary to drive freedom into the skulls of your fellow-Germans. Violence only creates bitterness, hatred, and reaction."[31]

Twenty-five years later, in a letter to Max Born, Einstein recalled this episode. Looking back, in the middle of World War II, he referred to their "illusion that we might truly be effective in turning those guys into honest democrats. How naive we both were as men of forty. To think of it makes me laugh. We did not realize, either of us, how much more is anchored in the spine than in the brain, and how much more firmly it is anchored."[32]

During the first few months, however, Einstein hailed the new republic unreservedly. He sympathized with the Germans in their harsh fate and suffered along with them, as if he now wished to accept as his homeland (albeit without any nationalism) the country which he had rejected in his youth. He even decided to use his trip to Switzerland, planned for December 1918, to make a detour to Paris "in order to request the Entente to save the hungry population here from starving to death. After so many lies it is difficult to help truth to prevail. But I think that I will be believed if I give them my word of honor."[33] This no doubt was too bold an idea. The journey to Paris did not materialize, though he did visit Switzerland.

About the middle of December, Einstein, together with Elsa, left Berlin for Zurich. His lecture contract had to be fulfilled and his divorce finalized. Naturally, he also wanted to see his sons, talk to his friends, and, even though it was winter, recuperate a little. He found a "radiant landscape and well-fed citizens who have nothing to fear."[34] Still, the transformations in Germany continued to occupy his mind: "But God knows I like the people with their worries, for whom the morrow is one of uncertainty. How will everything develop? I can hardly tear my thoughts away from the changed and still changing Berlin."

On December 23 the education council of the canton of Zurich completed its formalities and, with true Swiss caution, approved the lecture contract "by way of experiment for the current winter semester."[35] After Christmas the university remained closed until the end of January because of a shortage of coal. In the meantime, Einstein and Elsa went to Arosa and Lucerne, returning to Zurich at the beginning of February. Throughout the month he gave his two-hour lectures three times a week, twenty-four hours in all. There were forty-five registered students and 105 guests. As he wrote to Sommerfeld, he almost had to "drain his lungs and brain preaching, so that I now have a real hunger for rest."[36]

On February 14, his marriage to Mileva was dissolved by the district court, for "adultery," "character incompatibility," and other weighty reasons.[37] Custody of their sons was awarded to Mileva. Einstein undertook to pay 8,000 francs annually, to be accounted against the interest of the hypothetical Nobel Prize. Moreover, in line with Swiss custom at that time, Einstein, as the guilty party, was ordered not to marry for the next two years.

On his fortieth birthday, March 14, 1919, Einstein was back in Berlin. The day passed quietly, as he had wished: "Outside of Haberlandstr. 5 no one here was aware of the important event—thank God."[38] The political situation had calmed down, at least superficially, but Einstein's optimism had evaporated. He was disappointed both by developments in Germany and by the victorious powers' intransigent attitude toward Germany in the treaty of Versailles. "Those states,"[39] he wrote to Ehrenfest, "whose victory during the war I felt to be the incomparably lesser evil, are now proving an only slightly lesser evil. Then there are the exceedingly dishonest politics. Reaction with all vile deeds in a repulsive revolutionary disguise." Nevertheless, he felt content in Berlin. "Unless something quite terrible happens I certainly won't leave."[40]

Einstein's political sympathies were with the radical left. Even bolshevism on the Russian model, of whose inevitability he was for a time convinced, would not have been a reason for him to leave Berlin. Nothing could be worse than a return to the conditions of former imperial Germany. Toward the end of March he gave a lecture for which an admission fee was collected to benefit the Socialist Students' Union—needless to say, on the relativity principle.[41]

Just as important to him as the transformation of Germany was international reconciliation. To promote foreign confidence in Germany's young democracy he joined a committee which, without any official standing, relied on the personal engagement of its six members. Its purpose, as he explained to Lorentz, for whose collaboration or at least support he had asked, was "to subject to conscientious investigation all the accusations leveled against the German conduct of the war—accusations known abroad and regarded as proven—by means of German and if possible foreign official documents, and to publish the result over here."[42] To begin with, the atrocities committed by German troops in the French city of Lille were to be documented.

Einstein's participation on this committee cannot have been very intensive; he was too often away on his travels. Nor was the work entirely satisfying. The first edition of a brochure produced in September 1919 gave him "a real fright"[43]: because of a tactless introduction and sloppy documentation he feared that it "might do more harm than good." That edition was pulped, and a second version was published with Einstein's approval in January 1920. Even before it appeared, it was fiercely attacked, but Einstein reported to Lorentz that "gradually the conviction is gaining ground [in Germany] that grave crimes have been committed and that one is not free from guilt for the terrible hatred."[44]

In connection with the refusal of most foreign scientists to resume relations with their German colleagues, either individually or through international scientific organizations, Einstein thought that "it will do no harm if boycott by foreigners makes them realize how dependent they are." But the condemnation of the German professors in the (then by no means concluded) "paper war" seemed to him "too harsh, despite the horrible things that have happened. . . . Even *a priori* it is not credible that the inhabitants of an entire big country are inferior."[45] The Germans had a good advocate in Einstein, the more so as he was the only scholar in Berlin who could count on credibility and sympathy abroad. But for all that he did not wish to become a German citizen. He was a Swiss citizen and wanted to remain one.

In April and May Einstein was back in Zurich, both to visit his sons and his friends and because of his lecture contract. Interest in the lectures seems to have flagged considerably: only fifteen university students and twenty-two guests had registered, at a fee of 10 francs. The education council, thrifty as ever, thought this was not enough and concluded that "no demand exists at our university for the regular repetition of the lecture contract."[46] The faculty insisted on an extension and got its way. But even so, Einstein's second semester as a contract lecturer in Zurich was also his last—not so much because of the tightfistedness of the authorities as because of his own lack of time and interest.

Regardless of his earlier intentions—and the two-year ban on remarrying imposed on him by the divorce court in Zurich—Einstein walked into the Berlin registry office on June 2, 1919, and married his cousin Elsa.[47] Probably no one outside Haberlandstrasse 5 had any knowledge of this event. After all, it was only the legitimation of a long-existing situation. Elsa's daughters Ilse and Margot had already adopted the name Einstein and addressed their cousin familiarly as "Albert,"[48] though in front of others invariably as "Father Albert."[49] Probably Einstein had, even before the marriage, moved across the landing from his "interim"[50] apartment to Elsa's place.

Their apartment was on the fourth floor, the top floor, of an externally unimpressive building, though it had an ornate entrance hall with a doorman and an elevator—typical of the lifestyle of a well-situated Berlin family. The apartment had seven rooms,[51] the largest ones furnished as drawing room, library, and dining room in the rather gloomy style of the period, regarded as both elegant and comfortable. Philipp Frank, a frequent visitor, recalled Einstein living "among fine furniture, carpets, and pictures. . . . But as one stepped into the building one felt that Einstein would always remain a stranger in such a bourgeois household, a globetrotter resting for a moment, a Bohemian as a guest in a bourgeois home."[52]

Elsa was three years older than Einstein, an attractive, vivacious, efficient woman, not without a sense of social standing, and receptive to her husband's fame. Philipp Frank records that there was some criticism of her in Berlin's academic circles,[53] though this seems to have been somewhat inconsistent. Some people found fault with her "standard" generally; others accused her of being too protective of her husband, as though he were her personal property; still others found her too eager for fame. But she certainly tried to create an environment for Einstein that would improve his shaky health and promote his work.

Einstein had his own spartan study and bedroom, but the main advantage of their top-floor apartment was that they were soon able to rent two attics in the roof above and convert them into a study. This "turret room," with its sloping walls, was also of monastic simplicity: it had shelves crammed with books, journals, and offprints; a desk and a chair on a platform in front of the window; and on the walls prints of Isaac Newton and Michael Faraday. This was Einstein's realm, and it was not allowed to be tidied up—only careful dusting was permitted. Here he worked and received his visitors, and here also was the headquarters of the Kaiser Wilhelm Institute for Physics.

An American journalist who climbed up to Einstein's study in March 1921 was impressed by its simplicity and the large number of English books on the shelves, as well as by Newton's portrait on the wall.[54] Einstein wore "shabby trousers and a pullover"; the visitor could not recall a collar or any similar insignia of correct bourgeois attire. In such comfortable clothes, Einstein spent more time in his "turret room" than in the apartment.

Einstein was aware that his talent for married life was somewhat limited, and he found this confirmed again after his marriage to Elsa. Before long he wrote to Besso of "the hard test of patience which marriage invariably involves."[55] Even in a public statement he cast doubt on the pleasures of married life, when he rejected any kind of war "except the inevitable war with one's wife."[56] He was certainly "glad that my present wife understands nothing about scientific matters. In this respect my first wife was quite different."[57] When Jewish students in Princeton wished to know what he thought about marriage between Jews and non-Jews, he replied: "That's dangerous—but then any marriage is dangerous."[58]

It may be assumed that Elsa did not have an easy time with her husband. She consoled herself with her image of a genius who, in the nature of things, could not be perfect:

> One must not dissect him, otherwise one discovers "deficits." Any genius has those, or does one really think he is without fault in every respect? By no means; nature doesn't work that way. Where it is overbrimmingly prodigal, there it takes something away in another respect, and that means deficit features! He's got to be viewed as a "whole," he cannot be placed in one category or another. Otherwise one experiences disappointments. But the Lord has put into him so much that's beautiful, and I find him wonderful, even though life at his side is enervating and difficult, not only in this but in every respect.[59]

In one of his last letters, Einstein gave a melancholy summary of his two attempts as a married man. After the death of his lifelong friend Michele Besso (and shortly before his own) he wrote to Besso's son Vero and his wife: "But what I admired most in him as a person was the fact that he managed for many years to live with his wife not only in peace but in continuing harmony—an undertaking in which rather shamefully I failed twice."[60] This does not sound as if Einstein necessary sought the fault, if there was a fault, on the part of the women.

Soon after his marriage, Einstein was back in Switzerland in the summer of 1919, both for recuperation and to visit his mother, who was now incurably ill with cancer—and also to explore the offer from Zurich that he had turned down the year before. In this he was no doubt motivated by the rapid decline of the mark against the Swiss franc after the treaty of Versailles; he now saw no other way to meet his obligations to Mileva and his sons.

When rumors of Einstein's intentions reached Berlin, Max Planck again went into action, as he had done when Einstein was first appointed. "You can imagine how I feel about this,"[61] Planck wrote to Einstein, and at once promised help. "This matter, which is so important to our Academy and to German science generally, must not be measured by economic aspects; in other words: the Academy or the state must, and will, provide for you the money you need in order to live here, provided that's what you want." Fritz Haber, who could not disguise the fact that "the war years have moved us apart,"[62] also intervened, using personal, political, and even historical arguments.

Haber need not have exerted himself so much, since Einstein, who hardly ever let anyone influence him, had already placed his destiny in the hands of Max Planck. During those years in Berlin, and regardless of their political differences, Planck had become a father figure to Einstein, almost alongside Lorentz. Einstein's reply to Planck is lost, but to Ehrenfest he wrote that he had promised Planck "not to turn my back on Berlin until circumstances arise which will lead him to regard such a step as natural and correct."[63] When Einstein returned to Berlin, he had definitively turned down Zurich. "I am exceedingly pleased that you are staying with us,"[64] Haber welcomed him. "Quite exceedingly."

Planck and Haber meanwhile had been negotiating with the ministry and with Kommerzienrat Leopold Koppel and achieved a raise in Einstein's salaries by altogether 10,000 marks, partly from the exchequer and partly from Koppel's pocket. This was a considerable sum by German standards, but rather modest in terms of foreign currency, and just as Einstein had foreseen, the mark dropped faster than his salaries could be adjusted. But he kept his promise to Planck, until in the fateful year 1933 Planck himself described Einstein's resignation from the Prussian Academy as the "only possible way out."[65]

During the rest of the summer vacation Einstein "most enjoyably spent wonderful days sailing,"[66] probably on the boat of the surgeon Professor Moritz Katzenstein, with whom he was to spend many more summers on the waters of Mark Brandenburg.[67] Sailing became Einstein's favorite occupation—but his first season ended prematurely as he had "unfortunately, while on sailor's duty, got myself a problem (stomach), so that I have to spend a few days in bed again."[68]

Just then he received an especially long, temperamental letter from his friend Ehrenfest, informing him that in the Netherlands "we are suddenly, all of us, agreed that we have to tie you down in Leyden."[69] Ehrenfest was offering to "tie Einstein down" only very loosely, and with a golden chain. The Dutch maximum salary of 7,500 guilders[70] was to be Einstein's minimum; he would simply receive as much money as he needed for his children and for himself. "You'll have to work out the sum yourself." There would be no obligations attached to the professorship. He would be able to travel as much as he wished, so long as his residence was in Leyden or its neighborhood, "so that one can say: Einstein is in Leyden—Leyden has Einstein." With touching warmth Ehrenfest described to his "dear, dear Einstein" the sympathy and affection, and indeed love, with which he would be received in Leyden. In a typically Ehrenfest postscript he added: "It really is a nuisance that you should have any say in a matter which we are in a much better position to judge than yourself!" Without even waiting for a week for Einstein's reply, he sent another urgent letter to Berlin: "Here we have nothing but people who love *you* and not just your cerebral cortex."[71]

There were also such people in Berlin, however. This time Einstein had no great conflicts of conscience. Basically, he had made his decision, though he took great care over his explanation: "It would be doubly mean of me if, at this very moment when my political hopes are fulfilled, I were, perhaps for the sake of outward advantages, to turn my back on people who have surrounded me with love and friendship, and for whom my departure at this moment of perceived humiliation would be doubly painful. You cannot imagine with how much cordiality I am surrounded here; these are not only people who catch the drops of lard that I sweat out of my brain."[72] To leave Berlin without weighty reasons would be "tantamount to a vile breach of a promise given to Planck and also otherwise disloyal. I would blame myself afterwards." Einstein, too, concluded his letter with a typical addendum: "I feel like some relic in an abbey: one can't do anything with the old bone, and yet . . ."

Ehrenfest passed on Einstein's refusal to Lorentz, adding: "I felt deeply ashamed when I received this letter. But very soon I was filled with a warm, proud joy at this wonderful person. If only one could relieve him of his financial worries. The Nobel Prize?!"[73] But that was to take another three years.

Another wish of Ehrenfest's was more readily fulfilled—Einstein promised to visit his friends in the Netherlands, whom he had not seen for three years. Provided his "tyrannical belly"[74] permitted, he would come before the fall was over. Ehrenfest offered to meet all costs from a Leyden University fund, whether or not Einstein would give a lecture, and the Ehrenfest home would offer health, comfort, and tranquillity. After a visit to the Dutch embassy in Berlin toward the end of September, Einstein asked his friends[75] to intervene in The Hague to speed up his visa. Three weeks later, on October 19, he wrote to his mother: "I'm off to Leyden tomorrow morning."[76]

Those were two weeks of harmony and happiness. Einstein stayed with the Ehrenfests; they made music together; he played with the Ehrenfest children—by then four of them—and, of course, talked physics. Einstein was so delighted with the loving care he received there that, on his return, he sent them each a kiss, according to the "gender, tradition, and temperament" of the recipient, from the little boy Vassik to Tatyana's aunt.[77] "I have truly learned to regard you all as a piece of myself, and myself as belonging to you."

Ehrenfest, his hope of detaching Einstein from Berlin dashed, now thought up a second-best solution: a visiting professorship that would bring his friend to Leyden for three or four weeks every year.[78] Einstein jumped at this offer of a "comet existence in Leyden. If it comes off I'll certainly accept."[79] The following year Einstein became a visiting professor in Leyden.

Needless to say, physics was much discussed during Einstein's stay with Ehrenfest and Lorentz, especially with the astronomer Willem de Sitter.[80] Lorentz had discovered what was to be officially announced in London only at the beginning of November—that light behaved in a gravitational field exactly as Einstein had predicted and with the amount of deflection calculated by him.

That certainly was the climax of Einstein's visit. His long-cherished wish had now been fulfilled and his Dutch friends and colleagues were delighted to have with them the man who had predicted it all. What no one could have foreseen were the far-reaching changes this confirmation of his theory would produce in Einstein's life. To his fellow physicists he had long been the genius of the century; now he became the genius of the century in the eyes of the general public.

CHAPTER TWENTY-TWO

Confirmation of the Deflection of Light: "The Suddenly Famous Dr. Einstein"

THE WAR AND BAD WEATHER during the solar eclipse of August 21, 1914, had actually saved Einstein from bad news: while the deflection of light in the sun's gravitational field would have been confirmed, the amount observed would have disagreed with the figure he predicted. Not until November 1915, within the framework of the completed theory, did he arrive at the double value of 1.7 seconds of arc. A week before that, he had correctly calculated the precession of the perihelion of Mercury, which confirmed his belief in his theory even without proof of the deflection of light. Although his conviction was based primarily on the inherent logic of the theory, his wish to have it experimentally confirmed did not therefore diminish. He had been annoyed with the German astronomers; he had ensured reasonable working conditions for Freundlich; and he had jubilantly welcomed Freundlich's rather premature results on the red shift. There was one such result in 1919,[1] and Einstein at once sent an enthusiastic postcard to the observatory in Babelsberg: "Your new result is most gratifying. Personally, I am now convinced of the existence of the red shift."[2] However, to judge by the literature, Einstein and Freundlich were alone in that belief.

Red shift and deflection of light, it should be pointed out, are not equivalent effects of identical conditions. The red shift follows solely from the equivalence principle, whereas the deflection of light depends also on the specific form of the theory of gravitation and hence on the structure of space. To be really satisfied, Einstein therefore needed confirmation of both effects—and this is where, in the middle of the war, British astronomers entered the scene.

As we have seen, Willem de Sitter sent a copy of Einstein's great *Annalen* paper to England and then, in a series of three articles in the *Monthly Notices* of the Royal Astronomical Society, presented the new theory in England. Arthur Stanley Eddington, the young and highly promising Plumian Professor of Astronomy at Cambridge, later said that the copy of Einstein's article sent to him by de Sitter was the only original source available in Britain or America during the war, but Einstein himself had used Swiss channels to disseminate it and had asked Besso to send some of his papers to the astronomer Ludwig Silberstein in London.[3]

Although Eddington established himself as Einstein's prophet in the English-speaking world and, through a comprehensive report,[4] brought the theory to the attention of specialized circles, it was not he who initiated the solar eclipse expeditions. Indeed, he was so firmly convinced by the new theory that he thought any expedition unnecessary. The man who, in the middle of the war, called for the verification of Einstein's predictions was Sir Frank Dyson, Astronomer Royal and director of the Greenwich Observatory.[5] As early as March 1917, Dyson pointed out that the solar eclipse due on May 29, 1919, would provide a unique opportunity to observe the deflection of light, as it would occur in front of the Hyades, a cluster of bright stars in the constellation Taurus. The position on the ground was rather less favorable than in the sky, as the band of total eclipse ran through tropical regions, a few degrees below the equator. However, Dyson asked for and obtained £1,000 to mount two expeditions.

In the end Eddington was put in charge of one of them, but this was due to his pacifism—which would have pleased Einstein just as much as his commitment to the theory.[6] As a Quaker, Eddington was determined to refuse to serve with the armed forces. When in 1917 the call-up age was raised to thirty-five, the dons of Trinity College, Cambridge, where Eddington was a fellow, were anxious to avoid a repetition of the embarrassment that had previously, and for the same reason, been caused by Bertrand Russell. Strenuous efforts were begun to obtain a deferment for Eddington, on the ground that he was more valuable to the country as a scientist than as a soldier. The dons were successful, and Eddington's deferment came in a letter which he only needed to sign. But, as a man of conviction, Eddington added to his signature the remark that he would have refused military service anyway. This resulted in difficulties with the authorities, and the Astronomer Royal had to intervene once more. Thanks to his good connections with the Admiralty he got his way, but only on condition that Eddington concern himself with the preparations for the eclipse expeditions and, if the war ended in time, lead one of them.

Instrument makers were too busy with military orders even to consider developing new astrographic cameras, so existing equipment at Greenwich had to be adapted to the new tasks. The guns had fallen silent in time, and in February 1919 the two parties set out on their journey, with two months to spare for preparations on the spot. Eddington, along with E. T. Cottingham, had chosen the small volcanic island of Principe in the Gulf of Guinea off West Africa, then under Portuguese sovereignty. Their colleagues Charles Davidson and A. C. D. Crommelin crossed the Atlantic, to Sobral in northwestern Brazil.

In England, meanwhile, the problem of the deflection of light had acquired a patriotic aspect. Outlining some problems for future researchers, Newton, in the appendix of his book *Opticks*, had asked, as the first of sixteen "queries": "Do not Bodies act upon Light at a Distance, and by their Action bend its Rays; and is not this Action (*ceteris paribus*) strongest at the least Distance?"[7] In view of Newton's concept of light as a stream of minute corpuscles, this seemed a reasonable question. Gravitation of the sun or a planet would have to affect the path of light as it affected the path of a comet, provided that these corpuscles had mass—and corpuscles without mass were scarcely thinkable in Newton's day. As wave optics, which regarded light as an oscillation of the ether, became accepted in the nineteenth century, Newton's "query" came to be considered obsolete and was forgotten. But Eddington thought it worth closer investigation. On the strength of Einstein's $E = mc^2$ Eddington ascribed a mass to the electromagnetic radiation, applied Newtonian mechanics and Newton's law of gravitation for a corpuscle passing close to the sun, and arrived at a deflection of 0.87 seconds of arc[8]—the precise figure calculated by Einstein in 1911 from the principle of equivalence.

Eddington's procedure was not strictly legitimate: there was no such simple link, described by Newton's law of gravitation, between electromagnetic field and gravitation. Nevertheless, through a crazy coincidence, it produced a plausible result. Eddington called the figure he arrived at the "Newtonian value." Unfortunately, he did not trouble to show the fundamental differences between Einstein's line of argument, based on the equivalence principle, and his own calculation based on Newtonian theory—which yielded the same result only by sheer coincidence.

Barring the unforeseen, which can never be ruled out in a new experimental field, Eddington was confronted with three alternatives.[9] First, the light beam could travel as straight as an arrow, i.e., it would not be deflected at all by the sun's gravitational field. Second, its deflection could agree with the "Newtonian value" of 0.87 second of arc. Third, it could be deflected by 1.7 seconds of arc, the value postulated by the general theory of relativity. Small wonder that the deflection of light was being viewed in England as a kind of wager between Isaac Newton and Albert Einstein. Eddington himself, however, was inclined to regard Einstein's result as the only possible one.

After weeks of drought on Principe, the day of the solar eclipse started with clouds and rain. When the total eclipse began, Eddington could see the lunar disk, surrounded by the solar corona, only through clouds. But there was no alternative to running the observation program as planned and hoping for a few gaps in the clouds before the eclipse was over. One astronomer changed the plates and the other timed the exposure with a metronome. "There is a marvellous spectacle above . . ."[10] Eddington recalled. "We are conscious only of the weird half-light of the landscape and the hush of nature, broken by the calls of the observers, and the beat of the metronome ticking out the 302 seconds of totality." His work done, he cabled to London: "Through cloud, hopeful. Eddington."[11]

Eddington and Cottingham were not entirely unlucky. Of the sixteen plates exposed, most were useless because of the clouds. Toward the end of the eclipse, however, the clouds had broken up sufficiently for all the stars to be visible on at least one plate. For the sake of comparison the same region of the sky had been photographed at Greenwich in January, with the same instrument, at a time when the sun was in quite a different position. The only useful photograph of the eclipse was therefore compared in Principe with this reference photograph, a micrometer being used for the purpose. This was a delicate procedure, as a deviation of one second of arc represented no more than one-sixtieth millimeter on the plates, and such a minute difference might easily be lost in a multitude of errors, from distortions in the instrument caused by temperature to the swelling of the emulsion in the tropical climate. The one successful photograph "gave a definite displacement, in good agreement with Einstein's theory and disagreeing with the Newtonian prediction."[12]

This conclusion, however, could not be announced at once, because there were four plates with a special emulsion that could not safely be developed in the tropical heat and would therefore be examined only after the expedition's return to England. Besides, there was the other party in Sobral. They had been lucky with the weather, but they had to produce their comparison plates locally and therefore had to wait two months until the region of the Hyades could be photographed in nocturnal darkness, without the sun in the foreground. As a result, including the return journey and careful evaluation, it was another five months before the two groups faced the public with their joint result.

Einstein, the cause of these spectacular activities, was meantime in Berlin or in Zurich and could do nothing but wait. He would have been informed of the planned British expeditions not later than the end of the war. In a public lecture in Berlin in mid-April 1919, on the fundamentals of relativity theory, he mentioned these enterprises, and a report in *Vossische Zeitung* brought the news to a wider public.[13] On May 29, the day of the solar eclipse, the paper published a long article—under the headline "The Sun Will Bring It All Out,"[14] a well-known line by the German poet Adelbert von Chamisso—which explained the purpose of the expeditions to the equator. It was probably this article which Einstein, "for the further nourishment of Mama's already massive maternal pride,"[15] sent to his sister in Lucerne.

Einstein must have heard, from Lorentz or from Ehrenfest, about the observations' ultimate success despite the clouds, because he wrote to his mother in June: "It said in a Dutch paper that both expeditions obtained successful photographs of the solar eclipse, so that the result should be made known within 6 weeks."[16] The Dutch papers had based their stories on the first reports in the British journals *Nature* and *Observatory*. Not until July 21 did *Vossische Zeitung* pick up the news and carry a brief note for its German readers. Einstein had hoped that the results would be out in August, but he had to be patient a little longer.

Other news came from America in the summer of 1919, and that was rather disappointing.[17] The astronomer William Wallace Campbell, who had taken photographs during an eclipse on the west coast of the United States in 1918, let it be known that they did not allow any conclusion, one way or another. The reason for this failure was that Campbell had to manage with improvised equipment, as his astrographs, left behind in Russia in 1914, were still somewhere en route between Japan and the United States. In the course of these investigations some plates of earlier solar eclipses were examined, with the result "that there is no deflection of light in a strong gravitational field and the 'Einstein effect' does not exist."[18] This result was widely publicized and presented at conferences in the summer of 1919.

Examination of the red shift had also produced a result, and that too was negative. The spectroscopist Charles Edward St. John at Mount Wilson Solar Observatory in California had accurately measured especially sharp spectral lines of the sun and in consequence believed that he could rule out the effect predicted by the general theory of relativity. Aside from Einstein's enthusiasm for Freundlich's results—which were obtained from stars and were questioned by most astronomers—things were not going well with the experimental verification of the general theory of relativity. When Max Born asked Einstein what he would do if the predicted effect were not observed, "he replied with unshakable equanimity: 'I would be greatly surprised.' "[19]

In mid-September, with the results of the British expeditions long overdue and having heard nothing all summer, Einstein inquired in a very long letter to Ehrenfest, with deliberate casualness, about the state of affairs regarding the deflection of light: "Have you heard anything about the English solar eclipse expedition?"[20] The answer from Leyden came by telegram on September 22: "Eddington found star dislocation at solar rim provisional magnitude between nine tenths second and double—Lorentz."[21] Lorentz had received this information from a colleague at Leyden, just back from a meeting of the British Association for the Advancement of Science, held in Bournemouth from September 13 to 16. In a lecture that was scarcely noticed, Eddington had expressed himself very cautiously and vaguely,[22] as the evaluations had not by then been concluded. In fact, the data conveyed by Lorentz were not very accurate, certainly not accurate enough to distinguish between the "Newtonian value" of 0.87 second of arc and the double figure of 1.7 seconds.

If Einstein had any misgivings about that information as a full confirmation of his theory, he certainly did not take long to dismiss them. As soon as he received the telegram, he showed it to a doctoral student, but he remained unmoved by her jubilation: "I knew all the time that the theory was correct." But supposing his prediction had not been confirmed, or had even been refuted? "In that case I'd have to feel sorry for God, because the theory is correct."[23] To his mother he wrote a few days later: "Today joyous news. H. A. Lorentz cabled me that the English expeditions have really proved the deflection of light."[24] The same day, in another letter—to a physicist—he expressed a reservation: ". . . admittedly so far with slight accuracy 0.9–1.8 sec., whereas the theory demands 1.7 sec. No doubt the definitive evaluation will provide more accurate data."[25]

Nevertheless, he seems to have informed his colleagues in Berlin that his theory was now confirmed, because at the beginning of October Max Planck warmly congratulated him on the news "contained in Lorentz's telegram. Thus the close bond between beauty, truth, and reality has once again proved effective. You yourself have frequently observed that you had no doubt about the result, but it is a good thing that this fact has now been established beyond doubt also for others."[26]

Einstein had given the glad tidings not only to his mother and his colleagues, but also to journalists such as Alexander Moszkowski. This led to complications. A euphoric article in the *Berliner Tageblatt* of October 8, 1919, claimed that the deflection of light "agrees, within the range of error, with the magnitude predicted by Einstein. This is possible only if Einstein's fundamental framework, the general theory of relativity, represents the true constitution of the universe."[27] That wording was too bombastic and sensational for Einstein's taste. He therefore promptly sent a brief note—nine lines—to his friend Arnold Berliner, the editor of *Naturwissenschaften*, which immediately published it. In the note, Einstein stated, still rather boldly, that the British had "confirmed the deflection of light at the edge of the Sun, as demanded by the general theory of relativity," but otherwise endeavored to put the current situation in its proper perspective: "The provisionally established value lies between 0.9 and 1.8 seconds of arc. The theory demands 1.7."[28]

Not until his visit to Leyden was Einstein finally certain that his theory had been confirmed. On October 23, two days after his arrival, a letter came from England. "This evening," Einstein immediately wrote to Planck, "Hertzsprung showed me a letter from Arthur Eddington, according to which accurate measuring of the plates exactly yielded the theoretical value for the deflection of light. It is a mercy of fate that I was allowed to live to see this."[29] Two days later—a Saturday afternoon—Einstein was a guest at a meeting of the Amsterdam Academy, and he reported to his mother that "Lorentz spoke on general relativity and on the result of the English expeditions—naturally, in order to please me. The result is now definite and means an exact confirmation of my theory."[30] However, Lorentz had spoken informally and not publicly. The public announcement was to be reserved for an event in London.

▪ ▪ ▪

When Einstein returned to Berlin at the beginning of November, he found further congratulations waiting for him, even before he or anyone else on the European continent could have known that Eddington would in fact announce the value predicted by Einstein. These rather premature messages included one from Carl Stumpf—a psychologist at Berlin University and a member of the Prussian Academy—who lent a patriotic note to his commendation of his colleague's great efforts: "I feel a need to express to you my most sincere congratulations on the magnificent new successes of your gravitation theory. We all cordially share the excitement that must fill you, and are proud that, after the military and political collapse, German science was able to score such a victory."[31] Einstein may have been taken aback at being regarded as a German hero, and he replied to the old gentleman factually and modestly: "If indeed there were any great efforts at all, then these occurred several years ago. More recently I did not engage in any particular efforts and therefore do not, in this respect, deserve the sympathy that speaks from your kind lines."[32]

The most significant greeting came from his friend Zangger in Zurich, who thereby proved the "downright unfailing understanding for objective and psychological situations"[33] with which Einstein had always credited him: "Your confidence, your intellectual assurance, that the light must be curved round the Sun, for instance during the time you spent with us, was an overwhelming psychological experience for me. You were so sure that your assurance had a violent effect."[34]

Although Arthur Stanley Eddington was also sure—and indeed had approached the problem not with proper scientific skepticism but with a marked bias in favor of Einstein—he and his colleagues experienced a few ups and downs in their evaluation of the material.[35]

On one of the plates measured in Principe, and on the four plates developed in England, Eddington had found Einstein's value confirmed, albeit only within a rather large margin of error. Better photographs were expected from Sobral, but Crommelin and Davidson would not return from Brazil until the end of August. Their plates, taken with a large astrograph, were the first to be developed and, to

Eddington's surprise and indeed horror, fairly unambiguously showed not Einstein's value but only half his figure, i.e., the "Newtonian" value. This discrepancy was the reason for Eddington's vague formulation at the Bournemouth conference in mid-September and for the imprecise data which Lorentz in his telegram had passed on to Einstein on September 22. But the pictures on these plates were distorted, probably because of overheating of the heliostat, the system of mirrors by which the image of the sun was directed at the lens of the camera.

There remained seven photographs taken in Sobral with an auxiliary instrument—a telescope with a ten-centimeter lens. To evaluate these plates, a measuring instrument had first to be converted. This dragged on into October, but the evaluation time and again showed Einstein's value with absolute unambiguity and great accuracy. Eddington, who readily admitted that on this issue he did "not have an entirely open mind,"[36] could breathe again. He convinced himself and his colleagues that the astrograph plates with the "Newtonian" value had too many flaws even to be considered, and thus the Einsteinian value was ensured.

On Thursday, November 6, in London, a joint meeting of the Royal Society—the oldest and most venerable of British scientific institutions—and the Royal Astronomical Society was scheduled. It was held in Burlington House on Piccadilly, the Royal Society's home, and there was only one item on the agenda: notification and discussion of the results of the two solar eclipse expeditions. Everyone who was anyone in English physics and astronomy was present.

"The whole atmosphere of tense interest was exactly that of a Greek drama,"[37] Alfred North Whitehead, who had come specially from Cambridge, later recalled.

> We were the chorus, commenting on the decree of destiny in the unfolding development of a supreme incident. There was dramatic quality in the very staging:—the traditional ceremonial, and in the background the picture of Newton to remind us that the greatest of scientific generalizations was now, after more than two centuries, to receive its first modification. Nor was the per-

> sonal interest wanting: a great adventure in thought had at length come to safe shore.
>
> Let me remind you that the essence of dramatic tragedy is not unhappiness. It resides in the solemnity of the remorseless working of things. This inevitableness of destiny can only be illustrated in terms of human life. . . . The laws of physics are the decrees of fate.

Despite the drama perceived by Whitehead, the event also had the very British character of a wager between gentlemen: Newton versus Einstein, or, representing them, Sir Oliver Lodge versus Arthur Stanley Eddington.

The first speaker was the Astronomer Royal, Sir Frank Dyson, whose farsightedness, drive, and internationalist attitude had made the planning of expeditions possible in the midst of a war.[38] He described the purpose of the expeditions, their equipment, and the difficult evaluation procedures, and eventually announced: "After a careful study of the plates I am prepared to say that there can be no doubt that they confirm Einstein's prediction. A very definite result has been obtained that light is deflected in accordance with Einstein's law of gravitation."[39] Commerlin next explained the results from Sobral, and Eddington explained the results from Principe. The numerical values obtained were 1.98 ± 0.12 seconds of arc for the Sobral pictures and 1.61 ± 0.30 seconds of arc for the pictures from Principe (which had been impaired by clouds). Both results ruled out the "Newtonian" value; their mean value was almost exactly equal to Einstein's prediction of 1.74 seconds of arc.

The two presidents—J. J. Thompson for the Royal Society, and Alfred Fowler for the Royal Astronomical Society—supported the conclusions presented by Dyson and Eddington. In the discussion that followed, everyone waited to hear Sir Oliver Lodge, who had publicly wagered that there would be no deflection of light at all, or, if there was any, that it would be in line only with Newton's law of gravitation; and who in his books still adhered to the long-abandoned ether theory and had even calculated the mass of the ether at thousands of tons per cubic millimeter. Lodge wordlessly left the assembly. In the end it was mainly Ludwig Silberstein, the man to whom Einstein had sent some of his papers via Besso, who adopted a skeptical position. With some justification he pointed out that very accurate measurements of the spectral lines in sunlight had revealed no red shift, and that Einstein's theory should stand or fall on that criterion alone. With rather less justification, or none at all, he argued that the deflection of light could be explained by refraction in the solar atmosphere. With a fine sense of drama he gestured toward Newton's portrait: "We owe it to that great man to proceed very carefully in modifying or retouching his law of gravitation."

In the discussion, however—no doubt largely because of the authority of Dyson and Eddington—the view prevailed that, after the explanation of the precession of Mercury's perihelion, confirmation of the predicted deflection of light represented a confirmation of the general theory of relativity and hence also of Einstein's theory of gravitation. Sir Joseph John Thompson, a successor of Isaac Newton in the president's chair of the Royal Society, went so far as to proclaim it to be objective reality. He expressed his opinion—the majority opinion—in a pretty compliment not only to Einstein but also to his great predecessor: "This result is not an isolated one; it is part of a whole continent of scientific ideas. . . . This is the most important result obtained in connection with the theory of gravitation since Newton's day, and it is fitting that it should be announced at a meeting of the Society so closely connected with him. The result is one of the highest achievements of human thought."

This highest achievement of human thought also seemed to be the most incomprehensible. In the course of the discussion Thompson conceded that "no one can understand the new law of gravitation without a thorough knowledge of the theory of invariants and of the calculus of variations." There was a rumor in the air that only three persons really understood relativity theory. When the assembly ended, Silberstein, who regarded himself as one of the few experts on relativity theory, approached Eddington and paid him a compliment: surely Eddington was one of the three. Eddington was silent. "Don't be so modest, Eddington!" Silberstein urged him. "Not at all," Eddington replied. "I'm just wondering who the third one might be."[40]

■ ■ ■

"The Glorious Dead," "King's Call to His People," "Armistice Day Observance," "Two Minutes' Pause from Work"—these were the headlines in the London *Times* on November 7, 1919. On the same page, however, above a two-column article, there was also the headline "Revolution in Science—New Theory of the Universe—Newton's Ideas Overthrown." This was the report on the Royal Society's meeting the day before. Inserted into the text were the subtitles "Momentous Pronouncement" and "Space 'Warped.' " According to the article, J. J. Thompson had referred to "one of the most momentous, if not the most momentous, pronouncements of human thought." Einstein's theory, the article continued, meant that "our conceptions of the fabric of the universe must be fundamentally altered." At the same time, the newspaper was unable to disclose any details of "the famous physician Einstein"—there was no indication of his age, family, or first name. Nor was there any mention of the fact that he lived in Berlin, the capital of the recently defeated enemy.

Nevertheless, from then on articles on relativity began to appear frequently in the British press, especially in the *Times*. "All England is talking about your theory,"[41] Eddington wrote to him; in Cambridge he had given lectures to packed audiences, and hundreds more had vainly clamored for admission. A. F. Lindemann, the father of Nernst's former assistant, told Einstein that it was "amusing to read whose country's child you are being made out to be. . . . You are presented as a Pole, or a Swiss, etc., but most particularly as one who did not sign that unfortunate letter"—meaning the manifesto of the ninety-three German intellectuals in the first year of the war. Lindemann summed up the public fuss as being due to the fact "that your theory wrecks Newton's theory and that the world is no longer what we are accustomed to regard it as, and the whole Euclidian geometry had, as it were, gone to the devil, that space is bent, etc. etc. Naturally this has hurt our national feelings and thrown the world into great turmoil."[42] Eddington, on the other hand, was viewing the political effects of the fuss more positively: "For scientific relations between England and Germany this is the best thing that could have happened."[43]

Arnold Berliner, the editor of *Naturwissenschaften*, showed Einstein a letter from the physicist Robert W. Lawson, who had done research in Vienna and had heard Einstein lecture to the Naturforscher meeting there in September 1913. Lawson had been caught in Austria at the outbreak of war and had been interned, but he had been allowed to continue to work at the Radium Institute in Vienna. Lawson, now back in England, wrote: "Here the talk is practically of nothing but Einstein, and if he came over now I believe he would be fêted like a victorious general. The fact that the theory of a German was confirmed by observations by Englishmen has, as is becoming more obvious every day, brought the possibility of cooperation between these nations a lot nearer. Thus Einstein, quite apart from the high scientific value of his inspired theory, has done an inestimable service to humanity."[44]

Before the wave of tumult over relativity even reached Germany, it hit the American shore with an unimaginable roar. Nothing much was known of Einstein there. American physicists had shown no particular interest in relativity theory, regarding it, with American pragmatism, as a typical piece of German metaphysics. To the astronomers, the deflection of light was mainly an experimental challenge rather than a reason to concern themselves with the underlying theory. In fact, they knew very little about the theory. America lacked a prophet of Eddington's intelligence and persuasiveness who might help the new theory gain ground.

Thus until November 1919 Einstein had been totally unknown to the general public; after that date, there was scarcely anyone who had not heard of him. The index of *The New York Times* first lists his name on November 9,[45] although the compilers of the index had overlooked the fact that Einstein and his theory were actually first mentioned in the summer of 1918, in an account by W. W. Campbell of his observations of the solar eclipse.[46] Thereafter his name figures in the index every year. A brief report appeared on November 9, based essentially on the first article in the London *Times*, and a long article was featured on the following day, with six eye-catching headlines: "All Lights Askew in the Heavens," "Men of Science More or Less Agog at Results of Eclipse Observations," "Einstein's Theory Triumphs," "Stars Not Where They Seemed or Were Calculated to Be, but Nobody Need Worry," "A Book for 12 Wise Men," "No More in All the World Could Comprehend It, Said Einstein, When His Daring Publishers Accepted It."

As the reporter understood nothing of what he was writing about, he seized on the incomprehensibility of the subject, all the way through to the final paragraph, which had this to say about Einstein: "When he offered his last important work to the publishers, he warned them there were not more than twelve persons in the whole world who would understand it, but the publishers took the risk." Only Americans could have credited such a story, believing that *The New York Times*—"All the News That's Fit to Print"—could not possibly publish an invention.[47] The fictitious story, however, became established, perhaps because in the land of democracy, where everybody claimed the right to be heard about everything, such intellectual exclusivity was felt to be offensive.

In an editorial of November 11, *The New York Times* itself made this point. The fact that it had taken two presidents of Royal Societies "to give plausibility, or even thinkability, to the declaration that as light has weight space has limits" was not a good sign.

> It just doesn't, by definition, and that's the end of it—for common folk, however it may be for higher mathematicians. . . . Quite the most disturbing feature of the situation is the assumption that only men of wonderful learning have the ability, and therefore the right, to see what meaning there is in the fact that light, being subject to a turning from the straight path by a mass of matter like the sun, must itself have of matter at least the quality of weight. . . . It is difficult to avoid the suspicion that if the masters could do any more they wouldn't recoil from the task of putting what it means to them into words for the rest of us at least to mull over. If we give up, no harm would be done . . . but to have this surrender done for us—is, well, just a little irritating.

And so it continued. On November 13 an editorial writer had a consolation ready, and was, moreover, able to invoke the physicist who was then the most famous in America: "People . . . who have felt a bit resentful at being told that they couldn't possibly understand the new theory, even if it were explained to them ever so kindly and carefully, will feel a sort of satisfaction on noting that the soundness of the Einstein deduction has been questioned by R. A. Millikan." On November 16: "These gentlemen may be great astronomers, but they are sad logicians. Critical laymen have already objected that scientists who proclaim that space comes to an end somewhere are under some obligation to tell us what lies beyond it." On the same day an interview was published with Charles Poor, professor of celestial mechanics at Columbia University. This honest man saw the uncertain times, plagued by war, strikes, and bolshevik rebellion, as an expression of a deep-seated spiritual unrest that stopped at nothing and had penetrated even into science. This anxiety, sounding like a belated echo of comment on Galileo's propagation of the Copernican system, was discussed in an editorial on December 7, under the headline "Assaulting the Absolute": "The raising of blasphemous voices against time and space threw some [people] into a state of terror where they seemed to feel, for the first few days at least, that the foundations of all human thought have been undermined."

For a few months the readers of *The New York Times* may well have felt that something not accessible to everybody must be rejected as un-American. But gradually a different attitude emerged. The incomprehensible acquired a halo: it became numinous, and soon Einstein was included in this quasireligious concept. "To fictions such as time, space, mathematics, the universe, we should be all the kinder," declared an editorial in *The New York Times* on January 31, 1921. "Whatever else be fictional in a fictional Whole, we believe implicitly in Professor Einstein because of the subtlety of his characters and his plot and the limpidity of his dialogue."

"Einstein awoke in Berlin on the morning of November 7, 1919, to find himself famous,"[48] an English biographer wrote. This might have been true if he had been living in London, but in Berlin Einstein woke up and found nothing of the sort. Only a telegram from Lorentz informed him that day of the public announcement in London that his general theory of relativity had been confirmed, and this was no surprise to Einstein anyway.

The reason was that in Germany, unlike England or the United States, Einstein's name had long been familiar to the wider public—more so than any other scientist's. His fame in Germany had begun with his extraordinary invitation to the Prussian Academy, and it had steadily grown. In March 1918 the Philosophical Faculty of the University of Göttingen had awarded him the prize of the Otto Vahlbruck Foundation in Hamburg, a prize endowed with 11,000 marks, and this event was duly publicized in ten dailies and periodicals. (Einstein suspected that Hilbert had been the initiator, and thanked him for it.[49]) The University of Rostock had given Einstein an honorary doctorate—nothing special, but rather unusual for someone who, by the standards of the privy council, was still a young man. Perhaps the most striking indication of Einstein's public standing had been the fact that *Vossische Zeitung*, reviewing a brochure of lectures given at a festive meeting of the German Physical Society on the occasion of Planck's sixtieth birthday, headlined its article "A Physicist's View of the World: Professor Einstein on the Motivation of Research"[50]—even though lectures by Warburg, Sommerfeld, and Laue were also published, and even though, strictly speaking, Max Planck should have occupied center stage. All that was long before the tumult over relativity.

Also, the British expeditions had been extensively reported in the German press, though without emphasis on the assumed conflict between Einstein and Newton that was later highlighted by the British and American newspapers. Indeed, the confirmation of Einstein's prediction had been anticipated in October, not only in a spectacular article in *Berliner Tageblatt* on October 8, but a week later also in *Vossische Zeitung*. In consequence, the German press lacked the sensational element which, in England and in the United States, overnight catapulted the "suddenly famous Dr. Einstein"[51] into the spotlight.

Not until November 18 did *Vossische Zeitung* carry a report—and this was a perfectly sober one—on the Burlington House meeting of the Royal Society and the Royal Astronomical Society. On November 30 there followed an extensive and factual article by Erwin Freundlich, *Albert Einstein: On the Triumph of His Relativity Theory.* Freundlich began: "A scientific event of exceptional significance has not, in Germany, met with the attention its importance demands." For some time to come nothing else appeared in this major Berlin daily. *Frankfurter Zeitung* had asked Max Born for a contribution, which on November 23 appeared on the front page under the headline "*Space, Time, and Gravity.*" Einstein found it "excellent."[52] These were factual reports, in the best tradition of popularized science, without any sensationalism. *Naturwissenschaften*, the weekly journal for high-level scientific discussion, waited until the first issue of 1920, and even then confined itself to an unadorned account of the report in the British journal *Nature* on the Burlington House meeting of November 6.[53]

The London *Times*, meanwhile, had thought of inquiring at the source: Einstein was asked to explain his theory to the British public. He began by explaining why he was glad to grant that request:

> After the lamentable breach in the former international relations existing among men of science, it is with joy and gratefulness that I accept this opportunity of communication with English astronomers and physicists. It was in accordance with the high and proud tradition of English science that English scientific men should have given their time and labor, and that English institutions should have provided the material means, to test a theory that had been completed and published in the country of their enemies in the midst of war.[54]

This was followed by a very lucid, though perhaps not readily comprehensible, presentation of the two theories of relativity and their relation to traditional physics. With reference to the concept of "Einstein versus Newton"—an aspect that had been blown up in England—he concluded with an homage to Newton:

> No one must think that Newton's great creation can be overthrown in any real sense by this or any other theory. His clear

> and wide ideas will for ever retain their significance as the foundation on which our modern conceptions of physics have been built.

Reacting to reports which had meanwhile reached him from England, Einstein added to this conclusion a little postscript with certain undertones:

> The description of me and my circumstances in *The Times* shows an amusing feat of imagination on the part of the writer. By an application of the theory of relativity to the taste of readers, today in Germany I am called a German man of science, and in England I am represented as a Swiss Jew. If I come to be regarded as a *bête noire*, the descriptions will be reversed, and I shall become a Swiss Jew for the Germans and a German man of science for the English.

The Times, judging by an editorial in the same edition, was not amused. The tribute paid to the impartiality of English scientists had been "well-intended, if somewhat superfluous." As for his postscript, "We concede him his little jest. But we note that, in accordance with the general tenor of his theory, Dr. Einstein does not supply any absolute description of himself."

At about the same time, the Berlin correspondent of *The New York Times* called on the "suddenly famous Dr. Einstein." Naturally, he asked Einstein about the story of the "twelve men" who alone understood his theory, whereupon "the doctor laughed good-humoredly, but still insisted on the difficulty of making himself understood by laymen." Evidently the correspondent had his own difficulty, for in his report[55] the thought experiment underlying the general theory of relativity became an actual event—a real man who in 1915 fell from a real roof in Berlin, but, landing on a heap of refuse, survived. Einstein, who had seen the fall from his window, had rushed down to question the miraculously saved man about his sensations. The man had told him that during his fall he had not been aware of what was normally attributed to gravity, and—presto!—the key to the general theory of relativity was found. It remains an open question whether the journalist had put the story together from half-understood information, or whether Einstein himself had been fooling his visitor—which would not have been out of character.

Possibly in response to the garish publicity in the British and American press, *Berliner Illustrirte Zeitung* on December 14, 1919, took up the subject. On its front page it carried a magnificent portrait photograph of Einstein with the caption "A New Giant in World History: Albert Einstein, whose researches mean a complete overthrow of our views of nature and which rank as equal with the discoveries of Copernicus, Kepler, and Newton." Fame had at last caught up with Einstein at his place of work.

Even then he could not suspect the extent to which his life would be changed by this transformation from a genius among physicists to a public cult figure. A few weeks later, he knew. To him it was "dazzling misery."[56]

PART V

SPLENDOR AND BURDEN OF FAME

CHAPTER TWENTY-THREE

Relativity under the Spotlight

"THIS WORLD IS A CURIOUS MADHOUSE" was Einstein's reaction, in September 1920, to a situation at whose center, to his own surprise, he suddenly found himself. "At present every coachman and every waiter argues about whether or not the relativity theory is correct. A person's conviction on this point depends on the political party he belongs to."[1] That it had come to this was due to the political upheavals in Germany and to that newly emerged power of the twentieth century, the media. With their love of banner headlines and superlatives, the media had seized on Albert Einstein and did not let go of him. "Never before has anything of the kind been experienced," wrote Einstein's first biographer, trying to find words for the public enthusiasm and his own enchantment:

> A flood of amazement was sweeping the continent. Thousands of people who had never in their life worried about lightwaves or gravitation were seized by that flood and carried aloft, if not to comprehension then at least to a wish to comprehend. And they all understood enough to realize that from the intellectual work of a quiet scholar a message of salvation had emerged for the exploration of the universe.... No name was uttered as much during that time as that of this one man.... Here was a man who had reached out for the stars, a man in whose theory one had to penetrate to forget one's earthly troubles....
>
> Even the idea that there was a Copernicus walking among us had something elevating about it. Anyone paying homage to him had a sense of transcending space and time, and this homage was

an attractive feature of a period otherwise so poor in pleasant things.[2]

The emergence of an Einstein myth may have had something to do with a longing for peace and with the symbolism surrounding the stars. The starry sky had always been a mystery; and now there was a wise man who proclaimed new and scarcely believable news from the seat of the gods—news that, to the war-weary, must have sounded like a secularized Christmas message. The fact that a theory created in Berlin had been confirmed, at considerable expense and effort, by the British must have been widely seen as a token of real peace. Abroad it was being pointed out that Einstein had not been a signatory to the notorious "Manifesto of the Ninety-three," that he had rejected the war, and that he was Swiss and not German anyway. And to Germans he appeared as a personification of Kant's postulate of "the starry sky above and the moral law within." That his theory was not understood merely enhanced the mystery around him; had not the great prophets always "spoken in tongues"?

Once before, a physical discovery had made banner headlines—the discovery of X-rays by Wilhelm Conrad Röntgen. On New Year's Day 1896 he sent out the first offprints of his paper *On a New Kind of Rays* to his colleagues, along with photographs of X-rayed hands; on January 5 the new rays made the front page of the *Wiener Neue Presse* under the headline "A Sensational Discovery."[3] On January 7 the German and English papers carried the news, and on January 9 the Kaiser ordered a telegram to be sent to Röntgen: "If the report is proved true, I congratulate you from a full heart and praise God that this new triumph of science has been granted to our German fatherland." The Kaiser moreover requested "a report on your invention," and Röntgen immediately set off from Würzburg to Berlin, where on January 12 he gave a demonstration of his new rays at court. More than a thousand scientific publications about the rays and their application appeared before the year was out.

■ ■ ■

Matters were different with Einstein and his theory of relativity. The fact that space was warped and that light did not travel in a straight line was of no practical consequence. The strange structure of the world, as revealed at speeds approaching the velocity of light, was not perceptible at the comfortable speeds of normal life. Furthermore, the theory itself was beyond normal comprehension. Thus writers focused their attention on Einstein's personality.

"Ever since the announcement of the deflection of light a cult has been practiced with me, so that I feel like a graven image," Einstein complained within a few months of that event. "But that, too, will pass with God's help,"[4] he added—but here he was badly mistaken. The frenzy got worse. "I never understood," Einstein mused in retrospect in 1942, "why the theory of relativity with its concepts and problems so far removed from practical life should for so long have met with a lively, or indeed passionate, resonance among broad circles of the public. . . . What could have produced this great and persistent psychological effect? I never yet heard a truly convincing answer to this question."[5] Not an answer but a shrewd quip was supplied by Charlie Chaplin. He had invited Einstein to the premiere of *City Lights* in Los Angeles on January 31, 1931. As the crowds cheered them, Chaplin is reported to have observed: "They cheer me because they all understand me, and they cheer you because no one understands you."

Einstein never left any doubt that he disliked the "relativity circus"; at best it was a source of amusement to him, but more often it was an annoyance. Yet the media alone could not have kept the frenzy and occasional hysteria going if Einstein had not sometimes played along with them. Not that his fame had gone to his head—for that he was too intelligent and too remote from everyday life. Also, the personality cult conflicted with his democratic convictions. It was not, therefore, simply modesty when, in a newspaper article, he described it as "unjust, and indeed distasteful" if a few individuals are credited with

> superhuman powers of intellect and character. This has become my fate now, yet there is a grotesque contradiction between the capabilities and achievements people attribute to me and what I

> actually am and can do. Awareness of this curious state of affairs would be intolerable if it did not also hold a fine consolation: it is a welcome phenomenon in our supposedly materialist time that it makes heroes of men whose goals lie exclusively in the spiritual and moral domain.[6]

There was yet another consolation for Einstein: he was able to use his enhanced reputation to work for pacifism, democracy, and international understanding, and to address the fate of the Jews, which had lately come to his awareness.

He probably enjoyed contact with journalists as a pleasant change from the stiff, ceremonious behaviour of the world of learning. In addition to the articles which he himself wrote, there were interviews; and there were also photographs—so many, in fact, that he would occasionally give his occupation as "photographer's model."

Added to all this, at the end of the 1920s, were films. The new sound newsreels eagerly pounced on the man with the waving mane of hair, the pipe, and the violin case. The cameras were in position when, in September 1930, Einstein opened a radio exhibition in Berlin before a huge crowd. He urged "all those of you present and absent" not to forget men like Maxwell and Hertz, whose research had made radio possible. Many film clips of Einstein's visits to America show a man evidently enjoying the fuss made over him. For instance, on his arrival in the harbor of San Diego on December 31, 1930, he was greeted with rhythmical shouts of "Einstein! Einstein!" and led by a parade of cheerleaders as he descended the gangway to be greeted by mermaids reclining on flower-covered floats.[7] At the reporters' request he tossed his hat into the air; he reacted to moderately intelligent questions with "cheap jokes," and himself laughed loudest at them. Only later did he learn to use the media for his own purposes instead of letting himself be used.

Quite apart from the shrill "relativity circus," however, there was no denying that the most important physicist since Newton strode onto the scene not as a retiring scholar but as a militant intellectual like Voltaire or Zola. As a politically motivated intellectual, Einstein was certainly unique among natural scientists—an irritation to some colleagues, and a provocation to many Germans. Conflicts and malicious hostility were inevitable. He was not always satisfied with his own reactions: "The tragedy of my situation is that I am unable to muster even a fraction of self-assurance in order to act with 'dignity' the part assigned to me through no fault of my own."[8] During those first years of his fame he sometimes lost his composure, and at times even his temper.

The first row came at the beginning of 1920. Einstein's lectures on "Introduction to the Theory of Relativity" became a local attraction, something a visitor had to see, along with the palace and the Brandenburg Gate. Genuine students could barely find seats among all the sightseers. Once the curiosity of these sightseers was satisfied, after fifteen or thirty minutes, they would cause another disturbance by leaving before the end of the lecture. Students who had paid their tuition asked the principal to stop the traffic of illegitimate visitors. Only Einstein had nothing against the visitors; he even pleaded that anyone should be allowed to attend his lectures, and for the benefit of the merely curious he announced a short break to enable them to make an orderly withdrawal. This led to some correspondence, with the principal instructing Einstein on the university rules, and Einstein accepting his guidance.[9] On February 12, Einstein discussed the problem with his audience—both those who had been properly admitted and the others. Representatives of the students' committee thereupon proposed that registered students' fees should be refunded and Einstein should continue to run his class as a cycle of free public lectures, a Solomonic solution to which Einstein agreed.

There had been vehement arguments before this. Members of the students' committee complained that Einstein had referred to "outcasts of mankind,"[10] and, as a countermove, the left-wing *Vorwärts* spoke of "excesses by an anti-Semitic student mob."[11] The principal hastily offered an assurance that "with not a single word has there been any mention of anti-Semitism, Jewry, etc.";[12] the state secretary in the Ministry of Education telephoned Einstein; the ministry released a press statement; and *Berliner Tageblatt* summed up the events to the effect that "the protests were not of a political, and especially not of an anti-Semitic character."[13] Einstein partly but not entirely confirmed this in an official statement: "There can be no question of a scandal said to have taken place yesterday; nevertheless, a few remarks that were made testified to a certain animosity toward me. There were no anti-Semitic utterances as such, but their undertone could be so interpreted."[14]

The defeat in the war and the birth pangs of the republic represented a setting for coarse attacks on the Jews, who were blamed for everything. "There is strong anti-Semitism here and raging reaction, at least among the 'educated,' "[15] Einstein wrote, describing the situation in Germany toward the end of 1919. Was it this tense atmosphere that made Einstein sense anti-Semitism in incidents at the university that were often more comical than politically significant? Certainly, by going into print with his unsubstantiated suspicion, it was he himself who first linked his name publicly with anti-Semitism, even before the anti-Semites had discovered him as a target.

In any case, Berliners had other worries. A few weeks later, in mid-March, there was a radical right-wing putsch in Berlin, which forced the government to flee the city for several days. "Here we have confused conditions, corruption, and dictatorship of the sabre," Einstein reported to Zangger in Zurich. "The military are murdering with impunity, a sorry bunch. The barbarism is frightful."[16] When the putsch had collapsed, owing to a general strike, and the government had returned to Berlin, conditions were still far from normal: "The country is like a man who badly upset his stomach, but has not yet vomited enough."[17] In this atmosphere, poisoned by political extremism, some people saw a chance of making a name for themselves by attacking Einstein and relativity theory. And—Germany being Germany—these people even established a club.

The founder and mouthpiece of the Arbeitsgemeinschaft deutscher Naturforscher zur Erhaltung reiner Wissenschaft e. V.—the Working Party of German Scientists for the Preservation of a Pure Science (registered as an association)—was Paul Weyland, a graduate engineer with journalistic and political ambitions, initially a covert and later an open anti-Semite, a curious example of those dubious figures who in turbulent times appear in public life. According to Laue, Weyland was to be "classified as a crook."[18] He was probably financed by groups who preferred to keep in the background; he certainly had enough money to offer fat fees to physicists prepared to give public lectures against the theory of relativity.[19] At the beginning of August he opened a campaign in *Tägliche Rundschau* with an article describing the theory of relativity as "scientific mass suggestion" and "a big hoax," as well as accusing Einstein of plagiarism.[20] Weyland used the ensuing debate[21] to advertise his real objective: a public meeting on August 24, 1920, for which he had hired the great hall of the Philharmonic Society. Along with Weyland, the physicist Ernst Gehrcke was announced as a speaker.

Gehrcke was then an *Oberregierungsrat*, a senior official, at the Physical-Technical Reich Institute, and undoubtedly a good experimental physicist; he was the inventor of the Lummer-Gehrcke plate and a few other techniques useful in spectroscopy. But he did not understand Einstein's theoretical reflections and had turned his obtuseness into a militant virtue. Ever since 1911 he had written articles against relativity. By 1913, when the theory was still the exclusive domain of experts in the field, he believed that he had refuted it in the journal *Naturwissenschaften*, calling it "an interesting case of mass suggestion in physics, especially in the lands [where German is spoken]."[22] Max Born immediately wrote and sent in a retort,[23] but this had no effect on Gehrcke.

As no one believed his "refutations," Gehrcke in 1916 opened a second front. He unearthed some studies by the Pomeranian schoolmaster Paul Gerber, who, about the turn of the century, had tried to explain the perihelion movement of Mercury by procedures which were entirely arbitrary and were rightly rejected by astronomers. By reprinting Gerber's paper along with his own comment in *Annalen*[24] Gehrcke insinuated, first, that Einstein was a plagiarist and, second, that the problem of the perihelion precession had been solved before relativity theory. Einstein, who was normally not averse to a "paper war" if the intellectual level appealed to him, this time informed the editor of the journal that he "would not reply to Gehrcke's tasteless and superficial attacks, as every reasonable person can do so for himself."[25] However, he did reply a year later, when Gehrcke repeated his foolish arguments elsewhere.[26] Einstein's retort was concise and factual.[27] A sharper, crushing reproof came from the respected Munich astronomer Hugo Ritter von Seeliger, who pointed out that the Pomeranian schoolmaster had copied things which had long been known to every worker in the field, and that his so-called explanation was based on a crude mathematical mistake. Seeliger publicly declared that Gehrcke was "totally unacquainted with the whole issue, both in a factual and in a historical respect."[28]

Although probably no anti-Semite but only a passionately obtuse physicist, Gehrcke now found his platform, which the physicists had denied him, in Weyland's association.

The event staged by the Arbeitsgemeinschaft, however, was concerned less with physics than with a "link with anti-Semitic politics, as manifested already by the distribution of political smear sheets in the foyer of the hall,"[29] as noted by Max von Laue, who attended. The audience in Philharmonic Hall, which included Einstein, first heard a generally abstruse speech by Weyland and then abstruse physics from Gehrcke. Weyland denounced the relativity theory as a publicity stunt, described it as scientific dada, and defamed its author as a plagiarist. He was followed by Gehrcke, "and although he warmed up the old rubbish"—as Max von Laue recorded—"his calm and factual way of speaking was a relief after Weyland, who is the equal of the most unconscionable demagogue."[30] The "old rubbish," of course, was the paper by Gerber which Gehrcke had unearthed.

Reporting the Arbeitsgemeinschaft event, most of the liberal and left-wing Berlin papers deplored the polemical methods introduced into science by Weyland. Laue, Nernst, and Rubens in a joint statement condemned the fact that not only objections to Einstein's theory had been raised, but also "objections of a malicious character against his person as a scientist":

> It cannot be our task to discuss here in detail the unparalleled profound intellectual work which led Einstein to his theory of relativity. Surprising successes have already been achieved; further verification must of course be left to future research. What

> we do want to emphasize, and what was not touched upon in a single word yesterday, is that quite apart from Einstein's relativistic research, his other work already assures him of an immortal place in the history of our science. In consequence, his influence on the scientific life not only of Berlin, but of the whole of Germany, cannot be overestimated.
>
> Anyone who has the joy of being close to Einstein knows that he cannot be surpassed by anyone with regard to respect of other people's intellectual property, personal modesty, and dislike of publicity. We regard it as a demand of justice to voice this conviction of ours without any delay, the more so as no opportunity to do so was offered last night.

This declaration was carried by numerous papers on August 26, even by *Tägliche Rundschau*, Weyland's own mouthpiece.

Meanwhile Einstein himself had joined the fray. His response, under the headline "My Reply—About the Antirelativistic Association," appeared on the front page of *Berliner Tageblatt* on August 27. It was a searing polemic, angry and vicious: "I am well aware that the two speakers are unworthy of a reply by my pen, for I have good reason to believe that motives other than the search for truth underlie their enterprise." Nevertheless, in two long columns, he made mincemeat of Gehrcke and gave free vent to his anger: "If I were a German nationalist, with or without swastika, instead of a Jew with a libertarian international attitude . . ." The tone was justified by the circumstances, but, as Bertolt Brecht put it, hatred distorts one's features.[31] Two days later, in an interview, Einstein summed up his situation in Berlin vividly: "I feel like a man lying in a good bed, but plagued by bedbugs."[32]

Such rude words had not been heard before from members of the academy, and Einstein's best friends were horrified. Ehrenfest was unable to believe that "at least some of the phrases were written by your own hand."[33] Hedwig Born was worried about the possible consequences of the "maladroit reply" in the *Tageblatt:* "Anyone not knowing you might get a wrong picture of you. That, too, is painful."[34] Sommerfeld, who had followed "the Berlin campaign against you . . .

with real fury," summed up various people's opinion of the article to the effect that it was regarded as "not very happy and not really like yourself." He made no criticism, though, of Einstein's worst remark: "But the thing about the bedbugs was good."[35]

Immediately after the event at Philharmonic Hall Einstein had "for two days"[36] toyed with the idea of leaving Germany. He evidently made no secret of that intention, because virtually all the daily papers reported it toward the end of August. Sommerfeld implored him: "You must not leave Germany." The Ministry of Culture was getting nervous, and even the German chargé d'affaires in London reported to Berlin about reports in the English press: "The attacks on Prof. Einstein and the agitation against the well-known scientist are making a very bad impression over here. At the present moment in particular Prof. Einstein is a cultural factor of the first rank, as Einstein's name is known in the broadest circles. We should not drive out of Germany a man with whom we could make real cultural propaganda."[37] Einstein meanwhile had long recovered his "old phlegmatic mood" and could look at the comical side not only of the antirelativist club but also of his friends' concern. To the "dear Borns" he wrote: "Don't be too severe with me. Everyone must, from time to time, make a sacrifice on the altar of stupidity, to please the deity and mankind. And I did so thoroughly with my article. This is confirmed in a sense by the enormously approving letters of all my dear friends."[38]

Among the public there was no shortage of declarations of solidarity with Einstein—by artists such as Max Reinhardt and Stefan Zweig[39] and by the Social-Democrat minister of culture, Konrad Haenisch, who expressed the hope "that there is no truth in the rumors that, because of those ugly attacks, you wish to leave Berlin, which has been proud and will remain to be proud to count you, revered Herr Professor, among the greatest ornaments of our science."[40] Einstein replied at once "with a sense of sincere gratitude. Quite aside from the question whether I deserve so much benevolence and esteem, I have these days experienced that Berlin is the place to which I am most closely bound by personal and scientific relationships. I would only follow a call to a foreign country in the event that external conditions compel me to do so."[41]

The presiding secretary of the Prussian Academy, however, saw no reason to comply with a suggestion by the ministry "to come out in the quarrel which has now erupted among the public about the value of Einstein's theory of relativity."[42] Max Planck agreed: "It would mean doing too much honor to the dubious characters envious of our colleague's fame if we rolled out the heavy guns of the Academy against them. Besides—and this is the main thing—Einstein has no intention of leaving Berlin because of those goings-on."[43]

However, the affair was by no means settled. For one thing, the dubious characters did not let go, and for another, Einstein had committed an embarrassing faux pas in his article: he had quite violently attacked Philipp Lenard, the Nobel Prize laureate of 1905. It was Lenard's experiments on the photoelectric effect that had led the young Einstein toward the hypothesis of light quanta; and the two men had, by correspondence, assured each other of their esteem, indeed of their admiration—and these were not just polite phrases. Meanwhile, however, they had come into conflict over the general theory of relativity. In a *Jahrbuch* article in 1918, *On the Relativity Principle, the Ether, and Gravitation*, Lenard had raised objections based on what he described as "sound common sense." Before the year was out, Einstein replied to these and other objections in the form of a dialogue between a "critic" and a "relativist," written in a relaxed tone, with just a touch of condescension.[44]

In the *Berliner Tageblatt*, however, Einstein had been downright aggressive. He said that while he still admired Lenard "as a master of experimental physics," Lenard "has so far achieved nothing in theoretical physics, and his objections to the general theory of relativity are of such superficiality that until now I had thought it unnecessary to answer them in detail."[45] Einstein's rude attack had been provoked by the fact that reprints of Lenard's *Jahrbuch* article had been offered for sale in the foyer of Philharmonic Hall, and even more so because the Arbeitsgemeinschaft was advertising a further event with Lenard as the speaker. This, however, had been a unilateral action by Paul Weyland, who had put Lenard's name on the poster without asking him, relying on Lenard's nationalistic views.

Lenard, despite his resolute—and not always informed—opposition to the general theory of relativity, had invariably maintained an academic tone, and indeed in his article, according to Sommerfeld's impression, had referred to Einstein "very decently."[46] Thus he was outraged by the article in the *Berliner Tageblatt*, and rightly so. Sommerfeld, as chairman of the Physical Society, was anxious to limit the damage and asked Einstein "to write some conciliatory words to Lenard. . . . If you say to him that your defense had been aimed not at the learned critic but at Weyland's presumed comrade-in-arms, and that, if requested, you would state this publicly, his anger would probably abate."[47] Lenard insisted that Einstein "should retract [his remarks] just as publicly as he made them; otherwise the injustice done to me could not be undone—provided this is possible at all."[48] Einstein did not comply with Sommerfeld's request, and this attempt at mediation therefore collapsed. This did not bode well for the meeting of the Society of German Scientists and Physicians called for the end of September 1920.

The meeting, the first since the end of the war, was to be held in Bad Nauheim from September 19 to September 25. A joint session of the mathematical and physical sections was to be concerned with relativity theory. Einstein had suggested that there should also be "a discussion on the theory of relativity. There, anyone who dares face a scientific forum would be able to voice his objections."[49] This debate took place on September 23, following the lectures on relativity theory.

Alerted and worried by the incidents in Berlin, the managers of the congress had restricted access to the great Kursaal to registered participants and set up strict controls; they had even asked the police to be present. Einstein did not give a lecture, but he took a lively part in the discussions, especially on Hermann Weyl's report of his attempt to develop a unified theory of electricity and gravitation. The last speaker was the experimental physicist Ludwig Grebe of Bonn, who presented measurements of the red shift in the sun's gravitational field, which, to Einstein's delight, seemed to confirm his theoretical predictions. (As so often before, Einstein's joy was premature.)

At midday began the tensely expected general discussion. Max

Planck was in the chair, and the protagonists were Lenard and Einstein. But the dramatics feared by some and hoped for by others did not take place. As far as can be judged from a greatly abridged and perhaps editorially massaged report,[50] the rules of polite behavior were followed. "Excited slippages into personal attacks were swiftly damped down by Planck, and the audience was sparing in its expressions of rejection and agreement."[51] The big newspapers, which were not going to miss the occasion, had nothing unusual to report the following day. Not only Einstein, but Lenard as well, was evidently able to act on the podium as if the *Berliner Tageblatt* article had never been written.

From a scientific point of view, the discussion, as expected, was unproductive. Lenard did not accuse relativity theory of a conflict with experience or of logical flaws, nor did he take up Gehrcke's charge of plagiarism. Instead, he opened the debate with objections of a general character, such as that "the theory offends against the simple common sense of a scientist." He described an actually existing ether as indispensable and also said that thinking in graphic pictures was indispensable. Einstein replied to this last point that opinions on what was or was not graphic were by no means a matter of course: "I believe that physics is conceptual and not graphic. As an example of our changing views on what is graphic, I would remind you of the concept of the graphic in Galileo's mechanics in different periods."[52] But Einstein had put all this much better in his *Dialog*—not only his own defense of relativity theory but also the objections of its critics.

Planck was eventually able to wind up the fruitless debate with a joke: "As the theory of relativity has unfortunately not so far been able to prolong the absolute time available for our session, from nine to one o'clock, the meeting must now be adjourned."[53]

Einstein's opponents, who had hoped the meeting would end with a judgment against his theory, were disappointed. Paul Weyland, in a newspaper article, fulminated against the "throttling of the Einstein-adversaries."[54] But his "antirelativity" club soon petered out. Even Gehrcke before long saw Weyland as a "dubious type," and Lenard agreed: "Unfortunately Weyland really turned out to be a crook."[55]

The attempt to mount an organized opposition to Einstein and the theory of relativity had deservedly collapsed, and what was subsequently portrayed as a bitter struggle of anti-Semites against Einstein[56] was, on closer inspection, no more than a curious marginal phenomenon of those politically sensitive early years of the Weimar Republic. Before the Nauheim conference, the only instance of anti-Semitism was the distribution, recorded by Laue, of smear sheets in the foyer of the Philharmonic Hall. It was only after the Nauheim meeting that Lenard garnished his critique of relativity theory with the most distasteful excrescences of anti-Semitism.

Below the surface, however, emotions must have been running high. After the Nauheim conference Sommerfeld wrote to Elsa Einstein that he was glad "the whole crisis was settled with more or less propriety. The principal merit, of course, belongs to your husband, to his kindness and factual manner—qualities one cannot credit his opponent Lenard with."[57] Einstein's recollection was different. The meeting left a nasty taste in his mouth, partly because of his own behavior; he would "never again let myself be upset as in Nauheim. It is quite incomprehensible to me that, because of the bad company, I lost myself so deeply in humorlessness."[58] When the physicists held their own meeting in Jena the following year, Einstein stayed away, "because the bigwigs bored me far too much in Nauheim last year. I am trying altogether to be as independent of that mob as possible, and, what's more, I am succeeding."[59]

Even though Germany's leading physicists were proud to have the "new Copernicus" in their midst, they were by no means happy to see Einstein's name forever figuring in the newspapers. A typical remark was that of Felix Klein: "In his personal utterances Einstein is always so charming—quite unlike the foolish publicity circus that has been set in motion for him."[60]

The touchiness of scholars of the time toward anything that suggested publicity was also experienced, for instance, by Max Born. He had written a semipopular book on relativity theory and wanted to include a photograph of Einstein and a short biographical note. He sent this two-page tribute to Elsa Einstein, who, after reading it, felt that "I must give you a kiss for it. Please don't change a single word of it."[61] But when the book appeared, Laue immediately remonstrated with Born—and this was intended as the action of a friend. Laue argued "that he and many other colleagues would take umbrage at the photograph and the biography. This sort of thing does not belong in a scientific book, even if it is intended for a broader readership."[62] When Born thereupon asked the publisher, Ferdinand Springer, to dispense with the portrait and the biographical note, Springer felt that "personally neither you nor my firm need to make such a retreat."[63] Nevertheless, reprints appeared without these "personal additions" in order to avoid any suspicion of unscientific publicity.

No sooner was this little affair over, as well as the bigger affair of the Naturforscher assembly, than further trouble arose. The trade journal of the German publishers' association announced the forthcoming publication of a book with the ponderous title *Einstein—Insights into the World of His Ideas—Generally Comprehensible Observations on the Relativity Theory and a New World System—Based on Conversations between Einstein and Alexander Moszkowski.* The author was the journalist who, during the war, had invited Einstein to join a literary circle at the Hotel Bristol and who had written articles about the solar eclipse for the *Berliner Tageblatt.* His list of publications did not exactly suggest respectability, consisting as it did of a collection of jokes and occultist books.

Einstein's friends were horrified. Einstein himself being away on a trip, Freundlich tried to prevail on the author to withdraw the book—needless to say, in vain. Hedwig Born wrote Einstein a long letter, urgently asking him to prohibit publication of the book and painting a lurid picture of what would otherwise happen: "A new and even worse campaign will be unleashed, not only in Germany, no, everywhere, and the disgust of it will choke you. . . . The fact, as it will quite simply be seen, is that a man still in his early forties, i.e., still in his younger years, has authorized an unpleasant author to record his conversations. . . . Except for 4–5 of your friends, the book will mean a moral death sentence on you. It would provide the best confirmation of the accusations that you are blowing your own [horn]."[64] A few days later Max Born agreed with his wife: "You've got to shake off this M., otherwise Weyland will have won along the whole line, and Lenard and Gehrcke will triumph. . . . You should listen to and obey people with judgment (and not your wife)."[65] Einstein thought the verdict on Moszkowski was "too harsh," but he obeyed. "I have sent him a registered letter to the effect that his splendid *opus* must not be printed."[66]

Yet neither Moszkowski nor his publisher could be ordered about so easily. Einstein was advised to take legal action, but he rejected that idea "because it would merely magnify the scandal."[67] Altogether, Einstein did not regard the impending affair as nearly so dramatic as his friends did. "It is a matter of indifference to me, along with the shouting and the opinion of everyone. In point of fact, nothing can happen to me."[68] Back in Berlin for the winter, he awaited the inevitable with black humor: "The stinkpot of my friend Moszkowski has not yet burst open thanks to the slowness of the printing press," he described the situation to a worried Ehrenfest. "A pity I can't produce him for you in the flesh. You appreciate that sort of thing. Frau Born's crescendo has meanwhile developed into a fortissimo penetrante, so that I had to quieten her with gentle irony."[69] When the "splendid opus" eventually came out, Einstein was surprised at those who thought it necessary to read it and who were now boring him with critical or outraged reactions. "It is inconceivable to me," he reproved Zangger, "that you spent even a minute on Moszkowski; I haven't done so myself, so my soul's salvation is not in danger. I know that incense, too, is a commercial commodity, even if '*non olet*' is even less appropriate here than on other occasions."[70]

Fortunately, the book was not a catastrophe, nor did it trigger one. That "everything about Einstein is self-advertisement" had been claimed by his critics even before, and Einstein accepted the verdict with a shrug of his shoulders: "Just as for the man in the fairy tale everything he touched turned to gold, so for me it turns into media clamor—*suum cuique*."[71] The "media clamor" was shrill anyway, and Moszkowski's book, though repeatedly reprinted until 1921, added scarcely anything to it.

When Max Born in his old age picked up the book he had once so fiercely attacked, he found it "not as bad as I expected."[72] While no great biographer, Moszkowski seems to have presented Einstein's views on a whole range of subjects, from Rutherford's recent transformation of atomic nuclei to educational problems, with reasonable authenticity. As Moszkowski can have learned many of the details only form Einstein himself, and because it was written much earlier than most other biographies, his book, for all its flaws, remains an indispensable source.

Besides, Moszkowski's *Einstein* was just one title in flood of articles and publications on the theory of relativity—over fifty books and pamphlets in 1921 in Germany alone.

CHAPTER TWENTY-FOUR

"Traveler in Relativity"

When, in November 1919, Albert Einstein's fame spread throughout the world, the Prussians did not wish to be left behind. On November 26, only three weeks after the memorable meeting of the Royal Society in London, members of the state budget commission demanded that the Prussian government, in consultation with the Reich government, should "make available the necessary means to enable Germany to cooperate successfully with other nations on the development of the fundamental discoveries of Albert Einstein and to facilitate his own further research."[1] The sum of 150,000 marks was made available, for which Einstein was duly grateful, although, as he put it to the minister of culture, he could not quite "suppress some painful misgivings: at this moment of extreme hardship, would not a decision like this justly arouse bitter feelings among the public?"[2] With a sideswipe at Germany's astronomers, Einstein added that no special financing from the state would be necessary "if the observatories and astronomers of this country were to place a part of their equipment and their energy at the service of this cause."

Thus it was not Einstein but his faithful astronomer Erwin Freundlich who saw in this official offer a chance to combine the advancement of science with that of his own career. Einstein had made three predictions: the perihelion precession of Mercury, the deflection of light, and the red shift in the gravitational field of the sun. Freundlich had believed for some time that in order to confirm the third prediction—the red shift—a special tower telescope would have to be built, on the model of Mount Wilson in California, then the only one in the world, but if possible larger and with better instrumentation. This would cost more than could be raised from the state coffers, and Freundlich therefore persuaded Walther Nernst to canvas for donations from industry.

An appeal for an "Albert Einstein donation," drafted by Freundlich and approved by Nernst, referred to the patriotic "duty of honor of all those who care about Germany's position in the world of culture to raise whatever sums they can to enable at least one German astronomical observatory to verify the theory in direct cooperation with its creator."[3] The appeal was signed by all the important physicists and astronomers of the Prussian Academy, as well as by Adolf von Harnack, the president of the Kaiser Wilhelm Society. This seal of approval made it easy for Nernst and Fritz Haber, with their connections to the "Reich Association of German Industry," and more particularly the chemical industry, to obtain several hundred thousand marks, so that detailed planning of the "Einstein Tower" could start in the spring of 1920. The firms of Zeiss and Schott in Jena were prepared to deliver the equipment at cost, and the state provided a site on the Telegraphenberg in Potsdam, on the property of the Astrophysical Observatory. Erwin Freundlich became the scientific director.

The architect chosen was Erich Mendelsohn, a friend of Freundlich's. Inspired by Freundlich, Mendelsohn had sketched impressive solar observatories while still in the trenches, at a time when no one could tell if they would ever be constructed or how they would be paid for.[4] After considerable technical difficulties, the forty-six-foot tower was completed in 1922, though it took another two years for the sophisticated equipment to be installed. Mendelsohn's dramatic structure is regarded as the most important building of expressionist architecture. Einstein, however, did not like it;[5] his artistic taste was conservative anyway, and perhaps the structure suggested a rather oversized submarine conning tower. (It might have been more suitably named after an admiral than after a pacifist.) Einstein consoled himself with the thought that on the vast expanse of the Telegraphenberg the tower would be out of public view.

Although Einstein was appointed lifetime chairman of the board of the "Einstein Foundation," he badly neglected his duties, scarcely ever attended its meetings, and generally treated its business in an offhand manner. This was not, however, because of the architecture; rather, it was because his relations with Freundlich were gradually deteriorating and the tower, with its equipment, failed to come up to his expectations. Typically, Einstein had been certain at the laying of the foundation stone that the red shift would "eventually provide a brilliant confirmation of the theory; I never had a second's doubt of that."[6] But the enormous multiplicity of lines in the solar spectrum and the turbulent movements in the solar atmosphere were masking the red shift to such an extent that the separation of the effect predicted by Einstein proved impossible, even with this special telescope. It was of some use only for the exploration of the solar surface.

This was not Freundlich's fault. The actual confirmation would not be achieved until five years after Einstein's death—and not by observations of solar or stellar light, but under terrestrial conditions. The stunning precision of the Mössbauer effect, discovered in 1958, made it possible to demonstrate a minute difference in gravity between the ground and the top of the seventy-nine-foot Jefferson Tower of Harvard University. The result was what Einstein had always predicted: a "brilliant confirmation."

Einstein's letter to the minister of culture, thanking him for supporting his researches on relativity theory, also contained a very personal request, one which was more important to him at the end of 1919 than physics or his own fame. He wanted to bring his terminally ill mother, then in a private hospital in Lucerne, to Berlin for the last few months of her life. He therefore asked the minister of culture to drop a hint to the housing authority to ensure that an additional room at Haberlandstrasse 5, which had already been authorized, would in fact be made available for Pauline and a nurse. The request was not granted. When Pauline, by then paralyzed by progressive metastases, arrived in Berlin at the end of December, along with her daughter, Maja, and a woman doctor, she was accommodated in Einstein's study. His mother's suffering profoundly shocked Einstein: "One feels right into one's bones what ties of blood mean."[7] Pauline Einstein died toward the end of February 1920. "I know what it means to see one's mother in torturing agony; there is no consolation. We all have to bear such heavy blows, for they are indissolubly linked with life."[8]

There were also other worries, lesser ones but still absorbing much energy. Some of these were financial. With the collapse of the mark, it became increasingly difficult to raise the Swiss francs needed for the support of Mileva and their two sons. "It would be an enormous relief to me if my boys and Mileva could move to Germany next year,"[9] he wrote; he wanted to accommodate them near Karlsruhe, with a distant cousin who was the principal of a high school. Mileva rejected this plan—she did not want to be so close to relatives of her former husband—and a move to Freiburg was then proposed. But Mileva and the boys wished to stay in Switzerland. Einstein was understanding: "Maybe I can raise enough foreign money to let them stay in Zurich. This might have advantages for the future of my children, for the sake of which it would be right to overcome the difficulties."[10] His change of attitude may also, at least in part, have been due to the fact that his original hope for a democratic transformation of Germany had been disappointed.

Whatever promises he may have made to Planck and Haber—or to the minister of culture, Haenisch—about staying in Berlin, and however much he may have justified his refusal of offers from Zurich and Leyden by loyalty to his Berlin colleagues, Einstein had no real sense of being at home in Berlin. When Max Born consulted him about his own possible move from Frankfurt to Göttingen, Einstein revealed himself to his close friend as a homeless stranger, one "who leaves today and stays tomorrow—a potential itinerant." He did not feel

> qualified to give advice, being a person with no roots anywhere. My father's ashes lie in Milan. I buried my mother here [in Berlin] a few days ago. I myself have roamed about ceaselessly—a stranger everywhere. My children are in Switzerland under conditions which make it a complicated undertaking for me to see

> them. A person like myself regards it as an ideal to be *at home* somewhere with his near and dear ones; he has no right to advise you in this matter.[11]

From Columbia University in New York he received an offer, richly endowed, without any obligations, and with the assurance that for some time to come America would be "a quieter place for research than Germany"; he turned it down, with the explanation that "viewed objectively, it makes no difference in the end where I study and work."[12]

Einstein's unconditional loyalty was solely to a small circle of people to whom he felt attached through shared "scientific aspirations." This circle was significantly enriched when Niels Bohr came to Berlin from Copenhagen for a series of lectures in April 1920.

Einstein had admired Bohr ever since 1913, when Bohr published his first papers on the structure of atoms and an explanation of the spectral lines. Although he had not yet met the young physicist, his junior by six years, he had formed a good picture of him from Ehrenfest's accounts. Toward the end of 1919 he wrote to Ehrenfest in Leyden that he would now "engross himself in Bohr, seeing that you instilled in me a deep and warm interest in him. You made me realize that here is a man with a deep insight, one in whom major connections come to life."[13] In his old age, Einstein recalled how he had been in despair over the intricacies of radiation theory, and how Bohr had brought the first surprising insights, by ingeniously grafting "quantum conditions" onto the classical physics of the atom, which Rutherford had postulated as a miniature planetary system:

> It was as if the ground was being pulled away from under one's feet, without any solid place left to build upon. That this swaying basis was sufficient to enable a man like Bohr to discover the principal laws of the spectral lines and the electron shells of the atoms, along with what they meant for chemistry, seemed to me like a miracle—and still seems to me like a miracle. This is the peak of musicality in the field of ideas.[14]

On his visit, Bohr had brought butter and other nutritious things to Haberlandstrasse, "a magnificent present from Neutralia, where milk and honey still flow,"[15] as Einstein put it in his letter of thanks, to which Elsa added that her housewife's heart had "reveled at the sight of such delicacies." The talks between the two men must of course have been far more wonderful; they concerned the riddle of radiation and atomic mechanics. Although these conversations produced no hard results, let alone breakthroughs—the state of physics at the time was too confused for that—the scientists evidently derived great pleasure from their mutual understanding.

For Einstein Bohr's visit was another opportunity to put his gift for deep admiration into tender words: "Seldom in my life has a person given me such pleasure by his mere presence as you have," he wrote to Bohr after his departure. "I am now studying your great publications and—unless I happen to get stuck somewhere—have the pleasure of seeing before me your cheerful boyish face, smiling and explaining."[16] Two days later he reported to Ehrenfest: "Bohr was here, and I am just as enamored of him as you are. He is like a sensitive child and walks about this world in a kind of hypnosis."[17] Bohr was no less touched and, in slightly awkward German, replied: "It was to me one of my greatest experiences to have met you and to talk to you. You cannot imagine what a great inspiration it was for me to hear your views from you in person. . . . I will never forget our conversation on the way from Dahlem to your home."[18]

"Scientifically, I don't have much to show at the moment. My life is too hectic,"[19] Einstein wrote toward the end of 1919, when he found himself at the center of public interest. This state of affairs continued. Six months later he wrote: "As for work, there isn't much going on at the moment. I dissipate my strength, have a huge correspondence to deal with, to advise, to patronize, but am making no progress with the big problems."[20] These lamentations recall remarks he made at the age of twenty-seven, about the approach of "stationary and sterile old age."[21]

He really was publishing very little now—just one original paper in

1920, and this had nothing to do with the "big problems." But it was by no means unimportant. In fact, it marked the beginning of a development that would lead to a Nobel Prize only after his death.

The paper, *Propagation of Sound in Partially Dissociated Gases,*[22] submitted to the academy on April 8, 1920, contained a theoretical development of a proposal for determining the kinetics of fast chemical reactions by relaxation methods. Experimental investigation at Nernst's institute, however, failed because of the limitations of the techniques then available. Not until the 1950s were these ideas realized in Göttingen by Manfred Eigen, who was awarded the Nobel Prize for his work in 1967.[23]

Even in his wildest dreams, Einstein could hardly have expected the enormous productivity of his peak years—1915 to 1917—to continue. The idea of a "midlife crisis" is tempting and may have a grain of truth in it, but there were also objective reasons. The fruits of his prolonged "pregnancy period" since 1907 had been successfully harvested, and in a sense were used up. Time and again in the development of theoretical physics, we encounter such periods of stagnation, when even the greatest geniuses are unable to produce outstanding work. And around 1920, physics was in one of those periods as far as the "big problems" were concerned.

Moreover, the hardships of the war years and the postwar situation—which was not much better—along with personal worries and a protracted series of illnesses, also took their toll. Although he could once more describe himself as a "fairly robust fellow,"[24] Einstein still had to keep to a diet. Besides, the upheaval resulting from his fame was not exactly conducive to scientific work. At the same time, there is no doubt that the dissipation of his energies—which he himself diagnosed—would not have occurred if he had still been as enthralled by physics, and as obsessed by physics, as in earlier years.

But everything was different now. Whatever he did or wherever he went, he inevitably attracted public attention. Even going to a concert or a first night at a theater became a much-noticed event. His foreign trips no longer had a personal or scientific character but were politically significant, with a touch of the sensational. Although he himself viewed his role of "traveler in relativity"[25] with ironical detachment, he was certain of the attention of the German embassies wherever he went, and of the government in Berlin. This was not so much because, as was reported to him, "half the foreign ministry had been sitting over Born's introduction to relativity";[26] rather, it was because this universally celebrated "traveler in relativity" was a major cultural asset to a Germany still isolated scientifically as well as otherwise.

In May 1920 Einstein was again in Leyden. His visiting professorship had not yet been approved; however, Ehrenfest had arranged for the visit to be financed from a special fund and had also seen to it that, by way of consolation, Einstein was accepted into the Amsterdam Academy. In Leyden Einstein was, as always, happy; and his thanks to Ehrenfest were again effusive: "You are all so good and cordial to me, in a way I cannot explain to myself, pampered and overrated lout that I am. But I am grateful to you from the bottom of my heart, and I do appreciate it. For the two of us, moreover, it is strangely good to be more often together, because we are made for each other by nature."[27]

Unlike his earlier visits, however, this one included some official pomp. A lecture in the ancient Aula, *Space and Time in Modern Physics*, was given the character of a special tribute to Albert Einstein by the ancient academic ritual. The German minister was present, and on the following day he invited Einstein and some of the Leyden professors to lunch at the legation in The Hague. Afterward he reported to Berlin that the lecture had "made an exceptional impression on the crowded audience and was received with great enthusiasm. This strong impression was further heightened by Einstein's exceedingly modest behavior."[28] For Germany's diplomatic representative the most important aspect, of course, was that Einstein's activities "greatly contribute to bringing Germany's and Holland's scientific circles closer together."

No sooner was Einstein back in Berlin than his suitcases were packed again for a trip to Norway, at the invitation of the Students' Union of Oslo University. This time he was accompanied by his stepdaughter and secretary, Ilse—after a ruthless assessment of his possible traveling companions: "I would take with me only one of the women, either Elsa or Ilse. The latter is more suitable because she is healthier and more practical."[29] In Oslo Einstein gave three lectures, which earned him an honorary membership in the Students' Union, "a fresh and pleasant bunch of youngsters."[30] The Norwegian foreign minister turned up for the lectures, as of course did the German minister, who reported to Berlin: "Admiration for the scientist was extraordinary."[31] Einstein's sojourn ended with a sailing party on the fjord, followed by a picnic—according to Scandinavian custom, in formal dress, despite the "formidable heat."

On the return journey, Einstein first made a stopover at Copenhagen to consolidate his friendship with Niels Bohr and to give a lecture at the Astronomical Society. The previous day the press had duly featured the importance of the "most famous physicist of the present day," and after the lecture Denmark's leading daily, *Berlingske Tidende*, observed: "Einstein's work must become the common property of the entire civilized world and gain acceptance everywhere."[32] The German minister, in his report, volunteered this remark: "Although Einstein is Swiss by birth and said to be of Jewish extraction, his work is yet a link in the chain of German scientific research."[33] This was partly wrong and partly right, but it certainly was typical of the attitude of officialdom and the public mood in the Weimar Republic.

Einstein's homeward journey was via Kiel, where he gave a lecture at the university and paid a visit in passing to the Anschütz-Kaempfe gyrocompass factory to promote some joint patent projects. In Berlin, too, he had thought up "an amusing technical thing . . . in brotherly cooperation with Nernst."[34] This probably concerned a new cold-generating process, offered as Patent Nernst/Einstein to the firm of Borsig in 1922, though nothing much came of it.

He spent the summer of 1920 in Berlin, mostly sailing on the boat of his medical friend Moritz Katzenstein, ruffled only by the distasteful business of the "Antirelativity Company, Inc.," and the upheaval that followed his juicy reply in the *Berliner Tageblatt.* In September came the Nauheim meeting of the Naturforscher, which, though it ended with a debacle for the organized antirelativists, had nevertheless upset him. From Nauheim he proceeded to Stuttgart, where, toward the end of September, he "had to preach for the benefit of a *Volkssternwarte*,"[35] a publicly owned observatory; then he went on to Hechingen, his wife's birthplace, to visit their relatives. After that he met his sons during their autumn break in October, in the nearby township of Benzingen at the house of the priest Camillo Brandhuber—Switzerland had become too expensive for Einstein after the collapse of the mark. From Benzingen he returned direct to Leyden, where, as Ehrenfest reported to him, "those ink-shitters have at last got to the point of approving your professorship."[36]

On October 27, Einstein gave his inaugural lecture, again in a festive setting. This turned into a tightrope act, bringing confusion rather than clarity. Its very title, *The Ether and the Relativity Theory*,[37] was a tribute to the *genius loci*, the revered Hendrik Antoon Lorentz, who had loyally held on to his belief in the ether. Einstein, who had once dismissed the ether from physics as "superfluous," now, to everyone's surprise, declared that in the general theory of relativity "the ether concept had once more acquired a clear content," albeit one differing from Lorentz's. "The ether in the general theory of relativity is a medium which itself is bereft of all mechanical and kinetic properties, but which has a share in determining mechanical (and electromechanical) occurrences."[38]

This "medium" was represented by the gravity potentials and was therefore identical with the gravitational field—which makes one wonder what meaning Einstein's new terminology could have had, other than as a compliment to his father figure Lorentz. Besso, for one, instantly saw through the ploy: "You have endowed the word with its only possible meaning in the new sphere, to prevent the people who believe in it, especially Lorentz, from being further alarmed by apparent deviations—at any rate, something human. But also something humanly beautiful."[39] In physical terms, however, it was less beautiful, and indeed superfluous. Moreover, Einstein had not allowed for the fact that his stubborn opponents, like Lenard and Gehrcke, would hardly miss this opportunity to accuse him of inconsistency and self-contradiction. Einstein never used this formulation again.

The 2,000 Dutch guilders—Einstein's fee for the few weeks of his visiting professorship—were extremely welcome, even though they were no longer adequate to maintain Mileva and the two boys in Zurich. With the progressive collapse of the mark, hard currency had become a permanent problem.

Einstein's foreign earnings were no longer sent to Berlin but instead went to Leyden, where Ehrenfest acted as his trustee. As Einstein was not only violating German currency regulations but also concealing earned income from the tax authorities, the two men used a comical code when corresponding about these financial transactions. Thus Ehrenfest, reporting to Einstein the receipt of payments from England and the Netherlands, referred to the "results which you and I obtained here on the concentration of Au ions."[40] Evidently the two conspirators did not expect any official postal censor to have a scientific background; otherwise they would hardly have used such an obvious code as the chemical symbol for gold. Nor did Ehrenfest bother to conceal the name of Einstein's English publisher: "Thanks to the arrival of a new chemical from Methuen, and a chemical just now produced here locally, the concentration has risen to 6.7×10^{-1}. This is quite a useful value." Einstein understood, of course, that hidden behind the gold ions were 6,700 Dutch florins, which he could certainly use for his commitments in Zurich. "Your news about the high concentration of Au ions is most welcome, especially as the state of our researches here makes such a high concentration seem most desirable."[41]

In money matters Einstein was by no means the fool he has often been portrayed as in practical affairs. Although his personal needs were modest—often to the point of asceticism—in his early years he never had enough money for it to become a matter of indifference to him. He understood the value of money and was not averse to driving a hard bargain.

For his book for the general public, he had demanded an exorbitant share of the retail price: twenty percent. When his publisher, Vieweg in Braunschweig, complained that at a time of depreciation such a large share would drive up the price of the book, and proposed that Einstein content himself with the customary ten percent, Einstein persisted: "Not only the paper and printing costs, but everything the publishing house needs for its existence and the author for his life, has gone up proportionally and continues to do so."[42] He showed the same firmness with regard to his royalties from foreign editions, which always exceeded the customary rate. But he also made sure his translators were decently paid, and on one occasion actually proposed to renounce part of his own earnings in favor of the needy translator of the Czech edition of his book.[43]

Despite his careful management Einstein's financial worries were getting steadily worse. His Dutch colleagues therefore recommended some unusual steps. Lorentz drew his attention to the fact that "it is now actually possible to make money from the relativity theory."[44] The occasion was a competition sponsored by the American journal *Scientific American* for the best popular-scientific elucidation of the theory of relativity in no more than three thousand words. The winner would receive an award of $5,000, provided by a wealthy American. Lorentz believed that the journal "would have done better to ask you for an article and offer you the sum as a fee. Now I wish you had some small article, and we are certain the judges would be sensible enough to choose it. . . . It is not your fault that scientific achievements, such as the world is indebted to you for, do not free one from all money worries." Lorentz even offered to have an English translation made in Leyden should there be any difficulties with that in Berlin.

Einstein was already aware of the competition: he had been notified directly by *Scientific American.* He thanked Lorentz for his offer of help but added that he had "immediately decided not to participate in this competition. For one thing, I do not like dancing around the Golden Calf, and for another I have so little talent for that kind of dancing that I would hardly earn applause with it."[45]

Even without Einstein's participation, the competition attracted much attention. Along with a lot of patent nonsense, the journal received 275 articles that deserved to be taken seriously; seventeen were short-listed, including those by two competent scholars in Einstein's circle of friends—the philosopher Moritz Schlick and the astronomer Willem de Sitter at Leyden.[46] The prize went to an Englishman named Lyndon Bolton, a man unknown in scientific circles. He was an employee of the British Patent Office.

By the end of the year Einstein again felt that the only way out of his financial worries was to persuade Mileva to move with the boys to Germany. There "they could save something for days of hardship, whereas now everything is spent, and they live in almost depressed conditions, while I have difficulties seeing them. But they resist the idea, and there is no one in Zurich to explain the business to them properly."[47]

Another solution, though, was money from abroad. That was why Einstein was now prepared to accept invitations from American universities, at a high price. For a six-week lecture cycle he demanded a frighteningly high fee (he even frightened himself): about twice the annual salary of a top-grade scientist in the United States. "I have demanded $15,000 from Princeton and Wisconsin University. Probably it will scare them off. But if they do bite the bullet, I shall be buying economic independence for myself—and that's not a thing to sniff at."[48]

While awaiting the decisions from America, he began 1921 with a journey to Prague, where a decade earlier he had held his first professorship. To avoid attracting attention, he did not put up in a hotel but stayed with his successor Philipp Frank in Frank's temporary accommodation at the Physical Institute of the German university. Einstein slept on a sofa in what had once been his own study.

Prague was no longer an Austrian provincial city but the capital of the new Czechoslovak Republic, which had risen from the ruins of the Habsburg empire and was building a democracy under the wise leadership of its president, Thomas Masaryk—a development Einstein followed with sympathy. The Germans, no longer the rulers but a minority, regarded the great man as one of their own; one German daily welcomed him in comically mistaken terms: "Now the whole world will be able to see that an ethnic group that has produced a man like Einstein, the Sudeten German tribe, can never be oppressed."[49]

His evening lecture, at the scientific society Urania, took place in an atmosphere of sensational expectation. The room was "dangerously overcrowded," as Frank recorded. "Everybody wanted to see the world-famous man who had turned the laws of the universe upside down and proved the 'warping' of space. . . . But the audience was much too excited even to try to follow the lecture. They didn't want to understand, but to be present at an exciting event."[50] At a follow-up meeting with a smaller circle of notables, the guest was complimented in all kinds of speeches; but when the time came for Einstein to reply, he picked up his violin and played a Mozart sonata, because "that was probably pleasanter and easier to understand." The following evening there was a discussion meeting, again at the Urania, where the philosopher Oskar Kraus tried to convince not so much Einstein as the public of the "elementary absurdities" of relativity theory. As Philipp Frank was in the chair, the debate, thanks to his adroit handling, ended after a series of comical and by no means serious incidents.

The following day Einstein went on to Vienna, where he stayed with Felix Ehrenhaft, the professor of experimental physics. On January 10 Einstein gave a lecture at the institute, a strictly professional lecture for physicists only—but after that his peace was at an end. Ehrenhaft's house was besieged all day long by people who hoped to see the great man appear on the balcony, though Einstein was sitting inside, talking to Friedrich Adler, his old friend from Zurich, who had been released from prison and was now trying to organize an alliance of socialist parties.

For January 13 a major public lecture was scheduled in a concert hall seating three thousand. Before entering the sold-out hall, Einstein felt a little uneasy and asked Ehrenfest to stick close to him and to sit next to him on the platform. Even more so than in Prague, the audience was "in a strangely euphoric state, when it was a matter of indifference whether they understood anything so long as they were near a spot where miracles were performed."[51]

On the return journey Einstein had been urged to make a stopover in Munich to give a lecture there. Sommerfeld, alluding to his own efforts on behalf of relativity, had observed "that we have a greater claim on you than has any other place, as your relativity gospel took earlier and firmer root here than elsewhere."[52] But Einstein had already undertaken to give a lecture at the Technical University of Dresden, and thought it would be "too unnerving and too strenuous to squeeze in another stop in Munich; after all one's got to manage one's whole lifetime with the miserable bundle of nerves that nature had equipped one with for the journey."[53]

Back in Berlin, Einstein had to prepare a lecture he had been commissioned to give on Friedrichstag (Frederick's Day), the annual commemoration of the great reformer and patron of the Prussian Academy, on February 6. This lecture, *Geometry and Experience*, must have convinced the academicians, or those not previously convinced, that they had among their ranks a veritable master of scientific rhetoric, one who combined substance and attractive presentation to an outstanding degree.

A week later, he was once more on the move, this time on a confidential political mission to help Germany. On behalf of a group of pacifists from the New Fatherland League, which now was part of the left wing in the democratic spectrum of the new government, he left for Amsterdam on February 13, 1921, along with the diplomat Count Harry Kessler, to visit the central headquarters of the International Trade Union League. Kessler, with some surprise, recorded in his diary that Einstein was evidently traveling by sleeper for the first time and inspected everything with keen interest.[54]

The two emissaries were supposed to persuade the International Trade Union League to intervene at the Reparations Conference in London, due to begin on March 1, in favor of a solution which would enable the Germans to survive economically and which therefore would also be advantageous to international trade; if necessary, they were to resort to a general strike. The negotiations were conducted mainly by Count Kessler; they produced no result because the trade unionists, realistically assessing their limited power base, saw no chance of dissuading France from claiming exorbitant reparations. Einstein observed dryly: "The more impossible the conditions are, the more certain it is that they won't be realized."[55] But the conditions were harsh enough to ensure that the postwar chaos would not readily evolve into peace.

Further journeys were already planned. In the spring Einstein would assume his pleasant duties as visiting professor in Leyden and participate in the Solvay Congress in Brussels, the first since the end of the war. There he would give a paper on the confusion that had developed over the Einstein–de Haas effect. In the fall he would then make a big leap across the ocean. But things turned out otherwise.

To begin with, the American universities had, in February, declined Einstein's expensive services: "America isn't coming off because my demands were thought too high," he observed, though without regret. "I am happy not to have go there; surely this isn't an agreeable way of making money, and anything but a pleasure."[56] But a week later it turned out that the following spring he would go neither to Leyden, nor to Brussels, but to America after all—not as a "traveler in relativity" but in the service of Zionism.[57] Midway through his life, Einstein, the man without a home, had accepted an obligation which, next to physics, became the most important thing in his life: the knowledge that he belonged to Jewry.

CHAPTER TWENTY-FIVE

Jewry, Zionism, and a Trip to America

AT THE AGE OF SEVENTY-THREE EINSTEIN, reviewing his life, declared that "my relationship with Jewry had become my strongest human tie once I achieved complete clarity about our precarious position among the nations."[1] He began to develop this clarity at age thirty-five, when he came to Berlin, because it was only in the society of the imperial German Reich that "I discovered that I was a Jew, and this discovery was brought home to me by non-Jews rather than by Jews."[2] This is actually something of an overdramatization. Einstein had never forgotten his Jewish origins: his Christian environments had ensured that—more so in Germany, less so in Switzerland. Still, as a young man with bourgeois-liberal views and a belief in enlightenment, he had refused to acknowledge this. From his sixteenth year onward he had described himself as "without religious denomination." National and religious labels seemed to him throwbacks to an earlier age, labels for which he, a natural nonconformist, had no use.

His older friend Michele Besso, who was anything but an Orthodox Jew and, like Einstein, was married to a Gentile, seems to have repeatedly warned him in Bern that one's tie to one's origins is not easily broken, or perhaps is altogether unbreakable. Many years later Besso returned to their conversations on the subject: "It was, I believe, at the entrance to the Kleine Schanze in Bern—probably in 1908 or 1909—that I pointed out to you that your belief that you were detached from Israel could not be right."[3] At the time, Besso had been unable to convince his friend, and he had since been tortured by the idea that his own defense of Judaism and the Jewish family may have been responsible for the fact "that your family life took such a turn, and that I had to bring Mileva back from Berlin to Zurich."[4]

In Prague, similarly, Einstein had shown no interest in the Jewish tradition, the flourishing life of the Jewish community, or the lively discussions about Zionism. For the "comical scene" of his oath of office, as he wrote with some amusement, he "specially made use . . . of his re-assumed Jewish 'faith,' "[5] but this was only a formality. To his surprise, though, he discovered that the only colleague to whom he felt attracted in Prague, the mathematician Georg Pick, was also of Jewish extraction. "*Sangue non è acqua,*"[6] he said, expressing it in Italian because the German concept of "blood bonds" would have sounded too martial to him.

His experience in Prague may have struck a chord in him, for two years later—only five weeks after his arrival in Berlin—Einstein for the first time, and very decisively, avowed his Jewishness. The Academy in St. Petersburg invited him to visit Russia on the occasion of the 1914 solar eclipse, but Einstein rejected it emphatically: "It goes against the grain to travel without necessity to a country where my tribal companions were so brutally persecuted."[7] He had found, not the religion of his forefathers, but a sense of belonging to what he called his "tribe." As the Jewish religion was alien to him, and remained so, he did not refer to coreligionists: this was not the sense in which he belonged to the Jews. Instead, he declared his identity in a phrase that sounded odd even to Jewish ears—"tribal companions."

It was not only Gentiles who led Einstein to discover his Jewishness, but also—and perhaps more so—those Jews who believed in assimilation and integration with their environment, especially academics. While he was still in Bern, working at the Patent Office, he was shocked by the numerous Jewish privatdozents from wealthy families, who, despite being repeatedly bypassed, continued to aspire to a professorship as a token of social acceptance. To Einstein, this seemed undignified and subservient: "Why are these fellows, who make out very comfortably by private means, so anxious to land state-paid positions? Why all that humble tail-wagging to the state? Why does a fellow like Abraham not withdraw proudly? All this is rather advantageous to the public purse. Besides, it's easier to manage with dogs than with wolves."[8]

He resented the lack of pride, self-assurance, and solidarity among the successful Jews, and, to counteract this, resolved to acknowledge his own Jewishness as a profound obligation and to be proud of it:

> It was only when, at age 35, I got to Berlin that I understood the Jewish community of destiny, and that I felt a duty to oppose, as far as I could, the undignified demeanor of my Jewish colleagues. This was a purely emotional reaction and was not based on the fact that substantial portions of our spiritual inheritance may have passed down to me.[9]

He viewed the assimilationist aspirations of the Jewish circles in which he moved in Berlin—including his relatives and his cousin Elsa—as "mimicry": as an attempt, under pressure from the majority, to appear at least as German as the Germans, even though that majority was clearly not prepared to accept Jews as equal members of society. "I watched the undignified mimicry of valuable Jews, and the sight made my heart bleed."[10] His friend Fritz Haber represented an extreme—"the pitiable baptized Privy Councillor of yesterday and today,"[11] as Einstein once put it, though without actually naming Haber. With at least one eye to his career, Haber had had himself baptized a Protestant and, despite his outstanding intelligence, looked and acted like a caricature of a Prussian. Shortly after Einstein's arrival in Berlin, Haber appears to have advised him to have himself baptized a Protestant too: "Do this thing, so that you belong to us wholly and totally."[12]

It is a testimony to the deep respect which Haber and Einstein had for each other that their diametrically opposed views on the fate of the Jews, on Prussian patriotism, and on the war never seriously tested their friendship. Admittedly, they achieved this by largely avoiding these controversial issues. With other people, Einstein could react rather brutally if someone did not match up to his idea of a good Jew. "A pity you had yourself baptized," he once reproved a "tribal companion." "If I had not heard such favorable things about your character otherwise, I would be seriously concerned. As a rule it reflects a preponderance of self-interest over a sense of community."[13]

Most Jews in Germany preferred assimilation—absorption into the German culture, either with or without baptism. To them, religious belief was a private matter. Haber, as noted, was an extreme; but this applied also to Einstein's unbaptized Jewish colleagues, like Max Born. Even the war was hailed as a melting pot: numerous Jews, invoking their Maccabean tradition, threw themselves into battle, fought and suffered, and lost their lives along with their German fellow citizens.

To Einstein, Judaism was not a religion that one could join or leave, but a "community of destiny." The German-Jewish symbiosis, with which the Weimar Republic of the 1920s is in retrospect credited, did not exist for him, even though he was one of those who tried to personify it outwardly. During that allegedly productive phase, he gave this advice to Jewish students:

> We should be clearly aware of our otherness and draw our conclusions from it. There is no point in trying to convince the others of our emotional and intellectual equality by way of deduction, since the root of their behavior is not located in the cerebrum. Instead we should socially emancipate ourselves and essentially satisfy our social requirements ourselves. We should have our own student societies, and practice polite but consistent reserve toward non-Jews.[14]

In February 1919, when Einstein was the most famous physicist in Berlin (though not yet a legend), Kurt Blumenfeld and Felix Rosenblueth asked if they might visit him at Haberlandstrasse. The two were officials of the Zionist movement, and they were looking for prominent Jews who might be willing to support the establishment of a Jewish homeland in Palestine.[15] At first, Zionism was a somewhat remote issue to Einstein, both geographically and because of its nationalist component—the creation of a Jewish state. But when Blumenfeld explained that the concept of Zionism was intended to "give

Jews inner security," as well as "independence and inner freedom," Einstein recognized his own objectives. After some further discussion, he eventually assured Blumenfeld: "I am against nationalism but for the Jewish cause."[16] This was not then a popular attitude in Germany, least of all with influential assimilated Jews.

The struggle for specifically Jewish self-assurance was more important to Einstein, and above all more promising, than the struggle against anti-Semitism. That same year, 1919, he was asked to intervene on behalf of a Jewish mathematician whose application for a professorship had been rejected, presumably because of anti-Semitism. But Einstein did not wish to "act the bow-wow again, or else my barking won't be any use in more blatant cases." He reacted to anti-Semitism with a shrug of the shoulders, almost as if it were a natural law, and he was even prepared to turn the issue upside down: "Anti-Semitism must ultimately be understood as a thing based on real inherited characteristics, disagreeable though it often is to Jews. I could well imagine myself choosing a Jew for a companion if I had the choice."[17] This is not to say that he was indifferent to the fate of Jewish scholars; he merely thought it pointless for too many Jews to compete for the few available positions and thereby cause bad blood among the Germans. He believed that an improvement in the academic job market could come only from Jewish solidarity and self-help: "On the other hand, I would think it sensible for Jews themselves to collect money to provide support and teaching facilities for Jewish researchers outside the universities."[18]

Meanwhile the Zionists were about to establish a Hebrew University in Jerusalem. Hugo Bergmann, Einstein's acquaintance from Prague, who had gone to Palestine, approached him for support, addressing him—and this was before the hullabaloo over relativity—as "the greatest Jewish scientist."[19] Einstein instantly assured Bergmann of his interest in all matters relating to the new colony and "in particular the university to be established there."[20] He was therefore invited to a congress in Basel, in mid-January 1920, where the matter would be discussed. The fact that international fame had caught up with him by the time that date arrived made his participation even more valuable: "I believe that this enterprise deserves keen support. I am going there not because I consider myself a particular expert, but because my name, with its high currency after the English solar eclipse expeditions, may be useful to the cause by encouraging lukewarm tribal companions."[21]

After the war thousands of Jews had flooded into Berlin from the east, from Russia and Poland, fleeing hunger, hardship, and brutal persecution. In a district of wooden shacks behind Alexanderplatz a Jewish shtetl had come into existence, every bit as picturesque, poor, and overcrowded as any in Russia. For the integrated Berlin Jews these impoverished eastern Jews, with their Hasidic customs and religious rites, were embarrassing relatives with whom they wished to have nothing to do. Nationalist German circles were meanwhile demanding the immediate deportation of these people.

In this tricky situation, the newly famous physicist declared in a newspaper article his solidarity with the eastern Jews. Like most middle-class Berlin Jews, Einstein never visited the shtetl district, even though it was just fifteen minutes from the university and the academy; but there was no question in his mind that these eastern Jews were his "tribal companions." In his article, he emphasized the barbarism of the demanded expulsion of these poorest of the poor, warning that any such measure was unlikely to help Germany regain its lost moral credit abroad. He offered the eastern Jews hope that "in the newly developing Jewish Palestine they might find a true homeland as free sons of the Jewish people."[22] Blumenfeld's Zionist appeals had fallen on fertile ground; but to the upright Berlin Jews Einstein, for all their admiration for him, became a sort of black sheep whose actions were just as incomprehensible as his theories.

Soon there occurred an actual conflict with the Central-Verein Deutscher Staatsbürger jüdischen Glaubens (Central Association of German Citizens of the Jewish Faith). In April 1920 Einstein declined an invitation from this association to take part in a meeting on fighting anti-Semitism in academic circles. He had his doubts about the effectiveness of the enterprise, and used the occasion for lecturing the association on Judaism as he understood it:

First of all, anti-Semitism and the servile attitude among us Jews ourselves would have to be fought against by means of enlightenment. More dignity and independence in our own ranks! Not until we have the courage to see ourselves as a nation, not until we respect ourselves, can we acquire the respect of others; or rather, it will then come by itself. Anti-Semitism in the sense of a psychological phenomenon will exist so long as Jews come into contact with non-Jews—and what does it matter? Perhaps we owe it to anti-Semitism that we have survived as a race; I at least believe so.

When I read "of German citizens of the Jewish faith" I cannot avoid a wry smile. What is concealed behind that pretty name? What is Jewish *faith?* Is there a kind of non-faith through which one ceases to be a Jew? No. But behind that name there are two confessions of beautiful souls, to wit:

(1) I don't wish to have anything to do with my poor eastern-Jewish brethren.

(2) I do not wish to be seen as a child of my people, but merely as a member of a religious community.

Is that honest? Can an "Aryan" feel respect for such pussy-footing? I am neither a German citizen, nor is there in me anything that can be described as "Jewish faith." But I am happy to belong to the Jewish people, even though I don't regard them as the Chosen People. Why don't we just let the Goy keep his anti-Semitism, while we preserve our love for the likes of us?[23]

At least Einstein remembered just in time that the members of the Central-Verein were also "tribal companions." To heal any wounds the letter might cause, he added a request in conclusion: "Don't look too angry because of this avowal. It is not meant in a hostile or unfriendly manner." Needless to say, though, there were angry looks, even very angry looks, especially when an excerpt from the letter was published a few months later.[24] Only Zionists could agree with Einstein's views. But it was not so much his views as his fame that led Chaim Weizmann, the president of the World Zionist Organization, to ask Einstein to accompany him on a trip to America.

▪ ▪ ▪

On February 19, 1921, Kurt Blumenfeld came to Einstein, a telegram in his hand from the London headquarters of the World Zionist Organization, signed by Weizmann.[25] Einstein was to be persuaded to participate in America in a propaganda and fund-raising drive for the Jewish Development Fund "Keren Hajessod," especially for the benefit of the planned Hebrew University in Jerusalem. Initially, Einstein was not too pleased, but he yielded to Weizmann's request—which, to Blumenfeld's immense surprise, he seemed almost to regard as a command. Einstein insisted, however, that his wife must accompany him because of his state of health. When Blumenfeld immediately agreed, Einstein added the condition that both on board ship and in the hotels separate rooms be reserved for him and his wife, to enable him to work.

Einstein had no illusions about the role intended for him by the Zionist Organization: "Naturally, I am needed not for my abilities but solely for my name, from whose publicity value a substantial effect is expected among the rich tribal companions in Dollaria."[26] It certainly irritated him to "serve as bait and as something to boast about" but he accepted it in the interest of the cause, as "in other respects I do what I can for my tribal companions who are so meanly treated everywhere."[27] The departure was planned for mid-March. On the return trip he would make a stopover in England, in response to an invitation from the University of Manchester, which had also arrived in February.

Einstein very soon discovered that anything connected with this trip had political implications, mostly unpleasant ones. That was true even of his withdrawal from the Solvay Congress in Brussels, which he himself greatly regretted. German physicists had not been invited, apart from Einstein, who was regarded as Swiss or—as Ernest Rutherford put it—"for this purpose regarded as international."[28] As for his colleagues in Berlin, Einstein found himself between two stools because of his original acceptance of the invitation. Nernst was furious, viewing Einstein's behavior as a breach of solidarity; but Haber, like Einstein himself, had seen the conference as an opportunity to renew international relations, and now regretted Einstein's cancellation.

When the intended trip to the United States and England—the wartime enemies—on the invitation of the Zionist leadership in London, became known in Germany, opposition was universal, especially among Jews. Blumenfeld reported to Weizmann that Einstein's "main interest in our cause was due to his dislike of Jewish assimilation. . . . The trip has earned him outraged letters from German assimilants, which he has passed over with a smile."[29] Ehrenfest, however, understood that Einstein, because of his innermost convictions, could not avoid participating in the "Jerusalem drive into Dollaria";[30] he did not even hold Einstein's sudden cancellation of his activities as a visiting professor in Leyden against him. "You golden person, you angel," Einstein thanked him: "You don't berate me because of my Zionist escapades. Here there is considerable outrage, which, however, leaves me cold. Even the assimilated Jews are lamenting or berating me."[31]

Far more interesting and complex than the snap reactions of conservative Germans were the concerns of assimilated Jews like Fritz Haber. Haber felt that the painfully achieved status of Jews in Germany was threatened by Einstein's fraternization with the enemy and his involvement with a backward movement, Zionism. Their exchange of opinions on this issue, unyielding in substance yet almost caressing in tone, says as much about the two scientists as it does about the opposite poles of Jewish existence in the Weimar Republic.

In a four-page letter, Haber deployed the whole of his eloquence to remind Einstein of his obligations toward the German Jews: "All your deeds, so long as I have known you, have invariably sprung from the nobility of human nature and the goodness of your heart. . . . People's needs [and] deeper concerns have set upon you a deserved crown and invested your activity and passivity with an importance that, in the past, was possessed only by the actions of princes. What you do, you do not only with an effect on yourself." This was not just flattery—Haber's admiration for Einstein, eleven years his junior, was boundless. But it did have an objective: to dissuade Einstein from his trip.

Germans would regard Einstein's trip as treason to their cause,

Haber argued, and the damage would have to be borne by all the Jews in Germany:

> To the whole world you are today the most important of German Jews. If at this moment you demonstratively fraternize with the British and their friends, people in this country will see this as evidence of the disloyalty of the Jews. Such a lot of Jews went to war, have perished, or become impoverished without complaining, because they regarded it as their duty. Their lives and death have not liquidated anti-Semitism, but have degraded it into something hateful and undignified in the eyes of those who represent the dignity of this country. Do you wish to wipe out the gain of so much blood and suffering of German Jews by your behavior? . . . You will certainly sacrifice the narrow basis upon which the existence of academic teachers and students of the Jewish faith at German universities rests.[32]

Haber evidently had this urgent, beseeching letter taken to Haberlandstrasse by his chauffeur, because Einstein replied the same day, spontaneously and in the certainty of doing the right thing. He agreed with his "dear friend Haber" that the timing of the trip was not exactly favorable, in view of the victorious powers' tougher attitude regarding reparations and the peace treaty—which had emerged since he agreed to the trip. Still, he could not change his decision:

> Despite my internationalist beliefs I have always felt an obligation to stand up for my persecuted and morally oppressed tribal companions as far as is within my power. . . . Far more is involved, therefore, than an act of loyalty or one of disloyalty. Especially the establishment of a Jewish university fills me with particular joy, having recently seen countless instances of perfidious and loveless treatment of splendid young Jews, with attempts to cut off their chances of education.

Nor should he be accused of disloyalty to his German friends, as he had turned down other tempting offers from abroad: "This, incidentally, I did not from attachment to Germany, but from attachment to my dear German friends, of whom you are the most outstanding and most benevolent. Attachment to the political edifice of Germany would be unnatural for me as a pacifist." Fully realizing that most of his words would be deeply painful to Haber, he searched for a conciliatory conclusion: "Dear Haber! An acquaintance recently called me a 'wild beast.' The wild beast is fond of you and will call on you before its departure, provided this is possible in this turmoil."[33] Agreement was not possible. Haber might, at best, have consoled himself with the thought that Einstein would prove just as self-willed a partner for the Zionists as he was for the assimilated German Jews.

During the "turmoil" of the preparations for the trip, there were a few conversations with Blumenfeld, during which Einstein was to be instructed in the basics of Zionist policy. Blumenfeld must have found Einstein a difficult pupil. "Einstein, as you know, is no Zionist," he reported to Weizmann, "and I would ask you not to try to persuade him to join our organization." Any idea that Einstein might address American audiences had better be dismissed: "Please be very careful about this. Einstein is a bad speaker and sometimes in his naivety will say things that are unwelcome to us."[34] In fact, Einstein never became a member of any Zionist organization. But, as Blumenfeld realized, he was "always available . . . if we need him for a specific purpose."

Einstein's main reservation about Zionist politics was, and continued to be, his dislike of any kind of nationalism, including Jewish nationalism. While he felt himself to be part of the Jewish people, he could not see why a Jewish state was needed in Palestine. On the other hand, he "confidently believed that, owing to the small size and dependence of their colony in Palestine, they will be preserved from any power mania."[35] At that time some eighty thousand Jews were living in Palestine, fewer than in Berlin and far fewer than in New York, where Einstein was soon to arrive.

The journey began on March 21, 1921. The Einsteins went by train to the Netherlands and there boarded the steamship *Rotterdam*. The Zionist delegation led by Weizmann boarded the ship at Southampton.

15. With Peter Zeeman and Paul Ehrenfest at the University of Leyden, the Netherlands. In 1920 Einstein became a visiting professor at the university.

16. At a tea ceremony in Japan, 1922. During his journey, Einstein learned he had been awarded the 1921 Nobel Prize for Physics.

17. The "turret room" in the Einsteins' Berlin apartment, where the scientist worked and received visitors, 1927.

18. From left: Walther Nernst, Einstein, Max Planck, R. A. Millikan, and Max von Laue, Berlin, 1928. At the end of that year, Einstein finished his paper on the unified field theory. It was submitted to the Prussian Academy of Sciences on January 10, 1929, by Max Planck.

19. From left: Max Planck, British prime minister Ramsay MacDonald, Einstein, German minister of finance Hermann Dietrich, and Privy Councillor Schmitz of I. G. Farben, Berlin, 1931.

20. W. Adams, A. A. Michelson, Einstein, and R. A. Millikan at the California Institute of Technology, where the scientist was a research associate in 1931 and 1932.

21. During Einstein's visit to California, he received a handsome feathered headdress and the title "the Great Relative" from the Hopi.

22. The Einsteins were entertained by Charlie Chaplin in Los Angeles, 1932.

23. Speaking in Berlin's Philharmonic Hall, 1932.

24. With Albert, King of the Belgians, 1933. Einstein and "the Royals" enjoyed a warm friendship, and the scientist and Queen Elizabeth corresponded until his death.

25. With Peter Bergmann and Leopold Infeld at the Institute for Advanced Study, Princeton, 1937.

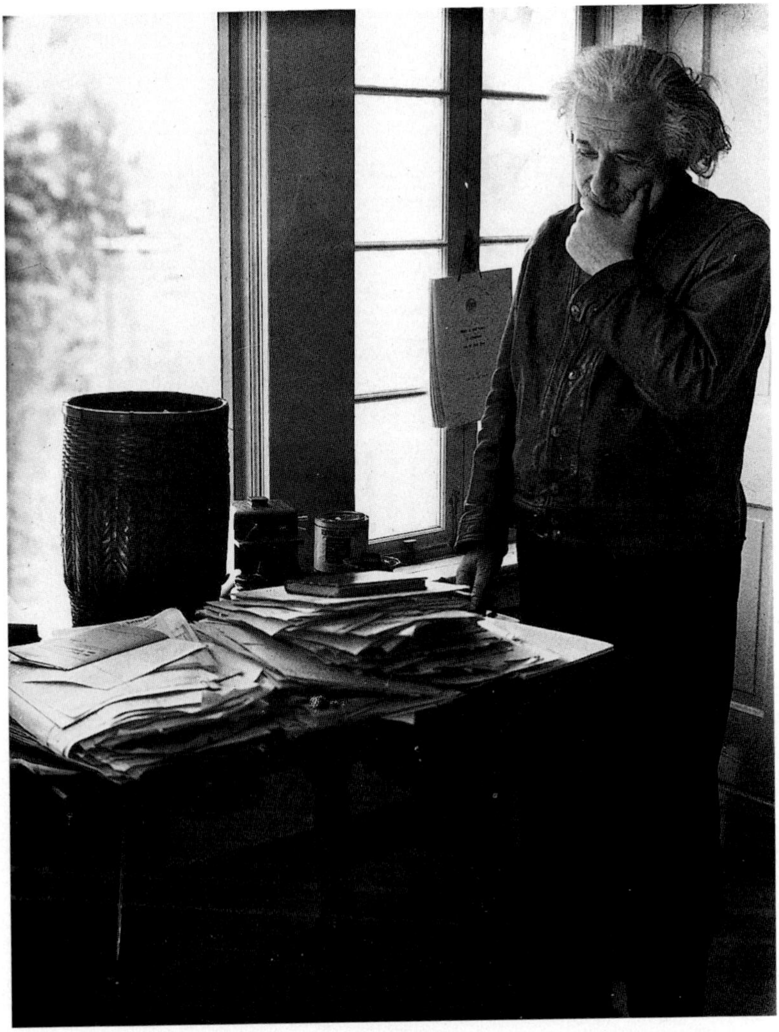

26. Einstein in his study at 112 Mercer Street, Princeton, 1937.

SECOND NEW

r News, Subscriptions
or Advertising
Call ATlantic 6100.

SATURDAY
MORNING,

Pittsburgh P

WORLD of OURS

-by-

Brooks Brokaw

R MEN ONLY

olumn is addressed
sively to men. It
out clothes. Lady
will gather nothing
cept perhaps the hint
n the field of fashion,
still the conqueror.

* * *

is a domain of human
which, during the past
seems to have been ex-
ppropriated and domi-
e female sex. Neverthe-
bout to prove (I think I
spite of the almost uni-
ndardization of men's
the strict unimaginative
aws laid down for their
clothes still do make
More important than
men have been able to
clothes fully express
nalities. In fact, such
us of the male sex in
that there is hardly a
n living today who can
tantly identified by a
ption of some particu-
of dress he is in the
aring.

* * *

CHAPLIN'S shoes for
e, are symbols of a per-
at all the world recog-
Charlie even without
them. From Shanghai
d you can festoon a the-
with a pair of sloppy
if you will, a flexible
ne, and battered plug
ou need not mention
me, or show his face,
e will know who is play-
The entire world fell a
... of the

Constantly Expressive—Einstei

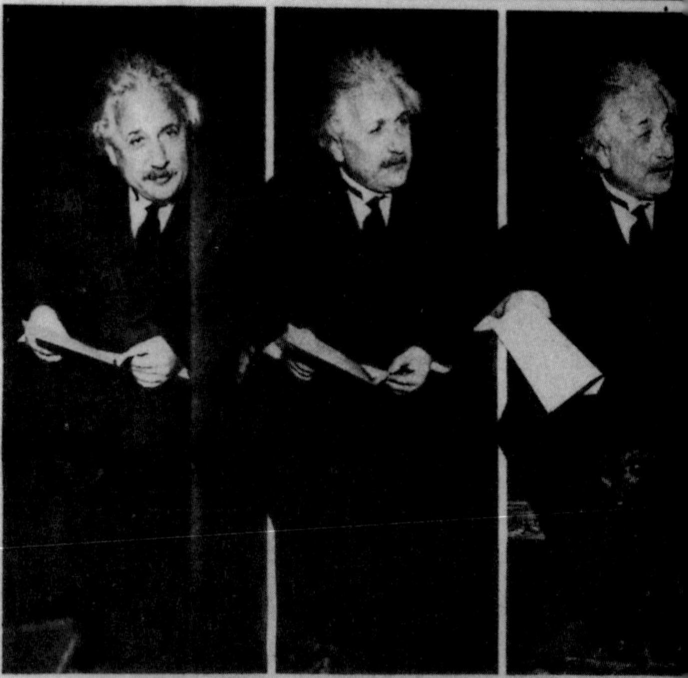

Professor Albert Einstein in a group of characteristic expressions, all moving slowly toward his famous smile in which his face actually lights up as he replies to the questions of his interviewers.

These portrait photographs, taken
newspapermen who crowded aroun
eagerly; next, getting a gleam of

CTION

st-Gazette

DECEMBER 29, 1934.

SPORTS, FINANCIAL, CLASSIFIED SECTION

rms Up to His Subject

—Post-Gazette Photos.

as he answered the 20 ow him first, listening o the question; third, explaining slowly and patiently; fourth, making his point clear; and last—the smile itself. He did not hesitate to say, "I don't know," frequently in reply to queries by his interviewers.

ATOM ENERGY HOPE IS SPIKED BY EINSTEIN

Efforts at Loosing Vast Force Is Called Fruitless.

SAVANT TALKS HERE

Now Indicates Doubt Of Relativity Theory He Made Famous.

Blind chance, or cause and effect—which ever you prefer—may run the universe!

Space may be infinite, or it may be finite, nobody knows!

It may be curved, or not curved, just as you please!

Whatever you decide, no one can contradict you, because no one has so far been able to prove any one of these contentions. Stil, you may be wrong, because some of the contentions may be proved in the future.

That is the contention of Prof. Albert Einstein.

But the "energy of the atom" is something else again. If you believe that man will someday be able to harness this boundless energy—to drive a great steamship across the ocean on a pint of water, for instance—then, according to Einstein, you are wrong now.

Energy of Atom.

The idea that man might some day utilize the atom's energy brought the only emphatic denial from the noted scientist yesterday when he was interviewed by a score of newspapermen at the house of Nathaniel Spear, near

27. In 1934, the *Pittsburgh Post-Gazette* reported that Einstein saw no practical applications for the "energy of the atom."

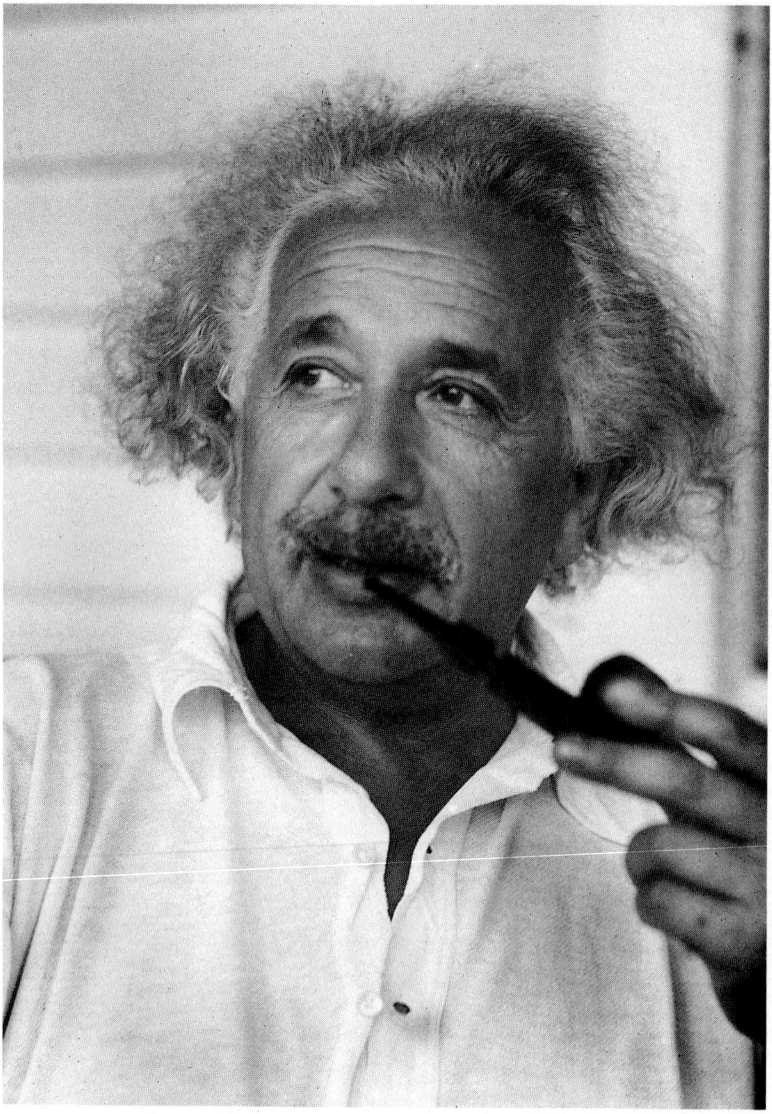

28. In 1938, in collaboration with Leopold Infeld, Einstein published *The Evolution of Physics*.

29. In 1939 and again in 1940, Einstein signed letters to President Franklin D. Roosevelt urging quick action on the part of the administration in organizing nuclear research in the United States.

30. With J. Robert Oppenheimer, then director of the Institute for Advanced Study, about 1947.

31. Honoring Einstein on his seventieth birthday, Princeton, 1949 (from left: H. P. Robertson, E. Wigner, H. Weyl, K. Gödel, I. I. Rabi, Einstein, R. Ladenburg, J. R. Oppenheimer, and G. M. Clemence).

32. With Israeli premier David Ben-Gurion, Princeton, 1951. The next year, following Chaim Weizmann's death, Ben-Gurion offered Einstein the presidency of Israel. Einstein was unable to accept.

33. With Niels Bohr, James Franck, and Isidor I. Rabi, 1954.

34. On NBC television, Einstein comments on the U.S. government's decision to develop the hydrogen bomb, 1952.

Weizmann was five years older than Einstein.[36] Born in a village in Belorussia, he had studied chemistry in Germany and Switzerland, taking his degree in Fribourg at the same time as Einstein left the Zurich Polytechnic with a teaching diploma. After a few years at the University of Geneva, Weizmann found a post at Manchester University and, as a result of some useful patents, became financially independent. During the war he developed important improvements in the manufacture of explosives. At first he worked for the British War Office under Lord Balfour; when Balfour took over the Foreign Office, Weizmann became his adviser on Palestine. His diplomatic masterstroke was his cooperation in the Balfour Declaration of November 2, 1917, which assured the Jewish people of a homeland in Palestine. In 1920 Weizmann became president of the World Zionist Organization. He was an intelligent man of worldwide interests, able to combine idealism with practical politics. Though capable of making speeches in many languages, he continued to prefer Yiddish, his mother tongue.

Shortly before his departure, Weizmann had received a surprising telegram from American Zionists, advising him to jettison his prominent traveling companion. Einstein's exorbitant demands for fees from a few universities had become known also outside academic circles, so that he was expected to prove of questionable value as a fund-raiser.[37] Weizmann, who had been unaware of Einstein's attempt to earn "economic freedom," instructed his American followers to pour oil on the troubled waters; on no account would he dispense with the most famous of all Jews. That Weizmann had made the right decision became apparent as soon as the *Rotterdam* entered New York harbor. "Einstein fever" had erupted.

An impressive welcoming committee, with the mayor of New York and the president of the city council, awaited Einstein; but droves of reporters, photographers, and even the odd film cameraman with massive equipment, stormed past them to board the *Rotterdam* in order to be the first to catch sight of the strange and wonderful man who had unhinged Newton's world. Einstein proved to be an exceedingly attractive and profitable media personality, even though he did not know any English. At one time, in Zurich in 1913, he had begun to learn the language, "slowly but thoroughly,"[38] but little of it seems to have stuck.

The first question put to him was how he would explain his relativity theory in a few sentences. Years of experience with reporters had prepared him for this kind of opening and—"not too seriously and more in fun"—he had his answer ready: "It used to be thought that if all things disappeared from the world, space and time would be left. According to the relativity theory, however, space and time disappear along with the things."[39] This kind of platitude pleased the reporters and, the following day, their readers. Actually, the only really witty remark seems to have been made by Weizmann. Asked whether he understood relativity theory, he is said to have replied: "During our crossing Einstein explained his theory to me every day, and on our arrival I realized that he really understands it."[40] Nevertheless, Einstein was the man of the moment, impressing the reporters with his "geniality, kindliness and interest in the little things of life."[41] This effect was heightened as, to the delight of the crowds, he walked down the gangplank with his pipe in one hand and his violin case in the other: "like an artist—a musician."

The motorcade made for City Hall, in front of which thousands of Jews cheered their idol. Inside, honorary citizenship was to have been bestowed on Einstein, but one of the city councilors had developed last-minute doubts whether this relativity business was not humbug after all, and it took a few days for these doubts to be assuaged and the ceremony to be restaged. Einstein was to be looked after by Felix Warburg, a wealthy partner of the banking firm of Kuhn, Loeb & Company, who had arranged comfortable accommodation at the fashionable Commodore Hotel. One of the first people to call on him there, to Einstein's great delight, was Max Talmud, his mentor from his schooldays in Munich; as Dr. Max Talmey he had become a successful physician in New York.[42]

Almost at once the fund-raising campaign began—sometimes with splendid dinners in exclusive hotels in a circle of wealthy Jews, but more often at meetings attended by the whole spectrum of American Jews, from small storekeepers to rich bankers and fashionable doctors, who wanted to hear Weizmann and see Einstein. "I had to let myself be paraded like a prize bull, and make a thousand speeches at big and small meetings," Einstein reported. Whether he really spoke much at the major meetings is questionable; Weizmann presumably had taken note of Blumenfeld's warning about Einstein's unpredictability as an orator. We do have a report of a meeting on April 12, attended by an audience of eight thousand: following Weizmann's address Einstein confined himself to such a short statement that he could not possibly make a false step—"Your leader, Dr. Weizmann, has spoken, and he spoke very well for all of us. Follow him, and you'll do the right thing. That's all I have to say."[43] At smaller meetings, especially at students' or medical associations, Einstein would appear on his own, particularly when these were concerned with donations for Hebrew University in Jerusalem, a topic on which there was no dissent between Einstein and Weizmann.

Weizmann, however, had come to the United States not only for fund-raising but also in order to get the local Zionist organization to adopt his political line. Louis Brandeis, the first Jew among the nine justices of the Supreme Court, and leader of the American Zionists, saw Jerusalem rather as Einstein did—as a cultural center of Jewry rather than the homeland of the Jewish people, let alone a Jewish state. For the difficult negotiations between Weizmann and Brandeis, or for the administrative work in connection with Keren Hajessod in the United States, Einstein's presence was not needed, and thus he found time to give lectures at a few universities, now for fees left to the discretion of his hosts.

On April 15 Einstein began a series of three lectures at Columbia University in New York, which had first invited him in 1912 and which, the previous year, had awarded him its Barnard Medal. Although Einstein spoke in German, the lecture hall was packed. The following week he lectured at City College of New York—which no doubt gave him particular pleasure, because this city institution enabled many impecunious Jews to study there. His lectures were translated into English paragraph by paragraph; and the young interpreter, who displayed great competence with both Einstein's language and his mathematics, soon acquired the reputation of being one of the few who understood relativity theory.

Einstein and Weizmann next traveled to Washington—Weizmann on political business and Einstein because he was to speak at the annual convention of the National Academy of Science. Despite the official invitation, the professor from Berlin got a rather frosty reception in the American capital. The Zionists had hoped he would receive an invitation to the White House, but this idea had been turned down by President Harding. When it was patiently explained that Einstein had not signed the Manifesto of Ninety-three, and that he was Swiss, traveling on a Swiss Confederation passport, the misgivings suddenly vanished. He was even allowed to accompany representatives of the American academy into the White House for a reception with the president. Nevertheless, the pattern of hospitality was below a politically significant level and was very different from the pomp and circumstance which, two weeks later, on May 20, would accompany Marie Curie's reception by President Harding.[44] But then Marie Curie was French.

There followed a few days at Princeton University, where Einstein received an honorary degree on May 9. The lessons of Washington had been duly learned, and Einstein was introduced from the outset as a Swiss, with a professorship in Leyden, the Netherlands. Naturally, his association with Berlin could not be totally concealed, but the official tribute to the new honorary doctor stressed his loyalty to moral standards, because "he refused to join with others in condoning the invasion of Belgium."[45]

John Hibben, the president of the university, hailed Einstein as a new Columbus of science, who sailed on his own through alien seas of thought. The ceremony suffered from problems with communication, and one newspaper described it as having been stage-managed by the Marx Brothers: "When Dr. Einstein was supposed to sit down he stood, he sat when he should have stood and finally when the smiling, confused scholar, . . . steered about by little tugs at his sleeve, was to receive his doctoral hood from President John Grier Hibben he turned his back in the confusion upon the president."[46]

That afternoon Einstein gave the first of his four lectures on the theory of relativity within the framework of the Stafford Little Lectures. As expected, the lecture hall, one of the biggest in the university, was filled to bursting, not only with students and faculty members, but also with sensation-seekers, many of whom had come from far afield. For those unable to follow German, a member of the physics department gave an English summary at the end of each lecture. But it was just as in Berlin: for the second lecture, the following day, there was plenty of room for anyone interested. These lectures were collected in a little book, Einstein's second, published first by Princeton University Press and then in England by Methuen—in order that it might contribute to the desirable "concentration of Au ions" in Leyden before being republished in Germany.[47]

During a reception in honor of the famous visitor, a rumor suddenly flew through the room that Professor Dayton C. Miller of Cleveland, repeating the Michelson-Morley experiment, had established a movement of the earth through the ether. Publication would follow shortly. Miller had constructed a substantially larger apparatus on Mount Wilson, in the hope that the ether drift could be observed there better than in the basement of a laboratory; and he believed he had succeeded in proving it, in a series of measurements in April 1921. If Miller's result was correct, the whole edifice of relativity would collapse and Einstein's lectures, instead of dealing with an established theory, would be just a beautiful, refuted dream. For reasons of scientific etiquette, Einstein could not very well declare Miller's result to be nonsense, so he produced one of his classical aperçus: "The Lord God is subtle, but malicious he is not."

While Miller's measurements are now of only historical interest—they were wrong—Einstein's remark has been carved in stone. The mathematician Oswald Veblen, who was present, was so impressed by Einstein's reference to God that he immediately jotted it down. Nine years later, when the new building of the mathematics department was completed, he recalled the words and had them engraved, in gothic script, in the chimneypiece of the Fine Hall common room. Along with Einstein's approval, Veblen received his interpretation: "Nature conceals her secrets by exaltedness, but not by cunning."[48]

▪ ▪ ▪

The relaxing and stimulating week in Princeton was followed by an exhausting tour with Weizmann through the Midwest. Even so, it left Einstein time for talks with colleagues. In Chicago, for the first time, he met Robert Andrews Millikan, then the foremost physicist in the United States, to whom Einstein was linked through the experimental side of the photoelectric effect. There was, of course, an obligatory lecture at the University of Chicago and even a side trip to Yerkes Observatory in Wisconsin.

Cleveland was the last stop. Weeks of media hype about the Jewish "new Columbus" had led most Jewish businesses to close for the day, and a large crowd had assembled at the railroad station to welcome the two men. A military band headed a motorcade of two hundred cars taking Einstein and Weizmann to their hotel. At the annual convention of the Zionist Organization, Weizmann succeeded, against Justice Brandeis, in getting his policy adopted. Einstein, meanwhile, managed to escape the turmoil and the intrigues and visit Professor Miller at Case Institute of Technology. They spent "one and a half hours in conversation about the ether drift experiments"[49] before Einstein returned to New York.

In New York, at the end of May, shortly before his departure for Europe, Einstein, rather exhausted but satisfied, drew up a balance sheet in a letter to Besso:

> I have two enormously tiring months behind me, but I also have the great satisfaction of having been of much use to the Zionist cause and of having ensured the foundation of the university. . . . It's a miracle I stood up to it. But now it's done and I'm left with the agreeable knowledge of having achieved something really good and of having battled bravely for the Jewish cause, regardless of all protests by Jews and non-Jews. Most of our tribal companions are more clever than courageous; that I've had ample opportunity to notice.[50]

It soon emerged, however, that as a fund-raiser Einstein had not been quite as successful as he had assumed. Instead of the expected $4 million to $5 million, only $750,000 had come in by the end of the year.[51]

Back in Berlin, Einstein told Ehrenfest about his newly confirmed Jewish identity: "Zionism really represents a new Jewish ideal, one that can give the Jewish people once more joy in its existence. . . . I am very glad I complied with Weizmann's invitation."[52] In the journal *Jüdische Rundschau* he described his strongest impression: "Not until I was in America did I discover the Jewish people. I had seen many Jews, but neither in Berlin nor elsewhere in Germany had I encountered the Jewish people. This Jewish people I saw in America came from Russia, from Poland, or generally from eastern Europe. These men and women have preserved a sound sense of nationality, not yet destroyed by isolation and dispersal."[53] He now had a concrete picture in mind when he spoke of his "tribe."

On May 30, Einstein and Elsa moved into their cabins aboard the British steamship *Celtic.* They were traveling on their own, as Weizmann's mission was not yet finished. On June 8 they arrived in England, the country which, by mounting the solar eclipse expedition, had in such an exemplary manner met Einstein's ideal of the international republic of scholarship. In that spirit, he had accepted the invitation from Manchester University to give the traditional Adamson Lecture.

England, however, did not consist exclusively of internationally minded, let alone pacifist, scientists like Eddington. A year earlier, Einstein was to have received the Gold Medal of the Royal Astronomical Society for 1920, in accordance with a majority vote of the membership. Einstein was informed of the honor and, despite some strident squabbling about the peace treaty among people involved with foreign policy, he had begun making preparations for his journey: "In the spring I'll be going to England to have a medal pressed into my hand and to have a look at the money business from the other side."[54] A few days later, though, he had received a letter from Eddington, who, with profound regret, had to cancel the invitation: the board's confirmation of the award had at the last moment been sabotaged by a chauvinistic lobby. Some British patriots evidently preferred not to award the medal at all, for the first time in thirty years, rather than hand it to someone from the enemy's capital.

Eddington asked Einstein not to take the embarrassing matter personally and regretted "that this promising beginning of a better international attitude has suffered a reverse from reaction. I am nevertheless certain that a better attitude is making progress."[55] This progress took time; it was 1926 before Einstein finally received the Gold Medal of the Royal Astronomical Society, after having been honored the year before, in 1925, with the incomparably more prestigious Copley Medal of the Royal Society.[56] Einstein took no offense; he simply acknowledged that England was much more difficult ground than the United States. Besides, he wanted to give a lecture in London.

For this reason, and because he knew virtually no English, he had asked Erwin Freundlich, whose mother was English, to act as his guide and interpreter. Freundlich took charge of Einstein immediately on his arrival in Liverpool. First of all, Einstein continued his new mission by addressing a Jewish students' association, appealing for support for the Hebrew University in Jerusalem. In the afternoon, in a festive setting, he gave the Adamson Lecture at the University of Manchester and afterward received an honorary degree. Einstein spoke in German, but this failing was more than compensated for by his other qualities. The *Manchester Guardian* remarked that "the excellence of his diction, together with the kindly twinkle which never ceased to shine from his eye through the sternest arguments, did not fail to make their impression upon the audience."[57]

For the next few days Einstein was Lord Haldane's guest in London. Haldane, the first Viscount Cloan, a fascinating man with exceptionally wide interests—which embraced the theory of relativity as well as reconciliation with Germany—was undoubtedly the ideal host for Einstein.

From his student days in Göttingen, Haldane had retained great sympathy for Germany. Despite his left-wing socialist links with the Fabian Society, he first became British Secretary for War; in this capacity he traveled to Berlin in 1912 in a sensational but unsuccessful mission to divert the two nations from their collision course. In 1915, because of his association with Germany, he was no longer acceptable as Lord Chancellor in the War Cabinet, and henceforth concentrated on his scientific studies. These resulted in a book, *The Reign of Relativity*. Although it misunderstood the philosophical implications of the theory of relativity, the book was displayed in bookstores along with Einstein's "generally comprehensible" little volume.

The invitation to London, in all its details, had been arranged while Einstein was in America; it included a lecture at King's College of the University of London. On June 10 Haldane met Einstein at the railroad station and drove with him straight to a meeting of the Royal Astronomical Society, the body which had first awarded him and then disallowed its Gold Medal. In an address Eddington recalled the British contribution to the confirmation and propagation of the general theory of relativity; and Einstein was able, in Burlington House, to see Newton's portrait, in front of which J. J. Thomson on November 6, 1919, had called relativity theory "one of the greatest achievements in the history of human thought." Only afterward did Haldane conduct his guest to his fine home in Queen Anne's Gate, St. James's Park.

In the evening Haldane gave a splendid dinner in Einstein's honor. The former Lord Chancellor would have liked to welcome Lloyd George, but as the prime minister wished to distance himself from a Berliner, the list of prominent guests was headed by the archbishop of Canterbury. Among the intellectual luminaries present were Harold Laski of the London School of Economics and George Bernard Shaw; and with two of the guests—Arthur Stanley Eddington and Alfred North Whitehead—Einstein could even talk physics. However, the conversation appears to have been dominated by the archbishop of Canterbury, who, despite intensive reading, had been unable to discover what relationship relativity theory had to religion; as a result, Einstein became the victim of small talk at a level where the fourth dimension blended into the spiritual.[58] The weekend was more restful. There were dinners with such lordships as Rothschild (the man of money) and Rayleigh (the man of the radiation law, whom Einstein had met at the Solvay Congress in Brussels).

On Monday, accompanied by Haldane, Einstein visited Westminster Abbey and, with a gift of flowers for the tomb of Isaac Newton, paid tribute to the giant upon whose shoulders he stood. This gesture was duly emphasized by Haldane in the afternoon, when at King's College he presided over Einstein's public lecture, in order to take the wind out of the sails of anyone who might feel hostility toward a speaker from Germany. The posters and tickets avoided any reference to Professor Einstein's origin, present residence, or academic posts, while pointing out that the proceeds would go to the Imperial War Relief Fund to mitigate the deplorable plight of war victims. Despite these shrewd measures, not one hand stirred at the beginning of the event to welcome the speaker with applause. But after his speech, given in German with humor, verve, and a warm tribute to Newton,[59] there was a standing ovation. Haldane and Einstein could be more than satisfied with this step toward German-English reconciliation.

In the course of this journey of nearly three months, Einstein had acted not only as a representative of Jewry, but also as an unofficial ambassador of the new Weimar Republic. Back in Berlin, he became a much-sought-after source of information. A report which he gave on June 30 at the invitation of the president of the German Red Cross, in an exclusive setting, was attended by Reich President Ebert and several cabinet ministers. But what Einstein said and wrote on this and similar occasions was received with indignation in America and largely canceled out the success of his trip.

An article which Einstein wrote for *Berliner Tageblatt*[60] about his impressions of America and which was reprinted by many newspapers, contained, along with enthusiastic admiration, some rather critical observations—for example, on Prohibition, the lack of taverns, the overvaluing of money, and political isolationism ("not worthy of that country"). *The New York Times* regarded Einstein's remarks as biting the hand that had fed him; but this may have been due more to American sensitivity at being lectured by the Old World than to the actual content of Einstein's article.

A more serious incident—in fact, a veritable disaster—was set off by an interview with a Dutch journalist, in which Einstein had spoken both recklessly and tactlessly.[61] Not only did he mock the Americans' excitement over a scientist whose work they did not understand; but he also accused American men, even though they were hardworking, of being nothing but "the toy dogs of the women, who spend the money . . . to wrap themselves in a fog of extravagance."[62] *The New York Times* followed its English version of the interview with an editorial in which Einstein's gaffes were attributed to his disappointment at "the failure of himself and his companion to make more than a partial success of the special mission for which they came to the United States." Three days later, after a flood of letters from outraged readers, there came another editorial, with the real reproof:

> Dr. Einstein will not be and should not be forgiven for the boorish ridicule of hospitable hosts who honored him because they believed the guarantors of his greatness in his own domain. That he is small out of that domain is a matter of no great consequence, however, for it is a peculiarity shared by many other specialists of like eminence, and in no degree reduces their value to the world.[63]

Einstein immediately declared that the interview had been wrongly reported and, by way of limiting the damage, said a few pleasant things to American reporters in Berlin. This was partially successful, but the matter of the "toy dogs" was not forgiven for a long time, much less forgotten. Once again Einstein had been taught a lesson on how to speak and how to keep silent while in the spotlight.

CHAPTER TWENTY-SIX

More Hustle, Long Journeys, a Lot of Politics, and a Little Physics

AFTER HIS RETURN FROM AMERICA, Einstein spent just one month in Berlin. Then he met his two sons and with them went vacationing in the small village of Wustrow on the Baltic coast, staying—very modestly—at the local bakery. The contrast with the grand style of his visits to America and England could hardly have been greater, but Einstein was happy with the simple life. Since his conflict with Mileva had been settled, he had enjoyed the company of his sons, even if they had not developed entirely along the lines he would have wished. Hans Albert, now seventeen, had "become a sound, independent fellow, assured, intelligent, modest." Tedel (Eduard), eleven years old, was "lively, roguish"; but both, to their father's regret, had a "somewhat mercantile spirit—without metaphysical needs."[1] Vacationing together evidently did all three of them some good: "They have turned out splendid," he exulted to Ehrenfest, "and we are *one* heart and *one* soul."[2]

In mid-August, straight from their vacation, Einstein and his sons went to Kiel to Hermann Anschütz-Kaempfe to sail on the *Förde* and to work seriously on a technical matter, a novel gyrocompass. Einstein must have been concerning himself intensively with this problem for some time, since by the end of 1920 Anschütz considered his work worth 20,000 marks—in cash, "because otherwise there might be questions of a tax nature."[3]

Because a mechanically mounted compass has certain inherent shortcomings, Anschütz had for some time toyed with the idea of a novel construction. The concept, to which Einstein contributed, was a sphere floating on a cushion of magnetic lines of force, without contact and without friction, held centered in a fixed position by the magnetic field. The arrangement of the electromagnetic fields was in itself a tricky matter. At first experiments were made with a ring of eight electromagnets grouped around the lower part of the sphere, until Einstein conceived the idea of using a single coil located within the sphere.

A great deal of development was necessary before the new compass was ready for marketing in 1925. The extensive correspondence between Einstein and Anschütz, as well as Anschütz's staff, testifies to the passion with which the former Patent Office clerk involved himself in this task. There was always a room ready for Einstein in Anschütz's house, so that Kiel at times became a second home for him—not only because of the gyrocompass but because of his love of sailing and his friendship with the head of the firm. After his first stay in Kiel he added the following remark to an otherwise technical letter: "The wonderful days in Kiel continue to revolve in my head."[4] The young Einsteins also sent Anschütz a polite thank-you letter.

In physics, too, Einstein once more turned to experiment. He was anxious to resolve the fundamental question whether electromagnetic radiation from an atom propagates as a spherical wave or unidirectionally, in the form of what was then called "needle radiation"—that is, as described in his quantum theory of emission formulated in 1916. "I have thought of a very interesting and quite simple experiment about the nature of light emission," he wrote to Born. "I hope I can carry it out soon."[5] Einstein believed that the propagation of light in dispersing media—media with a frequency-dependent index of refraction—must be different for spherical waves and for needle radiation. This difference should be observable in the radiation emitted by fast atoms, and should clarify the true nature of radiation.

This experiment occupied Einstein's attention for the rest of the year, even during his travels. In mid-October he went to Italy with Hans Albert. The occasion was an invitation from the university of Bologna, "where I am to give my lecture in Italian, those poor wretches!"[6] He saw Florence for the first time, but the most important news he had for Besso on a postcard was: "Interesting experiment on light emission in progress in Berlin."[7] Underneath he drew a sketch of the experimental setup. Via Zurich, Einstein went to Leyden, to discharge at least a minimum of his duties as a visiting professor, but also, and more importantly, to discuss with Ehrenfest the quantum problem and the experimental investigation of the nature of light.

Einstein did not have to do the experiment himself. After he had presented its theoretical basis to the Prussian Academy on December 8, 1921,[8] Emil Warburg, as its president, put the equipment of the Physical-Technical Reich Institute, as well as some efficient researchers, at the service of the project. Einstein worked out the details of the experiment with Hans Geiger and Walther Bothe, who performed the work swiftly and successfully. Einstein was delighted when the findings agreed with the calculations he had derived from the hypothesis of needle radiation. He was convinced that the traditional model of wave optics was now finished. "We now have definite proof that the undulation field has no real existence," he informed Born at the end of the year. He was once more euphoric: "This is my greatest scientific experience in years."[9]

On January 19, 1922, Einstein reported the successful outcome to the academy, but he withheld his lecture from publication because he had to acknowledge that objections being raised by Max von Laue were justified. Von Laue had been making these objections for some time, and had convincingly set them out the day before the session of the academy. According to Laue, the experiment itself was correct, but it was not conclusive, since the same result would be obtained for waves as for needle radiation. No distinction could, in consequence, be made between classical theory and quantum theory. Einstein, once again, was "a little wiser, but the poorer by one hope."[10] He reported to Born that he had "made a monumental booboo (experiment about light emission with canal rays). But one's got to live with that. Death alone stops one from making booboos."[11] To Ehrenfest he described his situation as follows: "I suppose it's a good thing that I have so much to distract me, else the quantum problem would have long got me into a lunatic asylum. . . . How miserable the theoretical physicist is in the face of nature, and in the face of his students."[12]

■ ■ ■

Distractions, as Einstein called them, were plentiful. The man who once had been merely a famous physicist was now a social and scientific institution and a new pattern was emerging for his life over the next ten years. He was no longer just a physicist, but a representative of science generally, as well as being a representative of the Weimar Republic and of Jewry—not necessarily in that order of priority. This earned him some hostility, but also a good deal of honor.

At the beginning of 1921 Einstein was accepted into the Pour le Mérite for Science and the Arts, the exclusive Order of Merit limited to thirty holders. Along with Einstein, the Göttingen mathematician Felix Klein, the painter Max Liebermann, and the poet Gerhart Hauptmann were inducted. At the age of forty-two, Einstein was by far the youngest member of this august circle. The Order of Merit was a great honor and fortunately involved little work, though it earned him a reproof from Nernst because, on public occasions, Einstein never wore his insignia.

Rather more work was involved in his membership in the senate of the Kaiser Wilhelm Society, to which he belonged until 1925. In addition there were countless commissions and boards that wanted to boast of having Einstein as a member. Most of these duties he tried to discharge conscientiously, even though deep down he had little interest in them. When Wilhelm Westphal, who was Einstein's junior by three years, resigned his membership on the board of the Einstein Foundation for professional reasons, Einstein, along with his official letter of thanks, regretting his departure, added a few handwritten lines: "As a private individual I am glad that from the sphere of cigar-smoking self-important gods you have found your way back to the sphere of persons active in the field of science."[13]

Nevertheless, when he was in Berlin, Einstein conscientiously and regularly joined the "self-important gods" in the academy, or, by preference, the Physical Colloquium. Young researchers, anxious to win their spurs in the exciting scientific atmosphere of Berlin, found him a practicing democrat, invariably friendly—even cordial—receptive, and helpful. He neutralized the inevitable halo of authority that surrounded by him by humor and self-mockery. In the colloquium he made no distinction between arguing with a student and a privy councilor. The students liked that, though the privy councilors liked it less. But then the privy councilors had to deal with the students, preparing them for a doctorate or *Habilitation*, while Einstein could confine himself to occasional lectures and rare seminars.

With the same relaxed informality, Einstein also moved in circles which, even after the end of imperial Germany, were still called "most exalted." In the case of bankers and industrialists this followed from their generous patronage of the Kaiser Wilhelm Society or the Einstein Foundation. With Walther Rathenau he often discussed political issues before Rathenau, in 1922, became foreign minister, and also during his brief period in office. Einstein had buried his bolshevik ideas since the first few months after the war and was prepared to give the government of the democratic center under Chancellor Wirth his conditional confidence: "After all, the men who now bear the burden of government are not responsible for the present difficult conditions; those responsible are the very people who now criticize most loudly."[14]

He also continued his Zionist work at the highest level by meeting the Soviet People's Commissar for Foreign Affairs, Georgiy Chicherin—who is said to have revered Einstein "like a god"—at his embassy on Unter den Linden to discuss with him the situation of Jews in the Soviet Union and ways of facilitating their emigration to Palestine.[15] What Einstein learned in that conversation merely confirmed his belief in the need for a Zionist movement.

His colleagues, who had been regarding Einstein as the greatest man in their profession, were often aghast when they read in the newspapers of his political activities or views. Some personal relationships broke under that strain.

Sommerfeld in Munich, for instance, was firmly convinced that an interview with Einstein on the front page of the Paris *Figaro* must be "a lie from start to finish" because it showed such a lack of good German behavior.[16] When Einstein read the article, though, he had to disappoint Sommerfeld. He had neither given nor authorized an interview, but the text nevertheless was based on some table talk he had had at a party in Berlin. "I recognize our conversation in the article. It states what I said, only illuminated by French Bengal light."[17] Sommerfeld learned in this way not only why Einstein had originally left Germany and not only that he had returned in 1914 on the condition that he would remain a Swiss citizen, but also his view that the Germans had played a disastrous role in the war and well deserved their defeat. As a postscript in his letter to Sommerfeld, Einstein wrote in the margin: "One should respect an honest person even when he holds different views from one's own." However, this plea proved vain. Sommerfeld did not reply, but Anschütz, who always spent the winter in Munich, reported to Einstein: "As for the way your letter hit Sommerfeld, a bomb is a feeble comparison. He gave it to me in utter despair, despairing of you and of mankind."[18] For the next few years their correspondence almost ceased; the few letters that were exchanged were restricted to trivialities, even in scientific matters.

On the other hand, the *Figaro* interview had a positive reception in France, and this may have induced Einstein's old friend Paul Langevin to renew an invitation to lecture at the Collège de France, originally envisaged for the autumn of 1914. Einstein had already declined invitations from the French League for Human Rights and from the Philosophical Society, mainly in order to avoid political complications, and thus found himself greatly embarrassed by Langevin's friendly gesture. He would have loved to accept it, but his German colleagues were very sensitive about this, and presumably the French were even more sensitive. French scholars were continuing the shooting war as a kind of paper war; they had not forgotten the unfortunate "Manifesto of the Ninety-three" and generally saw to it that the international boycott of German scientists and academic organizations was maintained.

In view of this unpromising situation, Einstein regretfully felt he had to decline the invitation: "I can think of nothing more beautiful that talking with yourself, Perrin, and Mme. Curie in a comfortable little room as in the past, and presenting the relativity theory to your students with a subjective note," he wrote at the end of his long letter, after carefully weighing the pros and cons. "But the great public and politics have long since got hold of me and tried to fit me to their purposes. . . . I would be certain to be asked also about my political views concerning Franco-German relations, and as I could not answer otherwise than truthfully, my answer would not win me sympathies either on this or on the other side of the Rhine. . . . Dear Langevin, it pains me that I cannot comply with your request, because I am fond of you."[19]

However, Einstein could not get rid of the feeling that, by declining the invitation, he had chosen the path of least resistance, whereas his French friend had taken a considerable risk in inviting him. A week later he informed Langevin that he had changed his mind, and why: "Further reflection and a chance conversation with Rathenau convinced me that, despite all the misgivings voiced in my letter, I should have accepted your invitation. In one's efforts gradually to remedy the misfortune of this war one should not be deflected by petty considerations; you and your colleagues have not allowed them to deflect you."[20]

In a carefully formulated letter Einstein informed his colleagues in the Prussian Academy of the invitation and of the reasons which had induced him to accept it, in particular that "this event is intended to serve the restoration of relations between German and French scholars."[21] In spite of all his good intentions, though, he expected that academic circles would take a dim view of his trip. "These circles, he says, are truly dreadful," Count Kessler recorded in his diary. "He is overcome by disgust when he thinks of them."[22]

Einstein arranged with Langevin that he would arrive on March 28. "I am not bringing my wife, because I think that we will be most comfortable alone. The simpler and less official everything is, the better."[23] Other arrangements included keeping his accommodation secret and avoiding private invitations. "Moreover, I want to have absolutely nothing to do with journalists. On the other hand, I would quite like to have a word with one or two politicians if the opportunity arises; maybe something can after all be achieved about the misfortune that is being carried into the world from your beautiful city."[24] Last but not least, he was looking forward to meeting his "good Solo" again, who functioned in Paris as his translator and publishing agent.

For his sake, too, Einstein had "shuffled off all kinds of things, to ensure that we have a little time for existence."[25]

On March 28 Einstein was met at the French frontier station by his friend Langevin and the astronomer Charles Nordmann.[26] When the train arrived at the Gare du Nord in Paris, they did not take the usual exit, in order to avoid the waiting journalists: like a trio of smugglers, they disappeared in the darkness across the tracks. The maneuver must have amused Einstein, but it made sense, since the outcome of his visit was still entirely uncertain. The French Physical Society had firmly refused to meet Einstein, and in the Académie des Sciences a debate on the subject had been ended when thirty of its members threatened to leave the hall in protest as soon as Einstein entered it.

When on Friday, March 31, Einstein began his lectures at the Collège de France the audience had been carefully selected and limited by the way tickets were issued. Paul Painlevé, a renowned mathematician who had also been minister of war and now held the high office of president of the Chamber of Deputies, had operated as a circumspect controller of the admission tickets. Madame Curie was present, as were Henri Bergson and even the Nobel laureate Charles Guillaume, whose objections to relativity theory not even Einstein could comprehend. No Germans, either from the embassy or journalists, had been invited.

Despite some misgivings—"if only my beak were better suited to French"[27]—Einstein lectured in the language of his hosts, with an accent and carefully, but clearly and simply. Langevin sat next to him, to help him, like a prompter, when he was searching for a word, or like a father attending his son's debut on the stage of the world. The subject matter was that of his Princeton lectures, which Solovine was just then translating into French but which had not yet been published.

After a weekend without public appearances, Einstein continued his lectures at the Collège de France, emerged successful from a debate with Guillaume, and on Thursday was the center of a discussion at the Philosophical Society. Press reports ranged from amicable to enthusiastic; attacks by nationalist papers were isolated exceptions.

As the German embassy summed it up in its report, Einstein simply was "a sensation which the intellectual snobbery of the capital would not do without."[28] In the end Einstein even risked an interview. He had also been able to talk to politicians, but only those who held no government office.

With the public his visit was such a success that the German embassy felt it had to warn Berlin not to assume that "Einstein's successful appearances in Paris were proof that in the field of science Germans could once more, untroubled, resume their former relations with French intellectual life and cultivate them on a personal basis."[29] Einstein himself appears to have been more optimistic. To his wife he reported with great satisfaction that she could "hardly imagine the sympathy with which I was met here. In political matters, too, I have found nothing but calm consideration of affairs and good will for understanding, incomparably better than I had expected. Tomorrow we're off by car to the war ruins."[30]

It was Charles Nordmann who, early in the morning on Monday, April 10, picked up the guest with a car. Along with Nordmann, Langevin, and Solovine, Einstein drove through a landscape laid waste, through wrecked towns and villages. Aghast, he stood in front of the fortifications and trenches in which, a few years before, millions of men had lost their lives. After a day of profound shock he boarded the train to Germany in the evening.

He made a detour via Kiel, where he successfully supported Anschütz in a patent suit before the district court and informed himself on progress with the work on the gyrocompass. Back in Berlin he wrote to Romain Rolland that he was "happy that my stay in Paris went so harmoniously, and I entertain the pleasant conviction of having contributed a little to the rapprochement of the minds."[31] He thanked Solovine for "unforgettable days, but damnably exhausting ones, they still stick in my nerves."[32]

When Einstein went to the academy meeting on April 20, he had the same experience as after his American trip. Most of the seats around him remained empty, since the modest attendance enabled the privy councilors to distance themselves from him.[33]

On the other hand, at a German-French friendship rally in the Reichstag on June 10, 1922, Einstein was fêted. This had been organized by the German Peace Cartel, an amalgamation of the New Fatherland League with other pacifist organizations. A like-minded delegation from France attended, led by Professor Victor Basch. "The demonstration was most impressive," noted Count Kessler, the principal speaker for the German side. "Basch, Einstein and I reaped tumultuous applause lasting several minutes."[34]

Two weeks later Einstein had to ask himself seriously if he could stay in Berlin any longer. The foreign minister, Walther Rathenau, had been shot and killed by a reactionary in the street on June 24. This was not the first incident of its kind. Since the end of the war there had been over three hundred murders "from the right" and twenty-two "from the left."[35] To Einstein, Rathenau's death was not only a political warning, but also a painful personal loss and a tragic confirmation of his own view that Rathenau should never have become foreign minister.

It was probably after Einstein's return from Paris that he and Blumenfeld had spent a long evening at Rathenau's home.[36] Blumenfeld had tried to win Rathenau over to Zionism, whereas Einstein would have been satisfied if Rathenau had given up his ministry. He gave his reasons in an obituary: "Given the attitude which a large section of the educated strata of Germany adopts toward the Jews, I believe that proud reserve on the part of Jews would be the natural thing in public life."[37] Einstein had never followed his own maxim, and Rathenau, who wished to achieve great things and perhaps even become a German Disraeli, was not prepared to follow it either. "Rathenau lacked psychological understanding of his own position," Einstein said on one occasion, explaining his misgivings to Blumenfeld. "If he were offered the post of Pope he'd be quite capable of accepting. And let me tell you, he wouldn't make a bad job of it either."[38] Rathenau in fact had been a good German foreign minister, anxious to guide Germany back, cautiously, into the family of nations. This was denounced as the servile policy of a Jew; hence his assassination.

▪ ▪ ▪

Einstein had reason to fear that he too was a target of the death squads. He therefore cut his lectures short and with his wife disappeared into the Anschütz workshops in Kiel—which in fact he had intended to do anyway,[39] in order to be available to Anschütz as an expert witness in a lawsuit, or, as Einstein put it himself, as a "bogeyman."

He now did what he had, in vain, advised Rathenau to do—he withdrew from areas of political conflict. On July 1 he resigned from the Commission for Intellectual Cooperation of the League of Nations. He had been invited to join that body only in May, along with H. A. Lorentz, Marie Curie, and Paul Painlevé, and he was the sole member of a nation excluded from the League. He had accepted after some hesitation, for reasons which were confirmed by the murder of Rathenau: "The situation here is such," he wrote to a friend in Geneva, "that a Jew does well to practice reserve in all public matters. I also have to admit that I have no desire to represent people who would certainly not choose me as their representative and with whose ideas . . . I do not agree."[40]

Next, in a letter to Planck, he canceled a firmly promised lecture at the Naturforscher convention in Leipzig at the end of September: "The fact is that I have been warned, by people to be taken entirely seriously (and independently by several of them), not to stay in Berlin in the immediate future, and especially against making a public appearance anywhere in Germany."[41] He even considered the idea, as he informed Marie Curie, of "resigning my post in the Academy and as Director of the Kaiser Wilhelm Institute as unobtrusively as possible, and to settle down as a private individual somewhere."[42] His preparations for this were already under way.

Anschütz thought it a "great sensation" when his guest confronted him with his intention of settling in Kiel and earning his livelihood, as in the days of the Patent Office, by practical work. "Einstein is weary of Berlin and everything connected with it in terms of visits and official matters, and *horribile scriptu* wants to go into technology. So he first of all asked me if I had any use for him, and if he could be of any value to my firm." To employ the world's greatest physicist in his factory was a prospect which, because of its historical dimension, "almost scared" him: "after all, it is no small matter to stand before all the world as a giant-eater. But I regard it as a very pleasant task to support Einstein in his escape from Berlin into the relative tranquillity here."[43]

Einstein's ingenious idea of ensuring the frictionless rotation of a gyrocompass by locating a coil inside the sphere had been patented by the Anschütz firm[44] on February 22, 1922, and represents his contribution to navigation. There was therefore no question that Anschütz would welcome Einstein "as a very valuable collaborator. He is so delighted with the gyrocompass and works so enthusiastically on all the tricky questions involved in this bold construction that I could not wish for anything better than being able to approach him with my problems at any time."[45] Einstein, for his part, was looking forward to his new life: "The prospect of a downright normal human existence in quietude, combined with the welcome chance of practical work in the factory, delights me. And then, the wonderful scenery, sailing—enviable."[46] Einstein had already considered buying a house, a splendid villa—though with a neglected garden—though he quickly dropped the idea because the Empress Auguste Viktoria had spent several years of her childhood in that house, and he was afraid that the people of Kiel would "regard the purchase of a property with such a history by a Jew as an act of provocation and take their revenge on me."[47]

But in any case, a few days later his intention to move was gone: he informed Anschütz that he would after all stay in Berlin. He used his wife as an excuse: she had a horror of any change and did not feel up to running a whole house any longer. In fact, as Elsa pointed out by way of correction, he had himself come to the conclusion that "this business of quietude [in Kiel] is an illusion. Nowhere can he submerge himself better than in Berlin; in a small town he would be exposed as on a platter."[48]

This hectic episode following Rathenau's murder did have an enjoyable postscript for Einstein. Anschütz, who was moving back to southern Germany and intended to be at his factory only periodically, had a house built on the bank of the Schwentine, near his factory. He would occupy its upper floor whenever he came to Kiel. The lower floor, with a fine view of the garden and the water, was equipped as a permanent refuge for Einstein, complete with a boat at the jetty outside his window, so that he might combine work on the gyrocompass with his hobby of sailing.

Meanwhile, despite the threatening situation, Einstein had returned to Berlin about the middle of July. "Here matters are in turmoil since the hideous murder of Rathenau," he wrote to Solovine. "I am being constantly warned, I have given up my lecture course and am officially absent, though I am really here. Anti-Semitism is very strong."[49] His caution was not of long duration. On August 1, the anniversary of the outbreak of war, he took part in a great antiwar rally in the Berlin Lustgarten. And he even spent a marvelous summer just outside the city.

Einstein had for some years toyed with the idea of buying a sailing boat and a weekend cottage somewhere on the water near Berlin.[50] Sailing and the Brandenburg lakes were to him the best things about Prussia. His money would probably never be sufficient for a country house, so he pursued his dream in a more modest way. He rented a small shack in the garden settlement of Boxfelde.

To Berliners, Boxfelde was "out in the country," though it was actually in the territory of the town of Spandau, which had just been incorporated into greater Berlin. Einstein's plot was on a picturesque bay formed by the Havel River, called Scharfe Lanke, where he could moor his sailboat. The plot was smaller than his drawing room in Haberlandstrasse, and the whole shack was barely the size of his study.[51] Einstein was fond of retiring to his "Spandau castle," where no one could disturb him. The fact that his wife could bear it for no more than two days at a time suited him fine.

Here he put up his sons in the summer of 1922. "The boys are here and reside in my Spandau castle," he reported to Anschütz, who would have much preferred to have him as his guest in Kiel. "I oscillate between my apartment in town and my castle, which has proved more watertight than my yacht."[52] His neighbors in the settlement remember him as a popular and peaceable weekend visitor, though not as a model gardener. He allowed the weeds to take over and the property to become so untidy that in September 1922 the district administration warned him that the plot would be relet unless it was put in order at once.[53] Einstein promised to do better and protested "that we too continue to have the greatest interest in renting the plot."[54]

In the course of the summer of 1922, Einstein allowed himself to be persuaded, mainly by letters from Marie Curie and by the visit of a representative from the Geneva headquarters of the League of Nations, to reconsider his withdrawal from the Commission for Intellectual Cooperation.[55] But he stayed away from the organization's constituent meeting in August.

Nor did he go to the Naturforscher convention in Leipzig, at which the association's centenary was to be celebrated. In his place Max von Laue gave the great keynote address on the theory of relativity. By staying away, Einstein saved himself an experience of intellectual poverty and malice. Unlike the Bad Nauheim meeting two years earlier, when the dispute about relativity theory was still conducted at the level of scientific etiquette, Leipzig became the scene of gross anti-Semitic boorishness.

The tone had been set by Philipp Lenard, when he prefaced the second edition of his foolish but not malicious pamphlet *Über Äther und Uräther* (*On Ether and Primal Ether*) with a "Reminder to German Scientists." In this he inveighed against "the confusion of concepts, hidden from those who knew nothing about race, surrounding Herr Einstein as a German scientist." He got worked up about the "well-known Jewish characteristic of immediately shifting factual problems to the field of personal quarrel," and he called for the cultivation of "sound German spirit." Then "the alien spirit will yield all by itself, the spirit which emerges everywhere as a dark power and which is so clearly seen in anything that relates to the 'relativity theory.' "[56] Lenard's followers distributed appeals outside the lecture rooms in Leipzig, objecting to the fact that the management of the congress had attached too much importance to relativity theory. They therefore called for a counterdemonstration.

This anti-Semitic activity was entirely in the spirit of an obscure politician in Munich, Adolf Hitler, who in the equally obscure *Völkischer Beobachter* had ranted: "Science, once our greatest pride, is today being taught by Hebrews, for whom . . . science is only a means toward a deliberate, systematic poisoning of our nation's soul and thus toward the triggering of the inner collapse of our nation."[57] However, the fact that this lunacy became official policy in 1933 should not lead one to conclude that it was already playing a major role in physics in the Weimar Republic of the 1920s.

Even far-right nationalist physicists, such as Wilhelm Wien, were too intelligent in their field to make themselves and their anti-Semitism look foolish by arguing against relativity. That is why Lenard remained alone with a small handful of quacks. He irritated even politically like-minded colleagues not only by his opposition to relativity theory but also because he felt that on the twenty-fifth anniversary of the discovery of X-rays, he should have been honored as their codiscoverer.

The organized struggle against relativity theory virtually collapsed with the Leipzig convention. Certainly not a single "antirelativist" was appointed to any professorship of physics until 1933—not even Johannes Stark, the Nobel laureate of 1919, who was soon to be Lenard's only ally of any scientific repute. That the author of the theory continued to get into conflicts was entirely due to politics, not to physics.

Einstein was spared the decision of whether he should or could stay on in Berlin. He was to go to Japan and would be away from Germany for the best part of six months. A Japanese publishing house had engaged him for what would now be called a publicity tour.[58] The connection between Einstein and the Kaizosha publishing house was established, without consulting Einstein, by Bertrand Russell. In 1921 Russell was in Japan in the service of this publishing house and its progressive left-wing periodicals. Asked who were the three most significant people alive—that is, who should be invited next—Russell named Einstein and Lenin, without offering a third name. As Lenin was otherwise engaged, the publishing house decided to approach Einstein.

A member of the firm, who was traveling in Europe at the time, was immediately ordered to Berlin, where, an appointment having been made by the Japanese embassy, he called at Haberlandstrasse.

Einstein, who had just returned from America, was again astonished at the interest people showed in him and his theory. During his triumphs in America he had felt "like a cheat, like a con-man, who didn't bring the people what they expected,"[59] he told Count Kessler. Nevertheless, after three conversations in September 1921, Einstein accepted the offer from Japan. He wanted to "see East Asia while the turbulence here continues, he might at least get something out of it."[60] He is unlikely to have regretted the fact that this trip was concerned more with commercial than with political aspects.

In January 1922 a formal contract was signed, under which Einstein would, during a six-week sojourn, give six scientific and six popular lectures in public. The fee, £2,000, was very generous. Even when £700 was deducted for travel expenses, the remaining £1,300 was an excellent deal—so much so that, given the shrinking value of the mark, he could afford to ask the academy to "suspend payment of my salary from October 1 for an indefinite period."[61] The period was "indefinite" because on his return journey he wanted to stop over in Palestine—this had long been an ardent wish of his—and then accept an invitation to Spain.

When Einstein and his wife left Berlin at the beginning of October, this was seen by many as a safety measure following Rathenau's assassination. That was not exactly true, as the contract had been signed earlier, but Einstein was very glad "to have the opportunity of a prolonged absence from Germany, which removed me from a temporarily enhanced danger without my having to do anything that might have been awkward for my German friends and colleagues."[62]

After a few days in Zurich and one day in Bern, the Einsteins boarded the Japanese steamship *Kitano Maru* in Marseilles. Einstein was pleased to find that the passengers were nearly all English and Japanese, "a quiet, refined company,"[63] who would not disturb him. On the six-week voyage he had intended to read several books and do a lot of work, but his stomach rebelled again; despite intensive treatment by a Japanese doctor, a fellow passenger, the condition was not permanently cured. Still, calls at Colombo, Singapore, Hong Kong, and Shanghai were of great touristic charm, even though he thought the poverty appalling and travel in a rickshaw pulled by a man embarrassing: "I felt acutely ashamed to be partly responsible for such a hideous treatment of human beings, but could do nothing about it. These beggars in royal shape pounce in droves on every stranger, until he has surrendered to them. They know how to beseech and to beg so one's heart is wrung by them."[64] His trip may have helped him escape the political upheavals, but not the hullabaloo. Sometimes, when the ship entered port, "*Deutschland, Deutschland über Alles*" was played in his honor—the wrong thing for a Swiss—and often he was caught up in the social events of German associations.

The most beautiful memory of the trip was Hong Kong, both for its scenery and because of its small Jewish community. Although most of the 120 Jews living there had immigrated from Arab countries, Einstein immediately saw them as his "tribal companions": "I am now fairly convinced that the Jewish race has more or less kept itself pure over the past 1,500 years, as the Jews from the Euphrates and Tigris countries are very similar to ours. A sense of belonging together also quite strong."[65]

On November 17 Einstein landed in Kobe; the next day he arrived in Tokyo. The Kaizosha publishing house had made careful preparations and had every reason to be satisfied with the publicity value of its guest. "His journey through Japan was like a triumphal progress,"[66] the German embassy reported. The enthusiasm eclipsed anything he had experienced in America. "At Einstein's arrival in Tokyo there was such a crowd of people at the station that the police could only helplessly watch the dangerous throng."[67]

The following day Einstein gave his first lecture, in front of an audience of two thousand. He spoke in German, interrupted by the translator—the physicist Jun Ishiwara. In 1910, Ishiwara had submitted the first Japanese paper on relativity theory; in 1912 he had worked with Arnold Sommerfeld in Munich and during that time had visited Einstein in Zurich. The procedure was slow, but Einstein is reported to have endured for five hours and to have spoken with great verve, perhaps because he knew that his listeners had paid an enormous entrance fee: 3 yen, enough to buy ten meals.

Einstein's host, Sanchiko Yamamoto, the head of the Kaizosha publishing house, systematically pursued his aim of turning Einstein's visit not only into an intellectual event but into solid profits: by the high cost of admission to the lectures; by a special issue in December of the popular-science journal *Kaizo*, which quickly sold out and was reprinted; and by increasing the press runs of the firm's other publications. The German ambassador recorded, not without pained surprise, that "the entire journey of the famous man has been mounted and executed as a commercial enterprise, and a rather profitable one at that. The contract, in so far as any terms have leaked out, even contained something humiliating for Einstein—he was not allowed to speak publicly outside the scheduled lectures! His learned words, converted into yen, flowed into Mr. Yamamoto's pockets."[68]

Yamamoto had shown some skill in planning Einstein's tour, with lectures and official engagements alternating with pleasant periods of rest, when the Einsteins, superbly cared for, could enjoy the scenery, the culture, and the people.[69] Thus the first lecture was followed by a day at the Imperial Academy, a visit to wonderful gardens, and a kabuki theater performance. The following day, at the traditional Chrysanthemum Feast in the imperial gardens, which celebrates the union between the imperial family and the people, Einstein was not just a guest but the center of attention. "Not the empress, not the prince regent or the imperial princes were the ones to hold court, but, instinctively or deliberately, everything revolved around Einstein." The German embassy has left us impressive accounts of "how the approximately 3,000 participants . . . because of Einstein totally forgot what the day signified. All eyes were turned on Einstein, everyone wanted at least to shake hands with the most famous man of the present day."[70]

This incomparable tribute was followed by the series of six scientific lectures in Tokyo, held, for reasons of fairness, in turn at the great universities of the city, before an audience of some 120 scholars with a knowledge of physics. Jun Ishiwara was invariably present, as was Ayao Kuwaki, who had visited Einstein in Bern in March 1909.

Meanwhile the press campaign continued in top gear, with true, half-true, and untrue stories; with poetical tributes to the great teacher; and with many caricatures. One factor that may have contributed to Einstein's great popularity was that the Japanese characters for "relativity principle" bore great similarity to those for "love" and "sex."[71] More serious matters seem to have occupied the cabinet council. The ministers of Her Imperial Majesty not only argued about whether a layman could hope to understand Einstein's lectures on relativity theory, but even conducted a sophisticated debate on what "to understand" actually meant.[72]

The two weeks in Tokyo were followed by a four-week tour: first to the north, where the visitor was welcomed in Sendai with a heroic poem, "To the Great Einstein"; and next to the south, to Nagoya and the ancient imperial city of Kyoto. Under his contract Einstein had to give only another four lectures, but the publisher Yamamoto persuaded him to give two more, always with the high admission fee of 3 yen and with a lot of publicity. At the University of Kyoto, moreover, Einstein allowed himself to be coaxed into giving an improvised lecture, on December 17, immediately before continuing his journey, on how he had come to create the theory of relativity. The world has reason to be grateful to the Kyoto professors and to his translator, Ishiwara, who took his speech down, as this lecture is one of the few statements in which Einstein described, if only in rough outline, not only the end result, but his struggle with the problems and the background to their solution.

The last stop on the tour was in Fukuoka, on the southern island of Kyushu. There he gave his final lecture, and at Christmas he played his violin at the YMCA. After a few relaxing days, including some at the home of Hayashi Miyake, the doctor who had attended him on board ship, Einstein had a rousing sendoff on December 29, as the "great teacher from the west."

Einstein was glad he had followed the "sirens of East Asia": "Japan was wonderful. Refined customs, lively interest in everything, intellectual naivety but good intelligence—a splendid people in a picturesque land."[73] Surprisingly, the "loner" recorded that in Japan he had "for the first time seen a healthy human society whose members are absorbed in it."[74] He kept a romantically tinted memory of both the land and its people. It was painfully reawakened on August 6, 1945: his train journey had taken him through Hiroshima.

Einstein had found Japan a fascinating experience, but primarily as a tourist. His visit to Palestine on the return trip, on the other hand, was a profoundly moving experience which marked his future life.

He disembarked at Port Said on February 1, 1923, and traveled by train to Lydda—present-day Lod—a small town between Tel Aviv, founded as recently as 1909, and Jerusalem, the three-thousand-year-old city of David. For the first few days the Einsteins were the guests of Sir Herbert Samuel (the future Lord Samuel), the high commissioner of the British-mandated territory of Palestine. Like Lord Haldane, Samuel belonged to the species of British "gentlemen philosophers" who had a passionate interest in relativity theory. Samuel would correspond with Einstein for many years to come, albeit not frequently. At the time, Samuel was the supreme authority in Palestine, and he accorded Einstein the honors of a state visitor, including a salvo of guns to salute his arrival at the high commissioner's residence.

The next day being the Sabbath, Einstein took a simple walk with Sir Herbert along the massive city walls built in the reign of Suleiman the Magnificent, allowed the magic of the Old City to act on him, and in the evening called on Hugo Bergmann, the "serious saint from Prague," who was attempting to build up a library. Not every sight gave Einstein the same kind of pleasure, as his diary reveals: "Then down to the Temple Wall (Wailing Wall), where dull-minded tribal companions are praying, faces turned to the wall, rocking their bodies forward and back. A pitiful sight of men with a past but without a future."[75] Orthodox Judaism with its fixation on Talmudic tradition remained alien to him. The men in their long black caftans, with beards and sidelocks, and with their big hats, were tribal companions—that was beyond any doubt—but whenever he encountered them he reacted with irritation.

Infinitely more attractive to him were the Jewish construction workers, artisans, and farmers he saw over the next few days. They were a living refutation of his initial skepticism, expressed in Berlin, about whether Jews still possessed those skills which were indispensable to the development of their colony. Now he watched Jews who, having come to Palestine from the ghettos of eastern Europe, without any training in practical occupations, were handling bricklayers' trowels as the most natural thing in the world, or were cultivating a soil made fertile only by great effort. Moreover, to Einstein's delight, they were practicing something like socialism—a foreman earned the same wage as his assistant.

It was this practical work, more than anything else, that made him believe in the future of a Jewish Palestine. One of his strongest impressions came from his visit to Tel Aviv, then more of a pioneer settlement than the great city it would one day become, but already called by the Jews their "Little Chicago": "What the Jews have achieved here in a few years arouses the greatest admiration. A modern Hebrew city raised out of the ground, with a lively commercial and intellectual life. An incredibly active people, our Jews!"[76] The honorary citizenship of Tel Aviv meant infinitely more to him than that of New York.

The most important event—which for Einstein perfectly combined emotion and intellect, his Jewishness and his "scientific aspirations"—was the ceremonial inauguration of Hebrew University. The honor had been reserved for him, as there could be no person more fit than the greatest Jewish scientist, who for many of his "tribal companions" was their greatest Jewish contemporary altogether. The university had become so dear to his heart that he had utilized the stopovers on his voyage to Japan to publicize it among Jewish communities, for instance in Singapore.[77] Now, on the afternoon of February 7, 1923, he was standing on Mount Scopus, from where Alexander the Great first saw the Great Jerusalem and where Titus rallied his Roman legions for the destruction of the Temple. In a provisional hall, draped with the insignia of the British Mandate, with flags of the Zionist movement, and with symbols of the twelve tribes of Israel, Einstein was to give the first lecture.

Menachem Ussishkin, the president of the Zionist Executive Council, who had been one of the party on the American trip, put the event in historic context:

> Three thousand years ago King Solomon built a house to the Lord of the world on Mount Moriah, and his first prayer in this house was that this should become a house of prayer for all peoples. Now, as we are building this house, we pray that it should become a house of science for the whole world. Professor Einstein, please step up on the stage which has been waiting for you these past two thousand years.[78]

There was probably no moment in Einstein's life when he regretted more bitterly that, during his Munich schooldays, he had, from "laziness and frivolity,"[79] neglected to learn Hebrew. He had no choice but to have the first sentence of his address translated into Hebrew for him and laboriously learn it by heart: "I, too, am happy to read my address in the country, whence the Torah and its light emanated to all the enlightened world, and in the house, which is ready to become a center of wisdom and science for all the peoples of the east."[80] With apologies that he could not give his address in the language of his people, he continued in French.

During his second week he toured the country in the company of Zionist officials, visiting schools, factories, and a congress of the Histadrut trade union—everywhere hailed and revered for his proud avowal of Judaism and his pride in the Jewish people. He had boundless admiration for the kibbutzim and their struggles against hunger, poverty, and malaria. While not seeing much of a future for the primitive-communist kibbutz experiment, he believed in building up a Jewish society:

> I greatly liked my tribal companions in Palestine, as farmers, as workers, and as citizens. The land, on the whole, is not very fertile. It will become a moral center, but it will not be able to absorb a major part of the Jewish people. On the other hand, I am convinced that colonization will succeed.[81]

After a stop at the picturesque Lake of Genezareth he returned to Jerusalem for one more day, gave a lecture in a crowded hall, and found himself confronted with a question which, unuttered, had accompanied him throughout Palestine: Would he come again, this time to stay? On the final evening he recorded in his diary: "They absolutely want to have me in Jerusalem and are assailing me in closed ranks on that question. My heart says yes, but my reason says no!"[82] He never returned, not even for a visit. He preferred to be attached to Zionism at a certain distance. But he readily accepted a role into which he had grown in Palestine, if not before. He whose people recognize no saints described himself, certainly with a touch of irony, as a "Jewish saint."[83]

The voyage continued to France. From there he did not return straight to Berlin, but first went to Spain for two weeks.[84] Barcelona, Madrid, Toledo, and Zaragoza were stages on his tour, during which academic lectures, addresses, and honors of every kind alternated in rapid succession. Einstein was admitted as a corresponding member of the academy, was given an honorary degree, and was received by the king. The German ambassador reported, without exaggeration, "that never in human memory has a foreign scholar met with such an enthusiastic and extraordinary reception in the Spanish capital."[85]

By the time he returned to Berlin in the middle of March, after an absence of just under six months, he had been awarded the Nobel Prize and, as a curious consequence of that distinction, had officially become a German and a Prussian.

PART VI

UNIFIED THEORY IN A TIME OUT OF JOINT

CHAPTER TWENTY-SEVEN

Einstein Receives the Nobel Prize and in Consequence Becomes a Prussian

In September 1922, when the tickets for the trip to Japan had long been ordered, Einstein received a letter from Svante Arrhenius, a prominent member of the Royal Swedish Academy of Sciences in Stockholm and chairman of the committee responsible for the Nobel Prize for Physics. Arrhenius had come to Leipzig to the Naturforscher convention, learned there about Einstein's travel plans, and with rather obvious hints endeavored to make him change his mind: "It will probably be very desirable for you to come to Stockholm in December, and if at that time you are in Japan that will be impossible."[1] In Leipzig Arrhenius had evidently informed Max von Laue about the state of affairs, which Laue promptly related to Einstein: "According to reliable news which I received yesterday, events may be taking place in November which would make your presence in Europe in December desirable."[2]

Einstein instantly realized that this concerned the awarding of the Nobel Prize on December 10, but even though Arrhenius, in the cryptic manner of academy members—who are sworn to secrecy on matters of the prize—hinted that Einstein's absence might jeopardize the vote in the academy, Einstein did not hesitate for a moment: "As my contract irrevocably ties me to Japan, I am quite unable to postpone the journey." He saluted his Swedish colleague "in the hope that the invitation envisaged for me will, as a result, only be postponed and not canceled."[3] His absence did not matter. While Einstein was somewhere between Hong King and Shanghai, en route to Japan, it was announced in Stockholm on November 9, 1922, that the Nobel Prize for Physics had been awarded to him.

We do not know when Einstein was informed of this honor; it certainly did not merit an entry in his travel diary. Nor do we know his spontaneous reaction to the apparent motivation behind the wording of the award: "in recognition of his merits for theoretical physics, more particularly for the discovery of the law of the photoelectric effect." Reference to an achievement dating back to 1905 saved the Swedish Academy from a delicate situation; by then the academy needed Einstein as a laureate more than he needed the prize—except, of course, for the money that went with it. The members of the academy had realized for a number of years that Einstein had to be given the prize; they were less clear about what he should receive it for.

Einstein had first been nominated for a Nobel Prize in 1910, by Wilhelm Ostwald, to whom he had once unsuccessfully applied for an assistant's post. Ostwald based his nomination on relativity theory. The committee for the physics prize, whose five members had to examine and evaluate the proposals before passing one of them on to the physical class—which in turn would submit it to the academy as a whole for a vote—recommended that Einstein's name should be set aside pending experimental confirmation of his theory.[4] This was not, at the time, unreasonable.

After 1912, Einstein's name was repeatedly put forward, not only for his relativity theory but also for his statistical work on the Brownian movement, and later also for the photoelectric effect.[5] After 1917 Einstein was being nominated by an ever-growing circle of those entitled to make nominations, but there was no coordinated campaign. From that year onward the nominations also listed the general theory of relativity and the perihelion of Mercury.

The theory of relativity gave the committee considerable problems, for a number of reasons. A principle or a theory certainly was not a "discovery" in the sense of Alfred Nobel's bequest. Nobel's directions, and the experimental positivism of the trend-setting physicists at the University of Uppsala, were significant. In 1908, the committee's proposal to award the prize to the theoretician Max Planck had been rejected by the physical class.[6] Because the committee was reluctant to trigger a similar conflict again, theoretical concepts were, for a long time, outside consideration. (Not until 1919 did Max Planck receive the prize for 1918; those who nominated him included Einstein.) Preference was given to "effects," such as those of Zeeman or Stark, and occasionally to practical matters such as an improvement in the intensity of a lighthouse beacon (1912) or a steel alloy (1920). Besides, there was no expert on the committee who might have explained the importance of relativity theory to his colleagues.

After the spectacular results of the solar eclipse in 1919, it became inevitable that relativity theory would be short-listed. In 1920, the chairman, Svante Arrhenius, personally undertook to prepare a detailed expertise for the committee. But Arrhenius, though a highly intelligent physical chemist (who had justly been honored with the prize for chemistry for 1903), was not quite up to this task. Uncritically, he listed all the objections found in the literature—not only entirely justified criticism but also the absurd statement by Ernst Gehrcke from Berlin that the precession of Mercury's perihelion had long been explained independently of relativity theory. As a result, a Nobel Prize went not to Einstein but to Charles Guillaume, for his stainless steel alloys.

The following year, Allvar Gullstrand was charged with the expertise on relativity theory. Gullstrand was a physiologist who had contributed so much to the understanding of the human eye as an optical system that in 1910 he could have received not only the prize for medicine—as in fact he did—but also the prize for physics. In the theory of relativity, however, he was no more than an eager dilettante, and his essay in consequence contained major misunderstandings.[7] The committee, confused, recommended that the awarding of the prize for 1921 be suspended until the following year.

In 1922 the number of nominations of Einstein had continued to increase. Again, the expertise was performed by Gullstrand, who did not discharge his task any better than the year before. However, one of the committee members, Carl Wilhelm Oseen, professor of physics at

Uppsala, came up with the idea of basing Einstein's nomination on his photoelectric effect, especially since Einstein's formula had by then been experimentally confirmed by Millikan. Oseen was therefore instructed to prepare a second expertise, on that subject. This turned out to be so brilliant that the committee recommended that the prize for physics for 1921 be awarded to Albert Einstein for his discovery of the law of the photoelectric effect. The proposal was adopted first by the physical class and, on November 9, by the plenary meeting of the Swedish Academy. The prize for 1922 went to Niels Bohr, "for his researches into the structure of atoms and the radiation emitted by them."

In judging the theory of relativity the committee did not exactly distinguish itself—though fortunately no one knew that at the time. Still, its rationale for choosing Einstein was by no means a mistake. His explanation of the photoelectric effect by the "heuristic viewpoint" of light quanta was certainly worth a Nobel Prize, and Arrhenius rightly pointed out in his citation that "an extensive literature has sprung up in this field, testifying to the exceptional value of this theory. Einstein's law has become the foundation of quantitative photochemistry, in the way that Faraday's law is that of electrochemistry."[8]

That the rather conservative Swedish Academy should have honored Einstein for the one paper which he himself had called "very revolutionary" in an amusing, ironical touch. All the same, the committee did show a sense of historical perspective by its successive awards to Max Planck in 1918, to Einstein and Bohr in 1922, and then to Millikan in 1923—a kind of recapitulation of the exciting story of early quantum theory.

Bohr immediately alluded to the symbolism of this sequence when, in his touchingly awkward German, he observed in his congratulatory letter to Einstein "that—quite apart from your great input into the world of human ideas—your fundamental contribution . . . should also outwardly be acknowledged before I myself was considered for such an honor."[9] Einstein, on board ship returning from Japan, replied: "Dear, or rather beloved, Bohr! Your cordial letter reached me shortly before my departure from Japan. I can say, without exaggeration, that it gave me just as much pleasure as the Nobel Prize. I find your fear that you might have got the prize before me particularly charming—that is real Bohr."[10]

On his return to Berlin, Einstein learned that some investigations ensuing from the award of the Nobel Prize had shown that he was a German and a Prussian. The question of Einstein's citizenship had arisen in Stockholm because a laureate was always accompanied, both at the presentation ceremony and at the succeeding festivities, by the diplomatic representative of his country. Indeed, should a laureate be unable to receive the prize in person, his place would actually be taken by his country's representative.

In Einstein's case, both the Swiss and the German minister claimed that privilege. In response to inquiries by the German minister, Rudolf Nadolny, the Berlin Academy cabled, sure of victory: "Einstein is a Reich German." His Swiss colleague was surprised, because Einstein invariably traveled on a Swiss passport, but withdrew gracefully.[11] An embarrassing moment arose when, on the day after the presentation, the Foreign Ministry in Berlin came through with a contradictory finding, to the effect that Einstein was after all a Swiss national. To avoid a diplomatic disaster, evidence had now to be found in Berlin that the German minister had in fact been Einstein's only legitimate representative.

The lawyers of the Berlin Academy put their heads together and discovered that on July 1, 1920, Einstein had taken an oath on the Reich Constitution and nine months later also the oath on the Prussian Constitution. Moreover, by taking up his post at the Prussian Academy he had become an "indirect state official" and hence a German citizen, even though he had not expressly given up any other nationality.[12] Upon his return Einstein immediately informed the Academy that he had expressly *kept* another nationality when he was invited in 1913: "As I attached importance to no change being made in my nationality, I made my acceptance of a possible invitation dependent on the fulfillment of this condition, which in fact was complied with. I have no doubt that this state of affairs can be verified from ministerial files; I also know that this state of affairs is known to my colleagues Haber and Nernst."[13]

In fact, Einstein had always been regarded as a Swiss national by the German authorities. Thus before his appointment to the board of the Physical-Technical Reich Institute the question had been raised whether a foreigner could be a member of that body, since the Reich Institute sometimes worked on confidential problems of relevance to the military. In the petition to the Kaiser, although these misgivings were played down, the point about his citizenship is clear: "Einstein is a citizen of Switzerland, but this fact should prove no obstacle to his invitation to the Board."[14]

Einstein, therefore, was so certain that he was Swiss and nothing else that he asked for his medal and scroll to be sent to him through the Swiss legation in Berlin. However, these had already been handed over to the German minister, Nadolny. The Nobel Foundation found a tactful way out by having them presented to Einstein, with Nadolny's agreement, by the Swedish minister in Berlin, Sten Ramel.

However, a prolonged search of the files and an extensive correspondence revealed nothing about an arrangement concerning Einstein's nationality dating to 1913 or 1914, let alone about any release from German nationality. Clearly the authorities, and Einstein himself, had simply forgotten to settle this issue definitively. The legal situation resulting from all this was explained to Einstein in June 1923 by a senior official of the Ministry of Education. Einstein evidently had not given up hope that something would emerge from the files, for he requested that his "possible Prussian nationality not be made public to outsiders."[15] Six months later, when nothing new had been discovered, Einstein finally, in a document for the Prussian Academy dated February 1924, declared that the relevant ministerial counselor in the Ministry of Education "firmly holds the view that my employment in the Academy had entailed acquisition of Prussian citizenship as nothing contradicting this view emerges from the files. I have no objection to this view."[16] Nor did anyone have any objection to the fact that Einstein retained his Swiss nationality alongside his German nationality. Nevertheless, in the wake of his Nobel Prize it had been definitively established that he was a German and a Prussian, and had been so since 1914.

The medal and scroll of the Nobel Prize were handed to Einstein in Berlin, but not the prize money, which, because of the German currency restrictions, he had asked to have transferred directly to Switzerland. His only obligation under the statutes was a lecture to be given by the laureate in Sweden. Arrhenius suggested to Einstein that he not wait until the following Nobel event in a gloomy Swedish December, but rather discharge his duty on the occasion of the Scandinavian Scientists' Convention in Götenborg in July. Although the laureate's lecture customarily dealt with the subject for which he had been honored, Arrhenius gave Einstein freedom of choice. However, "it is certain that one would be exceedingly grateful for a lecture on your theory of relativity."[17]

Thus it happened that even after his return to Berlin, Einstein continued to be on the move. He spent May as a visiting professor in Leyden and went to Sweden in July. On July 11 in Göteborg he gave his lecture, *Fundamental Ideas and Problems of the Theory of Relativity*, to an audience of two thousand which included the king.[18] He would have preferred to speak on his current reflections on a unified field theory, but he regretted "that my new theory on the essential unity of gravity and electromagnetism cannot be shaped in a popular manner."[19] He therefore presented unified field theory to a small circle of experts at the Technical University.

On his return journey he made a detour to Copenhagen to see Niels Bohr. His host met him at the railroad station. As always, the two were instantly engrossed in talk of physics, but this time they completely forgot the world around them: "We took the streetcar and talked so animatedly that we went much too far," Bohr related many years later. "We got off and traveled back, but again we rode too far. I don't know by how many stops, but we rode to and fro by streetcar, because Einstein at that time was really full of interest. I don't remember if his interest was more or less skeptical—but in any case we rode to and fro on the streetcar, and I can well imagine what the people thought about us."[20]

The sum of 120,000 Swedish crowns associated with the Nobel Prize—increased by interest accrued in the meantime, which led Einstein to thank Arrhenius for his "capitalist care"—was transferred after his lecture directly to Switzerland. What should have marked the end of all financial worries triggered a frightful row. Mileva and his sons were disappointed that the money (approximately 180,000 Swiss francs), in accordance with the divorce court ruling, was deposited in an inaccessible trust account, with only the interest at their disposal. Einstein complained that Hans Albert had "on the occasion of the arrangement of the N. Pr. written such an ugly and arrogant letter that I cannot meet with him this year."[21] The embittered father could do nothing but cancel their planned joined vacation. "And the wife also doesn't write like someone who knows that at times one has virtually given [her the shirt off one's back]."[22]

Sensible men like Heinrich Zangger and Hermann Anschütz intervened in these delicate family matters, acting as mediators, and managed at least to settle the dispute with Hans Albert. At the end of August 1923 Einstein with both sons was Hermann Anschütz's guest at Lautrach Castle in southern Germany, which Anschütz had purchased and made habitable. Subsequently the reconciled father spent two weeks in September with Hans Albert in Kiel, in the small flat Anschütz had provided for him as a permanent refuge. "I am again completely reconciled with Albert," Einstein reported from Kiel. "I am . . . in my hidey-hole here with him at the Anschütz factory, where we are having a wonderful time, and are able to make music together and sail."[23]

It took a lot longer for his anger with "the wife" to evaporate. Mileva had to write to Haber repeatedly before, through his intervention, Einstein stopped being annoyed with her. Eventually the Nobel Prize money was used to buy three multifamily houses on the Zürichberg; the rent income from them was to ensure the permanent financial security of Einstein's first wife.

▪ ▪ ▪

From Kiel, Einstein went directly to the physicists' convention in Bonn. This was the first time since the unfortunate event in Bad Nauheim three years earlier that he took part in a congress in Germany. In fact, he was participating even though the convention had been planned not purely as a scientific meeting but also as a patriotic event. Since it was taking place in the Rhineland, into which (like the Ruhr) French and Belgian troops had marched at the beginning of 1923 to ensure German reparation payments, the physicists intended their convention as a demonstration against the brutal policy of "mortgaged production capacity."

Because of that aggressive occupation, which represented the final factor in the economic catastrophe of galloping inflation, Einstein, immediately upon his return from Japan, had, for the second time, resigned from the League of Nations Commission for Intellectual Cooperation. Risking approval by the public and by false friends from the right, he proclaimed his conviction "that the League of Nations has neither the strength nor the good will to implement its great task. As a serious-minded pacifist I do not therefore consider it right to be in any way connected with it."[24]

As for the Solvay Conference of 1924, he did not even want to be invited, both because of the boycott of his German colleagues and because of the occupation: "If I took part in the congress," he wrote to Lorentz—justifying his refusal, but surely with a heavy heart—"I would in a sense aid and abet a course of action which I strongly and painfully feel to be an injustice." He also believed that "the French and the Belgians have over the past few years sinned so much that they no longer represent outraged innocence."[25]

With his tendency to sympathize with the underdog, Einstein now felt himself at one with the gagged Germans. Since he was prepared anyway to reconcile himself to the inevitable fact of his Prussian and German nationality, he went to the convention at Bonn; but he was unable to take its intention of being "a moral equivalent of a naval demonstration against France" quite seriously: "The wolves cannot slough off their skins, and one's got to howl with them in a comradely manner."[26]

Einstein had not registered to give a paper in Bonn, but he found the event "most interesting, and I've had my say now and again in the discussions."[27] His colleagues gratefully acknowledged not only his contributions to the discussions, but even more so the solidarity he showed by being present. The political animosities of earlier years had vanished, and so had the scientific controversies: "I am treated like a saint, though I don't feel too comfortable in that garb,"[28] he reported before the convention was even over.

Following the meeting in Bonn, Einstein spent two weeks with Ehrenfest in Leyden. As always, this was a very stimulating time—but not nearly long enough for his friend or for his own guest performance as a visiting professor. Soon, though, he was to meet his obligations with a longer stay of six weeks, as a result of what was happening in Berlin.

Einstein had been back in the capital for exactly three weeks when he received a warning that his life was in danger. On November 7, therefore, he took to his heels and returned to Leyden. A letter from Max Planck, addressed to Einstein in Leyden, is our only source—and then only in hints—for the "ghastly events" which had caused Einstein's precipitate departure. The coincidence of the dates, however, suggests that the threat of an attempt on his life might have been connected with the Beer Hall Putsch in Munich, where Adolf Hitler hoped, by marching on the Feldherrnhalle, to overthrow the republic.

Einstein had attracted the hatred of extreme right-wing circles not only by his fame but chiefly by his continuous commitment to the left of the political spectrum. Not only was he active in the League for Human Rights, which had developed from the New Fatherland League, but in 1921 he had taken part in the foundation of a Society of Friends of the New Russia, even becoming a member of its central committee.[29] In the autumn several newspapers reported that Einstein was planning a trip to the Soviet Union;[30] on October 6 *Berliner Tageblatt* announced that he had already left for Moscow, and at the beginning of November there was a story that he had spent three days in St. Petersburg.[31]

However, Einstein was actually in the Netherlands and in Berlin.

Neither in 1923 nor at any later date did he visit the Soviet Union; he preferred to show his sympathy for the communist experiment from a safe distance. It is uncertain whether Einstein had in fact ever toyed with the idea of a trip to the workers' paradise, or whether the newspaper stories were based only on the hopes of Soviet institutions. False or not, these reports of Einstein's flirtation with diabolical bolshevism were evidently sufficient for him to find himself once more, after the murder of Rathenau, a target of right-wing extremists.

Planck, while disapproving of his colleague's political activities, was nevertheless alarmed "that you've left and, as your wife tells me, have no wish to come back to us." He sincerely urged Einstein "not to take any step now that might make your return to Berlin impossible finally and for all time. No doubt a lot of tempting offers and invitations will come your way now, because foreign countries have long envied us for having this precious treasure. But please think also of those here who love and revere you, and don't let them suffer too much for the abject infamy of a vicious pack of dogs whom we must get under control."[32]

Planck involved the Berlin authorities, even initiating a scrutiny of the files of notorious psychopaths with an inclination to death threats, but he was unable to discover any concrete evidence of the origin of the threat to Einstein. He implored Einstein—directly and through letters to Ehrenfest and Lorentz—not to sever his relations with the Prussian Academy; he could do or not do whatever he thought necessary, but he should retain his official residence in Berlin and participate in the life of the academy to the extent of at least one lecture each year.[33]

Einstein does not seem to have viewed the affair in quite so dramatic a light as Planck. "Your kind letter to Ehrenfest gave me a great deal of pleasure," he wrote from Leyden on December 6. "For some time now there [has] no longer [been] any reason why I should persevere in this (really quite pleasurable) exile."[34] He intended to give a few more lectures to students in Leyden and return to Berlin shortly before Christmas. Then, abruptly, he referred to an enclosed paper—as "a sign of life"—for the *Proceedings of the Academy*, just as if the whole business had indeed been nothing but a "pleasurable exile."

There was clearly no room for fear in his personal makeup, and no sooner was he back in Berlin than he wrote to Michele Besso that he had "experienced a lot in the meantime. But outward experience remains on the surface, and the main thing is science."[35]

After three eventful and exciting years, Einstein was now anxious to steer his life into calmer waters, which would also be more conducive to reflection. To that end Berlin, for whatever reasons, seemed to him the most suitable place. The political situation in Germany had begun to calm down. With the introduction of the *Rentenmark* in the fall of 1923, and the resultant end to the inflation, his own modest monetary assets—and his wife's considerable assets—had melted away; but at least his foreign income in accounts in Leyden and New York had been saved from devaluation. The German economy showed signs of recovering, though the "golden 1920s" were to last no more than half a decade.

He was unwilling to embark on foreign journeys in the near future. He had explained to Weizmann that he would always be ready to support the Zionist cause, and more especially Hebrew University, with appeals, letters, or addresses, but only if he did not have to travel or attend congresses.[36] He declined an invitation from Millikan, who, as president of the California Institute of Technology in Pasadena, was planning to develop it into an elite institution. He also declined to tour South America—and he declined countless invitations to attend conferences and give lectures in Germany and elsewhere in Europe. He confined himself to a few trips to Leyden and Switzerland, and of course to his refuge at the Anschütz factory on the Kieler Förde. From there he wrote in May 1924: "Political conditions have calmed down, and, thank God, the far-too-numerous don't worry too much about me any longer, so that my life has become more tranquil and undisturbed."[37]

However, at the urging of Madame Curie and his revered H. A. Lorentz he again joined the Commission for Intellectual Cooperation and even attended its fourth session in Geneva, where on July 25, 1924, he was introduced to the distinguished circle with a eulogy by Henri Bergson. Einstein approached international understanding in an entirely pragmatic manner. His first concern was to standardize the terminology of physics and chemistry, and he supported the creation of an international meteorological office. "The League of Nations Commission was better than I had thought," he wrote after his stay in Geneva. "There's hope after all that things may get better in Europe."[38]

Berlin certainly had enough stimulation to offer Einstein, scientific as well as social. For several years, he had been considered an ornament to any salon or party in the worlds of politics, finance, and the arts. Typical of many was a banquet of bankers at Berlin's famous Kaiserhof hotel, even though Einstein felt a bit of an outsider there. "About a hundred 'prominent figures' from the world of politics, banking, and the intelligentsia," noted Count Kessler, with whom Einstein had spoken most of the time; "a mixture of capitalism and socialism, mostly on a Jewish basis."[39]

Einstein would often discuss political matters with Gustav Stresemann, the foreign minister; he was an acquaintance of Gerhart Hauptmann, the famous dramatist; he was a frequent guest at the house of Samuel Fischer, the publisher; and he indulged his love of music through a close friendship with Erich Kleiber, the conductor. Many artists considered it a privilege to paint the famous man's portrait, and Einstein fondly remembered his sittings at the apartment of Max Liebermann, whose incomparable sense of humor he admired even more than his skill as a portraitist: "I found that his picture was more like him than like me, but that actually improved it."[40]

Unlike the picture of the old man at Princeton with his chaotic mane of hair and his careless Chaplinesque attire, Einstein in midlife was an attractive, impressive man, whose features, eyes, speech, and mere presence aroused and indeed compelled attention. "One's first impression is that of astonishing youthfulness," wrote Charles Nordmann when he met the forty-three-year-old Einstein on his visit to Paris:

> Einstein is tall (about 1.70 m), with broad shoulders and a scarcely bent back. His head—the head in which the science of

> the world was newly created—instantly attracts lasting attention. . . . A little mustache, dark and very short, adorns a sensuous mouth, very red, rather big, with its corners betraying a permanent slight smile. But the strongest impression is that of stunning youthfulness, very romantic and at certain moments irresistibly reminiscent of the young Beethoven who, already marked by life, had once been handsome. And then suddenly laughter erupts and one is faced with a student.[41]

That such a man impressed women, and let himself be impressed by them, is hardly surprising. Vera Weizmann, who met him on the crossing to America, described him as "young, cheerful and given to flirtation."[42] Elsa Einstein could still smile at her Albertl's flirtations, but in 1923, after his return from Japan, she was by no means so sure of her husband. After his stepdaughter Ilse's marriage, another young woman, Betty Neumann, had become his secretary, and Einstein fell violently in love with her. Unlike his two marriages, this relationship aroused emotions that profoundly touched him. Nevertheless, this great love ended at the conclusion of 1924, when he wrote to "dear Betty" that he "must seek in the stars that which was denied him on earth."[43] This sounds like grand renunciation, but it was merely a variant of the justification he had given at the age of eighteen, when he parted from his first love, Marie Winteler.[44]

It was perhaps not only Einstein's detached observation of human folly, but also his renunciation of Betty Neumann, that led Count Kessler, a sensitive chronicler, to jot down: "The ironical *(narquois)* trait in Einstein's facial expression, the 'Pierrot lunaire' quality, the smiling and doleful skepticism that plays around his eyes, emerge ever more strongly." In Einstein's features Kessler believed he could discern that the man was "smiling not only at the surface, but at the roots, of human conceit."[45]

Perhaps a wish to put some distance between himself and his personal entanglements was a factor in Einstein's decision to leave Berlin for three months and undertake a major trip to South America. All he needed to do was reactivate invitations from scientific organizations in

Argentina, Uruguay, and Brazil. Three years earlier he had declined them, but now they suited him. At the beginning of March 1925 he embarked in Hamburg for a three-week voyage. He submitted to the traditional ceremony when the ship crossed the equator; he lectured the ship's officers on relativity theory; and every day in his cabin he played string quartets with some passengers and the leader of the ship's band.

His reception in Buenos Aires was exceedingly cordial, not only on the part of his Argentine hosts but also on the part of the German colony. This surprised him a little, as three years earlier the representatives of the same German colony had vigorously refused to invite or honor a "defeatist," who during the war had "made propaganda against Germany and therefore was a traitor to the Fatherland."[46] The fact that these people were now fêting him as an "exponent of German culture" gave rise to one of his sarcastic comments: "A funny lot, these Germans. To them I am a stinking flower, and yet they keep putting me into their buttonhole."[47]

Meanwhile, as expected, Einstein was boundlessly honored by the political and academic leaders of the countries he visited. He gave his lectures in the now familiar spectacular settings, and in a ceremonial session he was accepted as a foreign member of the Argentine Academy. "I was asked very stupid scientific questions, so that it was difficult to remain serious,"[48] he noted in his diary. But then he had not set out on this trip with any expectation of fruitful discussions about physics.

He even met a few relatives from the maternal, grain-trading side of his family, including his cousin Robert Koch, who had attended the cantonal school in Aarau with him: "We've grown old."[49] Among his "tribal companions" he solicited donations for Hebrew University in Jerusalem and was enthusiastically fêted by them, "because to them I'm a kind of symbol of cooperative activity among Jews. It gives me a lot of pleasure because I expect great things from the unification of Jews."[50]

After three weeks in Argentina, Einstein went on to Montevideo. Receptions by the president of the Republic of Uruguay and by notables in politics and science ensured a full program, which also included a cheerful evening with students, "guitar and singalong, me finally with my violin."[51] In Uruguay he met "genuine cordiality as rarely in my life," so much so that "with all that love I sometimes had to come up for air." He found Montevideo "a lot more human and pleasant" than Buenos Aires, and—as with his sympathies for Switzerland and the Netherlands—he attributed this to the country's small size. This led him to the conclusion: "The devil take the big states and their madness. I'd cut them all up into small ones if I had the power to do it."[52]

In conclusion, there was one more "big state," Brazil. Rio de Janeiro provided the customary festivities, not only within the circle of the Jewish community, but also at the German club Germania.

The German ministers in each of the countries he visited unanimously reported to Berlin that Einstein's "visit has been most useful for the German cause."[53] Einstein himself, returning to Berlin after a three-month absence, summed it up as "great fun without any real interest, but at least a few quiet weeks during the voyage."[54]

For two months Einstein again treated his colleagues in Berlin to his participation in academy meetings, before leaving them once more at the end of July to attend the League of Nations session in Geneva. In Zurich he looked in on Mileva and his sons; and he visited his old friend and frequent "lifesaver" Marcel Grossmann, who, he was shocked to find, had developed multiple sclerosis. From Switzerland he returned straight to Kiel, where he spent August sailing on the Förde and working on the gyrocompass at the Anschütz factory.

Epoch-making changes in physics were by then imminent. At one point, Einstein believed that he himself had accomplished a major step forward. On July 9, before his departure for Switzerland, he had submitted to the academy his *Unified Field Theory of Gravity and Electricity*, convinced of having "hit on the right solution."[55] It was not the first and not the last theory about which he spoke with such assurance. He had to admit to himself that with regard to the microcosm, the difficult question remained unresolved "whether this field theory is compatible with the existence of atoms and quanta"[56]; still, he felt sure that he was also on the right track to solving the riddle of the quanta.

At about the same time, a breakthrough toward the solution was becoming discernible, but along totally different and—to Einstein—surprising lines. The new quantum mechanics emerged: a scientific revolution if ever there was one. To Einstein's mind, though, it was a blind alley—certainly not a theory but a construct of, at most, temporary validity.

Einstein's future course is characterized by these two great themes: the struggle for a unified theory that would simultaneously solve the quantum problems in Einstein's sense, and his own never-ceasing criticism of quantum mechanics. It became a lonely road without lasting successes. His isolation increased over the years, but he bore it with a smile. Failure never made him despair.

The fact that Einstein pursued this road with unshakable perseverance and inexhaustible optimism to the end of his life had to do with a transformed concept of the cognition of nature, which evolved and became consolidated after the triumph of the general theory of relativity. Its origins can be found in wartime Berlin.

CHAPTER TWENTY-EIGHT

"The Marble Smile of Implacable Nature": The Search for the Unified Field Theory

BY THE TIME EINSTEIN GAVE his Nobel lecture on relativity theory in Göteborg in July 1923, he had long since turned his mind to new problems. Because of their complexity, these did not seem to him suitable for a public lecture, but he wanted to give his audience at least a small glimpse of what he was after. "The intellect seeking after an integrated theory," he declared, "cannot rest content with the assumption that there exist two distinct fields totally independent of each other by their nature."[1] He did not realize then that with these words he was describing what would be his passionate scientific quest to the end of his days: the search for a unified field theory of gravity and electromagnetism.

His longing was to remain unfulfilled. Yet, undeterred and undeflected, Einstein pondered this unification; over the years and decades he lost himself in ever more abstract reflections, studied advanced mathematics, and entangled himself in the most complex calculations. For two decades he never doubted that the problem could be solved. Eventually a moment came when he had to admit to himself that "most of my intellectual offspring end up very young in the graveyard of disappointed hopes."[2] But still he continued his search with unshakable optimism, inspired by a task which had become his destiny.

During the first decade of his quest, he still had a few scientists on his side, or at least interested colleagues; but with advancing age his isolation could no longer be overlooked—and it was made worse by his persistent critique of quantum mechanics. He was entirely aware of this, scoffed at himself as a "petrefact," but never ceased to "sing his lonely old song."[3] When he realized that he had very little time left, he had the latest calculations on the unified theory brought to his sickbed. He died the following night. The unfinished task had been with him to his final breath.

For most of his contemporaries, and even more so for the younger generation of physicists, this work was not only unfinished but, at least in its later phases, totally useless. Even devoted admirers of Einstein would not dispute that the progress of physics would not have suffered unduly if the indisputably greatest scientist among them had spent the final three decades of his life—roughly from 1926 on—sailing.

Of course, he still managed to produce an occasional fine study, the kind that would have been a respectable lifework for a less inspired researcher. Equally, his critique of quantum mechanics had its positive side in that its proponents were forced to make it stand up to his and similar reservations. But his great passion, his vision of a unified field theory, produced not only no result, but, above all, no consequences. It remained a foreign body, of scarcely any importance to physics.

How did it happen that Einstein—who, over two decades of overwhelming creativity, had transformed and enriched physics as no one else—in his middle years adopted a course which ended in a blind alley from which he never emerged? It is quite possible that a decline in creative vigor and perhaps even a touch of old-age stubbornness were factors—yet these factors are far less interesting than the question of Einstein's motives, which guided him and focused him on his goal.

The vision which induced Einstein to pursue his lonely road was a fundamental theory for the whole of physics, based on field theory. This would not only unify the two areas but also explain the existence and properties of elementary particles, and it would clearly establish natural constants such as elementary charge, velocity of light, and quantum of action. Ultimately, through deduction alone, a quantum theory was to emerge that would conform to Einstein's idea of an objective description of nature.

Einstein was thus aiming at what today is loosely called a "theory of everything." He was encouraged by the spectacular success of the general theory of relativity—a powerful temptation for a theoretical researcher to tackle this greatest of all tasks. His vision had taken shape as a realizable project before the new quantum mechanics; that he refused to abandon it when the radical transformation of physics by quantum mechanics was making his concept at best questionable, if not indeed obsolete, was seen by most of his colleagues as one of the reasons for his failure. The other reason was that, ever since the 1930s, new particles, new forces, and new fields had been identified, whose very existence suggested that Einstein's program was too narrowly conceived. Even if he had been able to complete it, it would not have become a "theory of everything," and certainly not a theory of the microcosm. Einstein therefore did not achieve his objective, and to this day it is impossible to see how he could have achieved it or whether it is achievable at all.

The world which Einstein intended to comprehend theoretically presented itself fairly simply. There were two fields, the electromagnetic and the gravitational. These were equal in their dependence on the location of their potentials and in their great reach. But they were very different in strength—with regard to strength, in fact, the ratio between them would be represented by a figure with forty zeros. Another difference between them was the fact that masses are subject to universally attractive gravity, i.e., they have only one sign; whereas electrical charges, being either positive or negative, both attract and repel one another.

Along with the fields, of course, there was solid matter, whose elementary building blocks were believed to be the electron and the (roughly two thousand times heavier) proton. Both occurred in countless multiplicity, though invariably with an absolutely identical mass and charge, which called for an explanation from fundamental principles. Atomic nuclei were then, in the absence of other possibilities, thought to be composed of protons and electrons, held together by strong electrical forces.

Within this orderly framework it must have seemed natural to begin the search for a "theory of everything" with a search for a unified field of which the two known fields would be specific aspects. Einstein was neither the first nor the last to undertake that task.

As early as 1912, when Einstein was deeply involved in the generalization of his theory of relativity, Gustav Mie, then a professor in Greifswald, Germany, made the first attempt at a comprehensive theory of matter, one which would simultaneously establish the "unity of the physical world picture." Proceeding from what mathematicians call a "nonlinear" extension of electrodynamics within the framework of the special theory of relativity, Mie hoped that the electron and the proton would result from his field equations as mathematically deducible "thought necessities." However, this hope was disappointed; nor was it fulfilled when David Hilbert, in his *Fundamentals of Physics*, integrated Mie's theory into the general theory of relativity.

We do not know when, despite the lack of any physical starting point, Einstein intensively took up the problem of a unified field theory. He may well have pondered the question in the early months of 1916, while working on his major article *The Fundamentals of the General Theory of Relativity*; there are some indications that he may have done so, but he certainly revealed nothing about it. By March 1918, however, he must have given it a lot of thought, because a letter to Hermann Weyl, complimenting him on his paper *Gravity and Electricity*, discloses that Einstein's struggle for unification had already begun: "And now you have actually given birth to the offspring that I was absolutely unable to produce—the construction of Maxwell's equations from $g_{\mu\nu}$."[4] This is a somewhat technical way of saying that Weyl had derived from the gravitational potentials a term that could be understood as the tensor of the electromagnetic field. Does this mean the problem had been solved even before Einstein was able to make a contribution to it?

Weyl was no stranger to Einstein. In 1913, when he was only twenty-nine and a privatdozent at Göttingen, Weyl had been invited to the Swiss Technical College in Zurich. There he had been able, from a colleague's proximity, to watch the genesis of the general theory of relativity. Thanks to his stupendous mathematical skills, Weyl soon succeeded in presenting Einstein's theory (completed in 1916) in a mathematically more elegant shape than its author had, by making use of a newly developed method called "infinitesimal parallel shifts" to extend Riemann's geometry, which before long gave rise to the mathematics of "affine connections." In the spring of 1918 Einstein, then sick in bed, read the galleys of Weyl's book *Space, Time, Matter* "with genuine enthusiasm. It is like a masterful symphony. . . . The concept of the work is magnificent."[5]

Shortly after writing this book, Weyl had completed his unified theory and sent it to Einstein with the request that he should submit it to the Prussian Academy for publication in its *Proceedings*. Einstein was again enthusiastic, but not for long, because he soon discovered a major flaw: in Weyl's presentation there were no invariant linear elements, which, in physical terms, meant that the yardsticks for space and time depended on their earlier history. Thus, for instance, hydrogen atoms which had traveled over different distances would have to exhibit shifts in their spectral lines—in total contrast to observation.

Four weeks later, therefore, Einstein's praise was enveloped in gentle irony: "Your argument is of wonderful homogeneity. Apart from agreeing with reality it is certainly a magnificent achievement of pure thought."[6] Weyl gloomily complained to Einstein "that you don't wish to know about this business at all. Naturally this worries me a lot, because experience has shown that one can rely on your intuition."[7] In fact, Weyl's theory could not be developed into anything useful; on the other hand, the "Eichin variance" he had used in it for the first time was subsequently to prove a powerful mathematical instrument in quantum mechanics and quantum field theory.

Despite his reservations Einstein supported publication of Weyl's paper, which, because of the paper shortage in the fourth year of war, had met with some opposition from the Berlin academicians. When it eventually appeared in *Proceedings*, Einstein recorded his criticism in a brief note at the end.[8]

Einstein's first intensive contact with unification was thus primarily critical. Nevertheless, his examination of Weyl's ideas left him with certain impressions to which he later returned. However, he did not yet believe that natural laws could be discovered by mathematical constructs alone.

First of all, Einstein intended to establish what the general theory of relativity could do to explain elementary particles. In the spring of 1919 he tried to demonstrate "that there are indications to support the idea that the building blocks of the electrical elementary structures forming the atoms are held together by gravitational forces."[9] These indications, however, vanished soon after publication.

All the more gratefully, then, he took up the proposal of the Königsberg mathematician Theodor Kaluza that the four-dimensional world be extended by a further dimension. This fifth dimension opened up new possibilities for the formulation of field theory. The objection that this was just a mathematical fiction, devoid of any relation to space or time, was softened by the fact that, in the end, it was once more restricted by meaningful supplementary conditions and that, in this way, contact was smoothly reestablished with the four-dimensional continuum of the real world. Einstein admired the formal unity achieved by this detour through the fifth dimension, submitted Kaluza's paper to the Prussian Academy in 1921, and himself published a paper on the five-dimensional method. For a while he even believed that this idea "smells to me more of reality than any other."[10] Very soon, however, he discovered that with this method too the electron remained a foreign body in the theory. Once more mathematics had proved elegant but unproductive in the physical sense.

Much the same was true of a proposal by Eddington, who had earned himself undying fame by proving the deflection of light and who now applied his mathematical skills to modifying and extending Weyl's theory in a way that would stand up to "Einstein's yardstick objections."[11] Soon, however, Einstein thought he saw a solution himself and, during his voyage to Japan, tried to develop it. On the way back he praised the voyage as "a wonderful existence for a ponderer—just as in a monastery. Plus the warmth of the Equator. Warm water lazily drips from the sky, spreading tranquillity and vegetation-like twilight."[12]

What he had put down on paper on board ship, entitled *On the General Theory of Relativity*, seemed to him so important that on arrival at Port Said he instantly posted it to Berlin, where Max Planck on his behalf submitted it to the Prussian Academy on February 15, 1923. This was a link between Eddington's formalism and Hamilton's principle, of which Einstein euphorically believed that it "leads to a theory almost free from arbitrary steps, one that conforms with what we know at present about gravity and electricity, and which unites both types of field in a truly perfect manner."[13] Back in Berlin, Einstein himself lectured at the academy about this concept in the spring of 1923 and supplemented his reflections with two more papers, though by then with the sobering realization that physically interesting results were refusing to come up. "Generally speaking, a rather resigned mood has taken hold of me concerning the whole problem," he reported to Weyl. "The mathematics is fine and good, but Nature is leading us a dance." Yet he would not give up. "The whole idea must be carried through and it is of a strange beauty; above it, however, hovers the marble smile of implacable Nature, which has given us more longing than intellect."[14]

In the summer of 1922, having skeptically reviewed all past attempts at a unified theory, Einstein wrote: "I believe that, in order to make any real progress, one would again have to find a general principle wrested from Nature."[15] He was then still hoping for inspiration from physics.

At times he had even looked for experimental approaches to the relationship between an electromagnetic field and gravity. In Zurich in 1913, probably unaware of Faraday's investigations more than half a century earlier, he had asked himself whether there could be a gravitational effect that was analogous to electromagnetic induction.[16] Sometime around 1922 he tried to get Walther Gerlach to take up experimentally the question whether moving matter produced a magnetic field. "What I had in mind was measurements along currents or waterfalls."[17] But Gerlach, who had just made a name for himself in the discovery of directional quantization—the "Stern-Gerlach experiment"—was unwilling to interrupt his academic career for what he thought a rather vague suggestion.

Einstein's refusal to be discouraged was due to the fact that from "wresting principles from Nature" he gradually drifted toward mathematics. This change in Einstein's way of thinking, with mathematics—formerly simply a tool of the physicist—becoming the source of cognition itself, distinguishes more than anything else the thoroughgoing physicist of his youth from the lonely seeker of his later years.

We have seen that as a young man Einstein viewed mathematical scholarship with almost amused skepticism. Not until his generalization of relativity theory did he acquire a proper respect for its subtleties. He was no doubt profoundly impressed by the fact that something like a prestabilized harmony emerged between Riemann's geometry, a subject originating from purely mathematical reflections, and the laws of gravity; but this did not then change his pragmatic attitude to mathematics.

Whenever he had an opportunity, he would explain to mathematicians that their abstract art, unless it paid due regard to facts, was pure speculation and not physics. "It seems to me that you greatly overestimate the value of formal points of view," he somewhat condescendingly, in 1917, lectured the *Praeceptor Mathematicae* at Göttingen, Felix Klein. Klein had examined certain transformations of Maxwell's equations, which Einstein had described as lacking any physical meaning. The formal points of view "are of course very valuable whenever an *already discovered* truth has to be definitively formulated, but they nearly always fail as a heuristic aid."[18] Einstein's letters and postcards to Weyl similarly reflect a persistent belief in the primacy of physics over mathematical speculation detached from experience.

In 1918 Einstein vigorously rejected the suggestion that in the general theory of relativity speculation had proved superior to empiricism. "But I believe that this development contains a different lesson, almost the opposite, namely that a theory, in order to deserve confidence, has to be based on generalizable facts. . . . Never has a truly useful and profound theory been discovered by pure speculation."[19] But by

1923 he was, nevertheless, about to turn toward the mathematical-speculative approach.

It was in his Nobel lecture that Einstein first clearly proclaimed mathematics to be the only signpost in the search for a unified theory: "Unfortunately we cannot in these efforts base ourselves on empirical facts as in the derivation of the theory of gravity (equality of gravitational and inert mass), but we are limited to the criterion of mathematical simplicity, which is not free from arbitrary aspects."[20] By finding the simplest differential equations that can be submitted to an affine connection, "we may hope to arrive at a generalization of the gravitational equations which would include also the laws of the electromagnetic field."

Einstein even believed that the paper he had written on his return voyage from Japan represented a partial fulfillment of that hope, though he had to add the reservation that he was not yet certain "whether the formal connection thereby gained can truly be regarded as a contribution to physics so long as it does not provide any new physical connections."[21] It did not provide them—but Einstein was not discouraged.

He drew his optimism and perseverance from his development of the general theory of relativity, which in his memory now appeared to him as a triumph of mathematical speculation. At the age of fifty he wrote: "The success of that experiment [in deriving] subtle natural laws from the conviction of the formal simplicity of the structure of reality, by a purely mental process, now encourages me to proceed along this speculative road, the dangers of which everyone who dares follow it should permanently keep before his eyes."[22] Einstein's misfortune was due not so much to the fact that he was not deterred by the dangers of mathematical speculation, but that he altogether lost sight of them. Thus in his Spencer Lecture, given in Oxford in 1933, he elevated mathematics to the "real creative principle."[23] A few years later he compressed a kind of intellectual minibiography into a single sentence: "Coming as I did from skeptical empiricism of [Mach's] type . . . the gravity problem turned me into a believing rationalist, i.e. into a person who seeks the only reliable source of truth in mathematical simplicity."[24]

At the very beginning of natural science based on mathematical laws, Galileo had enthusiastically declared that the "magnificent book of the universe" was "written in mathematical language . . . without an understanding of which one stumbles about as in a dark labyrinth."[25] Since then many researchers have experienced the well-nigh miraculous relationship between mathematics and physics. Indeed, a theoretical physicist without an awareness of that harmony or a sense of mathematical beauty and elegance would not get far. In fact, many a magnificent discovery has been theoretically anticipated by the conviction that an elegant mathematical structure must have its counterpart in physical reality. The most spectacular, though not the only, instance of this is the theoretical postulation of "antimatter" by P. A. M. Dirac.

But just as the theoretical physicist is helpless without mathematics, so mathematical speculation unrelated to reality remains void. Einstein allowed himself to be seduced into a belief that mathematical criteria were "the only reliable source of truth," and in his Spencer Lecture he said he believed it to be "true in a certain sense that pure thought is capable of comprehending reality."[26] This indicates a reckless overestimation of the possibility of understanding nature through mathematics alone—a mistake he would not have been capable of in his most productive years. Yet this faith, though ultimately unproductive, sustained him for decades in his search for the unified field theory.

As Einstein's own mathematical skills were not exceptional, ever since his work on the general theory of relativity, he had ensured that he would have the help of an outstanding mathematician, Jakob Grommer. Grommer is first mentioned as a coworker in *Cosmological Observations*;[27] his name stands alongside Einstein's at the head of some of his later papers, but most of the time it figures in the concluding acknowledgments.[28] The last mention is in a publication by Einstein of January 1929, which means that Grommer collaborated with Einstein for twelve years—longer than anyone else. But Grommer left virtually no traces in Einstein's correspondence, and the many people who knew Einstein in Berlin never mentioned Grommer. There was a reason for his shadowy existence.

Jakob Grommer came from an eastern Jewish family in Brest-Litovsk. We do not know the date of his birth, or even the year. It is reported that he was a studious disciple in the local yeshiva, intending to become a rabbi. One of the requirements for an aspiring rabbi at the time was that he should marry the daughter of an older rabbi. In this intention Grommer failed, not for any lack of Talmudic scholarship but because of the refusal of the older rabbi's daughter. The fact was that Grommer suffered from a disease of the lymphatic system, described picturesquely as elephantiasis because it leads to a shapeless enlargement of the extremities.

The failed rabbi, who until then had spoken only Yiddish and Hebrew, turned to mathematics, went to Göttingen, and soon produced an essay which so impressed David Hilbert that he wanted to accept it immediately as a doctoral thesis. There was a hitch, though: Grommer had never graduated from high school (a Yeshiva was no gymnasium). Still, Hilbert championed Grommer, and the (not exactly liberal-minded) faculty eventually decided to grant him a doctorate.[29]

It was probably Hilbert who drew Einstein's attention to this outstandingly gifted young Jewish mathematician. Grommer thus moved to Berlin to work with Einstein. This was not an official assistant's post; Einstein took care of Grommer's livelihood by sums from the budget of the Kaiser Wilhelm Institute for Physics, from an American foundation, and eventually from a fund provided for Einstein's researches by wealthy Berliners.[30]

In 1928 Grommer returned to his homeland. He became a professor in Minsk and a member of the Academy of the Belorussian Soviet Republic. He died in 1933.

Einstein did not long remain satisfied with his theory of 1923. With his next fundamental publication on unified field theory, in the summer of 1925, he officially announced that this theory "does not present the true solution of the problem."[31] But he cheerfully added good news: "After ceaseless searching over the past two years I now believe I have found the true solution."[32] This sequence—an unsentimental revocation followed by joyful announcement of the "true solution"—would characterize Einstein's later years.

The "true solution" of 1925 was also based on the mathematics of affine connections, but it took new paths. Einstein sought first of all the formally simplest term for the law of the gravitational field, and next the "most natural" generalization. In this extended framework—unlike in earlier theories—the fundamental tensor is no longer symmetrical. This Einstein regarded as an opportunity: in dividing the tensor he tried to identify its symmetrical component with gravity and its antisymmetrical component with the electromagnetic field. Admittedly, the reconstruction of the Maxwellian equations was possible only by recourse to a few artifices, and even then only for the limiting case of weak fields. That was all that this exceedingly compact paper (a mere six printed pages) had to offer—not much for a study entitled *Unified Field Theory of Gravity and Electricity*. The touchstone of the theory was whether the existence of the elementary particles could be derived from it; but in this regard Einstein had to confine himself to the statement that he had begun to address the problem in cooperation with Jakob Grommer.

Despite this shortcoming, Einstein was quite delighted with his new theory. During a "boring League of Nations meeting" he wrote to Besso that he saw in it "a splendid possibility that probably conforms with reality. . . . At least in the macroscopic range I don't doubt its correctness."[33] Only eight weeks later, his optimism had left him. "Now I have again great doubts about my work," he reported to Ehrenfest. Two days later, nothing was left: "My work of the last summer is no good."[34]

Nevertheless, this paper could have turned out to be one of Einstein's most spectacular achievements if only it had been correctly understood and interpreted: it could have provided nothing less than the theoretical basis for the existence of antimatter. But given the state of knowledge in 1925, this potential discovery not only went unrealized by Einstein; it actually became a major obstacle to him in his further reflections.

It is probable that in his investigations of the structure of general field theories Einstein had encountered a problem which he at first ignored or pushed aside, or at least did not pursue any further because it seemed to endanger his goal of a unified theory. However, in the fall of 1925 he presented it in a short paper entitled *Electron and General Theory of Relativity*.[35] It concerns the fundamental characteristics of such theories under mirror images of the space and time coordinates. Einstein demonstrated that in any relativist field theory in which gravity is represented by a symmetrical tensor and the electromagnetic field by an antisymmetrical tensor, the following applies because of invariance with regard to space and time mirroring: for every possible field corresponding to an elementary particle with positive charge there also exists a field describing a particle with negative charge but with identical rest mass. In later terminology this means that for every elementary particle of mass m and charge e there must exist an "antiparticle" of the same mass m but with a charge $-e$.

At the time, however, physicians knew of only two kinds of elementary particles—electrons as the carriers of negative charge and protons as the carriers of positive charge—and this knowledge was believed to be definitive and unalterable. However, since the mass of a proton is nearly two thousand times the mass of an electron, there was, in that limited framework, an asymmetry between negatively and positively charged elementary particles. Any theory postulating an equality of mass of negatively and positively charged particles was therefore in flagrant contradiction to what was then regarded as valid experience. Resigned, Einstein concluded that "the endeavor to amalgamate electrodynamics and the theory of gravity into a unity" in the light of his symmetry theorem was "no longer justified."[36]

If only Einstein had interpreted his theorem to mean that alongside the negative electron there must also exist an as yet undiscovered positively charged "antielectron" of identical mass, he would have predicted antimatter in an exceedingly elegant manner within the framework of "classical" physics. As it is, this achievement belongs to the younger physicist P. A. M. Dirac, who in 1930 deduced the existence of antimatter by linking the special theory of relativity to the new quantum mechanics. Two years later, when antimatter was first observed in cosmic radiation, in the form of the positively charged antielectron, the positron, Dirac's theory was triumphantly confirmed.

Einstein's paper, however, remained unnoticed, and he himself never referred to it in later years. It may be seen as a typical instance of a "premature" discovery, whose implications are not suspected even by its author. Instead, Einstein had to resign himself to the fact that the unified theory had receded into the distance and would require new approaches. Meanwhile, the new quantum mechanics provided a highly promising theory of the microcosm, but this was a theory which Einstein viewed with great skepticism on fundamental grounds.

CHAPTER TWENTY-NINE

The Problems of Quantum Theory

IN THE SUMMER OF 1925, a week after submitting his paper *Unified Field Theory* to the Prussian Academy, Einstein received a letter from Max Born in Göttingen, containing the first details of a newly completed study by Heisenberg: it "looks very mystical, but is certainly correct and profound."[1] With this stroke of genius the twenty-four-year-old Heisenberg, a student of Sommerfeld, Born, and Bohr, had sketched out the main features of a new quantum theory, illustrating it by two simple applications. Heisenberg's paper was the prelude to a time of intense, exciting creativity, which over the next two years led to the formulation of quantum mechanics as a full-fledged theory of the microcosm—the theory which has left its stamp on twentieth-century physics. But when Einstein read the paper, his reaction was as spontaneous as it was unambiguous: "Heisenberg has laid a big quantum egg. In Göttingen they believe in it (I don't)."[2] He persisted in his rejection: to the end of his life he did not believe in it.

Among the founders of quantum mechanics, Erwin Schrödinger was the only one who understood Einstein's skepticism. All the others—Einstein's coevals, like Bohr and Born; as well as the younger generation, including Heisenberg, Dirac, and Pauli—simply had to face, and eventually live with, the fact that Einstein never accepted the new quantum mechanics as a valid theory. Einstein himself had done more than anyone else to champion the "old" quantum theory, but in 1925 the reins were taken over by other, mostly younger, men, and Einstein's contributions to quantum mechanics were only critical. "Many of us think it a tragedy," Max Born wrote, "for him, who now has to travel his path in loneliness, and for us, who lack the master and standard-bearer."[3] Born, however, for all his close friendship with Einstein, was wrong here: Einstein never felt his loneliness to be a "tragedy," neither in his old age nor in his younger years, when he had been just as "alone" with his ideas about light quanta as he would be during the final three decades of his life.

In all things concerning quantum physics, Einstein had for twenty years been ahead of his time—right up to the first months of 1925, to the very threshold of the new quantum mechanics. This had begun while he was still at the Patent Office, with his concept of light quanta, which he himself described as "very revolutionary"; this concept was followed two years later, by the first quantum theory of solid bodies. After two more years, at the Naturforscher convention in Salzburg, he surprised his colleagues with his farsighted prediction "that the next phase of development in theoretical physics will bring us a theory of light that may be understood as a kind of fusion of the undulation and emission theories of light."[4] He meant that light must be neither continuous waves nor discrete energy quanta, but some "fusion" of the two—some third form, as yet unknown. However, even the most concentrated reflection did not bring him any closer to implementing his prediction.

Atomic physics entered an enormously productive phase in 1913, influenced by Niels Bohr and his model of the atom. Bohr's model, with its ad hoc discrete quantum states, ran counter to all ideas of "classical" physics, and Einstein reacted enthusiastically. Einstein was merely an observer, though, since just then he was totally immersed in the generalization of his theory of relativity. Only when he had accomplished that task did he return to his other great theme.

In 1916 Einstein published papers of incomparable elegance in which he described the emission and absorption of radiation in only the most general quantum concepts.[5] He succeeded in producing a "totally quantum-governed" justification of Planck's radiation formula; but more important to him was the fact that electromagnetic radiation was propagated not as a wave but as a stream of aimed particles, so-called "needle radiation," in which energy quanta were now also given a definite momentum—a second property, along with energy. As a result, the existence of light quanta was "as good as certain"[6] for Einstein. Two years later he confirmed his view: "But I no longer doubt the reality of quanta, even though I am still all alone with this conviction. This situation will persist so long as a mathematical theory is not successfully developed."[7]

The first indications were that the road to a mathematical theory of quanta would be long and difficult, because "chance" had for the first time entered the quantum processes. Direction and time of emission of a light quantum cannot be predicted; in a sense, a light quantum is left to itself to decide when and in what direction it exits from an atom. Einstein described this as "a weakness of the theory . . . that it leaves the time and direction of the elementary processes to 'chance.' "[8]

The discovery of what he put between quotation marks—chance—had made Einstein uneasy from the outset. "Chance" undermines causality and thus topples the framework of classical physics. Philosophers, loosely speaking, regard causality as the relationship between cause and effect; to physicists, however, it has a precise pragmatic meaning. From a given initial state, a system develops over time with unambiguous regularity, in such a way that all its future states are determined as solutions of differential equations in the space-time continuum. A light quantum whose future state comes about unpredictably through "spontaneous" emission was bound to be a foreign body in any classical theory.

Einstein would have loved to be able to take back this annoying discovery of chance and look for a causal description of quantum processes. "The business about causality irks me a lot," he wrote in 1920. "Is quantum light absorption and emission ever conceivable in terms of the demand for complete causality or does a statistical rest remain? I must confess that I lack the courage of conviction. But I am most reluctant to give up *complete* causality."[9] In fact, it was not long before he returned to a firm insistence on complete causality.

Over the next few years, Einstein thought a great deal about quantum problems, but he did not really become a member of the growing group of atomic physicists. Einstein would never have written, or even wanted to write, a book like Sommerfeld's *Atomic Structure and Spectral Lines*, first published in 1919, which immediately became the bible—and with its continually enlarged revisions became the chronicle—of the headlong development of the "old" quantum physics. He did not even wish to give lectures in this field, because, as he told some students in Zurich, "it is not my place to lecture on quantum theory. Much as I struggled with it, I hardly succeeded in achieving any real insight." . . . "Besides, I never bothered to gather together the many details and artifices of which the quantum theory at present consists, so that I couldn't give you a comprehensive overview."[10]

Einstein was not interested in a patchwork of details and artifices; he was fascinated by fundamental questions, such as the role of "chance" and the dual wave and the corpuscular aspects of radiation. Yet a decade after his lecture in Salzburg he was still on his own with this problem; all his colleagues regarded the wave theory of electromagnetic radiation as the final word. Even Niels Bohr, who was rethinking physics more and more radically, would not believe in light quanta. Indeed, Bohr amused his growing number of followers with the remark that if a telegram arrived from Einstein confirming the existence of light quanta, he—Bohr—would use that telegram as the strongest proof against them, because the information from Berlin would have been transmitted by waves and received in Copenhagen by an antenna. The universal rejection of light quanta was not even affected by the Nobel Prize which Einstein received in 1922 for his explanation of the photoelectric effect—an explanation based on light quanta.

In 1921, Einstein hoped that an experiment would prove that "needle radiation" was a stream of particle-like structures.[11] He soon had to admit that even with his plan for the experiment he had made a "monumental booboo." Nevertheless, in the end it was an experiment, not any theoretical argument, that helped the quantum concept to prevail.

At Washington University in St. Louis in 1923, the American physicist Arthur Holly Compton was investigating the diffraction of hard X-rays on electrons. Compton's observations were totally incompatible with the idea that X-rays were electromagnetic waves. On the other hand, his observations fit perfectly into a picture of particles bouncing off the electrons like billiard balls; the deflection of the particles was explained by the fact that they had clearly defined energy and momentum. Theoretically, the experiment could be interpreted entirely within the sacrosanct laws of conservation of energy and momentum.

The "Compton effect"—the name by which the phenomenon observed in this crucial experiment soon became known in the literature—by no means settled the issue, though. Rather, it exacerbated the problem by demonstrating, in Einstein's words, "that not only in regard to energy transfer, but also in regard to the impact effect, the radiation behaves as if it consisted of discrete energy projectiles."[12] Here we still have Einstein's "as if" conclusion, familiar from his "heuristic viewpoint" of March 1905.

Among physicists, however, Compton's experiment soon did bring about a change of attitude, and Einstein's light quanta at last became respectable. In 1926, these structures—with no rest mass, but possessing energy and momentum—were named "photons." A year later, physicists were considering photons every bit as real as electrons, without any "as if." No one would have blamed Einstein if he had basked in this belated triumph, but that was not his style.

Meanwhile, Einstein was feeling exceedingly unhappy about the kind of "fusion" of the wave and particle aspects of light demanded by quantum mechanics within Bohr's concept of complementarity. This was not what he had expected.

Ever since his years at the Patent Office, Einstein had been disturbed by a duality pervading the whole of physics. All field theories described the world as a continuum, reflected mathematically by partial differential equations; but at the atomic level both matter and radiation consisted of discrete building blocks—electrons and protons—and of light quanta. This dualism was intensified both by his own reflections and by the success of Bohr's model of the atom, with its discrete energy levels. In 1917, Einstein was seriously doubting that the well-tested classical tools of mathematical physics were still suitable in the new situation: "If the molecular view of matter is correct (useful), i.e. if a portion of the world has to be represented by a finite number of moving particles, then the continuum of present-day theory contains too great a multiplicity of possibilities." He suspected that the insufficient limitation of possible solutions to the continuum equations was the reason "why our present means of description fail in the face of the quantum theory. The question seems to me to be how it is possible to formulate statements on a discontinuum without resorting to a continuum (space-time). . . . For that unfortunately we still lack the mathematical form. How much painful effort I have already spent along these lines!"[13]

It seems, however, that he soon saw a way out of the problem—using the well-tested continuum theory he so loved. This was the idea of "overdetermination," a mathematical situation that arises when the number of equations is greater than the number of variables. In the general theory of relativity Einstein had found overdetermination useful at an important point: the transition from Riemann's geometry to the limiting case of the Euclidian world. For physics, the starting point in the ideal case would be a unified theory whose equations included at least those for the gravitational field and the electromagnetic field. Einstein was hoping, on the basis of this and some other conditions, to discover a suitable "overdetermined" system of equations that would permit only discrete solutions, which could then be identified with quantum conditions and elementary particles.

Implementation of this idea would have met his criteria for a complete theoretical comprehension of reality. Quantum phenomena would have been embedded in a natural way in the well-tested and indispensable continuum theory of classical physics and would have been deducible from it; all such calamities as chance and the dualism between particles and waves, or between continuum and discontinuum, would have been merely temporary obstacles during a transitional stage of theoretical confusion.

The vision of a "theory of everything" through "overdetermination" was not attainable; it did not play a major part in the theoreticians' discussions, and it played no part at all in the progress of physics. But it is a key to Einstein's concept of progress. Without this vision, it would be difficult to understand why he felt that his stubborn search for a unified field theory was also overwhelmingly important for quantum theory—and it would be even more difficult to understand why he never accepted quantum mechanics as the last word.

Younger physicists derided Einstein's attitude as intransigent or reactionary; but it was not the whim of a pigheaded elderly man—it was an attitude rooted in reasoning that left him no choice. He was a visionary who believed that he at least surmised the direction in which the "promised land" should be sought; and he must have regarded it as a kind of intellectual surrender that "chance," which he himself hoped to explain, had simply been elevated into a principle by the champions of quantum mechanics.

The first hint of a search for "overdetermination" appears abruptly in a letter to Max Born in January 1920, and it suggests that Einstein had been thinking about this for some time. "I continue to believe that one must look for such an overdetermination through differential equations and that the *solutions* no longer have continuum character. But how??"[14] The two question marks were certainly justified: easy as it was for Einstein to formulate the process, its execution evaded him. "I just can't manage," he complained a little later, "to give solid shape to my favorite idea of comprehending the quantum structure from overdetermination through differential equations."[15]

It took almost four years before Einstein considered publishing something on "overdetermination." He eventually wrote a paper in his "cheerful exile" in Leyden, where he had gone to escape the threats to his life in Berlin. In December 1923, Max Planck submitted it to the Prussian Academy on his behalf. However, this paper was more of a statement of hope than a result, as was reflected in its title, *Does the Field Theory Offer a Possibility of Solving the Quantum Problem?*[16]

Einstein sets out the problem in his introduction, concluding with a rhetorical question: can the quantum conditions of natural processes be "adequately described by a theory based on partial differential equations"? He continues optimistically: "Quite certainly; all we have to do is 'overdetermine' the field variables by equations."[17] This sounds simple enough but becomes exceedingly complex over the next few pages, and at times even opaque. Einstein readily admits that one vital deduction "is not as cogent as one might wish." Nor does he claim "that the equations set out by me really have any physical meaning." His paper, he remarks, will have achieved its purpose "if it induces mathematicians to cooperate and if it persuades them that the road here embarked on can be pursued and should certainly be thought through to the end."[18]

In fact, Einstein was presenting not a theory but a road—a road that led not so much to results as to further questions. A few weeks later he admitted the problems of his approach but also expressed the hopes he still held for it: "The mathematical aspect is enormously difficult, and the connection with what is accessible to experience is becoming less and less direct. Yet it is a logical possibility of accurately describing reality without a *sacrificium intellectus*."[19] No one, however, took up Einstein's suggestion, and he himself was unable to achieve any significant progress with his method of overdetermination.

While Einstein's endeavors with overdetermination were being largely ignored, Niels Bohr galvanized the world of atomic physics with reflections that propelled the crisis of the "old" quantum theory toward its climax. Only a few weeks after Einstein's paper, Bohr and his younger coworkers Hendrik Kramers and John Slater published a radiation theory without light quanta, in which anything concerned with quanta was fitted into the interaction of radiation and matter. However, this "rescue" of the description of radiation in space by the Maxwellian equations exacted a high price: the conservation laws for energy and momentum had to be abandoned for individual processes and were henceforth to be valid only statistically.

As always, Einstein was intensely interested in Bohr's ideas, but this time he was not enthusiastic: "That idea is an old acquaintance of mine, but I don't think he is a real fellow,"[20] he wrote, presumably referring to some reflections he had made, and rejected, while still at the Patent Office. Now he raised "a hundred objections" to Bohr's theory,[21] most of them highly technical, and he strongly criticized Bohr for having prematurely abandoned the laws of conservation and hence causality. "I would not like to be driven into abandoning strict causality without a great deal more opposition than has been shown so far. The idea that an electron exposed to a ray by *its own free decision* chooses the moment and the direction in which it wants to eject is intolerable to me. If that is so, I'd rather be a cobbler or a clerk in a gambling casino than a physicist."[22]

Bohr's theory of radiation without light quanta was actually only an intermezzo; it came to an early end as a result of a brilliant experiment by Bothe and Geiger, which proved that energy and momentum were conserved for each individual process. Meanwhile, Einstein had some considerable successes, moreover with papers on quantum theory. The fact that they were written "on the side"—his main interest was focused on the unified theory[23]—does not diminish their value. These were his last contributions of high creative power and would by themselves have assured him of a place in the pantheon of physics.

Toward the end of June 1924, Einstein received a letter from a young Indian physicist, Saryendra Nat Bose, who taught at the University of Dacca and had not so far called much attention to himself by his publications.[24] "Because we are all your disciples,"[25] Bose felt entitled to send Einstein an article—which represented a brilliant development of Einstein's ideas. Bose was able to derive Planck's radiation formula without recourse to classical electrodynamics, by treating radiation as a gas consisting of Einsteinian light quanta, similar to a gas consisting of "normal" molecules but with a modified way of counting. In the event that Einstein approved, Bose asked him to arrange for the paper to be published in *Zeitschrift für Physik*.

The editor of the *Philosophical Magazine* of the Royal Society in London had rejected the article, though Bose did not mention this in his letter. Einstein, however, immediately liked the "exceedingly interesting derivation"[26] by the unknown Indian, translated the article into German, and submitted it for publication without delay.[27] Einstein realized particularly—more clearly than the author himself—that Bose's novel statistical method of counting had implications far beyond the specific case of radiation. It contained the foundation of a new quantum statistics.

Ever since Boltzmann, physicists had been counting atoms as if, at least in principle, they could be numbered and individually identified—an illusion taken over from daily life and from classical physics. Bose's method put an end to this: identical light quanta, he argued, were not distinguishable, even in principle, and hence were devoid of any individuality. Einstein extended this idea to material structures such as atoms and molecules. This "loss of individuality" called for a new way of counting objects of the microcosm, one which differed essentially from the classical way. It led, as Einstein soon discovered, to a "hypothesis about an interaction between molecules of an as yet quite mysterious nature,"[28] with equally mysterious effects on observed phenomena.

Bose's paper was still with the printer when Einstein appeared at the Prussian Academy to present his own paper, *Quantum Theory of Single-Atom Gases*,[29] in which he applied Bose's method to material gas molecules, on the basis of a formal analogy between radiation and a gas. Six months later, Einstein continued this work in a *Second Treatise* in the conviction that "the analogy between quantum gas and molecular gas must be a complete one."[30] Three weeks after that, he supplied as a theoretical basis some additional "reflections on the quantum theory of ideal gases, now as free as possible from arbitrary hypotheses."[31]

Einstein's most important methodological result was his generalization of Bose's counting method into a quantum statistics valid equally for radiation and matter. It soon entered the literature as "Bose-Einstein statistics." What characterizes Bose-Einstein statistics is that particles are indistinguishable and any quantum level may be occupied by any number of particles.[32] In consequence, the particles—more so than would be possible in classical statistics—crowd into the lowest energy levels. This enabled Bose to offer a natural explanation of Planck's formula for quantum gas, and Einstein supplied an equally natural interpretation of the so-called third law of thermodynamics in an atomic or molecular gas, according to which entropy disappears at absolute zero.

The new statistics enabled Einstein to make remarkable predictions concerning the behavior of matter at extremely low temperatures, such as the disappearance of viscosity in liquefied gases. This "superfluidity" was subsequently discovered by Willem Hendrik Keesom in Leyden, in 1928. The concept of the "Bose-Einstein condensation" has proved useful to this day whenever matter in "degenerate" states has to be described.

From the analogy between quantum gas and molecular gas Einstein, in his *Second Treatise*, drew the far-reaching conclusion that a wave character must be assigned not only to light but also to matter, "by assigning to the gas, in an appropriate manner, a radiation process and by calculating its interference oscillations."[33] His mastery of fluctuation analysis was sufficient for him to put forward this astonishing claim; but for a further interpretation he had to have recourse to some ideas of Louis de Broglie, the younger brother of Maurice de Broglie, whom he had met at the Solvay conferences.

Louis de Broglie had submitted a doctoral thesis to Einstein's friend Paul Langevin in Paris in the spring of 1924, a thesis which went far beyond the customary scope. For each material particle, de Broglie postulated a simple relation between its momentum and its wavelength. This was something totally new, and a bold shot at the unknown, for until then particles had been regarded as compact concentrations of mass, which had absolutely nothing to do with waves. De Broglie's hypothesis could be justified only a posteriori, in that it permitted what Einstein later called a "very remarkable"[34] geometrical interpretation of Bohr's quantum conditions within the atom.

Langevin was so astonished at de Broglie's idea that he sent Einstein a carbon copy of the thesis and asked him for an unofficial second opinion. It must have struck a familiar chord in Einstein: he himself had discovered that relation between momentum and wavelength of a particle, though he had not published it because of lack of experimental evidence and because of difficulties with the energy and momentum theorem.[35] Langevin followed Einstein's advice: de Broglie received his doctorate, and five years later the Nobel Prize. Meanwhile, however, atomic physics, as if it were not entangled enough, had a new hypothesis which would not be easily integrated.

Einstein soon turned it to everybody's advantage. In his *Second Treatise* he was able to show that de Broglie's material waves corresponded exactly to those which had followed from his own fluctuation studies, with the result that the two arguments supported each other and jointly came close to certainty. In Einstein's cautious formulation, "it seems that an undulatory field is associated with every motion process, just as the optical undulatory field is associated with the movement of light quanta."[36]

Einstein was so fascinated by material waves that at the annual meeting of the Society of Scientists and Physicians in Innsbruck in September 1924 he suggested that experimenters might search for diffraction and interference phenomena in molecular beams.[37] This, however, was hopeless, because—as he himself demonstrated in his *Second Treatise* a few months later—their wavelengths are mostly considerably smaller than their molecular diameter, so that any such effect would not be susceptible to experimental verification. On the other hand, prospects were more favorable for slow electrons: by the summer of 1925 Walter Elsasser in Göttingen was able to show some first indications of such waves, and in 1927 the existence of material waves was experimentally confirmed by the diffraction of electrons on crystals.

With his work on the quantum theory of gases, Einstein had once more led the "old" quantum theory to a peak, bringing it close to the threshold of the new quantum mechanics, and in some respects even beyond. Thus the road he had taken in 1905 came almost full circle, but at a new level of understanding. Then, Einstein had suddenly and unexpectedly placed his quantum hypothesis alongside the wave theory of optics. Now, two decades later, he was able to connect material corpuscles with a wave field. The fact that immaterial light and material corpuscles both have characteristics of particles and waves no doubt appealed to his desire for "generalization."

The fact that particles and waves were still two separate phenomena, however, would not have appealed to Einstein. And this dualism of waves and particles was not resolved in the new quantum mechanics, for which in many respects Einstein had blazed the trail—instead, to his regret, duality was actually consolidated as a principle.

CHAPTER THIRTY

Critique of Quantum Mechanics

AFTER A QUARTER-CENTURY of improvised quantum theory, the breakthrough to quantum mechanics and its completion within a mere two years represented a collective though turbulent intellectual effort by a brilliant generation of physicists—a fascinating interplay of physics, mathematics, and cognitive theory.[1] Here, of course, this story can be presented only to the extent that it provides a background for Einstein's critical commentary and for what others saw as his lonely road.

We have already seen his reaction to Heisenberg's impressive opening in July 1925, the "quantum egg": unlike the physicists at Göttingen, he did not believe in it. Heisenberg's concepts were so radical that others, too, initially had difficulties with them—Niels Bohr, for one, and even Heisenberg himself. Everyone could understand that Heisenberg had unhesitatingly rejected the disparate patchwork of existing quantum theory, dismissing its models as, at best, accidentally correct. In its place he was proposing a fundamentally new quantum mechanics, which included only relations between observable magnitudes; thus there was no longer any talk of the "path" of an electron within an atom—it was appropriate to speak only of frequencies and amplitudes of radiation processes. This sounded like good positivist thinking, but it involved opaque physics and strange mathematics.

Heisenberg's most bizarre innovation was that the result of multiplying certain magnitudes depended on the order in which the multiplications were performed. Nothing of the kind had existed in physics before, even in the "old" quantum theory; but in Heisenberg's theory, "noninterchangeability" was an essential element, even though he was unable to state what mathematical structure was in fact involved. Max Born, a trained mathematician, soon discovered that his pupil's strange calculation amounted to matrix calculation, a method familiar to mathematicians. While Heisenberg traveled to Cambridge, England, and Copenhagen in the late summer, Born and Pascual Jordan transcribed his physics into correct mathematics, developing his assumption all the way to the first appearance of the "interchangeability" relation for the position q and the momentum p of a particle: $pq - qp = h/2\pi i$.

This "commutator" was also found—independently of Born and Jordan—by Paul Adrien Maurice Dirac at Cambridge, not only for position and momentum but for all magnitudes which physicists call "canonically conjugate." In his presentation of quantum mechanics, reduced to the most abstract structures, Dirac (who was then only twenty-three) was also able to express a system's motion equations with this new term. Thus the "commutator" became the signature of the new quantum mechanics and even mathematically marked the break with all traditional physics.

Over the following months Born, Heisenberg, and Jordan developed their famous "three-man paper" on what was now called "matrix mechanics." Along with an extended presentation of the foundations, it pointed the way to some noteworthy applications. Heisenberg presented the basic features and mathematical basis of the theory in an article completed shortly before Christmas. Meanwhile, Wolfgang Pauli had described the hydrogen atom by the new methods, much to the delight of Niels Bohr, whose scientific career was closely associated with this touchstone of atomic physics.

A first, improvised conference on quantum mechanics was organized in Leyden as a sidebar event on the fiftieth anniversary of H. A. Lorentz's doctorate on December 11, 1925. Einstein had come from Berlin, and Bohr from Copenhagen. Paul Ehrenfest, the host, proudly presented his two young students Samuel Goudsmit and Georg Uhlenbeck, who had just discovered a new property of the electron: torque, or "spin." Spin was almost as vital as mass and charge to understanding the electron, and it was as eagerly discussed in Leyden as the other aspects of the new quantum mechanics.[2]

For his return trip Bohr had chosen the route via Berlin, so that he could continue his conversations with Einstein on the train. To Bohr they were "a greater pleasure and more instructive than I can say."[3] Although we know nothing about the subject matter of these discussions, we may assume that Bohr's enthusiasm for the new physics would not have left Einstein unimpressed. When Einstein drew up a balance sheet for physics for 1925, he too regarded matrix mechanics as "the most interesting thing that theory has produced in recent times." His admiration, however, had a substantial admixture of mistrust: "A veritable witches' multiplication table, in which infinite determinants (matrices) take the place of cartesian coordinates. Exceedingly clever, and because of its great complexity safe against refutation as incorrect."[4]

Nevertheless, matrix mechanics must have exerted a strange fascination over Einstein over the next few months, because he reported to Born in Göttingen: "The Heisenberg-Born ideas are keeping everyone breathless, gripping the thinking and pondering of everybody interested in theory. Dull resignation has been replaced in us thick-blooded creatures by a unique tense expectation."[5] He was tactful enough to leave out the question whether the theory was true—or perhaps he was not clear in his own mind where the new development would lead.

Soon Einstein met the man who had aroused this "unique tense expectation," young Werner Heisenberg. He and Heisenberg had already exchanged a few letters. Thus Heisenberg records that Einstein had congratulated him on his theory as early as 1925 and had suggested a face-to-face discussion of its foundations, in a letter signed "in genuine admiration, yours, A. Einstein."[6] Even though quantum mechanics may have rather alarmed him, Einstein remembered his own younger years and was fond of youthful Hotspurs with unconventional ideas.

The opportunity for the desired conversation arose on April 28, 1926, after Heisenberg, en route to Copenhagen to assume a position as a *Dozent* with Bohr, had given a lecture in Berlin. Einstein asked

Heisenberg to accompany him to his home on Haberlandstrasse. What the two men had to say to each other was eventually reconstructed by Heisenberg—though not until four decades later, and then more in Heisenberg's voice than that of his host.[7]

"But surely you don't seriously believe"—Einstein said, opening his frontal attack on the central dogma of the new physics—"that one can include only observable quantities in a physical theory." Heisenberg retorted, as indeed Born and Bohr would have done, that he had followed the basic idea of relativity theory, in which "absolute" time, because unobservable, had been replaced by time measured by actual clocks and synchronization procedures. But Einstein did not want to understand this the way Heisenberg did; on the contrary, "Only the theory decides what can and what cannot be observed."[8] More generally, Einstein's did not approve of Heisenberg's talking about "what one knows about nature, instead of what nature really does. The physical sciences can only concern themselves with what nature really does."[9]

Einstein's disinclination to accept Heisenberg's arguments may also have been due to the fact that two weeks before this visit, he had already formed the impression that "the Born-Heisenberg business is probably not correct."[10] The reason was that an alternative had meanwhile emerged, which was a lot closer to his own way of thinking.

While matrix mechanics had reached a provisional goal with the "three-man paper," Erwin Schrödinger, entirely on his own, was searching for a totally different approach to the quantum riddle. Schrödinger, then thirty-eight, held the professorship in Zurich that had begun Einstein's academic career—but for Schrödinger's appointment, it had been upgraded to a full professorship. In the past, Schrödinger had only sporadically concerned himself with quantum theory, but this changed dramatically when Einstein's *Second Treatise* in 1925 drew his attention to the importance of de Broglie's material waves. Schrödinger made some unsuccessful experiments with a relativistic wave equation (which he ultimately rejected but which would later be rediscovered as the Klein-Gordon equation). Then, during the Christmas vacation of 1925, he discovered a nonrelativistic equation for a material wave field, from which he succeeded in calculating the quantum levels of the hydrogen atom by conventional methods.

On January 26, 1926, Schrödinger sent this trailblazing discovery to *Annalen*—and this was only the prelude to a creative explosion resulting in five papers altogether, written at roughly monthly intervals up to July 21, 1926. In these papers, he developed different aspects of the new wave mechanics. Three of these papers continued the title of the first: *Quantization as an Eigenvalue Problem*.

Schrödinger's first publication, at the beginning of April, was received enthusiastically. Here—unlike the abstract "witches' multiplication tables" of matrix mechanics—physicists found themselves on the familiar territory of partial differential equations; the discrete energy levels of the hydrogen atom emerged naturally and inevitably, rather like the nodes of a vibrating chord.

Max Planck had drawn Einstein's attention to Schrödinger's work "with justified enthusiasm,"[11] and Einstein studied the paper "with great interest." In fact, he found it a revelation: "Not such an infernal machine, but a clear idea—and logical in its application."[12]

Schrödinger's article was of major importance because, as its author noted, it offered information "on the relationship of the Heisenberg-Born-Jordan quantum mechanics to mine." Using modern functional analysis, he demonstrated in a mathematical tour de force that the two theories, while totally different in content and mathematics, were nevertheless mathematically equivalent to each other, once one had got down to the fundamental structures.[13] This explained why for all specific problems the two theories invariably yielded the same results. Instead of speaking of matrix *or* wave mechanics one could now, with every justification, speak of a single quantum mechanics.

Pleasing though this synthesis may have been, it was also rather odd. This quantum mechanics now took two shapes: one version was based on the visually comprehensible image of real material waves, while the other proscribed all images, declaring all visualizable models misleading and devoid of content.

Just as outspokenly as Schrödinger had criticized matrix mechanics, so its founders now reacted against wave mechanics. As for its alleged visualizability, Heisenberg curtly remarked: "I think it's rubbish."[14] And Pauli had a few strong words to say to "Dear Schrödinger" when a cutting reference by Schrödinger to "the local Zurich superstition" became known.[15]

The critics insisted that Schrödinger's wave functions could not possibly describe de Broglie's material waves, because they diverged too quickly, whereas an electron remained concentrated as a virtual point mass for all eternity. Besides, wave mechanics failed when applied to anything relating to quantum leaps. And in formal terms it was clear, especially in multielectron systems, that Schrödinger's waves did not even propagate in physical space, as one would expect material waves to do; they propagated only in the abstract structure of the configuration space. Einstein, though full of admiration, had not overlooked this difficulty. "We are all here fascinated by Schrödinger's new theory of quantum levels," he wrote in June 1926. But then he continued: "Strange as it is to introduce a field in the q space, the usefulness of the idea is quite astonishing."[16]

What, if anything, did Schrödinger's calculations mean? In his *Fourth Paper*, in June 1926, Schrödinger tried desperately to give his complex wave function a realistic interpretation. Meanwhile, though, Max Born cut the Gordian knot and advanced understanding by a vital step. In a brief note, described as "provisional," *On the Quantum Mechanics of Impact Processes*, he considered "only one interpretation possible": that the wave function was a measure of the probability of finding an electron at a definite level.[17] As a result, de Broglie's physically real material waves, which had been the starting point of Schrödinger's reflections, were to become purely abstract probability waves.

Born's mathematics was sketchy; and the correct version—that it is not the wave function but its absolute square that is proportional to probability—was added only at the proof stage, as a note. Still, he was already venturing out on an even more significant path: recommending that "determination be given up in the atomic world."[18]

Four weeks later a full-length version was completed, under the same title as the note, with correct mathematics and a precise definition of the concept of probability, which in quantum mechanics differs substantially from its meaning in a lottery or in classical physics. In a preamble to this paper, Born referred to his famous friend and colleague, taking up "a remark by Einstein on the relationship between wave field and light quanta; he said something to the effect that the waves merely served to show the way to corpuscular quanta, and in this context spoke of a 'ghost field.' "[19] Thus Einstein for the first time appeared in the role of midwife to the new physics—in this case even to its statistical interpretation.

In fact, Einstein in 1924 was so impressed by light quanta that he was inclined to assign them a higher degree of reality than light waves, contrary to the overwhelming evidence in favor of the wave theory of light. He never published these ideas, but he discussed them extensively with his colleagues. For instance, at the Innsbruck convention he told Pauli that "to his feeling something shadowy attaches to the undulatory character of light"; Einstein believed that the wave character of light was "increasingly produced as a secondary aspect and indirectly."[20]

According to Born, Einstein still assigned the role of "guide field" to Maxwell's now downgraded "ghost field": "It defines the probability of a light quantum, the carrier of energy and momentum, adopting a particular path, though the field as such possesses no energy and no momentum." In view of this analogy between light quanta and material particles, as worked out not least by Einstein himself, Born thought it reasonable "to regard the de Broglie–Schrödinger waves as a 'ghost field' or rather as a 'guide field.' "[21] Born then interpreted this "guide field" as a probability amplitude, which, while unfolding causally over time—in accordance with Schrödinger's equation—permits no more than probability statements about the future states of a system. He summed up this paradoxical situation in an elegantly formulated theorem of quantum mechanics: "The motion of particles follows probability laws, but probability itself propagates in conformity with the law of causality."[22]

After the publication of his paper, Born reported to Einstein that he was now very happy, because "my idea of interpreting Schrödinger's wave field as a 'ghost field' according to your understanding"[23] was proving ever more useful—though of course the field propagated not in ordinary space but in a configuration space. Einstein, however, vigorously objected to serving as a godfather to the statistical interpretation: "Quantum mechanics calls for a great deal of respect. But some inner voice tells me that this is not the true Jacob. The theory offers a lot, but it hardly brings us any closer to the Old Man's secret. For my part, at least, I am convinced that he doesn't throw dice."[24]

This disclaimer of Born's suggestion, in December 1926, was Einstein's first consistent rejection of the new physics. Of course, his reference to an "inner voice" was no argument but more a statement of faith; but precisely because of that it was a "hard blow"[25] to Born. The two friends did not then realize that this was the beginning of a lifelong dispute between them on a fundamental question: what physical knowledge actually meant and actually could achieve in the realm of atoms and quanta.

Quantum mechanics was soon able to address and solve many problems, but at the same time it gave rise to new and exceedingly complex questions. Thus conflicts invariably arose when an attempt was made to assign to a particle, at a given point in time, a definite position q and a definite momentum p. What had been a matter of course in classical physics—never even questioned—developed into a stumbling block in the microcosm. Pauli described it in these words: "One can view the world with the p eye and one can view it with the q eye, but if one tries to open both eyes together, one gets confused."[26] This paradox (like many others in quantum mechanics), proved impossible to resolve, and (as had occurred several times before) it was therefore elevated into a principle—in this case Heisenberg's "indeterminacy" or "uncertainty" principle, which explains why it is impossible to open "both eyes together."

The situation described by Pauli has its reason in reciprocal relations—the mathematical core of quantum mechanics. Their inevitable consequence is that the product of the fluctuation square of two canonically conjugate variables, such as position and impulse, is always greater than Planck's quantum: $\Delta q^* \; \Delta p \geq h/4\pi$. Heisenberg demonstrated by closely argued thought experiments that this mathematical statement describes the actual situation at every observation or measurement. If the position of an electron is accurately determined by irradiation with extremely shortwave light, then the electron inevitably undergoes a change in momentum in such a way that the product of both uncertainties can never be smaller than Planck's quantum of effect.

The time had now come for Heisenberg to draw far-reaching conclusions about the structure of natural laws: "In the precise formulation of the law of causality, 'If we accurately know the present, then we can calculate the future,' it is not the second clause that is wrong, but the first one. We cannot, as a matter of principle, gain knowledge of the present in all its determinants." Any hope of restoring the law of causality in whatever manner was described by Heisenberg as "unproductive and pointless." With massive finality he declared: "Because all experiments are subject to the laws of quantum mechanics . . . the invalidity of the law of causality is definitively established by quantum mechanics."[27]

In February 1927—before the uncertainty principle, and before Bohr's principle of complementarity—Einstein in a lecture to the Mathematical-Physical Department of Berlin University had declared that nature demanded "not quantum theory or wave theory, but nature demands from us a synthesis of the two concepts, although so far this has exceeded the intellectual powers of physicists."[28] This, incidentally, was the view he had first expressed at the Salzburg congress in 1909. But what he now read about uncertainty, about Heisenberg's definitive dismissal of causality, and about Bohr's complementarity seemed to him to be neither the "amalgamation" he had envisaged in 1909 nor a synthesis.

In March 1927, in an article on the two hundredth anniversary of Newton's death, Einstein referred to his many colleagues who had abandoned Newton's foundations of classical physics by declaring that "not only the differential law, but also the law of causality—previously the basic postulate of all natural science—had failed. Even the possibility of a space-time construct that could be unambiguously assigned to physical events is being disputed." Although Einstein conceded that the radical champions of quantum mechanics had not set forth on their road "without weighty arguments," he urged them to remember their great forerunner: "Who would have the temerity today to decide the question whether the law of causality and the differential law, these last two premises of Newton's way of looking at nature, must definitively be abandoned?"[29]

As Einstein was unable to discover any inherent error in quantum mechanics, he tried to keep the debate open by offering a causal alternative to the statistical interpretation. On May 15, 1927, he submitted to the Prussian Academy a paper with a cautiously questioning title: *Does Schrödinger's Wave Mechanics Determine the Motion of a System Completely or Only in a Statistical Sense?* In his first paragraph he announced that he would "demonstrate that Schrödinger's wave mechanics suggests that each solution of the wave equations should have motions of the system unambiguously assigned to it." But this is all that was printed of Einstein's only independent attempt at a quantum-mechanical theory, because its author withdrew it a few days later, no doubt for good reasons. Only the first page of the galley proof has survived in the files of the academy.[30]

It seems that after this fiasco Einstein determined to take the bull by the horns and directly challenge the central statements of quantum mechanics, to prove that it was only temporary and was in need of improvement. He evidently intended to do this at the Solvay Conference which had been called for October 1927 in Brussels.

Einstein had stayed away from the first two Solvay Conferences after the war—in 1921 because of his trip to America and in 1924 out of solidarity with his German colleagues, who had been excluded. Now, almost a decade after the end of the shooting war, the "paper war" of the scholars was also coming to an end. As a result, Hendrik A. Lorentz, in his traditional role of president, was once again able to invite German physicists to Brussels. The conference was to be about "electrons and photons," and the list of participants promised a summit of quantum-mechanics physicists.

The old masters Lorentz and Planck could hardly be blamed for having watched the headlong development of the past two years rather uncomprehendingly. The middle generation was represented by Bohr, Born, Ehrenfest, Einstein, and Schrödinger—the last two of these adopting a predominantly skeptical attitude. Finally, there were those who had earned quantum mechanics its reputation as "boys' physics": Pauli was only twenty-seven and Dirac and Heisenberg only twenty-five, but they were already highly experienced and highly conscious of having created the modern physics of the twentieth century.

Einstein, along with Madame Curie and Paul Langevin, was a member of the Scientific Committee. He had, moreover, been asked by Lorentz to deliver a report on the state of quantum mechanics. However, in June 1927 he withdrew his original acceptance:

> After much to and fro I have come to the conclusion that I am not competent to give such a report in a manner that would correspond to the actual state of affairs. The reason is that I have not been able to take such an intensive part in the modern development of the quantum theory as would have been necessary for this purpose. This is due partly to the fact that my receptivity is too small to fully follow the tempestuous development, and partly to the fact that I do not approve of the purely statistical interpretation upon which these new theories are based.[31]

Nevertheless, all those who had heard of Einstein's rejection of quantum mechanics through letters or rumors had been looking forward to his presentation as one of the highlights of the event.

As far as can be judged from the official report of the proceedings, Einstein did not take any major part in the discussions; whenever he did take part, he prefaced his remarks by the admission that he "had not penetrated deeply enough into the nature of quantum mechanics."[32] But argument was more lively at the club of the Fondation Universitaire and over meals in the hotel. Whereas Pauli and Heisenberg tended to disregard Einstein's objections, Bohr took them very seriously. His very thorough reconstruction of the discussions,[33] written with profound sympathy, shows that Einstein was not only disturbed by the statistical interpretation of elementary processes and the abandonment of causality, but also worried by a suspicion that quantum mechanics implied some novel instantaneous remote effects which would be in conflict with relativity.

These arguments between Bohr and Einstein have often, and rightly, been described as a struggle of Titans over the last riddles of the universe. For the participants, however, they were full of good humor, charm, and friendliness. Ehrenfest, who was always present and often talked to each of the two men separately—"Each night at 1 A.M. Bohr would come to my room to say *just one single word* to me until 3 o'clock"—on his return to Leyden gave his students a lively account of the congress. "It was wonderful for me to be present at the dialogues between Bohr and Einstein. Einstein, like a chess player, with ever new examples. A kind of *perpetuum mobile* of the second kind, intent on breaking through uncertainty. Bohr always, out of a cloud of philosophical smoke, seeking the tools for destroying one example after another. Einstein like a jack-in-a-box, popping out fresh every morning. Oh, it was delightful. But I am almost unreservedly pro Bohr contra Einstein. He now behaves toward Bohr exactly as the champions of absolute simultaneity had behaved toward him."[34]

While Einstein's coevals were saddened that the greatest among them would not allow himself to be convinced, the exponents of "boys' physics" just shrugged their shoulders, regarding his stance as "reactionary."[35] Einstein, however, proved his greatness, if only silently, by his nominations for the Nobel Prize. That he should have proposed Louis de Broglie for his material waves in 1928, along with the Americans Davisson and Germer for the experimental confirmation of de Broglie's prediction, was in line with his scientific sympathies. What is surprising is that in his nomination letter, anticipating the following year's prize, he suggested that the theoreticians Heisenberg and Schrödinger be considered. After prolonged reflection he came to the conclusion that "de Broglie should have priority, because his idea is certainly correct, whereas it is still problematical how much of the grandiosely conceived theories of the two last-named scientists will survive."[36] This curiously formulated proposal of Heisenberg and Schrödinger must have had the value of rarity in Stockholm.

The development and application of quantum mechanics released an undreamed-of explosion of knowledge, but Einstein did not participate in it. When he received the Max Planck Medal in June 1929, he recalled, with a touch of nostalgia, his unfulfilled dream of explaining quantum conditions through overdetermination of differential equations in the space-time continuum of a field theory: "This goal still stands unattained, and there is probably no expert to be found who shares my hope of getting to an understanding of reality by this means."[37] Using some arbitrary terminology, he criticized the structure of the laws of quantum mechanics as "subcausality" and clung to the hope that physics would eventually, by his road, arrive at "supercausality"—though he left his audience in the dark about the difference between this concept and ordinary causality.

Untroubled by these questions of principle, the quantum mechanics physicists were solving one problem after another, from the details of atomic spectra to the electronic theory of metals. Naturally, Einstein was impressed by these successes, and he publicly announced that he "greatly admired the achievements of the young generation of physicists grouped together under the name of quantum mechanics" and that he believed "in the profound truth contained in this theory, except that I think that its restriction to statistical laws will be a temporary one."[38] Shortly afterward, however, his judgment again became harsher; he was prepared to concede to quantum mechanics only the very inferior status "semiempirical," insisting that it was "not possible to get to the bottom of things by this semiempirical means."[39]

Although Einstein was in the minority with this judgment, he was not entirely alone. His most prominent ally was Erwin Schrödinger, who found the statistical interpretation so appalling that he sometimes regretted having created wave mechanics.

At the next Solvay Conference, in October 1930, Einstein was again present, "like a jack-in-the-box," with an exceedingly sophisticated thought experiment designed to refute uncertainty. Imagine a box filled with radiation. The box has a pinpoint shutter, whose opening and closing is controlled by a clock. By weighing the box before and after light emission the energy of the light could be accurately determined while the clock was accurately recording the time—contrary to the principle of the uncertainty of energy and time.

If Einstein intended to throw Bohr off stride with this thought experiment, he certainly succeeded, according to an eyewitness account: "To Bohr this was a heavy blow. At the moment he saw no solution. He was extremely unhappy all through the evening, walked from one person to another, trying to persuade them all that this could not be true, because if Einstein was right this would mean the end of physics. But he could think of no refutation. I will never forget the sight of the two opponents leaving the university club. Einstein, a majestic figure, walking calmly with a faint ironical smile, and Bohr trotting along by his side, extremely upset."[40]

After a sleepless night Bohr managed to refute Einstein with Einstein's own weapons. The uncertain elevation of the clock during the weighing operations must, according to the general theory of relativity, make the working of the clock inexact by the amount necessary for the uncertainty relation to be met for energy and time.[41] Unfortunately we have no record of Einstein's face when he was informed by Bohr that, of all things, he had overlooked relativity theory on a decisive point. This surprising reversal of fortune evidently left its mark on Einstein, because he never again tried to refute any statements of quantum mechanics.

He had even come to the conclusion that at least some aspects of quantum mechanics would survive. In 1931 he again proposed Schrödinger and Heisenberg for the Nobel Prize, this time without reservations and placing them at the top of his list: "the two men who above everyone else deserve the Nobel Prize for physics. I am convinced that this theory undoubtedly contains a piece of definitive truth."[42]

Even though Einstein now acknowledged the exceptional status of Heisenberg's achievement, this did not mean that he had made his peace with quantum mechanics, least of all with the interpretation by

Bohr and Heisenberg which came to be known as the "Copenhagen interpretation." In the future, however, his argument was not that quantum mechanics described individual processes incorrectly, but that it described them incompletely. The search for a complete description of nature would occupy him to his last breath.

CHAPTER THIRTY-ONE

Politics, Patents, Sickness, and a "Wonderful Egg"

After the exciting and exhausting first half of the 1920s, Einstein wanted nothing more than to "be left in peace,"[1] so he could devote himself to physics undisturbed. He never quite achieved that; but for a few years, after his return from South America in the summer of 1925, he got reasonably close to it. There were two tasks which he did not wish to give up: his membership on the board of Hebrew University in Jerusalem and on the Commission for Intellectual Cooperation of the League of Nations in Geneva. In neither of these bodies, though, was he always a comfortable partner; and he finally walked out on both of them in protest.

Closest to his heart, undoubtedly, was Hebrew University, which he had inaugurated in February 1923 with his address on Mount Scopus. The same year, while in Berlin, he accepted the editorship of the first volume of the new university's *Scripta Mathematica et Physica*, himself contributing a paper written jointly with Grommer, albeit not a very important one.[2] He also accepted membership in its academic council, consisting of nine scholars, together with his Berlin colleague August von Wassermann, the director of the Institute for Experimental Medicine of the Kaiser Wilhelm Society.

When in 1925 the direction of the university was entrusted to a board, Einstein was of course invited to be a member of it. However, at the time of its constituent meeting in Tel Aviv he was in South America, and it was only at its second meeting, in Munich in September 1925, that he realized that not everything in Jerusalem would be going the way he had hoped.

Einstein's idea had been an academically elite institution, certainly taking account of the special needs of the Jewish colonization of Palestine, but autonomous and solely dedicated to the highest scientific standards, based on the unity of research and teaching. This view met with only partial understanding on the part of American Jewry. Because the money came predominantly from them, they claimed the right to lay down the guidelines for the new university. They wanted to content themselves with a teaching institution at college level and to keep control of staffing—largely with an eye to finding prestigious positions for people from wealthy American Jewish families, with scholarly achievement not always the main consideration.

The president of the university was a former New York rabbi, Judah Magnes, who was a champion of American interests. Einstein was in vehement conflict with Magnes beginning with the meeting in Munich; and although both men were pacifists, there were exceedingly angry and implacable exchanges between them, including threats by Einstein to resign. Chaim Weizmann tried to mediate: he visited Einstein in Berlin and tried desperately to stop him from leaving the board on account of Magnes.[3] Weizmann, worldly-wise, was ready to settle for half a university rather than none at all; Einstein, unwilling to compromise, would rather have no university than half a university, or one he regarded as bad.[4]

Einstein's later summary of this controversy was exceedingly bitter, even considering his usual sharp tongue—possibly because it had been for him something like a disappointing love affair: "The bad thing about the business was that the good Felix Warburg, thanks to his financial authority, ensured that the romantic and incapable Magnes was made the director of the institute, a failed American rabbi, who, through his dilettantic political enterprises had become uncomfortable to his influential family in America, who very much hoped to dispatch him honorably to some exotic place. This ambitious and weak person surrounded himself with other morally inferior men, who did not allow any decent person to succeed there. . . . These people managed to poison the atmosphere there totally and to keep the level of the institution low."[5]

In 1928, Magnes's authority, originally confined to financial and administrative matters, was officially extended to appointments as well. At this point, Einstein felt that his principles of academic autonomy had been totally undermined and resigned from the board as well as from the academic council. The only favor he was prepared to do for Weizmann and Hebrew University was to resign quietly, without a public fuss.[6] In the hope of a better future he assured Selig Brodetzky, the vice-president of the board, that he would "never cease to regard the fate of the Jerusalem university as a matter close to my heart. . . . The main thing is that we all have the same goal, namely to serve the university. I hope that my method will contribute to that goal."[7]

As for the League of Nations Commission, Einstein already had one resignation and return behind him. From 1924 to 1927 he conscientiously attended the commission's regular meetings, which were always in July, even though he did not get much out of them. But as an "international person" he felt it his duty "to restore unity among the nations, so boundlessly destroyed by the world war, and to ensure that a better and more sincere understanding among nations renders impossible a repetition of the terrible disaster we have lived through. To cooperate toward that aim is, in my conviction, a duty which no one can shirk, no matter how great his achievements and in whatever field."[8] The wearying routine of the meetings was compensated for by his encounters with Marie Curie and, above all, H. A. Lorentz, who acted as chairman. Besides, he was able to make use of this opportunity for visits to Zurich or for a brief hike with his son Eduard.

Although nominated *ad personam*, Einstein rather felt himself to be a representative of Germany, which had been admitted to the League of Nations only in 1926. In the flush of solidarity he even, on one occasion, publicly called Germany "my own fatherland."[9] Privately, however, he liked the French "better than our lot"; and on the relations between the wartime enemies, who were still far from reconciled, he remarked that he would "probably not live to see these different worlds amalgamating. But it amuses me to watch them both, without feeling that I belong to the one or the other."[10]

Inevitably, he would occasionally get tired of diplomatic suavity and become involved in fierce quarrels. One such occasion was in January 1926 in Paris, when an "Institute for Intellectual Cooperation" was established in order to provide a permanent base for the commission. When Fascist Italy intended to delegate its minister of education, Alfredo Rocco, to the committee of the new institute, Einstein insisted that its members should be independent individuals. Proudly Einstein reported a "wild struggle with the Fascist minister of education, which the man will remember as long as he lives."[11] He even wanted to stand as a candidate against Rocco; but he was unable to prevent Rocco's election to the committee, because Mussolini threatened to walk out of the League of Nations.

Even apart from this incident Einstein had no illusions about the League. But now—unlike 1923, when he had demonstratively resigned from the commission—he praised the League of Nations as "the only great instrument for peace that we have, and while we should not hold back with our criticism of it, we have no right to deny it our cooperation."[12]

In 1927 Einstein supplemented his League of Nations work with membership in a "consultative committee of intellectual workers" under the International Labor Office in Geneva. But meanwhile he continued to be involved in something a more sensitive mind might have regarded as incompatible with those efforts for peace. This was the gyrocompass—not exclusively but predominantly a military piece of equipment. As many of his friends have recorded with astonishment, however, Einstein possessed "the curious gift of shaking off anything that is disagreeable to him, just like the water off a duck's back."[13]

Immediately after the sessions of the League of Nations Commission, Einstein had spent the whole of August 1925 in Kiel, where he went sailing with his son Hans Albert while supervising the gyrocompass program. To Anschütz he "could not be grateful enough for the wonderful hiding place you have created for me in Kiel."[14] In

October 1926, when Einstein again looked in on the Anschütz factory, tests on board a torpedo boat of the Reich navy had proved the superiority of the Anschütz compass over the older three-gimbal system, so that preparations were being made for serial production. The fact that it was mainly the navy that showed interest in the instrument did not seem to bother the pacifist Einstein any more than it bothered Anschütz.

Although Anschütz had remunerated Einstein generously in earlier years, Einstein's share in the gyrocompass project was now contractual: he was to receive three percent of the sales price of each instrument, and three percent of any revenues from licenses.[15] The contract was not with the Kiel firm, though, but with the Dutch firm Giro, a distribution company founded by Anschütz primarily to evade the ban imposed by the treaty of Versailles on exports of military articles.

From 1927 onward, the small German navy was being equipped with the new compass. In tests by the French and Italian navies, it proved so superior to all other systems that it became standard equipment for them—and for many other navies, including the Japanese. Even the British Sea Lords and the U.S. Navy were impressed by the test samples they had bought, although after weighing all considerations both eventually decided on domestic manufacturers. Proudly, the firm's chronicle recorded: "We were at the beginning of a strong upward trend, and when World War II broke out, the warships of all navies of any importance, except the Anglo-Saxon ones, went to sea with Anschütz gyrocompasses."[16]

Einstein also benefited, if only modestly and temporarily, from this development. From 1928 onward his share was transferred to him from the Koopmansbank in Amsterdam: initially just under $300 a year, and later $700 to $800—not vast wealth, but useful sums. In 1939, the first year of World War II, the money did not arrive, and Einstein, by then in Princeton, sent a reminder in January 1940. He was, however, informed that Giro had been in liquidation since 1938.[17] The parent firm in Kiel, whose owner had died in 1931, no longer needed a Dutch branch to circumvent armaments controls. Einstein, no longer receiving any payments from the German Reich, was at least spared any disquieting thoughts on the propriety of earning royalties from a device which guided German U-boats and Japanese aircraft carriers.

No sooner had the gyrocompass reached the stage of production than Einstein turned to another entirely practical problem: the "perfect"—in particular, the silent—refrigerator. In their home on Haberlandstrasse the Einsteins still used an old-fashioned icebox, possibly because in electric refrigerators the motor and compressor both made a lot of noise and because the coolant, which often leaked, was not entirely safe. Working with Leo Szilard—a brilliant and versatile young physicist who had come to Berlin from Budapest and established himself as a *Dozent* at the university—Einstein designed an original pump which was driven not mechanically but electromagnetically. A liquid metal (sodium, potassium, or a mixture of the two) is moved to and fro in a tube by an alternating electromagnetic field, which thus functions like a piston. The coolant is liquefied by periodic decompression, and the consequent evaporation produces the desired cold. This elegantly conceived pump would be inherently tight and, moreover, silent.

In November 1927 Szilard and Einstein jointly applied for their first patent for a novel refrigerator;[18] this was followed over the next two years by seven more patents, which concerned either details of the induction motor or ever-new variants to protect the original patent.[19] The basic idea was also registered in the United States, in Britain, and in the Netherlands, as well as at Einstein's former place of employment, the Patent Office in Bern, where his friend Michele Besso concerned himself with some "editorial criticism."[20]

Szilard meanwhile, at the research laboratory of AEG, was looking after the realization of this elegant idea. That cannot have been easy. Not only did the alkali metals have to be kept in a permanent liquid state through high temperatures; they were also highly reactive and corrosive, and therefore difficult to handle. Although prototypes were constructed,[21] they did not lead to a marketable product. The engineering was too demanding, and the market during the Great Depression was too unfavorable. Above all, conventional electric refrigerators were now quieter and safer, so that there was no further need for an alternative. If the two inventors had been hoping to make a fortune they were disappointed. They seem to have received only small fees from AEG.[22]

Not until much later did an application emerge for the Einstein-Szilard pump—not in refrigerators but in nuclear reactors. At Szilard's suggestion, the electromagnetic pump was developed at considerable expense after World War II for use in metal-cooled breeder reactors and melted sodium reactors, though again without great success.

In 1926, Einstein's work at the League of Nations work and his patents together gave him a justification for ending his favorite scientific appointment. "As a result of the League of Nations and several industrial matters in which I got involved I have so little time left that I cannot keep up my position in Leyden," he confessed to Ehrenfest, who had been waiting in vain for a more prolonged stay by Einstein, the visiting professor. Einstein added: "I have also become too infertile for this appointment to appear justified."[23]

The professors at Leyden put their heads together and came up with a wise decision which Ehrenfest conveyed to Einstein "in its jocular formulation": " 'Quite simply, Einstein becomes our emeritus professor.' This means: henceforward he is no longer expected to come to Leyden."[24] However, Einstein felt too young for a pension without duties, and so he proposed that his salary be paid to him only if he had actually earned it by his presence in Leyden; otherwise, it should be "used for the benefit of the institute or of young physicists."[25] This proposal, Ehrenfest felt certain, would "be received by all concerned with great jubilation."[26] The benefit fund must have grown to a respectable size. Einstein made frequent private visits and once, in February 1928, came as the official representative of the Prussian Academy for a commemoration of H. A. Lorentz, who had died at the age of seventy-five. But he made only a single stay as a visiting professor, over several weeks in 1930.

Needless to say, Leyden was not the only place that was interested in Einstein: invitations were arriving in vast numbers. But by then he had learned how to decline, and sometimes he even gave free rein to his impish moods. When it was suggested that he might participate in the musical inauguration of the First International Congress for Sexual Research by taking on one of the two violin parts in Brahms's string sextet (Op. 18), he replied: "Unfortunately I don't feel in a position, on the strength of either my sexual or my musical capacities, to comply with your kind invitation."[27]

Most of the time he expressed his regrets with less humor. Asked to support an event at the Berlin People's Observatory in Treptow Park, he grumbled: "Can you believe that I am tired of figuring everywhere as a bellwether with a halo. So count me out."[28] When he received an invitation from Reich Chancellor Wilhelm Marx, Einstein did not even decline in person but had his secretary write—admittedly, "most respectfully"—"that Herr Professor Einstein unfortunately cannot comply with the Herr Reich Chancellor's invitation for Nov. 30 as he has a previous engagement for that evening."[29] Perhaps he did not particularly like Marx, a Catholic Center politician; but anyway he now managed to keep unwelcome interruptions at arm's length.

Soon, however, there came an enforced rest, because of a serious illness. In March 1928 he had, as on a few earlier occasions, accepted an invitation from Wilhelm Meinhardt, the chairman of the board of Osram, who had a chalet at Zuoz in the Swiss Engadine. Einstein combined duty and pleasure by delivering the inaugural address at the university classes in Davos (also in Switzerland). These were intended to promote understanding by bringing together teachers and students from several nations. This event, the first congress of its kind, took place on March 18, 1928.[30] To help fund this nonprofit enterprise Einstein even agreed to be the violinist in a spontaneously formed string trio.[31] However, from what was planned as a restful recuperation in an Alpine winter landscape, Einstein returned to Berlin gravely sick.

During that time he had one of his courtroom assignments as an expert witness, this time in a patent suit between AEG and Siemens before the Reich Court. He went by train to Leipzig, submitted his expert opinion, and at once returned to the Engadine. Arriving in Zuoz at night, he trudged uphill through deep snow for several hundred yards to the Meinhardt chalet, carrying a heavy bag.[32] This effort led to a dramatic circulatory collapse, no doubt aggravated by years of neglecting his health. His friend Zangger was urgently contacted, and he, as Einstein later gratefully acknowledged, "devotedly cared for my corpse"[33] and arranged for careful transportation to Berlin. The patient described what he had felt like on that journey only when the danger was over: "But I was close to croaking, which of course one shouldn't put off unduly."[34]

His treatment in Berlin was taken over by Janos Plesch, a fashionable physician with the title of professor, who collected prominent patients the way some people collect postage stamps. Even before falling sick, Einstein had been Plesch's most precious trophy at his opulent stag parties, along with Fritz Haber, the pianist Artur Schnabel, the violinist Fritz Kreisler, and the diplomat Count Rantzau.[35] The exquisite food and wines served at these parties are reflected in the menus designed by another guest, the artist Max Slevogt.

Einstein's medical friends, the respectable professors Moritz Katzenstein and Rudolf Ehrmann, were clearly annoyed that he should have entrusted himself, in his condition, to such a flashy practitioner. "Ehrmann absolutely rejects Plesch, mainly on human grounds," Einstein informed Zangger, but he defended his choice of a doctor: "After all, one isn't such an unblemished angel oneself; therefore indulgence toward all other little pigs."[36]

Einstein was a very obedient patient, but not because of any confidence in Plesch's medical skill. On the contrary, as an experienced scientist Einstein informed Plesch that he "had always been convinced that our necessarily primitive thinking must inevitably prove inadequate to something as complex as a living organism, and that only patience and resignation, along with a healthy sense of humor and indifference to one's own existence, can help at all."[37]

He certainly needed these qualities, because for many weeks the bulletins from his sickbed continued to be depressing. "I was really feeling rather lousy, because my heart, despite ten weeks in bed, is still rather troublesome. Plesch has now diagnosed pericarditis, with an accumulation of fluid in the pericardium. . . . We'll wait for the result and see if Plesch is right."[38] In addition to strict bed rest, Plesch prescribed a salt-free diet and diuretics—treatments which took time. Not until the summer was Einstein sufficiently recovered to continue his convalescence on the Baltic.

He did not return to the island of Hiddenses, where he had sometimes gone in previous years, and where Berlin's fashionable intelligentsia enjoyed the simple life, often in the nude. Instead, he rented a house in Scharbeutz, then a sleepy little vacation place on the bay of Lübeck. "Here I am forced to just laze about under splendid beech trees on the Baltic,"[39] he wrote on a postcard to Ehrenfest. "We have been on the Baltic for some months now, and my vital spirits are strengthening again. Only here have I realized what an idiotic existence one leads in the city and how happy one can be in quietude and solitude. It is also wonderful for contemplation."[40] He was reading Spinoza's letters with "much pleasure": "He knew the liberating effect of rural remoteness."[41]

But his health continued to fluctuate. "I am a lot better already,"[42] he rejoiced after his arrival, but in September things again looked worse: "My heart is still rather slack."[43] His wife's worried assessment just before their return to Berlin was: "My husband has gained strength. But he has not recovered his former freshness and vigor. . . . He will have to content himself with living at a very leisurely pace."[44]

This, however, would not have been to his taste. Besides, he had plunged into an exciting scientific adventure, a new unified theory, in line with a decision he had made under the beeches of Scharbeutz: "I now believe less than ever in the essentially statistical nature of events, and have decided to use what little energy is left me in accordance with my own predilection, regardless of the present bustle around me."[45]

"In the tranquillity of my sickness I have laid a wonderful egg in the area of general relativity," he jubilantly announced toward the end of May, while still sick in bed. "Whether the bird hatching from it will be viable and of long life is still in the lap of the gods. Meanwhile I am blessing the sickness that has thus favored me."[46] He had thought up a new mathematics, not as an end in itself but as part of his great objective, a unified theory of gravity and electricity. To reach that goal he was prepared to make radical sacrifices. His earlier attempts had been designed so that the general theory of relativity was contained within the unified theory, or could be derived from it, but he now discarded this criterion. When he believed the "bird" to be viable, he declared that some substantial aspects of his finest theory "must be consigned to the junk room, despite their successes."[47]

As Einstein was feeling too weak to attend the meetings of the Prussian Academy, Max Planck submitted two papers on his behalf: a purely mathematical prelude on June 7 and the physical application a week later.[48]

Einstein's mathematical innovation was an amalgam of Riemann's geometry and its limiting case, Euclidian geometry. In Euclidian geometry the concept of parallelism is defined for all distances, but in Riemann's "warped" geometry it makes sense only for infinitely small distances. Einstein now had succeeded in implanting what he called "distant parallelism" into Riemann's geometry; this made it possible to compare the directions of finite line elements in curved space.[49]

In this space-time continuum with distant parallelism, Einstein obtained new kinds of tensors and invariants, from which he was able to construct physical concepts; but he did not yet succeed in deriving from his involved terms equations which corresponded to those for electromagnetic fields and for gravity. With these problems, he went to Scharbeutz on the Baltic in the summer, and only there did he discover a promising way of deriving the equations he needed.

Back in Berlin in the fall, there was a lot of work to be done before, around the turn of the year, he was once more euphoric: "But the best, that which I pondered about and calculated nearly all the days and half the nights, is now ready before me, compressed into 7 pages under the name of 'unified field theory.' This looks old-fashioned, and my dear colleagues will . . . at first stick their tongues out as far as they can. Because in these equations there is no Planck's *h*. But when they have clearly reached the limits of their statistical mania, they will remorsefully return to the space-time idea and then the equations will provide a starting point."[50] Elsa shared his enthusiasm in an abridged form: "He has lately worked magnificently and solved the problem which it has been his life's dream to solve."[51]

On January 10, 1929, the paper was submitted to the academy, again by Max Planck, because Einstein had every reason to avoid the attention of the press. Strange things had been appearing in the newspapers, even more bizarre than anything that had been reported during the excitement over relativity after the confirmation of the deflection of light.

It began with a report from Berlin carried in *The New York Times* on November 4, 1928, under the headline "Einstein on Verge of Great Discovery; Resents Intrusion." Whether the source of the story had been Einstein himself or gossip among his acquaintances remains uncertain; at any rate, though, he had learned his lesson. Ten days later the headline was "Einstein Reticent on New Work, Will Not 'Count Unlaid Eggs.' "[52] When Planck presented the paper to the academy on January 10, there was no end to the amazement that what was popularly called the "riddle of the universe" might have been solved by Einstein in so few pages. The Prussian state ministry, confused by the fuss in the newspapers, asked the academy for "authoritative information" but was told that the academy was not in the habit of commenting on scientific papers submitted to its meetings.[53] Telegrams demanding information arrived from all over the world, and a hundred curious journalists besieged the academy—but they were all told to wait for the routine publication of the paper on January 30. Wisely, the printing was set for the unusual number of a thousand.

During the weeks of tense anticipation Einstein was nowhere to be found. He had established himself comfortably at Janos Plesch's country residence, a large property on the west bank of the Havel at Gatow, which Plesch had purchased from a rich manufacturer of shoe polish.[54] In addition to the impressive villa (after World War II it was the residence of the British city commandant, and Queen Elizabeth II later stayed there during her visits to Berlin), there was also a neat pavilion, where Einstein could live whenever he chose. There he was on his own throughout the winter, cooking for himself—"like the hermits of old. And to one's surprise one notices how delightfully long the day is and how unnecessary a large part of all that busy and idle activity."[55]

That activity peaked on January 30. The thousand copies of Einstein's article were instantly sold out. New printings were hastily made—three of them, at a thousand copies each, a record for the academy's *Proceedings*. This had nothing to do with the scientific interest of the paper. Eddington reported that one copy had fetched up in the window of a London department store, all six pages pasted up next to each other, with "big crowds of people pushing forward to read it."[56] On February 1 the *New York Herald Tribune* printed a translation of Einstein's paper on the last page of its first section. More readily comprehensible to its readers than the new theory itself, no doubt, was the account of its transmission by telex, which was described as a great technical achievement. Because telex channels were restricted to sequences of numbers and letters, the *Tribune*'s Berlin correspondent had arranged with some physicists of Columbia University to encode the formulas. Einstein's text, complete with the coded equations, was keyed into the machine, in Berlin, and then the experts in New York decoded the chains of symbols and reconstructed the formulas—correctly, as far as one can judge.

Einstein could not understand the hullabaloo, and indeed got angry if any of his friends became involved in it—such as the philospher Hans Reichenbach, who had reported on "Einstein's New Theory" in *Vossische Zeitung* on January 25.[57] Meanwhile Einstein was himself writing a popular article, though not for the German papers. It appeared in the Sunday edition of *The New York Times* on February 3, 1929, and the following day in *The Times* of London.[58] No doubt this was partly due to attractive fees, but also partly to Einstein's more relaxed attitude toward any fuss made across the water.

Before the publication of his paper, Einstein had given his only interview on his new theory to an English paper, the *Daily Chronicle*. Its readers thus learned firsthand: "Now, but only now, we know that the force which moves electrons in their ellipses about the nuclei of atoms is the same force which moves our earth in its annual course about the sun, and is the same force which brings to us the light and heat which make life possible upon this planet."[59] This statement was more than premature. We do not "know" to this day; and even at the time, viewed dispassionately, the new theory was not even a tempest in a teapot—it was just a little breeze that soon died down.

▪ ▪ ▪

In his unified field theory Einstein had developed a set of equations which he was convinced correctly described the electromagnetic and gravitational fields. However, any connection not only with empirical knowledge but also with accepted theory, including his own general theory of relativity, was still something for the distant future: "A more thorough examination of the field equations will have to show whether the Riemann metric in conjunction with distant parallelism really supplies an adequate interpretation of the physical qualities of space."[60]

In the fall of 1929, he believed he had dealt with the remaining difficulties: "The latest results are so beautiful that I have every confidence in having found the natural field equations of such a variety."[61] But he also realized that he was alone in that confidence: "I have now completed the marvelous theory, to the lively mistrust and passionate rejection of my colleagues in the field."[62] This was still the case on December 12, when he submitted the "beautiful results" to the academy.[63]

In 1925, when Einstein first attempted a unified theory based on "affine connections," his colleagues—according to Born's recollection—regarded his "objective as attainable and highly important."[64] This time, they reacted first skeptically and then critically—not only because Planck's quantum of effect was missing but also because Einstein was abandoning several achievements of the general theory of relativity. Wolfgang Pauli, whom Ehrenfest called the "scourge of God" because of his sharp tongue, lived up to this title in a letter to Einstein. Pauli reminded Einstein of his own interpretation of the precession of Mercury's perihelion and of the deflection of light by the sun: "All this would seem to be lost in your extensive demolition of the general theory of relativity. But I stick to that fine theory, even if it is being betrayed by you. With your remark that you are still far from being able to assert the physical validity of the derived equations you have in effect silenced your critics among the physicists! All that is left to them now is to congratulate you (or had I better say: express their condolences?) on your having gone over to the pure mathematicians." The letter continued in this tone, and Pauli concluded by betting Einstein that "within a year, if not before, you will have given up that whole distant parallelism, just as earlier on you gave up the affine theory."[65]

Einstein thought this forceful letter "quite amusing . . . but a little superficial." Avoiding all specific questions, he reproved Pauli with Olympian kindness: "Only a man who is certain that he is viewing the unity of natural forces from the correct viewpoint is entitled to write as you did. I don't maintain at all that the road I have taken is necessarily the correct one. But I do maintain that intellectually it is the most natural road that has so far come to my knowledge. Until the mathematical consequences have been correctly thought through it is by no means justified to judge them dismissively. . . . Forget what you have said, and immerse yourself in the problem with such a mind as if you had just come down from the moon and still had to form a fresh opinion. And then don't say anything about it until at least three months have elapsed."[66]

Incidentally, Pauli would have lost his bet—not because he was wrong but because of his time limit. It actually took Einstein two years to give up his distant parallelism. He made no excuses for himself when, in January 1932, he admitted to Pauli: "So you were right after all, you rascal."[67]

CHAPTER THIRTY-TWO

Public and Private Affairs

Public excitement over Einstein's unified theory in January 1929 merged smoothly into the excitement over his fiftieth birthday, on March 14. The newspapers published enthusiastic tributes from all sides. These included the communists—the "revolutionary proletariat . . . salutes the great revolutionary of the natural sciences as a fellow fighter against the dark forces of ignorance, barbarism and reaction"[1]—and the popular bourgeois writer Emil Ludwig, who proclaimed: "He is like a magician and what hangs about him seems magic."[2]

There was even a slim volume in a limited edition of eight hundred copies, containing some of the doggerel Einstein was fond of writing; this was published by the Society of Friends of the Jewish Book.[3] It was adorned with a vignette of Einstein's portrait, designed by Harald Isenstein, a sculptor, who had also created a portrait bust of him, cast in bronze. This bust was purchased by the Prussian Ministry of Education and—as C. H. Becker, the minister of culture, informed Einstein by "state telegram" on his birthday—was to be erected at the Einstein Tower as "a lasting symbol of your great achievements."[4]

On his birthday, however, Einstein had disappeared. Once again he was at his comfortable hideout in Plesch's pavilion. He had left to Elsa the task of receiving well-wishers and dealing with basketfuls of mail, telegrams, and presents. There were countless eulogies, like the one from the Berlin city government which said that his name "would over the millennia be listed among those immortals upon whose scientific discoveries rests the concept that mankind has of the universe surrounding it."[5]

He thanked most of his well-wishers with verses—copied photographically—which he seemed to be able to shake out of his sleeve more easily than his formulas. To his colleagues, who had missed him at the weekly Colloquia, he added: "Relying on the saying that you can't keep a weed down, I have slowly recovered to the point where I may hope to take part again in our Wednesday mysteries."[6]

Among his personal letters of thanks, the most interesting is the one to Sigmund Freud, who had congratulated Einstein for being a "happy one." Einstein asked Freud: "Why do you emphasize happiness in my case? You, who got into the skin of so many people, and indeed of humanity, have had no opportunity to slip into mine."[7]

Among the many presents was one which gave rise to a grotesque comedy of errors. The well-connected Janos Plesch—who was acquainted with Gustav Böss, the mayor of Berlin—had suggested that the city of Berlin might give Einstein great pleasure by the gift of a house on the water. Plesch, as he himself pointed out, had acted "behind Einstein's back."[8] Böss, however, first made sure of Einstein's approval. He then got the approval of the city government, and he even believed he knew the exact property, presumably again through Plesch. Near Plesch's country residence the city had recently purchased the Neu Cladow estate on the Havel, with a small château and a large park. In the park, near the riverbank, was an elegant "nobleman's residence," and this jewel was to be given to Einstein. Although Einstein would not actually own it, he would have a lifelong right of residence. This was a generous birthday present—and an unusual one, without precedent in republican Berlin.

But when Elsa turned up to inspect the property, she was informed by an aristocratic gentleman that "she had no business on the estate."[9] The gentleman was entirely right: when the previous owners sold the property they had, in the contract, reserved for themselves a long-term right of residence. The city, therefore, had given away something which it may have owned but which it had no right to dispose of. An alternative was proposed, by which a substantial part of the grounds would be sliced off the estate, but this too failed in the face of the contract.

While the newspapers were publishing sarcastic commentaries on public incompetence, other properties were hurriedly proposed, all of them unsuitable. One was situated behind the stables of the Gatow estate, had no access to the water, and, as the journalists quickly discovered, "cannot be used at the moment because of a plague of midges and flies."[10] Another was next to a terminal where streetcars reversed direction with noisy screeching. Eventually it was suggested that Einstein should find a plot of land himself, which the city would buy and present to him—but that he would then have to have a house built on it and pay for that himself.

The newspapers got hold of this compromise proposal, and in consequence it came to the knowledge of some acquaintances of Einstein's, a family called Stern, who lived in Caputh, a village south of Potsdam at the meeting point of Lakes Templin and Schwielow. The Sterns were prepared to sell Einstein an unused part of their property, about one-third of an acre, on Waldstrasse. This was three minutes' walk from the water, but it was on a slight elevation at the edge of the forest, with a distant view of the Brandenburg lake scenery. Einstein knew Caputh from sailing, and he was agreeable.

Meanwhile, the city government had discovered that it was perhaps entitled to give away a right of residence, but not to make a purchase followed by a donation. In order to meet the budget requirements, therefore, a proposal was made at the city deputies' meeting on April 24 that "for the purpose of the donation of a property in Caputh as a gift to honor Prof. Einstein on the occasion of his fiftieth birthday, approximately 20,000 marks be made available from the resources of the land acquisition fund."[11] As a result, the project had moved into the political arena, and there it got stuck.

After Einstein had already applied to the Prussian Forestry Administration for the purchase of an adjoining strip of land, in order to round off the property to be given him as a gift, and indeed had submitted his building plans for the house in Caputh, the German Nationalist opposition in the city parliament exploited the situation to attack the Social-Democrat majority, insisted on a debate from which the public would be excluded, and delayed the decision—until, on May 14, the *Berliner Tageblatt* announced under a banner headline "Public Disgrace Complete—Einstein Declines."

The *Tageblatt*, usually well informed on matters concerning Einstein, reported that in a letter to the mayor of Berlin, Einstein had "with gentle sarcasm pointed out that life was much too short and that the affair of this gift to honor him had gone on too long for him now to be able to accept it."[12] It is possible that Einstein had meanwhile realized that the gift of a plot of land would seem rather strange in a republic, especially for a man with socialist leanings. The plan was strongly reminiscent of gifts of land by emperors and kings, even though it was on a considerably more modest scale than, for instance, Bismarck's estate, Sachsenwald. At any rate, Einstein decided to build his house on a plot paid for by himself, and he refused to change his mind on this point.

During the search for a plot, a young architect, Konrad Wachsmann, had attached himself to Einstein, hoping to profit from his client's fame. Wachsmann was employed by a timber construction firm in the Lausitz region; on the basis of its prefabrication methods he designed a country house with clean lines, somewhat suggesting the Bauhaus style.[13] Except for the foundations, it was to be built entirely of wood, but it was to be winter-proof and equipped with central heating.

While it was being built, the Einsteins spent their first summer in Caputh, in a rented and very sparsely furnished old house in an overgrown garden by the water, with its own mooring. Tied up to it was Einstein's most beautiful birthday present, paid for jointly by a group of his friends. It was a 215-square-foot dinghy cruiser, the *Tümmler* (*Porpoise*), a fine, comfortable mahogany-fitted vessel "given to me by high finance."[14] While Elsa, supervising the construction of the house, was getting on the architect's and the workers' nerves, Einstein enjoyed himself sailing on the Havel lakes and pondering the unified theory. Rural life pleased him enormously, and his health was improving: "My heart isn't bothering me anymore, so that I feel almost as before."[15] In August he even traveled to Zurich for a week for the World Zionist Congress—his first major trip since he had fallen ill eighteen months previously.

In September 1929 the Einsteins moved into their house. Elsa described it as "very artistic, very modern! Only four bedrooms, a very large living room, maid's room, and bathroom. The most modern central heating system and hot water in every corner."[16] This modernity, however, did not extend to the furnishings. Wachsmann had persuaded the famous Bauhaus artist Marcel Breuer to design the furniture—at a special price, of course, since such a famous client would be a good advertisement for the Bauhaus. But when Einstein saw some of the designs, he protested that he was "not going to sit on furniture that continually reminds me of a machine shop or a hospital operating room."[17] So nothing came of that idea; instead, Wachsmann—who was responsible for a number of elegant and efficient fittings—now had to watch his creation being completed by whatever pieces of furniture could be spared from the apartment on Haberlandstrasse. This reduced costs at a time when much of the Einsteins' savings had gone into the house—although naturally only from the account in Berlin, not from accounts abroad. Besides, Einstein was not in the least worried about disharmony in the interior decoration. In artistic matters he had the conservative tastes of any average middle-class person; the avant-garde had never appealed to him.

Once they had moved in, his new property and the lifestyle it represented exceeded all his expectations: "I like living in the new little wooden house enormously. Even though I am broke as a result. The sailing boat, the distant view, the solitary fall walks, the relative quiet; it is a paradise."[18] He left his refuge only for some meetings of the Prussian Academy, and now and then for some public occasion: "This evening I must radio Edison direct to America on his birthday—a beanfeast for the journalists. I'm going to Berlin specially."[19]

The official reason for Einstein's trip to Paris was that he was being awarded an honorary doctoral degree by the Sorbonne. The situation had changed since his last visit eight years previously. There were no more political problems; Einstein did not even mind being accommodated at the German embassy in the splendid Palais Beauharnais—

which made him something of a representative of the Weimar Republic. On the day after his arrival he gave the first of two lectures to physicists and mathematicians at the Institut Henri Poincaré, needless to say about his new unified field theory;[20] and in the evening he met with Maurice Solovine, his old friend from his days in Bern.

On November 9 came the solemn inauguration of the new semester in the Aula of the Sorbonne, when traditionally the honorary degrees were presented. As the German ambassador reported, Einstein "was honored by ovations lasting several minutes, which clearly characterized his rank with regard to the other honorary doctors, as well as the unique esteem he enjoys in the French world of science."[21] The solemn ceremony at the Sorbonne was followed by meetings of the Société Française de Philosophie and the Académie des Sciences. "Everywhere he was met with the greatest warmth and the most natural respect, and a conversation with him was universally regarded as a great honor."[22] Apart from the official events there were conversations with Louis de Broglie and Paul Langevin, as well as the mathematicians Elie Cartan and Jacques Hadamard. Only Marie Curie was missing: she was in America.

The newspapers had already reported that Einstein would be accepted into the Académie des Sciences as an *associé étranger*, and circumstances during his visits there all suggested that this was intended.[23] Einstein, however, objected to the fact that his friend Paul Langevin had still not been elected. Langevin had long been a fellow of the Royal Society and of many other learned societies; and recently he had succeeded Lorentz as president of the Institut Physique de Solvay and hence as the chief organizer of European physics. Out of loyalty to Langevin, then, Einstein initially declined this honor: "It would be exceedingly painful to me to be elected to this body while my revered friend Langevin is not a member of it. I am sure you will understand the feeling that motivates me in this request."[24]

The week's visit ended with a formal dinner at the residence of the German ambassador, to which the cream of the French intelligentsia had been invited. Back in Berlin, Einstein found his stay in Paris "wonderful, even though it represented the extreme stress that my shaky ego can still stand up to."[25] Months later he still recalled "the beautiful days in Paris, though I am happier with my relatively quiet existence here."[26]

In April 1930 Einstein moved back to his house in the country, accompanied by his wife and the indispensable maid. Now and again a secretary, Helen Dukas, came over from Berlin; she had been keeping Einstein's papers and correspondence in order since April 1928 and had almost become a member of the family. Helen Dukas, like Elsa, had been born in Hechingen; and Elsa had personally selected this young woman and had made sure she was reliable in every respect. The family also included Elsa's daughters, Margot and Ilse, and Ilse's husband, Rudolf Kayser, who had their own rooms in the house at Caputh and thus were able to spend weeks and even months there.

Then there was Dr. Walther Mayer, called "the calculator," a mathematician from Vienna and the author of a book on differential geometry, whom Einstein had brought to Berlin as his assistant toward the end of 1929. Their collaboration started promisingly and produced a joint paper[27] submitted to the Prussian Academy. Einstein greatly appreciated Mayer, describing him as "a splendid fellow who would have long had a professorship if he were not a Jew."[28] Mayer also lived in Caputh off and on—not in Einstein's house but nearby.

To Berliners, Caputh was really "at the back of beyond"—it was reached by train to Potsdam and from there by a rather infrequent bus—and Einstein did not even have a telephone installed there. Nevertheless, the quiet rustic existence of the summer while the house was being built did not last long. The happy homeowner had been rather too generous with his invitations, and everyone came: his sister Maja, his son Eduard, other relatives, friends, fellow physicists from Berlin and abroad, and a parade of prominent figures, from the celebrated poet Gerhart Hauptmann to the famous Indian poet and philosopher Rabindranath Tagore, an apostle of the mystical wisdom of the East.

Tagore, then on a world tour, did not want to miss visiting Einstein, but the villagers of Caputh must have been surprised to see him arrive—on July 14, 1930—in a flowing gown and with a large entourage. The entourage included two secretaries who busily wrote down what their master and his host had to say to each other on Einstein's terrace. A few weeks later, it appeared in print;[29] but there was no wonderful fruit from the tree of eastern or western wisdom—hence Einstein's comment: "The verbal dialogue with Tagore was a total failure because of communication difficulties, and should of course never have been published."[30] Despite this, it was reprinted many times.

Einstein was so fond of his rural existence that he stayed away from the physics Colloquium and even the weekly meetings of the academy, turning up only when Planck or he himself submitted a paper—and by now neither of these was a frequent event. "Apart from the many visitors who come and go, life is idyllic," Elsa said of a happy summer in Caputh. She was even enthusiastic about her husband's energy: "Albert is working as he has hardly ever worked before. Has thought up the most wonderful theory. It's getting more beautiful every day. If only it proves to be *true!!!*"[31] But her loyalty was of no use; the theory of distant parallelism had nothing more to yield, and even Einstein's collaboration with Walther Mayer produced nothing further that was suitable for publication.

To the Berliners Einstein was still the world's greatest scientist, and he became the subject of numerous anecdotes. For example, Janos Plesch was responsible for a story about something that occurred when he and Gustav Böss, the mayor of Berlin, were with Einstein at his country house. When the wind brought the pungent smell of sewage from a nearby sewage treatment plant, the mayor felt somewhat responsible for this unpleasantness and put the embarrassed question to Einstein whether this did not annoy him when he was staying in the country. Einstein's answer—"Now and again I return the compliment"[32]—soon made the rounds in Berlin; and Plesch said of Einstein, "He was not afraid of lasciviousness. One is at ease with him."[33]

Einstein was not only the subject of countless anecdotes but also an object of humor. These jokes were inspired in part by his increasingly conspicuous appearance. Comedians in nightclubs and cabarets often made cheap jokes about the supposedly unworldly professor, whose mane of hair and violin case made him look more like a down-at-heels musician than a scientist.

As a violinist Einstein was greatly feared. "Einstein's bowing was that of a lumberjack," a professional violinist recalled.[34] No doubt he had long ago given up regular practicing. Still, he could not bear to have his playing criticized, as the daughter of the publisher Samuel Fischer, who frequently accompanied him at the piano, informs us. She herself witnessed repeated outbursts of anger: once, playing the Bach concerto for two violins, he suddenly shouted at his partner, Eva Hauptmann—a good violinist who was the daughter-in-law of the poet—"Don't play so loud!" According to Tutti Fischer, Einstein got a lot more excited over musical disputes than scientific disputes.[35] Still, he actually performed publicly in the second movement of the Bach concerto, at a concert in the Oranienburger Strasse synagogue.[36] But as this was a benefit concert for the welfare and youth department of the Jewish community, perhaps the beauty of his tone was not very important.

Despite the relaxed atmosphere in Caputh and in the Einsteins' town apartment, their guests did not always feel comfortable. They could hardly miss the fact "that relations between him and his wife were inexplicably cool. Frau Einstein was there, and yet she was not there."[37] The atmospheric disturbances were due to the fact that Einstein, as his architect—among others—had observed, acted on women "as a magnet acts on iron filings"[38] and did not find this effect in the least disagreeable. After breaking off his relationship with Betty Neumann, he had clearly not contented himself, as he then wrote to her, with seeking his happiness in the stars. Although he remained married to Elsa, he confronted her, with characteristic directness, with his realization that he was not suited to the role of a faithful husband.

Since the fall of 1925, Einstein had frequently been seen in the company of a beautiful woman, not much younger than himself, a widow, elegant and exceedingly attractive. This woman, Toni Mendel, had an exquisite lifestyle, keeping a car and chauffeur, and living in a large villa on the Wannsee.[39] It had become Einstein's custom to spend many a day and many a night there, at one time even mooring his sailboat on the Wannsee. Einstein never made a secret of this relationship, and Toni Mendel seemed somehow to belong to his extended family, even if she was only just tolerated by his wife. Elsa Einstein showed little diplomacy in standing up to her husband; instead, she tried to bully him. On one occasion when Toni Mendel arrived to pick him up for an evening at the theater, there was a noisy row in the apartment while the car waited below—Elsa would not let him have the necessary petty cash for his night out.[40]

Einstein was not a man to worry about such scenes, and so Toni Mendel for many years remained his regular companion at concerts and at the theater, and even sailing at Caputh. But she was not the only one; as the housemaid recalled, the Herr Professor was "fond of looking at pretty women. He had a weakness for pretty women."[41] And there was no shortage of pretty women eager to be seen in public with a genius. This did not seem to affect his relationship with Toni Mendel, though, even when she left Berlin in 1932.

By then he had long had another woman friend, again with limousine and chauffeur, Estella Katzenellenbogen, who owned a chain of florist shops. But above all there was Margarete Lenbach, known as "the Austrian," a very pretty, youngish blond, whose presence Einstein's wife could not bear at all. Despite Elsa's attitude, "the Austrian" came to Caputh regularly once a week, and, again according to the housemaid's recollection, the Frau Professor on those days had no choice but to clear out and travel to Berlin early in the morning to do some shopping.[42]

Although Einstein cared little about the gossip in Berlin, and less about the outrage of the villagers at the female companions with whom he would go sailing or drop anchor among the reeds, he was anxious about his reputation with posterity. All the letters between him and Toni Mendel, even those in later years when she was living first in Zurich and then, like himself, in America, were destroyed at his request.[43] From none of the women, not even from "the Austrian," have any written traces of affection been preserved for us. He clearly intended these women, like his daughter, Lieserl, to vanish in the shadows of history.

▪ ▪ ▪

Einstein's desire to keep his private life private was clear to everyone who hoped to write about him—most of all his stepson-in-law Rudolf Kayser, an editor of *Neue Rundschau* and an advisor to the S. Fischer publishing house. Kayser was planning to publish a biography in connection with Einstein's fiftieth birthday.

The arguments used by Einstein to dissuade Kayser were probably the same which he used on David Reichinstein, another prospective biographer, some time later. Einstein thought it "in bad taste to have biographies or autobiographies published of persons still living." At the most he would accept accounts "which relegate personal aspects to the background." And finally, no doubt remembering the fuss about Moszkowski's book in 1921, he pointed out that "people of my personal circle are alienated as a result. They regard me as vain and publicity-seeking, and quite naturally so. Even though such views do not in themselves greatly concern me, they yet considerably complicate my life, as you can well imagine, and create a tense atmosphere, whereas I love a harmonious one more than anything."[44]

Having forbidden his son-in-law to publish in Germany, but not having ruled out publication abroad, Einstein suggested the same to Reichinstein. Although this was still "in bad taste," he conceded that "the authors really needed, or rather need, to make money, and that they cannot be expected to wait until I am dead."[45] However, when he had read parts of Reichinstein's manuscript he wrote to the author that "my good will is now definitively at an end. If you publish this manuscript anywhere and in any manner, then everything will be finished between us forever."[46] The book nevertheless appeared in Prague in 1934, much to Einstein's displeasure.[47]

Kayser's biographical sketch appeared in New York in 1930 under a pseudonym,[48] but with a short (and no doubt promotional) foreword by Einstein himself, confirming that the author was well acquainted with him. But Einstein's assertion that he had read the book and found all the facts accurate cannot be true, in view of several bad mistakes. Einstein probably avoided this book, as he did all subsequent books about him, following "a rule to which I have strictly adhered ever since my newspaper fame. In this way one doesn't get spoiled by praise or depressed by blame."[49]

Einstein himself often used the device he had suggested to his biographers: publishing abroad. When his article *The New Field Theory* appeared in *The New York Times* and *The Times* of London, but in no German paper, the reason may not only have been his fee—in hard currency—but also his concern that writing a newspaper article on his own work might have been regarded in Germany as a breach of scientific etiquette.

He chose this method also for a very personal text, which he had written in Berlin in 1920 but allowed to be published only in America.[50] Under the title *The World as I See It*, Einstein sketched out his thoughts on the "existential situation of us earth-dwellers," derived, at times even to the choice of words, from Schopenhauer. He listed his ideals as "goodness, beauty and truth" and praised "the mysterious . . . as the most beautiful thing we can experience." These are not profound or original insights, and if they have any significance at all, it is only that Einstein did not shrink from perpetuating the cultural values of a bourgeoisie that was scarcely capable of creating any new culture. These cozy ideas might have been set down by any professor or schoolmaster.

Einstein's resolute championship of social justice and democracy, especially the American version, would have been more unusual in academic circles. Now and again, though, a very German and very romantic cult of genius shows through. Einstein describes "personality" as the "really valuable quality: it alone creates things noble and sublime, while the herd as such remains dull in thought and dull in feeling." This kind of elitist contempt for the common man could have been voiced, or even written, by any reactionary academic—but not so the harsh words which Einstein had for the military: "If anyone can take pleasure in marching to music in line, dressed by the right, then I already despise him; he only received his brain in error, as his spine would be quite sufficient for him." If this had been printed in Germany, where a field marshal had for the second time been elected president, Einstein might well have found himself facing a criminal charge.

In a revealing and very intimate passage, which is not about how he sees the world but about how he sees himself, he explores a source of ambivalence: his "passionate sense of social justice" had always "strangely contrasted with a marked lack of a need to attach himself either to individuals or to human communities." He is a real "loner, who never belonged wholeheartedly to the state, the homeland, my circle of friends, or even my own family, but experienced with regard to all these ties a never abating feeling of outsiderness and a need for solitude. . . . One experiences clearly, but without regret, the limit of communication and consonance with other people."

When he was less steeped in Schopenhauer and speaking more freely and directly, he described his life in Berlin as that of "a Gypsy . . . an unattached person, who was fond of looking at the comical side of everything."[51] This sounds less bombastic, but it describes only one aspect of Einstein's fluctuation between his loner's need for independence and his growing urge to intervene in social and political matters—such as his pacifism and his appeals for resisting the draft.

Einstein had always been a pacifist, virtually by nature; but his attitude during World War I was essentially that of a private individual and did not go beyond that of a small circle of like-minded people. The fact that in the late 1920s he turned into a militant pacifist was due to his disappointment at the failure of the League of Nations to enforce disarmament and proscribe war. When the League achieved some minor successes, such as codifying the rules of warfare and banning the use of poison gas, Einstein responded with fierce public criticism and a call for the refusal of any kind of war service:

> To try to impose certain rules and restrictions on war seems to me quite pointless. War simply is no game and cannot be conducted according to the rules of a game. War must be opposed as such, and this can be done most effectively by the masses through an organized wholesale refusal of military service already in peacetime.[52]

This first declaration, at the beginning of 1928, was followed by a flood of statements, and by greetings and messages to members of pacifist organizations, with the result that Einstein soon rose to be the hero of militant international pacifism.

He was fully aware that with regard to this problem his feelings were "stronger than my reason."[53] His words, moreover, were often aggressive, implacable, and anything but peaceful. For him, the difficult problem of a "just war," discussed by international lawyers and philosophers from Grotius to Kant, was settled: military service, without exception, was preparation for organized murder. This is what he wrote in the "Golden Book of Peace" of the World Peace League in Geneva:

> No person has the right to call himself a Christian or a Jew so long as he is prepared, at the command of an authority, to engage in systematic murder or to allow himself to be misused in any way whatever in the service of such an enterprise or the preparation for it.[54]

Asked what should be done if war came, he had an unequivocal answer: "I would absolutely refuse any direct or indirect war service and endeavor to induce my friends to adopt the same attitude, moreover regardless of the judgment on the causes of war."[55] Such statements were, and no doubt would still be, regarded as scandalous among a broad cross-section of all nations. Einstein himself, as the political situation changed, was soon forced to revise his views.

Whatever the roots of Einstein's rigid pacifism, they had nothing to do with unconditional respect for human life. "In principle," he wrote in connection with the death sentence, in the midst of his pacifist commitments, "I would have no objection to the killing of worthless or even harmful individuals; I am against it only because I do not trust people, i.e., the courts. I appreciate more the quality than the quantity of human life."[56] This assessment of human life from the viewpoint of quality and quantity is the jargon of the "master race," and it is especially shocking coming from someone who regarded himself, and who was regarded by others, as a humanitarian pacifist. But anyone examining Einstein's attitudes will time and again come up against glaring contradictions. Einstein evidently was able to live with them, possibly because he did not even notice them.

In Germany all military matters continued to enjoy great respect, despite the horrors of the world war; but Einstein's call for a refusal of military service was largely academic there, since under the treaty of Versailles, Germany had only a small army and no conscription. Still, wherever a young man was arraigned in court for refusing to serve—whether in Finland or in Poland[57]—Einstein would raise his voice, writing to ministers and to military courts. On this issue he made no compromise: even the militia system of Switzerland—a country quite incapable of aggression, and a system to which he had paid his personal tribute ever since his forty-second birthday by paying his army tax—did not now escape his condemnation.[58] He even stuck to his pacifist principles when, in 1929, the security, and perhaps the very existence, of his "tribal companions" in Palestine was endangered.

Despite his ambivalent attitude toward Zionism and his anger at the way Hebrew University was developing, Einstein attended the sixteenth Zionist Congress in Zurich in August 1929. This event was the crowning culmination of Weizmann's efforts to unite all organizations involved in building up a Jewish Palestine under an umbrella Jewish Agency—regardless of any political differences between them, and naturally under his own presidency.

The speakers at the opening session—on Sunday, August 11—included the most prominent advocates of a Jewish homeland in Palestine, such as Felix Warburg and Lord Samuel (former British governor of Palestine), and of course also Albert Einstein. Typically, he sincerely welcomed "the courageous and efficient minority of those who call themselves Zionists" and then continued: "But we, the rest . . ."[59] The demonstration of Jewish unity did not leave Einstein unimpressed. In the evening at the Grand Hotel Dolder he wrote on the hotel's elegant notepaper: "This day the seed of Herzl and Weizmann has ripened in a wonderful way. Not one of those present remained unmoved." Weizmann wrote below, "Mille amitiés. Je t'embrasse,"[60] and kept this precious document to the end of his life. Soon, however, both men were to discover that the seed had ripened on contended ground.

No sooner had Einstein returned to Berlin than the newspapers reported serious attacks by Arabs on Jews in the Old City of Jerusalem. Excesses spread across Palestine like wildfire; hundreds of Jews were brutally done to death in Hebron and Safad before the troops of the British Mandate succeeded in quelling the revolt. Einstein had for a long time ignored the conflict between the long-settled Arabs and the Jewish immigrants, and he had underestimated it even after the outbreak of the revolt. Misreading the actual state of affairs, he tended to hold Britain's policy of "divide and rule" responsible for all the problems.

In a statement published first in England and then worldwide, Einstein sharply condemned the Arab marauders and the failure of the British Mandate; but unlike most Zionists, he called not for punitive action but for a fair settlement based on everyone's interests: "Jews do not wish to live in the land of their fathers under the protection of British bayonets. They come as friends of the kindred Arab nation."[61] When a speaker at a meeting in Berlin demanded that the Arabs be punished according to the law, Einstein reproached him for having "spoken like Mussolini" and shown "a lack of the spirit of conciliation."[62] When, finally, several Arabs who were found guilty of murder had been sentenced to death, Einstein, jointly with the Internationale of Military Service Refusers, approached the high commissioner of Palestine with a request for the sentences to be commuted to imprisonment.

Einstein also appealed to Weizmann to seek ways of "peaceful cooperation" with the Arabs; otherwise "we would have learned nothing from our 2000 years of suffering."[63] In an exchange of letters with the editor of an Arab journal in Jaffa, Einstein reiterated his conviction that "the two great Semitic peoples . . . have a great common future"[64] if only they could find a way to abolish mutual mistrust through sensible cooperation. His suggestion of a secret council of four Jews and four Arabs to represent and harmonize their interests vis-à-vis the British Mandate was, of course, both arbitrary and unrealistic.

▪ ▪ ▪

An important scientist who had passed his fiftieth birthday was expected by his colleagues not so much to produce exciting new discoveries as to be a dignified representative of their discipline and a wise organizer of research. Einstein had persistently evaded this second role while giving a dazzling performance in the first: as a representative of physics for the public. Newsreels suggest that he derived a good deal of pleasure from this role—for instance, in August 1930, when he opened the Seventh German Broadcast and Phono Exhibition at the foot of the radio tower in Berlin. Wearing an elegant suit, his mane of prematurely gray hair blown by the wind, he stepped up to the microphone-studded rostrum and, as his address was being carried by all German radio stations, began "Dear present and absent listeners."[65]

He did not overlook the opportunity to urge his audience of millions to show greater respect for science and technology: "When you listen to the radio, give a thought also to how men came to possess this wonderful instrument of information. Because the prime source of all technical achievement is the divine curiosity and the playful urge of the tinkering and pondering researcher, just as much as the constructive fantasy of the technical inventor." Citing a chain of scientific ancestors, he paid tribute to Oersted, Maxwell, and Hertz, and among the technological pioneers to Reis and Bell; but he also commemorated the "army of anonymous technicians, who so simplified the instruments of radio communications and adapted them to mass manufacture that they have become accessible to everybody." What C. P. Snow would later describe as the problem of the "two cultures" Einstein dealt with in a single brief remark: "They should all be ashamed, those who thoughtlessly make use of the miracles of science and technology, without understanding any more of them than the cow does of the botany of the plants it eats with enjoyment." However, his expectation that broadcasting would perform a unique function "in the sense of international conciliation" was not fulfilled; and less than three years later it was being used in Germany as a tool of the vilest propaganda.

▪ ▪ ▪

Since 1923 Einstein had not exercised his right to lecture at the university. Thus interest was all the greater when he did give an occasional lecture, not only because of his fame but because of his talents as an orator. "Among the 60 colleagues in the Academy most of them can, after due preparation, convey a more or less laborious instruction to a wider audience": this was how Fritz Haber characterized the average level of academic rhetoric when he thanked Einstein for having undertaken a public lecture. "But the combination of professional ability, skill of presentation, and personal freshness which is needed for shaking such a performance out of one's sleeve—that is found in you alone."[66]

In consequence, the hall was invariably crowded when Einstein lectured, no matter whether he spoke to the Mathematical-Physical Working Group of the university in Hall No. 122, accommodating just under a thousand, on *Theoretical and Experimental Aspects of the Question of the Generation of Light*,[67] or to the Kaiser Wilhelm Society in the Goethe Hall of the Harnack House in Dahlem, with a capacity of 450, on *Physical Space and the Ether Problem*.[68] For his lectures Einstein did not use prepared manuscripts; at most, he referred to a few notes on a scrap of paper.

A magnificent and well-documented instance of Einstein's skill as a speaker was his address in June 1920 at a festive meeting of the German Physical Society. To mark the fiftieth anniversary of Max Planck's doctorate the Max Planck Medal was presented for the first time, first to Planck himself and then, by Planck, to Einstein.[69] The way in which Einstein in his acceptance speech paid tribute to the achievements of his "revered master" as the basis of his own scientific development, the way he opposed his struggle with the quantum riddles to the already established statistical interpretation, and the way he finally thanked Planck for stimulating and supporting his own work and for setting an example of scientific integrity—all this marks an achievement of oratory that is rarely found in science.[70] And if Janos Plesch is to be believed, Einstein had jotted down a few words on the back of a shoemaker's bill just before the meeting, and then only because he feared that he might be overcome by this tribute to his great father figure.[71]

Einstein's admiration of the "master," however, stopped short of willingness to become Planck's successor at the university. Naturally enough, this most prestigious position had to be offered to Einstein first, because of his "unique position"; but "a confidential inquiry has revealed that Herr Einstein does not wish to see any change in his present relationship with the faculty, which fully satisfied him."[72] His comment on the prolonged search for a successor to Planck was: "Thank God I am aloof and don't have to participate more than I want to in this race of the intellects. Participation in it has always seemed to me a bad kind of slavery, no less than a hunger for money or power."[73] What mattered to him most was preserving his own independence. By then one could hardly have visualized him in the strict routine of faculty meetings and classroom teaching.

When Erwin Schrödinger was eventually appointed, Einstein had not only a new colleague but, soon, also a new friend. In many respects he felt that he had more in common with Schrödinger—who was Austrian—than with the stiff-necked Prussians.

While his work in Berlin represented an enormous enrichment of physics, Einstein had always been rather reluctant to concern himself with organizational matters. Throughout the two decades of his directorship no start was made on the construction of the Kaiser Wilhelm Institute for Physics (the institute was built only during the Nazi era, with money from the Rockefeller Foundation in America), and this was largely due to his own lack of interest.

As early as 1914, when—in connection with his appointment to the Prussian Academy—the foundation of a Kaiser Wilhelm Institute for Physics under his directorship was being considered, Einstein had wanted his salary to be settled independent of the Kaiser Wilhelm Institute. It might well turn out, he said, "that I am not the right person for the directorship of the institute and . . . the work connected with it, so that I might find myself induced to lay down that office."[74] Because of the war, the institute was not actually founded until 1917, and then only as a provisional expedient, without its own building or staff. It existed only in the attic of its director, Einstein, though it did have a fund for financing physical research.

After his trip to Japan, Einstein delegated its management to Max von Laue, and he wanted this arrangement to continue even when a proper institute was set up: "I would like it best if you were given the K.W. Institute," he wrote to his colleague, who was tired of "lecturing." "Then I would enjoy going there myself and taking part according to my abilities."[75] But Einstein did nothing toward the realization of such an institute.

Not until March 1929 did Einstein, jointly with colleagues from the board, submit a proposal for an institute of theoretical physics. This would concern itself also with topical experimental problems and in consequence would be rather expensive. The expenditure, however, would be more than justified because "theoretical physics has undergone a development unprecedented in the history of the discipline, a development so powerful and successful that we are almost embarrassed when looking for anything comparable in the entire history of science. It would be no exaggeration to say that theoretical physics . . . has over the past quarter-century become the center from which the most light and the strongest inspirations radiate for activity in other fields."[76] The following day the scientific staff of the Kaiser Wilhelm Institute in Dahlem were emboldened to propose to the president the establishment of an Institute for Theoretical Physics, in order to remedy the "lack of an intellectual center."[77]

Implementation of these plans was at first prevented by a shortage of money. This was remedied a year later, in April 1930, when the Rockefeller Foundation provided the colossal sum of $655,000 for building an Institute of Cell Physiology and an Institute for Physics.[78] The Institute of Cell Physiology, under Otto Warburg's energetic directorship, was inaugurated as early as 1931; but the Institute for Physics did not even have its questions of staffing settled by then.

The Americans, who had largely relied on Einstein's prestige as its director, were bitterly disappointed when they visited Berlin toward the end of 1930. Einstein had just left for America, and there was even a rumor that he had accepted a position at the California Institute of Technology. Friedrich Glum, the administrator of the Kaiser Wilhelm Society, with the title of director-general, was able to refute this rumor, though he admitted that Einstein was "erratic in his decisions" and a "difficult" personality, and that "with him one has to be prepared for anything." Glum was not even able to provide any information on Einstein's position in the institute: "Einstein could very well move into the institute and find it a suitable place for his activities, but it is also possible that he would prefer to stay in his own home and do his thinking there."[79]

Einstein in fact did prefer to do his thinking at home, so that Max Planck, who had become president of the society after Adolf von Harnack's death, had some difficulty applying the Rockefeller money and establishing the Institute. The American foundation, despite considerable misgivings, did not wish to revoke its promise, even after the National Socialist seizure of power, and so a Kaiser Wilhelm Institute for Physics eventually did come into being in 1936, under the directorship of Peter Debye, but this was at a time when Einstein had long since left Germany.

While his friends, and maybe he himself, believed that with the acquisition of his country house the "Gypsy" had settled down, Einstein was getting ready for another journey. The stars were showing him the way to California—in an entirely scientific sense. The astronomers of the Mount Wilson observatory, financed by the Carnegie Foundation and using a giant hundred-inch reflector telescope, come up with sensational observations and discoveries that cast a new light on the structure of the universe. Moreover, one of the guests in Caputh during that wonderful summer of 1930 had been Arthur Fleming, president of the board of the California Institute of Technology in Pasadena, which, under Millikan's presidency, had developed into a top-ranking university. When Fleming renewed Millikan's earlier unsuccessful invitations, Einstein now showed great interest, because of the Mount Wilson observatory and the theoretical processing of its astronomical discoveries by the fine physicists of Caltech, such as Richard Chase

Tolman and Paul Epstein. To cooperate with them, not just as a "table ornament," was so tempting to Einstein that he declared himself ready to spend the first two months of 1931 there as a "research associate."[80] A fee of $7,000 was agreed on; this was the annual salary of a senior professor. A more permanent association at an even higher fee was being considered.

Einstein's ready acceptance of this offer was connected with the universe, and possibly also with a probing search for a new orientation, but probably not with this earthly vale of tears or the political situation in Germany. Admittedly, the economic situation in Germany had further deteriorated since the fall of 1929, because of the worldwide depression, the reparation payments laid down in the Young Plan, and the consequential economy measures of Heinrich Brüning's government. The growing army of unemployed included many young scientists who, for lack of money, had lost their jobs. But no one, not even the politically attentive Einstein, suspected that the Weimar Republic was nearing its end.

Einstein certainly did not allow politics to mar his summer. In the Reichstag elections of September 14, a previously unimportant splinter group, the National Socialists—under their "Führer," Adolf Hitler—received over six million votes; and in consequence the worst kind of anti-Semitism established itself in the parliament as a political force. But in reply to a worried inquiry from the Berlin office of the Jewish Telegraph Agency "whether united Jewry is necessary for defense," Einstein confidently replied:

> For the time being I see the National Socialist movement only as a consequence of the momentarily desperate economic situation and as a childish disease of the Republic. Solidarity of Jews, I believe, is always called for, but any special reaction to the election results would be quite inappropriate.[81]

For one more year he would still be able to feel secure—until, sooner than some others, he realized that Germany was on a fatal course.

Before his trip to America's West Coast Einstein spent a few weeks traveling through Europe. First he took part in the Solvay Conference in Brussels from October 20 to October 25. Its real subject, the magnetic properties of matter, attracted him less than did the continuation of the debate on quantum mechanics, although he was unable to convert anyone to his point of view.[82]

This was followed by three days in London. Einstein had promised to participate in an event staged by two organizations dedicated to alleviating the hardships of eastern European Jews. Herbert Samuel, who had been Einstein's host as high commissioner of Palestine, was now England's postmaster general. He had invited Einstein to stay at his home, and Einstein had accepted with "great joy": "In this manner the business of a Jewish saint will not only be made easier for me, but will become a downright pleasure."[83]

Sir Herbert, with whom Einstein conversed mainly in French, arranged some pleasant dinner parties for him and conducted him to the Strangers' Gallery of the House of Commons. Even the charity dinner at the elegant Savoy Hotel, presided over by Lord Rothschild, turned out to be an agreeable occasion, especially thanks to George Bernard Shaw. Shaw, who was then seventy-four, made a spirited speech in front of the microphones and film cameras in honor of Einstein, praising him in the same breath as Ptolemy and Copernicus as a creator of universes.[84] Yet there were enough sparks of malicious wit in his address to make the guest of honor, as shown in the film clips, laugh heartily.

The following day, when he was Weizmann's guest, Einstein was still amused about the "monkey comedy" at the Savoy, when most of the guests were unsure whose hand they should shake first—Lord Rothschild's or the "Jewish saint's."[85] On the way to the station Einstein told his host, Sir Herbert, that in younger years he never dreamed that he would take part in the kind of public events for the sake of which he had come to England. He was not really suited to such matters: "Je ne suis pas 'practical.' "[86]

From London Einstein traveled to Zurich, but with a stopover in Brussels in order to visit the "royals" at Laeken Palace. The previous year, while visiting his favorite uncle Caesar Koch in Antwerp, Einstein had, on May 20, 1929, been invited to the palace by Queen Elizabeth of Belgium. Elizabeth, a member of the former Bavarian royal family of Wittelsbach, was musically minded; and Einstein had played a trio with her and a lady-in-waiting, taken tea with her, and tried to explain his physics to her.[87] At that time the king had been away on a trip, but he was now present. This second visit was described by Einstein in a lively letter: "I drove to the Royals at 3 o'clock, where I was received with touching cordiality. These two simple people are of a purity and goodness that is seldom to be found. First we chatted for an hour. Then came an English lady musician and we played quartets or trios (there was also a musical lady-in-waiting) for a few more hours, and very cheerfully. Then everyone left and I was alone with the Royals for dinner—no servants, vegetarian, spinach with fried egg and potatoes, just that. . . . I liked it there enormously and I am sure that the feeling is mutual."[88]

His impression was correct. An unusual friendship developed between Einstein and the "royals," which was to play an important part at certain turning points in his life. From the very start this relationship had nothing in common with the familiar convention of crowned heads occasionally cultivating intellectual giants. A shared view of the world war, during which King Albert had manfully stood by his tortured people, while Albert Einstein in Berlin had manfully distanced himself from German chauvinism, a shared interest in music and scholarship—all this undoubtedly promoted mutual understanding. But the friendship between Einstein and the "royals" went far beyond a harmony of interests. The letters which Einstein wrote to the "Dear Queen" to the end of his life—the king died in 1934—are touching testimony to a deep and enduring sympathy.

In Zurich Einstein was the guest of honor at the celebration of the seventy-fifth anniversary of the foundation of ETH, the Swiss Technical University. The honorary doctoral degree, awarded to the man rejected for an assistant's post years earlier, probably meant more to him than his many other doctorates, which he had long ceased to count. He stayed at the home of his "past one,"[89] had his old friend Besso visit him at Mileva's apartment, and enjoyed seeing his colleagues in Zurich. The only shadow cast on his stay was concern about his son Eduard, who was studying medicine, was highly intelligent, and had musical talent, but was showing serious symptoms of mental instability.

Back in Berlin, Einstein had three weeks left before his voyage to California. The trip to America was to have been a working visit, without much public fuss—but Einstein himself saw to it that the opposite happened. He could by now shrug off the excessive interest the press showed in him: "Any squeak I make becomes a trumpet solo."[90] But this did not stop him from sounding a vigorous solo himself. This was a lengthy article called *Religion and Science*, which on Sunday, November 9, took up the first four pages of *The New York Times Magazine*.

Einstein reproduced the critique of religion common among educated freethinkers, but then emphasized a higher level of "cosmic religiosity," devoid of all anthropocentric echoes and based on "the miraculous order that manifests itself in nature as well as in the world of ideas." From this he concluded not only that there was no conflict between science and cosmic religiosity, but that the latter was in fact "the strongest and noblest mainspring of scientific research."[91] Between the lines, but not to be overlooked, was a statement that, in this sense, Einstein was deeply religious.

Whereas this article remained deservedly unnoticed when it was republished in Germany,[92] it triggered violent controversies in America. Although there were many who recognized themselves in Einstein's "cosmic religiosity"—Reform Jews, enlightened Protestants, Quakers, and Unitarians—the orthodox of all faiths regarded it as atheism or "the sheerest kind of stupidity and nonsense."[93] At any rate, Einstein had again become a topic of interest in America at the right moment, and the vast number of telegraphed invitations and proposals which arrived at Haberlandstrasse from America made it likely that this would be a hectic trip.

CHAPTER THIRTY-THREE

Farewell to Berlin

THE LONG VOYAGE TO CALIFORNIA started in Antwerp on December 2, 1930, on board the *Belgenland*, a Belgian steamship on which Einstein was to travel several more times over the next few years. His party included his wife; his secretary, Helen Dukas, as "girl Friday"; and the "calculator," Walther Mayer, because Einstein intended to work during the voyage. In his luxury suite on the upper deck he felt "uncomfortable, like a con man and indirect exploiter"; compared with the elegantly restrained demeanor of the staff, he felt "odd with his peasant manners," and he invariably "dressed negligently, even for the sacred sacrament of dinner."[1] During the stopover at Southampton, he had an opportunity to admire the results of British education: "In England even the reporters are restrained. Honor where honor is due. A single 'No' is enough."[2] This would be very different in the New World.

Originally, Einstein had not intended to go ashore during the five days when the *Belgenland* was in New York harbor, in order to avoid attention. But *The New York Times* had already decided that it would be impossible for him to avoid the press, unless of course he had himself locked up in the purser's safe, "and even then there would be photos—of the safe."

As the *Belgenland* approached New York, there were "countless telegrams, so that the ship's radio operators were sweating,"[3] a foretaste of what was to come. The arrival in New York was "worse than the most fantastic expectation. Hordes of reporters came on board at Long Island, as well as the German Consul with his fat assistant

Schwartz. Plus an army of photographers who pounced on me like hungry wolves. The reporters asked exquisitely stupid questions, to which I replied with cheap jokes, which were enthusiastically received."[4] Everything was just as on his first visit ten years earlier, but possibly even more frenzied.

As Einstein was still not familiar with the English language, he spoke only German, even for his first greetings, which two broadcasting companies transmitted live from aboard. The manner in which he greeted American soil and the American people[5] was almost papal—if popes had then ever gone on tour.

In all this hullabaloo Einstein's wife proved a circumspect impresario, organizing the professor's appointments and making sure that for every photograph and every interview a small fee was paid—not into Einstein's pocket, but for the poor in Berlin and for draft refusers all over the world. In this respect he had every reason to be satisfied: "Thanks to Elsa's shrewd management I earned $1,000 for the charity box. By midday I was dead."[6]

Contrary to all plans, then, there were five exciting and exhausting days in New York while the *Belgenland* lay at anchor. Einstein was handed from one event to another and met the elite of the city as well as other celebrities passing through, such as Fritz Kreisler and Rabindranath Tagore—both of whom he already knew—and Arturo Toscanini, with whom he now shook hands for the first time. He was even able to see himself hewn in stone on the tympanum of Riverside Church above the Hudson, which was adorned with statues of great figures in world history, Einstein being the only living person among them. Some grotesque situations inevitably arose, on the lines of "Einstein escaping from reporters"; and it was only in his stateroom at night that he found any rest—the approach to his cabin was guarded by policemen.

In a festive ceremony with speeches by the mayor and the president of Columbia University, Nicholas Murray Butler, Einstein was made an honorary citizen of New York City. He himself made a speech at a Hanukkah celebration in Madison Square Garden, where he was claimed by the Zionists as one of their own.[7] His most controversial speech, however, was to a smaller audience at a meeting of the New History Society at the Ritz-Carlton Hotel on December 14. This was devoted to his great political passion, pacifism.

In his message of greeting from on board ship, Einstein had said that the Americans had the strength "to overcome the threatening specter of our era, militarism."[8] He now defined his own position and deplored the fact that "under the present military system any person can be compelled to commit murder in the name of his country." He also knew what was to be done about it: "uncompromising opposition"—that is, refusal of military service. "If even two percent of those called up declare that they will not serve, and simultaneously demand that all international conflicts be settled in a peaceful manner, governments would be powerless."[9] Finally, he called for the "creation of an international organization and of an international pacifist fund" to help those who found themselves in difficulties as a consequence of refusing to serve in the armed forces.

For a number of years this "two percent" speech became something of a Magna Carta of militant pacifism. It was extensively reprinted: in *The New York Times* for example, and also—excerpted—in Germany. An abridged version appeared under an outraged headline, "Einstein Begging for Military Service Objectors—Scientist's Unbelievable Publicity Methods in America"; and this was sent by the ministry to Planck as president and Glum as director-general of the Kaiser Wilhelm Society, with a request for information on "what attitude Herr Prof. Einstein adopts in the Kaiser Wilhelm Society."[10] In the United States, the speech did not meet with unanimous applause, but Einstein would have been pleased to note that many young Americans in the streets and on campuses were wearing buttons with the provocative slogan "Two Percent"—and everybody knew what that meant.

By the time the *Belgenland* left New York, on December 16, Einstein had had "to stand up to a trying amount of his fellow men's love,"[11] but he was probably highly satisfied with the way his visit had gone so far.

The voyage south and through the Panama Canal provided Einstein with unforgettable impressions of scenery and with some amusing folklore of Central American revolutions, which took place harmlessly but probably confirmed him in his dislike of using the term "revolution" in connection with science: "In Havana they were in the process of having a revolution while we were there, and in Panama shortly after our departure. Their president, a former fellow student from the Zurich Polytechnic, was deposed on this occasion."[12] He bore the interest of his fellow passengers with black humor. "Passengers becoming more importunate. Perpetual photographing," he noted in his diary. "The charity business with my autographs is booming. . . . They have gone crazy about me. How will it all end?" The voyage ended in San Diego on December 30, where the docking was attended by a spectacle suggesting that a reincarnated Columbus was about to step onto the shore of a new continent.[13] There was a four-hour show of the most garish American kind, with speeches and interviews, but it evidently gave him pleasure. Back home, his friends—watching his arrival in California in newsreels—feared that he was "totally immersed in the hullabaloo and razzmatazz of the Americans."[14] Hedwig Born, faintly irritated, wrote to him that it was "great fun to see and hear you in the newsreel. To see you (San Diego) presented with flower floats and beautiful mermaids, and suchlike! The world certainly has some amusing aspects. Even though these things look meshugge from outside, I still have the feeling that the good Lord knows what he is doing."[15]

In Pasadena, the Einsteins moved into a "small gingerbread cottage" in the immediate neighborhood of the Caltech campus. "Here in Pasadena it is like Paradise," he delightedly reported back to Berlin. "Always sunshine and clear air, gardens with palms and pepper trees and friendly people who smile at one and ask for autographs."[16]

During his first week, the famous man was immediately invited to Hollywood, where he watched a special screening of *All Quiet on the Western Front*, a film made from Erich Maria Remarque's novel; because of its realistic portrayal of men dying during World War I, the film was banned in Germany. Einstein declared the ban to be "a diplomatic defeat for our government."[17] He was the guest of Charlie

Chaplin, "who had set up in his home a Japanese theater, with genuine Japanese dances being performed by genuine Japanese girls. Chaplin is an enchanting person, just as in his film parts." On several occasions Einstein met the social critic Upton Sinclair: "He is in the doghouse here because he relentlessly lights up the dark side of the American bustle."[18] Over the next two months, though, Einstein also came to know the pleasant side of the American bustle, with brief excursions to fashionable places like Santa Barbara and Palm Springs; and Millikan invited him to go sailing on the Pacific.

Science, too, had its ceremonies. On January 15, Millikan gave a festive dinner at the Athenaeum, the elegant faculty club of Caltech.[19] Two hundred rich patrons of Caltech, in recognition of their donations, were invited to eat with the legendary Einstein, though he himself was more interested in talking to the physicists and astronomers who were present. In a brief after-dinner speech he thanked his colleagues for their work, without which his theory of relativity would "today be scarcely more than an interesting speculation": William Wallace Campbell, for his determination of the deflection of light in the sun's gravitational field; and Charles Edward St. John, for his efforts to prove the red shift.

For the first and only time, Einstein also met Albert Abraham Michelson, then seventy-nine and gravely ill. Einstein paid tribute to Michelson in this same after-dinner address, for his famous experiment "when I was still a little boy, hardly three feet high. It was you who led physicists onto new paths and by your wonderful experimental work even then prepared the road for the development of the relativity theory."[20] It was a pretty compliment, although Einstein avoided any reference to the part Michelson's experiment might have played, or not played, in the development of his own concepts.

Einstein also paid tribute to "the work of your wonderful observatory," which had been a major reason for his acceptance of the invitation to Pasadena: it had "led to a dynamic concept of the spatial structure of the universe, for which [Richard Chace] Tolman's work has provided an original and exceedingly clear theoretical expression."

■ ■ ■

As a result of that wonderful observatory on Mount Wilson, the structure of the universe now looked entirely different from what had been mapped out by Einstein in his pioneering cosmological study thirteen years earlier. Within the framework of the general theory of relativity, he had then described the universe as an unbounded but finite space, with a constant mean distribution of masses, unchanging over time, and, in retrospect, positively minute. Astronomers then knew of only one galaxy, our own Milky Way, and even that not very well. Einstein's description had been made possible only by his use of a "cosmological constant" in his equations, but it was nevertheless the first physically consistent model of the universe, and for the next five years it had been almost the only one.[21]

The first change in cosmology came through mathematical theorizing. In 1922 the brilliant Alexander Friedmann in St. Petersburg demonstrated that Einstein's field equations could be solved even without the "cosmological constant"; these solutions would be consistent with a universe with a spatially homogeneous distribution of masses, except that this universe would not be static but would change over time and space, either expanding or contracting.[22] The universe had been visualized by Einstein as without a beginning or an end; but Friedmann now gave it something like dynamics and a history. Einstein initially thought he could prove that Friedmann had made a mathematical mistake,[23] but he soon discovered a mistake in his own objections, retracted his criticism, and described Friedmann's paper as "clarifying."[24] There was then no question of deciding between the alternative concepts by astronomical observations.

All this changed in the course of the 1920s, thanks to Edwin Hubble's persistent work with his hundred-inch reflector on Mount Wilson. This new wonder of the world had been completed in 1919 and throughout three decades remained the most powerful telescope anywhere. For our knowledge of the universe, it had a significance comparable to that of Galileo's telescope for our knowledge of the solar system. Small patches of "nebulae" in the sky were resolved into individual stars and revealed themselves as galaxies similar to the Milky Way. Hubble determined the distance of the nearer galaxies and estimated the distance of the more remote galaxies, obtaining magnitudes of millions of light-years. He also measured the spectra of light from these distant worlds, in particular the red shift. According to Doppler's principle, the red shift had to be interpreted as an "escape velocity"; despite the use of the same term, it should not be confused with the very much smaller red shift in a gravitational field, as derived by Einstein from the equivalence principle.

By 1922, Hubble had collected sufficient data to venture what was probably the most important astronomical assertion of the twentieth century: that the galaxies are moving away from each other with escape velocities which increase with distance. In other words, the universe is expanding.[25] A simple calculation backward to the start of this expansion gave the age of the universe as approximately ten billion years. The expanding universe was evidently no longer described by Einstein's static cosmology, though it was possibly described by Friedmann's solutions.

Einstein, then deeply engrossed in "distant parallelism" as his variant of a unified theory, was reluctant to be distracted by cosmological models and initially merely took note of these sensational results. Soon, however, they seemed important enough for him to inspect them in person. He therefore often had himself driven up the twenty miles of hairpin turns to Mount Wilson, on a spur of the Sierra Nevada. On one occasion at least he was accompanied by movie cameras, which recorded him riding up the elevator to the observation platform and, for the benefit of the viewers, looking into something like an eyepiece.

Hubble and his colleague Milton L. Humason—who refined Hubble's spectroscopic measurements—showed their guest around their cathedral of astronomy, explaining their high-resolution spectrographs and presenting their evidence that the universe was expanding. Einstein was clearly impressed: he declared that the "cosmological constant" he had invented thirteen years earlier was superfluous and that a model of the universe based on Friedmann's solutions was appropriate.

Despite intensive discussions with the theoretician Tolman on how the new discoveries could best be represented through the general theory of relativity, no contribution to this set of problems came about in California. A paper, *On the Cosmological Problem of the General Theory of Relativity*,[26] submitted by Einstein to the Prussian Academy after his return to Berlin, hardly went beyond a report on the situation. The only paper he wrote in Pasadena, jointly with Tolman, was devoted to his hobbyhorse, the unanswered problems of quantum mechanics.[27]

Millikan, Einstein's host, had reacted with raised eyebrows to Einstein's "two percent" speech in New York. Thereafter, Einstein had exercised some moderation in political matters—not only because of Millikan's conservatism but also in order not to annoy the wealthy patrons on whose favor Caltech depended. Millikan therefore must have been rather irritated by an interview between Einstein and Upton Sinclair in a socialist weekly, and even more irritated by a speech Einstein made to students on the Caltech campus.

Einstein chose this stronghold of technological knowledge as a place to question the use of technology not only in war but also in peace, because technology had "turned men into slaves of the machine." He urged the students not to forget that they had a responsibility that went beyond their specialized fields: "Concern for man himself and his fate must always form the chief interest of all technical endeavors, concern for the great unsolved problems of the organization of labor and the distribution of goods—in order that the creations of our minds shall be a blessing and not a curse to mankind. Never forget this in the midst of your diagrams and equations."[28] Such language sounded very socialistic to conservative Californians. Millikan could hardly have found it helpful to his efforts to raise Caltech to the top of American science by associating Einstein permanently with it.

After two months Einstein had temporarily had enough of "this land of contrasts and surprises, where one in turns admires and shakes one's head. One feels that one is attached to the old Europe with its pains and hardships, and is glad to return there."[29] The homeward journey began by rail, across the continent. On a visit to an Indian reservation near the Grand Canyon on February 28 Einstein received from the Hopi not only a rich headdress but an amusingly punning title, "the Great Relative." In Chicago, where his train stopped for two hours, he was met by several hundred supporters of peace; to their great delight, he treated them to an abridged version of his "two percent" speech.[30]

The following morning, when the night train arrived in New York, where the steamship *Deutschland* was to sail at midnight, all hell once more broke loose, for the next sixteen hours. The German consul general recorded that "Einstein's personality, without any clearly recognizable reason, triggers outbursts of a kind of mass hysteria, not only among specially thus inclined groups of 'friends of peace' and the romantic dreamers of newly founded mystical religious communities, but also among relatively levelheaded circles, such as the American supporters of the Palestine program."[31]

Pacifist organizations claimed their hero immediately on his arrival in New York. Einstein invited them on board the *Deutschland* to a meeting restricted to four hundred persons and called for radical action: "The struggle against militarism will have a dramatic effect because it will create a conflict that will directly challenge our opponents."[32] These words set off such "a delirium of enthusiasm that numerous persons kissed Einstein's hands and clothing, and the poor man had eventually to be forcibly taken to his cabin in order to put an end to these demonstrations."[33] For the afternoon Einstein moved into a hotel, where he had to deal with an unending stream of journalists, visitors, and admirers.

The evening had been reserved for Einstein's other passion—Jewish development in Palestine. With funds running low, Weizmann had implored Einstein as early as the beginning of February to make himself available for an urgently needed fund-raising drive.[34] This was a request Einstein could not deny, and so he agreed to be the guest of honor and speaker at a great banquet given by the American Palestine Campaign at the Astor Hotel on the evening of his departure. The guests had to pay $100 each, but despite this high fee—in the midst of the Depression—the target figure of one thousand participants who wanted to see and hear Einstein was actually exceeded. He was celebrated as a "prince of the intellect," and the applause became an ovation when a telegram from President Hoover was read out. Hoover certainly had no sympathy for pacifism or socialism, but he could not avoid saluting the visitor: "My hope is that your visit to the United States has been as satisfying to you as it has been gratifying to the American people."[35] In his address, Einstein again appealed to the Jews to cooperate with the Arabs and pleaded for an arrangement on the model of the Swiss constitution.[36]

When Einstein returned to his ship shortly before midnight, pacifists were again lining the pier; as the ship pulled away, they chanted in unison, "No war forever."

On the stormy voyage, Einstein learned enough about conditions in Germany to view his return with mixed feelings. "In Germany everything is rocking, much worse than on this ship. But one is used to it and one cuts one's clothes according to one's cloth at the time. For the moment, at least, the Republic still stands."[37] But it was standing on very shaky feet.

In any case, Einstein did not intend to stay in Berlin long. In April he presented two papers to the Prussian Academy: his report on the cosmological problem;[38] and a paper written jointly with Walther Mayer during the crossing, on his "distant parallelism" theory,[39] which revealed that he would come no nearer to his great goal by this means. Then he took a month's leave to go to England, to give the Rhodes Lectures at Oxford.

This honor, accompanied by a fat fee, had first been offered to Einstein in 1927, on the initiative of Frederick Lindemann, director of the Clarendon Laboratory. For lack of time and because of a serious illness, he had to decline the invitation at that time. But Lindemann persisted, and in February he finally obtained Einstein's acceptance, from Pasadena.[40] It was a good decision, because Einstein spent a restful month in Oxford in May. He enjoyed the monastic life in Christ Church College, where Lindemann had made rooms available for him. His only obligation—and even that was interpreted loosely—was his participation in "the holy sacrament of the 'High Table.' " There was also the ceremonial awarding of an honorary doctorate, and the three lectures he gave at Rhodes House.[41] At these lectures, to begin with, the hall was overcrowded, but since some of the audience could not cope with Einstein's mathematics and others could not cope with his German, soon there remained only a small core of experts.

English reserve, as well as the fact that Oxford was used to eccentric geniuses, allowed Einstein to conduct his life according to his own taste and tempo, always discreetly looked after by Lindemann and Lindemann's servant and general factotum. No one was bothered by the fact that Einstein met frequently with pacifist student groups or that he discussed "the spread and effect of my speeches in America"[42] with representatives of the Internationale of War Service Refusers. And none of this was at the expense of physics, so Lindemann was more than content with his guest. "He threw himself into all the activities of Oxford science, attended the Colloquiums and meetings for discussion and proved so stimulating and thought-provoking that I am sure his visit will leave a permanent mark on the progress of our subject."[43]

Lindemann soon got his college to invite Einstein to spend a month in Oxford every year as a "research student"—the quaint terminology Christ Church uses for what other colleges call a "fellow."[44] Einstein accepted the five-year contract at an annual stipend of £400, not only because of his pleasant memories of his first visit, but also because the situation in Germany was becoming exacerbated.[45]

For the summer of 1931, Einstein again installed himself in his country retreat at Caputh, but not as a hermit. With a vast number of letters and statements he continued his pacifist campaign, reiterating that war could be prevented only by organized refusal to serve in the armed forces.[46] This commitment was in no way affected by his realization that Germany was on the way to becoming a dictatorship. Parliament had been dissolved, and the Brüning cabinet, governing by emergency decrees, already represented a quasi-dictatorship. The economic disaster, the Nazis' street fighting, and the weakness of the democratic parties boded ill for the future.

In this uncertain situation Einstein intended to ask Planck "to see to it that my German citizenship is rescinded. . . . Concern for the many people dependent on me, as well as a certain need for independence, compels me to take this step."[47] This letter was never sent; it was found, in its envelope, among Einstein's papers after his death.

Nevertheless, it shows that in the summer of 1931 Einstein was beginning to accustom himself to the idea that before long he would sever his ties with Germany.

If Einstein saw any hope of a turn for the better, he saw it on the left of the political spectrum. His sympathies were with the Social Democrats, but he was not afraid of contact with the communists. He readily signed appeals by the Rote Arbeiterhilfe (Workers' Red Help) and allowed himself to be used for propaganda purposes by the Communist International, the Comintern.[48] In the Marxist Workers' College "Masch," managed by the Communist Party, he gave a lecture on *What a Worker Should Know of the Theory of Relativity*[49]—not exactly a major problem of the class struggle. But he was not a dependable fellow traveler, because, for one thing, he urged that Stalin's mortal enemy Trotsky be given asylum in Germany.[50] He also refused to participate in peace congresses because of their preponderance of Soviet and communist supporters.[51]

Einstein's vacillation between "pink" and "red" testifies to his confusion in political matters. In 1930, he had signed an appeal against the Stalinist show trials in the Soviet Union. He saw the absurd charges against the "forty-eight vermin" as "either an act of desperation of a regime driven into a corner, or a mass psychosis, or a mixture of the two. . . . I am very sad that this development, which we had been watching with hopeful eyes, has now led to such atrocious events."[52] But a year later some pro-Soviet friends, of whom he had many, convinced him of the legitimacy and necessity of these trials, with the result that a communist propaganda sheet was able to print Einstein's recantation: "Today I most deeply regret that I then added my signature, because I have now lost confidence in the correctness of my views. Then I did not sufficiently realize that under the special conditions of the Soviet Union things are possible that, in the circumstances familiar to me, are totally unthinkable."[53] Whether it was instigated by Dimitri Marianoff—a journalist attached to the Soviet embassy with undefined responsibilities, who was married to Einstein's younger stepdaughter, Margot—or by Willi Münzenberg, the sinister propaganda chief of the Comintern[54] who had converted him, Einstein's statement reflects the limitations of his political judgment.

Einstein, incidentally, stuck to his view of the Moscow show trials. As Stalin's reign of terror got worse, not even his solidarity with the many accused and executed Jews, his "tribal companions," stopped him from trying to justify the trials. At the peak of the terror, in 1937, he wrote: "Indications are increasing that the Russian trials are no swindle, but that there is a conspiracy of those in whose eyes Stalin is a stupid reactionary who betrayed the idea of the Revolution. . . . At first I too was firmly convinced that these were instances of a dictator's arbitrary actions based on lies and swindle, but that was a deception."[55]

Einstein had returned to a threatened Germany in the knowledge that he would always have an acceptable—and superbly paid—fallback position in California. When in April 1931 he approached a senior official in the Prussian Ministry of Education with a request to establish an extraordinary professorship for his collaborator Walther Mayer, he frightened the official with the disclosure that he himself had received an offer from Pasadena, at an annual salary of $35,000. Unless a satisfactory solution was found for Mayer, who by then was forty-five, Einstein threatened, he "would otherwise have to go to Pasadena, because there the remuneration of Dr. Mayer would be no problem."[56]

In fact, whatever had been discussed in Pasadena, there can have been no firm contract but only an oral statement of intent by Arthur Fleming, the chairman of the Caltech board, who was inclined to act on his own. Over the summer, there had been exchanges of letters and telegrams between Pasadena and Berlin, but the figure mentioned in them was $20,000 for a ten-week stay at Caltech.[57] While Einstein was waiting for the contract to be mailed to him for his signature, Millikan, on a trip to Europe, visited him at Caputh and offered him a salary of $7,000 for his next visit, with a permanent arrangement to be settled the following year.

Einstein was irritated by this confusion in California and first went to give a lecture in Vienna, where Austrian officialdom "observed special reserve because he is a Jew and believed to be on the left politically."[58] After thinking the Pasadena offer over for a week, he wrote a grand refusal on October 19, informing both Fleming and Millikan that over the winter he would take a rest from these tiresome negotiations and seek out the sun of southern Europe. He intended to leave Berlin in any case. He informed his friend Besso that he would probably come to Switzerland in the winter "because things are getting uncomfortably hot for me here."[59]

Then, in a sudden change of mind, the reasons for which we do not know, Einstein after all accepted the offer from California on Millikan's terms; on November 14, he sent the signed contract back to Pasadena. A week later, he left Berlin, accompanied by his wife. He spent a few days in Belgium and Holland before embarking on the four-week voyage on December 2, this time on the American steamship *San Francisco*, which took him direct to California, sparing him the stress of a stopover in New York.

When the ship had left the coast of Europe behind, Einstein noted in his diary a decision of crucial importance for his future life: "Today I resolved in essence to give up my Berlin position. Hence a migrating bird for the rest of my life! Seagulls are still accompanying the ship, always on the wing. They are said to come with us as far as the Azores. These are my new colleagues, but, Heaven knows, more efficient than me."[60] He remained silent about his reasons, and also about his specific intentions. Presumably he had in mind shuttling between Pasadena and Oxford. As if to confirm the seriousness of his decision, he added: "I'm also learning English, but it won't stick in my elderly braincase."

Einstein arrived in Los Angeles shortly before the end of the year. This time, unlike the previous year, his arrival was almost normal. As a small compensation for the financial confusion, Einstein was able to move into Arthur Fleming's splendid accommodation at the Athenaeum. Among the colleagues with whom he met to work was Willem de Sitter, the Leyden astronomer—an expert on relativity, and like himself a visiting scientist. Jointly they produced a paper on an aspect of the expanding universe.[61] Einstein gave a few lectures on cosmological problems, and especially on a new variant of the unified theory, which he had worked out with Mayer.[62] But he did not give up his pacifist sermons; and, much to the displeasure of Millikan, he also meddled in American domestic problems such as racial discrimination.

Einstein's most important meeting, the one with the most far-reaching consequences, was with Abraham Flexner, a highly regarded scientific administrator. Flexner had a lot of money and a vague idea. Since the beginning of the century, he had exerted a major influence on academic teaching in America. Having first introduced a reform of colleges and then of medical schools, he was able, as secretary for fifteen years of the Rockefeller Foundation's "general education" board, to endow universities to the tune of more than $500 million—and, up to a point, to shape them in accordance with his criteria of academic excellence. The culmination of his career, thanks to a philanthropic donation of $5 million, was to be the establishment of an institute of higher or advanced study, where scholars, freed from university routine and duties, were to live solely for their research. Flexner had come to Pasadena to consult Millikan on the form of such an institution, and Millikan in turn referred him to his famous guest.

When Flexner and Einstein first met at the Athenaeum, only two conditions were attached to the new institution: at the wish of the donor family it was to be located somewhere in the state of New Jersey; and in view of its substantial but by no means unlimited finances, it should concentrate on theoretical disciplines. Einstein, who believed that scientific progress came from creative individuals rather than from organizational matters, supported Flexner's intention that, in contrast to the normal academic hustle and bustle, an informal enclave of scholars should be created—an ivory tower.[63] According to Flexner's records, nothing was said about Einstein's participation in the enterprise; but when the two men agreed to continue their talks in Oxford the following spring, both of them presumably entertained such an expectation.

When Einstein, in early March, once more boarded the *San Francisco* to return to Europe, he knew that he would return to Caltech the following year. But no permanent arrangements had yet been agreed on, and the idea of a permanent move to America did not greatly appeal to him anyway. As he explained to his friend Ehrenfest, who had asked him to look out for a possible position for him: "I must tell you quite frankly that in the long term I would prefer to be in Holland rather than in America, and that I am convinced that you would come to regret a change. Apart from the handful of really fine scholars, it is a boring and barren society that would soon make you shiver."[64]

Back in Berlin, Einstein at once went to the weekly meeting of the Prussian Academy, as he was anxious to submit a supplement to the paper whose first part he had submitted in October, before his departure for the United States. It again bore a promising title—*Unified Theory of Gravitation and Electricity*—but it no longer had anything to do with distant parallelism, which he had finally abandoned in the summer of 1931. As he later summed up, that "formerly rather interesting theory . . . simply does not lead to a description of the electromagnetic field. It took me a long time to realize this with certainty, because I was so fascinated by the naturalness of the theory."[65]

He did not mourn long over his three years of vain efforts with distant parallelism. Soon, he became fascinated with a five-dimensional formalism "which psychologically links up with Kaluza's well-known theory, while at the same time avoiding the extension of the physical continuum into one of five dimensions."[66] Each point of the four-dimensional space-time continuum was linked to five-dimensional vectors, by means of which a "fiver curvature" was achieved, from which the equations for the gravitational and the electromagnetic field derive through "rejuvenation." As at the beginning of every new theory, Einstein was enthusiastic; he wrote to Ehrenfest that this one "in my conviction definitively solves the problem in the microscopic area."[67] For the microscopic and atomic areas he already saw "an indication of a natural supplement that will perhaps supply the quantum laws."[68]

His colleagues, however, were not at all enthusiastic. The older ones kept silent; the younger ones were beginning to make jokes about Einstein.

After barely two weeks in Berlin, Einstein departed for England. He went first to Cambridge, where he gave a few lectures[69] and talked to Eddington, and then to Oxford for his sinecure as a "research student." Shortly before leaving Oxford, he was visited by Abraham Flexner, as had been arranged in Pasadena. Flexner's plan to locate the new institute at Princeton, by agreement with but totally independent of the university there, no doubt aroused Einstein's personal interest, which Flexner cannot have failed to notice. During a long walk in the grounds of the college, they again discovered that their ideas on the character of the new institution were largely identical, and Flexner summoned up his courage to put the vital question: "Professor Einstein, I would not presume to offer you a position in the new institute, but should you come to the conclusion that it could offer you those facilities which you treasure, then you would be welcome—on your own terms."[70]

No sooner was Einstein back at his country retreat of Caputh, after an appearance at the "Joint Peace Council" in Geneva toward the end of May,[71] than Flexner again appeared on the scene, on June 4. Now concrete terms were being discussed, such as a six-month stay each year, from the fall until roughly April; a position for Walther Mayer, independent of Einstein's; and Einstein's own salary and pension.[72]

Asked about his expectations for his salary, Einstein at first suggested $3,000. When he saw Flexner's puzzled expression, he added, "Or could I manage on less?"[73] Flexner, although he was an experienced administrator, took this as the modesty of a great scientist and told the story in that sense. Actually, it was a shrewd move on Einstein's part to lure Flexner into making an optimal offer without seeming greedy himself. The resulting offer was $10,000, with the institute paying Einstein's taxes and also Elsa's travel expenses—not a bad deal for five months.[74] Thus Einstein's idea of existence as a migratory bird had assumed solid shape, in a way he could not have imagined when he had jotted down the idea in his diary six months earlier. As, despite the rain, he saw his visitor to the last bus just before midnight, he assured Flexner: "I am full of enthusiasm for it."[75]

A few days later, when Flexner had confirmed the oral agreement in writing, Einstein cordially thanked him for the generous terms; indeed, he found the figures suggested for his and his wife's pension too high and proposed a reduction. On the other hand, he insisted that "Herr Mayer's appointment should be independent of mine. Otherwise I would feel that he would become unemployed on my death."[76] He concluded his written acceptance to come to Princeton in October

1933 with the assurance: "I am really delighted to be associated with you in such a wonderful purpose and am convinced that we shall be happy with each other."[77]

As political developments in Germany went from bad to worse, Einstein must have frequently thought of his contract with Flexner. Although in the spring Hindenburg once more won the presidential election, more than thirteen million votes had been cast for Hitler. Nevertheless, Einstein for the time being wanted to stay in Berlin. "He has wholly adapted to Caputh," his wife reported, "and keeps telling me that no one is going to make him leave. He knows no fear." But Elsa was worried and urged him "not to sign any appeals anymore and to live solely for his problems. He answered that . . . if I were as you want to have me, then I just wouldn't be Albert Einstein."[78]

Casting aside his wife's anxieties, Einstein, along with the artist Käthe Kollwitz and the writer Heinrich Mann, in a manifesto on the Reichstag elections in July, warned "that we are heading for the frightful danger of [becoming fascist]. This danger, in our opinion, can be removed only by the [cooperation] of the two great workers' parties in the electoral campaign."[79] During the campaign his name headed the posters calling for a united antifascist front between the Social Democrat and the Communist parties.[80] However, this did not come about, and in the July elections the Nazis became the strongest party, with 37 percent of the vote. The new Reich chancellor, von Papen, governed with a "cabinet of barons," dissolved the Reichstag, had the army drive Prussia's Social Democrat government from office, and appointed himself Reich commissioner for Prussia.

While many citizens were hoping that von Papen's "fundamentally new manner of government" would restore public order and, more especially, provide a shield against the National Socialists, Einstein revealed his gloomy expectations to his colleagues: "I certainly do not believe that a government by military force will prevent the coming revolution of the National Socialists. On the contrary, military government suppresses the will of the people. The people will look to a revolution from the right to protect them against the rule of the *Junkers* [the Prussian landed gentry] and army officers."[81]

▪ ▪ ▪

During that summer of political crisis, Einstein was involved in numerous political actions and discussions. The most impressive of these, even though it remained virtually unnoticed, was a public exchange of letters on the causes of war and on ways to prevent it, initiated by the Institute for Intellectual Cooperation. Einstein chose Sigmund Freud as his correspondent.

Freud and Einstein had repeatedly been bracketed when the most important living Jews were named. They first met in 1926, when Freud spent Christmas in Berlin with his son Ernst. There Einstein and his wife visited him. Describing the meeting and their two hours of conversation, Freud recorded: "He is serene, assured and courteous, understands as much of psychology as I do of physics, and so we had a very pleasant chat."[82] Einstein appreciated Freud more for his descriptive powers than for his psychoanalytical constructs, and preferred to "remain in the darkness of not-having-been-analyzed."[83]

When Einstein congratulated Freud on his seventy-fifth birthday, he added that every Tuesday evening, along with a female friend—presumably Toni Mendel—he read Freud's writings and could not sufficiently admire their "beauty and clarity. Except for Schopenhauer there is, for me, no one who can or could write like that." But as a "pachyderm," he nevertheless veered between "belief and disbelief."[84] A year later he thanked Freud "for many a beautiful hour which I owe to the reading of your works," and then continued: "I always find it amusing to observe that even people who regard themselves as 'unbelievers' with regard to your teachings, offer such little resistance to your ideas that they actually think and speak in your concepts the moment they let themselves go."[85]

In his open letter to Freud, Einstein did not invoke pacifism but instead, as "an individual free from emotions of a national nature,"[86] mapped out an order founded on peaceful arbitration of all conflicts between states by an institution that would have at its disposal the means to enforce this ideal of justice. "The road to international security lies via the unconditional surrender by states of some of their freedom of action, or sovereignty, and it seems unquestionable that there is no other road to such security."

▪ ▪ ▪

Einstein had kept quiet about his obligation to spend half the year in America; the authorities in Berlin and the Prussian Academy were taken by surprise by newspaper reports, at the end of August,[87] that the Institute for Advanced Study in Princeton would start functioning in the fall of 1933 and that Einstein was its most prominent acquisition.

Only in September, in response to a query, did Einstein feel inclined to inform the academy of his arrangement with Flexner, although he pointed out that he had already talked to Planck about it.[88] Nonchalantly, he left it to the Ministry of Education to decide "if under these new conditions a continuation of my employment at the Academy is at all possible, or desirable."[89] Planck probably intervened to ensure that Einstein was saved for the academy at least for the summer semester. Einstein on his own initiative proposed to the ministry a reduction by half of his annual salary—a gesture which, despite his prolonged and well-paid visits abroad over the past few years, he had not previously made.[90] When it was reported that he would move to the United States altogether, he corrected the reports: "I will not leave Germany. My permanent place of residence will continue to be Berlin."[91]

The California Institute of Technology, on the other hand, was immediately informed of Einstein's association with Flexner's new institute. In the circumstances, Einstein expected that Caltech would now do without his agreed visit to Pasadena during the coming winter.[92] Millikan was disappointed, believing that he had a right to Einstein's presence exclusively at Caltech, even though he still had not offered Einstein a long-term contract. However, he did not wish to do without the famous man and therefore confirmed his invitation for the winter. Millikan was hoping that Einstein might, in the future, divide his visits to America between Pasadena and Princeton. Einstein left this open as a vague possibility, but Flexner condescendingly rejected it.[93]

Much as Einstein was in demand, both Millikan and Flexner were continually worried about his political activity, which they not only personally disapproved of but had to defend, halfheartedly, to their wealthy patrons. Since Einstein's triumphal arrival in America in December 1930, irritation had been growing in the conservative camp about this strange professor, who did not confine himself to reporting amazing things about the universe but was raising his voice in pacifist and socialist speeches on very terrestrial—and, worse, domestic American—issues.

When Flexner proudly announced that he was about to inaugurate his Institute for Advanced Study with Einstein as its most prominent member, some conservative groups believed that this questionable foreigner represented a danger to the United States. The board of a "National Patriotic Council" felt it necessary to issue a warning against this "German bolshevist" with his dubious theories and scandalous opinions; and its female branch, the American Women's League, addressed a formal petition to the visa section of the State Department.[94] These patriotic women condemned Einstein's pacifist activities on sixteen pages, complaining—and here they were factually correct—that he supported communist associations such as the International Workers' Help, which was an organ of the Comintern. This, along with some rank nonsense about how relativity theory would undermine the church, the state, and science, culminated in a demand that Einstein should be prohibited from entering the United States of America.

The State Department sent the pamphlet of the patriotic women to the United States consulate in Berlin, where Einstein had in the past always received his visa without any problems, but where he was now to be questioned on the complaints. "Wouldn't it be funny if they didn't let me in?" Einstein scoffed to the Berlin reporter of *The New York Times*. "The whole world would be laughing at America."[95] And to make sure there would be something to laugh about, he improvised a sarcastic comment for the press:

> Never before have I been spurned so vigorously by the fair sex, or if this did ever happen, then not by so many at a time. But aren't they right, those vigilant women citizens? Why should one admit

> a person who devours hard-boiled capitalists with the same appetite and pleasure as the Minotaur monster in Crete devoured toothsome Greek virgins, and who moreover is mean enough to reject any kind of war, except the inevitable war with one's own wife? Listen therefore to your clever patriotic little women and remember that the Capitol of mighty Rome was once saved by the chatter of its loyal geese.[96]

The American consul, however, did not find the conflict at all amusing and summoned Einstein for a talk on December 5. When he had cautiously worked his way around to the question whether Einstein was a communist or an anarchist, Einstein—according to reports in the press—lost his patience and, in the form of an ultimatum, demanded a visa, which was promptly provided the following day.[97] According to recently released American government papers, however, Einstein did something different. He signed the declaration demanded of him, confirming that he was not a member of any radical organization.[98] Thus there was no further obstacle to his departure for California.

Nothing about Einstein's preparations for the journey had suggested a final farewell, and he had told the academy and his friends that he would be back in Berlin in April.[99] But he was haunted by dark premonitions. "Take a very good look at it," he said calmly to his wife as they locked up their villa in Caputh for the winter. "You will never see it again."[100]

On December 10, Einstein and his wife boarded a ship in Antwerp, which would again take them directly to California through the Panama Canal. To Millikan's relief, nothing spectacular accompanied their arrival in the port of Los Angeles. Einstein had evidently resolved to restrain his political impulses, possibly out of consideration for his hosts, but possibly also because his experience at the consulate in Berlin was giving him pause. After Einstein's first, exceedingly reserved, meeting with journalists, Millikan was very pleased with his guest's remarks, which "supplied no additional ammunition for those who are spreading grotesque and silly stories about his links with groups aiming to subvert American institutions and ideals."[101] Besides, the president of Caltech was anxious to shield his guest from the public—and perhaps to his own surprise, he succeeded in doing so.

Through an ironical turn of fate, Millikan had this time acquired Einstein's fee of $7,000 from the Oberlaender Trust in Philadelphia, a foundation of a family of German extraction, and in return had agreed that Einstein would make a speech that would be "helpful to American-German relations,"[102] to be broadcast by the National Broadcasting Corporation on January 23.

That evening Millikan gave a formal dinner at the Athenaeum; afterward, the guests moved in procession to the Pasadena Civic Auditorium, where Einstein was to speak at a symposium on "America and the World." In the pleasant style of an after-dinner talk, Einstein poked gentle fun at social taboos, beginning with dress, and criticized the use of negative labels such as "communist" in America, "Jew" among the right wing in Germany, and "bourgeois" in the Soviet Union. Millikan was satisfied with those irrelevant remarks, but not so *The New York Times*, which observed that Einstein's speech "had not thrown any new light on a dark situation."

Meanwhile, in Berlin, Adolf Hitler was getting ready to take over the government, and a week later he was appointed Reich chancellor. Einstein's premonitions had not deceived him. He was not to see his house in the country ever again, or Berlin, or Germany.

PART VII

THE PACIFIST AND THE BOMB

CHAPTER THIRTY-FOUR

Exile as Liberation

EXACTLY WHEN EINSTEIN, in California, learned about Hitler's seizure of power on January 30, 1933, or how he reacted to it, is difficult to establish. On February 2, he was still writing to Berlin, to the secretariat of the academy, about a reduction of his salary—just as if nothing had happened and his return in April was a matter of course. In his diary, which he kept only sporadically, there is no mention of Germany; there are only notes on his daily program: "Afternoon Tolman . . . about experimental work on cosmic rays. Evening Chaplin. Played Mozart quartets there. Fat lady whose occupation consists of making friends with all celebrities."[1]

However, four weeks after Hitler's assumption of power and two weeks before his own departure from Pasadena, Einstein had learned enough about the changes in Germany for his decision to be inevitable: "In view of Hitler I don't dare step on German soil," he wrote on February 27 to Margarete Lebach—whom he referred to as "the Austrian woman"—in Berlin. "I have already canceled my lecture at the Prussian Academy of Sciences."[2] The next day the Reichstag was in flames, and then came the first wave of brutal Nazi terrorism against left-wing politicians, intellectuals, and journalists.

The day before his departure, Einstein justified his decision not to return to Germany in a public statement that was widely reported: "As long as I have any choice in the matter, I shall live only in a country where civil liberty, tolerance, and equality of all citizens before the law prevail. . . . These conditions do not exist in Germany at the present time."[3] The journalist to whom he had handed this statement and defended it in an interview watched him walk across the Caltech campus after a seminar. Suddenly the ground shook under his feet: Los Angeles was experiencing the worst earthquake in its history. The reporter saw him walk calmly to his quarters at the Athenaeum, just as if nothing had happened.[4]

On March 11, Einstein and his wife left Pasadena by train, along the now familiar route across the continent. They spent March 14 in Chicago, where, in honor of his fifty-fourth birthday and for the benefit of Hebrew University in Jerusalem, a banquet had been arranged, at which eminent scientists—such as Arthur Compton and the governor of the state of Illinois—made speeches.

The next stop was New York, with receptions, rallies, and formal dinners every day, arranged by pacifist organizations to launch a volume of Einstein's pacifist speeches and writings,[5] and by friends of Hebrew University to collect donations. His views on the situation in Germany, which he expressed in speeches or interviews on these occasions, were snatched up by the press and disseminated. He resolutely condemned Hitler's government but was restrained in his references to the German people. He called on the civilized world to practice "moral intervention" against Nazism, in the hope that the German people would not be able to disregard foreign disapproval of the shameful treatment of pacifists, socialists, communists, and even liberals—but "it would be a great mistake to indulge in general anti-German agitation."[6]

When, on the day of his departure from New York, the newspapers reported that the Nazis, searching for weapons or other evidence, had broken into his house in Caputh, he described this action as "the result of the sudden takeover of police powers by the rabid mob of Nazi militia."[7] Even when he subsequently learned that these reports were false,[8] they confirmed him in abandoning any thought of returning to Germany. In the meantime he had spent a day in Princeton, where he met Oswald Veblen, a mathematician and a future colleague at the Institute for Advanced Study, and had looked around for a house—probably already with a permanent residence in mind.

During his crossing on the familiar *Belgenland*, Einstein had decided to sever his ties with Germany finally and completely. Immediately on his arrival in Antwerp, he had himself driven by car to Brussels, and at the German legation there he handed in his passport and declared his renunciation of German citizenship.[9] Before that, he had sent a letter to the Prussian Academy, the bitter necessity for which was beyond question but which must nevertheless have been painful for him to write:

> The conditions at present prevailing in Germany induce me to lay down herewith my position in the Prussian Academy of Sciences. Throughout 19 years the Academy provided me with the opportunity to devote myself to scientific work, free from any professional obligations. I realize the great measure of gratitude I owe to it. It is with reluctance that I leave its circle, also because of the stimulation and the beautiful personal relations which, during that long period of my membership, I enjoyed and always greatly appreciated. However, dependence on the Prussian government, entailed by my position, is something that, under the present circumstances, I feel to be intolerable.[10]

Although this may not have been his main consideration, Einstein hoped by this statement to save his colleagues in Berlin from running into problems with their new political masters. In this, however, he did not succeed. His departure from the Prussian Academy was to have a macabre epilogue.

For the Nazis, Einstein had always been one of their most hated figures. After their seizure of power, the German press was free to attack him with no holds barred. It is hardly surprising that *Völkischer Beobachter* published the most vile attacks on him, or even that the Nobel Prize laureates Philipp Lenard and Johannes Stark now thought the time had come for their distasteful anti-Semitism and their crazy idea of creating a "German physics." But even reputable papers were vying with each other to display their Nazi credentials. "Good News of Einstein—He Is Not Coming Back!" wrote one of them,[11] with Einstein's remarks in America providing the background for tirades of hatred.

Max Planck, uncertain how to react, on March 19 informed Einstein of his "profound distress" over "all kinds of rumors which have emerged in this unquiet and difficult time about your public and private statements of a political nature. I am in no position to judge their importance. But one thing I see very clearly—that these reports make it exceedingly difficult for all those who esteem and revere you to stand up for you." In a grotesque reversal of cause and effect, he held Einstein responsible for the fact "that as a result, the anyway difficult situation of your tribal companions and co-religionists here is in no way made easier but in fact made even more oppressed."[12] But there was worse to come.

On March 23, all the parties in the Reichstag, except the Social Democrats—the Communist deputies had already been taken to concentration camps or had gone underground—had voted in favor of an "empowering law" and thereby, within the meaning of the German constitution, legitimately installed Hitler's dictatorship. Thus legalized, the new rulers apparently had nothing more urgent to do than instruct the Prussian Academy to start formal disciplinary proceedings against Einstein, naturally with the objective of expelling him.[13]

Planck was on vacation when he learned of this intention, simultaneously learning from a newspaper report that Einstein had renounced his citizenship in Brussels and was resigning from the academy. With an obvious sense of relief he immediately wrote to Einstein—by registered letter to Ehrenfest's address in Leyden—"that this idea of yours seems to me the only way that would ensure for you an honorable severance of your relations with the Academy, and at the same time save your friends from an immeasurable amount of grief and pain."[14] This advice was not entirely unselfish: Planck was worried not so much about Einstein as about the reputation of the academy. That same day he wrote to the acting secretary in Berlin that "starting formal exclusion procedures against Einstein would bring me into the gravest conflicts of conscience. Even though in political matters a deep gulf divides me from him, I am, on the other hand, absolutely certain that in the history of centuries to come Einstein's name will be celebrated as one of the brightest stars that ever shone in our Academy."[15]

When Einstein's letter of resignation was received by the academy on March 30 and read into the minutes at its meeting, the issue of expulsion could have been settled, had it not been for the Nazis and their need for revenge. "The fury in the Ministry that he anticipated them by his resignation was indescribable,"[16] Max von Laue recalled. The least that was now being demanded of the academy was a sharp statement against Einstein, to be issued at once, by April 1. The reason was that for that day, a Saturday, the Nazi leadership had called for a "boycott of Jews," and having Einstein kicked out of the academy would have gone well with the posting of storm-trooper sentries outside Jewish shops. At the university and the state library, Jewish students, assistants, and *Dozenten* were prevented from entering and had their IDs and readers' tickets confiscated.

The only secretary of the academy then in Berlin, Ernst Heymann, a lawyer and Nazi sympathizer, felt that he had to add to these shameful excesses against Jewish citizens by issuing a statement on behalf of the academy, accusing Einstein of "atrocity propaganda" and concluding: "For this reason [the academy] has no cause to regret Einstein's resignation."[17]

Max von Laue was outraged at this insult and, at an extraordinary plenary meeting, attempted to get the academy to disavow Heymann's unauthorized action. However, that meeting of April 6, 1933, was for Laue "one of the most horrible experiences of my life."[18] Not one voice was raised in support of his motion; even Haber, who twenty years previously had passionately argued for Einstein's appointment to the academy, approved Heymann's formulation and, along with his colleagues, expressed his "thanks to him for his appropriate action."[19] The academy, which had included only one member of the Nazi Party and only one Nazi sympathizer, by this vote Nazified itself even before the government resorted to coercion.

Einstein was unwilling to let the libelous accusation of "atrocity propaganda" become attached to him; but, in an involved correspondence after his resignation, he was unable to induce his former colleagues to change their minds. They were loyally supporting the state, and as the Nazi regime had come to power lawfully, they were, with only one exception—Max von Laue—not even capable of opposition in spirit.[20]

In a personal letter to Planck, Einstein stated his view of his former colleagues' behavior:

> In particular I did not participate in any "atrocity propaganda." In favor of the Academy I assume that it made such a libelous statement only under external pressure. But even in that case it will scarcely redound to its credit, and many of the better men are probably already ashamed of it. . . .
>
> I should also point out that over all these years I have only promoted Germany's reputation, and that I never bothered that—especially over the past few years—the right-wing press systematically agitated against me, without anyone bothering to stand up for me. But now the war of extermination against my Jewish brethren has compelled me to throw the influence I have in the world into the balance in their favor.[21]

In his reply, Planck deplorably equated terror and persecution of left-wing sympathizers and Jews on the one hand with Einstein's pacifism on the other: "Two ideologies, which cannot coexist, have clashed here. I have no sympathy with the one or with the other. Yours, too, is alien to me, as you will remember from our conversations about the refusal of military service propagated by you."[22] By the time Planck wrote these lines, there had been not only the "boycott of Jews" but also a "law on the reinstallation of a professional civil service," which excluded Jews from state posts—and hence also from the universities and other research institutions such as the Kaiser Wilhelm Society.

The sharp differences between Einstein and Planck were no longer solely political but touched upon human rights as the most valuable achievement of European civilization. Nevertheless, this did not affect their personal relationship. Planck had expressed the confident hope "that despite the deep gulf that divides our political opinions, our personal amicable relations will never undergo any change."[23] And Einstein assured his father figure—the only one he had left after Lorentz's death—of his undying loyalty and reverence: "In spite of everything I am happy that you have met me in old friendship and that even the greatest stresses have failed to cloud our mutual relations. These continue in their ancient beauty and purity, regardless of what, in a manner of speaking, is happening further below."[24] If occasionally, speaking to someone else, he criticized Planck, his criticism was always rather mild: "Even as a Goy I wouldn't have remained President of the Academy and of the Kaiser Wilhelm Society under such conditions."[25]

Planck had stayed on in his posts in order to preserve the scientific institutions and steer them through the hazards of the moment. On the issue of Einstein, he was already thinking of future generations when he had the following read into the academy's record:

> Herr Einstein is not just one among many outstanding physicists, but Herr Einstein is the physicist through whose essays, published in our Academy, physical knowledge in this century has been deepened in a manner whose importance can only be measured against the achievements of Johannes Kepler or Isaac Newton.
>
> I feel it incumbent on me to say this, lest future generations should ever think that Herr Einstein's professional colleagues were unable fully to comprehend his importance to science.

It would have been fine if Planck had stopped there—but he added a badly conceived sentence, a semiofficial Nazi interpretation which undid his earlier efforts:

> It is therefore . . . greatly to be regretted that Herr Einstein through his political behavior himself rendered his continued membership in the Academy impossible.[26]

On the eve of this declaration, books had been piled up and burned opposite the university and within view of the academy. Writings by Sigmund Freud, Lion Feuchtwanger, Heinrich Mann, Kurt Tucholsky, and many others were consigned to the flames in a satanic spectacle.[27] But not even this macabre auto-da-fé made the mandarins of the academy see that it was not Einstein's behavior but the terror in Germany that had made his parting with the academy inevitable.

As Einstein had belonged not only to the Prussian Academy, but to countless scientific societies and associations, he requested his reliable friend Max von Laue "to see to it that my name is deleted from the lists of these organizations. . . . This is probably the right way to avoid new theatrical effects."[28] Particularly, he no longer wished to belong to the German Physical Society, whose president he had once been, or to the civil division of the famous Pour le Mérite Order.

On his arrival in Antwerp, Einstein had been welcomed by some Belgian colleagues, who offered him accommodations in an old country house in the neighborhood. But as he intended to stay in Belgium for the time being, he soon rented a vacation house, the Villa Savoyard, in the small seaside resort of Le Coq sur Mer, near Ostend. It was more modest than his house in Caputh, but magnificently situated among the dunes—an ideal refuge for him to think about the future and about his plans.

There, during the first days of April, news reached him that his bank accounts in Berlin had been confiscated. He did not regard this loss of 30,000 marks as tragic for very long: "I have now been promoted to the wicked beast in Germany and all my money has been taken. But I console myself with the thought that it would soon have gone anyway."[29] He did not need to accept the financial help offered him by his Dutch colleagues, "because I have been careful and made provision."[30] He had always deposited his foreign earnings in Leyden and in New York, so that, at least materially, he had no worries: "Personally, I was not really caught out, but practically all those were who are more or less close to me."[31]

His secretary, Helen Dukas, and Walther Mayer, "the calculator," arrived in Le Coq in April, thus completing the household-in-exile. His stepdaughter Margot and her husband Dimitri Marianoff had fled to Paris at the beginning of April, so that a search of Einstein's apartment on Haberlandstrasse, intended for Marianoff, produced no results.[32] Only his other stepdaughter, Ilse, and her husband, Rudolf Kayser, were still hanging on in Berlin, trying to save Einstein's papers, library, and furniture from being seized by the Nazis. Toward the end of May a squad of brownshirt storm troopers ransacked the apartment, picking up rugs, paintings, and a few other valuables.[33] Whatever was left, in particular his papers, was brought to France by sealed diplomatic bag, thanks to the help of the French ambassador André François-Poncet, and from there shipped to America.

■ ■ ■

As Einstein had for some time expected that he would have to leave Germany, the decision, made inevitable by Hitler's "national revolution," was in the end a kind of liberation. This was not the first time that he had stripped off his German nationality. He recalled being robbed of his childhood paradise in Munich, when he first decided he no longer wished to be a German, and he remembered "Papa" Winteler's mistrust of Switzerland's "large canton" in the north: "Once again the Germans are paying for Bismarck's disastrous educational efforts."[34]

In 1914 he had returned to the land of his birth with mixed feelings, not wishing to become a German citizen for a second time. That he became one nevertheless was due to bureaucratic accidents, but Einstein accepted it because he placed some hope in the new democracy. But even during the best periods of the Weimar Republic, he was aware of his precarious situation among Germans: "To them I am a stinking flower, and yet they put me in their buttonhole time and again."[35]

In contrast to his comments in America in March, he now judged the Germans more sharply—albeit, to avoid endangering his friends, only in private statements. His anger was directed mainly at the failure of the educated classes, especially the professors. He called on the foreign members of German learned institutions "not to go along with the fact that these societies accept without opposition the mortal struggle against liberal and Jewish intellectual workers. If appeals remain fruitless, I believe that another rupture of international relations would be entirely justified."[36]

Writing to Max Born, Einstein said: "You know that I never thought too favorably about the Germans (in the political and moral sense). But I have to confess that they did somewhat surprise me by their brutality—and their cowardice."[37] (Born, it may be noted, could have retained his professorship—there was an exemption for veterans who had served at the front—but he rejected this dubious favor.) Although Einstein realized that a lot of Germans were ashamed of their government and its criminal actions, he ruled out any pity or even sympathy for them: "I was there when for many years they nurtured the viper in their bosom, and when hell erupted they hid in their mouse holes. But they will soon feel on their own skins the consequences of their lack of a sense of duty."[38]

Fritz Haber, the baptized, overassimilated Jew (who, like Born, had served at the front), initially intended to stay. But in the end he could not find it in his heart to sack Jewish colleagues who did not enjoy his privileges; in fact, he considered going to Jerusalem. Einstein welcomed him in exile, not without some mockery: "I am delighted to know . . . that your former love for the blond beast has somewhat cooled off. Who would have thought that my dear Haber would turn up here as the champion of the Jewish, and indeed the Palestinian, cause." As Einstein was well aware of Haber's deep attachment to anything German, he made it clear that there was no place left in "Teutonia" for honest people, and that Haber should not regret his departure: "Surely there is no future in working for an intelligentsia that lies on its belly before common criminals and even, to some extent, sympathizes with those criminals. Me, they were unable to disappoint, because I never had any respect or sympathy for them—except for a few fine personalities (Planck 60 percent noble and Laue 100 percent)."[39]

A few months after the Nazis' seizure of power, when the Third Reich had just begun to reveal its perverse and diabolical character, and when many people still thought that things might not be too bad after all, Einstein had concluded the second German chapter of his biography, and this time for good: "I probably won't see the country of my birth again."[40] His parting was without nostalgia; and his solidarity henceforth was with the persecuted and expelled, particularly his "tribal companions": "[To me] the best thing always is contact with a few fine Jews. A few millennia of a civilized past mean something after all."[41]

If Einstein had needed any consolation during the first few weeks of his exile, he could have found it in a multitude of well-intentioned offers. He was asked to give some lectures in Brussels. In Oxford, Lindemann wanted to convert his "research studentship" into a professorship. From Madrid he had a prestigious invitation for the summer of

1934, which he accepted. He also accepted an offer from the Collège de France in Paris, which Langevin had arranged. Soon he complained: "I have more professorships than I have useful ideas in my brain. The devil shits on the big numbers."[42] And when he thought of his most important obligation, in Princeton, he feared: "When I have done this for a year I'll be dead."[43]

Abraham Flexner, who learned from the press of the offers that were pouring in to Einstein, may have been concerned that the ornament of his institution would squander his energies and neglect the Institute for Advanced Study. He therefore not only offered Einstein whatever assistance he needed, but also proposed that he come to Princeton not for the winter but for the whole academic year.[44] For a time Einstein hesitated, partly because he did not wish to give up Europe altogether and partly, perhaps, to ensure the best possible terms in Pasadena for his "calculator" Walther Mayer.

While Einstein realized that famous refugees such as Max Born would find suitable positions somewhere, he worried about the plight of the less famous: "My heart bleeds when I think of the young ones."[45] As first aid for Jewish university teachers, who had no chance of a livelihood in Germany, Einstein wanted "to try, together with a few friends, to set up a Jewish guest university for Jewish *Dozenten* and professors outside Germany (England?), to meet at least the most urgent requirements and provide a kind of intellectual asylum."[46] His liaison in England was Leo Szilard, who, immediately on his arrival there, had involved himself in organized assistance efforts. It was partly due to Szilard that an Academic Assistance Council was soon able to mitigate the worst hardships.

Over the first weeks of his exile, Einstein pursued his project of a refugee university with great zeal; he intended to "use all his influence to raise the money, the residence permits for those concerned, and a location. . . . This is the only way of helping in a dignified manner. It would also be a living disgrace for the Germans."[47] However, his commitment being greater than his organizational skills, he had to admit after a few weeks "that the difficulties are insuperable, and that such an enterprise would impair the efforts being made in the individual countries."[48] In fact, men like Lindemann and Rutherford were incomparably more efficient in organizing the Academic Assistance Council than Einstein could ever have been with his "refugee university."

One problem with the idea of a "refugee university" was the fact that there already was a Jewish university, the one in Jerusalem, which claimed to be for all Jews. Many Zionists, especially Chaim Weizmann, therefore believed that it was Einstein's duty to go to Jerusalem. But ten years previously Einstein had noted, with regard to moving to Palestine, that while his heart said yes, his reason said no.[49] Besides, there was his conflict with the administration of Hebrew University.

As early as March 1933, Einstein had brusquely declined an invitation from Weizmann. Even though, at Weizmann's request, he had rejoined the university's board in 1932 and had again raised money for it in America, the university to him was still "a real pigsty—nothing but charlatanism."[50] Without a "thorough purge" one could "not advise any decent person to go there."[51] Whether such fierce criticism was justified is an open question; but there is no question that Einstein, while calling for Jewish solidarity, was jeopardizing it. To Ehrenfest he reported that he was fighting for reform of the university "with a brutality that would amaze you."[52] Many of his "tribal companions" were not so much amazed as appalled.

In the past, Einstein had voiced his criticism of Hebrew University only privately, but he abandoned this restraint in the spring of 1933. He regretted publicly that "this university, upon which such great hopes were fixed, is unable to play the role in satisfying intellectual needs that should be expected of it at this critical period."[53] Also, for the first time he publicly announced that he had resigned from the board five years previously.

Although Weizmann was shocked by Einstein's behavior, he still wanted to lure the "Jewish saint" to Jerusalem. He again promised the reform demanded by Einstein, and even the establishment of a separate research institute. When Einstein got a news agency to inform Weizmann publicly that the latter knew very well "under what circumstances I am prepared to work for the Hebrew University,"[54] Weizmann twisted this statement around to mean that Einstein was ready to accept a professorship in Jerusalem. This was not so much diplomacy as blackmail, and Einstein complained about Weizmann, though not publicly: "He is an intelligent and charming man, but unfortunately a complete liar (a Jewish Alcibiades)."[55] Weizmann for his part felt that Einstein was "acquiring the psychology of a 'prima donna' who is beginning to lose her voice."[56]

One result of this polemic was that the board and a commission thoroughly examined the university. After two years of investigations and consultations, the responsibilities of Judah Magnes, whom Einstein saw as the root of all evil, were reduced to representative functions. Academic matters and staffing were now the business of a rector, or principal, a post to which Hugo Bergmann was appointed. Einstein knew Bergmann from Prague and esteemed him as a "serious saint"; Thus Hebrew University had set out on a new course which no longer put Einstein's loyalty to the test.

In addition to the news from Germany about the Nazi terror, Einstein had to face the fact that his younger son, Eduard, had developed schizophrenia. He therefore decided to postpone his trip to Oxford, originally planned for the end of May: "I wouldn't have a quiet minute in England," he apologized to Lindemann. "Although you are not a father yourself, I am sure you will understand."[57] He first had to give three lectures promised to the Belgian Franqui Foundation in Brussels before traveling through France to Zurich to see his son.[58]

It is a commonplace that sons of famous men—because of the attention and expectations to which they are subjected—are not to be envied; and it is also accepted that conflicts between parents leave a mark on their children. Einstein, who, in spite of being a "loner," was greatly attached to his sons, had tried hard, after his separation from Mileva, to perform his role as father. This had not always been easy for him, because of Mileva's "Medea syndrome" and because of the loyalty his sons had to their mother. Moreover, he had been a difficult father, who would stay out of touch for months on end and then would try forcibly to impose his will on his sons.

His older son, Hans Albert, had a taste of that when he intended to get married. Einstein opposed the marriage in a brutal way that far surpassed the scenes which his own mother had made about Mileva. Not only did he mobilize his friends, such as Anschütz and Zangger, to dissuade Hans Albert; he also had the medical history of the mother of his unwanted daughter-in-law investigated—after a hard life, she had had at one time undergone psychiatric treatment. Einstein was as firmly convinced of the hereditary nature of mental illness as he was of his unified field theory; for this reason, and because a sister of Mileva's had developed schizophrenia, he was greatly worried about the effect of mental illness on both sides of the family.[59] Hans Albert, however, who was his father's equal in strength of mind, married his fiancée as soon as he had passed the ETH diploma exam and was able to stand on his own feet. Two years later Einstein reported that his older son had "very disrespectfully promoted him to grandfather."[60] After that, relations between father and son, if not excessively warm, were free from conflict.

Eduard, the younger son, seems to have been less robust than Hans Albert. After a childhood marked by frequent illnesses, including tuberculosis at the age of thirteen, he developed into a sensitive boy, a good student, with considerable literary and musical talent, and certainly superior to his father as a poet.[61] While still at school he was fascinated by Sigmund Freud's writings, and after graduation he determined to study medicine. As a student he repeatedly visited his father in Berlin—for example, in March 1930 on Haberlandstrasse and in September in Caputh. "Albert is happy with him," recorded Elsa.[62]

Soon, however, signs of serious mental illness appeared. His studies were first neglected and subsequently became impossible, and in the fall of 1932 he had a major episode of schizophrenia and had to be taken to the Burghölzli psychiatric hospital. Zangger and Besso informed Einstein of this tragic development, urging him to look after his threatened son and take care of him: "That would be best for both."[63] Einstein, however, viewed his son's condition as more or less incurable and due to heredity on the boy's mother's side: "Everything unfortunately indicates that the grave heredity will [have] its decisive effect [on] him. I have seen it coming, slowly but irresistibly, ever since

Tedel's youth. The external occasions and influences play only a slight role in such cases compared to the secretory causes, which no one can get at."[64] And off he went to Pasadena.

Zangger initially thought that Eduard might make a full recovery and continue his studies, but this hope did not materialize. After painful months, which tested Mileva almost beyond the limits of her endurance, the young man again had to be admitted to Burghölzli in the spring of 1933. There is no written evidence about Einstein's feelings when he visited his son. No doubt he was profoundly shaken, and he certainly determined to make sure his son's future was financially secure. Beyond this, he may have acted on a view which he had once expressed to his friend Ehrenfest, when Ehrenfest's son Vassik (who had Down syndrome) had to be placed in an institution: "Valuable individuals must not be sacrificed to hopeless things, not even in this instance."[65] When Einstein left Eduard, it was a final farewell.

From Zurich Einstein went straight to England, where he arrived on June 1. The day after his arrival he was enthusiastically fêted as the guest of honor at Lord Rutherford's Robert Boyle Memorial Lecture—now not only as the greatest physicist since Newton but also as a symbol of moral integrity and opposition to dictatorship. With Rutherford, and particularly with Lindemann—who had toured Germany in April to inform himself on the situation of discharged researchers and recruit the best of the young physicists for his Clarendon Laboratory—Einstein discussed ways and means of assistance. But he avoided public political statements, realizing by then that they would only provoke the Nazis. Besides, in the atmosphere of creative quiet at Christ Church College, he wanted to prepare a few lectures.

On June 10, Einstein gave the Herbert Spencer Lecture at Rhodes House, a largely comprehensible exposition of his views: *On the Methods of Theoretical Physics.*[66] For the first time he spoke in English, although he was reading his lecture from a written translation of his German text. More significant than this linguistic premiere was the explanation of his research procedure, as it had matured over the past ten years and hardened into a credo. As there exists no logical road from statements of fact to theoretical conceptual systems, he explained to his audience that "the axiomatic foundation of physics cannot be derived from experience, but must be freely invented." Such invention must be guided by confidence "that nature represents the realization of the simplest that is mathematically thinkable," and by the hope that through pure mathematical construction "pure thought is capable of comprehending reality." Actually, this hope vastly overrates "pure thought" in matters of cognition of nature, but Einstein adhered to it, despite numerous disappointments, to the end of his life.

Two days later he gave the Deneke Lecture, this time in German and aided by just a few scraps of notes. He could not be persuaded to prepare this lecture—on the structure of physics—for publication. His next lecture, the George Gibson Lecture given at the University of Glasgow on June 20, was a fully worked-out text. His hosts had asked him to illustrate an important phase of his own scientific work; he complied with a fine reconstruction of his greatest triumph, the genesis of the general theory of relativity.[67]

Back at his refuge in Le Coq, Einstein, whether he wanted it or not, was once more caught up by politics, this time on the issue of his pacifism.

At the beginning of July Einstein was notified that "the husband of the second violinist would like to talk to you on an urgent matter."[68] It was obvious that this meant King Albert, and that the "urgent matter" concerned two young Belgians who were in prison for refusing to serve in the army. Pacifist sympathizers expected Einstein to intervene on their behalf, and the king wished to prevent that. A conversation in the gardens of Laeken Palace resulted in agreement between the two men on the essential aspects, as Einstein confirmed to the king in writing: for one thing, in view of developments in Germany, Belgium's army was purely an instrument of defense, and indeed was necessary for defense; for another, a foreigner enjoying the hospitality of a country should not intervene in matters such as refusal of military service. Einstein's former pacifist fervor now resulted merely in a request to the king to create alternative service for conscientious objectors instead of labeling them criminals.[69]

But Einstein was being pressed to defend the two imprisoned men, and he could no longer avoid a public statement on his change of mind. He provided one without delay. As Germany was "obviously working all out toward war," France and Belgium in particular were "in grave danger and absolutely dependent on their armed forces," he stated in a letter to be disseminated among pacifists. Moreover, he declared that although barely three years before he would "rather have been cut to pieces" than serve in the forces, "under today's conditions, if I were a Belgian, I would not refuse military service, but gladly take it upon me in the knowledge of serving European civilization."[70]

When in August 1933 this statement appeared in the papers in France, and subsequently in England and America, it had the effect of a bombshell among pacifists. Some were disappointed and others bitter about Einstein's volte-face, which many called an unscrupulous betrayal. "The antimilitarists are falling on me as on a wicked renegade," Einstein said, summing up their reactions. "Those fellows simply wear blinders and refuse to acknowledge the expulsion from 'Paradise.' "[71] Over the next few months and years, the former champion of militant pacifism would have to do a lot of explaining to his former comrades.

In countless statements Einstein argued that it was not that he had betrayed his convictions, but rather that the political situation had changed. "I am the same ardent pacifist as before," he insisted. "But I believe that the instrument of refusing military service can only be advocated again in Europe when the military threat to the democratic countries from the aggressively-minded dictatorships has ceased to exist."[72]

More sensitive criticism was directed not so much at Einstein's change of opinion as at the absolute assurance with which he always proclaimed his present views as the only correct ones. "Did he never consider the possibility of a situation like today's, when implementation of refusal to serve, as propagated by him, could become dangerous?" asked Romain Rolland, who, like Einstein, abhorred war but who had always regarded Einstein's "two percent" argument as deceptive and dangerous. For Rolland it now seemed that "Einstein, a genius in science, is weak, indecisive, and contradictory outside his own field. . . . His continuous change of opinion and the hesitation and discrepancy in his actions are worse than the inflexible obstinacy of a declared enemy."[73] Rolland's diagnosis proved wrong in that Einstein did not change his mind again: he would continue to insist that, against the German aggressor, military strength had to be the supreme law of "the nations which have stayed normal."

Much as Einstein's stay in Le Coq may have been like a vacation, the Belgian seaside resort had one major disadvantage—it was too close to Germany. It would have been foolhardy to ignore rumors that the Nazis were out for the life of their hated exile and had even put a price on his head.[74] The Belgian government had therefore detailed two police agents to protect the eminent guest, but Einstein found his continual supervision both comical and annoying, and hardly efficient.

Although the inhabitants of Le Coq had been asked to pretend ignorance when questioned about Einstein, any stranger asking the way to the Villa Savoyard was readily directed there. Naturally, the two detectives had no idea how to spot an assassin among harmless sightseers and legitimate visitors, with the result that many a genuine guest of the Einsteins was first detained in the dunes.[75] Elsa was living in a permanent state of excitement, so that Einstein, though at least pretending not to be afraid, decided at the beginning of September to leave the idyllic spot and the European continent altogether. Possibly, he was also reacting to the assassination of Theodor Lessing by Nazi agents in Marienbad (Mariánské Lázně) in Czechoslovakia on August 30, 1933.

On September 8, Einstein took the ferry to England. A few months earlier he had accepted an invitation from Commander Oliver Locker-Lampson, an officer, barrister, journalist, and member of Parliament. In this last capacity Locker-Lampson had induced the House of Commons to offer Einstein British nationality. In July, Einstein had already been Locker-Lampson's guest and, accompanied by him, had called on Winston Churchill, Austen Chamberlain, and Lloyd George. With

Churchill in particular he found agreement on the danger represented by Germany: "This is an eminently clever man, and I fully realized that these people have made good preparations and will act resolutely and soon."[76]

Einstein spent the last four weeks before his departure for America in "cheerful exile" at a vacation home belonging to Locker-Lampson on the Norfolk coast, to the north of London. He kept away from public controversies, but there was one important event he could not avoid. When he declared his readiness to speak at a joint public rally of the Academic Assistance Council, the Refugee Assistance Council, and other such organizations, the event was arranged in the Albert Hall, London's huge circular concert hall on the edge of Kensington Gardens.

On the evening of October 3, an audience of some ten thousand crowded into the rotunda. On the dais sat the great scholars—Lord Rutherford; Sir James Jeans, the physicist; and Sir William Beveridge, director of the London School of Economics. But most of those present had come to hear Einstein. He spoke in English, in a clear voice but with a strong accent. Without naming Germany, he depicted the danger emanating from that country; he thanked the English for having remained loyal to their tradition of justice and tolerance, and voiced the hope that it would be said in the future "that in our day the freedom and honor of this continent was saved by its western nations."[77] In conclusion, he perplexed his listeners by referring to the pensive quiet of lighthouses and lightships: "Would it not be possible to fill such posts with young people who wish to ponder scientific problems, especially those of a mathematical or physical nature?"[78]

Einstein made an overwhelming impression. The British press hailed the charismatic scientist as a "double symbol—a symbol of the mind traveling in the cold regions of space, and a symbol of the brave and generous outcast, but pure in heart and cheerful in spirit."[79]

A week later, in Southampton, Einstein boarded a steamship from Antwerp. His wife and Helen Dukas were already on board, and the party also included Walther Mayer. Einstein and his entourage were traveling on visitors' visas because the plan was still for only a six-month stay at Princeton. The following spring Einstein intended to be back in Europe. As the *Westernland* sailed down the Channel toward the Atlantic, he did not know that he was seeing the shores of the Old World for the last time. He was never to return.

CHAPTER THIRTY-FIVE

Princeton

EINSTEIN'S ARRIVAL IN AMERICA (after a crossing that was sometimes rough) had been organized by Flexner not as a triumphal progress but as a clandestine operation. As the *Westernland* approached New York harbor, a launch came alongside. Einstein and his party were taken to the Battery, on the southern tip of Manhattan, for speedy immigration processing, and he was then taken over by a trustee of the Institute for Advanced Study. While the mayor of New York was still waiting to welcome the world's most famous Jew with a parade at the Twenty-third Street pier—not entirely unselfishly, because the election campaign was in full swing and he needed the Jewish vote—Einstein had arrived by car in Princeton.

For the first few days, rooms had been booked at a small hotel, the Peacock Inn, where it was easier to shield Einstein from importunate curiosity than it would have been at the elegant Nassau Inn, the leading hotel in town. Within a few days Elsa had found an apartment in one of the houses, near the campus and opposite the Theological Seminary, which formed an elegant residential area.[1] After six months of wandering as a refugee, Einstein once more had a home, in a town that would be his refuge for the final two decades of his life. "Princeton is a wonderful piece of earth and at the same time an exceedingly amusing ceremonial backwater of tiny spindle-shanked semigods," he wrote, describing his experiences of the first four weeks, during which he had already found a rhythm for his life: "But it is possible, by offending against the *bon ton*, to ensure for oneself some splendid undisturbed existence, and that is what I am doing."[2]

Among America's elite universities, Princeton represents the dream of an academic ivory tower in its purest form—insulated from the rest of America, snobbish, arrogant, and with restrained wealth. The town is grouped around the all-dominating campus, and apart from the university there is virtually only the small trade associated with it, and the houses of affluent citizens, most of them alumni loyal to their alma mater, but making their money in New York, fifty miles distant, or elsewhere. The many scholars expelled from Germany who came to Princeton in the 1930s may have found that their acclimatization was made easier by reminders of Europe; and if they did not look too closely at the imitations of English Gothic and other styles, they might almost think themselves in Oxford or Cambridge.

The college was once, for many decades, a favorite for the education of "southern gentlemen," who were interested less in intellectual brilliance than in acquiring social polish. But around the turn of the century it developed an ambition to surpass the Old World even in academic standards; and thereafter Princeton established itself as a citadel of learning—and in some disciplines, such as mathematics, as a world leader. All this had been taken into consideration by Flexner when he established his Institute for Advanced Study at Princeton, and the university, despite some misgivings about this exclusive rival, readily assisted with its birth.

With no premises of its own, the institute's school of mathematics was offered hospitality at the university's Fine Hall. Fine Hall belonged to the department of mathematics, inaugurated only a few years previously; it was here that Einstein's remark, from his visit in 1921, was written on the chimneypiece: "The Lord God is subtle, but malicious he is not." Einstein was assigned an attractive corner room on the second floor, next to the rooms of his few colleagues at the institute, all of whom were mathematicians who had been snatched up by Princeton University—Oswald Veblen, as dean; James Alexander; and the brilliant John von Neumann, whom Einstein knew from Berlin. In January 1934 this high-powered circle was augmented by Hermann Weyl. Weyl had initially declined an invitation from the institute, hoping to continue the Göttingen tradition as Hilbert's successor—until Hitler quashed that hope. Weyl by then had given up his excursions into theoretical physics and was concentrating once more on pure mathematics. In consequence, "without prejudice to our continuing cordial personal relations,"[3] there was no scientific link between him and Einstein.

In addition to the four professors, the institute had over a dozen younger scientists, called "workers," who were hoping to benefit from this temporary collaboration with the "illustrious." Walther Mayer, thanks to Einstein's perseverance, had the exceptional status of an "associate." At the university, moreover, Einstein met a few acquaintances from Berlin, such as Professor Rudolf Ladenburg and the young theoretician Eugene Wigner. In March 1934 Erwin Schrödinger arrived, one of the few non-Jewish professors to have left Germany in the summer of 1933. The university offered him a prestigious professorship, but, having just received a Nobel Prize, Schrödinger declined, in the hope of a better-paid post at the institute. That, however, was not offered him, and in April he therefore returned to Europe.[4] Einstein would have liked to have Schrödinger as a colleague and even tried to speak for him, but he had lost all influence with Flexner as a result of a fierce conflict with him, culminating in a threat by Einstein to leave the institute.

What led to the controversy was Flexner's attempt to exert control, like a guardian, over his prominent employee's affairs. Partly through his own desires and partly through those of his conservative patrons, he felt it his duty to shield the institute and its members from publicity, and to this end every means, however dubious, seemed to him justified. Thus he had tried, by telegram, to forbid Einstein to take part in the rally at Albert Hall in London, but Locker-Lampson had informed Einstein of this only just before his departure for America.[5] On his arrival, Flexner warned Einstein against "irreponsible Nazi gangs in this country," to add weight to his urgent advice: "Your security in America depends on your silence and the rejection of all public appearances."[6]

Although Einstein himself felt a need for a quiet and retired life, he was unwilling to cut himself off totally, let alone to have his freedom curtailed by someone else. Even on such a trivial public occasion as a conversation with representatives of a student newspaper, Flexner intervened like an angry governess. When Einstein was to appear as a violinist at a charity concert in New York for the benefit of refugees, Flexner tried to foil the arrangement by making vicious telephone calls to the organizers, telling them he would "fire" Einstein. Flexner also regarded it as entirely normal to withhold any mail addressed to Einstein and to decline invitations on his behalf—he even declined an invitation to the White House, and this proved to be the last straw.

The invitation had been initiated immediately after Einstein's arrival by a friend of his, Rabbi Stephen Wise in New York. It was felt that a meeting between the famous exile and President Franklin Delano Roosevelt would draw attention to the plight of Jewish refugees from Germany. However, when the invitation arrived at the institute at the beginning of November, Flexner opened the letter and, without consulting Einstein, informed the president "that Professor Einstein had come to Princeton for the purpose of carrying on his scientific work in seclusion and that it was absolutely impossible to make an exception which would inevitably bring him into public notice."[7] Flexner added a reference to "irresponsible groups" and an assertion that he had, in agreement with Einstein, declined invitations even from scientific societies "in whose work he is really interested." No doubt the White House was meant to conclude that Einstein was not really interested in the work of the president.

When Einstein learned from Washington[8] that an invitation had been issued by the White House but had not reached him, the reason was obvious, and he did three things. He immediately assured Eleanor Roosevelt of his keen interest "in meeting the man who is addressing the greatest and most difficult problems of our era with gigantic energy."[9] He vented his fury at Flexner in a letter to Rabbi Wise, written from "Concentration camp, Princeton."[10] And he sent to the institute's trustees a long list of Flexner's arbitrary and tactless actions and gross misjudgments, concluding with a request that they ensure for him "security for undisturbed and dignified work, in such a way that there is no interference at every step of a kind that no self-respecting person can tolerate. If this were to be considered impracticable, I would propose that I discuss with you ways and means of severing my relations with your Institute in a dignified manner."[11] The threat was sufficient; after a "storm in the ivory tower," Flexner had to leave Einstein alone. Einstein was once more a free professor in a free country, but now without any influence on the running of the Institute for Advanced Study.

This bizarre episode was eventually followed by a visit to the president. On the evening of January 24, 1934, Einstein and his wife dined with the Roosevelts at the White House and spent the night in the Franklin Room. No record exists of the conversation on that evening. Einstein no doubt pointed to the danger emanating from Germany and to the plight of Jewish refugees, but the president preferred more harmless subjects like sailing and their acquaintance with the Belgian monarchs—about which Einstein reported to the queen in rhymed doggerel.[12] The visit certainly had no effect on bringing the fate of persecuted Jews to public attention.

"As far as possible I have followed the amicable advice to keep silent on political matters, though not from concern about my worn-out corpse but because I saw no chance of achieving anything useful."[13] This was one way Einstein justified his relative restraint in public statements. Another reason was that "as a German resident and state employee for many years, and as a Jew and a person robbed by the Nazis of his possessions, I would not be an objective judge in the eyes of the public. Altogether, in this struggle Jews should emerge as little as possible in public, because otherwise opposition to Hitlerite Germany might be labeled a Jewish affair and thus weakened."[14] He may also have considered that Americans, for all their open-mindedness, were not inclined to be lectured by guests, no matter how famous, in their own country.

Nevertheless, he still had some lessons for pacifists, even after the rise of Hitler. He still felt close to the American Friends of Peace, but his emphasis had shifted toward what was to become his passion in his later years—"world government." Increasingly, he argued in favor of ensuring peace "through the creation of an international organization embracing all major states . . . with a sufficiently strong executive power at its disposal."[15] In cautiously formulated statements he tried to persuade the Americans to give up their traditional isolationism and "to support in the near future the realization of an efficient international central power. The pursuit of international solidarity is today the best defense against fascism, which represents such a grave threat to our cultural life."[16]

Within a month of arriving at Princeton, Einstein was thinking about severing his only contractual link with Europe. He suggested to Lindemann "that under the present circumstances I am hardly justified in accepting a payment from Christ Church College for a number of years." Instead he proposed that "some other colleague in the field might benefit from that fine facility."[17] Although Lindemann tried to get him to visit Oxford, Einstein had made his summer plans by the end of March 1934: "If it is possible I will go into hiding somewhere in America over the summer. Why shouldn't an old guy have something like relative quiet once in a while?"[18]

Einstein stuck to that plan even when his stepdaughter Ilse fell seriously ill in Paris. In mid-May he saw his wife to the *Belgenland* in New York harbor and let her travel to Europe on her own. Elsa was only able to watch her daughter die. In August the ashes of Ilse—who had been only thirty-seven—were interred in the Netherlands, where Rudolf Kayser was living. Kayser was compiling a collection of his father-in-law's essays, which appeared in Amsterdam in German under the title *Mein Weltbild* (*The World as I See It*).

Einstein meanwhile had found a refuge for the summer, remote and suitable for sailing. Along with Walter Bucky, a radiologist who had treated Einstein's stepdaughters in Berlin, he rented a spacious, beautifully situated house on the coast of Rhode Island. Bucky had become an American citizen even before Hitler's seizure of power and was by then running a profitable practice. In the New World he became one of Einstein's closer friends; they also worked together on some patentable gadgets.

Einstein's health benefited greatly from sailing on a twenty-foot yacht, a compensation for his "fat boat" in Caputh, which had already been confiscated and resold by the German authorities as the property of an enemy of the nation and the state. The Atlantic coast, with its tides, forceful winds, and massive waves, was considerably more dangerous than the Havel lakes, and this probably suited him; he was, in a romantic way, fond of experiencing the play of primal elements when out sailing. He cheerfully dismissed other people's concerns about his safety; as he had done in Berlin, he refused to carry an auxiliary engine or life jackets: "If I have to drown, then let it be honestly." When Elsa returned from Europe, distressed and exhausted, she found her husband "in excellent shape. Nothing tragic really gets to him, he is in the happy position of being able to shuffle it off. That is also why he can work so well."[19] He did not return to Princeton until the beginning of October, as the second working period was starting at the institute.

The art historian Erwin Panofsky, who came from Aby Warburg's[20] Hamburg school and was one of the founding members of the institute's school of humanities, attributed the legendary reputation of the institute to the fact that "its members conduct their research activities openly and their teaching activities, as it were, secretly, whereas the contrary is true of many other learned institutions."[21] Many renowned researchers, regarding teaching and feedback from their students as indispensable, declined invitations to the institute; but this "ivory tower" setup was entirely to Einstein's taste. He did not wish to teach, even in secret. What he needed was time to think—and a collaborator to do his calculations.

His "calculator," Walther Mayer, however, sadly disappointed him. As late as the autumn of 1932 they had published a joint paper in the *Proceedings of the Prussian Academy*, in which they examined the spinors found by Dirac in terms of their usefulness for a unified theory.[22] These reflections were continued amid the dunes of Le Coq and published by the Amsterdam Academy.[23] In Princeton, however, Mayer participated in only one other paper on the same subject[24] before discovering that he was not, after all, the right "calculator" for Einstein's problems. Although he remained at the institute, he turned to independent research in pure mathematics. Flexner and Veblen continued to hold Einstein responsible for his "calculator," with the result that for some years he was not allowed to employ another assistant. However, he found some younger collaborators among the institute's "workers," who worked with him with great dedication.

In spite of some squabbles, the Institute for Advanced Study proved an ideal workplace for Einstein. His contract, originally for five-month stays over a period of five years, had been converted to that of "full professor," with a salary matching that of Oswald Veblen: $16,000 annually. This was about twice the salary of a professor at the university; indeed, Flexner's creation was said to be not only an institute for "advanced study" but also one for "advanced salaries." Einstein realized that he could not have hit on a better post and prepared to spend whatever years were left to him in Princeton. The "migrating bird for the rest of my life"[25] had come to rest. Europe now was only memories—some beautiful, some terrible.

"Sometimes I think back nostalgically to beautiful past hours; they tempt me to make a journey to Europe," he admitted to the queen of the Belgians in 1935, immediately noting one reason against it: "But so many obligations would await me there that I cannot summon the courage for such an undertaking."[26] He offered a similar reason for canceling Christ Church: "If I come to Oxford, I must also go to Paris and Madrid. But to undertake all this I lack the courage—and so I'll probably remain sitting here."[27] Naturally, he would also have had to go to Zurich—to his sick son and to Mileva—and no doubt he lacked the courage for that too.

He wasted no nostalgia on Germany, which in any case he would not have been able to enter. Although he wrote to his "dear old comrade" Laue "that the small circle of persons that used to be harmonically linked was really unique, and in such human purity would scarcely ever be encountered by me again," he also admitted "that I shed no tears for the wider circle; it was amusing for the unconcerned observer rather than lovable."[28] He did not specify who was included in his "small circle"—probably only Laue, Schrödinger, and Planck. Otherwise his "interest in the German business . . . [was] now only as a danger to the rest of the world."[29] That danger he saw approaching ineluctably. Since Hitler's rise to power, Einstein was convinced that Germany was heading for war, and he was astonished and bitter to see the Western nations doing little to avert that danger.

▪ ▪ ▪

Having decided to stay in Princeton, Einstein thought it best to become an American. As he had arrived on a visitor's visa, and as an application for immigration could be filed only at a consulate abroad, Einstein—with his wife, his stepdaughter Margot, and Helen Dukas—made a sea trip to Bermuda in May 1935. This brief and enjoyable excursion was to be his last stay outside the United States.

Summer in Princeton is hot, sultry, and humid. Einstein therefore spent it in Old Lyme on the estuary of the Connecticut River, where he indulged in veritable luxury. Proudly his wife described the rented vacation residence with "20 acres of land, meadows, fields, with all the summer glories, tennis court, swimming pool. . . . It is so elegant that for the first ten days we ate our meals at the servants' table, because we felt too grand in the grand dining room."[30] More important to her husband was a sailboat in which he made extensive trips in the wide estuary funnel of the Connecticut.

That August an opportunity arose to buy a house in Princeton, diagonally across the street from their apartment. It was clapboard in the restrained New England style, narrow-fronted with a small garden in front and a long narrow garden at the back. Inconspicuous and cozy, this house at 112 Mercer Street differed in no way from the average residences of the neighborhood, but because of its owner it became one of the best-known addresses in the world.

Einstein paid in cash and had enough money left for renovation and conversions. These were supervised by Elsa, who shuttled for the rest of the summer between Princeton and the elegant country seat in Connecticut. While Einstein went sailing, a back wall of the upper floor of the Princeton house was replaced by a large picture window, not exactly in style, but giving almost an effect of living outdoors; there was a view across wooded parkland to the Graduate College in the distance. This room became Einstein's study, and for the next two decades it was the place where he spent most of his time and where he felt happiest. The other rooms were largely furnished with pieces saved from Berlin, including the grand piano.

As soon as they moved in, Elsa had a foreboding that she would not enjoy their new home for long. A bad swelling had appeared near one eye, and it was diagnosed as a sign of serious circulatory and kidney problems. There followed a long winter of pain and medical treatments. Even a long summer vacation in the mild climate of the Adirondacks, at Saranac Lake in southern New York State, brought Elsa only moderate relief. During the next few months in Princeton Einstein was so worried about his wife that, as Elsa herself reported, "he went around miserable and depressed. I never thought he was so attached to me. That, too, helps."[31] She died in their house in Princeton on December 20, 1936.

The man whom Elsa had looked after for two decades as her "Albertl" adjusted quickly to the new situation. "I have got used extremely well to life here, I live like a bear in my den, and really feel more at home than ever before in my varied life," he wrote after a few weeks of being alone. "This bearishness has been further enhanced by the death of my woman comrade, who was better with other people than I am."[32] But of course he was not alone in his house. Helen Dukas took care not only of his mail and correspondence, his visitors and appointments, but also of the kitchen and household. Also living in the house was his stepdaughter Margot, who had meanwhile gotten a divorce. But the image of a bear was appropriate, because Einstein's deep-rooted sense of alienation became even more marked in Princeton: "I am not really becoming part of the human world here, for that I was too old when I arrived, and in point of fact it was no different in Berlin or in Switzerland. One is born a loner."[33] This attitude was probably intensified by the language problem. Although he wished to become an American, he never really came to grips with the language of his new country.

Einstein by then could read and understand English without difficulty, but he found writing and speaking very difficult. This did not matter at the institute, as many colleagues and most of his collaborators there had come from German-speaking countries, some of them without any knowledge of English. With the others, like the British-born Banesh Hoffmann, Einstein would speak English, but with a strong accent and a curious—because German—word order and stress.[34] This did not improve over the years. After more than a decade in America he confessed to Max Born, who refused to use German during the war: "But I cannot write English because of the insidious orthography. When I read I hear the word, but I cannot remember what it looks like on paper."[35]

It was not quite as bad as that. He was able to write simple notes in English, but he continued to write in German anything that was important to him and, if necessary, had it translated. He also preferred to read German rather than English and was astonished by the "funny thing that everything seems more plastic and alive when it appears in the old language."[36] Even though Germany had degenerated into "Barbaria," he felt at home only in German: he could formulate his ideas and express his feelings only in the words familiar to him from childhood, with the result that in the "daily struggle with English" only "the German stepmother-tongue is left as practicable, to my sincere regret."[37]

If Einstein felt fortunate "living here in Princeton in an island of destiny,"[38] this was largely due to the fact that the Princetonians respected his need for solitude. Of course they were interested when Einstein settled in their little town—this genius who had drawn so much attention by his wisdom about the universe and by his controversial political statements—but snobbishness as well as good manners ruled out excessive curiosity. Only from a distance were they amused by his mane of white hair (in the land of the crew cut); and from a distance they chuckled discreetly over his habit of licking an ice cream on Nassau Street on his way home from Fine Hall and were astonished by his utterly un-American long walks through the streets of Princeton.

Perhaps to bridge that remoteness, a whole collection of harmless anecdotes grew up around the famous man. It was amusing that the man who knew so much about the universe should lose his way while walking through Princeton and then forget his own telephone number, only to have Information refuse to give it to him because it was unlisted. The many stories of this kind[39] probably all had a grain of truth, but they probably said less about Einstein than they did about the way the citizens of Princeton reacted to the eccentric genius in their midst.

Einstein also led a rather marginal existence with regard to the growing colony of European refugees. He avoided John von Neumann's splendid parties, where science and politics were discussed into the small hours while liveried footmen served champagne, cocktails, and whiskey—just as he avoided the informal gatherings at the house of Hermann Weyl. With Thomas Mann, who was invited to the university in 1938 as a visiting professor for two years, he shared (as he did with all refugees) a revulsion for Nazi Germany, but no closer tie developed between them. The Manns lived only a few blocks away, in a grand house with servants, but this patrician setting for the intellect was no more to Einstein's taste than he was to the taste of the lady of the house: "He was very pleasant but not particularly stimulating. Einstein had something childlike about him, such big goggle-eyes. . . . Really an enormous specialized talent, but in ordinary life he was not a very impressive person."[40]

In New York he had a circle of friends to whom he was stimulating enough—above all the physician Gustav Bucky; the wealthy pharmaceuticals manufacturer Leon Watters; and the economist Otto Nathan, who had initially smoothed Einstein's way in Princeton in practical matters, had advised him on finances, and (jointly with Helen Dukas) had been chosen as his executor. In Princeton Einstein's contacts were less with the academic establishment than with outsiders, such as the philosopher Franz Oppenheim, the historian Erich von Kahler, and the Austrian poet Hermann Broch. Some of these, like Broch, he had helped to escape the Nazi terror.

Einstein was continuously aware of the plight of the exiles and of conditions in Germany, often in a family context. Soon after his arrival in America he had made arrangements for the admission of his wife's relatives, and in 1937 Hans Albert arrived in the United States with his wife and son. Two years later Einstein's sister, Maja, had to leave Italy. Her husband, Paul Winteler, was refused entry to the United States for reasons of health and remained in Geneva with the Bessos; but Maja moved into the house on Mercer Street in 1939.

In the course of helping his family, Einstein discovered how difficult immigration had become. An American consul would accept only an immigrant whose national quota was not yet filled and who, in addition, could produce an affidavit by a resident of the United States guaranteeing that the immigrant would not become a public charge. As in Germany, Einstein again criticized the lack of Jewish solidarity: "The main tragedy is that the sated Jews in the countries so far spared are, just as the German Jews once were, indulging in the foolish hope of being able to achieve their safety by silence or by patriotic gestures. For that reason they now sabotage the acceptance of German Jews, just as these previously sabotaged that of the Eastern Jews."[41] Besides, the American consuls saw to it that immigration was kept within limits.

Einstein did his best to help, with affidavits, gifts of money, and loans for travel expenses—sometimes without being asked. For example, in 1935 a young violinist in Berlin, Boris Schwarz, with whom Einstein had made music, received an inquiry from "Elsa Alberti" about whether he wanted to come to America. Soon Einstein's affidavit was at the Berlin consulate: Boris Schwarz could travel to New York. There he was accommodated by Einstein in a Jewish refugee home and given some fatherly advice on his first steps in this strange country.[42] Einstein also helped the children of his family physician in Berlin, Otto Juliusburger. The children came to America, thanks to Einstein's affidavits; and at the very last moment, the aged doctor and his wife were also able to leave Germany, again with Einstein's help.

After the annexation of Austria in March 1938, and the great waves of emigration which followed, the limitations of private help became obvious. Einstein therefore took the initiative and tried to start a program by which large organizations, such as the churches, the universities, and the Red Cross, would appeal to the "conscience of all well disposed people." He drafted a proclamation:

> No state is entitled to physically annihilate a section of its population living within its frontiers. We are determined, in every way open to us, to prevent innocent people being driven to their death, either by weapons or by systematic deprivation of all possibilities of livelihood. By its inhuman measures against German and Austrian Jews Germany has embarked on that road of annihi-

> lation and is using its military, political, and economic power toward the small states of eastern Europe to annihilate their Jewish populations in the same way.[43]

In spite of Einstein's prominence, though, this enterprise produced no results.

Referring to his personal assistance efforts, he reported in the summer of 1938 "that I have a contact agency for persecuted as well as intellectual eccentrics, and I can assure you that business is booming fantastically."[44] He believed that the problems of immigration could be overcome—but not the problem of employment in America. Job openings were still few after the Great Depression; there was anti-Semitism in the universities; and there was considerable reluctance to employ elderly men. Einstein felt driven to bitter sarcasm: "The best thing would be a generous attempt to feed our exiles over fifty in some cheap country in a kind of 'concentration camp.' But even that will remain a pious hope."[45]

In practical terms he had to point out that his own opportunities were limited. Although "as well known as a multicolored dog, I live very quietly and have virtually no human connections."[46] Soon, also, his financial resources were exhausted: "I can give no more affidavits, and would only jeopardize those already given if I issued new ones," he wrote after Hitler's occupation of Prague. "The few people I know who have money are also extended to the limit. The pressure on us from the poor people over there is such that one almost despairs in the face of so much misery and the slight possibilities of helping."[47] He saw no end to the misery; he only saw worse things approaching. "I wouldn't like to live if I didn't have my work."[48]

As in the past, it was physics that sustained him, even though he realized that in his field he had become an outsider and mainly "highly esteemed as an ancient labeled museum piece and curiosity."[48] However, "I am still working steadily, supported by a few adventurous young colleagues. I can still think, but my working energy has declined."[49]

CHAPTER THIRTY-SIX

Physical Reality and a Paradox, Relativity and Unified Theory

EINSTEIN WAS NOT ALONE in his mocking description of Princeton as an "amusing ceremonial backwater of tiny spindle-shanked semigods"; but very soon he was included by others among the curiosities of the ivory tower. "Princeton is a madhouse," observed the highly gifted J. Robert Oppenheimer about the Institute for Advanced Study in January 1935, with the self-assurance of youth—Oppenheimer was then thirty-one. "Its solipsistic luminaries shining in separate and helpless desolation. Einstein is completely cuckoo."[1] In this judgment, Oppenheimer—who, incidentally, became the director of that "madhouse" twelve years later—makes no claim to fairness, but still it reflects the fact that Einstein's stubborn criticism of quantum mechanics and his castle in the air, his unified theory, were getting on the nerves of the younger creative physicists. This was all the more true because Einstein refused to be diverted by newly discovered phenomena which opened undreamed-of new regions in physics.[2]

In 1932 the positron—the "antiparticle" of the electron—had been detected in cosmic radiation; this was the first representative of "antimatter." Precision measurements of radioactive decay inspired Wolfgang Pauli to postulate an entirely novel particle, the neutrino, with whose aid Enrico Fermi developed an effective theory of weak interactions. The discovery of the neutron as another building block of the atom opened the door, both theoretically and experimentally, to what for the first time could be called nuclear physics. And Einstein's former collaborator in Prague and Zurich, Otto Stern (before he had to abandon his chair in Hamburg in 1933 and emigrate to the United

States), demonstrated—by measuring the anomaly of the magnetic momentum of the proton—that even the nuclear building blocks must have an internal structure.

What excited the whole profession passed Einstein by: he noted it only sporadically, and it did not influence the problems he was concerned with. He was not unaware of this situation. In a moving obituary for his close friend Paul Ehrenfest, who committed suicide in September 1933, Einstein described "the increased difficulty that adaptation to new ideas inevitably imposes on a man of fifty." Indeed, he saw these difficulties as a major reason for Ehrenfest's suicide.[3]

Suicidal thoughts, needless to say, were alien to Einstein; with his robust psychological constitution he merely flung back at the quantum mechanists the accusation of reactionary unteachability: "The point is that all those fellows do not view the theory on the strength of the facts, but the facts on the strength of the theory only; they cannot escape from the conceptual net they have accepted, but can only daintily wriggle in it."[4] Einstein drew strength from his unshakable conviction that the new ideas and discoveries were only provisional and did not affect the search for the fundamental laws of physics: "I can derive only small pleasure from the great discoveries, because for the time being they do not seem to facilitate for me an understanding of the foundations," he said, in summary, after his first working period in Princeton—not without being somewhat ironical about his own efforts. "In consequence I feel like a kid who can't get the hang of the ABCs, even though, strangely enough, I still don't abandon hope. After all, one is dealing here with a sphinx, not with a willing streetwalker."[5]

If Einstein was chasing a sphinx, then the Institute for Advanced Study was the ideal place for him. Anyone of renown in mathematics was or had been a guest at the institute for a few months. The mathematicians were setting the tone, and they intended physics, even in its mathematical and theoretical aspects, to remain a marginal area. In consequence, for five years Einstein was the only professor in his field at the institute. But he did not regret this state of affairs, nor did he try to establish any close contacts with the physicists of the university.

On the rare occasions when he gave a seminar lecture, his colleagues were amazed by his style. Young John Archibald Wheeler, who was acquainted with the customary "retail" procedure of theoretical physicists—solving one equation after another—experienced with Einstein "for the first time how one can handle equations 'wholesale.' One counted the number of unknown quantities and the number of marginal conditions, and then compared these with the number of degrees of freedom. It was not so much a case of solving the equations as of deciding whether they even had a solution and whether that solution was the only possible one." Wheeler's lasting impression was "that Einstein went his way unwaveringly, unaffected by the great interest in nuclear physics that was then predominating in the United States."[6]

Despite the general shaking of heads over Einstein's obsession with a unified theory, there were a number of good physicists among the younger members of the institute who, unlike Oppenheimer, sought out and appreciated proximity to the great man: "The nice thing here is that I can work with young colleagues in the field."[7] Nowhere, in fact, did the "loner" Einstein have such regular and intensive contact with younger colleagues as in Princeton. They were not, formally, his assistants but came to him of their own accord, ignoring the widely accepted advice that for career reasons "it would be better not to work with Einstein."[8] Although this advice did reflect Einstein's strange position among physicists—more of a monument than a signpost, as Oppenheimer used to put it—none of them regretted having worked with him. It is true that none of them obtained a permanent post either at the institute or at the university (Einstein blamed this on the discreet anti-Semitism in Princeton), but all of them became professors.

Before long a routine developed for this cooperation. In the morning—not too early—Einstein would meet his coworkers in his room (Room 209 in Fine Hall) and discuss with them physical structures and mathematical approaches, the kind of thing Wheeler had described as "wholesale" physics. They separated about lunchtime. After lunch his collaborators engaged in the "retail" work, whose results would then be jointly discussed the following morning. Einstein was in the habit, after lunch and a siesta, of spending the afternoon in his cozy study on Mercer Street, where, with Helen Dukas's help, he dealt with his extensive correspondence, received occasional visitors, and mainly pursued his own thoughts.

The first joint publication required no mathematics, but its content was provocative. It dealt with quantum mechanics and had the effect of a bombshell, whose smoke has not entirely dispersed to this day. The problem raised in it, known as the "Einstein–Podolsky–Rosen paradox," is still being intensively discussed.[9] Actually, it is not so much a paradox as a logically impeccable conclusion which casts a spotlight on some peculiarities of quantum mechanics. The original idea came not from Einstein but from Nathan Rosen, a twenty-five-year-old American who had come to the institute in 1934 and had chosen to work with Einstein. Boris Podolsky, who was only seven years younger than Einstein, had worked at Caltech in Pasadena (where the two men had first met) and in 1934 was one of the few physicists at the institute. "For linguistic reasons the paper was written by Podolsky, after prolonged discussions," Einstein said, reporting on its genesis and its drawbacks. "But what I really wanted to say hasn't come out so well; instead, the main thing is, as it were, buried under learning."[10] Subsequently he returned to the "main thing" repeatedly on his own, and in German.[11]

The argument does not aim at proving mistakes or inner contradictions in quantum mechanics. By 1931, if not before, Einstein had accepted the new theory as efficient and free from contradictions, and even credited it with "undoubtedly a piece of definitive truth"[12]—but not the whole truth, let alone the definitive truth. Instead, hidden behind the title of the Einstein–Podolsky–Rosen (EPR) paper, *Can the Quantum Mechanical Description of Physical Reality Be Regarded as Complete?*, is an attempt to demonstrate that quantum mechanics represents only an incomplete description of physical reality and therefore is unable to get beyond the formulation of statistical regularities.

The rather vague concepts in the title are precisely defined in the opening paragraphs. A "complete theory" would have to satisfy the following condition: "Each element of physical reality must have a correspondence in physical theory." Physical reality is defined by a similarly pragmatic criterion: "If, without disturbing the system in any way, we can predict the value of a physical magnitude with certainty (i.e., with a probability of 1), then an element of physical reality exists that corresponds to that physical magnitude."[13] These preliminaries are intended to show, by means of a simple mental experiment, that the description of physical reality by quantum mechanics is incomplete. The thought experiment runs something like this:

Assume that two electrons E_1 and E_2 have collided and flown apart. When they are very distant from each other, the momentum of electron E_1 is measured; this, because of the conservation law, simultaneously establishes the momentum of E_2—without any disturbance to the E_2 system—at the moment of measurement, even if the two electrons are light-years apart. Thus the momentum of E_2 qualifies as an element of physical reality. But instead of measuring the momentum of E_1 one can measure its location, thereby obtaining the position of E_2; as this has been determined without any disturbance of E_2, the position of E_2 has therefore been established as a further element of physical reality.

The point of this argument is that according to quantum mechanics, if the momentum of a particle is known, then its position is necessarily uncertain, and hence not part of physical reality. This means that the reality of the momentum and the position of electron E_2 is "dependent on a measuring process at the first system, which in no way interferes with the second system. No reasonable definition of reality should permit this."[14] Hence the quantum mechanical description is incomplete, and a search for a complete description of physical reality is necessary and justified.

No sooner had the Einstein–Podolsky–Rosen paper been published, in May 1935, than alarm bells were ringing among the mandarins of quantum mechanics—not because they regarded the EPR paradox as a threat to their views but because of their irritation with Einstein. In Zurich, Wolfgang Pauli was furious: "Einstein has once again come out with a public comment on quantum mechanics. . . . As is well known, each time he does that is a disaster. . . . I would concede to him, though, that if a student in an early semester raised such objections to me, I would regard him as quite intelligent and promising."

To prevent confusion among less intelligent colleagues, Pauli suggested that Werner Heisenberg, the target of the ERP paper, should publish a rejoinder. He also considered, for "educational" reasons, "squandering paper and ink in order to formulate those facts demanded by quantum theory which cause Einstein particular intellectual difficulties."[15] He did not, though, do so, because Niels Bohr had already taken up the matter.

In Copenhagen the EPR publication seemed like "a bolt from the blue." When Bohr's collaborator Léon Rosenfeld reported on Einstein's arguments, "we dropped everything; we had to clear up such a misunderstanding at once. . . . Bohr, in great excitement, instantly began dictating the draft of a rejoinder."[16] But it took six weeks of work before the rejoinder was ready.[17] This did not try to prove a mistake, because there was, and is, nothing wrong with the situation set out by EPR. Instead, Bohr focused on imprecisions in the use of terms such as "physical reality" and "without any disturbance." His analysis made it once more clear that, because of the finite magnitude of the quantum of action, it was impossible—in contrast to classical physics—to speak of "physical reality" without including the measuring process in this reality. If that is borne in mind, the contradiction highlighted by EPR is only apparent, and quantum mechanics is a complete description of what physicists can discover about nature.

Einstein meanwhile was receiving letters from several colleagues who were quite certain that his argument was false, but who, to his amusement, adduced different reasons for their certainty.[18] Schrödinger was alone in sharing his delight "that you have . . . publicly caught dogmatic quantum mechanics by its throat."[19] Inspired by Einstein, Schrödinger on his part published a series of three articles arguing with orthodox quantum mechanics, including a paradox which has gained some popularity and is known as "Schrödinger's cat."[20] Einstein, however, reproved Schrödinger for publishing the articles in Germany: "In my opinion all well disposed individuals should break off relations while the scientists there tolerate this shameful regime."[21]

Of all the objections to the EPR paradox, Einstein thought that "Niels Bohr's view best meets the case."[22] However, if the quantum mechanical description, according to Bohr's analysis, was complete, then, in Einstein's opinion, there was the dilemma that the states of spatially separated systems were not independent of one another, i.e. that measurement on one system has instantaneous effects on a spatially distant system. Such "ghostlike remote effects" between separate systems—which could not, because of relativity theory, influence one another instantaneously—were something Einstein refused to accept, just as he refused to accept statistical regularities. "It seems hard to look into the cards of the Almighty. But I won't for one minute believe that he throws dice or uses 'telepathic' devices (as he is being credited with by the present quantum theory)."[23]

After lying dormant for a long time the EPR paradox has once more become interesting over the past two decades. It has been more precisely defined by theoretical analyses, and what was once a thought experiment has been performed as an actual experiment.[24] Accordingly, quantum mechanics supplies not only the correct but also the "complete" description. Most physicists now believe that the arguments about the EPR paradox always were much ado about almost nothing, but others, like Einstein, are irritated by the "ghostlike remote effect." The paradox and related recent discussion of fundamental concepts like "locality" and "separability" may highlight the circumstance that a satisfactory integration of the two great concepts of our century—quantum mechanics and the theory of relativity—is still missing.

Simultaneously with the debate about quantum mechanics, Einstein with Nathan Rosen embarked on a difficult attempt to derive the corpuscular structure of matter from a combination of the electromagnetic field and gravitation.[25] By means of a simple and clear transformation of the field equations they created mathematical elements, the so-called "Einstein–Rosen bridges," which they identified with the elementary particles. They succeeded in obtaining useful insights for a "bridge," and Einstein had great hopes of a development of their ideas. "Only the examination of a multi-bridge system will show whether this theoretical method can supply an explanation of the empirically proved mass identity of particles in nature, and whether it will agree with the facts so wonderfully comprehended by quantum mechanics."[26] But even the attempt to tackle the two-body problem[27] caused Einstein to bury his hopes again. To this day we do not know why electrons all have exactly the same mass.

The problem of gravitational waves, on the other hand, proved soluble, albeit after a few detours. In Newtonian theory, gravity is instantaneously present in space. No one ever asked how this force propagated; that question arose only with the general theory of relativity, under which gravitation propagates with the finite velocity of light. As early as during World War I, while he was sick in bed in Berlin, Einstein had deduced the existence of gravitational waves which, like electromagnetic waves, carried energy through space. Admittedly, he had then succeeded only in what is called the "first approximation."[28] Evidently he was no longer satisfied with that makeshift result and, jointly with Nathan Rosen, now looked for exact solutions of the field equations for gravitational waves. In doing so, he encountered a surprise: "Together with a young collaborator I arrived at the interesting result that gravitational waves do not exist, even though this had been regarded as certain according to the first approximation."[29] But when he tried to publish this result, Einstein learned something about the American scientific machinery that he did not like at all.

The editors of the *Physical Review*, the leading physics journal in the United States, returned the manuscript to the authors, along with a long opinion by a referee and requests for amendments. Einstein was outraged. Having articles examined by anonymous referees was the normal practice of the *Physical Review*, but Einstein refused to accept it.[30] That, at least, had been handled better in Germany. Einstein withdrew the paper, henceforward avoided the *Physical Review*, and published only in journals without referees.

Nevertheless, this incident had saved Einstein from publishing a mistake. In September 1936 he had still been complaining, "If only it wasn't so damn difficult to find exact solutions,"[31] but over the next few months he and Rosen succeeded in obtaining exact solutions for the field equations representing gravitational waves.[32] This remained a pioneering feat for a long time; only after Einstein's death were further solutions found. Despite an enormous technological input, however, gravitational waves have so far evaded experimental confirmation.

By the time the paper on gravitational waves appeared in 1937, in the *Journal of the Franklin Institute*, Nathan Rosen had left Princeton—his engagement there had not been renewed. Rosen, whose parents had immigrated from Russia, eventually went to the Soviet Union, with Einstein smoothing his way by a letter of recommendation to Vyacheslav Molotov, then chairman of the Council of People's Commissars. But new collaborators had appeared on the scene, as well as new problems.

Einstein's last major contribution to the development of the general theory of relativity was produced in collaboration with Banesh Hoffmann and Leopold Infeld. Hoffmann, who came from England, had taken his doctoral degree in Princeton, under Oswald Veblen, before becoming a member of the institute in 1935. Some time afterward, he became Einstein's collaborator. Leopold Infeld came to the institute from Poland in 1936, on a scholarship, after a few months in Edinburgh, Scotland, with Max Born, who had taken up a professorship there. Einstein proposed to Infeld and Hoffmann a joint investigation of motion in the general theory of relativity. In a paper published after two years of difficult work, they examined the problem of motion at a new level of generalization, at which the customary division into field equations on the one hand and the laws of motion on the other was overcome.

Newtonian celestial mechanics in its axiomatic structure consists of two clearly separate parts: the law of motion, and the law of gravity, which gives rise to the forces that keep the heavenly bodies in their orbits. The two parts stand alongside each other, unconnected. This separation into two strictly distinct sets of laws had not been overcome even by Einstein's initial treatment of motion in the general theory of relativity. However, by the early 1920s investigations by Lorentz, Eddington, and Levi-Civita suggested that in the general theory of relativity these two sets are not really separate. After two years' work, in 1938 Einstein, Hoffmann, and Infeld were able to show, in a voluminous publication,[33] that the field equations do in fact contain everything—not only gravitation, but also the movement of masses distributed in space. Thus the general theory of relativity now described not only space, time, and gravitation, but also, for the first time, the dynamics of matter.

In the same year as this publication, there also appeared, in April 1938, a book, *The Evolution of Physics*. This astonishing work, reflecting the history of a discipline through the eyes of its greatest representative, owed its genesis not to Einstein's desire to communicate, but to economics at the Institute for Advanced Study.

"Infeld is a splendid fellow. We've done a very pretty thing together," Einstein reported after six months of joint research, but despite his fervent support the institute would not extend Infeld's modest scholarship. "The Institute has treated him badly. But I'll help him prevail here."[34] Einstein, who felt he too had been badly treated by the refusal of a scholarship for his esteemed collaborator, wanted to defray the small sum from his own pocket. But Infeld, who was embarrassed by that suggestion, had an original idea: How about writing a book together for a wide readership? With Einstein as one of the two authors, this could not fail, so that Infeld's share in the proceeds would secure his livelihood. Einstein thought this was "not a silly idea at all. Not silly at all."[35] Infeld found a publisher, who paid him an advance, while Einstein planned the contents and the basic structure of the book. The book was a great success and Infeld's future in America was secure.

While Einstein was working on the motion problem with Hoffmann and Rosen, other collaborators arrived, with whom he could focus on his real passion, the unified theory. Peter Bergmann, born in Berlin, had taken his doctoral degree in Prague, under Philipp Frank, at the age of twenty-one; he arrived in Princeton in 1936 in order to work with Einstein. Valentin Bargmann was also a Berliner, but after Hitler's assumption of power he had gone to Zurich to complete his studies. He was a member of the institute from 1937 to 1946.

Alone and with his two young assistants Einstein pursued many approaches to unification. However, there were no breakthroughs and scarcely any publishable results. At times, as in the summer of 1938, he thought he was close to a solution: "This year, after twenty years of searching, I have found a promising field theory, one that is an entirely natural development of the relativist gravitation theory."[36] But this attempt, too, proved unsuccessful.

Even though Bergmann and Bargmann achieved only very slight success in terms of publications, they nevertheless—like Infeld and Hoffmann—felt enormously enriched by being able to share in the thoughts and the work of a genius. They noted with fascination how Einstein's thinking differed from conventional thought. He seemed to them less a strictly logical theoretician than a creative artist, full of imagination and frequently using arguments that would have been out of place in a scientific essay: "Einstein was motivated not by logic in the narrow meaning of this concept, but by a sense of beauty," one of his collaborators recalled. "In his work he was always looking for beauty."[37]

In his later years Einstein saw beauty in the laws of nature. He profoundly believed, with religious fervor, that simple laws existed, and that these could be discovered. Except for a brief phase during his adolescence, he never had any use for the personified God of the Judeo-Christian tradition. But even in his younger years, he saw God as the guarantor of the laws of nature. Initially this sounded like a playful formulation, but as he grew older the metaphor became a kind of heuristic principle: Einstein would attempt to slip into the role of the creator of the world and its laws. He surprised Banesh Hoffmann with his criterion: "When I am judging a theory, I ask myself whether, if I were God, I would have arranged the world in such a way."[38]

This belief in a lawful structure of the world gave him the strength and perseverance he needed throughout the decades when he was searching for a unified theory. He was capable of pursuing a theoretical concept, with great enthusiasm for months or even years at a stretch; but when grievous flaws emerged—which invariably happened in the end—he would drop it instantly at the moment of truth, without sentimentality or disappointment over the time and effort wasted. The following morning, or a few days later at the most, he would have taken up a new idea and would pursue that with the same enthusiasm.

To his assistants it seemed that they were working on field theories, whereas Einstein was always at the same time thinking of the foundations of quantum theory. When, in the course of their work on the motion problem, a system of four equations emerged instead of only three, the young people lost their courage because of the limited number of solutions. But Einstein said: "Oh, but that's beautiful." To his perplexed collaborators he explained: "We'll have an over-determined system, and then we can obtain quantum conditions analogous to Bohr's permitted orbits."[39]

He never wavered in his conviction that although the statistical laws of quantum mechanics were useful tools, they were not the foundations of physics. This attitude emerged regularly in his letters. In the summer of 1942 he explained it as follows: "As for myself, I am so old-fashioned and stubborn that I still do not believe that the Lord throws dice. Because if he had wanted to do that, he would have done it thoroughly, and not kept to a pattern in throwing dice. Gone the whole hog. In that case we wouldn't have to look for laws at all. It's true, everything points against the belief in total regularity. But I continue to search for just that. If what I find is no use in the end, then the fault is probably mine, not his."[40]

The fact that, with regard to science, things increasingly fell silent around him did not sadden Einstein. As he remarked in 1936, he was "living in the kind of solitude that is painful in one's youth but in one's more mature years is delicious."[41] He avoided talking at cross purposes by not talking at all. He even avoided Niels Bohr when Bohr came to the institute in January 1939 as a guest for a few months. Léon Rosenfeld, who accompanied Bohr as his assistant, experienced an abysmal silence falling on the conversation of the two great men, who once had liked nothing better than talking physics to each other:

> Einstein was only a shadow of himself. For days on end he kept himself shut up in his study and really spoke only to his two assistants, whose names, curiously, were Bergmann and Bargmann.

> Only once during those four months did he announce a lecture; it was about one of his countless attempts to establish a unified field theory. Bohr attended the lecture. In conclusion Einstein, fixing his eyes on Bohr, declared with emphasis that he had always hoped to be able to derive the quantum conditions in the way he had just described. During those four months Bohr and Einstein met only once at an afternoon reception, but their conversation did not go beyond banalities. Einstein allowed it to be clearly understood that he would rather avoid any discussion with Bohr. Bohr was profoundly unhappy about this.[42]

This episode is all the more remarkable because Bohr's stay in Princeton coincided with a rapid development in which a dramatic application of physics had moved into the realm of the feasible—the production of a nuclear chain reaction and, in consequence, of a bomb of almost unimaginable destructiveness. Einstein had been ignoring nuclear physics because it did not bring him closer to the "secret of the Old Man." But it was to catch up with him before long.

CHAPTER THIRTY-SEVEN

War, a Letter, and the Bomb

EINSTEIN SPENT THE SUMMER of 1939 on Long Island. As in the two previous years, he had rented a house at Nassau Point, at the eastern tip of the island, with its numerous bays and tongues of land where it faces the Atlantic. Nassau Point is on Great Peconic Bay, sheltered from the ocean waves and hence an ideal sailing area. He enjoyed his boat, made music with some neighbors, and despite the often oppressive heat, worked.

About the middle of July two old acquaintances turned up—Eugene Wigner, a professor at Princeton University; and Leo Szilard, who had found asylum as a visiting researcher at Columbia University in New York. They had set out on the drive to Peconic because the rare metal uranium had become a crucial subject over the past few months. It was not yet certain, but by no means impossible, that uranium might be a suitable starting material for bombs of gigantic explosive power. Concern that Nazi Germany might develop such a weapon had driven Szilard, in particular, into a state of great excitement. He had therefore planned a role for Einstein, not as an authority on physics but as a friend of the queen of Belgium. The reason was that the largest deposits of uranium ore then known were in the Belgian Congo, and the largest stocks of metallic uranium were owned by a Belgian concern. Einstein was to write a letter to his "dear queen" to induce the Belgian government to stop uranium sales to Germany.

Even though Einstein may not have followed the headlong advances of nuclear physics, he instantly realized the importance of the problem—its military and political significance, as well as its significance for science.[1] Releasing the energy locked up in the atom was an old theme that had been at the back of the mind of physicists since the discovery of radioactivity before the turn of the century. In fact, Einstein, in the final sentence of his 1905 paper in which he derived his formula $E = mc^2$, had referred to this possibility in the distant future.

Fifteen years after Ernest Rutherford's discovery of the atomic nucleus, when quite a lot had since been learned about the natural transmutation of radioactive nuclei, it was once more Rutherford who in 1920, while bombarding nitrogen with alpha rays emitted by radium, succeeded in observing the transmutation or fission of the nitrogen nuclei. What was called "artificial radioactivity" was excitedly hailed in the newspapers as a potentially vast source of energy, supported by the simple and very attractive calculation that one gram of matter might replace three thousand tons of coal.

Sober scientists rejected such visions because of the rarity and the enormous cost of radium, but Privy Councillor Planck declared: "We confidently believe that German science will now find a way."[2] Einstein, the discoverer of the miraculous formula for the conversion of matter into energy—which for some optimists was even then opening the doors to the future—commented as follows:

> It might be possible, and it is not even improbable, that novel sources of energy of enormous effectiveness will be opened up, but this idea has no direct support from the facts known to us so far. It is very difficult to make prophecies, but it is within the realm of the possible. . . . For the time being, however, these processes can only be observed with the most delicate equipment. This needs emphasizing, because otherwise people immediately lose their heads. But if this road leads on, especially in such a way that the rays released by the alpha particles are in turn able to produce the same effects, one cannot be at all sure that such a development may not progress rapidly.[3]

This was a prophetic statement, because of what Einstein said was the prerequisite for development: "that the rays released . . . are in turn able to produce the same effects." This was the basic concept of a chain reaction.

To his early biographer Moszkowski, Einstein had pointed out that such a development would be anything but desirable: "Assuming that it were possible to effect that immense energy release, we should merely find ourselves in an age compared to which our coal-black present would seem golden."[4] His objections concerned the difficulty of regulating such an energy source—fear of what was later to be called nuclear terrorism—and, of course, the technology of weapons: "All bombardments since the invention of firearms put together would be harmless child's play compared to its destructive effect."[5] Thus, most of what can be said about the use of nuclear energy had been stated by Einstein as early as 1920, at a time when details were still shrouded in the fog of the future. The prerequisite, as he realized, was the achievement of a chain reaction, and humankind would be better off if it did not succeed.

For a long time, such ideas were so speculative that they could not be published in serious scientific journals, but they were always in physicists' minds. A breakthrough came in 1932, when the neutron was discovered as a constituent of the atomic nucleus; before long, the neutron was used as a free particle for bombarding nuclei.[6] As early as 1934, Leo Szilard, always sparkling with intelligence and imagination, applied for a patent for a nuclear chain reaction based on neutron multiplication and had it deposited as a secret paper with the British Admiralty. Einstein, like practically every other physicist, remained skeptical, however; and as it turned out later, Szilard's chain reaction would have consumed more energy than it released.

Toward the end of 1935, Einstein had been invited to lend special luster to the annual meeting of the American Association for the Advancement of Science in Pittsburgh by giving the Willard J. Gibbs Memorial Lecture. He demonstrated to his audience how he had derived his famous formula on the equivalence of matter and energy.[7] Small wonder that the three dozen reporters covering a press conference on that subject wished to know whether the huge amounts of energy corresponding to his formula might be released by bombarding an atom. Einstein thought this about as unpromising "as firing at birds in the dark, in a neighborhood that has few birds."[8] The *Pittsburgh Post-Gazette* announced in a headline that Einstein had wrecked all hope of deriving energy from the atom.

Meanwhile, Enrico Fermi's team in Rome—using a superb experimental setup—had been bombarding heavy atomic nuclei, especially uranium, with neutrons. But Fermi was interested chiefly in the creation of transuranium elements, and so he failed to notice that uranium nuclei were split in the process. Not until the end of 1938 did Otto Hahn and his collaborator Fritz Strassmann make this discovery, at the Kaiser Wilhelm Institute in Berlin-Dahlem. Thanks to their precise radiochemical methods, they identified barium in the reaction products. Barium has a nucleus about half the weight of the uranium nucleus, and therefore nuclear fission was suggested.

In a letter to Lise Meitner, his colleague for many years, who had left the Kaiser Wilhelm Institute after the annexation of Austria and had fled to Sweden, Otto Hahn gave a detailed account of this result. Over Christmas, Lise Meitner discussed this sensational news with her nephew, Otto Frisch, who had been introduced to the theory of the atomic nucleus by Niels Bohr in Copenhagen. Together Frisch and Meitner wrote a brief notice for the British journal *Nature;* this was the first mention of splitting the atomic nucleus. Frisch also informed Bohr, who had developed a useful theoretical model of the nucleus and of chain reactions. Within five minutes, Bohr had understood everything, wondering merely why he had not thought of it himself.

Bohr was about to leave for America to spend several months at the Institute for Advanced Study in Princeton. On January 26, he spoke at a conference in Washington about Hahn and Strassmann's discovery and its interpretation by Frisch and Meitner. The result was that several physicists left the conference prematurely, rushing back to their laboratories to repeat the experiments. Everything was as Bohr had said, and, like Bohr, many of them may have wondered why they themselves had not thought of it.

A few days later, on a five-minute walk from the Princeton Club to Fine Hall, Bohr realized that only the isotope uranium-235 could be split—an isotope present in natural uranium to no more than 0.7 percent. This conclusion was a triumphant confirmation of the reliability of Bohr's nuclear model; but as separating isotopes is difficult, it was a setback for any practical applications.

On March 14, 1939, when *The New York Times* observed Einstein's sixtieth birthday with an extensive interview, Einstein said that the results obtained so far "do not justify the assumption of a practical utilization of the atomic energies released in the process. . . . However there is no single physicist with soul so poor who would allow this to affect his interest in this highly important subject."[9]

That same month, however, a crucial discovery was made almost simultaneously in France and America: in Paris by Frédéric Joliot-Curie and in New York by Enrico Fermi—who, after receiving the Nobel Prize in Stockholm on December 10, had not returned to fascist Italy but had taken up a professorship at Columbia University. Joliot-Curie and Fermi observed that when a uranium nucleus was split by a neutron, as a rule two neutrons were released, which in turn could each split another uranium nucleus. The chain reaction had moved into the realm of the possible, and Fermi, looking out his window, mused that a single fission bomb would destroy all that he could see of New York.

As feverish experimenting went on in the laboratories of Princeton, New York, and elsewhere, and as Niels Bohr with John Archibald Wheeler in his room at Fine Hall was working out the theoretical basis of the fission of uranium by neutrons, Einstein, only a few doors down the corridor, was deeply immersed in his work on the unified theory, unapproachable and even incommunicado. The fact that in mid-July he was nevertheless involved in the developing drama had—as mentioned earlier—nothing to do with physics but everything to do with his friendship with the "dear queen" in Brussels, and the Belgian uranium deposits.

When Szilard and Wigner talked to him on the terrace of his house on Peconic Bay, Einstein immediately declared himself ready to write a letter, though not to the queen but to the Belgian government. This letter was to be conveyed by the American State Department. Einstein made a draft on the spot and gave it to his visitors to take home with them. Back in New York and uncertain about the next step, Szilard consulted Dr. Gustav Stolper, a politically experienced émigré and a former member of the Reichstag. Stolper in turn directed Szilard, with his unusual problem, to Alexander Sachs, a banker who belonged to an unofficial circle of advisers to President Roosevelt. Sachs declared himself ready to submit an appropriate letter from Einstein to the president. By the time Szilard, now in the company of Edward Teller, next called on Sachs, the letter to Einstein's "dear queen" had turned into a letter to the president of the United States.

Einstein dictated another draft, on the basis of which Szilard, after consultation with Sachs, produced two English versions, a longer and a shorter one, as it was difficult to judge with how many words one could bother the president. In August 1939 Szilard sent both versions to Einstein, who immediately signed and returned them. The most important passages of the longer letter—the one that was eventually used—read as follows:

> Sir,
> Some recent work by E. Fermi and L. Szilard, which has been communicated to me in manuscript, leads me to expect that the element uranium may be turned into a new and important source of energy in the immediate future. Certain aspects of this situation seem to call for watchfulness and, if necessary, quick action on the part of the administration. I believe, therefore, that it is my duty to bring to your attention the following facts and recommendations.
>
> In the course of the last four months it has been made probable—through the work of Joliot in France as well as Fermi and Szilard in America—that it may become possible to set up nuclear chain reactions in a large mass of uranium, by which vast amounts of power and large quantities of new radium-like elements would be generated. Now it appears almost certain that this could be achieved in the immediate future.
>
> The new phenomenon would also lead to the construction of bombs, and it is conceivable—though much less certain—that extremely powerful bombs of this type may thus be constructed.

> A single bomb of this type, carried by boat or exploded in a port, might well destroy the whole port with some of the surrounding territory. However, such bombs might very well prove to be too heavy for transportation by air.

In the following paragraphs Einstein and Szilard made organizational proposals for a collaboration between government and physicists. They suggested the appointment of a special adviser for these questions, recommended the involvement of industrial laboratories, and did not omit to mention the need for financial support for the relevant research. They next referred to Germany and adduced a personal argument that might have been more persuasive to the president than all the physics: "I understand that Germany has actually stopped the sale of uranium from the Czechoslovakian mines which she has taken over. That she should take such early action might perhaps be understood on the ground that the son of the German Undersecretary of State, von Weizsäcker, is attached to the Kaiser Wilhelm Institute of Berlin, where some of the American work on uranium is now being repeated."[10]

What Einstein did not know, but may have suspected, was that his letter about uranium fission was not the first that a physicist had addressed to a government. As early as April 1924, Paul Harteck, who had taken up Otto Stern's former professorship in Hamburg, and his coworker Wilhelm Groth had recommended to the German Army Ordnance Department the use of a chain reaction in a bomb: "The country that exploits it first will have an incalculable advantage over the others."[11]

On September 1, 1939, Germany attacked Poland, and on September 3 World War II began.

Not until October 11 did Roosevelt find time to receive his friend and adviser Alexander Sachs. Sachs presented Einstein's letter, along with background material about physics in which Szilard summed up the hectic development of nuclear research in the first half of 1939 and its future prospects, together with a memorandum Sachs himself had written on the steps which the government should take. The president immediately understood the crux of the matter. "Alex," he interrupted him. "What you are after is to see that the Nazis don't blow us up." When Sachs confirmed this, Roosevelt summoned his secretary, General E. M. Watson, and instructed him: "This requires action."[12]

A few days later the president thanked Einstein "for your recent letter and the most interesting and important enclosure." He also informed Einstein of what had been done in the meantime: "I found this data of such import that I have convened a board consisting of the head of the Bureau of Standards and chosen representatives of the Army and Navy to thoroughly investigate the possibilities of your suggestion regarding the element of uranium."[13] The board met on October 21, with Enrico Fermi and the three members of the "Hungarian gang"—Szilard, Teller, and Wigner, who were all born in Budapest—providing the nuclear expertise, even though all four of them were foreigners. Like most committees and commissions, this one was not very farsighted: only $6,000 was approved for a whole year's research on uranium.

Everything was moving too slowly for Szilard, so he again turned to Einstein. On March 7, 1940, Einstein wrote a second letter, addressed to Sachs but intended for the president. In this he stressed the urgency of the problem in view of the German efforts, and also the need for secrecy. The president reacted by suggesting enlarging the board and including Einstein. Einstein, however, declined, rather brusquely and without giving a reason, possibly because he had not been invited to the first meetings. But he recommended the creation of a framework "that would facilitate the continuation of these and similar researches at a faster pace and on a greater scale than hitherto."[14] For the time being this letter marked the end of Einstein's contact with the government.

Einstein's interventions, including his first letter, had no appreciable effect on the course of events. Organization of nuclear research in the United States did not get beyond the stage of improvisation until the fall of 1941, when results achieved in England, in particular the so-called Frisch-Peierls Report, reached Washington through official channels. Eventually, on December 6, 1941, the secret "Manhattan Engineering District" was set up. This was the biggest technological and scientific enterprise the world had ever seen—had the world been allowed to see it. December 6, of course, was only one day before Japan's attack on the American Pacific Fleet at Pearl Harbor. Three days later, Germany declared war on the United States.

Roosevelt's decision to speed up the development of the bomb had contradictory consequences for Einstein. He was consulted as a physicist, but only briefly. At the beginning of December 1941, Vannevar Bush, the chief of the Office for Research and Development at the center of nuclear research, requested Einstein's assistance with a problem of isotope separation. Contact was made through official channels, via Frank Aydelotte, who in 1939 had succeeded Abraham Flexner as director of the Institute for Advanced Study. On December 19, Einstein handed Aydelotte the result of his reflections, which were passed on to Bush. In his covering letter Aydelotte remarked that Einstein had been "very much interested. . . . I very much hope that you will make use of him in any way that occurs to you, because I know how deep is his satisfaction at doing anything which might be useful in the national effort."[15]

This, however, did not happen, because to do useful work, Einstein would have had to be fully informed about the project. "I wish very much that I could place the whole thing before him," Bush wrote to Aydelotte, "but this is utterly impossible in view of the attitude of people here in Washington who have studied his whole history."[16] The FBI and Army Intelligence had come to the conclusion that Einstein represented a security risk: "In view of his radical background, this office would not recommend the employment of Dr. Einstein, on matters of a secret nature, without a very careful investigation, as it seems unlikely that a man of his background could, in such a short time, become a loyal American citizen."[17] The FBI and other secret agencies had never been informed of Einstein's letter to the president.

Thus the man who had drawn Roosevelt's attention to the atom bomb—and who, incidentally, had taken his oath as an American citizen on October 30, 1940—was spared having to decide to what extent he wished to take part in the development of this weapon. Judging by Aydelotte's letter, he would have gone a long way.

▪ ▪ ▪

Somewhat later, however, Einstein was able to contribute to military research. On May 16, 1943, a navy lieutenant heading a research team for conventional high explosives near Washington called on him. Asked whether he would work as an adviser to the navy generally, and in the area of high explosives in particular, Einstein reacted with considerable pleasure. "He felt very bad about being neglected. He had not been approached by anyone to do any war work," the lieutenant, who enlisted him on the spot, later recalled.[18] Einstein was delighted and looked forward "with great satisfaction" to working for the navy.[19] Aydelotte, after consulting Einstein, proposed the modest patriotic fee of $25 per day. Einstein, as he joked, was now in the navy without getting the prescribed haircut.

Every other week, or sometimes only once a month, Einstein was visited by the explosives experts, who brought him problems such as the optimal detonation of torpedoes. His solutions, arrived at by thought, were accurate: "Very expensive experiments performed much later showed that he had been right."[20] His visitors from Washington found him cheerful and more than happy to be able to contribute to the defeat of the Hitler regime. To an old friend he wrote about his "connection with the navy as a theoretical expert," hoping "that the war will be over before my activity on these lines has any consequences."[21] In fact, there is no record of any major consequences of his work on high explosives.

On the other hand, Einstein made some generous contributions to the war chest of the United States. The Book and Authors War Bond Committee, an organization which auctioned off manuscripts by famous authors and with the proceeds purchased war bonds, asked Einstein for the manuscript of his 1905 treatise on relativity. Einstein was unable to oblige; in his younger years he had always discarded his manuscripts once his work had appeared in print, saving only the more practical reprints. As a substitute, though, he offered his German draft of a paper he had recently completed with Valentin Bargmann, on bivector fields.[22] This was gratefully accepted.

Soon afterward, however, the university's chief librarian, acting as an emissary of the committee, suggested that Einstein might, for this good cause, copy his relativity paper again. Einstein was astonished but complied. "When others were doing so much for the war, this was the least he could do."[23] So Helen Dukas dictated and Einstein wrote. Now and again he pulled himself up and asked if he had really said that. When Helen Dukas confirmed it, he grumbled: "I could have put this much more simply."[24] His secretary did not record which passages he was referring to.

The auction took place on February 3 in Kansas City. The reconstructed manuscript on relativity was bought by an insurance company for $6.5 million, and the original manuscript about bivector fields brought $5 million. Both manuscripts were presented as a gift to the Library of Congress. Einstein, who did not think much of this manuscript fetishism, advised the economists to rethink their theories of value.

With America's entry into the war and the virtually simultaneous inception of the Manhattan Project, the university's physics department soon found itself depopulated. Even in the institute's "ivory tower," the quiet deepened, and several colleagues, such as John von Neumann, could now be contacted only through a post office box in Santa Fe. The institute had for some time been in a new building, Fuld Hall, located in a vast park in western Princeton, and reachable from Einstein's home on Mercer Street in a half hour's walk. There he had a spacious corner room with large windows on the ground floor, and next to it a smaller one for the assistant who had meanwhile been authorized for him.

By then Einstein even had a colleague who was a physicist. Wolfgang Pauli had joined the institute in 1940 because he had felt uneasy in Zurich, with Hitler at the door. "I am very glad that he is here and I have interesting conversations with him,"[25] Einstein reported of his sharp-tongued colleague. Despite their disagreements on quantum mechanics, they actually produced a joint publication, about relativity theory.[26]

In Einstein's search for the unified theory, of course. Wolfgang Pauli was more of a critic than a constructive coworker: "What God has put asunder, let no man join together," Pauli used to say. But Einstein was not discouraged: "I work like a man possessed, i.e., I ride my hobbyhorse like wild,"[27] he said, describing his mental state at the time. All this was reminiscent of his frenzy while he was wrestling with the generalization of the theory of relativity. Even on vacation he was sometimes "so excited . . . over difficulties and doubts in a matter very close to my heart, that I was unable to free myself day or night, or think of anything else."[28] Inevitably, disenchantment followed: "One thing is certain: the Almighty has not made it easy for us. While one is still young one doesn't realize it so clearly—fortunately."[29] But he was never downhearted.

A researcher of similar perseverance was the mathematician Kurt Gödel, who, after the annexation of Austria and an adventurous escape via Siberia and Japan, had arrived in Princeton in 1940 and obtained a post at the institute. Gödel had become famous for his work on the "decision problem" and was interested in the logical foundations of mathematics; he became not only a fascinating discussion partner for Einstein but also a close friend.

Einstein was fond of discussing philosophical aspects of scientific thought with Gödel and Pauli; and when Bertrand Russell came to Princeton in 1943—not as a professor but as an independent writer—these conversations were formalized: the four met regularly at Einstein's house one afternoon every week. There probably never was an intellectual gathering of higher caliber—but the discussions, at least to Bertrand Russell, were disappointing: "Although all three of them were Jews and exiles, and, in intention, cosmopolitan, I found that they all had a German bias towards metaphysics, and in spite of our utmost endeavors we never arrived at common premises from which to argue."[30] Einstein had no useful results to report, either.

On international politics and the war the four men hardly needed to talk, as they all had the same opinions and hopes—the earliest possible and most thorough victory over Germany. One exception, however, emerged when Bertrand Russell suggested that after the war the victors should help the Germans get back on to their feet and forget their crimes. Einstein indignantly rejected that idea.[31] With all the wrath he was capable of, he hoped "that the hair-raising crimes of the Germans would soon be revenged."[32] After Stalingrad, he was happy that "things are now going downhill for the Germans. . . . But it is too slow. What will remain of European life? There is nothing left but the consolation that—one is not there oneself. Is it possible that one has sunk so low that one can think like that?"[33]

When the Allies' landings in Normandy brought the defeat of Germany within reach, Einstein was thankful "that I have lived to see this turn of events, which looks like justice." Possibly as a reaction to Bertrand Russell's remark, he added: "But perhaps it would be more circumspect to die soon, so as not to be disappointed again, as in the years between 1918 and 1938."[34] For Einstein now, the criminal was no longer Hitler, or the leaders of the Nazi Party, or even the National Socialist movement, but the German nation in its totality. In his *Obituary for the Heroes of the Warsaw Ghetto* he wrote: "The Germans as a whole nation are responsible for these mass murders and must be punished for them as a nation, if there is any justice in the world. When they are wholly defeated and, as after the last war, bewail their fate, one should not allow oneself to be deceived a second time, but remember that they entirely deliberately used the humanity of the others for preparing their last and worst crime against mankind."[35] Many others felt like that at the time, but after the war most of them behaved in line with Bertrand Russell's prediction. Einstein, by contrast, never forgave the Germans.

There is no written evidence of how much Einstein knew during the war about the development of the atom bomb. This top secret matter was talked about only in hints, and not written about at all. But he must have realized that a gigantic effort was in progress, since virtually all nuclear physicists, as well as countless scientists from other disciplines, had vanished, their addresses unknown. Now and again one of them would turn up at Princeton for a few days, such as Niels Bohr on December 22, 1943. Bohr had escaped from occupied Denmark and was participating on the British side of the Manhattan Project as a kind of elder statesman. Bohr visited Einstein and Pauli, and although nothing is known about their conversation, Einstein was probably able to form a picture of the state of affairs.

In the fall of 1944, it seems that Otto Stern, his old collaborator from Prague and Zurich, informed Einstein at least about the impending success of the bomb project. Certainly they discussed the alarming postwar implications, because Einstein knew of Stern's great concern—which also became his own: "Then there'll be in every country a continuation of secret rearmament with technological means, which inevitably will lead to preventive wars (veritable wars of annihilation, worse in loss of life than the present one)."[36]

After a further talk with Otto Stern on December 11, Einstein in a letter to Niels Bohr explained how he thought the threat of catastrophe might be avoided. Influential scientists were to instruct the unsuspecting politicians in good time about the scale of the danger. "There is yourself, with your international contacts, A. Compton here, Lindemann in England, Kapitza and Joffe in Russia, etc. The idea is to induce them to prevail upon the political leaders in their countries, in order to achieve an internationalization of military power—a road that was some considerable time ago rejected as being too adventurous. But this radical step with all its far-reaching prerequisites concerning supranational government seems the only alternative to a secret technological arms race."[37] This letter already contains all the key aspects of Einstein's attitude to nuclear armament and postwar international politics.

Bohr had entertained similar ideas and had even submitted them to various government quarters, even to Roosevelt and Churchill. But he had met with no understanding and had merely exposed himself to the suspicion of being unreliable.[38] Einstein's letter, which reached him in Washington, therefore alarmed him. He immediately hastened to Princeton and made Einstein see that such an enterprise might have the worst possible consequences if anyone bound to official secrecy were to participate in it. Einstein understood and promised Bohr he would remain silent.

In March 1945, Leo Szilard appeared on the scene again. The final defeat of Germany was imminent and a Nazi "miracle weapon" need no longer be feared, but meanwhile the American bomb would be available in the summer. Szilard and a few colleagues were concerned about the state of information and the government's wisdom in using the atom bomb, as well as about the shape of the postwar world. Szilard, who wished to submit his views to the president, asked Einstein for a letter of introduction. This had to be formulated very cautiously, as Einstein was not supposed to know about Szilard's purpose or the intended recipient.[39] Roosevelt was never to see this letter. He died on April 12, 1945, and Einstein mourned "the loss of an old and dear friend."[40]

On August 6, shortly after eight o'clock local time, the Japanese city of Hiroshima was wiped out by the first uranium bomb. Einstein was again spending his vacation on Saranac Lake in the Adirondacks. Helen Dukas heard the news on the radio, and when she told Einstein he said: "*Oh, Weh*. And that's that."[41] On August 9 a plutonium bomb was dropped on Nagasaki. Japan surrendered; the war was over.

On August 11, 1945, Niels Bohr wrote in the London *Times:* "The formidable power of destruction which has come within the reach of man may become a mortal menace unless human society can adjust itself to the exigencies of the situation."[42] Einstein remained at his vacation home and kept silent. The U.S. government meanwhile had published the Smyth Report,[43] a semiofficial chronicle of the development of the atom bomb, written by a professor at Princeton University. One sentence of the report mentioned Einstein's letter to Roosevelt of August 1939, which was thereby moved from the sphere of secrecy into the limelight. Under the shadow of the new weapon of mass destruction, Einstein could no longer avoid the debate about the atom bomb and the new world order.

CHAPTER THIRTY-EIGHT

Between Bomb and Equations: "But the Equations Are for Eternity"

WITH THE DROPPING of the atom bomb and the first accounts of the gigantic effort that had led to its construction, physics and physicists were suddenly in the public spotlight. Because of the swift conclusion of the war, they were at first hailed as heroes who had saved the Americans an invasion of Japan, with its inevitably heavy loss of life. But once the victory celebrations died down, the physicists once more found themselves seen as terrifying sorcerers' apprentices—and some may even have seen themselves in that role.

While J. Robert Oppenheimer had become famous as the "father of the atom bomb," Einstein inadvertently found himself regarded as a kind of superfather because of the almost mythical authority he enjoyed among the public, because of his magical formula $E = mc^2$, and because of the fact, now declassified, that he had drawn the president's attention to the possibility of an atomic bomb in August 1939. Einstein, however, remained at Saranac Lake; not until mid-September did he make his first public comment on the new weapon of mass destruction, when a reporter for *The New York Times* tracked him down at his idyllic refuge.

Einstein by then knew what he had to do: he merely had to resume his prewar pacifism—modified by the debate, during the final year of the war, over the postwar order. In his opinion, as the reporter reproduced it, "The only salvation for civilization . . . lies in the creation of world government, with security of nations founded upon law. . . . As long as sovereign states continue to have separate armaments and armaments secrets, new world wars will be inevitable."[1]

Einstein's idea of world government would often later be described as naive; but during the final year of the war and for some time after the war ended, it was shared by more than just a small minority. Roosevelt and his successor, Truman, had, up to a point, embraced the idea by the establishment of the United Nations. However, after the war conflicts emerged between the victors. This was not a good climate for handing over one's atom bomb to some international organization. Most Americans and their politicians were convinced that the "secret of the bomb" must remain their monopoly for a long time if Soviet expansionism was to be curbed.

Although the existence of the atom bomb heightened distrust between the superpowers, for Einstein it confirmed the need for a world government. Back in Princeton, he signed a declaration, along with Thomas Mann and a dozen prominent Americans, which called for some rethinking: "The first atomic bomb destroyed more than Hiroshima. It also exploded our inherited, outdated political ideas."[2] They argued that the United Nations Charter, signed in San Francisco by fifty-one states on June 26, 1945, was not sufficient to guarantee peace and would prove a "tragic illusion." "We must aim at a Federal Constitution of the world, a working worldwide legal order, if we hope to prevent atomic war."

The atom bomb had made young physicists despair of their science, but they were hoping for clarifying and helpful words from Einstein. One of them wrote to him of the "cruel irony that one of the greatest and most joyous triumphs of the scientific intellect was to bring not spiritual uplift and greater life affirmation but disappointment and death. The final and unlimited confirmation of your formula $E = mc^2$ should have marked the beginning of an era of light. Instead we are staring in confusion into impenetrable darkness." There were plans for a national congress of scientists; even though this did not take place, Einstein would not have forgotten the anxious words with which he had been asked to participate: "We need you here to speak out on the social implications of the era of atomic energy. We feel instinctively that with you here we shall feel inspired to better master our future; we almost feel that it is your duty to help us and eventually the world, not only because yours is at once a humble and a powerful voice, but also because your own work, the mass-energy principle, has in a larger sense led us to this road that has two turnings—and we count on you to help us all travel along the correct path."[3]

In view of the many millions of human lives lost and the technological horrors perfected in World War II, Einstein did not see the atom bomb as a fundamentally new problem. Nor did he ever subsequently condemn the dropping of the atom bombs on the Japanese cities. He began a lengthy article, which was repeatedly reprinted, with the sobering statement:

> The release of atomic energy has not created a new problem. It has merely made more urgent the necessity of solving an existing one. One could say that it has affected us quantitatively, not qualitatively.[4]

His aim was and remained the abolition of war altogether, through the establishment of a world government which alone would have military means at its disposal, including nuclear weapons. Because of that aim, he was even able to see a positive aspect in the nuclear threat—"it may intimidate the human race into bringing order into its international affairs, which, without the pressure of fear, it would not do."[5]

At a Nobel memorial event on December 10 in New York, Einstein gave a speech in which he referred to the "responsibility," not to say the sense of guilt, felt by scientists, and also to his own initiative in the construction of the atom bomb:

> We helped create this new weapon in order to prevent the enemies of mankind from achieving it first; given the mentality of the Nazis, this could have brought about untold destruction as well as enslavement of the peoples of the world. This weapon was delivered into the hands of the American and the British nations in their role as trustees of all mankind, and as fighters for peace and liberty; but so far we have no guarantee of peace nor any of the freedoms promised by the Atlantic Charter. . . . The war is won—but the peace is not.

In this new situation Einstein called for "bold action and a radical change in our mentality. . . . Otherwise our civilization will be doomed."[6]

This warning had become necessary because relations between the great powers and, in consequence, the domestic climate in America had changed dramatically over a few months. The Soviet Union, the former ally, now became the enemy in the cold war.

In order to inform the public and the politicians about the bomb and its effects, a considerable number of the physicists of the Manhattan Project set up an Emergency Committee of Atomic Scientists. At the request of Leo Szilard, the driving force behind this enterprise, Einstein became its chairman,[7] not as a "table ornament" but as an eager fund-raiser and a committed propagandist. He signed a great number of appeals and proclamations, mostly drafted by others; and he frequently spoke on the radio or took part in broadcast discussions by telephone. In the fall of 1946, he invited prominent Americans to a meeting of the Emergency Committee at the Institute for Advanced Study in Princeton, complete with "luncheon" and tea with the director, Frank Aydelotte, and a relay of Einstein's address over a nationwide network, NBC.

Despite the initial publicity, though, the Emergency Committee failed to produce any visible effect on politics. Soon, also, differences emerged among its members; Einstein tried to mediate but could "not bring himself to agree with the views of certain of the committee members."[8] At the end of 1948 the committee suspended its activities—but not so Einstein, who anyway felt happier as a lone fighter than in the atmosphere of a committee where compromise was inevitable.

Despite Einstein's support for world government and his many pronouncements on the bomb, this commitment remained marginal: "If on occasion you see my name linked with political excursions," he wrote to his friend Besso at the peak of his activity, "don't think that I spend a lot of time on such matters, for it would be a pity to waste much strength on the arid soil of politics. But now and again comes a moment when one can't do anything else."[9] Once, walking to the

Institute of Advanced Study with his assistant, Munich-born Ernst Straus, he assessed his work on the Emergency Committee as follows: "Yes, that's how one has to divide one's time between politics and our equations. But our equations are much more important to me, because politics is for the present, while such an equation is for eternity."[10]

As for his own contribution to the terrifying new weapon, he felt he had to justify himself in the eyes of those who were hoping for support from him. "Had I known that the Germans would not succeed in developing the atom bomb, I would not have supported its construction," he told the magazine *Newsweek* in March 1947. He qualified the title of the three-page story—*Einstein, the Man Who Started It All*—by explaining that the military development of nuclear energy would have been not much different without his intervention at the time.[11] Most historians of the Manhattan Project would agree with this. Even without Einstein's letter, the bomb would have been built, and probably by no more than a week later. But this does not affect the issue of moral responsibility.

Einstein stuck to the justification that it was necessary at the time to outrace the Germans. But he certainly minimized his contribution when he later told the editor of the Japanese journal *Kaizo*: "My participation in the development of the atom bomb consisted of a single action—I signed a letter to Roosevelt";[12] and when, in a letter to Max von Laue in Germany, he repeated that "the business with the atom bomb and Roosevelt was limited to my signing a letter written by Szilard."[13] In fact, he had dictated the letter (in German), he had drafted and signed two further letters; and he would have been glad to contribute not only as a writer of letters but as a physicist, had he not been excluded as a security risk by "people in Washington." On the other hand, there is of course no doubt that Hitler had driven him to feel obliged to recommend the development of a weapon of mass destruction. "If I had known that these fears were groundless, I would not have taken part in opening that Pandora's box."[14]

Einstein totally rejected any accusation that because of his formula $E = mc^2$ he bore a special responsibility. "So you believe that poor me by discovering and publishing the relation between mass and energy has played a major part in bringing about our lamentable situation," he said with some irony to a historian who knew nothing of physics. To foresee the bomb at a time when he was still at the Patent Office would have been quite impossible, as its construction required a number of discoveries in nuclear physics which in 1905 could not even be surmised. "But even if this obstacle had not existed, it would surely have been ludicrous to keep silent about the consequences of the special relativity theory. . . . The theory owes its genesis to the endeavor to discover the properties of the 'luminiferous ether.' Not a hint of possible technological applications was in sight."[15]

It would never have occurred to Einstein to abandon or even restrict the exploration of nature because of possible problematical consequences. The search for the laws of nature was to him humanity's most noble pursuit. Anything else was part of the political and moral sphere, in which human beings so often failed. Even so, he occasionally speculated about the justification of this clear distinction:

> I believe that the terrible decline in man's ethical behavior is due primarily to the mechanization and depersonalization of our lives—a disastrous by-product of the development of the technological-scientific intellect. Nostra culpa! I see no way of dealing with this fatal shortcoming. Man cools more quickly than the planet he inhabits.[16]

When the nuclear race was on and the two superpowers faced each other with deadly threats—and in Korea even waged a kind of war by proxy—Einstein sometimes came close to despairing of humanity: "If all efforts are in vain and mankind ends in self-destruction, the universe will not shed a single tear over it."[17] After a conversation with him, Hermann Broch was impressed by what Einstein, despite all his pessimism, would still regret: "Mankind remains as idiotic as it always was, and it's no great pity; but that no one would then play Bach or Mozart any more—that is a pity."[18] Regardless of occasional attacks of melancholy, Einstein never abandoned hope, either in physics or in his political commitments: "After all, to despair makes even less sense than to strive for an unattainable goal."[19]

▪ ▪ ▪

One goal he was certainly *not* striving for, and one that he regarded as unattainable anyway, was any fundamental change in the German nation or any reconciliation with it. His anger at his former "step-fatherland" was so implacable that—in line with the Morgenthau Plan—he thought it "absolutely necessary" to "prevent permanently any substantial industrial power in Germany. . . . I am not out for revenge, but for the greatest possible safety from attacks by the Germans, a safety that cannot be expected to come from moral influence on them."[20] He even rejected humanitarian programs to help Germans suffering hardships in their destroyed country.

When James Franck initiated a constructive policy for the Americans as an occupying power, not for a "milder peace" but in order to prevent a "situation of spiritual and psychological decay," Einstein brusquely reproved him: "The Germans butchered millions of civilians according to a well prepared plan, in order to move into their place. If they had butchered you too, this would not have happened without some crocodile tears. They would do it again if only they were able to." Einstein felt that "not a trace of a sense of guilt or remorse is to be found among the Germans"; he therefore called Franck's initiative a "stinking business" and announced that, when a suitable occasion arose, he would "publicly speak out against it."[21] Hermann Broch, who had joined Franck's initiative and had cooperated in drafting the appeal, succeeded in dissuading Einstein only to the extent that he would "now desist from his threatened public protest against the program."[22]

The first person to discover that Einstein would literally have no more to do with Germany was Arnold Sommerfeld. Einstein had resigned from the Bavarian Academy in 1933, and Sommerfeld now wanted to reverse that. "I am therefore asking you," Sommerfeld wrote, totally misjudging Einstein's attitude, "to bury the hatchet and once more to accept membership in the Academy."[23] Although it was a "genuine pleasure" for Einstein to hear from his old colleague, he could not have refused him more sharply: "With the Germans having murdered my Jewish brethren in Europe, I do not wish to have anything more to do with Germans, not even with a relatively harmless Academy."[24]

It would be a mistake, though, to see in Einstein's attitude a hatred of everything German. In human matters as in physics, his "rational basis [was] confident belief in unlimited causality. 'I cannot hate him because he must act the way he does.' "[25] This viewpoint, taken over from Spinoza, acknowledged neither sin nor guilt, and it had helped Einstein to show toleration and patience for many human follies and trespasses. But now it led him, in his judgment on the Germans, to an implacable conclusion, beyond crime or punishment: If the Germans had acted the way they had to act, then they would do so again unless prevented. This, then, was his attitude—encountered by all those who approached him in an official capacity.

Otto Hahn's request that Einstein become a "foreign scientific member" of the Max Planck Society, founded in Göttingen as a successor to the Kaiser Wilhelm Society, was rejected by Einstein with the crushing argument: "The crimes of the Germans are really the most hideous that the history of the so-called civilized nations has to show. The attitude of the German intellectuals—viewed as a class—was no better than that of the mob. There is not even remorse or an honest desire to make good whatever, after the gigantic murdering, is left to be made good. In these circumstances I feel a deep aversion to being involved in a business that represents part of German public life, simply out of a need for cleanliness."[26] Even less did he wish to have anything to do with institutions of the Federal Republic born in 1949. Its president, Theodor Heuss, who, in connection with the reestablishment of the civilian class of the Pour le Mérite Order, approached Einstein as one of four surviving prewar members, was informed by Einstein that it was "evident that a proud Jew no longer wishes to be connected with any kind of German official event or institution."[27]

He did not even want the Germans to read his works. When his old publisher Vieweg wished to republish his slim "generally comprehensible" book "on the special and general theory of relativity," he was informed by Einstein: "After the mass murder committed by the Germans against my Jewish brethren I do not wish any publications of mine to appear in Germany."[28] Max Born, who intended to return to Germany after being pensioned in Edinburgh in 1953, received a rather tactless ticking off for "moving back into the country of the mass murderers of our tribal companions. . . . But then we know quite well that collective conscience is a very lousy little plant which tends to die just when it is most needed."[29]

At the same time, Einstein's attitude toward Germans as a whole never affected individuals who were his colleagues and friends—even those who, like Hahn and Sommerfeld, had remained steadfast under the Nazi regime. His only intensive exchange of letters was with Max von Laue, whom he had always and unqualifiedly regarded as a "really fine fellow." But in an obituary in 1942, Einstein paid tribute to Walther Nernst, who had died in 1941. Just as if there were no war, he gave homage to the liberal-mindedness of his sometime patron and colleague, and wistfully recollected an international spirit of science that existed no longer.[30]

For Max Planck, Einstein preserved a deep reverence. When Planck died on October 4, 1947, at the age of eighty-nine, Einstein, at a memorial meeting of the American Academy of Sciences, paid his respects to his father figure as a personification of the ideal of knowledge.[31] In a letter of condolence to Planck's widow, Einstein found moving words to describe what this man had meant to him personally: "It was a beautiful and fruitful time that I had the privilege of spending in his proximity. His gaze was fixed on eternal things, yet he took an active part in everything that belonged to the human and temporal sphere. . . . The hours which I was permitted to spend at your house, and the many conversations which I conducted face to face with that wonderful man, will remain among my most beautiful recollections for the rest of my life. This cannot be altered by the fact that a tragic fate tore us apart."[32]

Einstein had not actually gone to the memorial meeting for Max Planck in Washington but had arranged for his eulogy to be read out. The reason was his own deteriorating health. He now hardly ever left Princeton.

The three ladies at 112 Mercer Street were evidently unable to give the obstinate old man the same care that Elsa had given him. A series of illnesses began after the war, with marked physical deterioration, but the causes remained uncertain for a long time. When he was better again he drew up a humorous balance sheet for that period: "I had become markedly weak and looked like a specter. It turned out that for years I had been observing an unintentional starvation treatment which impaired the elementary balance of substances. With correctly dosed fattening-up I gained 15 pounds over four weeks, acquired a human appearance again, and the weakness was gone. Thus the Devil granted me one more reprieve."[33]

However, painful abdominal attacks then recurred, accompanied by vomiting and lasting several days. Rudolf Ehrmann, a specialist in stomach complaints who had treated Einstein in Berlin, suspected an intestinal ulcer. His patient, by then seventy, with stoical equanimity wondered "that this incredibly complex machinery is ever able to function, moreover for so many years! One only has to think of life to realize how lousily primitive our whole science is."[34] Eventually the surgeon Rudolf Nissen was called in. Nissen—who had met Einstein in Berlin, though not as a patient—diagnosed a cyst in the abdomen. After an "impressive council of doctors who are my friends,"[35] Einstein was admitted in December 1948 to Jewish Hospital in Brooklyn, New York, for surgical "medical repair." The surgeon found several adhesions of the intestine, which had apparently caused the attacks. The suspected cyst, however, turned out to be an aneurysm, an arteriosclerotic dilation of the major abdominal aorta, of an alarming size: it was as large as a grapefruit.[36] So far the wall of the aneurysm was firm, and surgical intervention, given the high risk associated with it at that time, was not indicated. After a month, the patient was discharged from the hospital.

Accompanied by Walter Bucky, Einstein spent a few weeks of convalescence in Florida. "One gets hungry, the old crate is working better, and one is developing something like a paunch," he reported. But he could not indulge in laziness, as he had promised to write a concluding reflection on the nineteen contributions to the volume devoted to him in the *Library of Living Philosophers.* "So I'm just writing, always dissatisfied with what I have written, yet unable to improve on it."[37] In the end he was, after all, satisfied at having once more defended "the good Lord against the suggestion that he continuously rolls dice."[38] When he returned to Princeton he was feeling better, but not well.

His sister Maja's health was considerably worse than his own. He had been living with Maja in cordial harmony; friends were astonished at how much the siblings with increasing years resembled each other in appearance, gestures, and facial expressions. Maja had intended to return to Switzerland at the end of the war, but she suffered a stroke from which she never recovered. Progressive arteriosclerosis immobilized her and kept her bedridden. Her brother was a loving companion during her prolonged suffering. "During the last few years I read to her every evening from the best books of the old and the new literature. Strangely enough, her intelligence had hardly suffered from the advanced illness, even though toward the end she could hardly talk audibly anymore." Maja died on June 25, 1951. "Now I miss her more than can be imagined."[39]

Mileva, Einstein's first wife, had died in Zurich in 1948. She had never had an easy life, and her son Eduard's schizophrenia upset her beyond endurance. Added to this were her own illnesses, as well as permanent financial problems—which Einstein had brought on by transfers of money and also by some complicated transactions, such as taking over the house on Huttenstrasse.

Eduard was at the Burghölzli psychiatric hospital in Zurich. The author Carl Seelig, who was already looking after the mentally ill poet Robert Walser, proved a kindly and helpful mentor to Eduard. Einstein could "not thank [him] enough for [his] sympathetic occupation with my sick son. He represents the virtually only human problem that remains unsolved. The rest have been solved, not by me but by the hand of death."[40]

Einstein, however, felt unable to keep up any contact with his son, even by letter: "There is a block behind it which I am unable to analyze fully. But one factor is that I believe I would be arousing painful feelings of various kinds in him if I made an appearance in whatever form."[41] Eduard Einstein survived his father by ten years. He died at the Burghölzli in 1965.

▪ ▪ ▪

It was after the war that the picture posterity has of Einstein began to take shape. Without Elsa's supervision, the indifference that suited him in external matters was gaining the upper hand. As early as 1942 he had described himself mockingly as "a kind of ancient figure known primarily by his non-use of socks and wheeled out on special occasions as a curiosity."[42] Now, marked by sickness and haggard, often dressed not just comfortably but grotesquely, he might have become a Chaplinesque tragicomic figure, had not his features and his deep, intelligent eyes enthralled every visitor.

And there were a lot of visitors. Old friends like Maurice Solovine and Max von Laue were cordially welcomed and stayed at his house. Prominent political figures, from the Indian prime minister Jawaharlal Nehru with his daughter, Indira, to David Ben-Gurion from Israel, made a point of driving to Princeton from New York or Washington for an afternoon in order to talk to the wise man. Of particular charm was a visit by the Juilliard String Quartet in the late fall of 1952. After an afternoon concert at the university the four young musicians came to Mercer Street and played Beethoven and Bartók, before surprising Einstein with a request to be allowed to make music with him. After prolonged protest—he had stopped playing the violin seven years previously—Einstein chose Mozart's great G Minor Quintet and took the second-violin part. Despite his unpracticed and fragile fingers, he played with a good sense of coordination and intonation, and with impressive concentration. Admittedly, concession had to be made to him in the matter of tempi, but the fact that the movements became slower and slower did not diminish the warm brilliance of Mozart's music and even intensified the happiness of the young virtuosi. At the moment of farewell, Einstein said he could not understand why the Juilliard Quartet was often criticized for its fast tempi.[43]

Some time later, on November 16, 1952, Einstein was offered an honor which moved him greatly, but which he could not accept. Chaim Weizmann, the first president of the young state of Israel, had died on November 9. Israel's premier, David Ben-Gurion, had taken up a suggestion, thrown out in the Israeli press, that the vacant post be offered to Einstein as the greatest living Jew and as an expression of

Israel's special link with scientific humanism. By telephone and telex he instructed Abba Eban, then Israel's ambassador in Washington, to offer Einstein the presidency.[44]

Before the ambassador could even prepare for this delicate task, he had a surprise: Einstein was on the telephone. Ben-Gurion's intention had been reported by the news media and Einstein had learned of it on the evening of November 16. "The little quiet household was much ruffled," reported a colleague from the institute who was visiting Einstein. " 'This is very awkward, very awkward,' the old gentleman was explaining, while walking up and down in a state of agitation which was very unusual with him."[45] He decided to call Washington at once, and in decisive terms he told Abba Eban that Ben-Gurion should dismiss the idea. He asked Eban to report to Israel that he was honored, but also that his refusal was unalterable.

Eban realized that he would be unable to make Einstein change his mind, but he would not accept a refusal of Israel's presidency over the telephone. He therefore sent his deputy to Princeton the following day. When he presented the official letter, Einstein had already drafted his reply:

> I am deeply moved by the offer from our state Israel, and at once saddened and ashamed that I cannot accept it. All my life I have dealt with objective matters, hence I lack both the natural aptitude and the experience to deal properly with people and to exercise official functions. For these reasons alone I would be unsuited to fulfill the duties of that high office, even if advancing age was not making increasing demands on my strength.
>
> I am the more distressed over these circumstances because my relationship to the Jewish people has become my strongest human bond, ever since I became fully aware of our precarious situation among the nations of the world.[46]

Abba Eban was deeply impressed by the fact that in the very first sentence Einstein referred to "our state Israel." Einstein had in fact wholeheartedly welcomed the foundation of the state, even though he regretted that it had not been possible without violence and although he continued to view coexistence between Jews and Arabs as a task still to be accomplished. To the editor of the Tel Aviv daily which had suggested Einstein for the presidency, he also justified his refusal by expressing his concern about "the difficult situation that might arise if the government or parliament took decisions which would bring me into a conflict of conscience, the more so as moral responsibility would not be eliminated by the circumstance that *de facto* one has no influence on events."[47]

To a friend, his justification was simpler: "The offer from my Israeli brethren moved me deeply. But I declined straight away with genuine regret. Although many a rebel has become a bigwig, I couldn't make myself do that."[48] While Ben-Gurion was awaiting Einstein's decision, he asked his assistant, the future president Yitzak Navon, over a cup of coffee: "Tell me what to do if he says yes! I've had to offer the post to him because it's impossible not to. But if he accepts we're in for trouble."[49]

At the Institute for Advanced Study, Einstein had been retired in the spring of 1946, but he had ensured that there would be no change in his salary or his working facilities: "I threatened that I would leave Princeton when I was pensioned off, and they didn't want that because of my popularity."[50]

In the fall of the same year, J. Robert Oppenheimer assumed the directorship of the institute and began to develop it into an outstanding center of research in physics. Einstein esteemed the new director as an "unusually capable man of many-sided education" but did not come into close contact with him, "perhaps partly because our scientific opinions are fairly diametrically different."[51] As all the younger physicists whom Oppenheimer brought to the institute had been brought up on quantum mechanics, they would not and could not concern themselves with Einstein's problems. The few who called on Einstein invariably felt enriched by their contact but were astonished to find him talking as if he had a direct line to God. The many who ignored him or smiled at him from a distance nevertheless felt, when they caught sight of him, a sense of reverence for the greatest living representative of their science, though from the past—almost as if Isaac Newton had appeared among them.

This was always the case when he attended a lecture, and it was so at a symposium by which the institute honored his seventieth birthday, in March 1949. Abraham Pais recalled that most of the three hundred participants had already taken their seats when Einstein appeared. There was a moment of respectful silence in the hall before the guests rose to their feet and applauded him.[52] Even Pauli (of the sharp tongue) was—again according to Abraham Pais—somewhat transformed in Einstein's presence and could not help feeling a touch of awe.

While his life's work was being celebrated as immortal, Einstein himself, two weeks after his birthday, summed it up differently:

> There isn't a single concept of which I am convinced that it will stand up, and I feel unsure if I am even on the right road. My contemporaries, however, see me as a heretic and a reactionary who has, as it were, outlived himself.
>
> This, of course, is due to fashion and shortsightedness, but the sense of inadequacy comes from within. Well—it probably can't be otherwise if one is critical and honest, and humor and modesty keep one on an even keel, in spite of outside influences.[53]

Sometimes he expressed himself more robustly about his contemporaries: "It is rather hard to find that we are still at the stage of infants in arms, and it is not surprising that the fellows are reluctant to admit it (myself too)."[54] But he did admit it to himself and indefatigably worked on his great task, the explanation of quantum theory on the basis of a unified theory resting on the general theory of relativity. Often he was hopeful that he had at least caught hold of a shred of the truth. "Maybe even my general relativity theory of the non-symmetrical field is the correct thing," he thought in 1952. "However, the mathematical difficulties of a comparison with experience are prohibitive for the time being. Be that as it may, we are certainly as far from a truly sensible theory (the dual nature of light quanta and particles) as fifty years ago."

Although he instructed his assistant that it was already a success "if one can compel nature to stick its tongue out at one,"[55] he kept raising his own criteria for a sensible theory. Not only would the electron, and hence in his view matter generally, be derived from such a theory as a logical necessity, but the magnitudes of natural constants like the velocity of light were no longer to be empirical facts but would be derived unambiguously from the theory.

In the past Einstein had tried to find out the intentions of the—metaphorically understood—Almighty and to discover how he created the world. Now he raised his sights; he wanted to discover if God actually had any choice in creating the world: "What really interests me is whether God could have made the world differently; in other words, whether the demand for logical simplicity leaves any freedom at all."[56]

For the sake of solving the ancient philosophical problem of contingency, at least within the framework of physics, Einstein was even prepared to give up the field theory and search for alternative physical and mathematical structures. Despite various attempts along those lines, nothing matured to the point of publication.

While he indefatigably continued to work on a unified field theory, he by no means rejected the idea that maybe everything was entirely different. In the summer of 1954, in his last letter to Besso, who, half a century earlier, had been with him at the genesis of the relativity theory, he said: "However, I regard it as entirely possible that physics cannot be based on the field concept, *i.e.* on continuous structures. In that case *nothing* remains of my entire castle in the air, including the gravitation theory."[57] In this harsh judgment, too, Einstein was alone.

Politics and secrecy moved into the ivory tower with J. Robert Oppenheimer. Military police kept a twenty-four-hour guard over the safe in which the "father of the atom bomb" deposited his secret papers for his conferences with the government on nuclear matters. While Einstein was immersed in thought in his fine street-level corner room, No. 109, one floor above him Oppenheimer held conferences with bomb-experienced colleagues—from Enrico Fermi, John von Neumann, and John Archibald Wheeler to Edward Teller—to discuss the construction of a "superbomb": a hydrogen bomb.

Einstein is unlikely to have spoken about the dangers of that enterprise with the bomb builders who came and went at the institute. But when the president of the United States announced the successful construction of the hydrogen bomb, Einstein spoke to the whole nation, for the first time on the new medium of television, in an address filmed in Princeton the day before. If this development succeeded, he said, "radioactive poisoning of the atmosphere and, hence, annihilation of all life on earth will have been brought within the range of what is technically possible. The weird aspect of this development lies in its apparently inexorable character. Each step appears as the inevitable consequence of the one that went before. And at the end, looming ever clearer, lies general annihilation."[58]

Reaction was overwhelming. The following day the *New York Post* ran a banner headline across the whole page: "Einstein Warns World: Outlaw H-Bomb or Perish!"[59] Papers throughout the world followed suit. It was all in vain. The arms race was running its course with the "apparent inevitability" diagnosed by Einstein.

Just as he did not give up in his scientific endeavors, so he continued his struggle against the nuclear arms race. Moreover, he intervened in many other controversies of American domestic policy; during the dark period of Senator Joseph McCarthy's witch-hunts he called for the defense of civil liberties and pleaded for civil disobedience. As during the Weimar Republic, he was now revered by the left and by liberals, and fiercely attacked from the right, which even demanded that he be stripped of his U.S. citizenship and deported.

When he saw how free expression of opinion was being muzzled in the universities and in public life, how intellectuals and scientists were persecuted, Einstein at seventy-five provoked the nation with the statement that as a young man in that situation he would "not try to become a scientist or scholar or teacher. I would rather choose to be a plumber or a peddler in the hope to find that a modest degree of independence is still available under present circumstances."[60] He was delighted when the Chicago Plumbers' and Sanitary Engineers' Union sent him a membership card.

Thanks to his science, however, Einstein had not only acquired the independence he was striving for, he had also made use of it to his final breath. In February 1955 Bertrand Russell approached him with a proposal to make it emphatically clear to the public and to the world's governments that in a nuclear war there would be neither victors nor vanquished, only total catastrophe.[61] In a protracted correspondence the two grand old men agreed on an appeal to be signed by scientists whose voices carried international weight. Russell bore the brunt of organizing this project and also drafted the definitive document, the Russell–Einstein Manifesto. Einstein signed it on April 11, 1955, and returned it to Bertrand Russell with a short covering note.

It was his last letter. By the time Russell held it in his hands, Einstein was no longer alive.[62]

CHAPTER THIRTY-NINE

"An Old Debt"

THE END CAME QUICKLY and surprisingly, even though it had long been expected. In the summer of 1950, Einstein's physicians had found that his aneurysm was getting bigger. From then on, Einstein knew that his time was limited. Calmly he awaited his death, hoping to "die gracefully."[1] On March 18 he had signed his will, which appointed Helen Dukas and Otto Nathan as his literary executors and instructed that all his manuscripts were to go to the Hebrew University in Jerusalem.

He wished to have the simplest possible funeral. He who once had laid flowers on Newton's grave, commemorating the giant on whose shoulders he stood, did not even want a gravestone. Replying to a bold question from a student—What would become of his house after his death?—he said with a roguish smile: "This house will certainly not become a place of pilgrimage, where pilgrims would come to view the bones of the saint."[2] That was in the fall of 1953.

In 1954, Einstein developed hemolytic anemia. In 1955, he was able to report that medical skill had allowed him to survive it. "The crate is again functioning, more or less, only the brain has gotten very rusty—the Devil altogether counts the years conscientiously, one's got to acknowledge that."[3] He was again able to go to the Institute for Advanced Study every morning and work with his assistant Bruria Kaufmann, a young woman physicist born in Palestine and trained in America.

Preparations were being made in Bern and Berlin to mark the

fiftieth anniversary of the theory of relativity. "Age and sickness make it impossible for me to participate in such events, and I have to confess also that this divine dispensation has something liberating about it for me," he wrote to Max von Laue. "If I have learned anything from a long life's ponderings it is that we are much further from a deeper insight into the elementary processes than most of our contemporaries believe, so that noisy celebrations are not much in line with the real state of affairs."[4]

For Einstein's seventy-sixth birthday, which was observed without any fuss, a physicist at the university had thought up a graceful present—an experimental demonstration of the equivalence principle[5] that was to remind him of the "happiest idea" of his life, half a century earlier at the Patent Office in Bern. The setup, involving a broomstick, a spring, and a sphere, was demonstrated with great ceremony and obvious enjoyment.

A month later, on April 11, Einstein signed the appeal drawn up by Bertrand Russell against the arms race, and in the afternoon he received the Israeli ambassador, Abba Eban, to discuss a planned radio address on the seventh anniversary of the establishment of Israel. None of his visitors suspected that death was close.

On Wednesday, April 13, strong pains set in. His family doctor feared a small perforation of the aneurysm. Walter Bucky and Rudolf Ehrmann, as well as other doctor friends, arrived from New York the following day to be with Einstein at the end. He resolutely rejected surgery. "I would like to go when I want to. To prolong life artificially is tasteless."[6] A few weeks earlier he had thought of death as "an old debt that one eventually pays. Yet instinctively one does everything possible to postpone this final settlement. Such is the game that nature plays with us. We may ourselves smile that we are that way, but we cannot free ourselves of the instinctive reaction to which we are all subject."[7] Now that he knew his time had come, he tried to smile and to be free from instinct, so as to "die gracefully."

On Friday he had to be hospitalized. His son Hans Albert came from California; Otto Nathan came from New York. On Sunday, however, Einstein's condition had improved so much that he asked for his calculations and for the draft of his broadcast for his "Israeli brethren." He was unable to finish it, though. Shortly after one o'clock on Monday morning, he became restless, spoke a few words in German which the night nurse could not understand, and died. The aneurysm had finally ruptured.

Twelve close friends assembled in the afternoon for a simple ceremony at the crematorium. Otto Nathan gave a short address and recited the epilogue to Schiller's *Glocke*, which Goethe had written for Schiller's funeral service. Einstein's ashes were scattered in an undisclosed place.

His stepdaughter Margot, who was in the same hospital and had been allowed to see him a few times, reported about his last few hours: "He . . . waited for his end as for an impending natural event. Fearless as he had been in life—so quietly and modestly he was facing death. He left this world without sentimentality and without regrets."[8]

Three weeks before his death, Einstein had commemorated his old friend Besso: "Now he has preceded me a little by parting from this strange world. This means nothing. To us believing physicists the distinction between past, present, and future has only the significance of a stubborn illusion."[9]

NOTES

Note: For works listed in the Bibliography, short citations are given here. Refer to the Bibliography also for abbreviations used.

1. Family

1. Maja Winteler-Einstein, *Albert Einstein—Beitrag für sein Lebensbild*, a 36-page manuscript, completed on February 15, 1924. Extracts published in CP1, p. LVI. The editor's notes, even more than Maja's manuscript, are a reliable source for many details of Einstein's family and the first years of his life.

2. Hans Eugen Specker (ed.), *Einstein und Ulm* (Ulm/Stuttgart, 1979). Contains numerous documents from the Ulm city archives. Einstein's birth certificate is reproduced on p. 61.

3. Einstein to Dr. Hans Mühsam, March 4, 1953, ETH, in *Helle Zeit—Dunkle Zeit*, p. 56.

4. CP1, p. LVI.

5. Einstein to Carlos Erlander, March 18, 1929, Ulm city archives.

6. Einstein to editor of *Ulmer Abendpost*, March 16, 1929, Ulm city archives.

7. Maja Winteler-Einstein, *Beitrag*, CP1, p. L.

8. Reiser, p. 28.

9. Ibid., p. 26.

10. Quoted in Hoffman/Dukas, p. 22. These sentences were excerpted by Helen Dukas; the original documents are in the Einstein Archives of Hebrew University in Jerusalem, in the so-called Family Folder. They are not at present accessible.

11. Maja Winteler-Einstein, CP1, p. LVII.

12. Einstein to Sybille Blinoff, May 21, 1954.

13. Ernst G. Straus, *Reminiscences*, in *Jerusalem Symposium*, p. 419; according to other accounts—e.g., Maja Winteler-Einstein, CP1, p. LVII—the repetition of the sentence was spoken softly. I regard the version related by Einstein himself to his assistant Straus to be the most plausible.

14. Maja Winteler-Einstein, CP1, p. LVII.

15. Ibid.

16. Erik H. Erikson, *Psychoanalytic Reflections*, in *Jerusalem Symposium*, p. 157.

17. Einstein to James Franck, quoted in Seelig (1952), p. 72.

18. *Autobiographisches*, p. 7. Einstein often jocularly referred to his autobiography as his "*Nekrolog*," his "Obituary."

19. Ibid.

20. Einstein's reply to a written

question at a press conference in New York on March 18, 1953, quoted in Seelig (1960), p. 399.

2. School

1. CP1, p. LIX, n. 42.

2. Reiser, p. 30.

3. Einstein in a draft letter to an unknown addressee, April 3, 1920.

4. Reiser, p. 30.

5. Einstein, draft letter, April 3, 1920.

6. Maja Winteler-Einstein, CP1, p. LIX.

7. Hoffman/Dukas, p. 27.

8. Reiser, p. 33.

9. Student register of elementary school on Blumenstrasse, CP1, p. LIX, n. 42.

10. Frank, p. 21; see also Reiser, p. 26.

11. Frank, pp. 24f.; see Moszkowski, p. 221: "With bitter sarcasm he told me the teachers had the character of sergeants—those later at the *Gymnasium* predominantly had the character of lieutenants."

12. E. A. Pariser, *Albert Einstein*, in *Umschau*, 1933, No. 10.

13. Dr. H. Wieleitner, *Albert Einstein am Münchner Luitpold-Gymnasium*, in *Münchner Neueste Nachrichten*, March 14, 1929.

14. E.g., Moszkowski, pp. 71ff. and 221ff.

15. Antonina Vallentin, *Das Drama Albert Einsteins—Eine Biographie* (Stuttgart, 1955), p. 16.

16. Moszkowski, p. 221.

17. Abraham A. Fraenkel, *Lebenskreise—Erinnerungen eines jüdischen Mathematikers* (Stuttgart, 1955), p. 16.

18. The syllabi are reproduced in CP1, Appendix B, pp. 346ff.

19. *Skizze*, p. 10.

20. Einstein to Sybille Bintoff, May 21, 1954.

21. Ibid.

22. Maja Winteler-Einstein, CP1, p. LIX.

23. Reiser, p. 28.

24. *Autobiographisches*, p. 2.

25. Maja Winteler-Einstein, CP1, p. LX.

26. Reiser, p. 29.

27. *Autobiographisches*, p. 1.

28. Ibid.

29. Ibid., p. 2.

30. Moszkowski, p. 222; see also Frank, p. 26, for a similar version of the same story.

31. Moszkowski, p. 223; also Maja Winteler-Einstein, CP1, p. LXI, and *Autobiographisches*, p. 4.

32. Max Talmey, *The Relativity Theory Simplified and the Formative Period of Its Inventor* (New York, 1932), p. 163. On his immigration to America, Max Talmud had anglicized his last name.

33. Whether Spieker's textbook is really the "sacred little geometry book" (see below), as Talmey claims, remains uncertain; for a "little" book it would seem too voluminous. According to Einstein's sister's account, the "sacred" book could equally well have been a volume of Adolf Sickenberger's *Leitfaden der elementaren Mathematik*, which was used at the Luitpold Gymnasium but may have been bought ahead of time.

34. Heinrich B. Lübsen, *Einleitung in die Infinitesimalrechnung (Differential- und Integralrechnung). Zum Selbstunterricht. Mit Rücksicht auf das Notwendigste und Wichtigste* (Leipzig, Friedrich Branstätter). Also intended for independent study was Lübsen's

Ausführliches Lehrbuch der Analysis as well as his *Ausführliches Lehrbuch der ebenen und sphärischen Trigonometrie.* All three volumes are now in the Einstein Archives. On the flyleaf they bear the signature "Jakob Einstein," which makes it exceedingly probable that Albert obtained these volumes not from Talmud but from his uncle's library. The *Einleitung in die Infinitesimalrechnung* has a few glosses in Einstein's own hand.

35. Talmey (see n. 32), p. 164.

36. *Autobiographisches*, p. 4.

37. Ibid., p. 3.

38. Bertrand Russell, *The Autobiography of Bertrand Russell* (London, 1967), p. 36.

39. *Autobiographisches*, p. 4.

40. Ibid., p. 2.

41. Ibid., p. 5.

42. Ibid.

43. Talmey (see n. 32), p. 164.

44. Fritz Genewein to Albert Einstein, Munich, October 20, 1924.

45. Alfred Einstein to Albert Einstein, Berlin, December 9, 1927.

46. Maja Winteler-Einstein, p. LVIII.

47. Einstein to Philipp Frank, draft of a letter, 1940.

48. Ibid.

49. Talmey (see n. 32), p. 165.

50. *Weltbild* (1934), p. 8 (written about 1930).

51. Reiser, p. 33; on Ruess, see also Moszkowski, p. 221.

52. Reiser, pp. 34f.

53. Einstein to Philipp Frank, draft of a letter, 1940.

54. Maja Winteler-Einstein, CP1, p. LXIII.

55. Einstein to Philipp Frank, draft of a letter, 1940.

56. See also Erich Kiesel, *Münchner Stadtanzeiger*, 1979, No. 22, p. 8.

57. For details see CP1, notes to pp. LI–LIII; and Pyenson, pp. 35–53.

58. The patents can be inspected at the German Patent Office in Munich; they are listed in Pyenson, p. 54.

59. Friedrich Uppenborn, *Die Versorgung von Städten mit elektrischem Strom. Nach Berichten elektrotechnischer Firmen über die von ihnen verwendeten Systeme* (Berlin/Munich, 1891); pp. 63–66 describe the Einstein three-conductor DC system.

60. G. N. Reinhart, in Alfred Kuhlo, *Geschichte der bayrischen Industrie* (Munich, 1926), p. 31.

61. Ibid.

62. Maja Winteler-Einstein, CP1, p. LIII.

63. In that respect the memorial plaque now affixed to Adlzreiterstrasse 12 does not entirely accord with the facts: Albert Einstein never lived in that building, but only on the site.

64. See n. 24.

65. Reiser, pp. 41f.; Frank, p. 31.

66. Officially, his release from school was dated December 29, 1894; Kgl. Luitpold-Gymnasium, annual report for 1895.

3. A "Child Prodigy"

1. Maja Winteler-Einstein, CP1, p. LXIV.

2. Ibid.

3. Ernesta Pelizza Marangoni, *Momenti pavesi nella vita di Alberto Einstein*, in *La Provincia Pavese*, May 14, 1955, pp. 1–3.

4. Quoted from Harry A. Cohen, *An Afternoon with Einstein*, in *Jewish Spectator*, January 1969, p. 16.

5. Marangoni, p. 1.

6. Einstein to Marie Winteler, April 21, 1886.

7. Einstein to Tullio Levi-Civita, March 26, 1915.

8. To Ernesta Marangoni, Princeton, August 16, 1946, published in Elena Sanesi, *Three Letters by Albert Einstein and Some Information on Einstein's Stay at Pavia*, in *Physis* 18 (1976), pp. 174–178. In the original the sentences read: "I mesi felici del mio soggiorno in Italia sono le piu belle ricordanze. . . . Giorni e settimane senza ansie e senza tensione."

9. Otto Neustätter to Einstein, March 12, 1929.

10. Jagdish Mehra, *Physikalische Blätter*, 27, 1971, pp. 385–91, with partial facsimile of the Einstein text which is still in the possession of Caesar Koch's heirs, in CP1, Doc. 5, p. 6.

11. Einstein to Caesar Koch, summer 1895, CP1, Doc. 6, pp. 9f.

12. Einstein, *Über die Untersuchung*, CP1, p. 6.

13. Clark, p. 17.

14. Leonhard Sohncke, *Die Umwälzung unserer Anschauungen vom Wesen der elektrischen Wirkungen*, in *Himmel und Erde. Illustrierte naturwissenschaftliche Monatsschrift* 3 (1891), pp. 157–72; CP1, p. 6, n. 14.

15. Albin Herzog to Gustav Maier, September 25, 1895, CP1, Doc. 7, pp. 12f.

16. Reiser, pp. 42f.

17. *Skizze*, p. 9.

18. For details of the requirements of the examination, see CP1, Appendix C; the individual examination results can no longer be found.

19. *Skizze*, p. 9.

20. Ibid.

21. See Pyenson, pp. 9ff., and CP1, pp. 11ff.

22. CP1, Doc. 8, p. 13, and Doc. 12, p. 18.

23. *Skizze*, pp. 9f.

24. On Jost Winteler, see the short biography in CP1, p. 388.

25. Gustav Maier to Jost Winteler, October 26, 1895, CP1, Doc. 9, p. 14.

26. Anna Winteler-Besso in Edgar Lüscher, *Schweizerische Lehrerzeitung* (1944), p. 623.

27. Hans Byland, *Aus Einsteins Jugendtagen. Ein Gedenkblatt*, in *Neue Bündner Zeitung*, February 7, 1928.

28. Central State Archives, Stuttgart, E 1516, Royal Württembergian List for the Danube Region of the Loss of Reich and State Citizenship through the Granting of Charters of Release in the Calendar Year 1896, No. 8; also Ulm City Archive, B122/53, No. 2: Emigrations and Renunciations of Citizenship.

29. It may therefore be assumed that Hermann Einstein set the application in motion in Pavia no later than September, and that, lacking a better description, he treated Albert's occasional help in the factory as his "occupation."

30. Einstein to Maja Winteler-Einstein, summer 1935, quoted in Seelig (1952), p. 22.

31. Einstein to Besso, February 16, 1936, *Speziali*, p. 308.

32. To Julius Katzenstein, California, December 27, 1931. Stargard Katalog 620, No. 423, June 1980.

33. CP1, Doc. 82, p. 269.

34. Frank, p. 34: "At the same time he left the Jewish religious community" is the presumed source of this assertion.

35. Einstein in a letter of April 15, 1954, quoted in *Helle Zeit—Dunkle Zeit*, pp. 57f.

36. Einstein to Marie Winteler, April 21, 1896, CP1, p. 21.

37. Einstein to Pauline Winteler, May 1897, CP1, Doc. 34, pp. 55f.

38. Quoted from CP1, p. 385.

39. Einstein to Julia Niggli, August 1899, CP1, Doc. 51, pp. 221f.

40. Antonina Vallentin, p. 9.

41. CP1, Doc. 11, p. 17.

42. Hermann Einstein to Jost Winteler, December 30, 1895, CP1, Doc. 14, p. 19.

43. CP1, Doc. 19, p. 23.

44. On all examination papers and results: CP1, pp. 23–42.

45. CP1, Doc. 22, p. 28, reproduces the original French text with all mistakes, and corrections by the teacher Jakob Hunzicker. Grade 3–4 seems exceedingly generous.

46. Albert Einstein to Heinrich Zangger, spring 1918, ETH; see *Briefe*, p. 18.

47. Albert Einstein to Mileva Marić, September(?) 18, 1899, CP1, Doc. 54, p. 230.

48. *Skizze*, p. 10.

49. Goethe, ed. Erich Trunz, vol. 14, p. 82 (Hamburg, 1960; Hamburg Edition).

4. "Vagabond and Loner": Student Days in Zurich

1. See *Gottfried Semper und die Mitte des 19. Jahrhunderts*, symposium at ETH, Zurich, in December 1974 (Basel, Stuttgart, 1976); and Gordon A. Craig, *Zürich im Zeitalter des Liberalismus 1830–1869* (Munich, 1988), pp. 201ff. Later additions toward the Rämistrasse have impaired the proportions of Semper's building, so that only the side facing Zurich can be regarded as original—and even that was somewhat affected by the university's new building next door, erected in 1914.

2. When Einstein began his studies, there were exactly 841 students at the Polytechnic; Program of the Polytechnic 1897b, pp. 35ff. All other figures from the same source.

3. Craig (see n. 1) p. 150.

4. *Autobiographisches*, p. 7.

5. *Skizze*, p. 11.

6. Ibid.

7. Ibid., p. 10.

8. Ibid.

9. See CP1, p. LIV, n. 21.

10. Kornprobst had collaborated closely with Jakob Einstein in Munich. Three patents dated 1889 and 1890 had been granted to "Einstein & Co and Sebastian Kornprobst." These were Nos. 53,546, 53,846, and 60,361 of the German Patent Office, Munich.

11. Ernesta Pelizza Marangoni, *Momenti pavesi nella vita di Alberto Einstein*, in *La Provincia Pavese*, May 14, 1955, p. 1.

12. Albert Einstein to Maja Einstein, Zurich, 1898; included in typescript of Maja Winteler-Einstein, *Beitrag*, p. 17; CP1, Doc. 38, p. 221.

13. Ibid.

14. Albert Einstein to Alfred Stern, Milan, May 3, 1901, CP1, Doc. 104, p. 296.

15. On this: Werner G. Zimmermann, *Albert Einstein in Zürich*, *Neue Zürcher Zeitung*, March 11, 1979.

16. Albert Einstein to Alfred Stern, Milan, May 3, 1901, CP1, Doc. 104, p. 296.

17. Imposition of fine No. 6619 of April 23/28, 1897, CP1, Doc. 33, p. 54.

18. *Skizze*, p. 11.

19. Ibid.

20. Seelig (1960), p. 55.

21. Einstein to Frau Anna Grossmann, Princeton, September 26, 1936.

22. *Skizze*, p. 11.

23. Einstein to Frau Anna Grossmann, Princeton, September 26, 1936.

24. Einstein to Besso, Princeton, March 6, 1952, and Einstein to Pauline Winteler, Zurich, May 1897, CP1, Doc. 34, p. 56.

25. Report on instrumental examination at the cantonal school, March 31, 1896, CP1, Doc. 17, p. 21.

26. See the report of Susanne Markwalder, in Seelig (1960), pp. 56–62.

27. Einstein to Pauline Winteler, Zurich, probably June 7, 1897, CP1, p. 57; the woman presumably was Selina Caprotti's mother.

28. Seelig (1960), p. 58.

29. Einstein to Julia Niggli, probably July 28, 1899, CP1, Doc. 48, p. 219.

30. For Mileva Marić, see the short biography in CP1, pp. 380f. The book by Desunka Trbuhović-Gjuvić, *Im Schatten Albert Einsteins* (Bern and Stuttgart, 1983), conveys a good impression of Mileva's origins, her determination to study, and her tragic life. As this book has become a kind of feminist anti-Einstein manifesto, it should be pointed out that most of the claims about Mileva's scientific work, particularly her alleged contributions to the special theory of relativity, as well as the conclusions drawn from them, are devoid of any foundation. See my article *Mutter der Relativitätstheorie* in *Die Zeit*, November 16, 1990.

31. Mileva Marić to Einstein, Heidelberg, end of October 1897, CP1, Doc. 36, pp. 58f. This is the earliest extant letter; at least one letter written by Einstein to Mileva Marić in the summer has been lost.

32. Ibid.

33. Einstein to Mileva Marić, Zurich, probably February 16, 1898, CP1, Doc. 39, p. 212.

34. Ibid.

35. Einstein's elaborations of Weber's lectures are reproduced in CP1, pp. 63–210, but are not very illuminating.

36. Einstein to Mileva Marić (see n. 33).

37. Einstein to Mileva Marić, Milan, probably September 28, 1899, CP1, p. 234.

38. CP1, Doc. 42, p. 214.

39. *Skizze*, p. 10.

40. Henry Crew, *Notes of Travel, Europe*, 1895. Entries for July 20, 21, and 22. A travel diary owned by the American Institute of Physics, quoted from Russell McCormmach, Editor's Foreword, p. XVI, in *HSPS*, 7, 1976, Princeton.

41. CP1, Doc. 28, p. 47.

42. Quoted in Seelig (1960), p. 65.

43. Ibid., p. 47.

44. Louis Kollros, *Erinnerungen eines Kommilitonen*, in *Helle Zeit—Dunkle Zeit*, p. 21.

45. Einstein 1901, pp. 513–23.

46. *Autobiographisches*, p. 7.

47. Ibid., p. 6.

48. Ibid., p. 7.

49. Ludwig Boltzmann, *Vorlesungen über Gastheorie* (Leipzig, 1896 and 1898).

50. See Einstein to Mileva Marić, September 1899, CP1, Doc. 54, p. 230.

51. *Skizze*, p. 10.

52. Max Wertheimer, *Productive Thinking* (New York/London, 1945), p. 218.

53. Ibid.

54. Einstein, *How I Created the Theory of Relativity*, in *Physics Today*, August 1982, pp. 45–47; here, p. 46. This text is based on a lecture given by Einstein at Kyoto University, Japan, on December 14, 1922. Einstein used no notes and spoke in German. Professor Ishiwara translated into Japanese and took notes in Japanese. This Japanese record was then translated into English by Yoshimasa A. Ono. The quotation therefore is a translation from German into Japanese, and from Japanese into English. Despite its questionable transfer through two translations, the Kyoto lecture is an early and (if only for that reason) indispensable source for reconstructing Einstein's thinking at the genesis of the theory of relativity.

55. Reiser, p. 52.

56. *Skizze*, p. 10.

57. Ibid.

58. Einstein to Mileva Marić, Zurich, probably spring 1898, CP1, Doc. 41, p. 213.

59. Paul Drude, *Physik des Äthers auf elektro-magnetischer Grundlage* (Stuttgart, 1894).

60. Probably Hermann von Helmholtz, *Wissenschaftliche Abhandlungen* (Leipzig, 1895); especially pp. 476–504: *The Principle of the Least Action in Electrodynamics*, a reprint of the eponymous article from *Annalen der Physik und Chemie*, 47, 1892, pp. 1–26.

61. Heinrich Hertz, *Untersuchungen über die Ausbreitung der elektrischen Kraft* (Leipzig, 1892); especially pp. 147–70: *The Forces of Electrical Oscillations, Treated According to Maxwell's Theory*; pp. 208–255: *On the Fundamental Equations of Electrodynamics for Bodies at Rest*; pp. 256–85: *On the Fundamental Equations of Electrodynamics for Moving Bodies*, all reprints of work published in journals in 1889 and 1890.

62. Einstein to Mileva Marić, Mettmenstetten, probably August 10, 1899, CP1, Doc. 52, p. 226.

63. Ibid.

64. Ibid.

65. Ibid., p. 227.

66. Hendrik Antoon Lorentz, *Versuch einer Theorie der elektrischen und optischen Erscheinungen in bewegten Körpern* (Leyden, 1895); according to Frank, p. 38, and Reiser, p. 49, Einstein also studied the popular textbook of August Föppl, *Einführung in die Maxwellsche Theorie der Elektrizität* (Leipzig, 1894).

67. *Verhandlungen der Gesellschaft deutscher Naturforscher und Ärzte*, Vol. 70, 2nd part, 1st half, p. 83; published in full in *Annalen der Physik und Chemie*, Vol. 65, No. 3 (1898), Appendix, p. I–XVII.

68. H. A. Lorentz, ibid., pp. 86ff.

69. Einstein to Mileva Marić, Milan, probably September 28, 1899, CP1, Doc. 57, p. 233.

70. Ibid. This clearly resolves the much disputed question whether and when Einstein learned of the Michelson-Morley experiment. Wien's survey of thirteen experiments on the motion of the ether lists the Michelson-Morley experiment among the ten "with negative result."

71. See n. 67.

72. Maja Winteler-Einstein, CP1, p. LV, n. 29.

73. Einstein to Maja Einstein, Spring 1899, CP1, Doc. 44, p. 215.

74. Einstein to Mileva Marić, March 1899, CP1, Doc. 45, p. 216.

75. Einstein to Mileva Marić,

Mettmenstetten, August 1899, CP1, Doc. 50, p. 220.

76. Ibid., p. 221.

77. Hans Byland, *Aus Einsteins Jugendtagen. Ein Gedenkblatt*, *Neue Bündner Zeitung*, February 7, 1928.

78. Einstein to Mileva Marić, Mettmenstetten, probably August 10, 1899, CP1, Doc. 52, p. 227.

79. Ibid.

80. Einstein to Mileva Marić, Milan, probably September 28, 1899, CP1, Doc. 57, p. 235.

81. Ibid.

82. The lectures for which Einstein enrolled at the Polytechnic are listed in CP1, Appendix E, pp. 362–69.

83. Reiser, p. 51.

84. This applies to ETH's Department XII to this day.

85. Einstein to Arnold Heim, Princeton, July 14, 1952.

86. See Julius Braunthal, *Victor und Friedrich Adler—Zwei Generationen Arbeiterbewegung* (Vienna, 1965), pp. 195ff.

87. Reiser, p. 50.

88. Friedrich Adler, quoted in Seelig (1960), p. 164.

89. The attempt to establish a close link between Einstein's relations with revolutionary fellow students and his subsequent research in physics, or even, by construing "isoemotional lines," to view them as the prerequisite of his innovations in physics—as in Lewis S. Feuer, *Einstein and the Generations of Science* (New Brunswick/London, 1982)—seem to me mistaken.

90. Einstein to Mileva Marić, Milan, probably October 10, 1899, CP1, Doc. 63, p. 238.

91. Ibid.

92. Mileva Marić to Helene Kaufler, Zurich, March 9, 1900, CP1, Doc. 63, p. 243.

93. From later letters—e.g., from one written by Einstein to Mileva Marić at the end of July 1900 (CP1, Doc. 68)—it emerges that serious intentions of marriage existed by that time.

94. Mileva Marić to Helene Kaufler, Zurich, March 9, 1900, CP1, Doc. 63, pp. 243f.

95. As diploma essays were not kept in the Polytechnic's archives, those by Einstein and Marić must be considered lost. Not even the subjects are known.

96. Einstein to Carl Seelig, Princeton, April 8, 1952, in ETH.

97. The results of the final diploma examination in Section VIA are in CP1, Doc. 67, p. 247.

98. Ibid. The decision not to award a diploma to Mileva Marić cannot readily be understood, as her grade average of 4.0 would, according to present practice, correspond to "satisfactory." I have been unable in the ETH archive to find any minimum average below which a diploma would be refused.

99. *Autobiographisches*, p. 8.

100. *Skizze*, p. 12. See also Chapter 5 of this book.

101. Einstein to Mileva Marić, Melchtal, probably August 1, 1900, CP1, Doc. 69, p. 251.

102. Einstein to Mileva Marić, Milan, probably September 6, 1900, CP1, Doc. 74, p. 257.

5. Looking for a Job

1. Einstein to Mileva Marić, Melchtal, probably July 29, 1900, CP1, Doc. 68, p. 248; the following quotations are also from that letter.

2. Gustav Robert Kirchhoff, *Vor-*

lesungen über mathematische Physik, Vol. I: *Mechanik* (Leipzig, 1897).

3. Einstein to Mileva Marić. Melchtal, probably August 1, 1900, CP1, Doc. 69, pp. 250f.

4. Einstein to Mileva Marić. Zurich, probably August 9, 1900, CP1, p. 253.

5. Einstein to Mileva Marić. Melchtal, probably August 6, 1900, CP1, Doc. 70, p. 251.

6. Einstein to Mileva Marić, Zurich, probably August 9, 1900, CP1, Doc. 71, p. 253.

7. Einstein to Mileva Marić, Milan, probably August 20, 1900, CP1, Doc. 73, p. 255.

8. Seelig (1960), p. 61.

9. Einstein to Mileva Marić, Milan, probably September 13, 1900, CP1, Doc. 75, p. 260.

10. Einstein to Mileva Marić, Zurich, probably August 9, 1900, CP1, Doc. 71, p. 253.

11. Seelig (1960), p. 48.

12. Einstein to Mileva Marić, Zurich, probably August 9, 1900, CP1, Doc. 71, p. 253.

13. Einstein to Mileva Marić, Zurich, probably August 14, 1900, CP1, Doc. 72, p. 255.

14. Einstein to Mileva Marić, Zurich, probably August 9, 1900, CP1, Doc. 71, p. 253.

15. Einstein to Mileva Marić, Milan, probably August 30, 1900, CP1, Doc. 74, p. 257.

16. Einstein to Mileva Marić, Milan, probably September 13, 1900, CP1, Doc. 75, p. 260.

17. Einstein to Mileva Marić, Milan, probably August 30, 1900, CP1, Doc. 74, p. 258.

18. Ibid.

19. Einstein to Mileva Marić, Zurich, probably August 9, 1900, CP1, Doc. 71, p. 253.

20. Einstein to Mileva Marić, Milan, probably September 19, 1900, CP1, Doc. 76, p. 261.

21. Einstein to Mileva Marić, Milan, probably October 3, 1900, CP1, Doc. 79, p. 266.

22. Einstein to Adolf Hurwitz, Milan, probably September 23, 1900, CP1, Doc. 77, p., 263.

23. Einstein to Adolf Hurwitz, Milan, September 26, 1900, CP1, Doc. 78, p. 264.

24. Einstein to Helene Kaufler, Zurich, October 11, 1900, CP1, Doc. 81, p. 268.

25. Einstein to Mileva Marić, Milan, probably October 3, 1900, CP1, Doc. 79, p. 267. Gustav H. Wiedemann was, until his death in 1899, editor of *Annalen der Physik und Chemie*, the most important journal of the exact sciences in the German-speaking area since its foundation by Poggendorff. At the time Einstein began to publish in *Annalen*, the journal was edited by Paul Drude, who deleted "chemistry" from its title. Because of Wiedemann's towering reputation as editor, the journal continued for a long time to be called *Wiedemanns Annalen*—as Einstein referred to it.

26. Einstein 1901.

27. Einstein described the interaction between two molecules of types i and j by a potential $P = P - c_i c_j P(r)$, with $c_i c_j$ being specific constants for the forces and $P(r)$ being a universal function. This was insufficient, if only because of the disparate sizes of the molecules.

28. Einstein to Marcel Grossmann, Milan, April 14, 1901, CP1, Doc. 100, p. 290.

29. Ibid.

30. Ibid., pp. 290f. See also A. v. Humboldt, *Kosmos*, Berlin (1845), Vol. 1, p. 6: "The most important result of profound physical research is therefore this: in the multiplicity to recognize the unity . . . remembering the exalted mission of Man, to comprehend the spirit of nature that lies concealed under the cover of the phenomena."

31. Einstein 1902b.

32. Einstein to Mileva Marić, Schaffhausen, November 28, 1901, CP1, Doc. 126, p. 321.

33. CP1, Doc. 132, p. 331.

34. Einstein to Johannes Stark, Bern, December 7, 1907, in Stark, p. 272.

35. Mileva Marić to Helene Savić, née Kaufler, Zurich, December 20, 1900, CP1, Doc. 85, p. 273.

36. Einstein to Otto Wiener, Zurich, March 9, 1901, CP1, Doc. 90, p. 277.

37. Einstein to Wilhelm Ostwald, Zurich, March 19, 1901, CP1, Doc. 92, p. 278.

38. Einstein to Wilhelm Ostwald, Milan, April 3, 1901, CP1, Doc. 95, p. 284, enclosing his address.

39. Hermann Einstein to Wilhelm Ostwald, Milan, April 13, 1901, CP1, Doc. 99, p. 289.

40. Ostwald had received the Nobel Prize for chemistry in 1909 and was therefore, according to the rules, entitled to propose recipients for the prize for chemistry and physics.

41. Pais, p. 506.

42. Einstein to Heike Kamerlingh Onnes, Milan, April 12, 1901, CP1, Doc. 98, p. 288.

43. Einstein to Professor Paalzow, Milan, April 12, 1901, in MPG.

44. Einstein to Mileva Marić, Milan, probably March 23, 1901, CP1, Doc. 93, p. 279.

45. Einstein to Mileva Marić, Milan, probably March 27, 1901, CP1, Doc. 94, p. 281.

46. Einstein to Marcel Grossmann, Milan, April 14, 1901, CP1, Doc. 100, p. 290.

47. The advertisement appeared in *Physikalische Zeitschrift*, Vol. 2, No. 25, March 23, 1901 (though the issue was presumably available a few days before its date of publication), and Riecke's rejection arrived in Milan on March 27, 1901. See also CP1, Doc. 93 and 94, pp. 279ff.

48. Einstein to Heinrich Zangger, undated, summer 1912; Weber had died in Zurich on May 24, 1912.

49. Einstein to Maja Winteler-Einstein, Berlin, 1918, published undated in *Briefe*, p. 16.

50. Einstein to Mileva Marić, Milan, probably March 27, 1901, CP1, Doc. 93, p. 280.

51. Einstein to Mileva Marić, probably Milan, April 4, 1901, CP1, Doc. 96, p. 285.

52. Einstein to Mileva Marić, Milan, April 15, 1901, CP1, Doc. 101, p. 292.

53. Einstein to Marcel Grossmann, Milan, April 14, 1901, CP1, Doc. 100, p. 290.

54. Einstein to Mileva Marić, Milan, April 15, 1901, CP1, Doc. 101, p. 291.

55. Einstein to Carl Seelig, Princeton, March 26, 1952, ETH. In this letter Einstein mistakenly referred to "nine years" of statelessness instead of the actual five years.

56. Maja Winteler-Einstein, *Beitrag*, typescript, p. 20.

57. On the procedure for acquiring citizenship, see CP1: Editorial Note, pp. 239ff., and the relevant documents.

58. Zurich Municipal Police Detective's report, V District, of July 4, 1900, CP1, Doc. 66, p. 246.

59. Minutes of the "Civic Section" of the Zurich City Council of December 14, 1900, CP1, p. 272.

60. Pathé Journal London, Libr. No. UN 142, I/1932.

61. Questionnaire for Civic Rights Applicants of the City of Zurich, completed toward the end of October 1900, CP1, Doc. 82, p. 269.

62. Report of the Swiss Information Bureau, Zurich, January 30, 1901, CP1, Doc. 88, p. 275.

63. Ibid.

64. Service Book, CP1, Doc. 91, p. 278.

65. Declaration released on June 2, 1931, by the League of Opponents of War, signed by, among others, Einstein; reproduced in part in *Frieden*, p. 154.

66. Letter from Einstein to a Swiss conscientious objector of August 15, 1931, in *Frieden*, p. 160.

67. Einstein to Gustav Wissler, Princeton, August 24, 1949, ETH.

68. *Einstein on His Theory*, in *The Times* (London), November 28, 1919.

69. Einstein to Mileva Marić, Milan, April 15, 1901, CP1, Doc. 101, p. 292.

70. Mileva Marić to Helene Savić née Kaufler, Zurich, undated, probably mid-May 1901, CP1, Doc. 109, p. 302.

71. Einstein to Mileva Marić. Winterthur, probably May 28, 1901, CP1, Doc. 111, p. 304.

72. Ibid.; the "wonderful paper" was by Philipp Lenard, *Erzeugung von Kathodenstrahlen durch ultraviolettes Licht*, in *Annalen der Physik*, 1900, pp. 359–75.

73. Einstein to Mileva Marić, Winterthur, probably June 4, 1901, CP1, Doc. 112, p. 306.

74. Einstein to Jost Winteler, Winterthur, probably July 8, 1901, CP1, Doc. 115, p. 310.

75. Ibid.

76. Ibid.

77. See also CP1, Doc. 113, 117, 118, and 120.

78. Einstein to Marcel Grossmann, Winterthur, probably September 6, 1901, CP1, Doc. 122, p. 315.

79. Einstein to Mileva Marić, Winterthur, probably June 4, 1901, CP1, Doc. 112, p. 306.

80. Einstein to Mileva Marić, Winterthur, probably July 7, 1901, CP1, Doc. 114, p. 308.

81. Einstein to Jost Winteler, Winterthur, probably July 8, 1901, CP1, Doc. 115, p. 310.

82. Records of Section VIA, July 26, 1901, ETH.

83. Einstein to Marcel Grossmann, Winterthur, probably September 6, 1901, CP1, Doc. 122, p. 315.

84. Seelig, p. 84.

85. Einstein to Mileva Marić, Schaffhausen, December 12, 1901, CP1, Doc. 127, p. 323.

86. Einstein to Mileva Marić, Schaffhausen, probably November 28, 1901, CP1, Doc. 126, p. 321.

87. Mileva Marić to Helene Savić, Neusatz, early December 1901, CP1, Doc. 125, p. 320.

88. Einstein to Mileva Marić, Schaffhausen, probably November 28, 1901, CP1, p. 322.

89. Einstein to Mileva Marić, Schaffhausen, December 28, 1901, CP1, Doc. 131, p. 330.

90. Einstein to Mileva Marić, Schaffhausen, probably November 28, 1901, CP1, Doc. 126, p. 322.

91. Einstein to Mileva Marić, Schaffhausen, probably December 17, 1901, CP1, Doc., 128, p. 326.

92. Friedrich Adler to Victor Adler, Zurich, June 19, 1908, in Julius Braunthal, *Victor und Friedrich Adler* (Vienna, 1965), p. 196.

93. Einstein to Mileva Marić, Schaffhausen, probably December 17, 1901, CP1, Doc. 128, p. 326.

94. Receipt of the university chancellery dated February 1, 1902, CP1, Doc. 132, p. 331.

95. Mileva Marić to Helene Savić, Neusatz, probably early December 1901, CP1, Doc. 125, p. 319.

96. Ibid., p. 320.

97. Einstein to Mileva Marić, Schaffhausen, probably December 12, 1901, CP1, Doc. 127, p. 322.

98. Ibid.

99. *Schweizerisches Bundesblatt* (Swiss official gazette), No. 50 of December 11, 1901, p. 1265.

100. Einstein to Mileva Marić, Bern, mid-February 1902, CP1, Doc. 137, p. 336.

101. CP1, Doc. 129, p. 327.

102. Ibid.

103. Einstein to Mileva Marić, Schaffhausen, December 19, 1901, CP1, Doc. 130, p. 328.

104. Ibid.

105. Einstein to Mileva Marić, Schaffhausen, December 28, 1901, CP1, Doc. 131, p. 329.

106. Einstein to Conrad Habicht, Bern, February 4, 1902, CP1, Doc. 133, p. 331.

107. Einstein to Mileva Marić, Milan, probably April 10, 1901, CP1, p. 286.

108. *Autobiographisches*, p. 19.

109. Ibid., p. 20.

110. Einstein to Mileva Marić, Milan, probably March 27, 1901, CP1, Doc. 94, p. 282.

111. Einstein to Mileva Marić, Schaffhausen, probably December 17, 1901, CP1, Doc. 128, p. 325.

6. Expert III Class

1. To Mileva Marić, Bern, probably February 4, 1902, CP1, p. 332.

2. Ibid.

3. *Anzeiger für die Stadt Bern*, February 5, 1902.

4. Flückiger, pp. 11, 12.

5. To Mileva Marić (see n. 1).

6. To Mileva Marić (see n. 1).

7. To Mileva Marić, Bern, probably February 17, 1902, CP1, p. 335.

8. Pauline Einstein to Pauline Winteler, Milan, February 20, [1902,] CP1, p. 336.

9. Talmey, pp. 166f.

10. To Mileva Marić, Bern, probably February 8, 1902, CP1, p. 334.

11. Ibid.

12. Maurice Solovine, in Solovine, p. V.

13. Ibid., p. VII.

14. To M. Besso, Princeton, March 6, 1972, in Besso, p. 464.

15. To C. Seelig, Princeton, April 20, 1952, ETH.

16. To Pauline Winteler, Zurich, May 1897, CP1, p. 56.

17. To C. Seelig, Princeton, April 20, 1952.

18. Einstein 1902a (*Über die thermodynamische Theorie der Potentialdifferenz zwischen Metallen und vollständig*

dissoziierten Lösungen ihrer Salze und über eine elektrische Methode zur Erforschung der Molecularkräfte [*On the Thermodynamic Theory of the Potential Difference Between Metals and Completely Dissociated Solutions of Their Salts and on an Electrical Method for Investigating Molecular Forces*], AdP, 8, 1902, pp. 798–814).

19. Ibid., p. 814.

20. To J. Stark, Bern, December 7, 1907, in Stark, p. 267.

21. *Skizze*, p. 12.

22. Justice and Police Department to the Federal Council, Bern, June 2, 1902, CP1, p. 338. His grading as "Expert III Class" corresponded, according to the career ranks at the time, to that of "Engineer II Class," mentioned in the advertisement, and does not therefore suggest any "reduction in rank."

23. To Mileva Marić, Winterthur, second half of May 1901, CP1, Doc. 110, p. 304.

24. To Frau Grossmann, Princeton, September 26, 1936.

25. To Hans Wohlwend, Bern, beginning of September 1902, Flückiger collection, excerpt published in *Der Bund*, March 2, 1985.

26. Ibid.

27. To Conrad Habicht, Bern, Summer 1905, in Seelig, p. 126.

28. *Skizze*, p. 12.

29. Ibid.

30. Ibid.

31. To Michele Besso, Berlin, December 12, 1919, in Besso, p. 148.

32. Flückiger, p. 57.

33. Ibid., p. 58.

34. Friedrich Haller to the Federal Council, Bern, January 27, 1906; Patent Office, Bern.

35. Negative judgment on patent application, December 11, 1907, Patent Office, Bern.

36. To Mileva Marić, Bern, probably February 17, 1902, CP1, p. 336.

37. To Emile Meyerson, Berlin, January 27, 1930.

38. Moszkowski, p. 227.

39. To Mileva Marić, Milan, March 23, 1901, CP1, p. 281.

40. Private communication from Helen Dukas to Abraham Pais, in Pais, p. 47.

41. Maja Winteler-Einstein, *Beitrag*, p. 23.

42. To C. Seelig, Princeton, May 5, 1952.

43. Sayen, p. 70.

44. To Mileva Marić, Schaffhausen, December 12, 1901, CP1, p. 324.

45. To M. Besso, Bern, January 22, 1903, in Besso, p. 3.

46. Einstein 1902b.

47. Einstein 1903.

48. *Autobiographisches*, p. 7.

49. To M. Grossmann, Winterthur, September 6, 1901, CP1, p. 315.

50. Einstein 1902b, p. 417.

51. W. Voigt, *Ludwig Boltzmann*, obituary in PhZ, Vol. 7, 1906, p. 49.

52. Einstein in fact had managed solely with the law of the preservation of energy and the Liouville theorem; see Einstein 1902b, p. 427.

53. Einstein. 1902b, p. 433.

54. To M. Besso, Bern, January 22, 1903, in Besso, p. 3.

55. To M. Besso, Bern, March 17, 1903, in Besso, p. 14.

56. Max Born in Schilpp, p. 85.

57. To M. Besso, Berlin, June 23, 1918, in Besso, p. 126.

58. Einstein 1916b, p. 481.

59. Hertz, *Über die mechanischen Grundlagen der Thermodynamik* in

AdP, Vol. 33, 1910, pp. 225–74 and pp. 537–52.

60. Einstein 1911c, p. 175.

61. Ibid.

62. J. Sauter, *Comment j'appris à connaître Einstein*, broadcast, August 6, 1955, in Flückiger, pp. 154ff.

63. Ibid.

64. To M. Besso, Princeton, July 13, 1953, in Besso, p. 471.

65. On Paul Gruner, see Flückiger, pp. 72ff.

66. Reglement über die Habilitation an der philosophischen Fakultät der Hochschule Bern von 1891.

67. To M. Besso, January 22, 1903, in Besso, p. 4.

68. To M. Besso, March 22, 1903, in ibid., p. 14.

69. Minutes of *Naturforschende Gesellschaft zu Bern*, State and University Library, Bern.

70. To F. Baltzer, Princeton, December 1936.

71. To Mileva Marić, Schaffhausen, December 17, 1901, CP1, p. 325.

72. Mileva Marić to Einstein, Stein am Rhein, November 13, 1901, CP1, p. 317.

73. To Mileva Einstein-Marić, Bern, end of August 1903.

74. To C. Seelig, Princeton, May 5, 1952, ETH.

75. Ibid.

76. Solovine, p. XII.

77. To Mileva Marić, Milan, March 27, 1901, CP1, p. 282.

78. The theory that Mileva Marić had a significant share in Einstein's work was propagated by Desunka Trbuhović-Gjuvić, *Im Schatten Albert Einsteins*. However, it rests on unprovable speculations; see n. 30 to Chapter 4.

79. To Conrad Habicht, Bern, April 14, 1904, in Seelig, p. 100.

80. Justice and Police Department to the Federal Council, Bern, March 10, 1904. Federal archives, Bern.

81. This may be deduced from the letter of the Justice and Police Department to the Federal Council, representing Einstein's development at the Patent Office. Federal archives, Bern.

82. To Vero and Bice Besso, Princeton, March 21, 1955, in Besso, p. 538.

83. Seelig, p. 120.

84. To M. Grossmann, Bern, probably April 3, 1904, in Seelig, p. 101.

85. Einstein 1911c, p. 175.

86. Einstein 1904, p. 360.

87. Ibid., p. 361.

88. Ibid., p. 362.

89. See Editorial Note, *Einstein's Reviews for the Beiblätter zu den Annalen der Physik*, CP2, pp. 109–111.

90. CP2, p. 332.

91. Ibid., p. 317.

92. To Conrad Habicht, end of May or beginning of June 1905, in Seelig, pp. 124f.

93. *Autobiographisches*, p. 12.

7. "Herr Doktor Einstein" and the Reality of Atoms

1. Tony Cawkell and Eugene Garfield, *Assessing Einstein's Impact on Science by Citation Analysis*, in *Einstein: The First Hundred Years* (London, 1980), p. 32.

2. Einstein 1905a.

3. Einstein 1905b.

4. Examples in n. 1, above.

5. E.g., Clark, p. 86.

6. The entry for Einstein in the *Dictionary of Scientific Biography* does

not mention the dissertation at all; neither does Born in his article in Schilpp; nor do the contributions to the Einstein symposia of 1979. Abraham Pais eventually remedied this shortcoming: Pais, pp. 88ff.

7. To M. Besso, Bern, January 22, 1903, in Besso, p. 4.

8. J. Sauter, *Erinnerungen 1955*, in Flückiger, p. 158.

9. Maja Winteler-Einstein, *Beitrag*, p. 23.

10. A. Kleiner, degree expertise, Zurich, July 24, 1905, STZ.

11. Expertise by A. Kleiner, July 24, 1905, STZ.

12. Expertise by H. Burkhardt, July 24, 1905, STZ (underlining by Burkhardt).

13. *Autobiographisches*, p. 19.

14. From Ludwig Boltzmann's inaugural lecture on natural philosophy, December 11, 1903, in Boltzmann, *Populäre Schriften* (Braunschweig, 1979).

15. L. Boltzmann, *Vorlesungen zur Gastheorie*, Part II (Leipzig, 1898), p. VI.

16. *Autobiographisches*, p. 18, jocularly referred to by Einstein as his "*Nekrolog,*" his "*Obituary.*"

17. To M. Besso, Bern, March 17, 1903, in Besso, p. 14.

18. Einstein 1906a, p. 289.

19. Einstein 1906a, p. 305.

20. This is probably the background of a well-known but scarcely plausible anecdote: "Laughing, Einstein later related that his dissertation was initially returned to him by Kleiner with the observation that it was too short. After he had inserted a single sentence it was tacitly accepted" (Seelig, p. 112). There is no support for this version. Einstein was presumably referring to the *publication* of his dissertation, though his subsequent addendum is rather longer than a single sentence.

21. Einstein 1906a, pp. 305f. In his dissertation Einstein had obtained $N=2.1 \times 10^{23}$. Using values from the 1905 new edition of the data compilation by Landolt and Börnstein he obtained, in his appendix, $N=4.15 \times 10^{23}$. In 1911, following correction of a calculation error, this became $N=6.56 \times 10^{23}$ (Einstein 1911a, p. 592). Nowadays 6.022×10^{23} is considered the "best" value.

22. For more detail, see *Einstein's Dissertation on the Determination of Molecular Dimensions*, CP2, p. 180.

23. To Ludwig Hopf, Zurich, December 1910–January 1911.

24. Einstein 1911d.

25. Einstein 1905b, p. 549.

26. On the Brownian movement, see Mary Jo Bye, *Molecular Reality* (London, New York, 1972) and the still very interesting book by Jean Perrin, *Les Atomes* (Paris, 1913).

27. Essentially, the shortcomings resulted because, before Einstein, it was believed that the velocity of the particles could be measured. This, however, is not possible; instead, one measures their mean displacement.

28. Einstein 1905b, p. 549.

29. Ibid., p. 560.

30. See introduction, Einstein 1906b, p. 371. The Einstein-Siedentopf correspondence is lost.

31. Einstein 1906b.

32. According to notes by Heinrich Zangger from the 1950s, this meeting took place in 1905 or 1906 (CP2, p. 217; the account in *Helle Zeit—Dunkle Zeit*, p. 42, to the effect that Zangger and Einstein had known

each other since 1902 does not seem to be correct).

33. Einstein 1907c.

34. To Jean Perrin, Zurich November 11, 1911.

35. Ibid.

36. Ibid.

37. Minutes No. 1038, *Naturforschende Gesellschaft Bern*, 1907.

38. Einstein 1908c.

39. Einstein 1906b, p. 343.

40. Einstein 1908a.

41. C. Röntgen to A. Einstein, Munich, September 18, 1906.

42. Einstein 1905b, p. 549.

43. Einstein 1915a (1925), p. 291.

44. Max Planck to the Prussian Ministry of Education, October 7, 1913, in Berlin, p. 100.

8. The "Very Revolutionary" Light Quanta

1. To Conrad Habicht, Bern, end of May or beginning of June 1905, in Seelig, pp. 124f.

2. Einstein 1905a.

3. E.g., in the *Introduction to the Critique of Judgment*, first version, section II; and in *Critique of Judgment*, §78, A.

4. Max Planck to Robert Williams Wood, October 7, 1931.

5. Gustav Kirchhoff, *Über das Verhältnis zwischen den Emissionsvermögen und den Absorptionsvermögen*, in Poggendorf's *Annalen der Physik und Chemie*, Vol. 160, 1860, p. 292.

6. Max Planck, *Wissenschaftliche Selbstbiographie* (Leipzig, 1948), pp. 22f.

7. Friedrich Paschen to Heinrich Kayser, February 8, 1898, in STPK, Darmstädter collection.

8. Max Planck to Robert Williams Wood, October 7, 1931.

9. Ibid.

10. Max Planck, *Zur Theorie der Energieverteilung im Normalspektrum* in VhDPG, Pt. 2, 1900, pp. 237–45.

11. Max Planck to Robert Williams Wood, October 11, 1931.

12. Planck, *Wissenschaftliche Selbstbiographie* (n. 6 above), p. 29.

13. To Mileva Marić, Milan, mid-March 1899, CP1, p. 216.

14. To Mileva Marić, April 4, 1901, CP1, p. 284.

15. To Mileva Marić, Milan, April 10, 1901, CP1, p. 287.

16. *Autobiographisches*, pp. 16f.

17. Ibid., p. 17.

18. To Mileva Marić, Winterthur, end of May 1901, CP1, p. 304.

19. To Julia Niggli, Zurich, July 28, 1899, CP1, p. 219.

20. To Mileva Marić, Mettmenstetten, probably August 10, 1899, CP1, p. 227.

21. Ibid., p. 229.

22. Einstein 1905a, p. 132.

23. Ibid., p. 133.

24. Ibid., p. 136.

25. Ibid., p. 143. Actually, what Einstein wrote was not $h\nu$, but $R\beta\nu/N$, with R being the gas constant and N the Avogadro number, $R/N=k$ and $\beta=h/k$. However, $h\nu$ is the form that later became customary and will be familiar to many readers. I hope my use of $h\nu$ will not be perceived as a violation of historical accuracy.

26. Pais, p. 387.

27. Einstein 1906c, p. 199.

28. Ibid.

29. Ibid., p. 202.

30. No details have come down to us on the history of publication of Ein-

stein's early papers. But it was just under three months from receipt of the paper to the date of publication—a little longer than usual but still entirely within the normal range.

31. Max von Laue to Albert Einstein, Berlin, June 2, 1906.

32. Max Planck to Albert Einstein, Berlin-Grunewald, July 6, 1907; this is the first extant letter of the Einstein-Planck correspondence. Their exchange of ideas in writing began no later than the summer of 1906, but earlier letters—like many later ones—have not survived.

33. Max Planck to H. A. Lorentz, Berlin, July 10, 1909. Lorentz's letters to Planck have been lost; in this quotation, Planck confirms Lorentz's opinion.

34. Nomination for the acceptance of Albert Einstein as a member of the Royal Prussian Academy of Sciences in Berlin, Berlin, June 12, 1913, signed by Planck, Nernst, Rubens, and Warburg, in Planck's handwriting; in Berlin, p. 96.

35. *Max Planck als Forscher*, in NW, Vol. 1, 1913, pp. 1077f. Einstein's remark here refers to the Wien function $f(\mu/T)$, which is equivalent to the Kirchhoff function, although it additionally depends on a variable μ/T.

36. To Philipp Lenard, Bern, November 16, 1905, cited in A. Kleinert and C. Schönbeck, *Lenard und Einstein*, in *Gesnerus*, Vol. 35, 1978, p. 319. This is the only extant letter from what presumably was not a very voluminous correspondence.

37. Rudolf Ladenburg, *Die neueren Forschungen über die durch Licht- und Röntgenstrahlen hervorgerufene Emission negativer Elektronen*, in *Jahrbuch der Radioaktivität und Elektronik*, Vol. 6, 1909, pp. 425–84.

38. R. A. Millikan, *Review of Modern Physics*, Vol. 21, 1949, p. 343.

39. R. A. Millikan, *Quantenbeziehungen beim photoelektrischen Effekt*, in PhZ, Vol. 17, 1916, pp. 217–21.

40. Ibid., p. 221.

41. Einstein 1909c, p. 188.

42. Einstein 1911i, German 1914, p. 359.

43. A good approximation for "specific heat" is $C = 6$ *cal/mol* × *degree.*

44. Cited in *Dictionary of Scientific Biography*, Dulong.

45. According to the equal distribution theorem the mean kinetic energy per degree of freedom is $^1/_2kT$; for a crystal this means $^3/_2NkT$ per mol, and with an equal amount for the potential energy a total of $3Nkt = 3RT$, hence $C = 3R = 6$ cal/mol × degree.

46. To Mileva Marić, Milan, March 23, 1901, CP1, p. 279.

47. Einstein 1907a.

48. This is the delta function introduced into quantum mechanics by Dirac. Einstein uses a form of that function on p. 183 of Einstein 1907a.

49. Einstein 1907a, pp. 183f.

50. AdP, 22, 1907, p. 800.

51. Einstein 1907a, p. 188.

52. Ibid.

53. W. Nernst, *Untersuchungen über die spezifische Wärme bei tiefen Temperaturen*, in SB, 1911, pp. 306–315.

54. W. Nernst to A. Schuster, March 17, 1910. Original in the archive of the Royal Society, London,

Schuster papers; copy MPG, Kangro collection.

9. Relative Movement: "My Life for Seven Years"

1. Kyoto lecture, December 14, 1922, in *Physics Today*, p. 46, August 1982. The following account is partly based on this text.

2. Ibid.

3. Joseph Sauter, *Comment j'ai appris à connaître Einstein*, in Flückiger, p. 156.

4. To C. Seelig, Princeton, March 11, 1951, in Seelig, p. 114. See also the Kyoto lecture (n. 1).

5. Einstein 1905d. Now and again I refer to "relativity theory" in contexts when this term did not yet exist.

6. R. S. Shankland, *Conversations with Albert Einstein*, in *American Journal of Physics*, Vol. 31, 1963, p. 56.

7. See the opening of the Kyoto lecture.

8. Galileo Galilei, *Dialogo*, Giornata seconda, 1632, p. 212.

9. Ibid., p. 213; Galileo's text of the crucial restriction, placed between brackets, equivalent to the determination of an inertial system, runs: "pur che il moto sia uniforme e non fluttuante in qua e in la."

10. Isaac Newton, *Philosophiae naturalis principia mathematica* (London, 1687), Corollary V.

11. Ibid.

12. I. Kant, *Kritik der reinen Vernunft*, B 40.

13. I. Kant, *Metaphysische Anfangsgründe der Naturwissenschaften*, A, p. 1.

14. The term "inertial system" was introduced by Ludwig Lange in 1885. See Lange's papers *Die geschichtliche Entwicklung des Bewegungsbegriffs* (Leipzig, 1886) and *Das Inertialsystem vor dem Forum der Naturforschung* (Leipzig, 1902).

15. The term "Galileo transformations" was introduced in 1909 by Philipp Frank: *Sitzungsberichte der Akademie der Wissenschaften zu Wien, Abt. IIa.*, 1909, p. 382.

16. Henri Poincaré in *La Science et l'hypothèse* (Paris, 1902). Poincaré's remark refers to Fizeau's experiment of 1851.

17. Heinrich Hertz, *Gesammelte Werke*, Vol. 1 (Leipzig, 1895), p. 339.

18. The fullest account of ether physics is still in Edmund Whittaker, *A History of the Theories of Aether and Electricity* (London, 1910); enlarged 2-vol. edition (London, 1953). See also Max Born, *Die Relativitätstheorie Einsteins* (1st ed., 1920; new ed., 1969). *Einstein's Theory of Relativity* (New York, 1962).

19. The velocity of light is given as 299,792,458 meters per second. Lately this value has been laid down by definition and is no longer measured. With a defined time standard, accurately realizable by atomic clocks, the meter is now a derived unit.

20. Albert Abraham Michelson, *Light Waves and Their Uses* (Chicago, 1907), p. 159.

21. *Späte Jahre*, p. 228.

22. According to Lorentz's own account in *Electromagnetic Phenomena*, in *Proceedings of the Academy of Science, Amsterdam*, Vol. 6, 1904, p. 810.

23. Henri Poincaré, *La Théorie de Lorentz et le principe de la réaction* in Johannes Bosscha (ed.), *Recueil de travaux offerts par les auteurs à H. A. Lorentz* (The Hague, 1900). Poincaré's contribution, pp. 252–78; the synchronization method is described on p. 272.

24. Henri Poincaré, *L'État actuel et l'avenir de la physique mathématique*, in *Bulletin des sciences mathématiques*, Vol. 28, 1904, pp. 302–24. German translation in *Physikalische Blätter*, Vol. 15, 1959, pp. 145–49 and pp. 193–201.

25. Ibid., p. 196 of the German translation.

26. Ibid., pp. 199f.

27. Henri Poincaré, *Sur la Dynamique de l'électron* in *Comptes rendus*, Vol. 140, 1905, pp. 1504–1508. An extensive version appeared in *Rendiconti del Circolo Matematico di Palermo*, a journal that was somewhat out of the way, but often chosen for this topic: *Rendiconti*, Vol. 21, 1906, pp. 129–175.

28. The paper, *Zur Elektrodynamik bewegter Körper*, does have four footnotes, but these simply provide more precision than the text. There are no bibliographical citations.

29. Schilpp, p. 1.

30. I. B. Cohen, *An Interview with Einstein*, in *Scientific American*, Vol. 93, July 1955, pp. 69–73.

31. In Schilpp, p. 20.

32. Ibid.

33. R. S. Shankland, *Conversations with Albert Einstein*, in *American Journal of Physics*, Vol. 31, 1963, p. 48.

34. In Schilpp, p. 6.

35. To Mileva Marić, Mettmenstetten, probably August 10, 1899, CP1, p. 225. Heinrich Hertz's theory used the concept of an ether participating in the motion of matter.

36. To Mileva Marić, Mettmenstetten, September 10, 1899, CP1, p. 230.

37. See the Kyoto lecture, Einstein 1922j; also Reiser, p. 52.

38. To Mileva Marić, Milan, September 28, 1899, CP1, 233.

39. To Mileva Marić, Milan, March 27, 1901, CP1, p. 282.

40. To M. Grossmann, Winterthur, early September 1901, CP1, p. 316.

41. To Mileva Marić, Schaffhausen, December 17, 1901, CP1, p. 325.

42. Ibid.

43. To Mileva Marić, Schaffhausen, December 19, 1901, CP1, p. 328.

44. To Mileva Marić, Schaffhausen, December 28, 1901, CP1, p. 330.

45. Ibid.

46. To M. Besso, Bern, January 1903, in Besso, p. 4.

47. To Mileva Marić, Mettmenstetten, August 10, 1899, CP1, p. 227.

48. Solovine, p. VIII.

49. Henri Poincaré, *La Science et l'hypothèse*.

50. Joseph Sauter, *Zur Interpretation der Maxwell'schen Gleichungen des elektromagnetischen Feldes in ruhenden isotropen Medien*, in AdP, Vol. 6, 1901, pp. 331–38.

51. Joseph Sauter, *Comment j'ai appris à connaître Einstein*, in Flückiger, p. 154.

52. To Carl Seelig, Princeton, February 19, 1955.

53. Ibid.

54. To J. Stark, Bern, September 25, 1907, in Stark, p. 269.

55. That Einstein was certainly able to use the City Library emerges from a letter to Gruner of February 11, 1908; Flückiger, p. 117.

56. Max Abraham, *Prinzipien der Dynamik des Elektrons*, in AdP, Vol. 10, 1903, pp. 105–179.

57. Max Abraham, *Zur Theorie der Strahlung und des Strahlungsdrucks*, in

AdP, Vol. 14, 1904, pp. 236–87. In §8 of his paper on relativity, Einstein calculated the radiation pressure on a mirror, obtaining a term which, in first approximation, was "in agreement with experience and with other theories." The only "other theory" was that of Abraham, who had in fact obtained the same result by other methods.

58. Wilhelm Wien, *Differentialgleichungen der Elektrodynamik für bewegte Körper*, in AdP, Vol. 13, 1904, pp. 641–62; Part II, pp. 663–68.

59. Wilhelm Wien, *Erwiderung auf die Kritik des Hrn. Abraham*, in AdP, Vol. 14, 1904, pp. 635–37.

60. Emil Cohn, *Über die Gleichungen des elektromagnetischen Feldes für bewegte Körper*, in AdP, Vol. 7, 1904, pp. 29–56.

61. Walter Kaufmann, *Die elektromagnetische Masse des Elektrons*, in PhZ, Vol. 4, 1902, pp. 54–57. Kaufmann published further papers on this subject in other journals, notably in *Göttinger Nachrichten*.

62. W. Wien, *Differentialgleichungen der Elektrodynamik für bewegte Körper*, in AdP, Vol. 13, 1904, p. 662.

63. To Robert W. Lawson, Berlin, January 22, 1920.

64. *A Brief Outline of the Development of the Theory of Relativity*, in *Nature*, Vol. 106, February 17, 1921, pp. 782–84.

65. The original is in the Pierpont Morgan Library in New York; a copy is in EA under file ref. 2–070.

66. Frank, p. 38, and Reiser, p. 49, agree that Einstein had read August Föppl's *Einführung in die Maxwell'sche Theorie der Elektricität* (Leipzig, 1894) as a student.

67. M. Besso to Einstein, Geneva, August 3, 1952, in Besso, p. 478.

68. Draft of the *Nature* article, beginning of 1920; emphasis is Einstein's.

69. To Paul Ehrenfest, Prague, April 25, 1912.

70. To Erika Oppenheim, Berlin, September 13, 1932.

71. In Schilpp, pp. 19f.

72. Ibid., p. 20.

73. Aurelius Augustinus, *Confessiones*, Lib. 11.

74. Isaac Newton, *Principia Mathematica*.

75. Unpublished manuscript: draft of a popular article from the 1920s, EA 2–069, p. 1.

76. Ernst Mach, *Die Mechanik in ihrer Entwicklung*, Chapter 2.6.2.

77. Henri Poincaré, *La Science et l'hypothèse*.

78. Henri Poincaré, *La Mesure du temps*, in *Revue de metaphysique et de morale*, Vol. 6, January 1898, pp. 1–13; also in *La Valeur de la science* (Paris, 1905).

79. Henri Poincaré, *La Mesure du temps*.

80. Henri Poincaré, *L'État actuel et l'avenir de la physique mathématique*, lecture in St. Louis, 1904.

81. Einstein 1907h.

82. Ibid., p. 413.

83. Draft for *Nature*, January 1920, footnote on p. 20.

84. Sound recording on disk of February 6, 1924, for the collection of voice records of the Prussian State Library, Berlin; transcription by Friedrich Herneck, *Über die deutsche Reichsangehörigkeit Albert Einsteins*, in NW, No. 2, 1961, p. 104.

85. Pais, p. 161.

86. Kyoto lecture (see n. 1).
87. In Einstein 1906d, p. 627.
88. H. Poincaré (see n. 80).
89. H. Poincaré, *La Valeur de la science*.
90. Joseph Sauter, *Comment j'ai appris à connaître Einstein*, in Flückiger, p. 158.

10. The Theory of Relativity: "A Modification of the Theory of Space and Time"

1. To Conrad Habicht, Bern, end of May or beginning of June 1905.
2. Einstein 1905d, p. 891.
3. An expression from his unpublished draft for *Nature*.
4. Einstein 1905d, p. 891.
5. Ibid. In his papers on relativity theory Einstein until 1907 denotes the velocity of light in the rather old-fashioned way, *V*, in line with Lorentz's *Versuch* of 1895. By 1905, however, the modern designation, *c*, was already customary, and I am therefore using it here.
6. Einstein 1905d, p. 892.
7. Einstein 1907h, p. 416.
8. The independence of the velocity of light of its source was first proved in 1913 by Willem de Sitter by means of precise measurements on binary stars.
9. Einstein 1905d, p. 892.
10. Schilpp, p. 20.
11. It soon became customary to derive the Lorentz transformations from the invariance of the wave equation, and from them in turn to derive the kinematic effects. This is how Einstein proceeded in his *Jahrbuch* article, and how Laue proceeded in the first textbook. There is, of course, nothing wrong with this procedure. But it starts with a wave equation to which a good deal of "theory" attaches, and therefore no longer makes it clear that kinematic effects as well as the Lorentz transformations derive directly from Einstein's principles.
12. Still to be recommended: Einstein 1917a; and Max Born, *Die Relativitätstheorie Einsteins* (1920, and many later editions). *Einstein's Theory of Relativity* (New York, 1962).
13. Einstein 1905d, p. 892.
14. Ibid., p. 893.
15. Ibid.
16. Ibid., p. 894.
17. This is due to the transitivity of the synchronism defined by Einstein: if *A* is synchronous with *B*, then *B* is also synchronous with *A*. If *A* is synchronous with *B* and *C*, then *B* is also synchronous with *C*.
18. Einstein 1905d, p. 895.
19. Ibid., p. 896.
20. Ibid.
21. Einstein 1907h, p. 417.
22. Einstein 1905d, p. 896.
23. H. Poincaré, *Sur la Dynamique de l'électron*, in *Comptes rendus*, Vol. 140, 1905, p. 1504. It should be noted that mathematically equivalent formulas had been derived from the wave equation by W. Voigt as early as 1877, but this attracted little attention. As Voigt's physical context was totally different, the name given by Poincaré has become accepted.
24. To C. Seelig, Princeton, February 19, 1955.
25. This concerns the implicitly performed factoring of $a(v) = P(v)(1 - v^2/C^2)$ on p. 285, which can be understood only in light of the result to be derived.

26. Mathematically, this means the invariance of the expression $x^2 + y^2 + z^2 = c^2t^2$ under Lorentz transformations.

27. H. Minkowski, *Raum und Zeit*, lecture at the meeting of German Scientists and Physicists in Cologne on September 21, 1908, in H. A. Lorentz, A. Einstein, H. Minkowski, *Das Relativitätsprinzip. Eine Sammlung von Abhandlungen*. With a contribution by H. Weyl and notes by A. Sommerfeld. Preface by O. Blumenthal (1913), reprinted (Stuttgart, 1982), p. 54.

28. Einstein 1905d, p. 903.

29. Ibid., pp. 904f. The expression here given for the time dilatation is obtained from the exact formula in good approximation if magnitudes of fourth and higher order are disregarded.

30. Ibid., p. 905. Einstein specifically refers to a "balance-wheel clock," since a pendulum clock is part of a physical system to which the Earth necessarily belongs and thus had to be excluded. (See the footnote, of uncertain origin, in the reprint of 1913, ed. by O. Blumenthal; see n. 27.)

31. Einstein 1911c, p. 13.

32. Paul Langevin, *L'Évolution de l'espace et du temps*, in *Scientia*, Vol. 10, 1911, pp. 31–54.

33. When clocks are carried around the globe in aircraft, the sum of two effects is in fact measured—the time dilatation according to the special relativity theory and the slowing down of clocks in the gravitational field according to the general theory of relativity.

34. Einstein 1905d, p. 903.

35. Ibid., p. 907.

36. Ibid.

37. Einstein still uses Heinrich Hertz's component notation; Lorentz and most of the authors in *Annalen* had by then gone over to the vector notation in use today.

38. Einstein 1905d, p. 910.

39. Ibid., p. 915.

40. Max Abraham, *Zur Theorie der Strahlung und des Strahlungsdrucks*, in AdP, Vol. 14, 1904, pp. 236–87.

41. Einstein 1905d, p. 915.

42. Einstein considers a "material point furnished with an electric charge *e*," which purely for convenience he calls an "electron." Naturally, his results are valid for any other charged particle, and some important partial results are valid even for uncharged "ponderable" matter.

43. Einstein defines "force" as it is defined in school—mass times acceleration. Planck's definition as a derivation of momentum from time is more appropriate for a mechanics with variable masses, as the theorems of the conservation of momentum and energy thereby acquire a simple form. See M. Planck, *Das Prinzip der Relativität und die Grundgleichungen der Mechanik*, in VhDPG, Vol. 8, 1906, pp. 136–41.

44. Einstein 1905d, p. 920.

45. Ibid., p. 921.

46. To Conrad Habicht, Bern, August-September 1905, in Seelig, p. 126.

47. Hasenöhrl, *Zur Theorie der Strahlung in bewegten Körpern*, in AdP, Vol. 15, 1904, pp. 344–77.

48. Einstein 1905e, p. 641.

49. Ibid.; symbols have been adjusted to later usage.

50. J. Precht, in AdP, Vol. 21, 1906, p. 599.

51. Einstein 1906d, p. 633.

52. Einstein 1907f.

53. Einstein 1907h, §11.

54. W. Braunbek, *Die empirische Bestimmung des Masse-Energie-Verhältnisses*, in ZfP, Vol. 107, pp. 1–11; here, p. 1. In this article Braunbek was not allowed to mention Einstein.

55. Einstein 1907h, p. 443.

11. Acceptance, Opposition, Tributes

1. Max Planck, *Wissenschaftliche Selbstbiographie* (Leipzig, 1948), p. 22.

2. To Ph. Lenard, Bern, November 16, 1905; see also A. Kleinert and C. Schönbeck, *Lenard und Einstein. Ihr Briefwechsel*, in *Gesnerus*, Vol. 35, 1978, pp. 318–33. Lenard's first letter to Einstein is lost.

3. Maja Winteler-Einstein, *Albert Einstein. Beitrag für sein Lebensbild*, p. 23 (see n. 1 to Chapter 1). I assume that Maja Einstein here gets something mixed up with an earlier episode, the more so as in 1905 she was living in Berlin; and she did not move near her brother in Bern until 1907. In 1908 Einstein was indeed disappointed that there was no reaction to his first attempt to generalize relativity theory. Maja Einstein's mention of Planck's request for "elucidation of a few dark points" certainly supports this assumption.

4. W. Kaufmann, *Über die Konstitution des Elektrons*, in SB, 1905, pp. 949–56.

5. To M. Solovine, Bern, May 3, 1906, in Solovine, p. 4.

6. Einstein 1913d, p. 1079.

7. Planck's first extant letter to Einstein is dated July 6, 1907.

8. See the Planck-Wien correspondence. Planck and Wien often discussed editorial problems concerning *Annalen*, STPK.

9. M. Planck, *Wissenschaftliche Selbstbiographie*, 1948, p. 31.

10. M. von Laue to C. Seelig, Fribourg, March 13, 1952, ETH.

11. Jakob Johann Laub to C. Seelig, Fribourg, September 11, 1959, ETH.

12. To Rudolf Ladenburg, Bern, December 20, 1907, in Stargard, Kat. 641, No. 420, March 1988.

13. A. Sommerfeld to H. A. Lorentz, Munich, December 26, 1907, in Lorentz Papers, Microfilm Reel 4, American Institute of Physics. I am obliged to Diana Barkan for drawing my attention to this letter.

14. To Max Born, Berlin, January 27, 1920, in Born, p. 45.

15. M. Planck, *Das Prinzip der Relativität und die Grundgleichungen der Mechanik*, in VhDPG, Vol. 8, 1906, pp. 136–41.

16. In Schilpp, p. 8.

17. Einstein 1905d, p. 921.

18. W. Kaufmann, *Über die Konstitution des Elektrons*, in AdP, Vol. 19, pp. 487–553.

19. Ibid.

20. H. A. Lorentz to Poincaré, Leyden, March 8, 1906.

21. Max Planck, *Die Kaufmannschen Messungen der Ablenkbarkeit von β-Strahlen in ihrer Bedeutung für die Dynamik der Elektronen*, lecture on September 19 in Stuttgart, in PhZ, Vol. 7, 1906, pp. 753–59.

22. W. Kaufmann, *Diskussionsbemerkung*, ibid., p. 760.

23. M. Planck, *Diskussionsbemerkung*, ibid.

24. Einstein 1907h, §10, pp. 436–39.

25. Ibid., p. 439.

26. To Marcel Grossmann, Milan, April 1, 1901.

27. A. H. Bucherer to Einstein, Bonn, September 7, 1908. Bucherer's results were published in *Messungen an Bequerel-Strahlen. Die experimentelle Bestätigung der Lorentz-Einsteinschen Theorie*, in PhZ, Vol. 9, 1908, pp. 755–62. Bucherer's data, however, were by no means as good as he himself believed; thus the controversy was not finally settled until a decade later.

28. A. H. Bucherer to A. Einstein, Bonn, September 9, 1908. Einstein's letters to Bucherer are lost.

29. Einstein 1907d.

30. Contribution to the discussion at the meeting of the Zurich Naturforschende Gesellschaft on January 16, 1911; Vol. 56, No. 4, p. VIII.

31. *Was ist Relativitätstheorie?* in *Weltbild*, p. 127, written toward the end of 1919.

32. Ibid., p. 128.

33. Ibid., p. 127.

34. Schilpp, p. 6.

35. M. Planck, *Die Kaufmannschen Messungen*, p. 756 (see n. 21).

36. Einstein 1911e.

37. For instance, in 1912e and 1913c.

38. Primarily in 1907e.

39. Max Planck, *Wissenschaftliche Selbstbiographie* (Leipzig 1948), p. 32.

40. Minkowski, *Raum und Zeit*, p. 60 (see n. 27 to Chapter 10).

41. To E. Zschimmer, Berlin, September 30, 1921.

42. See Schilpp, p. 12.

43. To C. Habicht, Bern, end of May or beginning of June 1905.

44. Isaac Newton to Robert Hooke, February 5, 1675 or 1676. For more detail, see Robert K. Merton, *On the Shoulders of Giants* (New York, 1965).

45. *Die hauptsächlichen Gedanken der Relativitätstheorie*, unpublished manuscript, c. 1920, EA 2089.

46. *New York Times*, April 4, 1921.

47. *Was ist Relativitätstheorie?* in *Weltbild*, p. 129.

48. *Grundgedanken und Probleme der Relativitätstheorie*. Nobel lecture in Göteborg on July 11, 1923, in *Les Prix Nobel* (Stockholm, 1923), p. 3.

49. Unpublished draft, quoted from M. Klein, *Einstein on Scientific Revolutions*, in *Vistas in Astronomy*, Vol. 17, 1975, p. 113.

50. Max von Laue to Jakob Laub, September 2, 1907, in Gerd Rosen Katalog, No. 35, No. 4578, auction of November 8, 1960.

51. Max Planck, *Acht Vorlesungen über Theoretische Physik, gehalten an der Columbia University in the City of New York im Frühjahr 1909* (Leipzig, 1910), pp. 117f.

52. *Sitzungsberichte der Naturforschenden Gesellschaft in Zürich*, meeting of January 16, 1911, No. 4 (quarterly), 1911.

53. See also H. Poincaré, *Dernières Pensées* (Paris, 1913).

54. To H. Zangger, November 15, 1911.

55. *Geometrie und Erfahrung*, lecture on "Frederick Day" at the Prussian Academy of Sciences in Berlin on January 21, 1921, published in *Weltbild*, p. 122.

56. To A. Sommerfeld, Berlin, January 28, 1922; in Sommerfeld, p. 99, with further data.

57. To M. Besso, Princeton, March 6, 1952, in Besso, p. 464.

58. To Max Born, Princeton, October 12, 1953.

59. To André Mercier, Princeton, November 9, 1953.

60. I. B. Cohen, *An Interview with Einstein*, in *Scientific American*, Vol. 93, July 1955, p. 69. See also Einstein's preface, written in 1953, to Stilman Drake's translation of the *Dialogo* of Galileo Galilei, Einstein 1953f.

61. Pais, p. 171.

62. H. A. Lorentz, *Theory of Electrons*, lectures at Columbia University, New York, Spring 1906 (Leipzig, 1909).

63. H. A. Lorentz, *Alte und neue Fragen der Physik*, in PhZ, Vol. 11, 1910, p. 1236.

64. Max Born, in Born, p. 72.

65. H. A. Lorentz, *Das Relativitätsprinzip* (Leipzig, 1914), p. 23.

66. Einstein 1914o.

67. To J. J. Laub, Bern, March 19, 1909.

68. R. A. Millikan, *Albert Einstein on His Seventieth Birthday*, in *Review of Modern Physics*, Vol. 21, 1949, p. 343.

69. M. von Laue, *Das Relativitätsprinzip* (Braunschweig, 1911), p. 14.

70. R. S. Shankland, *Conversations with Albert Einstein*, in *American Journal of Physics*, Vol. 31, 1963, p. 48.

71. Ibid., p. 55.

72. To Mileva Marić, Milan, September 28, 1899, CP1, p. 233; the "interesting paper" was W. Wien, *Über die Fragen, welche die translatorische Bewegung des Lichtäthers betreffen*, in AdP, Vol. 65, Appendix No. 3, pp. I–XVII.

73. To A. Sommerfeld, Bern, January 14, 1908.

74. For instance, in Einstein 1917a.

75. An exception here is the Kyoto lecture of 1922, when Einstein, referring to Michelson's null result, said, "This was the first path which led me to the special theory of relativity." Did J. Ishiwara, who took down the lecture given in German without notes and translated it into Japanese, overinterpret here?

76. To F. H. Davenport, Princeton, February 9, 1954.

77. To C. Seelig, Princeton, February 19, 1955.

78. Einstein 1916i, p. 103.

12. Expert II Class

1. To W. Dällenbach, Ahrenshoop, August 8, 1918.

2. Friedrich Haller to the Swiss Federal Council, Bern, January 17, 1906; Swiss Federal Office for Intellectual Property, Bern.

3. Flückiger, p. 68.

4. To L. Chavan, Prague, c. 1912, quoted in Flückiger, p. 66.

5. See Yardley Beers, *American Journal of Physics*, Vol. 46, 1978, p. 506.

6. Karl Biedermann to Carl Seelig, Bern, March 2, 1952.

7. The former Besenscheuerweg is now Tscharnerstrasse.

8. To M. Solovine, Bern, March 6, 1906, in Solovine, p. 5.

9. Ibid.

10. Ibid.

11. Mileva Einstein to Helene Savić-Kaufler, undated, December 1906, in D. Tribuhović-Gjuvić, *Im Schatten Albert Einsteins* (Bern, 1983), p. 82.

12. To Conrad Habicht, Bern, undated, September 1905, in Seelig, p. 126.

13. Ibid.

14. Mileva Einstein to Helene Savić-Kaufler (see n. 11), p. 82.

15. Einstein's collection of offprints is now at the Weizmann Institute, Rehovoth, Israel.

16. Maja Winteler-Einstein (see n. 1 to Chapter 1), p. 25.

17. Albert Einstein to J. J. Laub: "The photograph has at last arrived." Bern, undated, 1909; Laub to Albert Einstein, July 9, 1953, mentions a photograph of 1906.

18. To W. Wien, Bern, July 29, 1907. Only a few of the Wien-Einstein letters are preserved in the Wien papers of STPK; the subject of their correspondence was mainly the differences between phase and group velocity.

19. Ibid.

20. To J. Stark, Bern, December 14, 1907, in Stark, p. 227.

21. See his refusal of a post in Aachen, offered to him by J. Stark: "As I am a married man, the income from that post would not be large enough," Bern, April 6, 1909, in Stark, p. 278.

22. To the Director of Education, Bern, June 17, 1907, Cantonal Archive Bern, BB III 6, Vol. XV, 1907; see also Flückiger, p. 113.

23. Minutes of faculty meeting of October 28, 1907, Cantonal Archive, Bern.

24. To M. Besso, Princeton, March 6, 1952, in Besso, p. 464.

25. To M. Besso, Bern, March 17, 1903, in Besso, p. 14.

26. To M. Besso, Princeton, March 6, 1952, in Besso, p. 464.

27. B. G. Teubner to Einstein, Leipzig, October 3, 1907.

28. S. Hirzel to Einstein, Leipzig, November 3, 1908.

29. J. Stark, December 2, 1907, in Stark, p. 278.

30. To J. Stark, December 14, 1907, in Stark, p. 227.

31. To Eilhard Wiedemann, Bern, June 14, 1909.

32. Einstein 1917a.

33. Einstein 1922a.

34. To J. Stark, Bern, September 25, 1907, in Stark, p. 269.

35. Einstein 1907h.

36. To J. Stark, Bern, November 1, 1907, in Stark, p. 271.

37. Ibid.

38. Einstein 1907h.

39. Einstein 1923a, pp. 4f.

40. Ibid., p. 5.

41. To W. de Sitter, Berlin, November 4, 1916.

42. In Schilpp, p. 24.

43. See the letter to J. Stark of November 1, 1907; in it, Einstein makes no mention of the theory of gravitation.

44. To Conrad Habicht, Bern, December 24, 1907.

13. The New Copernicus: From "Bad Joke" to "Herr Professor"

1. J. J. Laub to Einstein, Würzburg, March 1, 1908, ETH.

2. To Arnold Sommerfeld, Bern, January 14, 1908, DMM; *Physikalische Blätter*, 40, No. 2, 1984, p. 29 (not in Sommerfeld). Sommerfeld's letters during the first few years of their correspondence are lost.

3. To Conrad Habicht, Bern, December 24, 1907.

4. To Marcel Grossmann, Bern, January 3, 1908, ETH; in Hoffmann/Dukas, p. 104.

5. Ibid.

6. Ibid.

7. To the Educational Council of the Canton of Zurich, Bern, January 20, 1908, STZ, U 84d. 2.

8. To Paul Gruner, Bern, February 11, 1908, in Flückiger, p. 117, in facsimile.

9. Flückiger, p. 119, gives further details of the *Habilitation* procedure.

10. Einstein 1909c, pp. 185–93.

11. To Paul Gruner, Bern, February 11, 1908, in Flückiger, p. 117.

12. Semester report of the University of Bern, SS 1908, in Flückiger, p. 121.

13. Lucien Chavan, Theorienhefte, Einstein-Haus, Bern.

14. Einstein and Laub, *Elektromagnetische Grundgleichungen für bewegte Körper* [*Electromagnetic Fundamental Equations for Moving Bodies*], in AdP, Vol. 26, 1908, pp. 532–40; *Corrections*, Vol. 27, 1908, p. 232.

15. Einstein and Laub, *Die im elektromagnetischen Felde auf ruhende Körper ausgeübten ponderomotorischen Kräfte* [*The Ponderomotoric Forces Exerted in the Electromagnetic Field on Bodies at Rest*], in AdP, Vol. 26, 1908, pp. 541–50; *Bemerkungen zu unserer Arbeit . . .* [*Observations on our Work . . .*], in AdP, Vol. 28, 1909, pp. 445ff.

16. To J. J. Laub, Bern, undated, probably Spring 1909; Katalog Gerd Rosen, Berlin, Auction No. 36, April 1961, p. 3.

17. To Walter Dällenbach, Ahrenshoop, August 8, 1918, quoted in Flückiger, p. 160. These problems have not been definitively solved to this day (personal communication from Jürgen Renn).

18. J. J. Laub to Einstein, Würzburg, May 18, 1908, ETH.

19. Einstein 1907b.

20. Einstein 1908a.

21. To Conrad Habicht, Bern, December 24, 1907.

22. Ibid.

23. To Arnold Sommerfeld, Bern, January 14, 1908, in *Physikalische Blätter*, 40, 1984, No. 2, p. 30 (not in Sommerfeld).

24. Einstein 1908a, p. 217.

25. J. J. Laub, *Albert Einstein und Albert Gockel*, in *Academia Freiburgensis*, 1962, pp. 30–33.

26. To J. J. Laub, Bern, July 30, 1908, ETH.

27. To J. J. Laub, undated, August 1908, in *Albert Einstein und Albert Gockel* (see n. 25), p. 31.

28. To Albert Gockel, Bern, December 3, 1908, STPK, Autogr. I/III.

29. Ibid.

30. Ibid.

31. H. A. Einstein in an interview with Bernard Mayes, BBC, in G. J. Whitrow (ed.), *Einstein: The Man and His Achievement* (London, 1966), p. 17.

32. To M. Solovine, Bern, December 3, 1908; Hans Albert's pet name "Buio" is Swiss for *Bub* (boy), in Solovine, p. 10, in facsimile.

33. Whitrow (see n. 31), p. 21.

34. See also a postcard from Einstein to the Habicht brothers from summer 1907 (undated): "Now in the course of July I could put both of you up because my sister is away."

35. Max Planck to Einstein, Axalp, September 8, 1908.

36. To J. Stark, Bern, February 22, 1908, in Stark, p. 274.

37. To J. Stark, Bern, December 14, 1908, in Stark, p. 277.

38. Hermann Minkowski, *Raum und Zeit* (see n. 27 to Chapter 10). For the genesis of Minkowski's concept of space-time see also Peter Louis Galison, *Minkowski's Space-Time: From Visual Thinking to the Absolute World*, in HSPS, 10, 1979, pp. 85ff.

39. Max Born, *Physik im Wandel meiner Zeit* (Braunschweig, 1958), p. 218.

40. Max Born, *Autobiography* (New York, 1978), p. 131.

41. Hermann Minkowski (see n. 38), p. 111.

42. J. J. Laub to Einstein, Würzburg, May 18, 1908, ETH.

43. Pais, p. 151.

44. Seelig, p. 46.

45. Einstein 1917d; quoted from Braunschweig, 1969 edition, p. 46.

46. Ibid., p. 48.

47. Louis Kollros, *Erinnerungen eines Kommilitonen*, in *Helle Zeit—Dunkle Zeit*, p. 25.

48. On the history of theoretical physics in the German-speaking countries, see the unique book by Christa Jungnickel and Russel McCormmach, *Intellectual Mastery of Nature—Theoretical Physics from Ohm to Einstein*. Vol. 2, *The New Mighty Theoretical Physics 1870–1925* (Chicago, 1986).

49. Rudolf G. Ardelt, *Friedrich Adler* (Vienna, 1984), pp. 156–94, gives a detailed account of Friedrich Adler's ambivalent attitude toward the Zurich professorship as well as the background of the establishment of the post.

50. Ibid., pp. 159ff.

51. Friedrich Adler to Victor Adler, Zurich, June 19, 1908, in ibid., pp. 163f.

52. To J. J. Laub, Bern, March 19, 1909, in ETH.

53. To M. Besso, Princeton, March 6, 1952, in Besso, p. 464.

54. Friedrich Adler to Victor Adler, Zurich, July 1, 1908, in Rudolf G. Ardelt, p. 164.

55. To. J. J. Laub, Bern, July 30, 1908, ETH.

56. To J. J. Laub, Bern, May 19, 1909, ETH.

57. Friedrich Adler to Victor Adler, Zurich, November 28, 1908, in R. G. Ardelt, p. 165.

58. To J. Ehrat, Bern, February 15, 1909, in Seelig, p. 155.

59. To J. J. Laub, Bern, May 19, 1909, ETH. This letter contains a detailed account of the professorship episode from Einstein's point of view.

60. To J. Ehrat, Bern, February 15, 1909, in Seelig, p. 155.

61. To A. Kleiner, Bern, February 25, 1909, in Universitätsarchiv Zurich.

62. Expert opinion by Alfred Kleiner, quoted in a letter of the Dean, Professor Otto Stoll, of March 4, 1909, to *Regierungsrat* Heinrich Ernst, Director of Cantonal Education in Zurich, pp. 7f., STZ, U.110b. 2, Einstein's personal file.

63. Ibid., p. 6.

64. The exceptionally gifted Walter Ritz died four months later, on July 7, 1909, at the age of only thirty-one.

65. For the reactions of heads of university departments and ministries to the appointment of Jewish physicists, see Jungnickel and McCormmach, pp. 286f (see n. 48).

66. See n. 62 above, p. 8.

67. Ibid., p. 9.

68. To J. J. Laub, Bern, May 19, 1909, ETH.

69. This emerges from the pro-

tocol of the Government Council of July 14, 1910, STZ.

70. To J. J. Laub, Bern, May 19, 1909, ETH.

71. To Carl Seelig, Princeton, Summer 1952, in Seelig, pp. 157ff.

72. Quoted from Seelig, p. 159.

73. Justice and Police Department to Swiss Federal Council, Bern, July 12, 1909, in Flückiger, p. 70.

74. To M. Solovine, Bern, March 18, 1909; in Solovine, p. 12.

75. To J. J. Laub, Bern, Monday, undated, probably June 1909 (not 1908 as assumed by Seelig, p. 147), ETH.

76. To H. A. Lorentz, Bern, undated, probably April 1909.

77. To J. J. Laub, Bern, Monday, undated, probably June 1909.

78. To J. J. Laub, Bern, May 19, 1909, ETH. Lorentz's early letters to Einstein are lost. On May 23, 1909, Einstein wrote Lorentz a seventeen-page letter on problems of radiation theory.

79. Ibid.

80. To J. Stark, Bern, July 31, 1909, in Stark, p. 279.

81. According to the report on the congress (*Verhandlungen der Gesellschaft Deutscher Naturforscher und Ärzte, 81. Versammlung in Salzburg,* Leipzig 1910) and the program of the "well attended" physical section (PhZ, 10, November 10, 1909, No. 22, p. 777), those present included Max Born, Paul S. Epstein, James Franck, Philipp Frank, Albert Gockel, Otto Hahn, Friedrich Hasenöhrl, Ludwig Hopf (later an assistant at Zurich University), Rudolf Ladenburg, Anton Lampa, Max von Laue, Lise Meitner, Gustav Mie, Max Planck, Heinrich Rubens, Clemens Schaefer, Arnold Sommerfeld, Johannes Stark, Wilhelm Wie, and Woldemar Voigt.

82. PhZ, 10, No. 22, p. 777.

83. The invitation is thought to have been brought by Rudolf Ladenburg on a visit to Bern in the summer of 1908 (Clark, p. 93); I have found no evidence of this. Nor do the extant fragments of the correspondence between Planck and Einstein contain anything about an invitation for a lecture. However, since Planck presided over the afternoon meeting on September 21 and was therefore responsible for its program, it seems very likely that the suggestion that Einstein should give one of the keynote addresses did come from Planck.

84. Max Born, *Physik im Wandel meiner Zeit* (Braunschweig, 1958), p. 193.

85. Einstein 1909e.

86. Ibid., p. 820.

87. Ibid., p. 817.

88. Ibid., p. 820.

89. Ibid.

90. Max Planck's contribution to the discussion, ibid., p. 825.

91. Max Born, *Die Relativitätstheorie Einsteins* (Berlin, 1920), p. 238 (only in this first edition).

92. Wolfgang Pauli, *Einsteins Beitrag zur Quantentheorie,* in Schilpp, p. 78.

14. Professor in Zurich

1. To Lucien Chavan, Zurich, October 19, 1909.

2. Friedrich Adler to Victor Adler, Zurich, October 28, 1909, in R. G. Ardelt, p. 166 (see n. 49 to Chapter 13).

3. To J. J. Laub, Zurich, December 31, 1909, ETH.

4. Ibid.

5. To M. Besso, Zurich, November 17, 1909, in Besso, p. 16.

6. To A. Sommerfeld, Zurich, January 19, 1910, DMM, *Physikalische Blätter*, 40, No. 2, 1984, p. 32 (not in Sommerfeld).

7. To Eilhard Wiedemann, Bern, July 14, 1909, DMM.

8. Hans Tanner, in Seelig, p. 171.

9. Max Fisch, in Seelig, p. 170.

10. To M. Besso (see n. 5).

11. To J. J. Laub (see n. 3).

12. To J. J. Laub, Zurich, March 16, 1910, ETH.

13. Ibid.

14. Seelig, p. 188.

15. To C. Seelig, Princeton, August 20, 1952, ETH.

16. Einstein 1911b. This paper begins: "My colleague Professor Zangger drew my attention to an important remark . . ."

17. Hans Tanner, in Seelig, p. 170.

18. George Hevesy, interview with T. S. Kuhn and E. Segre of May 25, 1962, in *Sources for History of Quantum Physics;* quoted from T. S. Kuhn, *Black-Body Theory and the Quantum Discontinuity*, p. 215 (Oxford, 1978).

19. To J. J. Laub, Zurich, March 16, 1910, ETH.

20. Following obvious modifications of Einstein's theory, Peter Debye (AdP, Vol. 39, p. 789), as well as Max Born and Theodor von Karman (PhZ, Vol. 13, p. 297), developed formulas which correctly describe the course of specific heat during approach to absolute zero.

21. In Seelig, p. 177.

22. To A. Sommerfeld, Zurich, January 19, 1910, DMM; in *Physikalische Blätter*, 40, 1984, p. 32 (not in Sommerfeld).

23. To J. J. Laub, Zurich, March 16, 1909, ETH.

24. To Emil Fischer, Zurich, November 5, 1910, STPK.

25. To J. J. Laub, Zurich, December 31, 1909, ETH.

26. To J. J. Laub, Zurich, March 16, 1910, ETH.

27. To A. Sommerfeld, Zurich, July 10, 1910, DMM; in *Physikalische Blätter*, 40, 1984, p. 33 (not in Sommerfeld).

28. Ibid.

29. Armin Hermann, in Sommerfeld, p. 23.

30. To J. J. Laub, Zurich, Saturday, undated, July 1910, ETH.

31. Ibid.

32. Einstein 1910a and 1910b.

33. W. Pauli, *Einsteins Beitrag zur Quantentheorie*, in Schilpp, p. 79.

34. To J. J. Laub, Zurich, November 4, 1910, ETH.

35. To J. J. Laub, Zurich, November 11, 1910, ETH.

36. See the letters to Sommerfeld of January 19, 1910; and to Laub of summer 1910, undated.

37. Seelig, p. 101; the lecture was given on May 7, 1910. In a letter to his mother of April 29, 1910, Einstein says: "In a few days I give a lecture to the convention of Swiss scientists, for which I haven't prepared at all."

38. Carl Gustav Jung to Carl Seelig, February 25, 1952, ETH.

39. Ibid.

40. To Conrad Habicht, Zurich, December 14, 1910.

41. To Conrad Habicht, Zurich, undated, February-March 1910.

42. C. Habicht and P. Habicht, *Elektrostatischer Potentialmultiplikator nach A. Einstein*, in PhZ, 11, 1910, pp. 532–35.

43. Ibid., p. 535.

44. To M. Besso, Prague, February 4, 1912, in Besso, p. 47.

45. To M. Besso, Prague, December 26, 1911, in Besso, p. 42.

46. To Conrad Habicht, Princeton, August 15, 1948.

47. *Marian von Smoluchowski*, AdP, Vol. 25, 1908, pp. 205–26. See Armin Teske, *Marian Smoluchowski—Leben und Werk*, Polska Akademia Nauk (Warsaw, 1977), pp. 215–32.

48. *Marian von Smoluchowski*, in NW, Vol. 5, 1917, p. 737.

49. To J. J. Laub, Zurich, Saturday, undated, Summer 1910, ETH.

50. Einstein 1910c.

51. Ibid., p. 1294.

52. To M. Grossmann, Milan, April 14, 1901, in CP1, Doc. 100, p. 290.

53. Quoted in Teske (see n. 47), p. 219.

54. Ibid., p. 231.

55. Einstein 1917g.

56. His invitation to Prague is described in the following accounts:

(a) Andreas Kleinert, *Anton Lampa und Albert Einstein—Die Neubesetzung der physikalischen Lehrstühle an der deutschen Universität Prag*, in *Gesnerus*, 32, 1975, pp. 285–92. (b) Jan Havranek, *Albert Einstein's Appointment to Professor in Prague*, in *Acta Universitatis Carolinae—Hist. Univ. Car. Pragensis*, 8, 1977, pp. 109–12. (c) Jozsef Illy, *Einstein in Prague*, in *Isis*, 70, 1979, pp. 76–84.

57. Friedrich Adler to Victor Adler, Zurich, March 30, 1910, in Adler-Archiv, Vienna.

58. To Pauline Einstein, Zurich, April 29, 1910.

59. Friedrich Adler to Victor Adler, Zurich, April 29, 1910, in Adler-Archiv, Vienna.

60. Illy, p. 77 (see n. 56).

61. This expertise is lost; Andreas Kleinert has reconstructed it from oral tradition (Kleinert, p. 288 and n. 9, p. 291; see n. 56). This reconstruction is in line with Planck's form of words three years later, when he proposed Einstein's election to the Prussian Academy of Sciences.

62. Max Planck, *Acht Vorlesungen über Theoretische Physik, gehalten an der Columbia University in the City of New York im Frühjahr 1909* (Leipzig, 1910).

63. To A. Sommerfeld, Zurich, July 10, 1910, in *Physikalische Blätter*, 40, 1984, p. 33 (not in Sommerfeld).

64. To. J. J. Laub, Zurich, Saturday, undated, summer 1910.

65. Kleinert, p. 288 (see n. 56).

66. Count Karl Stürgkh to Emperor Francis Joseph, Vienna, December 16, 1910; quoted in Kleinert, p. 289.

67. Petititon of fifteen students to the Directorate of Education, Zurich, June 23, 1910, STZ, U 110b.2 (44).

68. Ibid.

69. Protocol of Governmental Council, July 14, 1910, STZ, U 110b.2.

70. Ibid.

71. Frank, p. 136.

72. Illy, p. 77 (see n. 56).

73. Friedrich Adler to Victor Adler, Zurich, September 23, 1910, in Adler-Archiv, Vienna.

74. Illy, p. 78 (see n. 56).

75. Frank, p. 137; Hoffmann/Dukas, p. 114.

76. To Paul Ehrenfest, Prague, April 25, 1912.

77. To Carl Schröter, Zurich,

December 12, 1910, ETH. Schröter was professor of botany at the Polytechnic.

78. Illy, p. 78 (see n. 56).

79. Protocol of Governmental Council, Zurich, February 10, 1911, STZ.

80. Havranek, p. 106 (see n. 56).

81. A. Kleiner to a colleague (no longer identifiable), Zurich, January 18, 1911, STZ.

82. Emil Fischer to A. Einstein, Berlin, November 1, 1910, in Fischer Papers, Bancroft Library, University of California, Berkeley.

83. To Emil Fischer, Zurich, November 5, 1910, STPK.

84. On Emil Fischer and Franz Oppenheim as treasurer of the German Chemical Society, see Emil Fischer, *Aus meinem Leben* (Berlin, 1922).

85. Franz Oppenheim to Emil Fischer, Berlin-Wannsee, November 9, 1911, in Fischer Papers, Bancroft Library, Berkeley.

86. To Emil Fischer, Zurich, November 5, 1910, STPK.

87. To H. A. Lorentz, Zurich, January 27, 1911.

88. To Friedrich Adler, Basel, February 8, 1911, ETH.

89. To H. A. Lorentz, Zurich, February 15, 1911.

90. To H. A. Lorentz, Prague, November 23, 1911.

91. Graveside oration for H. A. Lorentz, 1928, in *Weltbild*, p. 27.

92. *H. A. Lorentz als Schöpfer und Persönlichkeit*, written in 1953, in *Weltbild*, p. 31.

93. Seelig, pp. 177ff.

94. To A. Stern, Zurich, undated, end 1910, ETH.

15. Full Professor in Prague—But Not for Long

1. To M. Grossmann, Prague, March 1911, undated (postmarked March 27, 1911).

2. The address was Třebízkého ulice 1215, now Lesnická ulice 7; there is a bust on the facade to commemorate its first and most famous resident.

3. See also a letter to Besso, end of September 1911, in Besso, p. 30, and a letter from Besso, October 29, 1911, ibid., p. 37.

4. To H. Zangger, Prague, undated, spring 1911.

5. To M. Grossmann, Prague, end of March 1911.

6. Ibid.

7. To H. Zangger, Prague, undated, spring 1911.

8. To M. Besso, Prague, May 13, 1911, in Besso, p. 19.

9. To L. Chavan, Prague, July 6, 1911, ETH.

10. Otto Stern, in an interview with Res Jost, recorded on November 25, 1961; a transcript is at ETH; I am obliged to Professor Jost for permitting me to quote from it.

11. To M. Besso, Prague, May 13, 1911, in Besso, p. 19.

12. The natural science institutes of the Philosophical Faculty are on Weinberggasse 3, now Viničná ulice 3.

13. To M. Besso, Prague, May 13, 1911, in Besso, p. 19.

14. Frank, p. 145.

15. To M. Grossmann, Prague, end of March 1911, ETH.

16. To A. Stern, Prague, March 17, 1912, ETH.

17. To L. Chavan, Prague, July 6, 1911, ETH.

18. To J. J. Laub. Prague, August 10, 1911, ETH.

19. To M. Besso, Prague, February 4, 1912, in Besso, p. 45.

20. To J. J. Laub, Prague, August 10, 1911, ETH.

21. To H. Zangger, Prague, September 20, 1911.

22. *Prager Tagblatt*, January 22, 1911, quoted from Illy, p. 79 (see n. 56 to Chapter 14).

23. Ibid.

24. Gerhard Kowalewski, *Bestand und Wandel* (Munich, 1950), p. 237.

25. Hugo Bergmann, *Personal Remembrances of Albert Einstein*, in *Boston Studies in the Philosophy of Science*, Vol. XIII, p. 390.

26. Max Brod, *Streitbares Leben* (Munich, 1969), p. 201.

27. Ibid., p. 202.

28. Frank, pp. 152ff.

29. To Hedwig Born, Berlin, September 8, 1916, in Born, p. 21.

30. Frank, p. 143.

31. To M. Besso, Prague, May 13, 1911, in Besso, p. 19.

32. Ibid., p. 20.

33. Ibid.

34. To Walther Nernst, Prague, June 20, 1911, quoted in Jean Pelseneer, *Le premier Conseil de Physique*, unpublished manuscript, Institut Solvay, Brussels.

35. Ernest Solvay, *Sur l'Etablissement des Principes Fondamentaux de la Gravito-Matérialique* (Brussels, 1911).

36. See also Jagdish Mahra, *The Solvay Conference on Physics* (Dordrecht/Boston, 1975), p. 4.

37. Max Planck to Walther Nernst, Berlin, June 11, 1910, in J. Pelseneer (see n. 34).

38. Ibid.

39. To M. Besso, Prague, May 13, 1911, in Besso, p. 19.

40. To M. Besso, Prague, September 11, 1911, in Besso, p. 26.

41. To M. Besso, Prague, October 21, 1911, in Besso, p. 32.

42. Ibid.

43. To H. Zangger, Prague, undated, beginning November 1911, ETH; in *Helle Zeit—Dunkle Zeit*, p. 43.

44. To H. Zangger, Prague, November 16, 1911, ETH; in *Helle Zeit—Dunkle Zeit*, p. 43. The "three languages" were, of course, French, German, and English.

45. To M. Besso, Prague, December 26, 1911, in Besso, p. 40.

46. In *Proceedings of the Conseil Solvay 1911*, German version in *Abhandlungen der Bunsen-Gesellschaft*, No. 7 (Halle, 1914), pp. 339f.

47. To H. Zangger, Prague, November 16, 1911, ETH; in *Helle Zeit—Dunkle Zeit*, p. 43.

48. To M. Besso, Prague, December 26, 1911, in Besso, p. 40.

49. To H. Zangger, Prague, undated, beginning of November 1911, ETH; in *Helle Zeit—Dunkle Zeit*, p. 42.

50. Quoted in the Earl of Birkenhead, *The Professor between Two Worlds* (Cambridge, 1962), p. 42.

51. Clark, p. 185.

52. To H. Zangger, Prague, undated, beginning of November 1911, ETH; in *Helle Zeit—Dunkle Zeit*, p. 43.

53. Details of the correspondence between Einstein and Julius are in Clark, pp. 187ff.

54. To H. Zangger, Prague, September 20, 1911.

55. H. Zangger to Ludwig Forrer, Zurich, October 9, 1911, in Swiss Federal Archive, Bern.

56. To H. Zangger, Prague, September 20, 1911.

57. To H. Zangger, Prague, November 15, 1911.

58. To M. Grossmann, Prague, November 18, 1911, ETH; in Seelig, p. 225. Grossmann's letter, to which the present letter is the answer, is lost.

59. H. A. Lorentz to Einstein, Leyden, December 8, 1911.

60. To M. Grossmann, Prague, December 10, 1911, ETH.

61. M. Grossmann to Einstein, Zurich, December 12, 1911, ETH.

62. H. Zangger to Rudolf Gnehm, Zurich, December 14, 1911, ETH.

63. Marie Curie to Pierre Weiss, Paris, November 17, 1911, ETH; Poincaré to Pierre Weiss, pp. 228ff., in Seelig, p. 228.

64. A. Sommerfeld, quoted from Seelig, p. 230.

65. Seelig, pp. 231f.

66. Minutes of the Swiss Education Council for 1912, meeting of January 22, 1912, in Swiss Federal Archive, Bern.

67. To the Federal President of Switzerland, Prague, February 2, 1912, ETH.

68. To Alfred Stern, Prague, February 2, 1912, ETH. The "bear cubs" are his two sons.

69. To H. Zangger, undated, spring 1912.

70. George Peagram to Einstein, New York, January 9, 1912.

71. To George Peagram, Prague, January 29, 1912.

72. To H. Zangger, Prague, January 27, 1912.

73. To H. Zangger, Prague, February 19, 1912.

74. *In memoriam Paul Ehrenfest* (1934), in *Späte Jahre*, p. 204.

75. Ehrenfest's diaries and notebooks are at the Museum Boerhaave in Leyden. Martin J. Klein made use of them for his—in many respects remarkable—biography, *Paul Ehrenfest*, Vol. 1: *The Making of a Theoretical Physicist* (Amsterdam, 1970); here pp. 174ff. The second volume has not so far been published. The quotation here is from Karl von Mayenn, *Einsteins Dialog mit den Kollegen*, in *Berlin Symposion*, pp. 464ff.

76. To M. Besso, Prague, March 3, 1912 (not in Besso). In Einstein's correspondence there is no reference to "talking shop."

77. W. Nernst to Emil Fischer, Berlin, July 30, 1910, STPK, Darmstaedter Collection F 2e 1898 (5).

78. To Mileva Marić, Milan, March 23, 1901, in CP1, p. 281.

79. Einstein and Mileva lived apart after 1914; they were divorced in 1919.

80. H. A. Einstein in G. J. Whitrow, *Einstein—The Man and His Achievement* (London, 1967), p. 20.

81. To M. Besso, Prague, March 3, 1912 (not in Besso).

82. To M. Besso, Prague, December 21, 1911, in Besso, p. 32.

83. To M. Besso, Zurich, end of February 1914, in Besso, p. 32.

84. To C. Seelig, Princeton, May 5, 1952, ETH.

85. See also a letter to Annelie Meyer-Schmid of December 17, 1926.

86. To C. Seelig, Princeton, April 20, 1952, ETH.

87. To C. Seelig, Princeton, May 5, 1952, ETH.

88. To Elsa Löwenthal, Prague, Tuesday, April 30, 1912. Elsa Löwenthal's letters to Einstein were destroyed by him at her request, presumably to deny Mileva a point of attack. Elsa Löwenthal kept Einstein's letters—of which this is the first—in a tied-up folder; on a card attached to it she had written: "Especially beautiful letters from better days."

89. To Elsa Löwenthal, Prague, May 7, 1912.

90. To Elsa Löwenthal, Prague, May 21, 1912.

91. Otto Stern in an interview with Res Jost (see n. 10).

92. Ibid.

93. Einstein 1912a and 1912d. *Thermodynamische Begründung des photochemischen Äquivalentgesetzes [Thermodynamic Justification of the Law of Photochemical Equivalence]*, in AdP, 37, p. 832, and 38, p. 881 (1912).

94. Johannes Stark, *Über die Anwendung des Planckschen Elementargesetzes auf photochemische Reaktionen. Bemerkung zur Mitteilung des Hr. Einstein [On the Application of Planck's Elementary Law to Photochemical Reactions. Observation on Einstein's Paper]*, in AdP, 38, 1912, p. 467.

95. Einstein 1912d.

96. Havranek, p. 108 (see n. 56 to Chapter 14).

97. Frank, p. 170.

98. Ibid.

99. *Prager Tageblatt* of August 5, 1912, quoted in Illy, p. 84 (see n. 56 to Chapter 14).

100. To C. Habicht, Prague, February 12, 1912 ETH; in Seelig, p. 221.

101. Illy, p. 84.

102. See also John Stachel, *The Genesis of General Relativity*, in *Berlin Symposion*, p. 432.

16. Toward the General Theory of Relativity

1. Einstein in his Kyoto lecture of December 14, 1922, in Einstein 1922j.

2. Ibid.

3. *Dictionary of Scientific Biography*, Vol. 10, p. 61.

4. *Grundgedanken und Methoden der Relativitätstheorie in ihrer Entwicklung dargestellt [Fundamental Ideas and Methods of the Theory of Relativity Presented in Their Development]*, unpublished manuscript of 1920, quoted from Pais, p. 175.

5. Ibid.

6. Henri Poincaré, *Sur la Dynamique de l'électron*, in *Rendiconti del Circolo Matematico di Palermo*, Vol. 21, 1906, pp. 129ff. and 166ff.

7. Hermann Minkowski, *Grundgleichungen der elektromagnetischen Vorgänge in bewegten Körpern*, in *Göttinger Nachrichten*, 1908, pp. 53ff.

8. See also Einstein's letter to J. Stark of September 25, 1907, according to which Einstein was acquainted with only five papers on relativity theory, none of which dealt with gravitation theory; in Stark, p. 269.

9. Einstein 1933d, pp. 135f.

10. Ibid.

11. "Inertial mass" m_i appears as inertial resistance to acceleration in Newton's law of force, his so-called Second Axiom: $K = m_i b$. "Gravitational mass" m_g is that of gravitation $K = m_g M_{g2}/r^2$. Hence it follows for free fall, with $b = M_{g2}/r^2$, that $m_i = m_g$.

12. Einstein 1917d, p. 54.

13. Einstein 1970h, p. 454.

14. Ibid., p. 456.

15. Ibid., p. 458.

16. Ibid., p. 459.
17. Ibid., p. 461.
18. Ibid., p. 462.
19. J. Stark to Einstein, Greifswald, October 4, 1907, in Stark, p. 270.
20. To J. Stark, Bern, November 1, 1907, in Stark, p. 271.
21. Einstein 1933d, p. 136.
22. This emerges from the "Corrections" written toward the end of February 1908, in *Jahrbuch*, Vol. 5, pp. 98f.
23. To Conrad Habicht, Bern, December 24, 1907, in ETH and Seelig, pp. 127f.
24. *Helle Zeit—Dunkle Zeit*, p. 14.
25. Einstein 1911g, p. 898.
26. Ibid., p. 906.
27. Ibid., p. 908.
28. To E. Freundlich, Prague, September 1, 1911.
29. To E. Freundlich, Prague, January 8, 1912.
30. To E. Freundlich, Prague, September 1, 1911.
31. To J. J. Laub, Prague, August 10, 1911, ETH.
32. To L. Hopf, undated, Prague, December 1911 or January 1912.
33. Einstein 1912b and 1912c.
34. To M. Besso, Prague, March 26, 1912 (not in Besso).
35. F. A. Lindemann to his father, Brussels, November 4, 1911; reprinted in the Earl of Birkenhead, *The Professor between Two Worlds* (Cambridge, 1962), p. 42.
36. Quoted in Constance Reid, *Hilbert* (Berlin, 1970), p. 112.
37. Interview typescript, p. 11, ETH.
38. See also Hugo Bergmann, *Personal Remembrances of Albert Einstein* in *Boston Studies*, Vol. XIII, p. 389; and Gerhard Kowalewski, *Bestand und Wandel* (Munich, 1950), p. 238.
39. Max von Laue, *Das Relativitätsprinzip* (Braunschweig, 1911).
40. To Alfred Kleiner, Prague, April 1, 1912, ETH.
41. AdP, 38, 1912, p. 356.
42. Kyoto lecture, December 14, 1922, Einstein 1922j.
43. To L. Hopf, Prague, June 13, 1912.
44. To L. Hopf, Zurich, August 16, 1912.
45. Louis Kollros, *Erinnerungen eines Kommilitonen*, in *Helle Zeit—Dunkle Zeit*, p. 27.
46. *Skizze*, p. 15.
47. See also Einstein's account in the Kyoto lecture, Einstein 1922j, p. 47; and the accounts of Pais and Straus, in Pais, p. 213.
48. To A. Sommerfeld, Zurich, October 29, 1912, in Sommerfeld, p. 26.
49. A. Sommerfeld to D. Hilbert, Munich, November 1, 1912, in Sommerfeld, p. 27.
50. *Skizze*, p. 16.
51. To P. Ehrenfest, Zurich, May 28, 1913.
52. To Ernst Mach, Zurich, undated, summer 1913.
53. To P. Ehrenfest, Zurich, May 28, 1912.
54. Einstein 1913c.
55. John Norton, *How Einstein Found His Field Equations: 1912–1915*, in HSPS, Vol. 14, Part 2, 1984, pp. 253–315.
56. Einstein 1933d, in *Weltbild*, p. 138.
57. Max Born, *Besprechung von "Entwurf einer verallgemeinerten Relativitätstheorie und einer Theorie der*

Gravitation," in NW, Vol. 2, 1914, pp. 448f.

58. To H. A. Lorentz, Zurich, August 14, 1913.

59. Einstein 1913f.

60. Einstein 1913g.

61. Ibid., p. 1251.

62. Ibid., p. 1255.

63. Ibid., p. 1262.

64. Conference report in PhZ, Vol. 14, 1913, p. 1073.

65. Ibid.

66. Discussion on Einstein's lecture, ibid., p. 1263.

67. To E. Freundlich, Zurich, undated, January 1914.

68. To M. Besso, Zurich, undated, end of 1913.

69. PhZ, Vol. 14, 1913, p. 1262.

70. To E. Freundlich, Zurich, undated, August 1913.

71. To E. Freundlich, Zurich, undated, 1913.

72. To E. Freundlich, Zurich, undated, August 1913.

73. To George Hale, Zurich, October 14, 1913.

74. To E. Freundlich, Zurich, December 7, 1913.

75. Ibid.

76. To E. Freundlich, Zurich, undated, beginning of 1914.

77. Einstein 1914i.

78. To M. Besso, Zurich, undated, beginning of March 1914, in Besso, p. 52.

79. Einstein 1914a.

80. To H. Zangger, Zurich, March 10, 1914.

81. To M. Besso, Zurich, undated, beginning of March 1914, in Besso, p. 53.

82. To H. Zangger, Zurich, March 10, 1914.

17. From Zurich to Berlin

1. The address was Hofstrasse 116. The details of the apartment are taken from a tenancy dispute, Zurich District Court, B XII Zch 63 14 32. STZ.

2. To L. Hopf, Prague, June 12, 1912.

3. To A. Kleiner, Prague, April 10, 1912.

4. To M. von Laue, Prague, June 10, 1912.

5. To L. Hopf, Prague, June 12, 1912.

6. This tradition continued. Because of the war, the post of "extraordinary" professor at first remained vacant after Laue's departure. In 1917 it was upgraded to an ordinary professorship for Ernst Schrödinger, to ensure that he would stay longer than his predecessor. Schrödinger received the Nobel Prize in 1933.

7. Otto Stern, typescript, p. 11, ETH.

8. Ibid.

9. M. von Laue to C. Seelig, Berlin, March 13, 1952, ETH.

10. To M. Besso, Zurich, undated, end of 1913, in Besso, p. 50.

11. To M. Solovine, Zurich, undated, March 1913, in Solovine, p. 14.

12. Einstein 1913b.

13. Einstein 1913c, p. 226.

14. To Marie Curie, Zurich, April 3, 1913.

15. Quoted in Eve Curie, *Madame Curie;* English translation by Vincent Sheean (London/Toronto, 1938), p. 296.

16. Quoted in Martin J. Klein, *Paul Ehrenfest* (Amsterdam, 1972), p. 295.

17. M. von Laue to Carl Seelig, Berlin, March 13, 1952, ETH.

18. Otto Stern interview, typescript, p. 12, ETH.

19. Jagdish Mehra, *The Solvay Conference on Physics* (Boston, 1975), p. XXIII.

20. F. Haber to Hugo Krüss, Prussian Ministry of Education, Pontresina, January 4, 1913.

21. See Berlin, pp. 7f; and Documents 1–6, pp. 96–100.

22. Berlin, p. 96.

23. Ibid., p. 97.

24. Ibid., p. 99.

25. Ibid., p. 97.

26. Seelig, p. 245.

27. W. Nernst to F. A. Lindemann, Berlin, August 18, 1913.

28. To Elsa Löwenthal, Zurich, undated, about July 14, 1913.

29. To J. J. Laub, Zurich, July 22, 1913, ETH, in Seelig, p. 245.

30. *Max Planck als Forscher*, in NW, Vol. I, November 1913, pp. 1077–99.

31. To M. Besso, Zurich, undated, end of 1913, in Besso, p. 50.

32. Ibid.

33. Fritz K. Ringer, *Die Gelehrten* (Stuttgart, 1983), p. 43.

34. Friedrich Schmidt-Ott, *Erlebtes und Erstrebtes* (Wiesbaden, 1952), p. 119.

35. To J. J. Laub, Zurich, July 22, 1913, ETH, in Seelig, p. 245.

36. To H. A. Lorentz, Zurich, August 14, 1913.

37. To P. Ehrenfest, Zurich, undated, end of 1913.

38. Hugo Bergmann, *Personal Remembrances of Albert Einstein*, in *Boston Studies*, XIII, p. 390.

39. Aurel Stodola to Einstein, Zurich, May 12, 1919.

40. To H. Zangger, Berlin, July 7, 1915.

41. To Elsa Löwenthal, Zurich, undated, after March 14, 1913.

42. To Elsa Löwenthal, Zurich, undated, about July 14, 1913.

43. To Elsa Löwenthal, Zurich, undated, mid-August 1913.

44. To Elsa Löwenthal, Zurich, October 10, 1913.

45. To Elsa Löwenthal, Zurich, undated, mid-October 1913.

46. To Elsa Löwenthal, Zurich, undated, end of November 1913.

47. To Elsa Löwenthal, Zurich, undated, mid-August 1913.

48. To Elsa Löwenthal, Zurich, undated, end of November 1913.

49. To Elsa Löwenthal, Zurich, undated, mid-December 1913.

50. To Elsa Löwenthal, Zurich, undated, end of December 1913.

51. To Elsa Löwenthal, Zurich, undated, beginning of December 1913.

52. To Elsa Löwenthal, Zurich, undated, end of December 1913.

53. The address was Ehrenbergstrasse 33.

54. Berlin, Doc. 7: Letter from Presiding Secretary of the Academy, Gustav Roethe, to Einstein, November 22, p. 101.

55. To the Royal Prussian Academy of Sciences, Zurich, December 7, 1913, in Berlin, Doc. 8, p. 101.

56. Otto Stern interview, typescript, p. 13, ETH.

57. Berlin, Doc. 22–32, pp. 112–20.

58. Einstein 1914e.

59. Louis Kollros, *Erinnerungen eines Kommilitonen*, in *Helle Zeit—Dunkle Zeit*, p. 30.

60. To P. Ehrenfest, Berlin, April 10, 1914.

61. Ibid.

62. To A. Hurwitz, Berlin, May 4, 1914.

63. Ibid.

64. Ibid.

65. To P. Ehrenfest, Berlin, May 25, 1914.

66. To P. Ehrenfest, Berlin, April 10, 1914.

67. M. Besso to Einstein, Bern, January 17, 1928, in Besso, p. 238. In this letter Besso recalls that "I had to bring Mileva back from Berlin to Zurich."

68. Personal information from Helen Dukas to A. Pais, in Pais, p. 242.

69. To C. Seelig, May 5, 1952.

70. H. A. Einstein in G. J. Whitrow, *Einstein* (London, 1967), p. 20.

71. Einstein 1914f.

72. Einstein 1914k.

73. *Erwiderung des Sekretars Max Planck auf die Antrittsrede Einsteins* [*Reply by Secretary Max Planck to Einstein's Inaugural Address*], in SB, Vol. 2, pp. 742–44.

18. "In a Madhouse": A Pacifist in Prussia

1. To P. Ehrenfest, Berlin, August 19, 1914.

2. Ibid.

3. Foreword to Johan Hannak, *Emanuel Lasker* (Princeton, 1952), p. X.

4. Max Planck, *Rektoratsrede am 3. August zum Stiftungsfest der Berliner Universität,* in *Deutsche Hochschulstimmen,* No. 33, 1914.

5. See also Kurt Mendelssohn, *Walther Nernst und seine Zeit* (Weinheim, 1976), pp. 112f.

6. "*Aufruf an die Kulturwelt*" ["Appeal to the Cultured World"], in *Vossische Zeitung* of October 4, 1914, p. 7. Repeatedly reproduced in collections of World War I documents, e.g., in *Aufrufe und Reden deutscher Professoren im Ersten Weltkrieg* (Stuttgart, 1975), pp. 47-49, though without the list of signatories.

7. Georg Friedrich Nicolai, *Die Biologie des Krieges* (Zurich, 1919), p. 12.

8. Ibid., p. 13, "*Aufruf an die Europäer*" ["Appeal to Europeans"].

9. Ibid., p. 13.

10. Ibid., p. 14.

11. To P. Ehrenfest, Berlin, undated, end of 1914, in Frieden, p. 20.

12. Berlin, p. 10.

13. Letter from the Reich Chancellor to Einstein, Berlin, January 16, 1917, in Berlin, p. 160.

14. To H. Zangger, no source, undated, Berlin, February 1915.

15. Romain Rolland, *La Conscience de l'Europe,* entry of September 16, 1915. This is Rolland's diary of the war years, 1914–1918.

16. Franziska Baumgartner-Tramer, *Erinnerungen an Einstein,* in *Der Bund,* Bern, July 10, 1955; reprinted in Frieden, p. 27.

17. To Romain Rolland, Berlin, August 23, 1915, in Frieden, p. 31.

18. To P. Ehrenfest, Berlin, August 23, 1915, in Frieden, p. 30.

19. To H. A. Lorentz, Berlin, August 2, 1915, in Frieden, p. 29.

20. To H. A. Lorentz, Berlin, September 23, 1915.

21. To H. A. Lorentz, Berlin, August 2, 1915, in Frieden, p. 30.

22. On science in World War I, see also John Ziman, *The Forces of Knowledge* (Cambridge, England, 1976), pp. 302ff.

23. *Nachruf auf Otto Sackur* [*Obituary for Otto Sackur*], in PhZ, 16, 1915, pp. 113ff.

24. To P. Ehrenfest, Berlin, August 19, 1914.

25. The address was Wittelsbacher Strasse 13, corner of Konstanzer Strasse; the building was destroyed in World War II.

26. Rudolf Jakob Humm, *Tagebuch*, in Seelig, p. 259.

27. To M. Besso, Berlin, February 15, 1915 (not in Besso).

28. To P. Ehrenfest, Berlin, August 19, 1914.

29. See also E. F. Freundlich's application to the academy of December 7, 1913; and notes in Berlin, pp. 164ff.

30. J. Earman and C. Glymor, *Relativity and Eclipses*, in HSPS, 11, 1980, pp. 49ff.

31. Ibid.

32. J. Crelinstein, *William Wallace Campbell and the "Einstein Problem,"* in HSPS, 14, 1983, pp. 1ff.

33. Ibid.

34. Einstein 1914l, p. 1030.

35. Ibid., p. 1030.

36. Ibid., p. 1079.

37. Ibid., p. 1080.

38. To P. Straneo, Berlin, January 7, 1915.

39. To T. Levi-Civita, Berlin, March 5, 1915.

40. To T. Levi-Civita, Berlin, March 26, 1915.

41. To M. Besso, Berlin, February 15, 1915, in Besso, p. 58.

42. Flückiger, p. 172.

43. Ibid.

44. Einstein, 1915d, p. 157.

45. Ibid., p. 170.

46. De Haas found a somewhat greater value in Leyden, $g = 1.2$; see also *Proceedings of the Royal Academy of Sciences, Amsterdam,* 1916, pp. 1281ff.

47. Einstein 1916g.

48. To M. Besso, Berlin, February 15, 1915, in Besso, p. 58.

49. Einstein 1915d.

50. See also Peter Galison, *How Experiments End* (Chicago, 1987), pp. 34–74, for an extensive account; and Pais, pp. 247–52, for a more compact one.

51. Minutes of the Meeting of the Board of Trustees of the Physical-Technical Reich Institute of March 8 and 9, 1922, in Berlin, p. 161.

52. To H. Zangger, Berlin, May 28, 1915.

53. To Romain Rolland, Berlin, March 22, 1915; in Frieden, pp. 31f., where the date is erroneously given as May 22, 1915.

54. R. Rolland to Einstein, Geneva, March 22, 1915.

55. To A. Sommerfeld, Sellin (Rügen), July 15, 1915, in Sommerfeld, p. 30.

56. A. Sommerfeld to W. Wien, December 25, 1914, STPK.

57. To H. Zangger, Berlin, July 7, 1915.

58. R. Rolland, *La Conscience de l'Europe* (see n. 15), Vol. 1, pp. 696ff.

59. Ibid.

60. *Meine Meinung über den Krieg,* undated manuscript, end of October or beginning of November 1915, STPK; with the excision of two paragraphs published in *Das Land Goethes 1914–1916—Ein vaterländisches Gedenkbuch* [*The Country of Goethe 1914–1916—A Patriotic Album*], published by Berliner Goethebund (Berlin, 1916.)

61. To Berliner Goethebund, Berlin, November 11, 1915, STPK.

62. To A. Sommerfeld, Berlin, November 28, 1915, in Sommerfeld, p. 32.

19. "The Greatest Satisfaction of My Life": The Completion of the General Theory of Relativity

1. To Walter Dällenbach, Berlin, March 31, 1915.

2. Ibid.

3. To A. Sommerfeld, Sellin (Rügen), July 15, 1915, in Sommerfeld, p. 30.

4. To W. J. de Haas, no place, undated, end of August 1915.

5. Erwin Freundlich, *Über die Gravitationsverschiebung der Spektrallinien bei Fixsternen [On the Gravitational Shift of the Spectral Lines of Fixed Stars]*, in PhZ, XVI, 1915, pp. 115–17.

6. To M. Besso, Berlin, February 15, 1915, in Besso, p. 57.

7. To W. Dällenbach, Berlin, May 31, 1915.

8. To D. Hilbert, Berlin, November 7, 1915; in this letter Einstein writes that "four weeks ago" he realized the faulty nature of his earlier derivations.

9. To H. A. Lorentz, Berlin, January 1, 1916.

10. To A. Sommerfeld, Berlin, November 28, 1915, in Sommerfeld, p. 32.

11. See also John Norton, *How Einstein Found His Field Equations: 1912–1915*, in HSPS, Vol. 14, Part 2, 1983, pp. 253ff.

12. See also Einstein's letter to Conrad Habicht of December 24, 1907, quoted in Chapter 16 (see n. 23 to Chapter 16).

13. Einstein 1915g, p. 799.

14. See also John Norton (see n. 11).

15. Einstein 1915i, p. 779.

16. Einstein 1915h.

17. Actually, it is only a coincidence that the deflection of light by the curvature of space has the same value as that derived from the equivalence principle, so that the result appears to be multiplied by 2.

18. To A. Sommerfeld, Berlin, December 9, 1915, in Sommerfeld, p. 37.

19. To P. Ehrenfest, Berlin, January 16, 1916.

20. Quoted from Pais, p. 256.

21. This is one example that can stand for many others: P. A. M. Dirac, *Methods in Theoretical Physics*, in *From a Life of Physics* (Trieste, 1968).

22. To A. Sommerfeld, Berlin, November 28, 1915, in Sommerfeld, p. 33.

23. Einstein 1915i, p. 847.

24. To H. Zangger, no place (Berlin), undated, "Friday" (presumably December 3, 1915).

25. To A. Sommerfeld, Berlin, December 9, 1915, in Sommerfeld, p. 37.

26. To M. Besso, Berlin, December 10, 1915, in Besso, p. 60.

27. To M. Besso, Berlin, December 21, 1915, in Besso, p. 61.

28. To P. Ehrenfest, Berlin, December 26, 1915.

29. To H. Zangger, no place (Berlin), undated (presumably December 3, 1915); see also H. A. Medicus, *A Comment on the Relations between Hilbert and Einstein*, in *American Journal of Physics*, 52, 1984, p. 206.

30. D. Hilbert, *Die Grundlagen der Physik* in *Nachrichten von der Königlichen Gesellschaft der Wissenschaften zu Göttingen, Mathematisch-physikalische Klasse*, 1915, pp. 395–407.

31. On Einstein and Hilbert, see the following studies: (a) Jagdish Mehra, *Einstein, Hilbert, and the Theory of Gravitation* (Dordrecht and Boston, 1974); (b) J. Earman and C. Glymour,

Einstein and Hilbert: Two Months in the History of General Relativity, in *Archive for the History of the Exact Sciences*, 19, 1978, pp. 291ff., (c) Pais, pp. 261–65.

32. F. Klein to W. Pauli, Göttingen, March 8, 1921, in Pauli, p. 27.

33. F. Klein to W. Pauli, Göttingen, May 8, 1921, in Pauli, p. 31.

34. Hilbert, p. 395.

35. Pais, p. 265. Hilbert's letter of apology, which Einstein mentioned to Straus, seems to be lost. Possibly Einstein did not keep it because he did not wish to be reminded of an unpleasant episode.

36. To D. Hilbert, Berlin, December 10, 1915.

37. See also Constance Reid, *Hilbert* (New York and Heidelberg, 1970), p. 141.

38. Rudolf Jakob Humm, in Seelig, pp. 260ff.

39. F. Klein to W. Pauli, Göttingen, May 8, 1921, in Pauli, p. 30.

40. To M. Besso, Berlin, January 3, 1916, in Besso, p. 63.

41. To H. A. Lorentz, Berlin, January 17, 1916.

42. Ibid.

43. *Die Grundlage der allgemeinen Relativitätstheorie*, in AdP, 49, 1916, pp. 769–822. Published in Einstein, Lorentz, and Minkowski, *Das Relativitätsprinzip*, 3rd ed. (1919 et seq. English translation, 1920. French translation by Solovine, 1933).

44. Max Born, *Einsteins Theorie der Gravitation und der allgemeinen Relativität*, in PhZ, 17, 1916, pp. 51–59.

45. To M. Born, Berlin, February 27, 1916, in Born, p. 20.

46. Erwin Freundlich, *Die Grundlagen der Einsteinschen Gravitationstheorie* (Berlin, 1916).

47. Seelig, p. 257.

48. To M. Besso, Berlin, August 28, 1918, in Besso, p. 138.

49. To Otto Naumann, *Min. Direktor* in Prussian Ministry of Education, Berlin, December 7, 1915, in Berlin, pp. 167f.

50. Hermann Struve to Otto Naumann, Berlin-Babelsberg, December 20, 1915, in Berlin, p. 170.

51. To A. Sommerfeld, Berlin, November 28, 1915, in Sommerfeld, p. 36.

52. Ibid.

53. H. Struve to O. Naumann, Berlin-Babelsberg, December 20, 1915, in Berlin, p. 169.

54. To A. Sommerfeld, Berlin, February 2, 1916, in Sommerfeld, p. 39.

55. To M. Besso, Berlin, December 21, 1915, in Besso, p. 61.

56. To Hermann Weyl, Berlin, November 23, 1916.

57. Karl Schwarzschild, *Über das Gravitationsfeld eines Massenpunktes nach der Einsteinschen Theorie*, in SB, 1916, pp. 189–96.

58. Karl Schwarzschild, *Über das Gravitationsfeld einer Kugel aus inkompressibler Flüssigkeit nach der Einsteinschen Theorie*, in SB, 1916, pp. 424–34.

59. Einstein 1916c.

60. Einstein 1918a, p. 154.

61. On this test and other experimental tests of the general theory of relativity, see the precise and comprehensible presentation by Clifford M. Will, *Was Einstein Right?* (New York, 1987).

62. To M. Besso, Berlin, May 14, 1916, in Besso, p. 69.

63. Einstein 1913c, p. 228.

64. To P. Ehrenfest, Berlin, February 4, 1917.

65. Einstein 1917a, p. 144.

66. Ibid.

67. Ibid., p. 151.

68. *Grundzüge der Relativitätstheorie*, Appendix 1 (Braunschweig, 1956), p. 111. Amplified version of Einstein 1921f.

69. George Gamow, *My World Line* (New York, 1970), p. 150: "he remarked that the introduction of the cosmological term was the biggest blunder he ever made in his life."

70. Leopold Infeld, in Schilpp, p. 343.

71. To A. Sommerfeld, Berlin, February 8, 1916, in Sommerfeld, p. 40.

72. To A. Sommerfeld, Berlin, August 3, 1916, in Sommerfeld, p. 41.

73. Einstein 1916c, p. 696.

74. To M. Besso, Berlin, undated, presumably mid-July 1916, in Besso, pp. 78f.

75. Einstein 1916j, p. 319.

76. Ibid., p. 322.

77. Einstein 1916k; this is the issue published in memory of Professor Alfred Kleiner, who had died on July 3, 1916. Identical text in PhZ, 1917, p. 121–28.

78. Ibid., p. 127.

79. To M. Besso, Berlin, September 6, 1916, in Besso, p. 82.

80. Einstein 1917h, p. 128.

81. To M. Besso, Berlin, March 9, 1917, in Besso, p. 103.

20. Wartime in Berlin

1. To W. Dällenbach, Berlin, May 31, 1915.

2. To M. Besso, Berlin, December 21, 1915, in Besso, p. 61.

3. To M. Besso, Berlin, July 14, 1916 (not in Besso).

4. To M. Besso, Berlin, July 21, 1916 (not in Besso).

5. To H. Zangger, Berlin, July 25, 1916.

6. To M. Besso, Berlin, September 6, 1916, in Besso, p. 81.

7. To M. Besso, Berlin, August 24, 1916, in Besso, p. 80.

8. To P. Ehrenfest, Berlin, August 25, 1916, and M. Besso, Berlin, September 6, 1916.

9. P. Ehrenfest, *Fragmentarischer Entwurf aus dem Jahre 1928*, probably for a memorial address on the death of H. A. Lorentz, quoted in Martin Klein, *Paul Ehrenfest* (Amsterdam, 1970), pp. 303f.

10. To M. Besso, October 31, 1916.

11. To P. Ehrenfest, Berlin, October 18, 1916.

12. To P. Ehrenfest, Berlin, October 24, 1916.

13. To M. Besso, Berlin, October 31, 1916, in Besso, p. 84.

14. To H. A. Lorentz, Berlin, November 13, 1916, in Frieden, p. 36.

15. Wien papers, STPK, and letter from A. Sommerfeld to W. Wien of December 25, 1914, DMM.

16. To H. Zangger, Berlin, undated, probably April 1917.

17. To M. Besso, Berlin, March 9, 1917, in Besso, p. 103.

18. To H. A. Lorentz, April 3, 1917, in Frieden, p. 38.

19. To P. Ehrenfest, Berlin, February 4, 1917.

20. To H. Zangger, Berlin, December 6, 1917.

21. Einstein 1916m.

22. Ibid. p. 509.

23. There is, incomprehensibly, no literature on Hermann Anschütz-Kaempfe, his fascinating life, or his

inventions. I base myself on the archive of the firm of Anschütz & Co. in Kiel and on the publication prepared for the firm's fiftieth anniversary in 1955. I am grateful to Dipl.-Ing. Bernhardt Schell for his help.

24. Patent expertise of August 7, 1915. Einstein later recalled that he had been motivated to conduct his gyromagnetic experiments by "technical expert opinions on the gyrocompass" (letter to Emile Meyerson of January 27, 1930), but these experiments began toward the end of 1914. Thus he must have been commissioned to give an expert opinion soon after his arrival in Berlin in 1914.

25. To M. Besso, Berlin, May 14, 1916, in Besso, p. 69.

26. To R. Rolland, Lucerne, August 22, 1917, in Frieden, p. 38.

27. Adler's assassination of Stürgkh, his trial, and its political significance are described in Julius Braunthal, *Victor and Friedrich Adler* (Vienna, 1965), pp. 230–51.

28. To Katya Adler, Berlin, February 20, 1917.

29. To Friedrich Adler, Berlin, April 13, 1917.

30. To H. Zangger, Berlin, undated, end of April 1917.

31. To M. Besso, Berlin, April 29, 1917, in Besso, p. 105.

32. M. Besso to Einstein, Zurich, May 5, 1917, in Besso p. 110.

33. The full text is in *Vor dem Ausnahmegericht—eine Dokumentensammlung* (Vienna, 1923).

34. Friedrich Adler to Katya Adler, February 17, 1917, in Braunthal (see n. 27), p. 248.

35. To M. Besso, Berlin, April 29, 1917, in Besso, p. 106.

36. Frank, p. 289; Philipp Frank had been sent Adler's treatise by the court with a request for an expert opinion.

37. To M. Besso, May 13, 1917, in Besso, p. 114.

38. *Friedrich Adler als Physiker. Eine Unterredung mit A. Einstein*, in *Vossische Zeitung*, May 23, 1917, morning ed., p. 2.

39. To Friedrich Adler, quoted from Braunthal (see n. 27).

40. Friedrich Adler, *Ortszeit, Systemzeit, Zonenzeit und das ausgezeichnete Bezugssystem der Elektrodynamik—Eine Untersuchung über die Lorentzsche und Einsteinsche Kinematik* (Vienna, 1920).

41. To C. Seelig, Princeton, April 8, 1952, ETH.

42. To M. Besso, March 9, 1917, in Besso, p. 103.

43. To P. Ehrenfest, Berlin, February 14, 1917.

44. To M. Besso, Berlin, March 9, 1917, in Besso, p. 102.

45. Mileva Marić to Helene Savić, Zurich, end of May 1901, in CP1, Doc. 109, p. 303.

46. To M. Besso, Berlin, May 8, 1917, in Besso, p. 113.

47. To M. Besso, Berlin, May 13, 1917, in Besso, p. 114.

48. See also Pschyrembel, *Klinisches Wörterbuch*, Berlin (1982), p. 1113. "Scrofulosis" is now only a historical term.

49. To M. Besso, Berlin, March 9, 1917, in Besso, p. 102.

50. Ibid.

51. To H. Zangger, Berlin, undated, end of April 1917.

52. Data on Einstein's lectures during the war years are taken from the listings in *Physikalische Zeitschrift.*

53. To P. Ehrenfest, Berlin, June 3, 1917, in Frieden, p. 38.

54. Hedwig Born, *Albert Einstein ganz privat*. Repeatedly published: e.g., in *Helle Zeit—Dunkle Zeit*, pp. 35ff., here p. 36.

55. On Hans Mühsam and Einstein, see *Helle Zeit—Dunkle Zeit*, pp. 48–58.

56. Moszkowski, pp. 16f.

57. Max Wertheimer, *Productive Thinking* (New York and London, 1945), p. 158.

58. To M. Besso, Berlin, May 13, 1917, in Besso, p. 114.

59. To M. Besso, Berlin, June 24, 1917, in Besso, p. 117.

60. To M. Besso, Benzingen, September 3, 1917, in Besso, p. 119.

61. To M. Besso, Berlin, September 22, 1917, in Besso, p. 121.

62. A. von Harnack to Einstein, Berlin, September 12, 1917, MPG. A few documents on the foundation of the Kaiser Wilhelm Institute for Physics are in Berlin, pp. 147ff.

63. To W. von Siemens, Berlin, December 18, 1917, MPG.

64. To W. von Siemens, Berlin, January 4, 1918, in Berlin, pp. 150f.

65. Activity report of the Kaiser Wilhelm Institute for Physics for the period from April 1, 1919, to March 31, 1920, in Berlin, p. 152.

66. To Max Planck, Berlin, undated, beginning July 1918, on an application by Debye, passed on to Planck, MPG.

67. To H. Zangger, Berlin, December 6, 1917.

68. Ibid.

69. To M. Besso, Berlin, January 5, 1917, in Besso, p. 124.

70. Ibid.

71. To A. Sommerfeld, Berlin, undated, beginning of March 1918, in Sommerfeld, p. 48.

72. *Motiv des Forschens*, in *Weltbild*, pp. 107–10, under the title (changed by the editor) *Prinzipien der Forschung*.

73. To H. Zangger, Berlin, December 6, 1917.

74. To H. A. Lorentz, Berlin, December 18, 1917.

75. Max Planck to Einstein, Berlin, December 29, 1917.

76. To G. Nicolai, Berlin, February 28, 1917.

77. High Command in the Marches. Chief of Staff to Police President of Berlin, along with a "blacklist," Berlin, January 20, 1918, in Berlin, pp. 198f.

78. To R. Rolland, Lucerne, August 22, 1917, in Frieden, p. 39.

79. Circular to Colleagues, April 1918; in the Hilbert file of EA, 13 115; further data are taken from a covering letter by Einstein to Hilbert.

80. To D. Hilbert, Berlin, end of April 1918.

81. D. Hilbert to Einstein, Göttingen, April 27, 1918.

82. D. Hilbert to Einstein, Göttingen, May 1, 1918.

83. To D. Hilbert, Berlin, May 24, 1918.

21. Postwar Chaos and Revolution

1. To M. Besso, Berlin, January 5, 1918, in Besso, p. 124.

2. To Hedwig Born, Berlin, February 8, 1918, in Born, p. 23.

3. To D. Hilbert, Berlin, undated, end of April 1918.

4. To M. Besso, Berlin, June 23, 1918, in Besso, pp. 126f.

5. To P. Ehrenfest, Berlin, June 5, 1918.

6. Rudolf Ehrmann, in *Helle Zeit—Dunkle Zeit*, pp. 58ff.

7. To M. Besso, Berlin, September 22, 1917, in Besso, p. 121.

8. Ibid.

9. To M. Born, Ahrenshoop, undated, July 1918, in Born, p. 26.

10. Ibid.

11. To M. Besso, Ahrenshoop, August 20, 1918, in Besso, p. 133.

12. To M. Born, Ahrenshoop, August 2, 1918, in Born, p. 27.

13. To M. Besso, Ahrenshoop, August 20, 1918, in Besso, p. 133.

14. Ibid.

15. To M. Besso, Berlin, September 8, 1918, in Besso, pp. 139f.

16. Ibid.

17. Ibid.

18. Protocol of the Education Council of Canton Zurich of December 23, 1918, excerpt, ETH; quotations from a letter from Einstein (undated) to the education authority, received on December 20, 1918.

19. To M. Besso, Ahrenshoop, July 29, 1918, in Besso, p. 130.

20. To M. Besso, Berlin, undated, December 1916, in Besso, p. 98.

21. Currency exchange statistics of the German Bundesbank, Frankfurt.

22. To M. Besso, Berlin, January 5, 1918, in Besso, p. 124.

23. To M. Besso, Berlin, June 23, 1918, in Besso, p. 126.

24. Notebook for the relativity class WS 1918/19, EA.

25. To Maja and Paul Winteler, Berlin, November 11, 1918, in Frieden, p. 44.

26. To Pauline Einstein, Berlin, November 11, 1918, in Frieden, p. 43.

27. To M. Besso, Berlin, December 4, 1918, in Besso, p. 145.

28. Accounts of this event are given in Born, pp. 202ff.; and in Frieden, pp. 44f.

29. As remembered by Max Born, in Born, p. 205.

30. M. Born, in Born, p. 206.

31. Manuscript of a speech intended for the revolutionary students but probably not made, November 1918, in Frieden, p. 45.

32. To M. Born, Princeton, September 7, 1944, in Born, p. 202.

33. To H. A. Lorentz, Berlin, December 6, 1918, in Frieden, p. 48.

34. To M. Born, Arosa, January 19, 1919, in Born, p. 28.

35. Protocol of the Education Council of Canton Zurich, December 23, 1918, ETH.

36. To A. Sommerfeld, Zurich, March 5, 1919, in Sommerfeld, p. 55 (where the date is mistakenly given as January 5, 1919).

37. District Court, Zurich, B XII Zch. 6314.43.

38. To Pauline Einstein, Berlin, undated, March 1919.

39. To P. Ehrenfest, Berlin, March 22, 1919, in Frieden, p. 48.

40. To Pauline Einstein, Berlin, undated, March 1919.

41. Ibid.

42. To H. A. Lorentz, April 26, 1919, in Frieden, p. 51. A. Fölsing does not know who the other five members of the committee were, but in his letter to Lorentz Einstein vouched for their "absolutely pure and earnest attitude and sense of justice."

43. To H. A. Lorentz, Berlin, September 21, 1919, in Frieden, p. 53.

44. To H. A. Lorentz, Berlin, March 18, 1920, in Frieden, p. 54.

45. To H. A. Lorentz, Berlin, September 21, 1919, in Frieden, p. 53.

46. Protocol of the Education Council of Canton Zurich, July 8, 1919, ETH.

47. Register of marriages, Berlin-Wilmersdorf, No. 623/1919.

48. Thus in letter from Ilse Einstein: "Albert."

49. Friedrich Herneck, *Über die deutsche Reichsangehörigkeit Albert Einsteins*, in NW, No. 2, 1961, p. 95.

50. To H. Zangger, Berlin, December 6, 1917.

51. There is a detailed description of the apartment at Haberlandstrasse 5 in Herneck (see n. 49), pp. 28ff.

52. Frank, p. 219.

53. Ibid.

54. Elias Tobenkin, *Interview with Einstein*, in *New York Evening Post*, March 26, 1921.

55. To M. Besso, New York, May 28, 1921, in Besso, p. 163.

56. *Antwort an amerikanische Frauen*, 1932, in Weltbild, p. 45.

57. In a conversation with Esther Salaman, spring 1924, in Seelig, p. 316.

58. Sayen, p. 70.

59. Elsa Einstein to Herman Struck and wife, Berlin, undated, 1929, MPG.

60. To Vero and Bice Besso, Princeton, March 21, 1955, in Besso, p. 537.

61. M. Planck to Einstein, Berlin, July 20, 1919.

62. F. Haber to Einstein, Berlin, undated, July 1919.

63. To P. Ehrenfest, Berlin, September 12, 1919.

64. F. Haber to Einstein, no place, undated, "Saturday," August 1919.

65. M. Planck to H. von Ficker, Munich, March 31, 1933, in Berlin, p. 245.

66. To M. Born, Berlin, September 1, 1919, in Born, p. 33.

67. *Nachruf auf Moritz Katzenstein*, in *Helle Zeit—Dunkle Zeit*, p. 46.

68. To M. Born, Berlin, September 1, 1919, in Born, p. 33.

69. P. Ehrenfest to Einstein, Leyden, September 2, 1919.

70. In 1919, 7,500 Dutch guilders represented more than 50,000 marks; more important to Einstein would have been the fact that they represented 15,000 Swiss francs. Currency statistics of the German Bundesbank, Frankfurt.

71. P. Ehrenfest to Einstein, Leyden, September 8, 1919.

72. To P. Ehrenfest, Berlin, September 12, 1919.

73. P. Ehrenfest to H. A. Lorentz, Leyden, September 21, 1919.

74. To P. Ehrenfest, Berlin, September 12, 1919.

75. To P. Ehrenfest, Berlin, September 28, 1919.

76. To Pauline Einstein, Berlin, October 19, 1919.

77. To P. Ehrenfest, Berlin, November 9, 1919.

78. P. Ehrenfest to Einstein, Leyden, November 24, 1919.

79. To P. Ehrenfest, Berlin, December 4, 1919.

80. Pierre Kerszberg, *The Inverted Universe: The Einstein–de Sitter Controversy and the Rise of the Relativistic Cosmology* (Oxford, 1989), gives a full account of the discussions between Einstein and de Sitter.

22. Confirmation of the Deflection of Light: "The Suddenly Famous Dr. Einstein"

1. E. Freundlich, *Zur Prüfung der allgemeinen Relativitätstheorie*, in NW, 7, 1919, pp. 629ff.

2. To E. Freundlich, Berlin, March 29, 1919.

3. See also letter to M. Besso, Berlin, May 8, 1917, in Besso, p. 112.

4. A. S. Eddington, *Report on the Relativity Theory of Gravitation* (London, 1918).

5. See also John Earman and Clark Glymour, *Relativity and Eclipses,* in HSPS, 11, 1980, pp. 49–85.

6. Subramanyan Chandrasekhar, *Verifying the Theory of Relativity,* in *Bulletin of the Atomic Scientists,* June 1975, pp. 17–22.

7. Isaac Newton, *Opticks,* 1704.

8. A. S. Eddington, *Report on the Relativity Theory of Gravitation,* p. 56; *Space, Time, and Gravitation* (London, 1920).

9. F. Dyson, A. S. Eddington, and C. Davidson, *A Determination of the Deflection of Light by the Sun's Gravitational Field,* in *Philosophical Transactions of the Royal Society,* 220, 1920, pp. 291–333.

10. Eddington, *Space, Time, and Gravitation,* p. 115.

11. *Nature,* June 5, 1919; *Observatory,* June 1919.

12. Eddington, *Space, Time, and Gravitation,* p. 116.

13. *Vossische Zeitung,* April 15, 1919.

14. Kurt Joel, *Die Sonne bringt es an den Tag,* in *Vossische Zeitung,* May 29, morning edition.

15. To Maja Winteler-Einstein, Berlin, end of May or beginning of June 1919.

16. To Pauline Einstein, Berlin, June 11, 1919.

17. See also Jeffrey Crelinsten, *William Wallace Campbell and the "Einstein Problem,"* in HSPS, 14, 1983, pp. 1–91, especially pp. 28–32.

18. Ibid., p. 42.

19. Max Born, *Die Relativitätstheorie Einsteins* (Berlin, 1920), p. 238.

20. To P. Ehrenfest, Berlin, September 12, 1919.

21. H. A. Lorentz to Einstein, telegram, September 22, 1919.

22. H. A. Lorentz to Einstein, October 7, 1919.

23. Ilse Rosenthal-Schneider, *Erinnerungen an Gespräche mit Einstein,* manuscript, July 23, 1957, p. 2; also Ilse Rosenthal-Schneider, *Begegnungen mit Einstein, von Laue, Planck* (Braunschweig, 1988), p. 60. Rosenthal-Schneider believes she remembers a cable from Eddington, but there was no such cable: her recollection no doubt refers to Lorentz's telegram.

24. To Pauline Einstein, Berlin, September 27, 1919.

25. To Walter Dällenbach, Berlin, September 27, 1919.

26. M. Planck to Einstein, Berlin, October 4, 1919.

27. Alexander Moszkowski, *Die Sonne bracht' es an den Tag,* in *Berliner Tageblatt,* October 8, 1919.

28. Einstein 1919d. This note was undoubtedly motivated by Moszkowski's article, not because Einstein was particularly excited (Pais, p. 303).

29. To M. Planck, Leyden, October 23, 1919. Postcard, MPG.

30. To Pauline Einstein, Leyden, undated, October 26, 1919.

31. C. Stumpf to Einstein, Berlin, October 22, 1919.

32. To C. Stumpf, Berlin, November 3, 1919.

33. To C. Seelig, Princeton, August 20, 1952, in *Helle Zeit—Dunkle Zeit,* p. 45.

34. H. Zangger to Einstein, Zurich, October 22, 1919.

35. See also F. Dyson, A. S. Eddington, and C. Davidson, *A Determination of the Deflection of Light by the Sun's Gravitational Field, from Observations Made at the Total Eclipse of May 29, 1919*, in *Philosophical Transactions of the Royal Society*, 220, 1920, p. 220, the "official" report. Eddington, *Space, Time, and Gravitation*, pp. 110–22.

36. Ibid., p. 116.

37. A. N. Whitehead, *Science and the Modern World* (London, 1926), p. 13.

38. According to the reports in *Observatory*, 42, 1919, pp. 389–98; in *Nature*, November 13; and in London *Times*, November 7, 1919.

39. *Observatory* (see n. 38), p. 391.

40. According to S. Chandrasekhar (see n. 6), Eddington related this story in Cambridge.

41. A. S. Eddington to Einstein, Cambridge, December 1, 1919.

42. A. F. Lindemann to Einstein, Sidholme, November 23, 1919.

43. See n. 41.

44. R. W. Lawson to Arnold Berliner, mid-November 1919, in Frieden, p. 46.

45. Pais, p. 312.

46. W. W. Campbell, *Clouds Fall Away from Solar Eclipse*, in *New York Times*, June 10, 1918.

47. E.g. Pais, p. 313, or Stanley Goldberg, *Understanding Relativity* (Boston, 1984), p. 312. It is simply inconceivable how, before the publication of the *New York Times* article of November 10, 1919, an American journalist could have spoken to Einstein. Even if Einstein, in connection with the publication of his great *Annalen* article as a separate brochure by Barth in Leipzig in 1916, had said anything of the kind, it is not clear how this could have come to the knowledge of Americans, quite apart from the fact that the implied irony would have been lost. Eddington believed that the source of this story was his conversation with Silberstein after the meeting on November 6, 1919.

48. Clark, p. 295.

49. To D. Hilbert, no place, undated, beginning April 1918.

50. *Vossische Zeitung*, July 23, 1918.

51. *New York Times*, November 18, 1919.

52. To Max Born, December 9, 1919, in Born, p. 38.

53. NW, 8, p. 20, issue of January 2, 1920.

54. *Was ist Relativitätstheorie?* in *Weltbild*, p. 127. English translation as *Einstein on His Theory* in London *Times*, November 28, 1919.

55. *New York Times*, December 3, 1919.

56. To Ludwig Hopf, Berlin, February 2, 1920.

23. Relativity under the Spotlight

1. To Marcel Grossmann, Berlin, September 12, 1920.

2. Moszkowski, pp. 26f.

3. See O. Glaser, *Wilhelm Conrad Röntgen und die Geschichte der Röntgenstrahlen* (Berlin, 1959). All data on the events of 1896 are taken from that book.

4. To H. Zangger, undated, Berlin, beginning of 1920.

5. Preface written in 1942 for Philipp Frank's biography, but not published until 1979.

6. *Berliner Tageblatt*, July 7, 1921, p. 2. Also in *Weltbild*, p. 42.

7. E.g., Pathé Journal, 1, 1931.

8. To Marcel Grossmann, Berlin, September 12, 1920.

9. Eduard Meyer, Rector of Berlin University, to Einstein, Berlin, February 12, 1920.

10. Eduard Meyer to Einstein, Berlin, February 13, 1920.

11. E.g., in *Vorwärts*, February 13, 1920.

12. Eduard Meyer to Ministry of Culture, Berlin, February 13, 1920.

13. Events in Professor Einstein's class. Statement by the Ministry of Culture as well as further comment in *Berliner Tageblatt* of February 14, 1920.

14. *Vorwärts*, February 14, 1920.

15. To P. Ehrenfest, Berlin, December 4, 1919.

16. To H. Zangger, Berlin, March 26, 1920.

17. To Aurel Stodola, Berlin, March 31, 1920.

18. Max von Laue to A. Sommerfeld, Berlin, August 23, 1920.

19. *Berliner Tageblatt*, September 4, 1920.

20. Paul Weyland, *Einsteins Relativitätstheorie—eine wissenschaftliche Massensuggestion*, in *Tägliche Rundschau*, August 6, 1920.

21. Ibid. A retort from Max von Laue appeared in *Tägliche Rundschau* on August 11, whereupon Weyland reacted on August 14 and 16.

22. E. Gehrcke, *Die gegen die Relativitätstheorie erhobenen Einwände*, in NW, Vol. 1, 1913, p. 62–66.

23. Max Born, *Zum Relativitätsprinzip: Entgegnungen auf Herrn Gehrckes Artikel*, in NW, Vol. 1, pp. 92–94.

24. E. Gehrcke, *Zur Kritik und Geschichte der neueren Gravitationstheorien*, in AdP, Vol. 51, 1916, pp. 119–24.

25. To W. Wien, Berlin, October 17, 1916.

26. E. Gehrcke, *Über den Äther*, in VhDPG, Vol. 20, 1918, pp. 165–99.

27. Einstein 1918i.

28. H. von Seeliger, *Bemerkung zu dem Aufsatz des Herrn Gehrcke "Über den Äther,"* in VhDPG, Vol. 20, 1918, p. 262. The fact is that soon after the discovery of Mercury's perihelion anomaly, calculations were made with a velocity-dependent gravitational potential, analogous to Weber's electrodynamic potential. Such a potential results in a precession of the perihelion, and by a suitable adjustment of a constant it is possible to "force" the correct amount. However, this procedure has no explanatory value, since there is no such thing as a velocity-dependent gravitational potential.

29. Max von Laue to A. Sommerfeld, Berlin, August 25, 1920.

30. Ibid.

31. Bertolt Brecht, *An die Nachgeborenen*, in Bertolt Brecht, *Gesammelte Gedichte* (Frankfurt, 1976), p. 722.

32. *Der Kampf gegen Einstein*, in *Vossische Zeitung*, August 29, 1920.

33. Paul Ehrenfest to Einstein, Leyden, September 2, 1920.

34. Hedwig Born to Einstein, Frankfurt, September 8, 1920, in Born, p. 58.

35. A. Sommerfeld to Einstein, Munich, September 3, 1920, in Sommerfeld, pp. 67f.

36. To A. Sommerfeld, Berlin, September 6, 1920, in Sommerfeld, p. 69.

37. German Chargé d'Affaires in

London, F. Sthamer, to German Foreign Ministry, London, September 2, 1920, in Berlin, p. 206.

38. To Hedwig and Max Born, Berlin, September 9, 1920, in Born, p. 59.

39. *Berliner Tageblatt*, August 31, 1920.

40. K. Haenisch to Einstein, Berlin, September 6, 1920. This "private letter" was handed to the press by the Ministry of Culture; it was published by *Tägliche Rundschau*, September 7, 1920.

41. To K. Haenisch, Minister of Culture, Berlin, September 8, 1920, in Berlin, p. 204.

42. G. Roethe, Managing Secretary, to M. Planck, Berlin, September 10, 1920, in Berlin, p. 205.

43. M. Planck to G. Roethe, September 14, 1920, in Berlin, 206.

44. Einstein 1918l.

45. *Meine Antwort*, in *Berliner Tageblatt*, August 27, 1920.

46. A. Sommerfeld to Einstein, Munich, September 11, 1920, in Sommerfeld, p. 71.

47. Ibid.

48. P. Lenard to A. Sommerfeld, Heidelberg, September 14, 1920, in A. Kleinert and D. Schönbeck, *Lenard und Einstein*, in *Gesnerus*, Vol. 35, 1978, p. 330.

49. *Meine Antwort* (see n. 45).

50. *Allgemeine Diskussion über Relativitätstheorie*, in PhZ, Vol. 21, 1920, pp. 666–68.

51. Walther Gerlach, *Erinnerungen an Albert Einstein*, in P. C. Aichelburg and R. Sexl (eds.), *Albert Einstein—Sein Einfluß auf Physik, Philosophie und Politik* (Braunschweig, 1979).

52. PhZ, Vol. 21, 1920, p. 666. A report in *Berliner Tageblatt* reproduces this passage identically as to substance but in somewhat different words. The reporter notes "laughter."

53. Frank, p. 275; *Berliner Tageblatt*, September 24, 1920.

54. Paul Weyland in *Deutsche Zeitung*, September 26, 1920.

55. E. Gehrcke to P. Lenard, February 3, 1921, with a marginal gloss by Lenard.

56. E.g., Clark, pp. 316-27; inspired by this also: Alan D. Beyerchen, *Wissenschaft unter Hitler* (Cologne, 1980), pp. 124ff. Born, too, in 1969 wrote that Lenard directed "sharp, vicious attacks on Einstein, with undisguised anti-Semitic tendency." This was evidently a projection backward from Lenard's behavior after the Nauheim conference.

57. A. Sommerfeld to Elsa Einstein, Munich, October 7, 1920, in Sommerfeld, p. 72.

58. To Max Born, no place, undated, Leyden, October 1920, in Born, p. 67.

59. To H. Zangger, no place, undated, autumn 1921.

60. F. Klein to W. Pauli, Göttingen, March 8, 1921, in Pauli, p. 27.

61. Elsa Einstein to Max Born, Berlin, June 26, 1920, in Born papers, STPK.

62. Max Born, in Born, p. 69.

63. F. Springer to Max Born, September 7, 1920, in Born papers, STPK.

64. Hedwig Born to Einstein, Leipzig, October 7, 1920, in Born, pp. 63ff.

65. Max Born to Einstein, Frankfurt, October 13, 1920, in Born, pp. 65ff.

66. To Max Born, Benzingen, Post

Veringenstadt, October 11, 1920, in Born, p. 66.

67. To Max Born, Leyden, October 26, 1920, in Born, p. 67.

68. Ibid.

69. To P. Ehrenfest, Berlin, November 26, 1920.

70. To H. Zangger, Berlin, March 14, 1921.

71. To Max Born, Berlin, September 9, 1920, in Born, p. 59.

72. See Born, p. 69.

24. "Traveler in Relativity"

1. In Berlin, p. 175.

2. To K. Haenisch, Berlin, December 6, 1919, in Berlin, p. 176.

3. Appeal for the Albert Einstein Donation, in Berlin, p. 177.

4. See the catalogue of the Erich Mendelsohn exhibit, *Ideen, Bauten, Projekte* (Berlin, 1987).

5. See M. Grüning, *Der Wachsmann Report. Auskünfte eines Architekten*, Berlin 1985, p. 403.

6. To M. Besso, Berlin, July 26, 1920, p. 152.

7. To H. Zangger, undated, Berlin, beginning March 1920.

8. To Hedwig Born, Christiania, June 18, 1920, in Born, p. 53. The date given by Born is a misreading.

9. To M. Besso, Berlin, December 12, 1919, in Besso, p. 148.

10. To M. Besso, Berlin, January 6, 1920, in Besso, p. 150.

11. To Max Born, Berlin, March 3, 1920, in Born, p. 48.

12. To N. M. Butler, President of Columbia University, Berlin, April 11, 1923.

13. To P. Ehrenfest, Berlin, November 9, 1919.

14. *Autobiographisches*, p. 17.

15. To N. Bohr, Berlin, May 2, 1920.

16. Ibid.

17. To P. Ehrenfest, Berlin, May 4, 1920.

18. N. Bohr to Einstein, June 24, 1920.

19. To M. Besso, Berlin, December 12, 1919, in Besso, p. 148.

20. To M. Besso, Berlin, July 26, 1920, in Besso, p. 153.

21. To M. Solovine, Bern, May 3, 1906, in Solovine, p. 4.

22. Einstein 1920d.

23. See Manfred Eigen, *Die "unmeßbar" schnellen Reaktionen*, in *Les Prix Nobel en 1967*, pp. 151–80.

24. To A. Sommerfeld, Berlin, January 4, 1921, in Sommerfeld, p. 77.

25. To F. Haber, Hechingen, undated, September 1920.

26. To M. Born, Berlin, January 31, 1921, in Born, p. 78.

27. To P. Ehrenfest, Berlin, June 6, 1920.

28. German Legation in The Hague to Ministry of Foreign Affairs, May 25, 1920, in Berlin, p. 225.

29. To F. Haber, undated, Hechingen, September 1920.

30. To M. Born, Christiania, June 18, 1920, in Born, p. 53; the date given by Born is a misreading.

31. German Legation in Christiania to Ministry of Foreign Affairs, June 22, 1920, in Berlin, p. 226.

32. According to the report of German Legation in Copenhagen to Ministry of Foreign Affairs, June 26, 1920, in Berlin, p. 226.

33. Ibid.

34. To M. Besso, September 12, 1919, in Besso, p. 148.

35. To M. Born, Berlin, September 9, in Born, p. 60.

36. P. Ehrenfest to Einstein, Leyden, July 25, 1920.

37. Einstein 1920f. The version of this lecture published by Springer in Berlin erroneously gives its date as May 5, 1920.

38. Einstein 1920f, p. 12.

39. M. Besso to Einstein, Bern, December 24, 1920, in Besso, p. 160.

40. P. Ehrenfest to Einstein, Leyden, January 17, 1922.

41. To P. Ehrenfest, Berlin, February 12, 1922.

42. To F. Vieweg, Berlin, January 9, 1922.

43. To F. Vieweg, Berlin, February 15, 1922.

44. H. A. Lorentz to Einstein, Leyden, September 10, 1920.

45. To H. A. Lorentz, Hechingen, undated, end of September 1920.

46. See Klaus Hentschel, *Zwei vergessene Texte Moritz Schlicks*, in *Gesnerus*, Vol. 31, 1988, pp. 300–311.

47. To H. Zangger, undated, beginning of January 1921.

48. To P. Ehrenfest, Berlin, November 26, 1920.

49. Frank, p. 283.

50. Ibid., p. 285.

51. Ibid., p. 290.

52. A. Sommerfeld to Einstein, Munich, December 18, 1920, in Sommerfeld, p. 73.

53. To A. Sommerfeld, Berlin, December 20, 1920, in Sommerfeld, p. 75.

54. Count Harry Kessler, *Tagebücher*, Frankfurt, 1982, pp. 249ff. Kessler gives a detailed account of the purpose and course of the journey.

55. To M. Born, Berlin, January 31, 1921, in Born, p. 79.

56. To P. Ehrenfest, Berlin, February 12, 1921.

57. Kurt Blumenfeld to Chaim Weizmann, Berlin, February 20, 1921.

25. Jewry, Zionism, and a Trip to America

1. To Abba Eban, Princeton, November 18, 1952; this letter is Einstein's reply to the offer to accept the presidency of the state of Israel.

2. Willy Hellpach, fall 1929, published in *Weltbild*, p. 104.

3. M. Besso to Einstein, Geneva, October 15, 1939, in Besso, p. 346.

4. M. Besso to Einstein, Geneva, July 23, 1927.

5. To H. Zangger, Prague, August 24, 1911.

6. "Blood is not water": to Ludwig Hopf, Prague, June 12, 1912.

7. To P. P. Lazarev, Berlin, May 16, 1914, quoted in Kirsten and Treder, *Albert Einstein 1879–1955*, in Winter and Jarosch (eds.), *Wegbereiter der deutsch-slawischen Wechselseitigkeit* (Berlin, 1983), p. 354.

8. To J. J. Laub, Bern, May 19, 1909.

9. To Julius Katzenstein, California, December 27, 1931.

10. See n. 2.

11. *Antisemitismus und akademische Jugend* (mid-1920s), in *Weltbild*, p. 94.

12. Frank, p. 251.

13. To Robert Eisler, Berlin, January 31, 1925.

14. *Antisemitismus und akademische Jugend*, p. 94.

15. Kurt Blumenfeld, *Erlebte Judenfrage* (Stuttgart, 1962), pp. 126ff.; *Einsteins Beziehung zum Zionismus und zu Israel*, in *Allgemeine Wochenzeitung der Juden in Deutschland*, September 16, 1955; published in *Helle Zeit—Dunkle Zeit*, pp. 74ff. Blumenfeld in *Erlebte*

Judenfrage dates his first visit to Einstein as 1920; this is incorrect.

16. According to Blumenfeld, in *Helle Zeit—Dunkle Zeit*, p. 76.

17. To Max Born, Berlin, November 9, 1919, in Born, p. 36.

18. Ibid.

19. H. Bergmann to Einstein, October 22, 1919.

20. To H. Bergmann, Berlin, November 5, 1919.

21. To M. Besso, Berlin, December 12, 1919, in Besso, p. 148.

22. *Berliner Tageblatt*, December 30, 1919, morning edition.

23. To Central-Verein Deutscher Staatsbürger Jüdischen Glaubens, Berlin, April 5, 1920. Published in David Reichenstein, *Albert Einstein. Sein Leben und seine Weltanschauung* (Berlin, 1932), pp. 144ff.; the text excerpt in Frank, p. 250, is not accurately reproduced.

24. News of the Jewish Press Office, Zurich, September 21, 1920, in *Jüdische Rundschau*, December 10, 1920.

25. K. Blumenfeld to Chaim Weizmann, Berlin, February 20, 1921, in Kurt Blumenfeld, *Kampf um den Zionismus. Briefe aus fünf Jahrzehnten* (Stuttgart, 1976), p. 65. In two autobiographical recollections Blumenfeld gives March 10, 1921, as the date of his first talk with Einstein; this is an error.

26. To F. Haber, Berlin, March 9, 1921.

27. To M. Solovine, March 8, 1921, in Solovine, p. 26.

28. E. Rutherford to B. Boltwood, February 28, 1921, in L. Badash (ed.), *Rutherford and Boltwood: Letters on Radioactivity* (New Haven, 1969), p. 142.

29. K. Blumenfeld to Chaim Weizmann, Berlin, March 16, 1921.

30. P. Ehrenfest to Einstein, Leyden, February 28, 1921.

31. To P. Ehrenfest, Berlin, March 8, 1921.

32. F. Haber to Einstein, Berlin, March 9, 1921.

33. To F. Haber, Berlin, March 9, 1921.

34. K. Blumenfeld to Chaim Weizmann, Berlin, March 16, 1921.

35. To M. Solovine, March 16, 1921, in Solovine, p. 30.

36. On Weizmann see Norman Rose, *Chaim Weizmann: A Biography* (New York, 1986).

37. Sayen, p. 47.

38. To M. Besso, Zurich, undated, end of 1913, in Besso, p. 50.

39. Frank, p. 296.

40. Seelig, p. 136.

41. *New York Times*, April 3, 1921.

42. Max Talmey, *The Relativity Theory Simplified* (New York, 1932), p. 174.

43. *New York Times*, April 13, 1921.

44. Eve Curie, *Madame Curie*, trans. Vincent Sheean (London, 1970), pp. 343–50. During her almost simultaneous visit to America Marie Curie appears to have enjoyed the same public interest as Einstein.

45. Citation for honorary degree, Princeton University, May 9, 1921.

46. *Philadelphia Evening Bulletin*, May 10, 1921.

47. Einstein 1922a.

48. To Oswald Veblen, Berlin, April 30, 1930.

49. D. C. Miller to T. C. Mendenhall, Cleveland, June 2, 1921, in *American Institute of Physics*.

50. To M. Besso, New York, May 28, 1921, in Besso, p. 163.

51. Norman Rose, *Weizmann* (New York, 1986), p. 214.

52. To P. Ehrenfest, Berlin, June 18, 1921.

53. Einstein, *About Zionism: Speeches and Letters* (New York, 1931), pp. 48f.

54. To M. Born, Berlin, January 27, 1920.

55. A. S. Eddington to Einstein, Cambridge, England, January 9, 1920.

56. It may be useful to point out here that the Royal Society in England has the status of a national academy, whereas societies such as the Royal Astronomical Society are comparable to specialist associations in other countries.

57. Clark, p. 335.

58. For a detailed account of the small talk, see Clark, pp. 338ff.

59. *Über Relativitätstheorie*, published in excerpt in *Weltbild*, p. 131.

60. *Meine ersten Eindrücke in Nordamerika*, in *Berliner Tageblatt*, July 7, 1921. Published in *Weltbild*, p. 41.

61. This interview first appeared in *Nieuwe Rotterdamsche Courant;* then, at the beginning of July 1921, in German papers; and, following a cabled report from Germany, in *The New York Times* on July 8. The present author has seen only the version published in *The New York Times*.

62. *New York Times*, July 8, 1921.

63. *New York Times*, July 12, 1921.

26. More Hustle, Long Journeys, a Lot of Politics, and a Little Physics

1. To H. Zangger, undated, Berlin, beginning of 1921.

2. To P. Ehrenfest, Berlin, September 1, 1921.

3. H. Anschütz-Kaempfe to Einstein, Munich, December 28, 1920.

4. To H. Anschütz-Kaempfe, Berlin, September 14, 1921.

5. To M. Born, Berlin, August 22, 1921, in Born, p. 86.

6. To M. Besso, Florence, October 20, 1921, in Besso, p. 170.

7. Ibid.

8. Einstein 1921c.

9. To M. Born, Berlin, December 30, 1921, in Born, p. 96.

10. To A. Sommerfeld, January 28, 1922, in Sommerfeld, p. 99.

11. To M. Born, undated, beginning of May 1922, in Born, p. 103.

12. To P. Ehrenfest, Berlin, March 15, 1922.

13. To W. Westphal, undated, Berlin, c. 1924.

14. To A. Sommerfeld, January 14, 1922, in Sommerfeld, p. 97.

15. See Kurt Blumenfeld, *Erlebte Judenfrage*, pp. 133 ff. Blumenfeld mistakenly refers to "Ambassador Chicherin."

16. A. Sommerfeld to Einstein, Munich, January 11, 1922, in Sommerfeld, p. 96. The interview was published in *Le Figaro*, Paris, on October 13, 1921.

17. To A. Sommerfeld, Berlin, January 28, 1922, in Sommerfeld, p. 99.

18. H. Anschütz-Kaempfe to Einstein, Munich, February 3, 1922.

19. To Paul Langevin, Berlin, February 27, 1922.

20. To P. Langevin, Berlin, March 6, 1922.

21. In Berlin, p. 210.

22. Count Harry Kessler, *Tagebücher*, p. 289.

23. To P. Langevin, March 13, 1922.

24. To P. Langevin, March 23, 1922.

25. To M. Solovine, Berlin, March 22, 1922, in Solovine, p. 38.

26. Detailed accounts of Einstein's visit to Paris are given in Charles Nordmann, *Einstein expose et discute sa théorie*, in *Revue des deux Mondes*, Vol. IX, 1922, pp. 129–66; and *L'Illustration*, Paris, April 15, 1922. A good survey in Michel Biezunski, *Einstein à Paris*, in *La Recherche*, Vol. 13, Paris, 1982, pp. 502ff.

27. To M. Solovine, Berlin, March 14, 1922, in Solovine, p. 36.

28. Berlin, p. 227.

29. Ibid.

30. To Elsa Einstein, Paris, April 9, 1922, in *Frieden*, p. 68.

31. To R. Rolland, Berlin, April 19, 1922.

32. To M. Solovine, Berlin, April 20, 1922, in Solovine, p. 40.

33. Frank, p. 314. Also Einstein to von Laue, Princeton, April 24, 1950.

34. Count Harry Kessler, *Tagebücher*, p. 333. Einstein's address is published in *Frieden*, pp. 70ff.

35. Emil Gumbel, *Vier Jahre politischer Mord* (Berlin, 1922; new ed., Heidelberg, 1980).

36. See K. Blumenfeld, *Erlebte Judenfrage*, pp. 142ff.

37. *In memoriam Walther Rathenau*, in *Neue Rundschau*, Berlin 1922, p. 815.

38. K. Blumenfeld to Einstein, Jerusalem, March 20, 1955. In this letter Blumenfeld quotes what Einstein had said to him in 1921.

39. To H. Anschütz-Kaempfe Berlin, July 1, 1922; and Anschütz-Kaempfe to Einstein, July 2, 1922.

40. To Pierre Comert, Berlin, July 4, 1922, in *Frieden*, p. 78.

41. To M. Planck, Kiel, July 6, 1922.

42. To Marie Curie, Kiel, July 11, 1922.

43. H. Anschütz-Kaempfe to A. Sommerfeld, Kiel, July 12, 1922.

44. German Reich Patent No. 394 667 of the Anschütz company, especially section 4.

45. See n. 43.

46. To H. Anschütz-Kaempfe, Berlin, July 12, 1922.

47. Ibid.

48. A. and E. Einstein to H. Anschütz-Kaempfe, Berlin, July 16, 1922.

49. To M. Solovine, Berlin, July 16, 1922, in Solovine, p. 42.

50. To M. Born, Berlin, September 9, 1919, in Born, p. 59.

51. Records of the Boxfelde gardening association before 1945 no longer exist. The author can therefore only make an assumption about the duration of his rental agreement. Frau Elfi Fillgis of Berlin, to whom the author is obliged for some information on "Einstein in Boxfelde," thinks that he had rented the plot from 1919 to 1928. In Einstein's correspondence, however, the plot has left unambiguous traces only in 1922. It was at Burgunderweg 2.

52. To H. Anschütz-Kaempfe, Berlin, July 25, 1922.

53. District Administration Spandau to Einstein, September 12, 1922.

54. To District Administration Spandau, Berlin, September 23, 1922.

55. Einstein's fluctuating relations with the Commission for Intellectual Cooperation are described in detail in Clark, pp. 431ff.

56. P. Lenard, *Über Äther und Uräther*, 2nd ed. (Leipzig, 1922), pp. 3 and 9.

57. Adolf Hitler, *Sämtliche Aufzeichnungen 1905–24*, E. Jäckel with A. Kuhn (eds.) (Stuttgart, 1980), p. 268. Hitler's article dates from January 3, 1921.

58. See Tsutomu Kaneko, *Einstein's Impact on Japanese Intellectuals*, in Thomas F. Glick (ed.), *The Comparativce Reception of Relativity* (Boston, 1987), pp. 351ff.

59. Count Harry Kessler, *Tagebücher*, p. 289.

60. Ibid.

61. To Prussian Academy of Sciences, Berlin, July 12.

62. To Wilhelm Solf, German embassy, Tokyo; Miyajima, December 20, 1922, in Berlin, p. 231.

63. *Reisetagebuch* [*Travel Diary*], October 6, 1922.

64. Ibid., Colombo, October 28, 1922.

65. Ibid., Hong Kong, November 9, 1922.

66. German embassy to German Foreign Office in Berlin, Tokyo, January 3, 1923, p. 230.

67. Ibid.

68. Ibid.

69. The schedule of Einstein's journey is given in Kenji Sugimoto, *Albert Einstein—Eine kommentierte Bildbiographie* (Gräfelfing, 1987), p. 78.

70. German embassy to Foreign Office in Berlin, Tokyo, January 3, 1923.

71. Kaneko (see n. 58), p. 363.

72. The discussions about relativity in the cabinet council were reported not only in a Japanese paper, but also, in English tradition, in *Japan Weekly Chronicle*, published in Kaneko; and Clark, p. 369.

73. To M. Solovine, Berlin, Whitsun 1923, in Solovine, p. 42.

74. To M. Besso, Kiel, May 24, 1924, in Besso, p. 202.

75. *Reisetagebuch*, February 3, 1923.

76. Ibid., February 8, 1923.

77. *Singapore Times*, November 3, 1922.

78. Quoted from the catalogue of the Einstein Exhibition 1879–1979 (Jerusalem, 1979).

79. To Heinrich Friedmann, Berlin, March 12, 1929.

80. See n. 78.

81. To M. Solovine, Berlin, Whitsun 1923, in Solovine, p. 44.

82. *Reisetagebuch*, February 13, 1923.

83. To M. Besso, Berlin, December 25, 1925, in Besso, p. 215.

84. Einstein's sojourn in Spain is described in detail in Thomas F. Glick, *Einstein in Spain* (Princeton, 1988).

85. German embassy, Madrid, to German Foreign Ministry, Madrid, March 19, 1923, in Berlin, p. 232.

27. Einstein Receives the Nobel Prize

1. S. Arrhenius to Einstein, Leipzig, undated (probably September 18, 1922).

2. M. von Laue to Einstein, Leipzig, September 18, 1922.

3. To S. Arrhenius, Berlin, September 20, 1922.

4. On the decision making of the Royal Swedish Academy of Sciences in the matter of the Nobel Prizes for physics and chemistry, see Elisabeth

Crawford, *The Beginnings of the Nobel Institution* (Cambridge, 1984).

5. A detailed report of the nominators and their motivation is given in Pais, pp. 505ff.

6. Elisabeth Crawford, pp. 132ff.

7. Gullstrand, for instance, believed that the precession of Mercury's perihelion was an experiment; see Pais, p. 507.

8. S. Arrhenius in *Les Prix Nobel en 1921–1922* (Stockholm, 1923), p. 63.

9. Niels Bohr to Einstein, Copenhagen, November 11, 1911 (draft), Niels Bohr Archive, Copenhagen.

10. To Niels Bohr, on board *Haruna Maru* near Singapore, January 10, 1923.

11. Report of the German legation in Stockholm to the Foreign Ministry, Stockholm, December 12, 1922, in Berlin, p. 113.

12. Report of the Academy to Ministry of Education, Berlin, January 13, 1923; protocol of academy meeting on January 18, 1923, and letter from the presiding secretary of the academy to Einstein of February 15, 1923. Official concern with Einstein's citizenship produced countless documents in the academy, in the embassies, and in the ministries; of these only thirty have survived. The seven most important, including those listed above, are published in Berlin, pp. 113ff.

13. To Presiding Secretary of the Academy, G. Roethe, Berlin, March 24.

14. K. Helferich, State Secretary in the Reich Office of the Interior, to Kaiser Wilhelm II, Berlin, December 22, 1916, in Berlin, p. 159.

15. Minute of Ministerial Councillor von Rottenburg, Berlin, June 19, 1923, in Berlin, p. 117.

16. To Prussian Academy of Sciences, Berlin, February 7, 1924, in Berlin, p. 118.

17. S. Arrhenius to Einstein, March 17, 1923.

18. Einstein 1923c.

19. To S. Arrhenius, Berlin, March 23, 1923.

20. Niels Bohr, interview of July 12, 1961, Niels Bohr Archive, Copenhagen.

21. To P. Ehrenfest, Berlin, July 20, 1923.

22. To H. Anschütz-Kaempfe, Berlin, July 26, 1923.

23. To P. Ehrenfest, Kiel, September 12, 1923.

24. *Vossische Zeitung*, Berlin, March 22, 1923.

25. To H. A. Lorentz, Berlin, August 16, 1923.

26. To H. Anschütz-Kaempfe, Kiel, undated (September 9, 1923).

27. To Betty Neumann, Bonn, September 21, 1923, MPG.

28. Ibid.

29. See Berlin, pp. 215ff.

30. *Deutsche Allgemeine Zeitung*, September 15, 1923; *Vossische Zeitung* No. 359.

31. *Kieler Zeitung*, November 2, 1923.

32. M. Planck to Einstein in Leyden, Berlin, November 10, 1923.

33. M. Planck to P. Ehrenfest, November 30, 1923; and to H. A. Lorentz, December 5, 1923.

34. To M. Planck, Leyden, December 6, 1923, MPG.

35. To M. Besso, Berlin, January 5, 1924, in Besso, p. 197.

36. To C. Weizmann, Berlin, October 27, 1923.

37. To M. Besso, Kiel, May 24, 1924, in Besso, p. 202.

38. To M. Solovine, Berlin, October 30, 1924, in Solovine, p. 48.

39. Count Harry Kessler, *Tagebücher*, p. 414 (entry of December 17, 1924).

40. To C. Seelig, Princeton, April 20, 1952. The picture has been lost.

41. Charles Nordmann, in *L'Illustration*, Paris, April 15, 1922.

42. Vera Weizmann, *The Impossible Takes Longer* (London, 1967), p. 103.

43. Pais, p. 320. The correspondence between Einstein and Betty Neumann is not accessible.

44. See letter to Pauline Winteler, Zurich, May 1897, in CP1, p. 55.

45. Count Harry Kessler, *Tagebücher*, p. 414 (entry of December 18, 1924).

46. Report from German embassy in Buenos Aires to Foreign Ministry, September 26, 1922, in Berlin, p. 228.

47. *Reisetagebuch*, April 17, 1925.

48. Ibid., April 16, 1925.

49. Ibid., March 30, 1925.

50. To M. Besso, Berlin, June 5, 1925, in Besso, p. 204.

51. *Reisetagebuch*, April 22, 1925.

52. Ibid., April 22 and 30.

53. Berlin, pp. 234ff.

54. To M. Besso, Berlin, June 5, 1925, in Besso, p. 404.

55. Einstein 1925c, p. 414.

56. To M. Besso, Geneva, July 28, 1925, in Besso, p. 209.

28. "The Marble Smile of Implacable Nature": The Search for the Unified Field Theory

1. Einstein 1923e, p. 9.

2. To H. Zangger, Princeton, February 27, 1938.

3. To Niels Bohr, Princeton, April 4, 1949.

4. To H. Weyl, Berlin, March 8, 1918.

5. Ibid.

6. To H. Weyl, Berlin, April 8, 1918.

7. H. Weyl to Einstein, Zurich, December 10, 1918.

8. Einstein 1918d.

9. Einstein 1919a, p. 349.

10. To H. Weyl, Berlin, June 6, 1922.

11. Ibid.

12. To Niels Bohr, near Singapore, January 10, 1923.

13. Einstein 1923a, p. 32.

14. To H. Weyl, Berlin, May 26, 1923.

15. Ibid.

16. Einstein 1913e.

17. Walther Gerlach, *Erinnerungen an Albert Einstein*, in Aichelburg and Sexl (eds.) *Albert Einstein* (Braunschweig, 1979).

18. To F. Klein, Berlin, December 12, 1917.

19. To M. Besso, Berlin, August 28, 1918, in Besso, p. 138.

20. Einstein 1923e, p. 9.

21. Ibid., p. 10.

22. Einstein 1929f, p. 127.

23. *Zur Methodik der Theoretischen Physik*, in *Weltbild*, p. 117.

24. To C. Lanczos, Princeton, January 24, 1938.

25. Galileo Galilei, *Il Saggiatore*, L. Sosio (ed.) (Milan, 1965), p. 200.

26. Einstein 1923e, p. 9.

27. Einstein 1917a.

28. Einstein 1925c.

29. Constance Reid, *David Hilbert* (Berlin, Heidelberg, New York, 1970), p. 131.

30. Berlin, p. 154; Einstein 1925c; Einstein 1929a.

31. Einstein 1925c, p. 414.

32. Ibid.

33. To M. Besso, Geneva, July 28, 1925, in Besso, p. 209.

34. To P. Ehrenfest, Kiel, September 18 and 20, 1925.

35. Einstein 1925d.

36. Ibid., p. 334.

29. The Problems of Quantum Theory

1. Max Born to Einstein, Göttingen, July 15, 1925, in Born, p. 121. The paper referred to by Born is W. Heisenberg's *Über quantentheoretische Umdeutung kinematischer und mechanischer Beziehungen* [*On Quantum-Theoretical Reinterpretation of Kinematic and Mechanical Relations*], in ZfP, Vol. 33, pp. 879–93.

2. To P. Ehrenfest, Kiel, September 30, 1925.

3. Max Born, *Einstein's Statistical Theories*, in Schilpp, p. 84.

4. Einstein 1909c, pp. 482f.

5. Einstein 1916j, 1916k, and 1917h.

6. To M. Besso, Berlin, September 6, 1916, in Besso, p. 82.

7. To M. Besso, Ahrenshoop, July 29, 1918, in Besso, p. 130.

8. Einstein 1916j, p. 23.

9. To M. Born, Berlin, January 27, 1920, in Born, p. 44.

10. To W. Dällenbach, undated, no place (probably Zurich), January 1919, in Stargard Catalogue, No. 588, February 1969.

11. Einstein 1921e.

12. Einstein 1924b.

13. To W. Dällenbach, undated, no place, (Berlin, 1917).

14. To M. Born, Berlin, January 20, 1920, in Born, p. 43.

15. To M. Born, Berlin, March 3, 1920, in Born, p. 49.

16. Einstein 1923k.

17. Ibid., p. 360.

18. Ibid., p. 361.

19. To M. Besso, Berlin, January 5, 1924, in Besso, p. 197.

20. To P. Ehrenfest, Berlin, May 31, 1924.

21. W. Heisenberg to W. Pauli, Göttingen, July 8, 1924.

22. To M. Born, Berlin, April 19, 1924, in Born, p. 118.

23. Esther Salaman, *Memoirs of Einstein*, in *Encounter*, April 1979, p. 22.

24. On the relationship between Bose and Einstein, see William Blanpied, *Einstein as a Guru? The Case of Bose*, in M. Goldsmith, A. McKay, and J. Woudhuysen, *Einstein: The First Hundred Years* (London, 1980), pp. 64ff.

25. S. N. Bose to Einstein, Dacca, June 15, 1924.

26. Einstein 1924i, p. 261.

27. S. N. Bose, *Planck's Law and Light Quanta Hypothesis*, in ZfP, Vol. 26, 1924, p. 178–81.

28. Einstein 1925a, p. 5.

29. Einstein 1924i.

30. Einstein 1925a, p. 3.

31. Einstein 1925b, p. 18.

32. A statistics in which every state can be occupied by no more than one particle was developed in 1926; it is called "Fermi-Dirac statistics." To Wolfgang Pauli we owe a profound relation between spin and statistics, as well as the symmetry characteristics of the wave functions of quantum-mechanical systems.

33. Einstein 1925a, p. 9.

34. Ibid., footnote on p. 9.

35. This is the recollection of I. I. Rabi and E. Wigner, in *Princeton Symposium*, pp. 471f.

36. Einstein 1925a, p. 10.
37. W. Pauli in Schilpp, p. 80.

30. Critique of Quantum Mechanics

1. Probably the best reconstruction of the history of quantum mechanics is B. L. van der Waerden, *Sources of Quantum Mechanics* (Amsterdam, 1967). This book contains all the essential original studies, with instructive introductions and interpretations. For the sources of these studies, see van der Waerden.

2. See Samuel A. Goudsmit, *It Might Well Be Spin*; George E. Uhlenbeck, *Personal Reminiscences*. Both articles in *Physics Today*, June 1976.

3. N. Bohr to P. Ehrenfest, Copenhagen, December 22, 1926.

4. To M. Besso, Berlin, March 7, 1926, in Besso, p. 216.

5. To Hedwig Born, Berlin, March 7, 1926, in Born, p. 127.

6. C. F. von Weizsäcker, *Zeit und Wissen* (Munich, 1982), p. 781. Letters from Einstein to Heisenberg are all lost; five letters from Heisenberg to Einstein are preserved in the Einstein Archive; the first dates from November 30, 1925, and is a reply to a letter from Einstein.

7. W. Heisenberg, *Der Teil und das Ganze* (Munich, 1969), pp. 90–100.

8. Ibid., pp. 91f.

9. Ibid., p. 98.

10. To P. Ehrenfest, Berlin, April 12, 1926.

11. To E. Schrödinger, Berlin, April 16, 1926.

12. To P. Ehrenfest, Berlin, April 12, 1926.

13. Matrix mechanics and wave mechanics are different representations of the abstract theory of operators in "Hilbert space," a subject on which Jordan submitted a comprehensive study in November 1928.

14. Heisenberg to W. Pauli, Copenhagen, June 8, 1926, in Pais, p. 328.

15. W. Pauli to E. Schrödinger, Hamburg, November 22, 1926, in Pauli, p. 357.

16. To Paul Epstein, Berlin, June 10, 1926.

17. M. Born, *On the Quantum Mechanics of Impact Processes*, in ZfP, Vol. 37, p. 863. Received on June 25, 1926. In the correction at the proof stage Born actually noted that the probability is proportional to the square of the wave function. He should have written "absolute square."

18. Ibid., p. 864.

19. Ibid., p. 804.

20. See Pauli's letter to Bohr, Hamburg, October 2, 1924, in Pauli, p. 164.

21. Born, see n. 17.

22. Ibid., p. 834.

23. M. Born to Einstein, Göttingen, November 30, 1926. This letter is not contained in the published correspondence.

24. To M. Born, Berlin, December 4, 1926, in Born, p. 129.

25. Born, p. 130.

26. W. Pauli to W. Heisenberg, Hamburg, October 10, 1926, in Pauli, p. 347.

27. W. Heisenberg, *Über den anschaulichen Inhalt der quantentheoretischen Kinematik und Mechanik* [*On the Visualizable Content of Quantum-Theory Kinematics and Mechanics*], in ZfP, Vol. 43, 1927, pp. 172–98.

28. Einstein 1927d.

29. Einstein 1927g, p. 276.

30. The preserved page of the proof is reproduced in facsimile in Berlin, p. 129. The manuscript of the paper is in the Einstein Archive.

31. To H. A. Lorentz, Berlin, June 17, 1927, quoted in Pais, pp. 431f.

32. Inst. Int. de Physique Solvay, *Rapport et discussion du 5ᵉ Conseil* (Paris, 1928), pp. 253ff.

33. See *Diskussion mit Einstein* . . . [*Discussion with Einstein* . . .], in Schilpp, pp. 124ff.

34. P. Ehrenfest to Goudsmit, Uhlenbeck, and Dieke, November 3, 1927.

35. W. Pauli to H. Weyl, Zurich, July 1, 1928, in Pauli, p. 506.

36. To Nobel Committee for Physics, September 25, 1928. The Swedish Academy essentially followed Einstein's proposal: the 1929 prize went to L. de Broglie, whereas Davisson was not honored until 1937.

37. Einstein 1929g.

38. Ibid.

39. To W. Pauli, Berlin, December 24, 1924.

40. Léon Rosenfeld in *Proceedings of the 14th Solvay Conference* (New York, 1968), p. 232, quoted in Pais, p. 454.

41. Bohr, in Schilpp, pp. 152ff.

42. To Nobel Committee for Physics, September 30, 1931. No prize was awarded in 1932; in 1933 Heisenberg received the prize for 1932; the prize for 1933 was awarded to Schrödinger and Dirac.

31. Politics, Patents, Sickness

1. To P. Ehrenfest, Berlin, July 12, 1914.

2. Einstein 1923i.

3. C. Weizmann to Einstein, July 9, 1926.

4. To C. Weizmann, Berlin, January 8, 1928.

5. To Fritz Haber, Le Coq-sur-mer, August 9, 1933.

6. To C. Weizmann, Berlin, June 14 and 20, 1928,

7. Quoted in Selig Brodetzky, *Memoirs: From Ghetto to Israel* (London, 1960), p. 130.

8. *Neue Zürcher Zeitung*, November 20, 1927; quoted in *Frieden*, p. 105.

9. Ibid.

10. To Elsa Einstein, Geneva, July 1927.

11. To Elsa Einstein, Paris, January 17, 1926, quoted in *Frieden*, p. 97.

12. *Neue Zürcher Zeitung*, November 20, 1927; quoted in *Frieden*, p. 106.

13. Janos Plesch, *Janos. Ein Arzt erzählt sein Leben* (Munich, 1949), p. 140.

14. To H. Anschütz-Kaempfe, Berlin, August 31, 1926.

15. Contract between Giro and Albert Einstein about the gyrocompass, Kiel-Neumühlen, October 11, 1926. The contract referred to German Reich Patent 394 667, especially to the construction of the coil described in subsection 4. Under this contract Einstein received royalties only for instruments sold abroad.

16. D. Bludau, *Anschütz & Co. 1905–1935* (Kiel, 1955), p. 46.

17. Einstein's correspondence with Giro, starting in 1928. Last letters: Einstein to Giro, Princeton, January 23, 1940; and Giro to Einstein, Amsterdam, February 27, 1940.

18. German Reich Patent 563 403 of November 13, 1927. L. Szilard and A. Einstein, *Kältemaschine*. These and all later patents were granted in Germany in 1934, when Einstein and Szilard had long since left the country.

19. The patents are listed in Berlin, vol. 2, p. 290.

20. See M. Besso to Einstein, Bern, May 1, 1930, in Besso, p. 256.

21. L. Szilard to Einstein, Berlin, September 27, 1930.

22. L. Szilard to Einstein, Berlin, October 12, 1929.

23. Quoted in Pais, p. 495.

24. P. Ehrenfest to Einstein, Leyden, January 11, 1927.

25. To P. Ehrenfest, Berlin, January 11, 1927.

26. P. Ehrenfest to Einstein, Leyden, January 13, 1927.

27. To K. Singer, Berlin, August 16, 1926.

28. To F. S. Archenold, Berlin, October 14, 1926.

29. To Office of the Reich Chancellery, Berlin, November 28, 1926.

30. Einstein 1930h. Introductory remarks on the Davos university courses before his lecture on March 18, 1928, in *Weltbild*, pp. 25–27.

31. *Weltbild*, p. 188.

32. Plesch (see n. 13), p. 148.

33. To H. Zangger, undated, Berlin, c. May 1928.

34. To M. Besso, Gatow (near Berlin), January 5, 1929, in Besso, p. 241.

35. Plesch (see n. 13), pp. 110ff.

36. To H. Zangger (see n. 33).

37. Plesch (see n. 13), p. 138.

38. To H. Zangger (see n. 33).

39. To P. Ehrenfest, Scharbeutz, July 12, 1928.

40. To P. Ehrenfest, Scharbeutz, August 28, 1928.

41. To L. Szilard, Scharbeutz, September 15, 1928.

42. To P. Ehrenfest, Scharbeutz, July 12, 1928.

43. To L. Szilard, Scharbeutz, September 15, 1928.

44. Elsa Einstein to Hermann Struck, Scharbeutz, September 27, 1928.

45. To P. Ehrenfest, Scharbeutz, August 28, 1928.

46. To H. Zangger, undated, no place (Berlin, end of May 1928).

47. To M. Besso, Gatow, January 5, 1928, in Besso, p. 240.

48. Einstein 1928c and 1928d.

49. A similar mathematics had already been developed by the French mathematician Elie Cartan. According to Cartan's recollection, he had even discussed it with Einstein in Berlin in 1922. See R. Debever, *Elie Cartan—Albert Einstein: Letters on Absolute Parallelism* (Princeton, 1979).

50. To M. Besso, Gatow, January 5, 1929, in Besso, p. 240.

51. Elsa Einstein to Hermann Struck, Berlin, December 27, 1928.

52. Quoted in Pais, p. 346.

53. Berlin, pp. 135ff.

54. *Wie sich die Haus-Geschichte der Villa Lemm belebt* [*How the Domestic Story of the Villa Lemm Comes to Life*], in *Der Tagesspiegel*, Berlin, June 7, 1987.

55. To M. Besso, Gatow, January 5, 1929, in Besso, p. 240.

56. A. S. Eddington to Einstein, February 11, 1929.

57. On the quarrel between Einstein and Reichenbach, see Klaus Hentschel, *Interpretationen* (Basel, Boston, and Berlin, 1990), pp. 192ff.

58. Einstein 1929c.

59. Einstein 1929b.

60. Einstein 1929a, p. 7.

61. To P. Ehrenfest, Caputh, September 24, 1929.

62. To Maja Winteler-Einstein, Caputh, October 22, 1929.

63. Einstein 1930a.

64. Max Born, in Born, p. 126.

65. W. Pauli to Einstein, Zurich, December 19, 1929, in Pauli, p. 527.

66. To W. Pauli, Berlin, December 24, 1929.

67. To W. Pauli, Pasadena, January 22, 1932.

32. Public and Private Affairs

1. L. Lenz, *Relativitätstheorie und dialektischer Marxismus*, in *Rote Fahne*, March 14, 1929.

2. E. Ludwig, *Zum 50. Geburtstag*, in *Berliner Tageblatt*, March 14, 1929.

3. *Gelegentliches. Zu Einsteins 50. Geburtstag*, 32 pp., Soncino, Gesellschaft der Freunde des jüdischen Buches (Berlin, 1929).

4. Berlin, p. 221.

5. Berlin city government and deputies to Einstein, March 14, 1929.

6. To "Dear Colleagues" (R. Ladenburg, H. Kopfermann, etc.), Berlin, April 9, 1929. Ladenburg papers, Stargard Catalogue 1989, No. 456.

7. To Sigmund Freud, Berlin, March 22, 1929. See also Freud to Einstein, November 23, 1930, in Ernest Jones, *Sigmund Freud*, Vol. 3 (Munich, 1984, p. 186).

8. Janos Plesch, in *Janos. Ein Arzt erzählt sein Leben* (Munich, 1949), p. 154.

9. *Einsteins Geburtstagsgeschenk*, in *Berliner Tageblatt*, April 17, 1929.

10. *Die Cladower Fußangel*, in *Berliner Tageblatt*, April 22, 1929.

11. Berlin, p. 221.

12. Einstein's letter declining the gift is lost. According to the recollection of the architect, Wachsmann, it was written on May 11. See Michael Grüning, *Ein Haus für Albert Einstein* (Berlin, 1990), pp. 122ff.

13. Exceedingly detailed, but not always accurate, information is given by Michael Grüning (see n. 12).

14. To Maja Winteler-Einstein, Berlin, November 19, 1929.

15. To Maja Winteler-Einstein, Caputh, August 19, 1929.

16. Elsa Einstein to Maja Winteler-Einstein, Caputh, August 19, 1929.

17. Wachsmann, quoted in M. Grüning (see n. 12).

18. To Maja Winteler-Einstein, Caputh, October 22, 1929.

19. Ibid.

20. Einstein 1930a.

21. Report by German Embassy, Paris, to Foreign Ministry in Berlin, November 22, 1922; in Berlin, p. 235.

22. Ibid.

23. For instance, *Berliner Börsenzeitung*, November 11, 1929, *Le Journal*, November 13, 1929.

24. To Emile Borel, Berlin, November 19, 1929.

25. To Maja Winteler-Einstein, Berlin, November 18, 1929.

26. To M. Solovine, Berlin, March 4, 1930, in Solovine, p. 56.

27. Einstein 1930c.

28. To M. Solovine, Berlin, March 4, 1930, in Solovine, p. 56.

29. *Tagore Talks with Einstein*, in *Asia*, Vol. 31, pp. 138–42.

30. To R. Rolland, Berlin, October 10, 1930.

31. Elsa Einstein to Hedwig Born, Caputh, September 13, 1930.

32. Plesch (see n. 8), p. 145.

33. Ibid., p. 146.

34. Walter Friedrich, in Herneck, *Einstein privat* (Berlin, 1978), p. 129.

35. Brigitte B. Fischer, *Sie schrieben mir* (Zurich and Stuttgart, 1978), p. 23.

36. The concert was on January 29, 1930. Einstein's partner was the renowned violinist Alfred Lewandowski.

37. Eugene Wigner, *Ich war Einsteins jüngerer Freund*, in Grüning (see n. 12), p. 529.

38. K. Wachsmann, in Grüning (see n. 12), p. 158.

39. Friedrich Herneck, *Einstein privat* (Berlin, 1978), pp. 146f.

40. Hertha Waldow, ibid., p. 44.

41. Ibid., p. 48.

42. Ibid., p. 124.

43. Ibid., pp. 146f.

44. To David Reichinstein, Oxford, April 2, 1932.

45. Ibid.

46. To David Reichinstein, Caputh, May 26, 1932.

47. David Reichinstein, *Albert Einstein, sein Lebensbild und seine Weltanschauung* (Prague, 1934).

48. Anton Reiser, *Albert Einstein: A Biographical Portrait* (New York, 1930).

49. To Antonina Luchaire-Vallentin, Princeton, November 10, 1953.

50. A. Einstein, *The World as I See It*, in *Forum and Century*, Vol. 84, 1931, pp. 193–94. Also in *Living Philosophies* (New York, 1931), pp. 3–7. Quoted here from the German original, *Weltbild*, pp. 7–10.

51. To Rudolf Kallir, Princeton, February 28, 1952.

52. Statement made to a working group on poison gas warfare in Geneva, Berlin, January 4, 1928, quoted in *Frieden*, p. 109. This collection of documents contains ample material on Einstein's pacifist commitment between 1928 and 1933.

53. To Jacques Hadamard, Caputh, September 29, 1929.

54. *Frieden*, p. 110.

55. Ibid., p. 113.

56. *Briefe*, p. 81; written November 4, 1931, in Berlin.

57. *Frieden*, pp. 144f., 151.

58. Ibid., p. 160.

59. Quoted from the report on the inaugural speeches at the first session, *Bericht des 16. Zionistischen Kongresses* (Zurich, 1929), p. 578.

60. August 11, 1929. Original in the Weizmann Archive, Rehovot.

61. *The Problem of the Jews in Palestine*, in *Manchester Guardian*, October 12, 1929.

62. Selig Brodetzky, *Memoirs: From Ghetto to Israel* (London, 1960), p. 137.

63. To Chaim Weizmann, Berlin, November 25, 1929.

64. To Azmi El Nashasili, editor of the journal *Falastin*, Berlin, January 16, 1930, and March 15, 1930; see *Brief an einen Araber*, in *Weltbild*, p. 102.

65. Film and sound record of Einstein's opening address at the Radio Exhibition on August 22, 1930, in German Radio Archives, Frankfurt. Transcription: Einstein 1930p.

66. F. Haber to Einstein, Berlin, November 17, 1930.

67. Lecture on February 23, 1927; Einstein 1927d.

68. Lecture arranged by the Kaiser Wilhelm Society for invited guests and "numerous representatives of the press" on December 11, 1929.

69. The Max Planck Medal has been awarded by the German Physical Society every two years since 1929 for outstanding achievements in theoretical physics. It is one of the most highly regarded distinctions.

70. Einstein 1929g.

71. Plesch (see n. 8), p. 143.

72. Report by the Philosophical Faculty of Berlin University to the Ministry of Culture, Berlin, December 4, 1926, in Berlin, p. 134.

73. To P. Ehrenfest, Berlin, May 5, 1927.

74. To M. Planck, Berlin, July 7, 1914.

75. To M. von Laue, Berlin, June 12, 1926.

76. Petition to the President of the Kaiser Wilhelm Society, signed by Einstein, Haber, von Laue, Nernst, Paschen, Planck, and Warburg; MPG.

77. Suggestion to the president of the Kaiser Wilhelm Society concerning the Kaiser Wilhelm Institute for Physics, Berlin, March 6, 1929; signed by Rudolf Ladenburg, Michael Polanyi, Otto Hahn, Lise Meitner, and others; MPG.

78. See Kristie Macrakis, *Wissenschaftsförderung der Rockefeller-Stiftung im "Dritten Reich,"* in *Geschichte und Gesellschaft*, Vol. 12, pp. 348ff.

79. L. W. Jones, deputy director for natural sciences of the Rockefeller Foundation, in memorandum of January 2 and 5, 1931. Quoted in Macrakis (see n. 78), pp. 356f.

80. See Arthur Fleming's report to Allan Balch, December 5, 1934, Millikan papers, Caltech, Pasadena.

81. To Jewish Telegraph Agency, Caputh, September 18, 1930.

82. See Chapter 30.

83. To Herbert Samuel, Berlin, September 25, 1930.

84. Excerpts from G. B. Shaw's address to Einstein on October 28, 1930, in *Weltbild*, pp. 184f.

85. Vera Weizmann, *The Impossible Takes Longer* (London, 1967), p. 113.

86. Note by Herbert Samuel of October 30, 1930.

87. Clark, p. 511, according to diary of Queen Elizabeth of the Belgians.

88. To Elsa Einstein, Zurich, c. November 1, 1930, in *Briefe*, p. 46.

89. To M. Besso, Zurich, November 1, 1930, in Besso, p. 263.

90. To P. Ehrenfest, Berlin, March 21, 1930.

91. Quoted in *Weltbild*, pp. 15ff.

92. *Berliner Tageblatt*, November 11, 1930.

93. *New York Times*, November 10, 14, and 16, 1930.

33. Farewell to Berlin

1. *Reisetagebuch*, December 2 and 3, 1930.

2. Ibid., December 3, 1930.

3. Ibid., December 10, 1930.

4. Ibid., December 11, 1930.

5. One of the two greetings is published in *Frieden*, pp. 132–34.

6. *Reisetagebuch*, December 11, 1930.

7. Short summary based on Einstein's travel diary and reports in the *New York Times*.

8. *Frieden*, p. 133.

9. The original German text of the "two percent" speech is not extant, but there are five English translations. The text here is retranslated from *Frieden*, p. 132.

10. *Der Tag*, Berlin, December 16,

1930. The article with the ministry's request for information and Planck's and Glum's initials is in the MPG Archive.

11. To Lebach family, Pasadena, January 16, 1931.

12. Ibid.

13. Chapter 27.

14. Max Born to Einstein, Göttingen, February 22, 1931, in Born, p. 152.

15. Hedwig Born to Einstein, Göttingen, February 22, 1931, in Born, p. 154.

16. See n. 11.

17. *Frieden*, p. 137.

18. See n. 11.

19. See the report *Professor Einstein at the California Institute of Technology*, in *Science*, Vol. 73, 1931, pp. 375–81.

20. The original German text of Einstein's after-dinner speech in Pasadena on January 15, 1931, is published as *Dr. Einstein's Address*, in *Proceedings of the American Philosophical Society*, Vol. 93, 1949, p. 544.

21. Actually, there was also a model of the universe developed by Willem de Sitter on the basis of Einstein's equations, with a vanishing mass density. This will not be considered here.

22. Alexander A. Friedmann, *Über die Krümmung des Raumes*, in ZfP, Vol. 10, pp. 377–86.

23. Einstein 1922i.

24. Einstein 1923l.

25. See Edmund Hubble, *The Realm of Nebulae* (New Haven, 1936; reprinted New York, 1958), which is still interesting.

26. Einstein 1931a.

27. Einstein 1931d.

28. *Frieden*, pp. 138f.

29. To Elizabeth, Queen of the Belgians, Pasadena, February 9, 1931.

30. Report of the German consulate–general, Chicago, March 6, 1931, in Berlin, p. 236.

31. Report of the German consulate–general in New York, March 21, 1931, in Berlin, pp. 237ff.

32. *Frieden*, p. 140.

33. Berlin, p. 238.

34. Telegram from Weizmann to Einstein, London, February 4, 1931.

35. *New York Times*, March 5, 1931, in *Frieden*, p. 140.

36. *New York Times*, March 5, 1931.

37. To Maja Winteler-Einstein, on board the *Deutschland*, mid–March 1931.

38. Einstein 1931a.

39. Einstein 1931b.

40. See the detailed account in Clark, pp. 529ff.

41. No text of the Rhodes Lectures is extant. Brief reports were published in *Nature*, Vol. 127, pp. 764, 790, 826. The first lecture dealt with relativity theory, the second with cosmology, and the third with Einstein's unified theory.

42. Diary, May 26, 1931.

43. F. C. Lindemann to Lord Lothian, June 27, 1931; quoted in Clark, p. 534.

44. F. C. Lindemann to Einstein, June 29, 1931.

45. To F. C. Lindemann, Caputh, July 15, 1931.

46. For further details, see the volume of documents *Einstein über den Frieden*.

47. To M. Planck, Caputh, July 17, 1931.

48. Frederick S. Litten, *Einstein and the Noulens Affair*, in the *British*

Journal of the History of Science, Vol. 24, 1991, pp. 465ff.

49. Marxistische Arbeiterschule "Masch" in Berlin, program for the first term of the sixth year. Einstein's lecture is announced in the program for October 25. The text is not extant.

50. To Rudolf Hilferding, then Reich minister of finance, Berlin, April 1929.

51. See the protracted correspondence with Henri Barbusse.

52. To Käthe Kollwitz, Caputh, October 10, 1930. The author has been unable to find the protest signed by Einstein.

53. *Das Neue Rußland*, Issue 8/9, 1931, p. 40, in Berlin, p. 222.

54. On Münzenberg's contacts with Einstein, see Babette Gross, *Willi Münzenberg*, series published by *Vierteljahreshefte für Zeitgeschichte* (Stuttgart, 1967), p. 234.

55. To Max Born, undated, no place (Princeton, February or March 1937), in Born, p. 179.

56. Minute of W. Windelband, *Ministerialrat* in the Prussian Ministry of Education, Berlin, April 11, 1931; also Windelband to Einstein, Berlin, April 15, 1930, in Berlin, pp. 139f.

57. See Clark, pp. 536ff.

58. Report of the German legation to the German Foreign Ministry, Vienna, October 15, 1931, in Berlin, p. 240.

59. To M. Besso, Caputh, October 30, 1931, in Besso, p. 276.

60. *Reisetagebuch*, December 6, 1931.

61. Einstein 1932b.

62. Einstein 1931c.

63. Abraham Flexner, *I Remember* (New York, 1940), pp. 381ff.

64. To P. Ehrenfest, April 3, 1932, Rotterdam, on board the *San Francisco*.

65. To Herbert E. Salzer, Princeton, September 13, 1938.

66. Einstein 1931c, p. 541.

67. To P. Ehrenfest, Caputh, September 17, 1931.

68. To P. Ehrenfest, Caputh, October 2, 1931.

69. In Cambridge Einstein gave the Rouse Ball Lectures on Mathematics.

70. Flexner (see n. 63), p. 383.

71. See *Frieden*, pp. 183ff.

72. Flexner (see n. 63), p. 384.

73. A. Flexner in *The New York Times*, April 19, 1955.

74. A Flexner, Memorandum of agreement with Einstein of June 6, 1932.

75. Flexner (see n. 63), p. 385.

76. To A. Flexner, Caputh, June 8, 1932.

77. Ibid.

78. Elsa Einstein to Antonina Luchaire-Vallentin, Caputh, June 6, 1932.

79. Appeal to Theodor Leipart, Ernst Thälmann, and Otto Wels from Heinrich Mann, Käthe Kollwitz, and Albert Einstein, Berlin, June 17, 1932.

80. Facsimile in Berlin, p. 235.

81. To Philipp Frank in Caputh in the summer of 1932, in Frank, p. 363.

82. S. Freud to S. Ferenczi, January 2, 1927. Quoted in Ernest Jones, *Sigmund Freud*, 3 Vols. (Munich, 1984), p. 160.

83. Draft of a reply to the request of a psychotherapist intending to write a book based on the psychoanalysis of important people, in *Letters*, p. 35.

84. To S. Freud, Caputh, April 29, 1931.

85. To S. Freud, Caputh, July 30, 1930.

86. To S. Freud, Caputh, July 20, 1930. The open correspondence

between Einstein and Freud on the question "Why War?" is published in *Frieden*, pp. 204–19.

87. *Vossische Zeitung*, August 25, 1932.

88. To Prussian Academy, September 6, 1932.

89. To Prussian Ministry of Education, Caputh, September 13, 1932.

90. See Letter from Ministry of Education to the Academy, December 24, 1924, in Berlin, p. 212.

91. *New York Times*, October 16, 1932.

92. Elsa Einstein to R. Millikan, Caputh, June 22, 1932.

93. Millikan to Flexner, July 25, 1932; Flexner to Millikan, July 29, 1932; Elsa Einstein to Millikan, August 13, 1932.

94. Mrs. Randolph Frothingham, President of the American Women's League, to Head of the Visa Section of the State Department, November 19 and 22, 1932. Section 1 of the FBI file on Albert Einstein, in Sayen, pp. 6–7.

95. *New York Times*, December 4, 1932.

96. Statement for Associated Press, Berlin, December 3, 1932. Published in *Weltbild*, p. 45.

97. *New York Times*, December 7, 1932.

98. Brief biography of Einstein in the files of the U.S. Army relating to the security clearance of physicists for the atomic bomb project in 1940. See R. A. Schwartz, *Einstein and the War Department*, in *Isis*, Vol. 80, 1989, pp. 281–84.

99. To M. Solovine, Caputh, November 20, 1032, in Solovine, p. 64.

100. Frank, pp. 363f.

101. R. A. Millikan to Wilbur K. Thomas, Oberlaender Trust, January 16, 1933.

102. Ibid.

34. Exile as Liberation

1. *Reisetagebuch*, February 7, 1933.

2. *Frieden*, p. 227. There is no record in the Prussian Academy's papers of Einstein's planned lecture or of its cancellation.

3. *New York World Telegram*, March 11, 1932.

4. Evelyn Seely, in *New York World Telegram*, March 11, 1932.

5. *The Fight against War*, Alfred Lief, ed. (New York, 1933).

6. *New York Times*, March 18, 1933.

7. *Frieden*, p. 229.

8. F. Herneck, *A. E. und das politische Schicksal seines Sommerhauses in Caputh*, in Herneck, *Einstein und sein Weltbild* (Berlin, 1976), pp. 263ff.

9. Actually, according to German law at the time, surrender of one's citizenship, as intended by Einstein, was not possible. Einstein was later officially declared to have forfeited his citizenship, in 1934.

10. To Prussian Academy of Sciences, SS *Belgenland*, March 28, 1933, in Berlin, p. 246.

11. *Berliner Lokalanzeiger*, March 17, 1933.

12. Max Planck to Einstein, Berlin, March 19, 1933, in Armin Hermann, *Max Planck* (Reinbek, 1973), pp. 77f.

13. Reich Commissioner Rust to the Prussian Academy, March 29, 1933, in Berlin, p. 245.

14. Max Planck to Einstein, Munich, March 31, 1931, in Hermann (see n. 12), p. 78.

15. Max Planck to Heinrich von

Ficker, Munich, March 31, 1933, in Berlin, p. 245.

16. M. von Laue to H. von Ficker, Göttingen, July 11, 1947, in Berlin, p. 273.

17. Berlin, p. 248.

18. M. von Laue to H. von Ficker, Göttingen, July 11, 1947, in Berlin, p. 274.

19. Minutes of the Extraordinary Session of the Plenum of the Prussian Academy of Sciences on April 6, 1933, in Berlin, p. 251. Nernst and Planck did not attend that meeting. At first Gottlieb Haberland had also demanded that the press statement "must not remain uncontradicted in its brusqueness," but later he did not support Laue at the meeting.

20. All important data on the dispute between Einstein and the Academy are in Berlin, pp. 243–75.

21. To M. Planck, Le Coq-sur-Mer, April 6, 1933, in *Frieden*, p. 232.

22. M. Planck to Einstein, Taormina, April 13, 1933.

23. M. Planck to Einstein, Munich, March 31, 1933.

24. To M. Planck, Le Coq-sur-Mer, April 6, 1933, in *Frieden*, p. 235.

25. To Ludwig Silberstein, Watch Hill, Rhode Island, September 20, 1934. Planck of course was secretary of the academy, which had no president; this mistake was made by Einstein more than once.

26. Minutes of the plenary meeting of the Prussian Academy on May 11, 1933, in Berlin, p. 267.

27. Despite plausible assertions, there is no evidence that Einstein's books were thrown into the flames at the book burning. Perhaps his two little volumes were too slim for the purpose.

28. To M. von Laue, Oxford, June 7, 1933.

29. To M. Born, Oxford, May 20, 1933, in Born, p. 160.

30. To M. Planck, Le Coq-sur-Mer, April 6, 1933.

31. To M. Besso, Le Coq-sur-Mer, May 5, 1933, in Besso, p. 294.

32. F. Herneck, *Einstein privat* (Berlin, 1978), pp. 150ff.

33. Ibid., pp. 152f.

34. To M. Besso, Le Coq-sur-Mer, May 5, 1933, in Besso, p. 294.

35. *Reisetagebuch*, Buenos Aires, April 17, 1925.

36. To P. Ehrenfest, Le Coq-sur-Mer, May 1, 1933.

37. To M. Born, Oxford, May 30, 1933, in Born, p. 160.

38. To P. Ehrenfest, Le Coq-sur-Mer, May 19, 1933.

39. To F. Haber, Le Coq-sur-Mer, August 8, 1933.

40. To F. Lindemann, Le Coq-sur-Mer, May 1, 1933.

41. To F. Haber, Le Coq-sur-Mer, August 8, 1933.

42. To M. Solovine, Le Coq-sur-Mer, April 23, 1933, in Solovine, p. 66.

43. To P. Langevin, Oxford, June 4, 1933.

44. A. Flexner to Einstein, April 15, 1933.

45. To M. Born, Oxford, May 30, 1933, in Born, p. 159.

46. To M. Solovine, Le Coq-sur-Mer, April 23, 1933.

47. To P. Ehrenfest, Le Coq-sur-Mer, May 2, 1933.

48. To M. Born, Oxford, May 30, 1933, in Born, p. 160.

49. *Reisetagebuch*, Jerusalem, February 13, 1923.

50. To M. Born, Oxford, May 30, 1933, in Born, p. 159.

51. To P. Ehrenfest, Oxford, June 14, 1933.

52. Ibid.

53. *Jewish Chronicle*, April 8, 1933.

54. Jewish Telegraph Agency, July 1, 1933.

55. To F. Haber, Le Coq-sur-Mer, August 8, 1933.

56. Quoted in Norman Rose, *Chaim Weizmann* (New York, 1986), p. 297.

57. To F. Lindemann, Le Coq-sur-Mer, May 9, 1933.

58. To Eduard Einstein, March 1931.

59. See the correspondence with Anschütz and, especially, with Zangger during 1926 and 1927.

60. To L. Hopf, quoted in Seelig, p. 339.

61. See Eduard Rübel (ed.), *Eduard Einstein* (Bern and Stuttgart, 1986). These "reminiscences by former classmates" contain many poems and aphorisms of Eduard Einstein.

62. Elsa Einstein to Hedwig Born, Caputh, September 18, 1930.

63. H. Zangger to Einstein, Zurich, October 1932.

64. To M. Besso, Caputh, October 21, 1933.

65. To P. Ehrenfest, undated, no place (Berlin, August 1932).

66. Einstein 1933c. German original text in *Weltbild*, pp. 113–19.

67. Einstein 1933d. German original text in *Weltbild*, pp. 134–38.

68. Quoted in *Frieden*, p. 242.

69. To King Albert of the Belgians, Le Coq-sur-Mer, July 14, 1933, in *Frieden*, p. 243.

70. To Alfred Nahon, Le Coq-sur-Mer, July 20, 1933, in *Frieden*, p. 245.

71. To Helen Dukas, September 10, 1933.

72. Statement of April 4, 1934, in *Frieden*, p. 165.

73. Romain Rolland, diary entry of September 1933, quoted in *Frieden*, p. 149.

74. There is no written evidence of assassination plans or of a price on Einstein's head.

75. Frank, p. 386.

76. To Elsa Einstein, July 21, 1933, in *Frieden*, p. 253.

77. Quoted from the original German draft of the speech, in *Frieden*, pp. 253ff.

78. There seems to be no reliable record of Einstein's actual words in the Royal Albert Hall. The German draft does not contain the "lighthouse" passage; the present author is quoting from Clark, p. 611, who based his account on reports in the press.

79. *New Statesman*, October 21, 1933, p. 481.

35. Princeton

1. The address was 2 Library Place.

2. To Elizabeth, Queen of the Belgians, Princeton, November 20, 1933.

3. H. Weyl to C. Seelig, Zurich, May 19, 1952.

4. See Walter Moore, *Schrödinger* (Cambridge, England, 1989), p. 292–94.

5. To Board of Trustees of the Institute of Advanced Study; German draft, undated, no place (Princeton, end of November 1933). Einstein lists Flexner's misdemeanors over five pages.

6. A. Flexner to Einstein, October 13, 1933.

7. A. Flexner to Franklin D. Roosevelt, Princeton, November 3, 1933.

8. Through a letter from Henry

Morgenthau, then under secretary of the Treasury.

9. To Eleanor Roosevelt, Princeton, November 21, 1933.

10. The author of this book has not seen this letter; it is quoted in Sayen, p. 66.

11. To Board of Trustees of the Institute for Advanced Study (see n. 5).

12. The verses for the Queen of the Belgians are published in *Letters*, p. 48.

13. To Elizabeth, Queen of the Belgians (see n. 2).

14. To Ludwig Silberstein, Watch Hill, Rhode Island, September 20, 1934.

15. Message to the Anti-War Committee at New York University, March 22, 1934, in *Frieden*, p. 264.

16. Message to the American League against War and Fascism, May 26, 1934, in *Frieden*, p. 267.

17. To F. Lindemann, Princeton, November 21, 1933.

18. To Max Born, Princeton, March 22, 1934, in Born, p. 170.

19. Elsa Einstein to Antonina Vallentin, Princeton, October 26, 1934.

20. Aby Warburg was the founder of the famous Warburg Cultural-History Library in Hamburg. After Hitler's rise to power the library with its more than 100,000 volumes was transferred to London, where the newly established Warburg Institute subsequently became part of the University of London.

21. Erwin Panofsky, *Sinn und Deutung in der bildenden Kunst* (Cologne, 1975), p. 379.

22. Einstein 1932b.

23. Einstein 1933a and 1933b.

24. Einstein 1934a.

25. Diary entry of December 6, 1931.

26. To Elizabeth, Queen of the Belgians, Princeton, February 16, 1935.

27. To F. Lindemann, Princeton, January 22, 1935.

28. To M. von Laue, Princeton, March 23, 1934.

29. To F. Lindemann (see n. 26).

30. Elsa Einstein to Antonina Vallentin, in Vallentin, *Das Drama Albert Einsteins* (Stuttgart, 1955), p. 224.

31. Antonina Vallentin (see n. 30), p. 227.

32. To M. Born, Princeton, undated (February 1937), in Born, p. 177.

33. To Otto Juliusburger, Princeton, September 28, 1937.

34. Banesh Hoffmann in *Princeton Symposion*, p. 477. In 1990 WNET in New York broadcast a television program about Einstein, which included a film clip of an address of 1950, which Einstein read from a written English translation. To help present-day Americans understand him, the passage had to have English subtitles.

35. To M. Born, Princeton, September 7, 1944, in Born, p. 202.

36. To Louis de Broglie, Princeton, February 8, 1954.

37. To Jean Bequerel, Princeton, August 16, 1951.

38. To Elizabeth, Queen of the Belgians, Princeton, March 20, 1936.

39. Many of the anecdotes circulating in Princeton about Einstein are recounted in Sayen.

40. Katia Mann, *Meine ungeschriebenen Memoiren* (Frankfurt, 1974), p. 122.

41. To M. Born, Princeton, March 22, 1934, in Born, p. 169.

42. See the account by Boris Schwarz, with excerpts from Einstein's letters, in *Jerusalem Symposium*, p. 409.

43. Memorandum of April 9, 1938, in *Frieden*, p. 293.

44. To Tatjana Ehrenfest, Peconic, August 4, 1938.

45. To W. Pauli, Peconic, September 1936, in Pauli, vol. 2, p. 601.

46. To M. Besso, Peconic, August 8, 1938, in Besso, p. 321.

47. To M. Besso, Princeton, October 10, 1938, in Besso, p. 330.

48. Ibid.

49. To M. Solovine, Princeton, April, 10, 1938, in Solovine, p. 70.

50. Ibid.

36. Physical Reality and a Paradox, Relativity and Unified Theory

1. J. R. Oppenheimer to Frank Oppenheimer, January 11, 1935, in Alice Kimball Smith and Charles Weiner (eds.), *Robert Oppenheimer, Letters and Recollections* (Cambridge, Mass., 1980), p. 190.

2. The best presentation of the new physics is Abraham Pais, *Inward Bound* (Oxford, 1986).

3. Einstein 1934d. A sympathetic account of Ehrenfest's tragic end is given by Hendrik Casimir, *Haphazard Reality* (New York, 1983), pp. 147–59.

4. To E. Schrödinger, Old Lyme, Conn., August 8, 1935.

5. To M. von Laue, Princeton, March 23, 1934.

6. J. A. Wheeler, *Erinnerungen*, in A. P. French (ed.), *Albert Einstein—Wirkung und Nachwirkung* (Braunschweig, 1985), p. 85.

7. To M. Besso, Princeton, June 9, 1937, in Besso, p. 313.

8. Leopold Infeld, *Leben mit Einstein* (Vienna, 1969), p. 52.

9. Einstein 1935a. For topical discussions of the Einstein-Podolsky-Rosen paradox, see J. S. Bell, *Speakable and Unspeakable in Quantum Mechanics* (Cambridge, England, 1987). A simple and—despite the questionable title—appropriate account of the various points of view is P. C. W. Davies and J. R. Brown, *The Ghost in the Atom* (Cambridge, England, 1986).

10. To E. Schrödinger, Old Lyme, Conn., June 19, 1935.

11. Einstein 1936a, 1948a, and 1949a.

12. To Nobel Committee for Physics, Caputh, September 30, 1931.

13. Einstein 1935a, p. 777.

14. Ibid., p. 780.

15. W. Pauli to W. Heisenberg, Zurich, June 15, 1935.

16. Léon Rosenfeld, in S. Rozenthal (ed.), *Niels Bohr* (Amsterdam, 1967), p. 128.

17. Niels Bohr, *Can Quantum-Mechanical Description of Physical Reality Be Considered Complete?* in *Physical Review*, 48, 1935, pp. 696–702.

18. Banesh Hoffmann, in G. J. Whitrow, *Einstein—The Man and His Achievement* (New York, 1975), p. 79.

19. E. Schrödinger to Einstein, Oxford, June 7, 1935.

20. E. Schrödinger, *Die gegenwärtige Situation in der Quantenmechanik*, in NW, Vol. 23, 1935, pp. 807–12, 823–28, 844–49.

21. To E. Schrödinger, undated, no place (Old Lyme, Conn., August 1935).

22. Einstein 1949a, p. 506.

23. To Cornelius Lanczos, Princeton, March 21, 1942.

24. See n. 9.

25. Einstein 1935b.

26. Einstein 1936a, in *Späte Jahre*, p. 304.

27. Einstein 1936b.

28. Einstein 1918a.

29. To M. Born, undated, no place (Knollwood, Saranac Lake, New York, end of August or beginning of September 1936); in Born, p. 171.

30. To J. T. Tate, Knollwood, Saranac Lake, July 27, 1936.

31. To M. Born (see n. 29).

32. Einstein 1937a.

33. Einstein 1938a.

34. To M. Born, undated, no place (February or March 1937), in Born, p. 179.

35. L. Infeld, *Leben mit Einstein* (Vienna, 1969), pp. 77ff., relates in detail the genesis of *Die Evolution der Physik* from his own angle.

36. To M. Besso, Peconic, August 8, 1938, in Besso, p. 321.

37. Banesh Hoffmann, in *Princeton Symposion*, p. 476.

38. Ibid.

39. Ibid., p. 477.

40. To Fritz Reiche, Knollwood, Saranac Lake, August 15, 1942.

41. George Schreiber (ed.), *Portraits and Self-Portraits* (Boston, 1936); German original text in *Späte Jahre*, p. 13.

42. Léon Rosenfeld to Friedrich Herneck, letter from 1962, published in F. Herneck, *Einstein und sein Weltbild* (Berlin, 1976), p. 280.

37. War, a Letter, and the Bomb

1. After Szilard's mention of the chain reaction, Einstein is said to have exclaimed: "And I never thought of that!" as in *Frieden*, p. 306. The present author does not consider this story plausible.

2. *Berliner Tageblatt*, July 25, 1920.

3. *Berliner Tageblatt*, July 25, 1920. A major article by Hans Dominik, *Auf der Suche nach dem Kohleersatz*, and *Die Urteile der deutschen Gelehrten: Einstein, Haber, Planck, Nernst und Rubens.* Moszkowski's book *Einstein* (Berlin, 1921) also contains a lengthy passage in which Einstein discusses the replacement of the coal economy by nuclear energy.

4. Moszkowski, p. 45.

5. Ibid., p. 46.

6. The history of the atom bomb has been set out in detail in so many books that the present author dispenses with bibliographical references for the short outline over the next few pages. A detailed account to be recommended is Richard Rhodes, *The Making of the Atomic Bomb* (New York, 1987).

7. Einstein 1935c.

8. *Literary Digest*, January 12, 1936.

9. Original German text of Einstein's reply in *Frieden*, p. 306.

10. To Franklin D. Roosevelt, Peconic, August 2, 1939. A copy of the English text is in the Einstein Archive.

11. Quoted in David Irving, *The Virus House* (London, 1967), p. 34.

12. Quoted in R. Rhodes (see n. 6), p. 314. Rhodes quotes Sachs's testimony before a Senate committee in the fall of 1934.

13. F. D. Roosevelt to Einstein, October 19, 1939, in *Einstein on Peace*, p. 297.

14. To Lyman J. Briggs, Princeton, April 25, 1940.

15. F. Aydelotte to V. Bush, December 19, 1941, in Clark, 684.

16. V. Bush to F. Aydelotte, December 30, 1941, in Clark, p. 684.

17. Quoted in Richard A. Schwartz, *Einstein and the War Department*, in *Isis*, Vol. 80, pp. 281–84. Schwartz gives a detailed description of the documents.

18. Stephen Brunauer, *Einstein and the U.S. Navy*, in B. H. Davis and W. P. Hettinger (eds.), *Heterogeneous Catalysis* (Washington, D.C., 1983), pp. 217–26.

19. To Stephen Brunauer, Princeton, May 17, 1943, in Brunauer (see n. 18), p. 220.

20. Ibid., p. 222.

21. To Otto Juliusburger, Princeton, September 13, 1943.

22. Einstein 1944a.

23. Julius Boyd to Dorothy Pratt, November 22, 1943, in Sayen, p. 149.

24. Pais, p. 147.

25. To E. Schrödinger, Princeton, February 12, 1943.

26. Einstein 1944a.

27. To Lina Kocherthaler, undated, no place (Princeton, 1943).

28. To Otto Juliusburger, undated, no place (Saranac Lake, summer 1942).

29. To E. Schrödinger, Princeton, Easter 1943.

30. Bertrand Russell, *Autobiography*, Vol. II (London, 1971), p. 224.

31. Bertrand Russell, in G. J. Whitrow (ed.), *Einstein—The Man and His Achievement* (New York, 1975), p. 90.

32. To Otto Juliusburger, Princeton, January 13, 1943.

33. To Otto Juliusburger, Princeton, March 31, 1943.

34. To Otto Juliusburger, Knollwood, Saranac Lake, September 6, 1943.

35. In *Späte Jahre*, p. 254.

36. To Niels Bohr, Princeton, December 12, 1944.

37. Ibid.

38. See Abraham Pais, *Niels Bohr's Times* (Oxford, 1991), p. 502.

39. To F. D. Roosevelt, Princeton, March 25, 1945, in *Frieden*, p. 318.

40. In *Frieden*, p. 345.

41. Helen Dukas, in Sayen, p. 151.

42. Quoted in Pais (see n. 38), p. 504.

43. Henry De Wolf Smyth, *Atomic Energy for Military Purposes*, U.S. Government Printing Office, September 1945. A little later this was also published by Princeton University Press.

38. Between Bomb and Equations: "But the Equations Are for Eternity"

1. *New York Times*, September 15, 1945.

2. This manifesto by Einstein, Thomas Mann, and others was published as a letter to the editor in *The New York Times* on October 10, 1945.

3. Daniel Posin to Einstein, Cambridge, Mass., October 21, 1945, in *Einstein on Peace*, p. 342.

4. *Atomic War or Peace*, in *Atlantic Monthly*, November 1945; frequently reprinted, e.g., *New York Times*, October 27, 1945.

5. *Atlantic Monthly*, November 1945.

6. *Der Krieg ist gewonnen—nicht aber der Friede*; first published in the *New York Times*, December 11, 1945; German original text, drafted jointly with Erich von Kahler, in *Späte Jahre*, pp. 133f.

7. Along with Einstein as chairman, the board of the Emergency Committee included Harold C. Urey

as his deputy, Hans A. Bethe, Torfin R. Hogness, Philip M. Morse, Linus Pauling, Leo Szilard, and Victor Weisskopf.

8. Harrison Brown, executive vice-chairman of the Emergency Committee, in *Einstein on Peace*, p. 505.

9. To M. Besso, Princeton, April 21, 1946, in Besso, p. 376.

10. Ernst Straus, in *Helle Zeit—Dunkle Zeit*, p. 71.

11. *Einstein, the Man Who Started It All*, in *Newsweek*, March 10, 1947.

12. To Katusu Hara, editor of *Kaizo* (Tokyo), Princeton, September 20, 1952.

13. To M. von Laue, Princeton, March 19, 1955.

14. Ibid.

15. To Jules Isaac, Princeton, February 28, 1955.

16. To Otto Juliusburger, Princeton, April 11, 1946.

17. To M. Solovine, Princeton, May 7, 1946, in Solovine, p. 120.

18. Hermann Broch to Hannah Arendt, July 24, 1950, in Hermann Broch, *Kommentierte Werkausgabe*, P. M. Lützeler (ed.), Vol. 13/3 (Frankfurt, 1981), p. 484.

19. To P. von Schöneich, Princeton, July 20, 1946.

20. To James Franck, Princeton, December 6, 1945.

21. To James Franck, Princeton, December 30, 1945.

22. Hermann Broch to James Franck, January 30, 1946, in Broch (see n. 18), p. 57.

23. A. Sommerfeld to Einstein, Munich, October 27, 1946, in Sommerfeld, p. 120.

24. To A. Sommerfeld, Princeton, December 14, 1945, in Sommerfeld, p. 121.

25. To M. Besso, Princeton, January 6, 1948, in Besso, p. 392.

26. To Otto Hahn, Princeton, January 28, 1949.

27. To Theodor Heuss, Princeton, January 16, 1951.

28. To Verlag Vieweg, Braunschweig, Princeton, March 25, 1947.

29. To M. Born, Princeton, October 12, 1953, in Born, p. 266.

30. Einstein 1942c.

31. The ceremony for Max Planck was held in Washington on April 23; Einstein's eulogy was read in German and English. In *Späte Jahre*, p. 220; *Out of My Later Years*, p. 229.

32. To Marga Planck, Princeton, October 1947, quoted in A. Hermann, *Max Planck* (Reinbek, 1973), p. 127.

33. To E. Schrödinger, Princeton, January 27, 1947.

34. To Erich Mühsam, Princeton, April 24, 1949.

35. To Erich Mühsam, Princeton, September 19, 1949.

36. R. Nissen to C. Seelig, June 29, 1955, in ETH and Seelig, p. 415.

37. To Otto Nathan, Florida, February 17, 1949.

38. To M. Besso, November 30, 1949, in Besso, p. 424.

39. To Lina Kocherthaler, Princeton, July 27, 1951.

40. To C. Seelig, Princeton, April 20, 1952, ETH.

41. To C. Seelig, Princeton, January 4, 1954.

42. To Erich Mühsam, Princeton, spring 1942, in *Helle Zeit—Dunkle Zeit*, p. 50.

43. Robert Mann, leader of the Juilliard Quartet, in *Princeton Symposium*, p. 527.

44. Abba Eban, *An Autobiography* (New York, 1977), pp. 166ff.

45. David Mitrany, quoted in Pais, p. 481.

46. To Abba Eban, Princeton, November 18, 1952, in *Frieden*, p. 570.

47. To Azriel Carlebach, Princeton, November 21, 1952, in *Frieden*, p. 571.

48. To Josef Scharl, Princeton, November 24, 1952, in *Frieden*, p. 571.

49. Yitzak Navon in *Jerusalem Symposium*, p. 295.

50. To E. Schrödinger, Princeton, May 20, 1946.

51. To C. Seelig, Princeton, August 12, 1954, ETH.

52. Pais, p. 5.

53. To M. Solovine, Princeton, March 28, 1949, in Solovine, p. 94.

54. To E. Schrödinger, December 22, 1950.

55. Ernst Straus, in *Helle Zeit—Dunkle Zeit*, p. 75.

56. Ibid., p. 72.

57. To M. Besso, Princeton, August 10, 1954, in Besso, p. 527.

58. Television address on February 12, 1950. Quoted in *Einstein on Peace*, p. 521.

59. "Einstein Warns World: Outlaw H-Bomb or Perish," *New York Post*, February 13, 1950.

60. *The Reporter*, November 18, 1954.

61. B. Russell to Einstein, February 11, 1955. The entire correspondence on the Einstein-Russell Manifesto is in *Frieden*, pp. 617–31.

62. B. Russell, in *Frieden*, Preface, p. 17.

39. "An Old Debt"

1. To Erich Mühsam, Princeton, summer 1946, in *Helle Zeit—Dunkle Zeit*, p. 50.

2. J. A. Wheeler, *Erinnerungen . . .*, in A. P. French (ed.), *Albert Einstein—Wirkung und Nachwirkung* (Braunschweig, 1985), p. 86.

3. To M. Solovine, Princeton, February 27, 1955.

4. To M. von Laue, Princeton, February 3, 1955.

5. A description of the experiment on the equivalence principle is given in I. B. Cohen, *An Interview with Einstein*, in *Scientific American*, Vol. 93, July 1955, p. 73.

6. H. Dukas to A. Pais, April 30, 1955, in Pais, p. 482.

7. To Gertrud Warschauer, Princeton, February 5, 1955.

8. Margot Einstein to Hedwig Born, April 1955, in Born, p. 309.

9. To Vero and Bice Besso, Princeton, March 21, 1955, in Besso, p. 538.

BIBLIOGRAPHY AND ABBREVIATIONS

To attempt a comprehensive bibliography of the by now enormous literature on Albert Einstein would go beyond the scope of this book. I therefore restrict myself to listing those books which were indispensable to me. These include, above all, publications of Einstein's original texts and letters, as well as some collections of documents. Titles not listed below but used sporadically are detailed in the endnotes.

So far, there is no collection of Einstein's most important scientific publications. (Only in the former Soviet Union is such a collection said to exist, translated into Russian.) The monumental project of the *Collected Papers*, published by Princeton University Press, will fill this gap, though its completion lies in the distant future. A bibliography of Einstein's publications may therefore prove helpful; most of them are in fact available in university libraries or in the libraries of sufficiently old physics institutes.

General Literature

CP1, CP2

The Collected Papers of Albert Einstein, ed. Johan Stachel et al., Princeton University Press, Vol. 1, 1987; Vol. 2, 1989.

Autobiographical Writings by Albert Einstein

Autobiographisches: in Paul Arthur Schilpp (ed.), *Albert Einstein als Philosoph und Naturforscher* [*Albert Einstein as a Philosopher and Scientist*] (Braunschweig, 1979). What he himself called his "Obituary," written in

1946; a kind of scientific autobiography. English: *Autobiographical Notes: A Centennial Edition*, trans. Paul Arthur Schilpp (La Salle and Chicago, Open Court Publ. Co., 1979).

Skizze: Albert Einstein, *Autobiographische Skizze*, in Carl Seelig (ed.), *Helle Zeit—Dunkle Zeit. In memoriam Albert Einstein* [*Bright Times—Dark Times*] (Zurich, 1956; Braunschweig, 1986). Written in 1954; a kind of postscript to his "Obituary."

Weltbild: Albert Einstein, *Mein Weltbild* [*The World as I See It*] (Amsterdam, 1934; enlarged new ed., ed. Carl Seelig (Zurich, 1953). *Ideas and Opinions*, based on *Mein Weltbild*, ed. Carl Seelig, and on other sources, new trans. and revisions by Sonja Bergmann (New York, Crown, 1954).

Späte Jahre: Albert Einstein, *Aus meinen späten Jahren* (Stuttgart, 1979); English: *Out of My Later Years* (Westport, Conn., Greenwood, 1970).

Einstein's Correspondence

Besso: *Albert Einstein—Michele Besso, Correspondance 1903–1955*, ed. Pierre Speziali (Paris, 1972).

Born: Albert Einstein—Max Born, *Briefwechsel 1916–1935*, ed. with commentary by Max Born (Munich, 1969).

D. Lohmeyer and B. Schell (eds.), *Einstein, Anschütz und der Kieler Kreiselkompaß* [*Einstein, Anschütz, and the Kiel Gyrocompass*] (Heide-in-Holstein, 1992). Contains the correspondence between Einstein and Hermann Anschütz-Kaempfe.

Pauli: Wolfgang Pauli, *Wissenschaftlicher Briefwechsel*, Vol. 1, ed. A. Hermann, K. von Meyenn, and V. F. Weisskopf (New York, Heidelberg, Berlin, 1979); Vol. 2, ed. K. von Meyenn (New York, Heidelberg, Berlin, 1984).

Solovine: Albert Einstein, *Lettres à Maurice Solovine*, ed. M. Solovine (Paris, 1956).

Sommerfeld: Albert Einstein—Arnold Sommerfeld, *Briefwechsel*, ed. by Armin Hermann (Basel and Stuttgart, 1968).

Stark: *Albert Einstein und Johannes Stark. Briefwechsel und Verhältnis der beiden Nobelpreisträger*, ed. Armin Hermann, in *Sudhoffs Archiv*, Vol. 50, 1966, pp. 267–85.

Collections of Edited Texts by Albert Einstein

Letters: Albert Einstein, *Briefe* [*Letters*], eds. Helen Dukas and Banesh Hoffmann (Zurich, 1981).

Frieden: Albert Einstein, *Über den Frieden*, eds. Otto Nathan and Heinz Norden (Bern, 1975). English: *Einstein on Peace*, eds. Otto Nathan and Heinz Norden (New York, Simon and Schuster, 1960).

Collections of Documents

Berlin: *Albert Einstein in Berlin 1913–1933, Darstellungen und Dokumente*, eds. Christa Kirsten and Hans-Jürgen Treder, 2 vols. (Berlin, 1979).

Ulm: *Einstein und Ulm*, Ulm City Archive, ed. Hans Eugen Specker (Ulm and Stuttgart, 1979).

Biographies

Clark: Ronald W. Clark, *Einstein—The Life and Times* (New York, 1971). Quotations are cited from the American paperback edition (New York, 1972).

Flückiger: Max Flückiger, *Albert Einstein in Bern* (Bern, 1972).

Frank: Philipp Frank, *Einstein—His Life and His Times* (New York, 1949). Einstein's foreword, written in 1942, is contained only in the German edition, *Einstein—Sein Leben und seine Zeit* (Braunschweig, 1979).

Hoffman/Dukas: Banesh Hoffmann and Helen Dukas, *Einstein—Schöpfer und Rebell* [*Einstein, Creator and Rebel*] (Zurich, 1976). Quotations are cited from the Frankfurt edition of 1978.

Moszkowski: Alexander Moszkowski, *Einstein—Einblicke in seine Gedankenwelt* [*Insights into His World of Ideas*] (Hamburg and Berlin, 1921).

Pais: Abraham Pais, *Subtle Is the Lord . . .* (Oxford and New York, 1982).

Pyenson: Lewis Pyenson, *The Young Einstein* (Bristol and Boston, England, 1985).

Reiser: Anton Reiser (pseudonym of Rudolf Kayser), *Albert Einstein: A Biographical Portrait* (New York, 1930).

Sayen, Jamie Sayen, *Einstein in America* (New York, 1985).

Seelig: Carl Seelig, *Albert Einstein und die Schweiz* [*Albert Einstein and Switzerland*] (Zurich, 1952). An enlarged edition is *Albert Einstein. Leben und Werk eines Genies unserer Zeit* [*Life and Work of a Genius of Our Time*] (Zurich, 1960).

Collections of Articles

P. C. Aichelburg and R. Sexl (eds.), *Albert Einstein. Sein Einfluß auf Physik, Philosophie und Politik* [*His Impact on Physics, Philosophy, and Politics*] (Braunschweig, 1979).

Berlin Symposium, *Einstein-Symposion Berlin*, eds. H. Nelkowski et al. (Berlin, Heidelberg, New York, 1979).

A. P. French, *Einstein—A Centenary Volume* (Cambridge, Mass., 1979).

Helle Zeit—Dunkle Zeit: Carl Seelig (ed.), *Helle Zeit—Dunkle Zeit. In memoriam Albert Einstein* (Zurich, 1956; reprint, Braunschweig, 1986).

Jerusalem Symposium: *Albert Einstein, Historical and Cultural Perspectives: Centennial Symposium in Jerusalem*, eds. Gerald Holton and Yehuda Elkana (Princeton, 1982).

Princeton Symposion: *Some Strangeness in the Proportion: A Centennial Symposium to Celebrate the Achievement of Albert Einstein*, ed. Harry Wolf (Reading, Mass., 1980).

Schilpp: P. A. Schilpp (ed.), *Albert Einstein als Philosoph und Naturforscher* [*Albert Einstein as Philosopher and Scientist*] (Braunschweig, 1979; original American edition, Evanston, 1949.)

Journals and Series

AdP: *Annalen der Physik*, fourth series, from 1901.

HSPS: *Historical Studies in the Physical Sciences.*

NW: *Die Naturwissenschaften.*

SB: *Sitzungsberichte der Preußischen Akademie der Wissenschaften zu Berlin.*

VhDPG: *Verhandlungen der Deutschen Physikalischen Gesellschaft.*

ZfP: *Zeitschrift für Physik.*

Archives

DMM: Bibliothek des Deutschen Museums München.

EA: Einstein Archiv, Jerusalem; copies in Mudd Manuscript Library, Princeton, and in Science and Engineering Library, Boston University.

ETH: Einstein-Sammlung der ETH Zürich.

MPG: Archiv der Max-Planck-Gesellschaft, Berlin.

STPK: Staatsbibliothek Preußischer Kulturbesitz Berlin.

STZ: Staatsarchiv des Kanton Zürich.

Scientific Publications by Albert Einstein

The publications are arranged chronologically by year of publication. Only in the few cases where the date of completion or submission to a journal is significant has this been indicated.

The list of scientific publications is complete. Writings of a more general character are listed selectively. Newspaper articles, etc., are listed only if they are primarily concerned with scientific matters.

1901 *Folgerungen aus den Capillaritätserscheinungen* [*Conclusions from Capillarity Phenomena*], AdP, 4, 1901, pp. 513–23.

1902 a *Über die thermodynamische Theorie der Potentialdifferenz zwischen Metallen und vollständig dissoziierten Lösungen ihrer Salze und über eine elektrische Methode zur Erforschung der Molecularkräfte* [*On the Thermodynamic Theory of the Difference of Potential between Metals and Completely Dissociated Solutions of Their Salts, and on an Electrical Method for the Exploration of Molecular Forces*], AdP, 8, 1902, pp. 798–814.

1902 b *Kinetische Theorie des Wärmegleichgewichts und des zweiten Hauptsatzes der Thermodynamik* [*Kinetic Theory of the Thermal Equilibrium and of the Second Law of Thermodynamics*], AdP, 8, 1902, pp. 417–33.

1903 *Eine Theorie der Grundlagen der Thermodynamik* [*A Theory of the Foundations of Thermodynamics*], AdP, 11, 1903, pp. 170–87.

1904 *Zur allgemeinen molekularen Theorie der Wärme* [*On the General Molecular Theory of Heat*], AdP, 14, 1904, pp. 354–62.

1905 a *Über einen die Erzeugung und Umwandlung des Lichtes betreffenden heuristischen Standpunkt* [*On a Heuristic Viewpoint Concerning the Generation and Transformation of Light*], AdP, 17, 1905, pp. 132–84. Completed March 17, received March 18, 1905.

1905 b *Eine neue Bestimmung der Moleküldimensionen* [*On a New Determination of Molecular Dimensions*], doctoral thesis, completed April 30, 1905, printed by K. J. Wyss, Bern, 1905. See Supplement 1906a and Correction 1911d.

1905 c *Über die von der molekulartheoretischen Theorie der Wärme geforderte Bewegung von in ruhenden Flüssigkeiten suspendierten Teilchen* [*On the Movement of Particles Suspended in Fluids at Rest, as Postulated by the Molecular Theory of Heat*], AdP, 17, 1905, pp. 549–60. Completed May 1905, received May 11, 1905.

1905 d *Zur Elektrodynamik bewegter Körper* [*On the Electrodynamics of Moving Bodies*], AdP, 17, 1905, pp. 891–921. Completed June 1905, received June 30, 1905.

1905 e *Ist die Trägheit eines Körpers von seinem Energieinhalt abhängig?* [*Does the Inertia of a Body Depend on its Energy Content?*], AdP, 18, 1905, pp. 639–41. Received September 27, 1905.

1905 f Twenty-five reviews in *Beiblätter zu den Annalen der Physik*, AdP, 29, 1905.

1906 a *Eine neue Bestimmung der Moleküldimensionen* [*A New Determination of Molecular Dimensions*], AdP, 19, 1906, pp. 289–305. Slightly revised version of 1905c, received August 19, 1905, with a "Supplement," pp. 305–306, of January 1906. Correction 1911d.

1906 b *Zur Theorie der Brownschen Bewegung* [*On the Theory of the Brownian Movement*], AdP 19, 1906, pp. 371–81. Received December 19, 1905.

1906 c *Zur Theorie der Lichterzeugung und Lichtabsorption* [*On the Theory of the Generation and Absorption of Light*], AdP, 20, 1906, pp. 199–206.

1906 d *Das Prinzip von der Erhaltung der Schwerpunktbewegung und die Trägheit der Energie* [*The Principle of the Conservation of the Motion of the Center of Gravity and the Inertia of Energy*], AdP, 20, 1906, pp. 627–33.

1906 e *Rezension von Max Planck: Vorlesungen über die Theorie der Wärmestrahlung* [Review of *Max Planck: Lectures on the Theory of Heat Radiation*], Leipzig 1906, Beiblätter zu AdP, 30, 1906, pp. 764–66.

1906 f *Über eine Methode zur Bestimmung des Verhältnisses der transversalen und longitudinalen Masse des Elektrons* [*On a Method for the Determination of the Ratio between the Transversal and Longitudinal Mass of the Electron*], AdP, 21, 1906, pp. 583–86.

1907 a *Die Plancksche Theorie der Strahlung und die Theorie der spezifischen Wärme* [*Planck's Theory of Radiation and the Theory of Specific Heat*], AdP, 22, 1907, pp. 180–90; Correction AdP, 22, 1907, p. 800.

1907 b *Über die Gültigkeitsgrenzen des Satzes vom thermodynamischen Gleichgewicht und über die Möglichkeit einer neuen Bestimmung der Elementarquanta* [*On the Limits of Validity of the Theorem of the Thermodynamic Equilibrium and on the Possibility of a New Determination of the Elementary Quanta*], AdP, 22, 1907, pp. 569–72.

1907 c *Theoretische Bemerkungen über die Brownsche Bewegung* [*Theoretical Observations on the Brownian Movement*], *Zeitschrift für Elektrochemie und angewandte physikalische Chemie*, 13, 1907, pp. 41–42.

1907 d *Über die Möglichkeit einer neuen Prüfung des Relativitätsprinzips* [*On the Possibility of a New Test of the Relativity Principle*], AdP, 23, 1907, pp. 197–98.

1907 e *Bemerkungen zu der Notiz des Hrn. Paul Ehrenfest: "Die Transfor-*

mation deformierbarer Elektronen und der Flächensatz" [*Observations on Herr Paul Ehrenfest's Communication "The Transformation of Deformable Electrons and the Surface Theorem"*], AdP, 23, 1907, pp. 206–8.

1907 f *Über die vom Relativitätsprinzip geforderte Trägheit der Energie* [*On the Inertia of Energy Postulated by the Relativity Principle*], AdP, 23, 1907, pp. 371–84.

1907 g *Rezension von J. J. Weyrauch: Grundriss der Wärmetheorie*, Band 2 [*Review of J. J. Weyrauch: Outline of Heat Theory*, Vol. 2], Beiblätter zu AdP, 31, 1907, pp. 777–78.

1907 h *Über das Relativitätsprinzip und die aus demselben gezogenen Folgerungen* [*On the Relativity Principle and the Conclusions Drawn from It*], *Jahrbuch der Radioaktivität und Elektronik*, 4, 1907, pp. 411–62. Actually published on January 22, 1908. Corrections in *Jahrbuch*, 5, 1908, pp. 98–99.

1908 a *Eine neue elektrostatische Methode zur Messung kleiner Elektrizitätsmengen* [*A New Electrostatic Method of Measuring Small Amounts of Electricity*]. PhZ, 9, 1908, pp. 216–17.

1908 b *Elementare Theorie der Brownschen Bewegung* [*Elementary Theory of the Brownian Movement*], *Zeitschrift für Elektrochemie und angewandte physikalische Chemie*, 14, 1908, pp. 235–39.

1908 c *Über die elektromagnetischen Grundgleichungen für bewegte Körper* [*On the Electromagnetic Fundamental Equations for Moving Bodies*], with Jakob Laub, AdP, 26, 1908, pp. 532–40; Correction, AdP, 27, 1908, p. 232.

1908 d *Über die im elektromagnetischen Feld auf ruhende Körper ausgeübten ponderomotorischen Kräfte* [*On the Ponderomotoric Forces Exerted in the Electromagnetic Field on Bodies at Rest*], with Jakob Laub, AdP, 26, 1908, pp. 541–50.

1909 a *Bemerkungen zu unserer Arbeit: "Über die elektromagnetischen Grundgleichungen für bewegte Körper"* [*Observations on Our Treatise "On the Electromagnetic Fundamental Equations for Moving Bodies"*], with Jakob Laub. Received December 19, 1908. "Supplement" received January 19, 1909, AdP, 28, 1909, pp. 445–47.

1909 b *Bemerkungen zu der Arbeit von D. Mirimanoff "Über die Grundgleichungen . . .* [*Observations on the Treatise by D. Mirimanoff "On the Fundamental Equations . . ."*], AdP, 28, 1908, pp. 885–88.

1909 c *Zum gegenwärtigen Stand des Strahlungsproblems* [*On the Present State of the Radiation Problem*], PhZ, 10, 1909, pp. 185–93.

1909 d *Zum gegenwärtigen Stand des Strahlungsproblems* [*On the Present

State of the Radiation Problem], with Walter Ritz, PhZ, 10, 1909, p. 323.

1909 c *Über die Entwicklung unserer Anschuungen über das Wesen und die Konstitution der Strahlung* [*On the Development of Our Views on the Nature and Constitution of Radiation*], lecture, Salzburg 1909, VhDPG, 11, 1909, pp. 482–500. Quoted from the reprint in PhZ, 10, 1909, pp. 817–25.

1910 a *Über einen Satz der Wahrscheinlichkeitsrechnung und seine Anwendung auf die Strahlungstheorie* [*On a Theorem of Probability Calculus and Its Application to the Theory of Radiation*], with Ludwig Hopf, AdP, 33, 1910, pp. 1096–1104.

1910 b *Statistische Untersuchung der Bewegung eines Resonators in einem Strahlungsfeld* [*Statistical Investigation of the Motion of a Resonator in a Radiation Field*], with Ludwig Hopf, ADP, 33, 1910, pp. 1105–15.

1910 c *Theorie der Opaleszenz von homogenen Flüssigkeiten und Flüssigkeitsgemischen in der Nähe des kritischen Zustandes* [*Theory of Opalescence of Homogeneous Liquids and Mixtures of Liquids in the Proximity of the Critical State*], AdP, 33, 1910, pp. 1275–98.

1910 d *Le Principe de relativité et ses consequences dans la physique moderne* [*The Relativity Principle and Its Consequences in Modern Physics*], *Archives des sciences physiques et naturelles*, 29, 1910, pp. 5–28, 125–44.

1910 e *Théorie des quantités lumineuses et la question de la localisation de l'énergie électromagnétique* [*Theory of Light Quanta and the Question of the Location of Electromagnetic Energy*], *Archives des sciences physiques et naturelles*, 29, 1910, pp. 525–28.

1910 f *Sur les Forces pondéromotrices qui agissent sur les conducteurs ferromagnétiques disposés dans un champs magnétique et parcourus par un courant* [*On the Ponderomotoric Forces Acting upon Ferromagnetic Conductors Placed in a Magnetic Field with a Current Passing through Them*], *Archives des sciences physiques et naturelles*, 30, 1910, pp. 323–24.

1911 a *Bemerkung zu dem Gesetz von Eötvös* [*Observation on Eötvös's Law*], AdP, 34, 1911, pp. 165–69.

1911 b *Eine Beziehung zwischen dem elastischen Verhalten und der spezifischen Wärme bei festen Körpern mit einatomigem Molekül* [*A Relation between the Elastic Behavior and the Specific Heat of Solids with Single-Atom Molecules*], AdP, 34, 1910, pp. 170–74.

1911 c *Bemerkungen zu den P. Hertzschen Arbeiten: "Über die mechanischen Grundlagen der Thermodynamik"* [*Observations on P. Hertz's Treatises "On the Mechanical Foundations of Thermodynamics"*], AdP, 34, 1911, pp. 175–76.

1911 d *Berichtigung zu meiner Arbeit: "Eine neue Bestimmung der Moleküldimensionen"* [*Correction to My Paper "A New Determination of Molecular Dimensions"*], AdP, 14, 1911, pp. 591–92.

1911 e *Die Relativitäts-Theorie* [*The Theory of Relativity*], *Vierteljahresschrift der Naturforschenden Gesellschaft Zürich*, 56, 1911, pp. 1–14.

1911 f *Elementare Betrachtungen über die thermische Molekularbewegung in festen Körpern* [*Elementary Observations on Thermal Molecular Motion in Solids*], AdP, 35, 1911, pp. 679–94.

1911 g *Über den Einfluß der Schwerkraft auf die Ausbreitung des Lichtes* [*On the Effect of Gravity on the Propagation of Light*], AdP, 35, 1911, pp. 898–908.

1911 h *Zum Ehrenfestschen Paradoxon* [*On the Ehrenfest Paradox*], PhZ, 12, 1911, pp. 509–10.

1911 i *L'État actuel du problème des chaleurs spécifiques* [*The Present State of the Specific Heat Problem*], lecture at Solvay Conference, 1911, in Paul Langevin and Maurice de Broglie (eds.), *La Théorie du rayonnemente et des quanta. Rapports et discussion de la réunion à Bruxelles, du 30 octobre au 3 novembre 1911*, Paris 1912, pp. 407–35.

1912 a *Thermodynamische Begründung des photochemischen Äquivalenzgesetzes* [*Thermodynamic Justification of the Photochemical Equivalence Law*], AdP, 37, 1912, pp. 832–38.

1912 b *Lichtgeschwindigkeit und Statik des Gravitationsfeldes* [*Velocity of Light and Statics of the Gravitational Field*], AdP, 38, 1912, pp. 443–58.

1912 c *Theorie des statischen Gravitationsfeldes* [*Theory of the Static Gravitational Field*], AdP, 38, 1912, pp. 443–58.

1912 d *Antwort auf eine Bemerkung von J. Stark: Anwendung des Planckschen Elementargesetzes* [*Reply to an Observation by J. Stark: Application of Planck's Elementary Law*], AdP 38, 1912, p. 888.

1912 e *Relativität und Gravitation: Erwiderung auf eine Bemerkung von M. Abraham* [*Relativity and Gravitation: Reply to an Observation by M. Abraham*], AdP, 38, 1912, pp. 1059–64.

1912 f *Bemerkung zu Abrahams Auseinandersetzung: Nochmals Relativität und Gravitation* [*Observation on Abraham's Argument: Once More Relativity and Gravitation*], AdP, 39, 1912, p. 704.

1913 a *Einige Argumente für die Annahme einer molekularen Agitation beim absoluten Nullpunkt* [*A Few Arguments in Favor of the Assumption of Molecular Agitation at Absolute Zero*], with Otto Stern, AdP, 40, 1913, pp. 551–60.

1913 b *Déduction thermodynamique de la loi de l'équivalence photochimique* [*Thermodynamic Deduction of the Law of Photochemical Equivalence*], lecture given on March 27, 1913, in Paris at annual meeting of the Société Française de Physique, *Journal de Physique*, 3, 1913, pp. 277–82.

1913 c *Entwurf einer verallgemeinerten Relativitätstheorie und eine Theorie der Gravitation* [*Draft of a Generalized Theory of Relativity and a Theory of Gravitation*]; I. Physical Part by Albert Einstein, II. Mathematical Part by Marcel Grossmann, *Zeitschrift für Mathematik und Physik*, 62, 1913, pp. 225–61. Also published as a separate brochure (Leipzig, 1913).

1913 d *Max Planck als Forscher* [*Max Planck as a Researcher*], NW, 1, 1913, pp. 1077–79.

1913 e *Gibt es eine Gravitationswirkung, die der elektromagnetischen Induktionswirkung analog ist?* [*Is There a Gravitational Effect Analogous to the Electromagnetic Induction Effect?*], *Vierteljahresschrift für gerichtliche Medizin*, 44, 1913, pp. 37–40.

1913 f *Physikalische Grundlagen einer Gravitationstheorie* [*Physical Foundations of a Gravitation Theory*], *Vierteljahresschrift der Naturforschenden Gesellschaft Zürich*, 58, 1913, pp. 284–90. Lecture given to that society on September 9, 1913.

1913 g *Zum gegenwärtigen Stande des Strahlungsproblems* [*On the Present State of the Radiation Problem*], PhZ, 14, 1913, pp. 1249–66. Lecture given to annual meeting of the Society of German Scientists and Physicians in Vienna on September 21, 1913.

1914 a *Nordströmsche Gravitationstheorie vom Standpunkt des allgemeinen Differentialkalküls* [*Nordström's Gravitation Theory from the Viewpoint of General Differential Calculus*], with A. D. Fokker, AdP, 44, 1914, pp. 321–28.

1914 b *Bemerkung zu P. Harzers Abhandlung: "Die Mitführung des Lichtes in Glas und die Aberration"* [*Observation on P. Harzer's Treatise "The Conduction and Aberration of Light in Glass"*], *Astronomische Nachrichten*, 199, 1914, pp. 8–10.

1914 c *Antwort of eine Replik P. Harzers* [*Reply to a Rejoinder by P. Harzer*], *Astronomische Nachrichten*, 199, 1914, pp. 330–64.

1914 d *Zum gegenwärtigen Stande des Problems der spezifischen Wärme* [*On the Present State of the Problem of Specific Heat*], *Abhandlungen der Deutschen Bunsengesellschaft*, 7, 1914, pp. 350–64.

1914 e *Zur Theorie der Gravitation* [*On the Theory of Gravitation*], *Vierteljahresschrift der Naturforschenden Gesellschaft Zürich*, 59, 1914, pp. 4–6.

1914 f *Das Relativitätsprinzip* [*The Relativity Principle*], *Vossische Zeitung*, April 26, 1914, pp. 33–34.

1914 g *Nachträgliche Antwort auf eine Frage von Reissner* [*Supplementary Answer to a Question by Reissner*], PhZ, 15, 1914, pp. 108–10.

1914 h *Prinzipielles zur verallgemeinerten Relativitätstheorie und Gravitationstheorie* [*Fundamental Remarks on the Generalized Theory of Relativity and the Gravitation Theory*], PhZ, 15, 1914, pp. 176–80.

1914 i *Kovarianzeigenschaften der Feldgleichungen der auf die verallgemeinerte Relativitätstheorie gegründeten Gravitationstheorie* [*Covariance Characteristics of the Field Equations of the Gravitation Theory Based on the Generalized Theory of Relativity*], with Marcel Grossmann, *Zeitschrift für Mathematik und Physik*, 63, 1914, pp. 215–25.

1914 j *Zum Relativitätsproblem* [*On the Relativity Problem*], *Scientia* (Bologna), 15, 1914, pp. 337–48.

1914 k Antrittsrede am 2. Juli 1914 [Inaugural address on July 2, 1914], SB II, pp. 739–42, along with Max Planck's reply, pp. 742–44.

1914 l *Formale Grundlage der allgemeinen Relativitätstheorie* [*Formal Basis of the General Theory of Relativity*], SB II, 1914, pp. 1030–85.

1914 m *Physikalische Grundlagen und leitende Gedanken für eine Gravitationstheorie* [*Physical Foundations and Guiding Ideas for a Theory of Gravitation*], *Verhandlungen der Schweizerischen Naturforschenden Gesellschaft*, 96, Part 2, 1914, p. 146.

1914 n *Besprechung von A. Brill: Das Relativitätsprinzip* [Review of A. Brill: *The Relativity Principle*], NW, 2, 1914, p. 1018.

1914 o *Besprechung von H. A. Lorentz: Das Relativitätsprinzip* [Review of H. A. Lorentz: *The Relativity Principle*], NW, 2, 1914, p. 1018.

1915 a *Theoretische Atomistik* [*Theoretical Atomistics*], in Paul Hinneberg (ed.), *Die Kultur der Gegenwart*, Part 3, Vol. 1, E. Lehrer: *Die Physik*, 1915, pp. 251–53; quoted from 1925 edition, pp. 281–94.

1915 b *Die Relativitätstheorie* [*The Theory of Relativity*], in *Die Physik* (see 1915a), pp. 703–13; quoted from 1925 edition, pp. 783–97.

1915 c *Antwort auf eine Abhandlung M. von Laues: Ein Satz der Wahrscheinlichkeitsrechnung und seine Anwendung auf die Strahlungstheorie* [*Reply to a Treatise by M. von Laue: A Theorem of Probability Calculus and Its Application to the Theory of Radiation*], AdP, 47, 1915, pp. 879–85.

1915 d *Experimenteller Nachweis der Ampèreschen Molekularströme* [*Experimental Proof of Ampère's Molecular Currents*], with W. J. de Haas, VhDPG, 17, 1915, pp. 152–70; Correction, p. 203.

1915 e *Experimenteller Nachweis der Ampèreschen Molekularströme* [*Experimental Proof of Ampère's Molecular Currents*], with W. J. de Haas, NW, 3, 1915, pp. 237–38.

1915 f *Grundgedanken der allgemeinen Relativitätstheorie und Anwendung dieser Theorie in der Astronomie* [*Fundamental Ideas of the General Theory of Relativity and Application of That Theory in Astronomy*], SB I, 1915, p. 315 (Summary).

1915 g *Zur allgemeinen Relativitätstheorie* [*On the General Theory of Relativity*], SB II, 1915, pp. 778–86, 799–801.

1915 h *Erklärung der Perihelbewegung des Merkur aus der allgemeinen Relativitätstheorie* [*Explanation of the Perihelion Precession of Mercury According to the General Theory of Relativity*], SB II, 1915, pp. 831–39.

1915 i *Feldgleichungen der Gravitation* [*Field Equations of Gravitation*], SB II, 1915, pp. 844–47.

1916 a *Eine neue formale Deutung der Maxwellschen Feldgleichungen der Elektrodynamik* [*A New Formal Interpretation of the Maxwellian Field Equations of Electrodynamics*], SB I, 1916, pp. 184–88.

1916 b *Über einige anschauliche Überlegungen auf dem Gebiet der Relativitätstheorie* [*On Some Graphic Reflections in the Area of the Relativity Theory*], SB I, 1916, p. 423 (Summary).

1916 c *Näherungsweise Integration der Feldgleichungen der Gravitation* [*Approximative Integration of the Field Equations of Gravitation*], SB I, 1916, pp. 688–96.

1916 d Gedächtnisrede auf Karl Schwarzschild [Commemorative Oration for Karl Schwarzschild], SB I, 1916, pp. 768–70.

1916 e *Hamiltonsches Prinzip und Allgemeine Relativitätstheorie* [*Hamiltonian Principle and General Theory of Relativity*], SB II, 1916, pp. 1111–16.

1916 f Vorwort zu E. Freundlich: *Grundlagen der Einsteinschen Gravitationstheorie* [Preface to E. Freundlich: *Fundamentals of Einstein's Theory of Gravitation*], Berlin 1916.

1916 g *Einfaches Experiment zum Nachweis der Ampèreschen Molekularströme* [*Simple Experiment for the Demonstration of Ampère's Molecular Currents*], VhDPG, 18, 1916, pp. 173–77.

1916 h *Grundlage der allgemeinen Relativitätstheorie* [*Foundation of the General Theory of Relativity*], AdP, 49, 1916, pp. 769–822. Also published as a separate brochure (Leipzig, 1916).

1916 i *Über Kottlers Abhandlung: "Einsteins Äquivalenzhypothese und die Gravitation"* [*On Kottler's Treatise "Einstein's Equivalence Hypothesis and Gravitation"*], AdP, 51, 1916, pp. 639–42.

1916 j *Strahlungsemission und Absorption nach der Quantentheorie* [*Emission and Absorption of Radiation According to the Quantum Theory*], VhDPG, 18, 1916, pp. 318–23.

1916 k *Quantentheorie der Strahlung* [*Quantum Theory of Radiation*], *Mitteilungen der Physikalischen Gesellschaft Zürich*, 16, 1916, pp. 47–62.

1916 l Besprechung von H. A. Lorentz, *Théories statistiques en thermodynamique* [Review of H. A. Lorentz, *Statistical Theories in Thermodynamics*], NW, 4, 1916, pp. 480–81.

1916 m *Elementare Theorie der Wasserwellen und des Fluges* [*Elementary Theory of Water Waves and of Flight*], NW, 4, 1916, pp. 509–10.

1916 n *Ernst Mach*, PhZ, 17, 1916, pp. 101–4.

1917 a *Kosmologische Betrachtungen zur Allgemeinen Relativitätstheorie* [*Cosmological Observations on the General Theory of Relativity*], SB I, 1917, pp. 142–52.

1917 b *Eine Ableitung des Theorems von Jacobi* [*A Derivation of Jacobi's Theorem*], SB I, 1917, pp. 606–8.

1917 c *Zum Quantensatz von Sommerfeld und Epstein* [*On the Quantum Theorem of Sommerfeld and Epstein*], VhDPG, 19, 1917, pp. 82–92.

1917 d *Über die spezielle und allgemeine Relativitätstheorie, gemeinverständlich* [*On the Special and General Theory of Relativity, Generally Comprehensible*] (Braunschweig, 1917). Numerous translations and new enlarged editions.

1917 e *Friedrich Adler als Physiker* [*Friedrich Adler as a Physicist*], *Vossische Zeitung*, No. 259, May 23, 1917.

1917 f *Besprechung von H. v. Helmholtz: Zwei Vorträge über Goethe* [Review of H. von Helmholtz: *Two Lectures on Goethe*], NW, 5, 1917, p. 675.

1917 g *Marian v. Smoluchowski*, NW, 5, 1917, pp. 737–38.

1917 h *Quantentheorie der Strahlung* [*Quantum Theory of Radiation*], PhZ, 18, 1917, pp. 121–28.

1918 a *Über Gravitationswellen* [*On Gravitational Waves*], SB I, 1918, pp. 154–67.

1918 b *Kritisches zu einer von Hrn. de Sitter gegebenen Lösung der Gravitationsgleichungen* [*Critical Remarks on Herr de Sitter's Solution of the Gravitational Equations*], SB I, 1918, pp. 270–72.

1918 c *Der Energiesatz in der Allgemeinen Relativitätstheorie* [*The Energy*

Theorem in the General Theory of Relativity], SB I, 1918, pp. 448–59.

1918 d *Kommentar zu H. Weyl: Gravitation und Elektrizität* [*Commentary on H. Weyl: Gravitation and Electricity*], SB I, 1918, p. 478.

1918 e *Über eine von Levi-Civita und Weyl gefundene Vereinfachgung der Riemannschen Theorie der Krümmung und über die hieran sich knüpfende Weylsche Theorie der Gravitation und Elektrizität* [*On a Simplification, Found by Levi-Civita and Weyl, of Riemann's Theory of Curvature and on Weyl's Resultant Theory of Gravitation and Electricity*], SB I, 1918, p. 615 (Summary).

1918 f *Motiv des Forschens* [*Motivation of Research*], in *Zu Max Plancks 60. Geburstag* [*On Max Plank's Sixtieth Birthday*] (Karlsruhe, 1917). Reprinted in *Mein Weltbild* under the title *Prinzipien der Forschung* [*Principles of Research*].

1918 g *Prinzipielles zur allgemeinen Relativitätstheorie* [*Fundamental Remarks on the General Theory of Relativity*], AdP, 55, 1918, pp. 251–54.

1918 h *Lassen sich Brechungsexponenten der Köroper für Röntgenstrahlen experimentell ermitteln?* [*Can Refraction Exponents of Bodies with Regard to X-rays Be Experimentally Established?*], VhDPG, 20, 1918, pp. 86–87.

1918 i *Bemerkung zu Gehrckes Notiz: Über den Äther* [*Observation on Gehrcke's Communication: On the Ether*], VhDPG, 20, 1918, p. 261.

1918 k *Rezension von H. Weyl: Raum, Zeit, Materie* [Review of H. Weyl: *Space, Time, Matter*], NW, 6, 1918, p. 373.

1918 l *Dialog über Einwände gegen die Relativitätstheorie* [*Dialogue on Objections to the Relativity Theory*], NW, 6, 1918, pp. 697–702.

1918 m *Notiz zu Schrödingers Arbeit: Energiekomponenten des Gravitationsfeldes* [Note on Schrödinger's Paper *Energy Components of the Gravitational Field*], PhZ, 19, 1918, pp. 115–16.

1918 n *Bemerkung zu Schrödingers Notiz: Lösungssystem der allgemein kovarianten Gravitationsgleichungen* [*Observation on Schrödinger's Communication: System of Solutions of Generally Covariant Gravitational Equations*], PhZ, 19, 1918, pp. 165–66.

1919 a *Spielen Gravitationsfelder im Aufbau der materiellen Elementarteilchen eine wesentliche Rolle?* [*Do Gravitational Fields Play a Significant Part in the Structure of Material Elementary Particles?*], SB I, 1919, pp. 349–56.

1919 b *Bemerkungen über periodische Schwankungen der Mondlänge, welche bisher nach der Newtonschen Mechanik nicht erklärbar schienen [Observations on Periodic Fluctuations of the Lunar Length, Which Did Not Previously Seem Explicable by Newtonian Mechanics]*, SB I, 1919, pp. 433–36; commentaries on this in SB II, 1919, p. 711.

1919 c *Über eine Veranschaulichung der Verhältnisse im sphärischen Raum, ferner über die Feldgleichungen der Allgemeinen Relativitätstheorie vom Standpunkt des kosmologischen Problems und des Problems der Konstitution der Materie [On a Graphic Presentation of Relations in Spherical Space, also on the Field Equations of the General Theory of Relativity from the Viewpoint of the Cosmological Problem and the Problem of the Constitution of Matter]*, SB I, 1919, p. 463 (only a short summary, essentially a report on 1919a).

1919 d *Prüfung der allgemeinen Relativitätstheorie [Testing of the General Theory of Relativity]*, NW, 7, 1919, p. 776.

1919 e *My Theory, Times* (London), November 28, 1919.

1919 f *Leo Arons als Physiker [Leo Arons as a Physicist], Sozialistische Monatshefte*, 52, 1919, pp. 1055–56.

1919 g *Induktion und Deduktion in der Physik [Induction and Deduction in Physics], Berliner Tageblatt*, December 25, 1919, 4th supplement, p. 1.

1920 a *Das Trägheitsmoment des Wasserstoffmoleküls [The Moment of Inertia of the Hydrogen Molecule]*, SB, 1920, p. 65 (summary only).

1920 b *Schallausbreitung in teilweise dissoziierten Gasen [Propagation of Sound in Partially Dissociated Gases]*, SB, 1920, pp. 380–85.

1920 c *Meine Antwort. Über die relativitätstheoretische G.m.b.H.* [My Reply. *On the Relativity-Theory Limited Liability Company*], *Berliner Tageblatt*, August 27, 1920, pp. 1–2.

1920 d *Bemerkung zur Abhandlung von W. R. Heß: Theorie der Viscosität heterogener Systeme [Observation on W. R. Hess's Treatise: Theory of Heterogeneous Systems], Kolloizidzeitschrift*, 27, 1920, p. 137.

1920 e *Äther und Relativitätstheorie [Ether and Relativity Theory]*, inaugural address for Einstein's visiting professorship in Leyden (Berlin, 1920).

1920 f *Inwiefern läßt sich die moderne Gravitationstheorie ohne die Relativität begründen? [How Can the Modern Theory of Gravitation Be Explained without Relativity?]*, NW, 8, pp. 1010–11.

1921 a *Geometrie und Erfahrung [Geometry and Experience]*, festive lecture on the anniversary of Frederick II, given at the Prussian Academy, SB I, 1921, pp. 123–30. Enlarged version as a separate brochure (Berlin, 1921). Also in *Weltbild.*

1921 b *Einfache Anwendung des Newtonschen Gravitationsgesetzes auf die kugelförmigen Sternhaufen* [*Simple Application of Newton's Law of Gravitation to the Spherical Star Clusters*], *Festschrift* on the tenth anniversary of the Kaiser Wilhelm Society (Berlin, 1921), pp. 50–52.

1921 c *A Brief Outline of the Development of the Theory of Relativity*, *Nature*, 106, 1921, pp. 782–84.

1921 d *Eine naheliegende Ergänzung des Fundaments der allgemeinen Relativitätstheorie* [*An Obvious Supplementation of the Foundations of the General Theory of Relativity*], SB I, 1921, pp. 261–64.

1921 e *Ein den Elementarprozeß der Lichtemission betreffendes Experiment* [*An Experiment Concerning the Elementary Process of Light Emission*], SB I, 1921, pp. 882–83.

1921 f *The Meaning of Relativity: Four Lectures Delivered at Princeton University, May 1921.* Trans. by E. P. Adams (Princeton, 1921); numerous translations and new editions.

1922 a *Experiment betreffend die Gültigkeitsgrenze der Undulationstheorie* [*Experiment Concerning the Limit of Validity of the Undulation Theory*], SB, 1922, p. 4 (summary only).

1922 b *Theorie der Lichtfortpflanzung in dispergierenden Medien* [*Theory of the Propagation of Light in Dispersing Media*], SB, 1922, pp. 18–22.

1922 c *Theoretische Bemerkungen zur Supraleitung der Metalle* [*Theoretical Observations on the Superconductivity of Metals*], in *Gedenkboek aangeboden aan H. Kamerlingh Onnes* [*Commemorative Book Offered to H. Kammerlingh Onnes*] (Leyden, 1922), pp. 429–35.

1922 d *Bemerkung zur Seletyschen Arbeit: Beiträge zum kosmologischen Problem* [*Observation on Selety's Paper: Contributions to the Cosmological Problem*], AdP, 69, 1922, pp. 436–38.

1922 e *Rezension von W. Pauli: Relativitätstheorie* [*Review of W. Pauli: Relativity Theory*], NW, 10, 1922, pp. 184–85.

1922 f *Emil Warburg als Forscher* [*Emil Warburg as a Researcher*], NW, 10, 1922, pp. 823–28.

1922 g *Bemerkung zu der Abhandlung von E. Trefftz: Das statische Gravitationsfeld zweier Massenpunkte in der Einsteinschen Theorie* [*Observation on E. Trefftz's Paper: The Static Gravitational Field of Two Point Masses in Einstein's Theory*], SB, 1922, pp. 448–49.

1922 h *Quantentheoretische Bemerkung zum Experiment von Stern und Gerlach* [*Quantum Theoretical Observation on the Experiment by Stern and Gerlach*], with Paul Ehrenfest, ZfP, 11, 1922, pp. 31–34.

1922 i *Bemerkung zu der Arbeit von A. Friedmann: Über die Krümmung des*

Raumes [*Observation on the Paper by A. Friedmann: On the Curvature of Space*], ZfP, 11, 1922, p. 326.

1922 j *Wie ich die Relativitätstheorie entdeckte* [*How I Created the Theory of Relativity*], lecture at University of Kyoto, Japan, on December 14, 1922. Einstein spoke off the cuff; simultaneous notes by the Japanese translator Yon Ishiwara were translated into English by Y. A. Ono; *Physics Today*, August 1932, pp. 45ff.

1923 a *Zur Allgemeinen Relativitätstheorie* [*On the General Theory of Relativity*], SB, 1923, pp. 32–38.

1923 b *Bemerkung zu meiner Arbeit "Zur Allgemeinen Relativitätstheorie"* [*Observation on My Treatise "On the General Theory of Relativity"*], SB, 1923, pp. 76–77.

1923 c *Zur affinen Feldtheorie* [*On the Affine Field Theory*], SB, 1923, pp. 137–40.

1923 d *Theory of the Affine Field, Nature*, 112, 1923, pp. 448–49.

1923 e *Grundgedanken und Probleme zur Relativitätstheorie* [*Fundamental Ideas and Problems of the Theory of Relativity*], Nobel lecture given in Göteborg on June 11, 1923, 10 pp., Imprimerie Royale, Stockholm, 1923. Also in *Les Prix Nobel en 1921-1922* (Stockholm, 1923).

1923 f *Quantentheorie des Strahlungsgleichgewichts* [*Quantum Theory of the Radiation Equilibrium*], with Paul Ehrenfest, ZfP, 19, 1923, pp. 301–306.

1923 g *Bemerkung zu der Notiz von W. Anderson: Neue Erklärung des kontinuierlichen Koronaspektrums* [*Observation on the Communication by W. Anderson: A New Explanation of the Continuous Spectrum of the Solar Corona*], *Astrologische Nachrichten*, 219, 1923, p. 19.

1923 h *Experimentelle Bestimmung der Kanalweite von Filtern* [*Experimental Determination of the Channel Diameter of Filters*], with Hans Mühsam, *Deutsche Medizinische Wochenschrift*, 49, 1923, pp. 1012–13.

1923 i *Beweis für die Nichtexistenz eines überall regulären zentrisch symmetrischen Feldes nach der Feldtheorie von Kaluza* [*Proof of the Nonexistence of a Universally Regular Centrally Symmetrical Field According to Kaluza's Field Theory*], with Jakob Grommer, *Scripta Mathematica et Physica*, Hebrew University, Jerusalem, Vol. 1, No. 7, 1923. German and Hebrew text.

1923 j *Bietet die Feldtheorie Möglichkeiten für die Lösung des Quantenproblems?* [*Does the Field Theory Offer Possibilities of Solving the Quanta Problem?*], SB, 1923, pp. 359–64.

1923 k *Berichtigung* [Correction] to 1923i, ZfP, 16, 1923, p. 228.

1924 a Geleitwort zu Lucretius: *De Rerum natura* [Prefatory note to Lucretius: *De Rerum natura*], in Latin and German, ed. H. Diels, Vol. 2 (Berlin, 1924).

1924 b *Das Komptonsche Experiment* [*The Compton Experiment*], *Berliner Tageblatt*, April 20, 1924, 1st supplement, p. 1.

1924 c *Zum hundertjährigen Gedenktag von Lord Kelvins Geburt* [*On the Centenary of Lord Kelvin's Birth*], NW, 12, 1924, pp. 601–2.

1924 d *Über den Äther* [*On the Ether*], *Verhandlungen der schweizerischen Naturforschenden Gesellschaft*, 105, 1924, pp. 85–93.

1924 e *Rezension von J. Winternitz: Relativitätstheorie und Erkenntnislehre. Eine Untersuchung über erkenntnistheoretische Grundlagen der Einsteinschen Theorie und die Begründung ihrer Ergebnisse für die allgemeinen Grundlagen des Naturerkennens* [Review of J. Winternitz: *Relativity Theory and Cognition Theory. An Examination of the Cognition-Theory Basis of Einstein's Theory and Justification of Its Results for the General Foundations of Knowledge of Nature*] (Leipzig, 1923). *Deutsche Literaturzeitung für die Kritik der internationalen Wissenschaft*, 1, 1924, col. 20–22.

1924 f *Rezension von A. C. Elsbach: Kant und Einstein. Untersuchungen über das Verhältnis der modernen Erkenntnistheorie zur Relativitätstheorie* [Review of A. C. Elsbach: *Kant and Einstein, Examinations of the Relationship between Modern Cognition Theory and Relativity Theory*] (Berlin and Leipzig, 1924). *Deutsche Literaturzeitung für Kritik der internationalen Wissenschaft*, 1, 1924, col. 1685–92.

1924 g *Theorie der Radiometerkräfte* [*Theory of the Radiometric Forces*], ZfP, 27, 1924, pp. 1–6.

1924 h *Anmerkung zur Arbeit von Bose: Wärmegleichgewicht im Strahlungsfeld bei Anwesenheit von Materie* [*Note on the Paper by Bose: Thermal Equilibrium in the Radiation Field in the Presence of Matter*], ZfP, 27, 1924, pp. 392–93.

1924 i *Quantentheorie des einatomigen idealen Gases* [*Quantum Theory of Single-Atom Ideal Gases*], SB, 1924, pp. 261–67.

1925 a *Quantentheorie des einatomigen idealen Gases* [*Quantum Theory of Single-Atom Ideal Gases*], second treatise, SB, 1925, pp. 3–14.

1925 b *Quantentheorie des einatomigen idealen Gases* [*Quantum theory of Single-Atom Ideal Gases*], SB, 1925, pp. 18–25.

1925 c *Einheitliche Feldtheorie von Gravitation und Elektrizität* [*Unified Field Theory of Gravitation and Electricity*], SB, 1925, pp. 414–19.

1925 d *Elektron und allgemeine Relativitätstheorie* [*Electron and General Theory of Relativity*], *Physica*, 5, 1925, pp. 330–34.

1925 e *Bemerkung zu P. Jordans Abhandlung: Theorie der Quantenstrahlung* [*Observation on P. Jordan's Treatise: Theory of Quantum Radiation*], ZfP, 31, 1925, pp. 784–85.

1925 f *Eddingtons Theorie und Hamiltonsches Prinzip* [*Eddington's Theory and the Hamiltonian Principle*], Appendix to German edition of A. S. Eddington, *Relativitätstheorie in mathematischer Behandlung* (Berlin, 1925).

1926 a *Über die Ursache der Mäanderbildung bei Flußläufen* [*On the Cause of Meander Formation in River Courses*], SB, 1926, p. 1 (summary only of a lecture on January 7). Text in NW, 14, 1925, pp. 223–24. Also in *Weltbild.*

1926 b *Über die Anwendung einer von Rainisch gefundenen Spaltung des Riemannschen Krümmungstensors in der Theorie des Gravitationsfeldes* [*On the Application of a Splitting, Discovered by Rainisch, of Riemann's Tensor of Curvature in the Theory of the Gravitational Field*], SB, 1926, p. 1 (summary only).

1926 c *Meine Theorie und Millers Versuche* [*My Theory and Miller's Experiments*], *Vossische Zeitung*, January 19, 1926.

1926 d *W. H. Julius, 1860–1925, Astrophysical Journal*, 63, 1926, pp. 196–98.

1926 e *Vorschlag zu einem die Natur des elementaren Strahlungs-Emissionsprozesses betreffendes Experiment* [*Suggestion for an Experiment Concerning the Nature of the Elementary Radiation Emission Process*], NW, 14, 1926, pp. 300–301.

1926 f *Interferenzeigenschaften des durch Kanalstrahlen emittierten Lichtes* [*Interference Characteristics of the Light Emitted by Canal Rays*], SB, 1926, pp. 334–40.

1927 a *Allgemeine Relativitätstheorie und Bewegungsgesetz* [*General Theory of Relativity and Law of Motion*], with Jakob Grommer, SB, 1927, pp. 2–13.

1927 b *Zu Kaluzas Theorie des Zusammenhanges von Gravitation und Elektrizität* [*On Kaluza's Theory of the Connection between Gravitation and Electricity*], 1st communication, SB, 1927, pp. 23–35; 2nd communication, SB, pp. 26–30.

1927 c *Einfluß der Erdbewegung auf die Lichtgeschwindigkeit relativ zur Erde* [*Effect of the Earth's Motion on the Velocity of Light Relative to the Earth*], *Forschungen und Fortschritte*, 3, 1927, p. 36–37.

1927 d *Theoretisches und Experimentelles zur Frage der Lichtentstehung* [*Theoretical and Experimental Aspects of the Problem of the Genesis of Light*], report on a lecture to the Mathematical Physical Work Team of Berlin University on February 23, 1927, *Zeitschrift für angewandte Chemie*, 40, 1927, p. 546.

1927 e *Formale Beziehung des Riemannschen Krümmungstensors zu den Feldgleichungen der Gravitation* [*Formal Relationship of Riemann's Tensor of Curvature to the Field Equations of Gravitation*], *Mathematische Annalen*, 97, 1927, pp. 99–103. Full treatment of 1926 b.

1927 f *Isaac Newton*, *Manchester Guardian*, March 10, 1927.

1927 g *Newtons Mechanik und ihr Einfluß auf die Gestaltung der theoretischen Physik* [*Newton's Mechanics and Its Influence on the Shaping of Theoretical Physics*], NW, 15, 1927, pp. 273–76. Also in *Weltbild*.

1927 h *Isaac Newton*, letter to the Royal Society, London, on the bicentenary of Newton's death, *Nature*, 119, 1927, p. 467.

1927 i *Establishment of an International Bureau of Meteorology*, report of a commission of the International Committee for Intellectual Cooperation, signed by M. Curie, A. Einstein, and H. A. Lorentz, *Science*, 65, 1927, pp. 415–17.

1927 j *Allgemeine Relativitätstheorie und Bewegungsgesetze* [*General Theory of Relativity and Laws of Motion*], SB, 1927, pp. 235–45.

1928 a *H. A. Lorentz*, *Mathematische naturwissenschaftliche Blätter*, 22, 1928, pp. 24–25. Also in *Weltbild*.

1928 b *À Propos de "La Déduction relativiste" de E. M. Meyerson* [*On the "Relativistic Deduction" of E. M. Meyerson*], *Revue Philosophique de France*, 106, 1928, pp. 161–66.

1928 c *Riemann-Geometrie mit Aufrechterhaltung des Fernparallelismus* [*Riemann Geometry with Preservation of Distant Parallelism*], SB, 1928, pp. 217–21.

1928 d *Neue Möglichkeiten für eine einheitliche Feldtheorie von Gravitation und Elektrizität* [*New Possibilities of a Unified Field Theory of Gravitation and Electricity*], SB, 1928, pp. 224–27.

1929 a *Zur einheitlichen Feldtheorie* [*On the Unified Field Theory*], SB, 1929, pp. 2–7.

1929 b *Einstein's New Theory*, interview, *Daily Chronicle*, January 26, 1929. Excerpts in *Nature*, 123, 1929, pp. 2–7.

1929 c *The New Field Theory*, *New York Times*, February 3, 1929; *Times* (London), February 4, 1929.

1929 d *Einheitliche Interpretation von Gravitation und Elektrizität* [*Unified

Interpretation of Gravitation and Electricity], SB, 1929, p. 102 (summary only).
1929 e *Einheitliche Feldtheorie und Hamiltonsches Prinzip* [*Unified Field Theory and Hamilton's Principle*], SB, 1929, pp. 156–9.
1929 f *Über den gegenwärtigen Stand der Feldtheorie* [*On the Present State of Field Theory*], *Festschrift* for A. Stodola (Zurich, 1929), pp. 126–32.
1929 g *Ansprache an Max Planck* [*Address to Max Planck*] on the inaugural award of the Planck Medal of the German Physical Society, on June 28, 1929, *Forschungen und Fortschritte*, 5, pp. 248–49.
1929 h *Space-Time, Encyclopaedia Britannica*, 14th ed., Vol. 21, 1929, pp. 105–8.
1929 i *Sur la Théorie synthétique des champs* [*On the Unified Field Theory*], with Théophile de Donder, *Revue générale de l'electricité*, 25, 1929, pp. 35–39.
1929 j *Simon Newcomb*, Science, 69, 1929, p. 249.
1929 k Sesión especial de la Academia [Special session of the Academy], April 16, 1925. *Anales de la Sociedad científica Argentina*, 107, 1929, pp. 337–47.
1930 a *Théorie unitaire du champ physique [Unified Physical Field Theory], Annales de l'Institut Henri Poincaré*, 1, 1930, pp. 1–24.
1930 b *Die Kompatibilität der Feldgleichungen in der einheitlichen Feldtheorie* [*The Compatibility of Field Equations in the Unified Field Theory*], SB, 1930, pp. 18–23.
1930 c *Zwei strenge statistische Lösungen der Feldgleichungen der einheitlichen Feldtheorie* [*Two Strict Statistical Solutions of the Field Equations of the Unified Field Theory*], with Walther Mayer, SB, 1930, pp. 110–20.
1930 d *Über die Fortschritte der einheitlichen Feldtheorie* [*On Progress Made by the Unified Field Theory*], SB, 1930, p. 102 (summary only).
1930 e *Zur Theorie der Räume mit Riemann-Metrik und Fernparallelismus* [*On the Theory of Spaces with Riemann Metrics and Distant Parallelism*], SB, 1930, pp. 401–2.
1930 f *Über die statistischen Eigenschaften der Strahlung* [*On the Statistical Characteristics of Radiation*], SB, 1930, p. 543 (summary only).
1930 g *Auf die Riemann-Metrik und den Fernparallelismus gegründete einheitliche Feldtheorie* [*Unified Field Theory Based on Riemann Metrics and Distant Parallelism*], *Mathematische Annalen*, 102, 1930, pp. 685–97.

1930 h *Raum, Äther und Feld in der Physik* [*Space, Ether, and Field in Physics*], *Forum Philosophicum*, 1, 1930, pp. 173–80.

1930 i *Raum-, Feld- und Ätherproblem in der Physik* [*The Problem of Space, Field, and Ether in Physics*], World Power Conference, Berlin Transactions, 1930, pp. 1–5.

1930 j *Rezension von S. Weinberg: Erkenntnistheorie* [Review of S. Weinberg: *Cognition Theory*], NW, 18, 1930, p. 536.

1930 k *Über Keppler*, *Frankfurter Zeitung*, November 9, 1930. Also in *Weltbild.*

1930 l *Religion und Wissenschaft* [*Religion and Science*], *New York Times*, November 9, 1930, *Berliner Tageblatt*, November 11, 1930.

1930 m *Concept of Space*, *Nature*, 125, 1930, pp. 897–98.

1930 n *Über den gegenwärtigen Stand der Relativitätstheorie* [*On the Present State of the Theory of Relativity*], *Yale University Library Gazette*, 6, 1930, pp. 3–6.

1930 o *Das Raum-Zeit-Problem* [*The Space-Time Problem*], Koralle, 5, 1930, pp. 486–88.

1930 p *Rede zur Funkausstellung* [*Speech at the Broadcasting Exhibition*], Berlin, August 22, 1930. Transcription from the sound recording by F. Herneck in NW, 48, 1930, p. 33.

1930 q Begleitwort [Introduction] to David Reichinstein, *Grenzflächenvorgänge in der unbelebten und belebten Natur* [*Boundary Processes in Inanimate and Animate Nature*] (Leipzig, 1930).

1931 a *Zum kosmologischen Problem der Allgemeinen Relativitätstheorie* [*On the Cosmological Problem of the General Theory of Relativity*], SB, 1931, pp. 235–37.

1931 b *Systematische Untersuchungen über kompatible Feldgleichungen, welche in einem Riemannschen Raume mit Fernparallelismus gesetzt werden können* [*Systematic Investigations of Compatible Field Equations Capable of Being Stated in a Riemann Space with Distant Parallelism*], with Walther Mayer, SB, 1931, pp. 257–65.

1931 c *Einheitliche Theorie von Gravitation und Elektrizität* [*Unified Theory of Gravitation and Electricity*], with Walther Mayer, SB, 1931, pp. 541–57.

1931 d *Knowledge of Past and Future in Quantum Mechanics*, with R. C. Tolman and B. Podolsky, *Physical Review*, 37, 1931, pp. 780–81.

1931 e Foreword to R. de Villamil, *Newton, the Man* (London, 1931).

1931 f Foreword to Sir Isaac Newton, *Opticks* (New York, 1931).

1931 g *Maxwell's Influence on the Development of the Conception of Physical*

Reality, in J. C. Maxwell, *A Commemorative Volume* (Cambridge, England, 1931), pp. 66–73. Also in *Weltbild*.

1931 h *Thomas Alva Edison, 1847–1931, Science*, 74, 1931, pp. 404–5.

1931 i Gedenkrede auf Albert A. Michelson [Memorial Address for Albert A. Michelson], *Zeitschrift für angewandte Chemie*, 44, 1931, p. 658.

1932 a *Einheitliche Theorie von Gravitation und Elektrizität* [*Unified Theory of Gravitation and Electricity*], with Walther Mayer, 2nd treatise (Continuation of 1931c), SB, 1932, pp. 130–37.

1932 b *Semi-Vektoren und Spinoren* [*Semivectors and Spinors*], with Walther Mayer, SB, 1932, pp. 522–50.

1932 c *On the Relation Between the Expansion and the Mean Density of the Universe*, with Willem de Sitter, *Proceedings of the National Academy of Science*, 18, 1932, pp. 213–14.

1932 d *Zu Dr. Berliners 70. Geburtstag* [*On Dr. Berliner's Seventieth Birthday*], NW, 20, 1932, p. 913.

1932 e *Gegenwärtiger Stand der Relativitätstheorie* [*Present State of the Relativity Theory*], *Die Quelle*, 82, 1932, pp. 440–42.

1932 f *Unbestimmtheitsrelation* [*Uncertainty Relation*], report of a colloquium at Berlin University on November 4, 1931, *Zeitschrift für angewandte Chemie*, 45, 1932, p. 23.

1932 g Prologue and Epilogue, in Max Planck (ed.), *Where Is Science Going?* (New York, 1932).

1933 a *Dirac-Gleichungen für Semi-Vektoren* [*Dirac Equations for Semivectors*], with Walther Mayer, *Proc. Akademie van wetenschappen*, Amsterdam, 36, II, 1933, pp. 497–502.

1933 b *Spaltung der natürlichsten Feldgleichungen für Semi-Vektoren in Spinorgleichungen vom Diracschen Typus* [*Splitting of the Most Natural Field Equations for Semivectors into Spinor Equations of the Dirac Type*], with Walther Mayer, *Proc. Akademie van wetenschappen*, Amsterdam, 36, II, 1933, pp. 615–19.

1933 c *On the Methods of Theoretical Physics*, Spence Lecture at Oxford University, on June 10, 1933 (Oxford, 1933).

1933 d *Origins of the General Theory of Relativity*, Gibson Lecture at Glasgow University on June 20, 1933, *Glasgow University Publications* No. 20.

1934 a *Darstellung der Semi-Vektoren als gewöhnliche Vektoren von besonderem Differentiationscharakter* [*Presentation of Semivectors as Vectors of a Particular Differentiation Character*], with Walther Mayer, *Annals of Mathematics*, 35, 1934, pp. 104–10.

1934 b *Rezension von R. Tolman: Relativity, Thermodynamics, and Cosmology* [Review], *Science*, 80, 1934, p. 358.

1934 c Introduction to L. Infeld, *The World in Modern Science* (London, 1934).

1934 d Nachruf auf Paul Ehrenfest [Obituary for Paul Ehrenfest], *Almanak van het Leidensche Studentencorps*, 1934. Also in *Späte Jahre.*

1935 a *Can Quantum-Mechanical Description Be Considered Complete?* with B. Podolsky and N. Rosen, *Physical Review*, 47, 1935, pp. 777–80.

1935 b *The Particle Problem in the General Theory of Relativity*, with N. Rosen, *Physical Review*, 48, 1935, pp. 73–77.

1935 c *Elementary Derivation of the Equivalence of Mass and Energy*, J. W. Gibbs lecture given to AAAS on December 28, 1934, *Bulletin of the American Mathematical Society*, 41, 1935, pp. 223–30.

1936 a *Physik und Realität [Physics and Reality]*, *Journal of the Franklin Institute*, 221, 1936, pp. 313–47. Also in *Späte Jahre*, pp. 63–106.

1936 b *Two-Body Problem in General Relativity Theory*, with Nathan Rosen, *Physical Review*, 49, 1936, pp. 404–5.

1936 c *Lens-like Action of a Star by Deviation of Light in the Gravitational field*, *Science*, 84, 1936, pp. 506–7.

1937 a *On Gravitational Waves*, with Nathan Rosen, *Journal of the Franklin Institute*, 223, 1937, pp. 43–54.

1938 a *Gravitational Equations and the Problems of Motion*, with Leopold Infeld and Banesh Hoffmann, *Annals of Mathematics*, 39, 1938, pp. 65–100.

1938 b *Generalization of Kaluza's Theory of Electricity*, with Peter Bergmann, *Annals of Mathematics*, 39, 1938, pp. 683–701.

1938 c *The Evolution of Physics: The Growth of Ideas from Early Concepts to Relativity and Quanta*, with Leopold Infeld (New York, 1938).

1939 a *Stationary Systems with Spherical Symmetry Consisting of Many Gravitating Masses*, *Annals of Mathematics*, 40, 1939, pp. 922–36.

1939 b *Our Goal*, lecture at Princeton Theological Seminary; Part 1 in *Späte Jahre*, pp. 37–40.

1940 a *Science and Religion*, Conference on Science, Philosophy and Religion (New York, 1940); Part 2 in *Späte Jahre*, pp. 41–47.

1940 b *Gravitational Equations and the Problems of Motion*, Part II, with Leopold Infeld, *Annals of Mathematics*, 41, 1940, pp. 455–64.

1940 c *Considerations Concerning the Fundamentals of Theoretical Physics*, *Science*, 91, 1940, pp. 487–92. Lecture to the American Scientific Congress, Washington, May 28, 1940.

1941 a *Five-Dimensional Representation of Gravitation and Electricity*, with Valentin Bargmann and Peter Bergmann, in *Theodore von Karman Anniversary Volume* (Pasadena, 1941), pp. 212–25.

1941 b Foreword to Peter G. Bergmann, *Introduction to the Theory of Relativity* (New York, 1941).

1942 a *Demonstration of the Nonexistence of Gravitational Fields with a Nonvanishing Total Mass Free of Singularities*, *Revista*, Tucuman Universidad Nacional, 2, 1942, pp. 11–16.

1942 b *The Common Language of Science*, *Advancement of Science*, 2, 1942, p. 109.

1942 c *The Work and Personality of Walther Nernst*, *Scientific Monthly*, 54, 1942, pp. 195–96.

1943 a *Nonexistence of Regular Stationary Solutions of Relativistic Field Equations*, with Wolfgang Pauli, *Annals of Mathematics*, 44, 1942, pp. 131–37.

1944 a *Bivector Fields I*, with Valentin Bargmann, *Annals of Mathematics*, 45, 1944, pp. 1–14.

1944 b *Bivector Fields II*, *Annals of Mathematics*, 45, 1944, pp. 15–23.

1944 c *Remarks on Bertrand Russell's Theory of Knowledge*, in *The Philosophy of Bertrand Russell* (Evanston, 1944), pp. 277–91 (Library of Living Philosophers, Vol. 5).

1945 a *On the Cosmological Problem*, *American Scholar*, 14, 1945, pp. 137–56; Corrections, p. 269.

1945 b *Generalization of the Relativistic Theory of Gravitation*, *Annals of Mathematics*, 46, 1945, pp. 578–84.

1945 c *Influence of the Expansion of Space on the Gravitation Fields Surrounding the Individual Stars*, with Ernst G. Straus, *Review of Modern Physics*, 17, 1945, pp. 120–24.

1945 d *A Testimonial from Prof. Einstein*, pp. 142–43, in Jacques S. Hadamard, *An Essay on the Psychology of Invention in the Mathematical Field* (Princeton, 1945).

1945 e *The Meaning of Relativity*, enlarged ed. of 1921f.

1946 a *Generalization of the Relativistic Theory of Gravitation*, *Annals of Mathematics*, 47, 1946, pp. 731–41.

1946 b *Elementary Derivation of the Equivalence of Mass and Energy*, *Technion Journal*, 5, 1946, pp. 16–17.

1946 c *Autobiographisches*, in Schilpp (Evanston, 1949), pp. 1–32.

1947 a *The Problem of Space, Ether, and the Field in Physics*, in Saxe, Commins, and Linscott (eds.), *Man and the Universe* (New York, 1947), pp. 471–82.

1947 b *Paul Langevin*, *La Pensée*, 12, 1947, pp. 13–14.

1948 a *Quantenmechanik und Wirklichkeit* [*Quantum Mechanics and Reality*], *Dialectica*, 2, 1948, pp. 320–24.

1948 b *Generalized Theory of Gravitation*, *Review of Modern Physics*, 20, 1948, pp. 35–39.

1948 c *Max Planck*, contribution to the memorial ceremony for Planck at the National Academy of Science, April 1948.

1949 a *Bemerkungen zu den in diesem Bande vereinigten Arbeiten* [*Observations on the Papers Collected in this Volume*], in Schilpp (Evanston, 1949), pp. 493–511.

1949 b *Motion of Particles in General Relativity Theory*, with L. Infeld, *Canadian Journal of Mathematics*, 3, 1949, pp. 209–14.

1950 a *Generalized Theory of Relativity*, Appendix II to 3rd edition of 1921f (Princeton, 1950).

1950 b *On the Generalized Theory of Gravitation*, *Scientific American*, 182, April 1950, pp. 13–17.

1950 c *The Bianchi Identities in the Generalized Theory of Gravitation*, *Canadian Journal of Mathematics*, 4, 1950, pp. 120–28.

1950 d Introduction to Philipp Frank, *Relativity—A Richer Truth* (Boston, 1950).

1953 a *Reply to Criticism of a Recent Unified Field Theory*, *Physical Review*, 89, 1953, p. 321.

1953 b Appendix II: *Relativity Theory of the Nonsymmetric Field*, in 4th edition of 1921f (Princeton, 1953).

1953 c *Elementare Überlegungen zur Interpretation der Grundlage der Quantenmechanik* [*Elementary Reflections on the Interpretation of the Foundations of Quantum Mechanics*], in *Scientific Papers Presented to Max Born* (New York, 1953).

1953 d *Louis de Broglie*, in *Louis de Broglie, Physicien et Penseur* (Paris, 1953).

1953 e *H. A. Lorentz als Schöpfer und Persönlichkeit* [*H. A. Lorentz as a Creator and Personality*], written for the centenary celebrations of Lorentz's birth, in *Weltbild*.

1953 f Vorwort [Preface] to Galileo Galilei, *Dialogue Concerning the Two Chief World Systems, Ptolemaic and Copernican*, trans. Stilman Drake (University of California Press, 1953).

1954 a *Algebraic Properties of the Field in the Relativistic Theory of the Asymmetric Field* (with Bruria Kaufmann), *Annals of Mathematics*, 59, 1954, pp. 230–44.

1955 a Revised version of 1953b.

1955 b *Erinnerungen—Souvenirs [Recollections*], *Schweizerische Hochschulzeitung*, 28, 1955, pp. 145–53. Reprinted as *Autobiographische Skizze in Helle Zeit—Dunkle Zeit*, pp. 9–17.

CHRONOLOGY

1876 *August 8:* Hermann Einstein and Pauline Koch are married in the synagogue of Cannstatt near Stuttgart, Germany.

1879 *March 14:* Albert Einstein is born in Ulm, the first child of Hermann and Pauline Einstein.

1880 *June 21:* The Einstein family moves to Munich. Hermann Einstein and his brother Jakob establish the electrical-engineering firm Einstein & Cie.

1881 *November 18:* Einstein's sister, Maria (Maja), is born.

1884 The first "miracle": the compass.

1885 *March 31:* The family moves to Rengerweg 14 (now Adlzreiterstrasse) in Munich's Sendling district, in the immediate neighborhood of the Einstein & Cie. factory.

Einstein had been having private tuition since 1894; on October 1 he joins the second grade of the Petersschule, a Catholic primary school.

He receives Jewish religious instruction at home.

He is taught the violin.

1888 *October 1:* Having passed the entry examination, Einstein is accepted at the Luitpold Gymnasium.

1889 *Fall:* The twenty-one-year-old medical student Max Talmud regularly visits the Einsteins; for the next five years he is Albert's friend and mentor. Talmud acquaints Albert with scientific and philosophical writings, including Kant's *Critique of Pure Reason.*

1890 The second "miracle": The "sacred little geometry book."

1892 Einstein is not "bar mitzvah" and therefore, in the rabbinical sense, is not a member of the Jewish community.

1894 *June:* Einstein & Cie. go into liquidation; the family moves to northern Italy (Milan, Pavia, and back to Milan). Einstein remains in Munich under the care of distant relatives.

December 29: Einstein leaves the Luitpold Gymnasium without graduating and joins his parents in Milan.

1895 Einstein sends an essay, *On the Investigation of the State of the Ether in the Magnetic Field*, to his uncle Caesar Koch in Belgium. He visits relatives in Genoa and prepares for the entry exam for the Swiss Polytechnic in Zurich (after 1911 the Swiss Technological University, ETH).

October 8–14: Although he is two years under the minimum age, he is admitted for the exam. His performance in mathematics and physics is outstanding. However, as his knowledge of other subjects is inadequate, he fails the exam.

October 26: Einstein is accepted for the trade department of the cantonal school in Aarau, Switzerland. He lives with the family of Jost Winteler, a teacher at that school.

1896 *January 28:* Release from Württembergian, and hence from German, nationality; for the next five years Einstein is stateless.

September: School graduation examinations. On October 3 Einstein graduates and thus is entitled to study at the Polytechnic.

October: Beginning of study at the Zurich Polytechnic, in the "School for Specialized Subject Teachers." Fellow students include Marcel Grossmann and Mileva Marić.

1898 *October:* Intermediate diploma examination.

1900 *Spring:* Diploma paper on heat conduction; oral and written exams. On *July 28* Einstein receives his diploma as a mathematics teacher.

October: Although he does not get the hoped-for post of assistant at the Polytechnic, Einstein starts work on a doctoral thesis.

December 13: He submits his first publication (on capillarity) to *Annalen der Physik*.

1901 *February 21:* Swiss citizenship.

Spring: Einstein applies unsuccessfully for assistant's posts in Germany, the Netherlands, and Italy.

May–July: Temporary teacher at the Technical College in Winterthur, Switzerland.

September: Teacher at a private school in Schaffhausen, Switzerland.
November: Einstein submits his thesis to the University of Zurich.
December 18: He applies for a post at the Swiss Patent Office—the "Swiss Office for Intellectual Property"—in Bern.

1902 "Lieserl," Mileva Marić and Einstein's daughter, is born in Novi Sad (then Hungary, after World War I Yugoslavia).
February: Einstein withdraws his thesis; he moves to Bern and in an advertisement offers private lessons.
June 16: Einstein is engaged as an "Expert III Class" at the Patent Office in Bern. He starts work on June 23. His salary is 3,500 Swiss francs.
October 10: His father dies in Milan.

1903 *January 6:* Einstein and Mileva Marić are married in Bern; their daughter remains in Novi Sad.
April: With Maurice Solovine and Conrad Habicht, Einstein founds the Akademie Olympia.
May: Member of the Naturforschende Gesellschaft zu Bern.
September: The Einsteins' daughter, Lieserl, is apparently adopted by strangers.

1904 *May 14:* Their son Hans Albert is born in Bern.
September 16: Einstein obtains a permanent position at the Patent Office; his salary rises to 3,900 Swiss francs.

1905 *March 17:* Einstein completes his paper on the light quanta hypothesis.
April 30: He completes a paper, *On a New Determination of Molecular Dimensions;* toward the end of July it is accepted as a doctoral thesis by the University of Zurich.
May 11: His paper on the Brownian movement is received by *Annalen*.
June 30: His paper *On the Electrodynamics of Moving Bodies*, the special theory of relativity, is received by *Annalen* and published on September 28.
September 27: A supplement to the special theory of relativity is received by *Annalen;* this contains the formula $E = mc^2$.

1906 *January 15:* After various formalities Einstein becomes "Herr Doktor."

April 1: Promoted to "Expert II Class" with a salary of 4,500 Swiss francs.

November 9: His paper on specific heat, the explanation of the quantum theory for solids, is received by *Annalen.*

1907 *June 17:* Application for *Habilitation* at Bern University; the application is rejected on October 28 until Einstein submits an original *Habilitation* thesis.

October–November: Einstein discovers the equivalence principle as the guiding idea on the road to the general theory of relativity. On December 4 he completes his great comprehensive treatise on the relativity principle.

1908 *February 28:* After submission of a *Habilitation* thesis and a trial lecture Einstein becomes a privatdozent at Bern University.

April: Johann Jakob Laub arrives in Bern for three weeks of joint work. Throughout the year, intensive experiments with the "little machine."

1909 *May 7:* Einstein is appointed extraordinary professor of theoretical physics at Zurich University; his salary is 4,500 Swiss francs. He starts his professorship on October 15.

July 9: Einstein receives his first honorary degree, from the University of Geneva.

September 20: Einstein is the guest of honor at the annual meeting of the Gesellschaft Deutscher Naturforscher und Ärzte in Salzburg; on September 21 he gives a major lecture on radiation theory. He makes the acquaintance of the leading German physicists.

October 15: Move to Zurich; beginning of his teaching activity.

1910 *April 21:* Einstein is nominated for the post of full professor of theoretical physics at the German University in Prague.

July 14: Zurich University increases his salary to 5,500 Swiss francs in order to retain him.

July 28: The Einsteins' second son, Eduard, is born.

September 24: Journey to Vienna for professional negotiations; Einstein visits Ernst Mach and Victor Adler.

November 1: Einstein receives a bursary of 5,000 marks annually over three years from an anonymous donor in Germany (Franz Oppenheim).

1911 *January 6:* Emperor Francis Joseph appoints Einstein a full professor in Prague as of April 1, 1911.
February 10: Einstein travels to Leyden to give a lecture and meets Hendrik Antoon Lorentz and Heike Kamerlingh Onnes.
April 3: Einstein and his family arrive in Prague.
August 23: Einstein is sworn in.
August 24: Negotiations begin about an appointment in Utrecht.
September: Heinrich Zangger visits Einstein in Prague; a post at ETH in Zurich is discussed.
September 25: Einstein makes the acquaintance of Fritz Haber at the annual meeting of the Society of German Scientists and Physicians in Karlsruhe.
October 30–November 4: He participates at the First Solvay Congress in Brussels, Belgium; he speaks on the quantum theory of solids. Afterward he visits Utrecht.
December 19–25: Einstein negotiates in Zurich about a post at ETH.

1912 *January 30:* Einstein is appointed professor of theoretical physics at ETH, Zurich; his salary is 11,000 Swiss francs.
April 15–22: Einstein travels to Berlin and there meets all the leading physicists; first ideas about a post in Berlin. On this visit Einstein discovers more than just family feelings for his cousin Elsa Löwenthal, née Einstein.
July 25: Move to Zurich.
August: Beginning of collaboration with Marcel Grossmann on the general theory of relativity.

1913 *March 27:* Lecture in Paris about the law of photochemical equivalence.
July 3: The physical-mathematical class of the Royal Prussian Academy of Sciences nominates Einstein a member. Max Planck and Walther Nernst travel to Zurich in mid-July to propose a move to Berlin. Einstein accepts.
August: Hike through the Engadine with Marie Curie.
September: Vacation in Novi Sad with Mileva's parents. On September 21 lecture on gravitation theory to the annual meeting of the Society of German Scientists and Physicians in Vienna; subsequently a journey to Berlin.

November 12: Kaiser Wilhelm II confirms Einstein's election to the Academy. Einstein accepts on December 7. His salary is 12,900 marks.

1914 Einstein leaves Zurich and visits Antwerp and Leyden. He arrives in Berlin at the beginning of April. Mileva and their sons follow in mid-April. At the end of June Mileva and Albert Einstein separate; Mileva returns to Zurich with the sons.

July 2: Einstein's inaugural address at the Prussian Academy. Planck makes the response.

August 1: Einstein reacts to the outbreak of World War I as a determined pacifist; for the first time he concerns himself with political problems.

November: He joins the New Fatherland League, whose goal is the creation of a "United States of Europe." He signs a "Manifesto to Europeans" drafted by Georg Nicolai.

1915 *January:* At the Physical-Technical Reich Institute, Einstein—jointly with J. W. de Haas—begins experimental investigation of the gyromagnetic effect (Einstein–de Haas effect).

End of June: Einstein pays a week's visit to Göttingen, staying with David Hilbert and giving six lectures on the general theory of relativity.

November: In four lectures to the Prussian Academy (the first on November 4, the last on November 25) Einstein formulates the completion of the general theory of relativity.

1916 *March 20: The Foundations of the General Theory of Relativity*, the first systematic representation, is completed. The article is published in *Annalen der Physik* and also as a separate brochure.

May 5: Einstein becomes president of the German Physical Society.

June: First treatise on gravitational waves.

July: Renewed work on quantum theory, resulting in three papers—on spontaneous and induced emission and absorption; on a new derivation of Planck's radiation formula on that basis; and on the particle concept of the photon.

December: Einstein completes *On the Special and General Theory of Relativity, Generally Comprehensible*, his best-known book.

December 30: The Kaiser, at "Supreme Headquarters," decrees that Einstein be invited to join the Board of the Physical-Technical Reich Institute.

1917 At the beginning of the year Einstein falls ill. Liver complaints, stomach problems, and duodenal ulcers follow one another. His recovery takes about four years. His cousin Elsa looks after him.

February: Within the framework of the general theory of relativity, Einstein maps out the first model of the universe; it contains a "cosmological constant" in order to guarantee the stability of a spatially bounded universe.

September: Einstein moves into his cousin Elsa's apartment at Haberlandstrasse 5 in Berlin-Schöneberg.

October 1: The Kaiser Wilhelm Institute for Physics, originally to have been established in 1914, starts operating under Einstein's management. It confines itself to promoting research in physics and astronomy.

1918 *February:* Second paper on gravitational waves; it contains the quadrupole formula.

August: Einstein declines a joint invitation from ETH and Zurich University on exceedingly favorable terms.

November 9: The German Reich surrenders; the Kaiser abdicates; the Republic is proclaimed. Einstein is delighted and commits himself to left-wing democratic goals.

1919 *January:* Sojourn in Switzerland for several months. Einstein lectures on relativity theory at Zurich University.

February 14: Divorce from Mileva; custody of the children is adjudicated to Mileva.

Spring: Discussions with Blumenfeld about Zionism; Einstein supports Zionist ideals but does not join any Zionist organization.

June 2: He marries his cousin Elsa in Berlin.

September 22: Einstein learns that evaluation of observations made by two British expeditions during the solar eclipse of May 29 has confirmed the deflection of light in the gravitational field of the sun, as he had predicted from the general theory of relativity.

November 6: At a ceremonial meeting of the Royal Society and the Royal Astronomical Society in London, the confirmation of Einstein's prediction is officially announced. The following day the Einstein legend starts.

November 26: Members of the Prussian Diet demand generous support for Einstein's research. This develops into the "Einstein donation," the funding for construction of the "Einstein Tower."

1920 *February:* Niels Bohr visits Einstein in Berlin.

Pauline Einstein dies at her son's home in Berlin after a serious illness.

June: Journey to Norway and Denmark to give lectures.

August 24: Public rally at the Berlin Philharmonic Hall against relativity theory. Einstein reacts very fiercely in a newspaper article published three days later.

September 23: Argumentative discussion with Philipp Lenard at the Naturforscher meeting in Bad Nauheim.

October 27: Inaugural lecture as visiting professor in Leyden.

December 31: Elected to the Pour le Mérite Order as its youngest member.

1921 *January:* Keenly attended lectures in Prague and Vienna.

April 2–May 30: First visit to the United States: fund-raising for Hebrew University in Jerusalem; lectures in Princeton, published in book form.

June: Breaks his return journey in England. Lectures in Manchester and London.

1922 *January:* Einstein submits his first paper on unified field theory to the Prussian Academy.

March 28–April 10: Journey to Paris. Lectures at Collège de France; visit to World War I battlefields.

April: Toward the end of the month Einstein becomes a member of the League of Nations Commission for Intellectual Cooperation.

June 24: The German foreign minister, Walther Rathenau, is assassinated. Einstein cancels lectures and public appearances.

October 8: Einstein undertakes a trip to Japan as the guest of a publisher of periodicals.

November 9: The 1921 Nobel Prize is awarded to Einstein; official news reaches him during his journey.

November 17–December 29: Stay in Japan; academic and popular lectures.

1923 *February 2–14:* On his return journey, visit to Palestine. Einstein becomes the first person to be given the freedom of the city

of Tel Aviv; he lays the cornerstone of Hebrew University in Jerusalem.
March: Resigns from the League of Nations commission, largely because of the occupation of the Rhineland by France and Belgium.
June: Founding member and presidium member of the Association of Friends of the New Russia.
July: Journey to Sweden and Denmark; in Göteborg he delivers his Nobel lecture.
December: In a paper submitted to the Prussian Academy, Einstein discusses *Possibilities of Solving the Quantum Problem*, essentially by marginal conditions for the field equation of the general theory of relativity.

1924 *June:* Einstein rejoins the League of Nations commission.
December: The "Einstein Tower," including its instrumentation, is completed. He becomes life chairman of the board of trustees of the Einstein Institute.

1925 He publishes *Quantum Theory of Single Atom Ideal Gases;* he discovers an argument for the wave character of matter. He formulates the Bose-Einstein statistics and, toward the end of 1925, publishes his discovery of the Bose-Einstein condensation. These are his last discoveries of indisputably major importance.
April–June: Journey to South America.
September: Einstein becomes a member of the administrative council of Hebrew University in Jerusalem, having been editor of the publications of its Institute for Physics since 1924.

1926 Formulation of quantum mechanics by Heisenberg, Schrödinger, Born, etc. Einstein voices his misgivings.

1927 *October:* The Solvay Congress in Brussels marks the beginning of his intensive dispute with Niels Bohr about the foundations of quantum mechanics.

1928 *February:* In Davos Einstein develops a serious heart condition. He has to stay in bed for four months; his recovery takes about a year.

1929 *June 28:* Max Planck receives the first and Einstein the second Max Planck Medal, instituted on the occasion of Planck's seventieth birthday.
October: Solvay Conference in Brussels. Visit to the Belgian royal

family; beginning of his friendship and lifelong correspondence with the "dear queen."

1930 Intensive commitment to pacifism.

December: Journey to the United States for research at the California Institute of Technology in Pasadena. Return in March 1931.

1931 "Student" of Christ Church College, Oxford, a kind of visiting professorship for four weeks each year.

December: Another journey to the United States, to Caltech in Pasadena.

1932 Political engagement for the preservation of the Weimar Republic.

July: At the suggestion of the League of Nations and its Institute for Intellectual Cooperation, correspondence with Sigmund Freud on "Why War?" Published in 1933.

August: Appointment to the Institute for Advanced Study in Princeton, then being established, as of October 1933. Einstein intended to spend half his year in Princeton and the other half in Berlin.

December: Journey to the United States, to Caltech in Pasadena. His return was planned for March 1933.

1933 *January 30:* Nazis seize power in Germany.

March 10: Before his departure for Europe, Einstein publicly declares that he will not return to Germany.

March 28: Einstein announces his resignation from the Prussian Academy of Sciences. He terminates all links with official German institutions. He remains in Belgium.

June: He delivers the Spencer Lecture in Oxford.

September: He stays in England until he has to leave for the United States.

October 3: Speech at the Royal Albert Hall in favor of an assistance fund.

October 17: Together with his wife, his secretary Helen Dukas and his assistant Walther Mayer, Einstein arrives in the United States and goes to Princeton.

1934 *Mein Weltbild* (*The World as I See It*), a collection of his articles, is published in Amsterdam.

His stepdaughter Ilse Kayser-Einstein dies in Paris; his other stepdaughter, Margot, joins the Einsteins in Princeton.

1935 *May 15:* The Einstein-Podolsky-Rosen paradox is published; Bohr's rejoinder marks a certain conclusion of the debate on the foundations of quantum mechanics.

August: Purchase of a house, 112 Mercer Street in Princeton.

1936 Work on the motion problem with Banesh Hoffmann and Leopold Infeld.

December 20: Elsa Einstein dies in Princeton.

1937 Einstein's son Hans Albert, with his family, comes to the United States.

Work with Valentin Bargmann and Peter Bergmann on the unified theory.

1938 Publication of *The Evolution of Physics* with Leopold Infeld.

1939 His sister, Maja, joins him in Princeton, where she remains until her death.

August 2: Einstein signs a letter to President Franklin D. Roosevelt, referring to the possibility of making an atom bomb and to the danger that the Germans might construct such a weapon.

September 1: German attack on Poland; beginning of World War II.

1940 *March 7:* Second letter intended for President Roosevelt, on the atom bomb.

October 1: Einstein becomes an American citizen; he retains his Swiss citizenship.

1941 *November 6:* Start of the "Manhattan Project" for the development of the atom bomb. Einstein, because he is viewed as a security risk, takes no part in it.

December 7: Japanese attack on Pearl Harbor; United States enters the war.

1943 *May 31:* Einstein becomes an adviser on high explosives to the U.S. Navy; his daily pay is $25.

1944 *February 3:* Einstein's handwritten copy of his paper *On the Electrodynamics of Moving Bodies* is sold at auction in Kansas City, the proceeds going to the Book and Author War Bond Committee; it fetches $6 million.

1945 In a letter to President Roosevelt, Einstein tries to enable Leo Szilard and other physicists to state publicly their anxieties about the atom bomb.

August 6: The first atom bomb is dropped on Hiroshima.

September: Through the publication of the Smyth Report, Einstein's letter of 1939 to Roosevelt becomes public knowledge.

December 10: At a Nobel ceremony in New York Einstein delivers a widely noted speech: "The war is won, but peace is not."

1946 *May 23:* Einstein assumes the chair of the Emergency Committee of Atomic Scientists.

December: He publicly champions a world government.

1947 Intensive commitment to arms control and world government.

1948 *August 4:* Mileva Marić dies in Zurich.

December: Einstein has an abdominal operation; an aneurysm of the major abdominal aorta is diagnosed.

1949 Recuperation in Florida.

Work on the final observations for the Einstein volume in the Library of Living Philosophers.

1950 *March 18:* In his will, Einstein stipulates that his papers be turned over to Hebrew University in Jerusalem.

1951 His sister, Maja, dies in Princeton.

1952 *November:* After Chaim Weizmann's death Einstein is offered the presidency of Israel; Einstein declines.

1953 *May 16:* Open letter to W. Frauenglass in defense of civil rights against the McCarthy committee, causing fierce controversy.

1954 *April:* Einstein supports J. Robert Oppenheimer on the issue of Oppenheimer's "national reliability."

1955 *March 15:* Michele Besso dies in Geneva.

April 11: Letter to Bertrand Russell on the "Einstein-Russell Manifesto" on disarmament.

April 13: The aneurysm ruptures.

April 18: Einstein dies. His body is cremated the same day. After a simple ceremony, his ashes are scattered at an undisclosed location.

INDEX